P9-DHQ-884

Microwave and Millimeter-Wave Mixers

OTHER IEEE PRESS BOOKS

A Century of Honors, The First One-Hundred Years of Award Winners, Honorary Members, Past Presidents, and Fellows of the Institute
MOS Switched-Capacitor Filters: Analysis and Design, *Edited by G. S. Moschtyz*
Distributed Computing: Concepts and Implementations, *Edited by P. L. McEntire, J. G. O'Reilly, and R. E. Larson*
Engineers & Electrons, *By J. D. Ryder and D. G. Fink*
Land-Mobile Communications Engineering, *Edited by D. Bodson, G. F. McClure, and S. R. McConoughey*
Frequency Stability: Fundamentals and Measurement, *Edited by V. F. Kroupa*
Electronic Displays, *Edited by H. I. Refioglu*
Spread-Spectrum Communications, *Edited by C. E. Cook, F. W. Ellersick, L. B. Milstein, and D. L. Schilling*
Color Television, *Edited by T. Rzeszewski*
Advanced Microprocessors, *Edited by A. Gupta and H. D. Toong*
Biological Effects of Electromagnetic Radiation, *Edited by J. M. Osepchuk*
Engineering Contributions to Biophysical Electrocardiography, *Edited by T. C. Pilkington and R. Plonsey*
The World of Large Scale Systems, *Edited by J. D. Palmer and R. Saeks*
Electronic Switching: Digital Central Systems of the World, *Edited by A. E. Joel, Jr.*
A Guide for Writing Better Technical Papers, *Edited by C. Harkins and D. L. Plung*
Low-Noise Microwave Transistors and Amplifiers, *Edited by H. Fukui*
Digital MOS Integrated Circuits, *Edited by M. I. Elmasry*
Geometric Theory of Diffraction, *Edited by R. C. Hansen*
Modern Active Filter Design, *Edited by R. Schaumann, M. A. Soderstrand, and K. R. Laker*
Adjustable Speed AC Drive Systems, *Edited by B. K. Bose*
Optical Fiber Technology, II, *Edited by C. K. Kao*
Protective Relaying for Power Systems, *Edited by S. H. Horowitz*
Analog MOS Integrated Circuits, *Edited by P. R. Gray, D. A. Hodges, and R. W. Brodersen*
Interference Analysis of Communication Systems, *Edited by P. Stavroulakis*
Integrated Injection Logic, *Edited by J. E. Smith*
Sensory Aids for the Hearing Impaired, *Edited by H. Levitt, J. M. Pickett, and R. A. Houde*
Data Conversion Integrated Circuits, *Edited by D. J. Dooley*
Semiconductor Injection Lasers, *Edited by J. K. Butler*
Satellite Communications, *Edited by H. J. Van Trees*
Frequency-Response Methods in Control Systems, *Edited by A. G. J. MacFarlane*
Programs for Digital Signal Processing, *Edited by the Digital Signal Processing Committee, IEEE*
Automatic Speech & Speaker Recognition, *Edited by N. R. Dixon and T. B. Martin*
Speech Analysis, *Edited by R. W. Schafer and J. D. Markel*
The Engineer in Transition to Management, *I. Gray*
Multidimensional Systems: Theory & Applications, *Edited by N. K. Bose*
Analog Integrated Circuits, *Edited by A. B. Grebene*
Integrated-Circuit Operational Amplifiers, *Edited by R. G. Meyer*
Modern Spectrum Analysis, *Edited by D. G. Childers*
Digital Image Processing for Remote Sensing, *Edited by R. Bernstein*
Reflector Antennas, *Edited by A. W. Love*
Phase-Locked Loops & Their Application, *Edited by W. C. Lindsey and M. K. Simon*
Digital Signal Computers and Processors, *Edited by A. C. Salazar*
Systems Engineering: Methodology and Applications, *Edited by A. P. Sage*
Modern Crystal and Mechanical Filters, *Edited by D. F. Sheahan and R. A. Johnson*
Electrical Noise: Fundamentals and Sources, *Edited by M. S. Gupta*
Computer Methods in Image Analysis, *Edited by J. K. Aggarwal, R. O. Duda, and A. Rosenfeld*
Microprocessors: Fundamentals and Applications, *Edited by W. C. Lin*
Machine Recognition of Patterns, *Edited by A. K. Agrawala*
Turning Points in American Electrical History, *Edited by J. E. Brittain*
Charge-Coupled Devices: Technology and Applications, *Edited by R. Melen and D. Buss*
Spread Spectrum Techniques, *Edited by R. C. Dixon*
Electronic Switching: Central Office Systems of the World, *Edited by A. E. Joel, Jr.*
Electromagnetic Horn Antennas, *Edited by A. W. Love*
Waveform Quantization and Coding, *Edited by N. S. Jayant*

Microwave and Millimeter-Wave Mixers

Edited by
Erik L. Kollberg
Department of Electron Physics and Onsala Space Observatory
Chalmers University of Technology

A volume in the IEEE PRESS Selected Reprint Series, prepared under the sponsorship of the IEEE Microwave Theory and Techniques Society.

The Institute of Electrical and Electronics Engineers, Inc., New York

PRINTED IN THE UNITED STATES OF AMERICA

IEEE Order Number: PC01735

Library of Congress Cataloging in Publication Data
Main entry under title:

Microwave & millimeter-wave mixers.

(IEEE Press selected reprint series)
"Prepared under the sponsorship of the IEEE Microwave Theory and Techniques Society."
"PC01735"—T.p. verso.
Includes bibliographical references and indexes.
1. Microwave devices—Addresses, essays, lectures.
2. Mixing circuits—Addresses, essays, lectures.
I. Kollberg, Erik L. II. IEEE Microwave Theory and Techniques Society. III. Title: Microwave and millimeter-wave mixers.
TK7876.M523 1984 621.381'32 84-10887

ISBN 0-87942-179-7

Contents

Introduction

Overall View of Microwave and Millimeter-Wave Mixers

IN a superheterodyne receiver an incoming signal at a microwave frequency f_s is converted to a low frequency f_{IF} typically in the megahertz range. This is accomplished by utilizing the properties of a nonlinear impedance element which is simultaneously interacting with the *signal* (f_s) and a (usually) much stronger signal at f_{LO} generated by a *local oscillator.* Fourier theory tells us that the resulting current waveform can be described as a summation of frequency components $nf_{LO} \pm mf_s$. Since the desired frequency component, the *intermediate frequency* $f_{IF} = |f_{LO} - f_s|$, is much lower than f_s it is much easier to amplify and analyze. The particular device accomplishing the frequency down-conversion is referred to as the mixer. The first mixer circuits were described already in 1924 by Armstrong [1] and independently in 1925 by Schottky [2]. It is important to remember that in the frequency conversion process, information about the frequency, amplitude, and phase of the original signal is preserved. Therefore, we find the mixer as a component in almost all kinds of microwave systems where sensitivity and simplicity are required, i.e., in various communication systems, interferometric systems, phased arrays, radars of all kinds, and instruments such as spectrum analyzers, etc.

The frequency conversion may be performed in any current or voltage dependent impedance. In *resistive mixers,* a nonlinear resistance is used, while in *parametric devices* a nonlinear capacitance or inductance will do the job. In practice, one may have a combination of resistive and parametric frequency conversion. For example, the Schottky diode is modeled as a voltage dependent resistance in parallel with a voltage dependent capacitance. The theoretical *classical* conversion efficiency of a *purely resistive* mixer is always less than one [3]; i.e., less power is delivered at the IF frequency than is brought to the mixer at the signal frequency. However, Tucker discovered in 1979 that if the nonlinearity is strong enough, *quantum effects* become important [4], and in fact conversion gain larger than unity is possible. This phenomenon has nothing to do with *parametric effects* which can also cause conversion gain larger than one. In parametric down-converters a negative resistance is associated with the conversion gain and is achieved only when the idler frequency ($2\ f_p - f_s$) is properly terminated [5]. Such purely parametric frequency down-converters with gain are not used in practice because of the intricate problem of designing them for stable optimum performance.

Several types of components have been used as nonlinear impedance elements in mixer applications. However, the main part of the present book will be devoted to mixers using semiconductor diodes as the nonlinear element in predominantly *resistive* mixers. The obvious reason is that those mixers are by far the most important ones and used in the overwhelming number of practical applications. The RF properties of the Schottky diode are similar to those of the p–n diode, but in the Schottky diode, which is a majority carrier device, it is easier to achieve a low shunt capacitance (C_o) and series resistance (R_s). This is important since it is necessary to use a diode with a cutoff frequency $f_c = (2\pi R_s C_o)^{-1}$ much higher than the signal frequency, in order not to lose a substantial amount of power in the series resistance. Schottky diodes with cutoff frequencies of several THz have been described in the literature. The noise properties of the Schottky diode are determined partly by shot noise, related to electrons passing the barrier, and partly by thermal and excess noise, related to carrier transport in the nondepleted part of the semiconductor. The noise from the diode can be decreased by as much as a factor of four by cooling the diode to liquid nitrogen temperatures (77 K) or lower. In particular, for millimeter-wave frequencies where low noise amplifiers do not yet exist, cooled mixers are quite important. Single sideband receiver noise factors (see Part I) approaching 1.5 dB have been obtained at such high frequencies as 100 GHz. A comparable uncooled system has a noise figure of somewhat less than 6 dB. At still higher frequencies, the noise temperature increases approximately with a quadratic law, yielding a 16 dB noise figure ($\approx$10 000 K noise temperature) for the uncooled mixer at 1000 GHz and considerably less, cooled. For microwave frequencies, uncooled mixer receivers can have noise figures below 4 dB at *X* band and somewhat less at lower frequencies.

In the early sixties, the tunnel diode was a promising device for amplifiers and for mixers. Since the tunnel diode offers a nonlinear *negative* resistance, conversion gain can be achieved. However, this implies that stable operation is again a problem and since the sensitivity does not seem to improve much over ordinary present-day Schottky mixers, the tunnel diode mixer is, today, of minor importance. Mixing can also be performed in a Schottky barrier FET and conversion gain can be realized in this device as well [6]. However, since the FET has three terminals it is a more convenient and stable device than the tunnel diode mixer. Concerning the noise properties, the FET mixer seems to offer no great advantage over the Schottky mixer and is, in general, noisier than the FET amplifiers. The reason seems to be related to the fact that a MESFET suffers from $1/f$ noise, which is particularly annoying below a few hundred megahertz. Although normal single-gate FET's are successfully used in many mixer applications, dual-gate FET's have an advantage since the local oscillator circuit and the signal circuit can be separated in a simple way. Perhaps further improvements will be realized in the future with more specialized mixer FET's, since no real effort in this direction has been pursued so far. As for the Schottky diode, cooling will significantly improve the noise properties of the FET mixer. Another interesting and recently (1982) developed nonlinear impedance element for mixer applications is the so-called planar-doped GaAs diode [10]. It can be tailored so that the *I–V* characteristic becomes symmetrical, a property which will simplify the design of subharmonically pumped mixers. They will also possibly offer improved performance, since the troublesome parasitic inductances present when two separate diodes are coupled in antiparallel are avoided.

Self-oscillating mixers have been developed using FET's, Gunn diodes, and Josephson junctions. The self-oscillating FET mixer has a much higher sensitivity than the Gunn mixer, which however is a simpler and less expensive component. The Josephson junction self-oscillating mixer is a much more extreme type of device and will perhaps have some importance in the future for submillimeter frequencies.

Very low noise temperatures can be achieved using liquid helium cooled ($\leq$4 K) superconducting elements. Historically, Josephson junctions were tried first and some successful mixers were constructed. Since the junction behaves as a nonlinear inductance element, the mixer is in fact a parametric down-converter. Gain has been demonstrated, but the very low noise temperatures hoped for were not achieved in practice and problems in connection with saturation were recognized. More promising seems to be the quasiparticle mixing mode which can be described in terms of purely nonlinear resistance mixing. However, due to the exceedingly strong nonlinearity, quantum effects become important and gain can be achieved [4]. Gain (of the order 4 dB) has been demonstrated at 35 GHz and typical mixer noise temperatures of about 10 K have been reported. The indium-antimonide (InSb) bolometer mixer is also an extremely low noise mixer device. However, due to the long thermal time constant of the elements, the IF frequency is limited to a few megahertz, which restricts the usefulness to applications like radio astronomy for submillimeter wavelengths. Finally, the super Schottky diode mixer should be mentioned. This element is a Schottky diode where the metal contact is made from a superconductor. The noise temperature of this diode is much lower than for the ordinary cooled Schottky diode, but the low cutoff frequency seems to restrict its possible practical use to a few tens of gigahertz.

During the last few years, the rapid development of low noise GaAs MESFET amplifiers [6],[7] has somewhat lessened the importance of sophisticated mixer designs for achieving lowest possible noise at frequencies below about 50 GHz. However, the best mixer performances are not that far behind the FET's, in particular for the higher microwave frequencies. Hence, if simplicity and price are important and sensitivity is not a major problem, a diode mixer is the obvious choice. The situation for low noise mixers further improves above about 50 GHz, where FET amplifiers are not yet competitive. For millimeter-wave frequencies (say about 100 GHz), the diode mixer will remain the most practical front end device for many years to come. If the ultimate sensitivity is required, as it is, for instance, in radio astronomy, superconducting quasiparticle mixers may be an alternative to diode mixers above about 30 GHz. For microwave frequencies (<30 GHz), a cooled FET amplifier followed by a Schottky mixer should be recommended.

Some Basic Definitions Concerning Mixer Sensitivity

Before we proceed to discuss some specific problems concerning mixers, we should halt and consider some important definitions related to their ability to detect faint signals. *The equivalent noise temperature* of a complete receiver system (T_{syst}) is a measure of the system's sensitivity. The lower the noise temperature the better the sensitivity. Remembering that the mixer in general responds at all frequencies where $f_s = nf_{LO} \pm f_{IF}$, we can express the total *system noise* temperature as

$$T_{syst} = T_A L_S \cdot \sum_{n=1}^{\infty} \left(\frac{1}{L_{n+}} + \frac{1}{L_{n-}}\right) + L_S \cdot T_{IF} + T_{MXR} \quad (1)$$

where L_S is the conversion loss at the signal frequency, L_{n+} and L_{n-} are the conversion loss for the sidebands $nf_{LO} \pm f_{IF}$, respectively, T_{IF} is the noise temperature of the intermediate frequency amplifier, and T_{MXR} is the equivalent noise of the mixer itself. T_A is the input (antenna) noise temperature, here assumed to be the same for all sidebands. L_S is equal to L_{1-} for a lower sideband mixer and to L_{1+} for an upper sideband mixer. For a harmonic mixer one of the harmonic sidebands is used; i.e., L_S may be equal to L_{2+} or say L_{2-}. In spectrum analyzers for example, harmonic mixers using higher order harmonic sidebands are commonly used. In a double sideband (DSB) mixer receiver, both the upper and the lower sidebands are supposed to contain useful information and the receiver noise temperature is then defined as

$$T_{syst}\,(DSB) = \frac{1}{1 + \frac{L_S}{L_i}} T_{syst}\,(SSB) \quad (2)$$

where L_i is the conversion loss at the image frequency ($2f_{LO} - f_s$ for an ordinary mixer). Consequently, T_{syst} (DSB) $= 0.5 \cdot T_{syst}$ (SSB) if $L_i = L_S$.

Notice that in (1) and (2) the background noise contribution T_A is included. When we talk about *receiver* noise temperatures we should not include T_A, i.e.,

$$T_{Rec}\,(SSB) = L_S \cdot T_{IF} + T_{MXR} \qquad (3)$$

$$T_{Rec}\,(DSB) = \frac{1}{1+\dfrac{L_S}{L_i}} \cdot T_{Rec}\,(SSB). \qquad (4)$$

The noise figure is commonly expressed as

$$F = \frac{T_{Rec} + T_o}{T_o} = L_S\,(N_R + F_{IF}^{-1}) \qquad (5)$$

where $T_o = 290$ K and

$$F_{IF} = \frac{T_{IF} + T_o}{T_o} \qquad (6)$$

$$N_R = \frac{T_{MXR} + T_o}{L_S \cdot T_o}. \qquad (7)$$

N_R is referred to as the noise ratio of the mixer diode and is usually of the order of 1–2 for ordinary room temperature Schottky diode mixers. In the next paragraph we will discuss the conversion loss problem, and after that the origin of T_{MXR}.

Calculation of Conversion Loss and Noise in Mixers

The basic theory for calculating diode mixer conversion loss was first studied in depth by Torrey and Whitmer [8] in 1948. However, the complexity of the nonlinear problem of the pumped diode for a long time made it necessary to simplify the analysis by assuming idealized switching of the diode or a sinusoidal local oscillator voltage across an exponential resistive diode [3]. The latter approach has been used by many authors with some success in the understanding of the behavior of the mixer. The early papers did not consider the fact that there is also a pumped nonlinear capacitance, which further complicates the problem. It was not until powerful computers became available that methods for solving the complete nonlinear problem were developed and applied to the single-ended diode mixer. It should be pointed out that although the diode mixer performs mainly as a resistive mixer, the pumped capacitance implies that parametric frequency conversion will also take place. In extreme cases, conversion gain due to parametric effects may even be obtained, a situation that one should probably try to avoid in practice since this will cause stability problems.

The Fourier expansion of the exact current and voltage waveform implies that an infinite number of frequency components $nf_{LO} \pm mf_s$ are generated in the mixing process. In actual calculations, it is of course possible to truncate the Fourier expansions and still obtain quite an accurate prediction of the mixer properties. First of all, if the signal power is very much smaller than the power of the local oscillator, it is sufficient to consider terms where $m = 1$. For the local oscillator then, it may, in some cases, be necessary to account for up to 6 harmonics in a diode mixer, i.e., $n \leq 6$. For a basic understanding of the mixer some of those papers dealing with simplified models are quite informative and helpful. The effects of higher order sideband terminations of resistive mixers are discussed in some detail in the important work by Saleh [3]. In short, one can conclude that the terminations at the harmonic sidebands are quite important for the conversion properties, not only because signal power may be dissipated, but also because of the intricate interference between waves being reflected and reconverted to the signal and IF port, respectively. If the harmonic terminations all are purely reactive they will dissipate no power, and zero decibel conversion loss may be approached theoretically. In summary, we can conclude that mainly the mixer embedding circuits together with the methods of pumping are responsible for the deterioration in performance. However, the diode parasitics, i.e., the capacitance and the series resistance are quite important at microwave frequencies and, in particular, at millimeter-wave frequencies. Essentially it is advantageous to have diodes with low shunt capacitances and low series resistances. In the superconducting quasiparticle mixer there is virtually no series resistance, and as mentioned above the resistive nonlinearity is so strong that quantum effects may produce conversion gain.

When the signal power approaches the same power level as the local oscillator, the mixer becomes saturated, i.e., the output IF power no longer increases linearly with the input power. This means also that the signal is "pumping" the diode and that intermodulation products such as $mf_s \pm nf_{LO}$ will be created in the mixer.

Calculation of noise in mixers is rather difficult independently of which type of nonlinear element used. However, some of the basic principles are essentially the same for any mixer, and it is therefore at least educational to discuss briefly the properties of the Schottky diode mixer, where noise generated at the harmonic sidebands $nf_{LO} \pm f_{IF}$ will be converted to the IF frequency and add to the total noise. Usually one considers two types of noise source for the diode mixer viz. thermal noise generated in lossy parts of the mixer circuit including the series resistance of the diode, and shot noise generated when electrons pass the barrier in the diode. There are independent noise sources related to each harmonic sideband and all of them are coherently modulated by the local oscillator. This phenomenon has been discussed by various authors. Held and Kerr have in their work described the shot noise as well as the thermal noise and scattering noise (antenna noise) by a noise correlation matrix which is compatible with the formulation of the small-signal mixer analysis based on the Torrey and Whitmer theory [8]. Consequently, the situation is again complicated and it seems that a completely straightforward way to describe the noise behavior of a mixer is hard to find. For example, Held and Kerr find that the shot noise observable from a dc-biased diode ($\approx T_o/2$) can be almost zero although the conversion loss is of the order 5 dB. If the quantum limit is approached, i.e., kT_{rec} is of the order hf_s (possible in superconducting mixers), special care must be taken concerning quantum effects. An excellent review of this problem has been published recently by Caves [9].

Mixer Configurations

In the simplest mixer configuration only one diode (or any other nonlinear element) is used. In the single-ended mixer the local oscillator power and the signal are fed to the diode in the same way, e.g., through the same waveguide. The local oscillator should be injected into the waveguide through a narrow-band filter to make the local oscillator circuit invisible to the signal. Many millimeter-wave mixers ($f_s > 50$ GHz) are built in this way, since they are not too difficult to fabricate in a rectangular waveguide configuration. However, microwave mixers can certainly also be designed using more than a single diode, and for microwave frequencies two or four diode configurations are quite common. Such a mixer is usually realized by using planar microstrip or slot line techniques. Various symmetry properties in the mixing process as well as in the circuitry, which can be identified when two or more diodes are used, have proved to be of substantial advantage in particular for designing mixers with broad bandwidth and large tunability.

The balanced mixer is perhaps the best known multidiode configuration. In this mixer the signal couples, with the same phase, to two diodes, while the local oscillator couples out-phase (or vice versa). One advantage of using this design is that noise originating from the local oscillator and converted to the IF will be in antiphase with the signal. Consequently, the noise can be made to cancel at the IF while the signal is recovered. Another important advantage is that the signal and local oscillator circuits can be isolated from each other due to the symmetry properties of the circuit. Therefore, the design is inherently broad-band, since no narrow-band filtering is required.

In most applications a single sideband mixer response is desirable. To achieve a true single sideband mixer performance, it is necessary to prevent the mixer from responding to signals at the image frequency ($2f_{LO}-f_s$) in particular, and if possible also to signals at the harmonic sideband frequencies ($nf_{LO} \pm f_{IF}$). The general name for mixers of this type is *image rejection mixers.* Image rejection can of course be realized by using a filter or a diplexer at the input port of a single-ended mixer. However, keeping in mind that the signal is actually generating power at the image frequency, it is also possible to *recover* at least some of this power and convert it to IF power yielding an *enhanced* mixer performance. Consequently, mixers that can do this trick are called *image recover* or *image enhanced* mixers. In practice, there are various ways of realizing this improvement. An obvious way is to use a filter in front of the mixer, which will cause a reactive termination at the image frequency. By a proper choice of image reactance, the conversion loss may be minimized. Another way to recover the image power is to use two mixers connected in such a way that the image power from one mixer is converted to the IF with an optimum phase in the other mixer (and vice versa). In his book [3], Saleh has discussed extensively the different mixer configurations using four diodes. By proper choice of symmetry it is possible to make, e.g., all even harmonic sideband currents cancel, yielding a more efficient mixing. A great number of practical designs using up to eight diodes have been described in the literature, and some of them are reproduced in the present volume. The designs utilize techniques based on rectangular waveguide circuits, microstrip circuits, fin-line configurations, and sometimes even a mixture of all three techniques. For *X* band mixers conversion loss numbers as low as 2 dB (excluding some circuit loss) have been reported, while at *K* band the lowest reported conversion loss numbers are around 3 dB.

It is perhaps appropriate at this point to remind the reader of two things. First, although the conversion loss (L_s) of the mixer is of prime importance for optimum mixer performance, the inherent mixer noise (T_{MXR}) must also be kept low by an appropriate choice of embedding circuit impedances. Second, it is not only the embedding circuit impedances at the sideband frequencies $nf_{LO} \pm f_{IF}$ that are important. The impedances at the harmonics of the local oscillator frequency (nf_{LO}) are also of utmost importance for obtaining an optimum pumped diode waveform, which determines the frequency conversion ability of the diode itself.

Another family of mixers are the harmonic diode mixers, where the LO frequency is of the order half the signal frequency, i.e., $f_{IF} = 2f_{LO}-f_s$. It is certainly possible to build a single diode mixer that may work in this way. However, a harmonic mixer using a single diode will perform worse in practice than a fundamental mixer. To optimize such a mixer it is important to control such frequency components as $f_{LO} \pm f_{IF}$ and perhaps also $3f_{LO} \pm f_{IF}$. This can be done by using suitable filters. In an elegant design, one uses two diodes with opposite polarity in parallel. Together they form a nonlinear element with a symmetrical *I–V* characteristic, which when pumped at f_{LO} will act as if the impedance is modulated with $2f_{LO}$. It will be possible to replace the diode pair with planar doped diodes [10] which have an inherent symmetrical *I–V* characteristic. This new device should make it easier to construct subharmonically pumped mixers.

For short millimeter- and submillimeter-wave frequencies, open mixer structures are used with a single diode coupled directly to an antenna or quasioptical system. Although the mechanical problems in constructing these mixers can be solved satisfactorily, the conversion loss and noise temperature of submillimeter-wave mixers, in particular, become unsatisfactorily high. This is largely due to the fact that the diode impedance becomes too low (the diode capacitance) and that it is difficult to meet the circuit requirements for optimum termination of all the important frequency components involved in the mixing process. For long millimeter wavelengths monolithic mixers have been constructed, where a diode, an IF circuit, and an antenna structure all have been manufactured on the same GaAs substrate. This is of course an extremely interesting technique having a lot of promise for the future.

Editorial Comments

The present collection of mixer papers is divided into ten major parts containing a total of 71 papers. Aside from these papers, it is recommended that the reader consult the reference list at the end of this Introduction, which refers to some important books or sections of books relating to mixer theory and technology. There is extensive literature concerned with the practical implementation of mixers and it is of course possible to include only a minor number of the many important papers. Thus, the reader is also urged to consult the reference lists of the individual papers of this volume for suggestions on further readings.

References

[1]. E. H. Armstrong, "The superheterodyne—Its origin, development, and some recent improvements," *Proc. IRE,* vol. 12, pp. 540–552, 1924.

[2] W. Schottky, "Über den Ursprung des Superheterodyngedankens," *Elek. Nachrichtentech.,* vol. 2, pp. 454–456, 1925.

[3] A. A. M. Saleh, *Theory of Resistive Mixers.* Cambridge, MA: M.I.T. Press.

[4] J. R. Tucker, "Quantum limited detection in tunnel junction mixers," *IEEE J. Quantum Electron.,* vol. QE-15, pp. 1234–1258, Nov. 1979.

[5] P. Penfield and R. P. Rafuse, *Varactor Applications.* Cambridge, MA: M.I.T. Press.

[6] R. S. Pengelly, *Microwave Field-Effect Transistors—Theory, Design, and Applications.* Chichester, England: Research Studies, 1982.

[7] H. Fukui, *Low-noise Microwave Transistors and Amplifiers.* New York: IEEE Press, 1981.

[8] H. C. Torrey and C. A. Whitmer, *Crystal Rectifiers* (M.I.T. Rad. Lab. Series, vol. 15). New York: McGraw-Hill, 1948.

[9] C. M. Caves, "Quantum limits of noise in linear amplifiers," *Phys. Rev. D,* vol. 26, pp. 1817–1839, 1982.

[10] R. J. Malik, T. R. AuCoin, R. L. Ross, K. Board, C. E. C. Wood, L. F. Eastman, "Planar-doped barriers in GaAs by molecular beam epitaxy," *Electron. Lett.,* vol. 16, p. 836, 1980.

Part I
Basic Mixer Theory

THE purpose of this part is to give the reader a basic understanding of mixers. All of the papers except for the last one (by Tucker), which concerns certain quantum effects, concentrate on diode mixers. More exact approaches for analyzing diode mixers will be found in the next part. Superconducting mixers will be discussed in some detail in Part VIII.

Several textbooks discuss one aspect or another concerning mixers. One important book on conversion properties of various types of diode mixers is by Saleh [1]. Using simple models he sorts out a number of important properties of mixers. The first paper of this part is an often quoted paper by Barber. The theory of mixers is briefly reviewed and results are presented in graphical form convenient to the designer. Emphasis is placed on the pulse duty ratio rather than on the magnitude of the generally nonsinusoidal LO voltage as a parameter for estimation of mixer performance. In the following paper, by Liechti, a more detailed analysis including capacitance variations is presented. It is shown that in a resistive mixer with equal termination of signal and image ports, the addition of a nonlinear capacitance can lower the conversion loss substantially. The next three papers discuss the influence of the diode capacitance. Mania and Stracca analyze the conversion loss for the image port either open-circuited or short-circuited and for various combinations of diode capacitance and series resistance. In the fourth paper McColl investigates how to choose the Schottky diode parameters for optimum performance at high frequencies. The analysis focuses on the competing requirements of impedance matching the diode to its embedding circuit and the finite dynamic range of the nonlinear resistance. Günes *et al.* find that a diode pumped with a sinusoidal voltage (Y mixer) will give a lower noise than when pumped with a sinusoidal current (Z mixer).

The theory of shot noise in mixers has been examined by several authors. Both the book by van der Ziel [2] and the paper by Kim in Part X of this volume are worthwhile reading as an introduction to the noise problem. Of fundamental interest is the paper by Kerr in this part. He analyzes the noise of a purely resistive exponential diode mixer and shows how the noise properties can be described in terms of a simple attenuator noise model. The analysis, however, is only valid if the nonlinear capacitance and series resistance are negligible. For the exact analysis, the reader should consult Kerr's papers in the next part. Kelly's paper, the sixth in this part, is also of great pedagogic value and is recommended for anyone who wants to "understand" where the signal power disappears in a mixer (see also Hine's paper in the next part). In the following two papers, by Korolkiewicz and Kulesza, the effect of source resistance and diode capacitance on broad-band balanced mixers is discussed.

The quantum theory of resistive mixers has been described by Tucker [3]. Quantum effects may be important if the I–V characteristic of the mixing element is curving considerably with a voltage interval related to the photon voltage $\Delta v = hf_s/e$ (h is the Planck's constant and e the charge of an electron). In the last paper of this section Tucker makes the astonishing prediction that conversion efficiencies greater than unity can be realized in a purely resistive mixer. This is a result of fundamental importance and basic for the success of the quasiparticle mixer described in Part VIII.

References

[1] A. A. M. Saleh, *Theory of Resistive Mixers.* Cambridge, MA: M.I.T. Press, 1971.

[2] A. van der Ziel, *Noise: Sources, Characterization and Measurement.* Englewood Cliffs, NJ: Prentice-Hall, 1970.

[3] J. R. Tucker, "Quantum limited detection in tunnel junction mixers," *IEEE J. Quantum Electron.*, vol. QE-15, pp. 1234–1258, Nov. 1979.

Noise Figure and Conversion Loss of the Schottky Barrier Mixer Diode

MARK R. BARBER, MEMBER, IEEE

Abstract—**The theory of mixer operation is briefly reviewed and the results presented in graphical form convenient to the designer. In particular, the minimum noise figure, conversion loss, and the source and output impedances are plotted as functions of the pulse duty ratio of the diode current. Emphasis is placed on the pulse duty ratio as a more fundamental parameter for defining mixer operation than the magnitude of the diode voltage which is generally nonsinusoidal.**

It is shown that Schottky diode mixers should exhibit single sideband noise figures as low as 3 dB at *X* band when used in conjunction with 1.5 dB noise figure IF amplifiers, provided the diodes have cutoff frequencies higher than 500 GHz.

I. Introduction

CONSIDERABLE interest has recently been shown in the use of the epitaxial Schottky barrier (ESBAR) diode as a microwave frequency down-converter. For the first time it has become possible to surpass the performance of the redoubtable point-contact diodes, through the use of photoresist techniques to achieve small areas and epitaxial material to achieve low series resistance. The new diodes are also more reproducible, have much lower reverse current leakage, lower $1/f$ noise, and can be designed for much higher dynamic range. Fig. 1 compares two Schottky diode *I–V* characteristics with those of two well-known point contacts. It should be noted that Schottky characteristics can be exponential over a range seven decades of current thus making it feasible to carry out accurate mathematical computations in closed form. Herein it is shown that Schottky diodes should exhibit overall calculated noise figures as low as 3 dB at *X* band when the image is open circuited and the following IF amplifier has a 1.5 dB noise figure.

II. Mixer Noise and Gain Formulas

The theory for calculating mixer gain has been established for a number of years and has been applied to mixers using point-contact diodes by Herold *et al.*,[1] Torrey and Whitmer,[2] and Strum.[3] The gain and impedance formulas used in this paper are identical to those used by previous workers (except for minor corrections and additions) but the paper differs in that it considers the noise theory and also breaks away from the assumption of a sinusoidal voltage across the nonlinear diode conductance.

Manuscript received October 19, 1966; revised May 9, 1967.

The author is with Bell Telephone Laboratories, Inc., Murray Hill, N. J. 07971

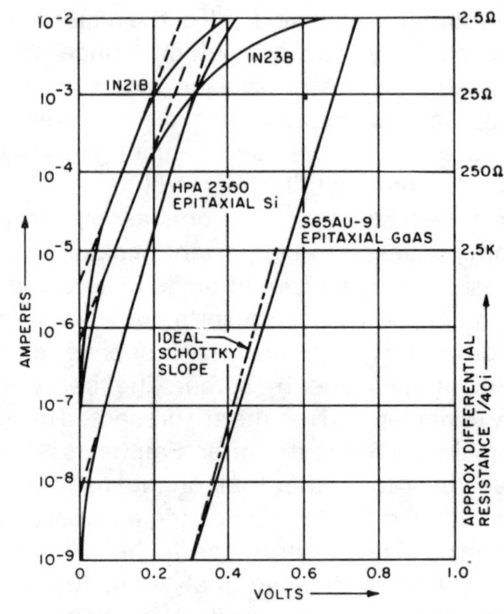

Fig. 1. Point contact and Schottky barrier *I–V* characteristics. The quantity i_0 in (3) for a particular diode is given by the intercept of the straight dashed line with the *y* axis. The value of α can be determined from its slope

$$\alpha = \frac{\Delta\,(\log_{10} i)}{\Delta V} \cdot 2.303.$$

The basic noise theory for diode mixers has been developed by Strutt,[4] Messenger and McCoy,[5] and summarized by Kim.[6] This theory neglects thermal noise generated in the diode series resistance and expresses the result in terms of ratios of the Fourier coefficients of conductance g and equivalent shot-noise current I_{eq} where these quantities are time varying due to the application of some arbitrary local oscillator voltage waveform across the diode.

$$g = g_0 + \sum_{n=1}^{\infty} 2g_{cn} \cos n\omega t. \tag{1}$$

$$I_{eq} = I_{e0} + \sum_{n=1}^{\infty} 2I_{en} \cos n\omega t. \tag{2}$$

The theory also neglects the effects of diode parasitic reactances, transit time phenomena, and conversion from frequencies near the higher harmonics of the local oscillator. It should be noted that in the range of voltages where the Schottky diode mixes most effectively the junction capacitance changes by almost 100 percent and only the average component can be tuned out exactly by the source admittance. The fundamental component will cause some reactive

Reprinted from *IEEE Trans. Microwave Theory Tech.*, vol. MTT-15, pp. 629–635, Nov. 1967.

mixing and may or may not be a contributor to the conversion loss and/or gain. A general treatment of conversion by means of a nonlinear admittance has been considered by Edwards.[7]

Formulas for conversion loss, minimum noise figure, optimum source impedance, and the corresponding output impedance are presented in the Appendix. Three values of conductance as seen by the diode at the image frequency are considered, namely, a short circuit, an open circuit, and a conductance equal to the signal source. The open-circuited image gives noise and loss figures approximately half a decibel lower than those obtainable with a short circuit.

III. Application of Theory to the Schottky Barrier Diode

At low frequencies the equation relating current and voltage in the Schottky barrier diode is

$$i = i_0(e^{\alpha V} - 1).$$

The effects of diode parasitic reactances which influence the validity of this equation at high frequencies are discussed in Section IV.

The differential conductance is therefore given by

$$g = \frac{di}{dv} = \alpha i_0 e^{\alpha V}$$
$$= \alpha(i + i_0) \cong \alpha i \qquad i \gg i_0. \tag{3}$$

Also, the mean-square shot-noise current is given by

$$\overline{i_{sh}^2} = 2qI_{eq}\Delta f$$

where

$$I_{eq} = i + 2i_0$$
$$\cong i \qquad i \gg i_0. \tag{4}$$

Since g and I_{eq} are simply related through the constant α, it is easy to show using (1) and (2) that

$$\frac{g_{c1}}{g_0} = \frac{I_{e1}}{I_{e0}},$$
$$\frac{g_{c2}}{g_0} = \frac{I_{e2}}{I_{e0}}. \tag{5}$$

Equations (5) permit a simplification of the noise theory for the Schottky diode, hence, all of the mixer formulas listed in the Appendix can be expressed as functions of the two ratios g_{c1}/g_0 and g_{c2}/g_0 which must approach unity if mixer loss and noise figure are to become small.

Throughout this paper it will be assumed that the peak forward current (i) greatly exceeds the reverse saturation current (i_0) and that the current versus voltage characteristic is purely exponential. From Fig. 1 it can be seen that (3), (4) and therefore (5) will become inaccurate when the maximum forward voltage is less than 0.1 volts thereby introducing errors in the noise calculations; this case is of little interest, however, because of the high noise figure. This conclusion is empirical and will depend on the care with which the junction is made, the material, and the junction area.

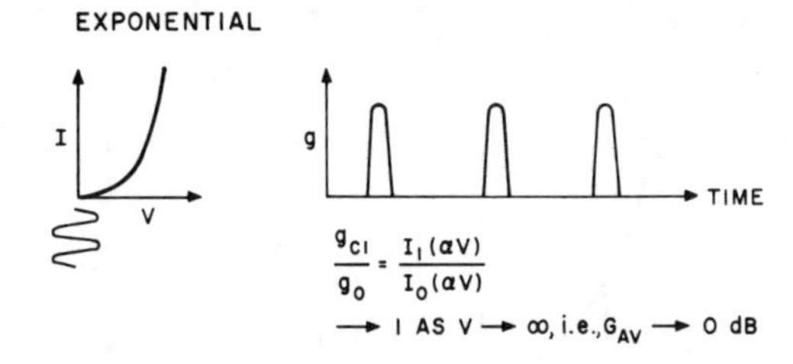

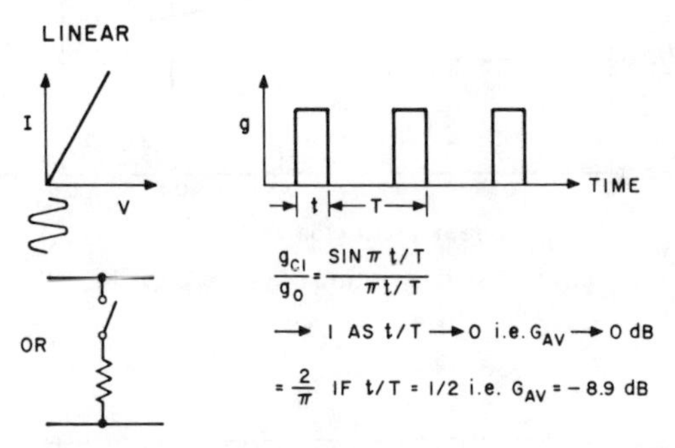

Fig. 2. Exponential and linear mixer conductance waveforms illustrating the definition of an effective pulse duty ratio. (G_{av} = available conversion gain.)

Case I. Sinusoidal Junction Voltage

If a sinusoidal oscillator voltage is assumed, which is certainly not the general case, it is possible to express V as

$$V = V_0 + V_1 \cos \omega t \tag{6}$$

where V_0 is the dc bias voltage. Substitution of this expression in (3) then defines the conductance waveshape

$$g(t) = \alpha i_0 e^{\alpha V_0} e^{\alpha V_1 \cos \omega t}$$
$$= \alpha i_0 e^{\alpha V_0}[I_0(\alpha V_1) + 2I_1(\alpha V_1) \cos \omega t$$
$$+ 2I_2(\alpha V_1) \cos 2\omega t + \cdots] \tag{7}$$

where I_0, I_1, I_2, etc. are modified Bessel functions with argument (αV_1). Hence from (1) and (7),

$$\frac{g_{c1}}{g_0} = \frac{I_1(\alpha V_1)}{I_0(\alpha V_1)},$$
$$\frac{g_{c2}}{g_0} = \frac{I_2(\alpha V_1)}{I_0(\alpha V_1)}. \tag{8}$$

The mixer gain and noise formulas can therefore be expressed as functions of the single variable (αV_1). Fig. 2 illustrates the conductance waveform which consists of a series of narrow pulses. The figure also shows the waveform for a conductance in series with an ideal switch. By determining the values of peak sinusoidal voltage and rectangular pulse duty ratio which give identical values of the ratio g_{c1}/g_0 in the two cases, it can be shown that these pairs of parameters will also give values of the ratio g_{c2}/g_0 for the two cases which are equal to within 3 percent or better for pulse duty ratios below 30 percent. It is therefore possible to define an effective rectangular conductance pulse duty ratio PDR

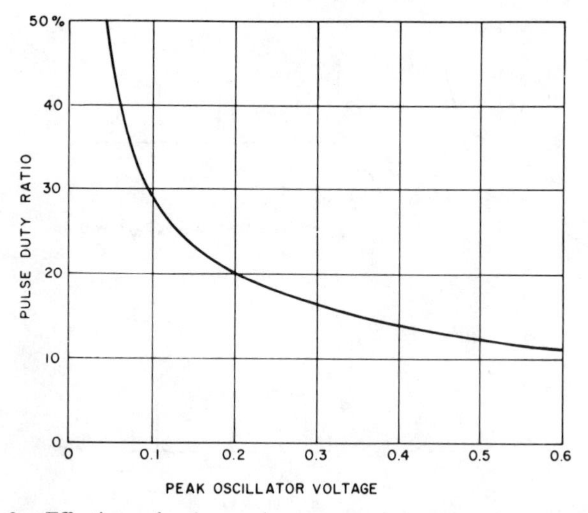

Fig. 3. Effective pulse duty ratio versus peak local-oscillator voltage.

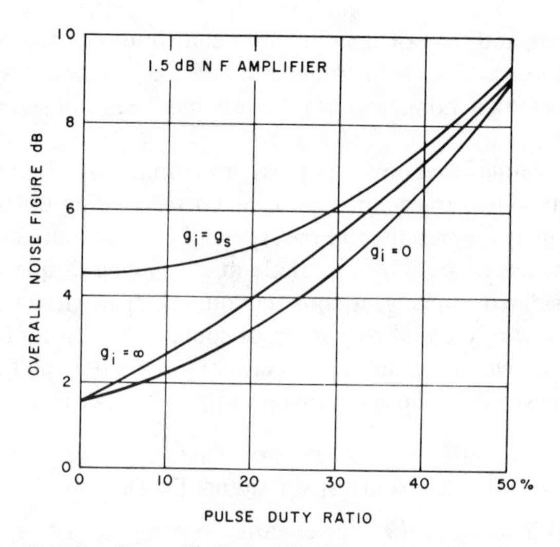

Fig. 6. Computed overall mixer noise figure including a 1.5 dB noise figure IF amplifier. This figure has been derived from Fig. 5 by means of the well-known cascade noise formula.

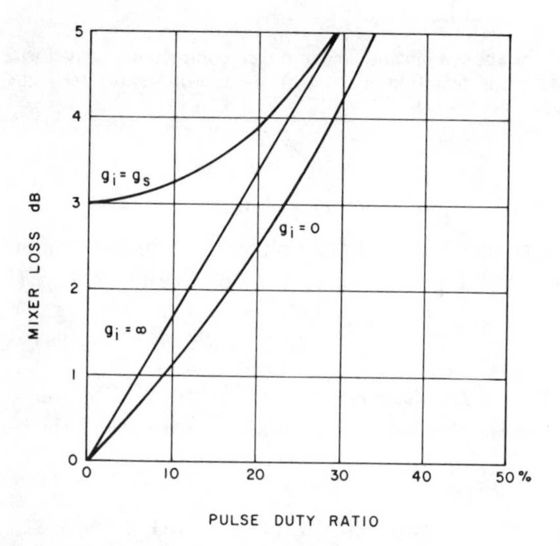

Fig. 4. Computed mixer conversion loss applicable to diodes with no series resistance.

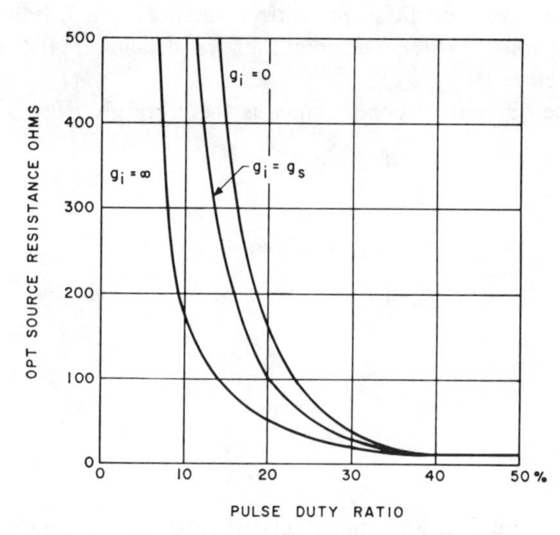

Fig. 7. Computed optimum source resistance for fixed $I_{max} = 10$ mA.

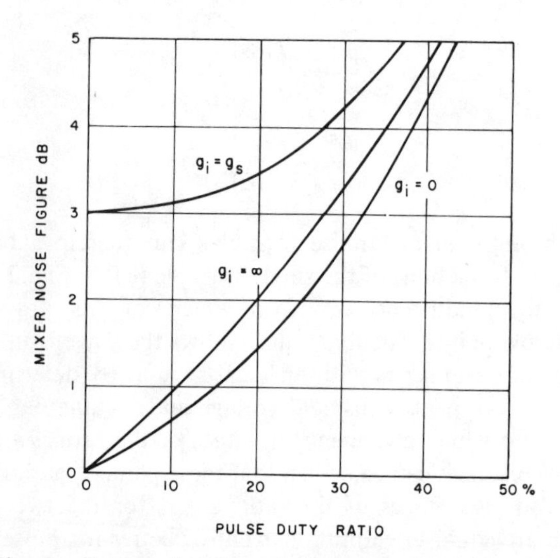

Fig. 5. Computed mixer noise figure applicable to diodes with no series resistance.

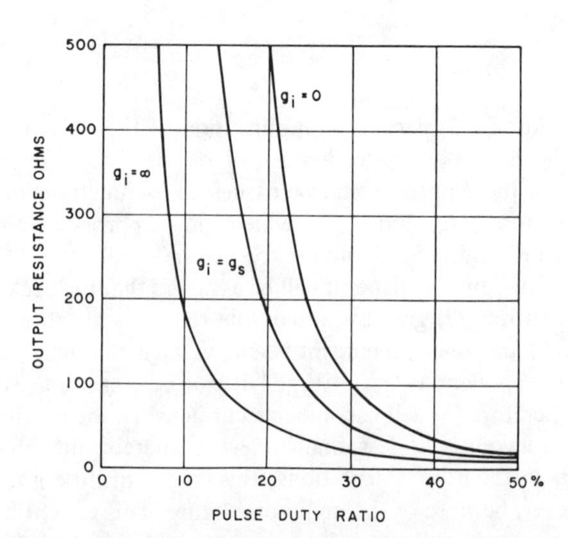

Fig. 8. Computed output resistance for fixed $I_{max} = 10$ mA and optimized source resistance.

$=t/T$ (see Fig. 3) which would result in mixer performance identical to that obtainable from an exponential device pumped by a specified sinusoidal voltage (V_1).

The pulse duty ratio is a very useful single parameter for defining mixer performance and can be shown to be equal to the width of the conductance pulse at 0.22 g_{max} relative to the interpulse period (T). Fig. 2 shows that the pulse duty ratio should be as small as possible if mixer loss and noise figure are to be reduced, a limit is set, however, by the source and output conductances which ultimately become unreasonably small.

Case II. Nonsinusoidal Junction Voltage

It has been found through low-frequency diode simulation studies, discussed in Section VI, that the conductance versus time waveform is always similar in shape to that occurring when a sinusoidal voltage exists across the Schottky diode junction. Typical conductance and voltage waveforms are shown in Fig. 13 for a case in which the voltage is quite nonsinusoidal. In such cases the effective pulse duty ratio adequately defines the conductance ratios and hence mixer performance. On the other hand, the peak local oscillator voltage or functions derived from that voltage, which were commonly used in the past,[2],[3],[8] have now ceased to be useful parameters.

Computed Results

Figs. 4 through 8 show the computed values of mixer loss, noise figure, and impedances plotted as functions of pulse duty ratio for three values of conductance (g_i) at the image frequency as seen by the diode.[1] In every case the source conductance (g_s) at the signal frequency has been chosen to minimize the noise figure. This condition also results in an input impedance match for all three cases of image loading.[2]

The source and output impedances have been computed assuming that the maximum forward current is 10 mA. It will be clear from Section IV that this is a reasonable assumption, however, for other values of maximum forward current the resistances should be multiplied by $10/I_{max}$ in milliamperes.

The rectified dc current is a useful quantity for checking mixer operation and can be calculated using (3) and (7):

$$I_{dc} = I_{max} \cdot e^{-\alpha V_1} \cdot I_0(\alpha V_1).$$

If $e^{-\alpha V_1} \cdot I_0(\alpha V_1)$ is plotted as a function of pulse duty ratio using Fig. 3, the result is found to be a straight line through the origin, hence

$$I_{dc} \cong 0.70 I_{max} \cdot (\text{PDR}). \tag{9}$$

[1] The curve of loss versus pulse duty ratio for the case of $g_i = \infty$ has been presented previously by Strum.[8]

[2] From Figs. 4 and 5 and the Appendix it can be seen that the mixer noise ratio is less than unity.

The local oscillator power absorbed by the diode can also be calculated if a sinusoidal voltage is assumed and is generally less than half a milliwatt.[9]

IV. Effect of Series Resistance and Junction Capacitance

When the diode has finite series resistance r_s, then both the forward and the reverse impedance at microwave frequencies are affected. In the forward direction, increasing the pump voltage causes the junction conductance to rise until it finally reaches $g_{max} = 1/r_s$. Beyond this point further drive only widens the effective relative conduction interval t/T (Fig. 9), which reduces the conversion conductance g_{c1}/g_0.

If the forward excursion is arbitrarily restricted to the point where the diode conductance rises to $g_{max}/5$ ($\cong$10 mA for the GaAs ESBAR diode in Fig. 1) the only way to reduce the loss and noise figure with increased sinusoidal pumping (smaller pulse duty ratio) is to use reverse bias. It should be emphasized, however, that if the voltage swings too far in the negative direction the loss will also increase rapidly due to the series resistance and junction capacitance which lead to a residual diode conductance $g_{min} = r_s \omega^2 c_j^2$ as shown in Fig. 9. This effect reduces the conversion conductance which now becomes

$$\frac{g_{c1}}{g_0} = \frac{\sin \dfrac{\pi t}{T}}{\pi\left(\dfrac{t}{T} + \dfrac{g_{min}}{g_{max}}\right)}$$

and, if the reduction amounts to 1 percent, the conversion loss of a typical mixer will have increased from about 3 to 3.3 dB. This increased loss can be avoided if

$$\text{PDR} > 100 \frac{g_{min}}{g_{max}} \cong 500 \left(\frac{f}{f_{c0}}\right)^2. \tag{10}$$

If a diode is to be operated at 10 GHz with an overall noise figure of 3 dB, including a 1.5 dB IF amplifier, a glance at Fig. 6 shows that the pulse duty ratio must be less than 20 percent and hence from (10) the cutoff frequency must be greater than 500 GHz.

Note that large voltage excursions require the largest possible ratio of reverse to forward diode resistance (f_{co}/f). In general at microwave frequencies, this switching ratio is far lower than the low-frequency switching ratio (r_{rev}/r_s) and it is mainly for this reason that the capacitance of mixer diodes has always been kept very low. From this point of view mixer diodes should have very high cutoff frequencies; values in excess of 500 GHz have been obtained in the best GaAs varactors made by J. C. Irvin of Bell Telephone Labs.

The concept of residual diode conductance used above is very approximate since the junction capacitance is, in actuality, a time-varying quantity and the rules for changing from impedance to admittance are much more complicated than the one used here. This section obviously over-

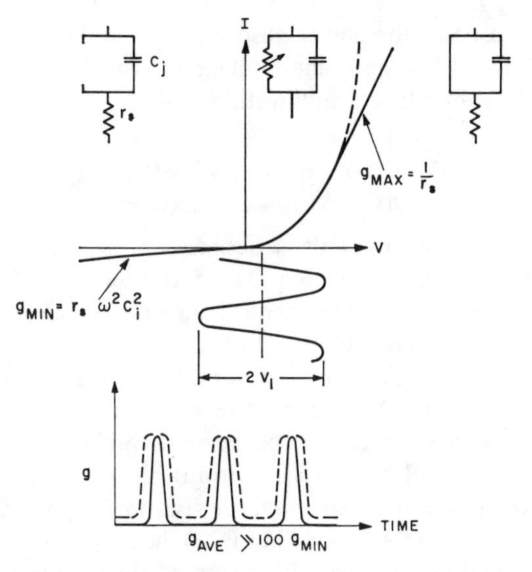

Fig. 9. Effect of diode parasitics on conductance versus time waveform.

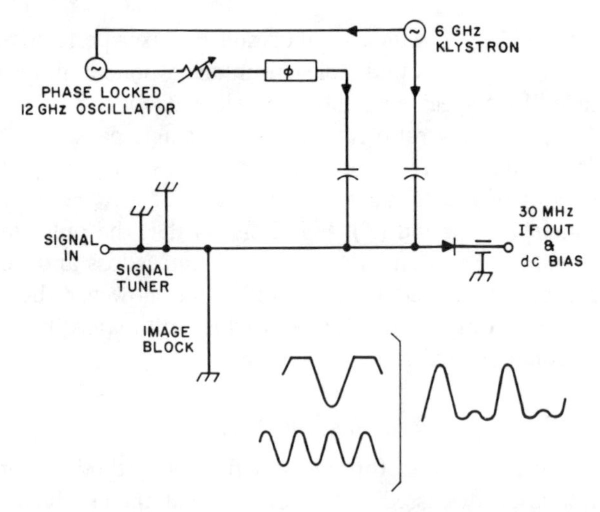

Fig. 10. Mixer circuit with inset illustrating principle of pump waveform improvement. The fundamental voltage will be truncated as shown when the diode resistance falls below that of the pump source.

simplifies the actual situation and has been included simply to illustrate some effects on mixer performance which can be caused by parasitic reactances and changes in pulse duty ratio. The thermal noise from the series resistance could also be important, especially at low-noise figures and whenever the junction resistance becomes comparable to r_s.

V. Experimental Results

A series of measurements at 6 GHz have been carried out using the coaxial circuit shown in Fig. 10. The pump waveform could be modified by adding phase locked second harmonic in an attempt to reduce the conductance pulse duty ratio; the principle is illustrated in the Fig. 10 inset and was described originally by Strum.[8] The added complexity of this scheme was found to be unwarranted however, since the overall noise figure improvement was only 0.2 dB. The image block was positioned so that the reactive susceptance seen by the diode junction at the image frequency was intermediate between a short and an open circuit.

Some typical results without pump waveform improvement are plotted in Fig. 11 using GaAs Schottky barrier diodes having series inductances of the order of 0.5 nH and a variety of junction capacitances. It was found that the overall noise figure could be minimized by successive adjustments of bias, pump power, source impedance, and a π network between the mixer output and the 30 MHz IF amplifier; regardless of the order in which the adjustments were carried out, the same set of final conditions was always obtained.

For each diode it was found that the measured values of minimum noise figure, source impedance, output impedance, and the dc current corresponded to the same pulse duty ratio on each of the theoretical curves of Figs. 6 through 8. For example a diode with 0.46 pF junction capacitance exhibited a minimum noise figure of 5.1 dB and thus from Fig. 6 must have operated with a pulse duty ratio of 30 percent. The measured 35 ohms source impedance, 80 ohms output impedance, and 1.4 mA rectified current are in good agreement with Figs. 7 and 8 and equation (9) assuming an image susceptance intermediate between a short and an open circuit. In addition it was found that noise figures could not be lowered when attempts were made to reduce pulse duty ratios below 23 percent by increasing the reverse bias. This may have been due to the limitation imposed by (10) or to the effects of diode parasitic reactances described in Section VI.

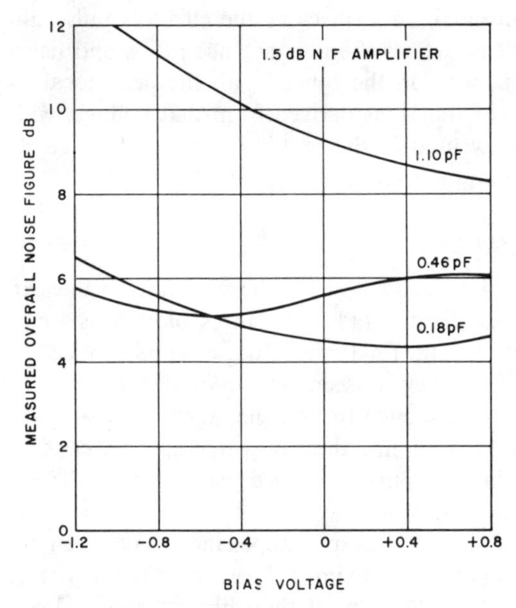

Fig. 11. Measured overall noise figure for conversion from 6 GHz to 30 MHz.

The variations in noise figure behavior with forward bias for different diodes can be seen in Fig. 11. This behavior is attributed to the effects of diode parasitic reactances and is discussed in Section VI.

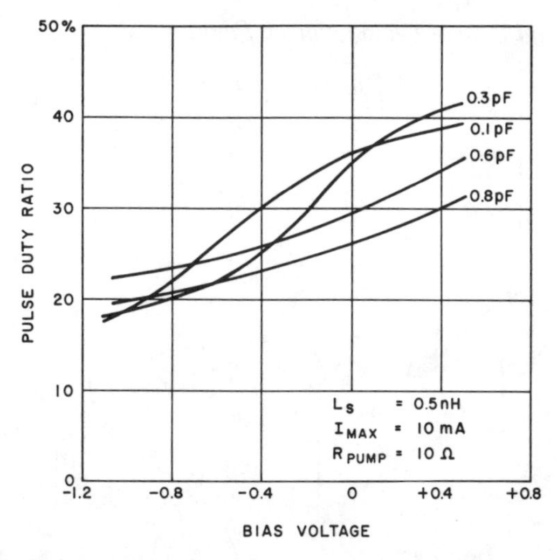

Fig. 12. Pulse duty ratio at 6 GHz measured using a simulated diode at 730 kHz.

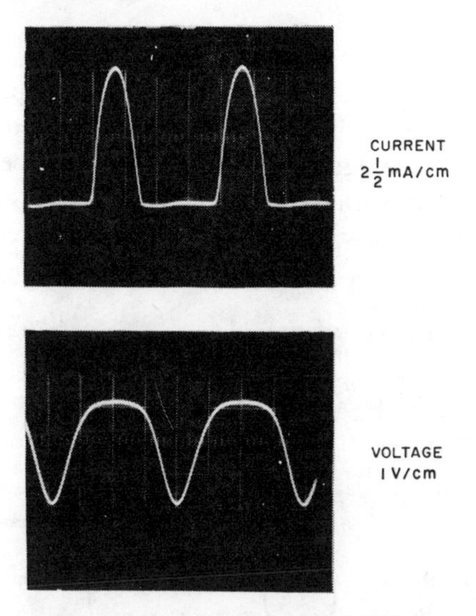

Fig. 13. Current and voltage waveforms at 6 GHz measured using a simulated diode at 730 kHz. $L_S=0.5$ nH, $C_j=0.3$ pF, $R_{\text{pump}}=10$ ohms, $I_{\max}=10$ mA, PDR=35 percent, and $V_0=0$ volts.

VI. Low-Frequency Diode Simulation Studies

The 6 GHz voltage and current waveforms occurring at the Schottky junction can be studied at 730 kHz if the fixed series inductance and variable junction capacitance are scaled up 8200 times. The variation of junction capacitance with voltage obeys an inverse square-root law and can be simulated using one or more 12 voH breakdown Zener diodes, e.g., the Western Electric 426P whose capacitance is ~800 pF at −0.7 volts and becomes infinite at +0.3 volts. Fortunately the Zener diode capacitance increases more rapidly than its forward conductance which can therefore be neglected. The Zener and Schottky diodes must be separately biased with a constant difference (~0.7 volts) to ensure that the Zener capacitance becomes infinite at the appropriate bias on the Schottky diode which is ~+1.0 volts for the epitaxial GaAs units. Fig. 12 shows measured pulse duty ratios versus bias voltage for a diode with 0.5 nH of inductance when the maximum forward current reaches 10 mA. As expected from Fig. 3 the pulse duty ratio generally falls with reverse bias and increased voltage swing but significant differences occur with forward bias; in particular the simulated 0.1 and 0.3 pF diodes behave in a manner similar to the 0.18 and 0.46 pF diodes at 6 GHz shown in Fig. 11 when the limitations due to finite cutoff frequency [equation (10)] are excluded.

It was found that a reduction of inductance to 0.1 nH produced smaller pulse duty ratios similar to those for the 0.8 pF diode, for a wide range of diode capacitance. Tuning out the inductance with a series capacitor provided a negligible one percent reduction in pulse duty ratio.

The pulse duty ratio also varied with the resistance of the pump source and reached a maximum at 25 ohms which was 5 percent higher than for the 10 ohm source results shown in Fig. 12. Lowest pulse duty ratios occurred with low and high values of source resistance, e.g, 10 and 100 ohms; the use of a low resistance value has been advocated previously by Strum.[8]

It should be emphasized at this point that the external embedding impedance for the diode in the low-frequency simulation is probably very different in the microwave experimental setup. It is unlikely, for example, that the changes in pulse duty ratio produced by changes in diode inductance are real. Actually, the inductance becomes part of the external embedding circuit for the junction and can be tuned out at any number of single microwave frequencies, for example, all of the harmonics of the local oscillator.

When the Schottky diode was driven to forward currents in excess of 10 mA the pulse duty ratio increased at a rate of 1 percent per milliamp probably due to the effects of series resistance as discussed in connection with Fig. 9. Fig. 13 shows typical current and voltage waveforms measured in the simulated model. It should be reiterated that the current pulse retains its typical shape even when the voltage waveform becomes highly nonsinusoidal.

VII. Conclusions

It appears that the pulse duty ratio of the Schottky diode current waveform is a more fundamental parameter for defining mixer operation than the magnitude of an erroneously assumed sinusoidal junction voltage.

Future efforts should be directed at lowering the present 23 percent pulse duty ratio limit which would require diodes with cutoff frequencies in excess of 500 GHz. In addition, diodes should have series inductances smaller than 0.5 nH which would make the choice of junction capacitance between 0.2 and 0.6 pF relatively unimportant.

It may also be desirable to use pump sources with resistances appreciably greater or less than 25 ohms and to break away from the common procedure of matching the pump to the diode.

Following these criteria and designing the mixers with open circuit image impedances, it should be possible to obtain X-band SSB noise figures of 3 dB with reasonable output resistances (250 ohms) driving 1.5 dB IF amplifiers.

APPENDIX

The following equations were used to compute Figs. 4 through 8. The origins of the equations were discussed in Section II. In the figures, pulse duty ratio has been used as the independent variable in place of g_{c1}/g_0 and g_{c2}/g_0 in the equations. This new variable can be determined from (8) and Fig. 3.

Image Conductance Infinite

$$L = \left(\frac{g_0}{g_{c1}}\right)^2 \left[1 + \left\{1 - \left(\frac{g_{c1}}{g_0}\right)^2\right\}^{1/2}\right]^2$$

$$F = \frac{1}{2} + \frac{L}{2}$$

$$\left(\frac{g_{\text{source}}}{g_0}\right)^2_{\text{opt}} = \left(\frac{g_{\text{out}}}{g_0}\right)^2 = 1 - \left(\frac{g_{c1}}{g_0}\right)^2$$

Image Conductance Equal to Zero

$$L = \left[1 + \left\{\frac{1 + \dfrac{g_{c2}}{g_0} - 2\left(\dfrac{g_{c1}}{g_0}\right)^2}{\left(1 - \left(\dfrac{g_{c1}}{g_0}\right)^2\right)\left(1 + \dfrac{g_{c2}}{g_0}\right)}\right\}^{1/2}\right]^2 \cdot \frac{\left(1 - \left(\dfrac{g_{c1}}{g_0}\right)^2\right)\left(1 + \dfrac{g_{c2}}{g_0}\right)}{\left(\dfrac{g_{c1}}{g_0}\right)^2\left(1 - \dfrac{g_{c2}}{g_0}\right)}$$

$$F = \frac{1}{2} + \frac{L}{2}$$

$$\left(\frac{g_{\text{source}}}{g_0}\right)^2_{\text{opt}} = \frac{\left(1 - \dfrac{g_{c2}}{g_0}\right)^2}{1 - \left(\dfrac{g_{c1}}{g_0}\right)^2} \cdot \left(1 + \frac{g_{c2}}{g_0}\right)\left(1 - 2\left(\frac{g_{c1}}{g_0}\right)^2 + \frac{g_{c2}}{g_0}\right)$$

$$\left(\frac{g_{\text{out}}}{g_0}\right)^2 = \frac{1 - \left(\dfrac{g_{c1}}{g_0}\right)^2}{1 + \dfrac{g_{c2}}{g_0}} \cdot \left[1 - 2\left(\frac{g_{c1}}{g_0}\right)^2 + \frac{g_{c2}}{g_0}\right]$$

Image and Source Conductances Equal

$$L = \left[1 + \left\{\frac{1 + \dfrac{g_{c2}}{g_0} - 2\left(\dfrac{g_{c1}}{g_0}\right)^2}{1 + \dfrac{g_{c2}}{g_0}}\right\}^{1/2}\right]^2 \cdot \frac{1 + \dfrac{g_{c2}}{g_0}}{\left(\dfrac{g_{c1}}{g_0}\right)^2}$$

$$F = 1 + \frac{L}{2}$$

$$\left(\frac{g_{\text{source}}}{g_0}\right)^2_{\text{opt}} = \left(\frac{g_{c1}}{g_0}\right)^2 \cdot \left(1 + \frac{g_{c2}}{g_0}\right) \cdot \left[\left(\frac{g_0}{g_{c1}}\right)^2 + \frac{g_0 g_{c2}}{g_{c1}^2} - 2\right]$$

$$\left(\frac{g_{\text{out}}}{g_0}\right)^2 = 1 - \frac{2\left(\dfrac{g_{c1}}{g_0}\right)^2}{1 + \dfrac{g_{c2}}{g_0}}$$

ACKNOWLEDGMENT

The author wishes to thank R. M. Ryder, W. J. Bertram, R. E. Fisher, I. Tatsuguchi and V. J. Glinsky for many helpful comments in connection with this work. Thanks are also due to R. D. Tibbetts for assisting with the measurements.

REFERENCES

[1] E. W. Herold, R. R. Bush, and W. R. Ferris, "Conversion loss of diode mixers having image-frequency impedance," *Proc. IRE*, vol. 33, pp. 603–609, September 1945.

[2] H. C. Torrey and C. A. Whitmer, *Crystal Rectifiers*, M.I.T. Radiation Lab. Ser., vol. 15. New York: McGraw-Hill, 1948.

[3] P. D. Strum, "Some aspects of crystal mixer performance," *Proc. IRE*, vol. 41, pp. 875–889, July 1953.

[4] M. J. O. Strutt, "Noise figure reduction in mixer stages," *Proc. IRE*, vol. 34, pp. 942–950, December 1946.

[5] G. C. Messenger and C. T. McCoy, "Theory and operation of crystal diodes as mixers," *Proc. IRE*, vol. 45, pp. 1269–1283, September 1957.

[6] C. S. Kim, "Tunnel-diode converter analysis," *IRE Trans. Electron Devices*, vol. ED-8, pp. 394–405, September 1961.

[7] C. F. Edwards, "Frequency conversion by means of a nonlinear admittance," *Bell Sys. Tech. J.*, vol. 35, pp. 1403–1416, November 1956.

[8] P. D. Strum, "Some aspects of the performance of mixer crystals," Airborne Instruments Lab., Melville, L. I., N. Y., Tech. Memo. 172-TM-3, May 1950.

[9] M. R. Barber and R. M. Ryder, "Ultimate noise figure and conversion loss of the Schottky barrier mixer diode," *Internat'l. Microwave, Symp. Digest*, pp. 13–17, May 1966.

Down-Converters Using Schottky-Barrier Diodes

CHARLES A. LIECHTI, MEMBER, IEEE

Abstract—The analysis of noise in down-converters, in which signal and image ports are equally terminated, is extended to include the effect of a nonlinear junction capacitance. The theory is applied to Schottky-barrier diodes, which are represented by a constant resistance in series with a parallel combination of a nonlinear conductance and nonlinear capacitance pumped in phase. Using numerical optimization procedures, the minimum overall receiver noise figure, the associated conversion loss, and the required optimum external terminations are calculated and compared with experimental results. It is shown that in a resistive mixer with equal termination of signal and image ports, the addition of a nonlinear capacitance can lower the conversion loss substantially; but the overall noise figure remains nearly invariant. A conversion loss as low as 3.5 dB was measured (3.3 dB calculated) for a silicon Schottky-barrier diode operated at 3 GHz, driven by 0.1 mW LO power, and without dc bias. The overall noise figure, including an IF amplifier noise figure of 1.5 dB, however, stays at 5.7 dB (5.5 dB calculated).

I. Introduction

THE question of whether or not a nonlinear capacitance in shunt with a nonlinear conductance would lower the conversion loss in mixers, where both elements are pumped in phase, has received considerable attention [1], [6]–[9]. Edwards [6] investigated the properties of down-converters with conjugate-matched signal and IF ports and short-circuited image port. He finds that for large frequency ratios, the addition of a nonlinear capacitance cannot lower the conversion loss by more than 1 dB. On the other hand, Torrey, Whitmer [1], and Macpherson [7] showed that for idealized down-converters with equally terminated signal and image ports operating conditions exist where the signal source impedance has a positive real part, the IF output resistance approaches infinity, and the conversion gain is arbitrarily large. It will be shown in this paper that equally terminated down-converters using Schottky-barrier diodes can theoretically be expected to have net conversion gain even at high frequency ratios. Under practically realizable operating conditions, it is demonstrated theoretically and experimentally that the effect of the nonlinear capacitance can lower the conversion loss as much as 1 dB.

The generation of shot noise in diodes pumped by a local oscillator was first described by Strutt [10]. Kim [11] has presented a more general analysis which includes the effect of series resistance and constant junction capacitance as well as termination of the image port by a general admittance. However, noise properties of mixers with image termination have been studied only for the simplified case of a constant junction capacitance. In addition, external impedances have not been optimized for minimum receiver noise figure.

In this paper the analysis of noise in down-converters, in which signal and image ports are equally terminated, is extended to include the effect of the nonlinear capacitance. The analysis is applied to Schottky-barrier diodes. The diode chip is represented by an equivalent circuit consisting of a constant resistance in series with a parallel combination of a nonlinear conductance and nonlinear capacitance, pumped in phase. Numerical procedures are used to vary the external signal impedance until the minimum overall receiver noise figure is reached. The resulting noise figure, the associated conversion loss, and the required external port terminations are discussed and compared with experimental data.

Manuscript received April 6, 1970; revised June 12, 1970.
The author is with Hewlett-Packard Associates, Palo Alto, Calif.

Reprinted from *IEEE Trans. Electron Devices,* vol. ED-17, pp. 975–983, Nov. 1970.

II. Numerical Results

A. Introduction

Mixer operation has been studied for an epitaxial silicon Schottky-barrier diode. The equivalent circuit for the diode chip is shown in Fig. 1. For the small-signal junction conductance g_j standard formulas, described in the literature [1], [5], and [12], were used. The voltage dependence of the forward biased junction capacitance $C_j(v)$ is given by [12]

$$C_j = \frac{C_{j0}}{\left(1 - \frac{v}{V_{df}}\right)^{1/2}} \qquad 0 \le v < V_{df}. \tag{1}$$

In the reverse bias region (v: negative), $C_j(v)$ can be approximated over a limited range by[1]

$$C_j = \frac{C_{j0}}{\left(1 - \frac{v}{V_{dr}}\right)^{P}} \qquad V_r \le v \le 0. \tag{2}$$

Here, C_{j0} stands for the capacitance at zero volts. V_{df}, V_{dr}, and p are parameters appropriately chosen to approximate the measured $C_j(v)$ relationship, and V_r defines the limit of the voltage range in which the approximation is valid. The series resistance R_s is treated as a constant. Parameters for the investigated diode are listed in Table I.

Theory of frequency conversion in nonlinear admittance mixers is extensively documented in the literature [6]–[9]. However, expressions for noise figure have only been derived for the simplified case of constant junction capacitance [11]. An extension of the theory for shot noise and thermal noise, which includes the nonlinear properties of the capacitance yields the following single-sideband noise factor F_m of a converter, which is terminated by Y_a at the signal and image port (Appendix I).

$$F_m = 2 + \frac{1}{|Y_a'|^2 \cdot \text{Real}\left(\frac{1}{Y_a}\right)} \Bigg\{ nG_0 + 2R_s|Y_a'|^2 + \frac{R_{\beta\beta}^2}{|Z_{\beta\alpha}|^2}\left[\frac{n}{2}G_0 + \frac{R_s}{(R_s + R_\beta)^2}\right] + 2nG_1R_{\beta\beta}\frac{\text{Real}\,(Z_{\beta\alpha})}{|Z_{\beta\alpha}|^2} + nG_2\,\text{Real}\left(\frac{Z_{\beta\alpha}}{Z_{\beta\alpha}^*}\right)\Bigg\}. \tag{3}$$

[1] For a uniformly doped semiconductor, (1) would be valid over the entire reverse voltage range. However, in an optimized microwave mixer diode with low series resistance, the low impurity density is uniform only in a small layer next to the metal contact. At greater depths the impurity concentration increases rapidly to a very high level, corresponding to low resistivity material. Therefore, (1) does not hold in the reverse direction where the depletion layer extends into regions with increasing impurity density. In (2), the values of V_{dr} and p are chosen for best approximation of the measured $C_j(v)$ curve.

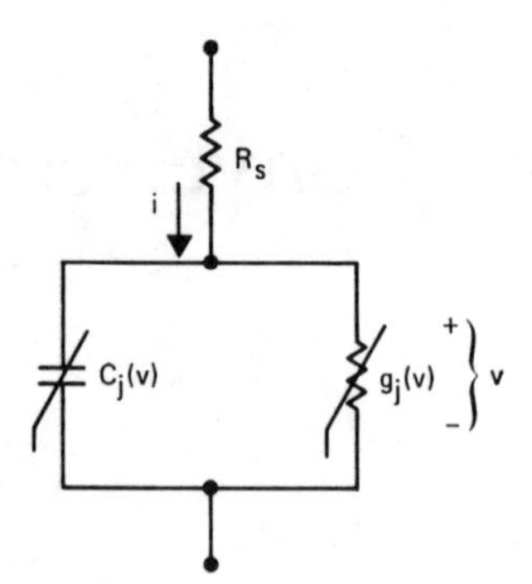

Fig. 1. Equivalent circuit of Schottky-barrier diode chip.

TABLE I
Characteristics at 25°C of the Schottky-Barrier Diode Investigated (HP 5082–2511)

Parameter	Symbol	Measured Conditions	Value	Unit
Diode ideality factor	n		1.04	—
Saturation current	I_s		9.64	nA
Junction capacitance	C_{j0}	0 volts	0.560	pF
"Forward diffusion voltage"	V_{df}		0.510	volts
"Reverse diffusion voltage"	V_{dr}		0.200	volts
Exponent	p		0.233	—
Approximation limit	V_r		−5.0	volts
Series resistance	R_s	10 mA	3.40	ohms

Here, Y_a' is the admittance of Y_a in series with R_s

$$Y_a' = \frac{1}{R_s + \frac{1}{Y_a}}. \tag{4}$$

R_β is the IF output resistance.

$$R_\beta = R_s + \frac{1}{G_{\beta\beta} - 2\frac{G_{\beta\alpha}\,\text{Real}\,[Y_{\alpha\beta}(Y_{\alpha\alpha}^* - Y_{\alpha\gamma}^* + Y_a'^*)]}{|Y_{\alpha\alpha} + Y_a'|^2 - |Y_{\alpha\gamma}|^2}}. \tag{5}$$

The matrix elements $Z_{\beta\alpha}$, $R_{\beta\beta}$, and the determinant D are

$$Z_{\beta\alpha} = -\frac{G_{\beta\alpha}}{D}(Y_{\alpha\alpha}^* - Y_{\alpha\gamma}^* + Y_a'^*) \tag{6}$$

$$R_{\beta\beta} = \frac{1}{D}[|Y_{\alpha\alpha} + Y_a'|^2 - |Y_{\alpha\gamma}|^2] \tag{7}$$

$$D = \left(G_{\beta\beta} + \frac{1}{R_s + R_\beta}\right)(|Y_{\alpha\alpha} + Y_a'|^2 - |Y_{\alpha\gamma}|^2) - 2G_{\beta\alpha}\,\text{Real}\,[Y_{\alpha\beta}(Y_{\alpha\alpha}^* - Y_{\alpha\gamma}^* + Y_a'^*)]. \tag{8}$$

n is the diode ideality factor and the elements of the mixer admittance matrix $Y_{ik} = G_{ik} + jB_{ik}$ are linearly related to the Fourier coefficients G_m and C_m ($m = 0, 1, 2$) of the pumped junction conductance and capacitance [1].

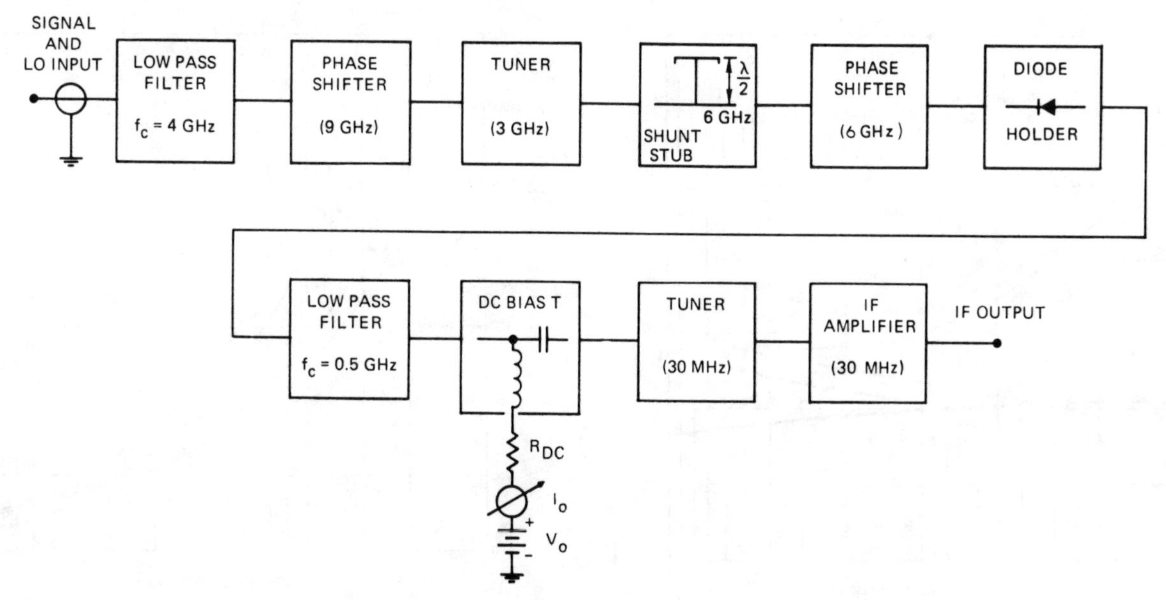

Fig. 2. Block diagram of the down-converter (description in footnote 5).

Mixer operation has been studied under the assumption that a pure sinusoidal 3-GHz local-oscillator voltage is applied to the junction.[2] Diode equivalent circuit parameters and LO drive characteristics were fed to a FORTRAN program which computed the Fourier coefficients of junction conductance g_j and capacitance C_j, optimized external termination of the signal port Y_a [3] for minimum overall receiver noise figure F_r,[4] and printed out the resulting conversion loss L_c, single-sideband noise figure F_r, IF output resistance R_β, optimum signal (image) admittance Y_a, peak LO current I_{peak}, and rectified dc current I_0. The parameters have been computed for the following cases:

1) idealized Schottky-barrier diode (parameters as stated in Table I, except that $R_s=0$ and $C_j=0$);
2) diode with constant junction capacitance (see Table I, except that $C_j=C_0$, $C_1=0$, $C_2=0$);
2') diode with constant junction capacitance (same as 2), except that thermal noise generated in R_s is neglected);
3) real diode with nonlinear capacitance (all parameters listed in Table I apply).

The parameters are presented in Figs. 3 to 7. In addition, values measured on diode type Hewlett-Packard 5082-2511 are shown. The block diagram of the mixer used is illustrated in Fig. 2. Conditions for measurement of diode parameters were:

intermediate frequency $f_{IF}=30$ MHz;

dc load resistance $R_{dc}<10$ ohms;

no dc bias voltage applied; $V_0=0$;

at each LO power level, matching networks at the RF and IF ports were tuned for lowest noise figure;

the second harmonic (6 GHz) and third harmonic (9 GHz) of the LO were short circuited at the chip terminals;[5]

all measurements were performed at room temperature (25°C);

accuracy of measurement:

conversion loss	±0.3 dB
noise figure	±0.7 dB
impedance	±5 percent.

B. Converter Characteristics in Absence of dc Bias

First, converter parameters in the absence of any dc bias voltage will be discussed. Results are plotted versus rectified dc current in Figs. 3 to 7.

Fig. 3 shows conversion loss for the idealized, the constant capacitance, and the nonlinear capacitance diodes. The loss of the idealized diode is only 0.5 dB higher than the theoretical limit (3 dB). The slight decrease in conversion loss, as incident LO power increases,

[2] The reason for this assumption is the following. In practice, distortions of the LO voltage waveform that broaden the current pulse width (e.g., clamping) have to be prevented in order to keep the conversion loss as small as possible [5]. This can be done by short circuiting the harmonic currents generated in the junction. Because true short circuits can be provided only at the chip terminals, the assumption of a purely sinusoidal voltage is only an approximation though acceptable, as long as the minimum value of junction resistance reached during an RF cycle is not much less than the series resistance. In absence of dc bias, this condition is met in the lower half of the LO power range investigated where most of the interest for comparison of the constant and nonlinear capacitance diode is concentrated. In addition, the up-converted small-signal currents closely spaced above and below the LO harmonics are also short circuited as postulated in the theoretical model.

[3] Signal and image-port are equally terminated.

[4] Including an IF amplifier noise figure of 1.5 dB.

[5] In the experiment, this was accomplished in the following way. The susceptances which have to be synthesized across the package terminals to short circuit the chip at 6 GHz and 9 GHz were calculated on basis of the known package parasitics. Then, these susceptances were realized by proper adjustment of line stretchers (Fig. 2) which control the reflection phase of the short-circuited shunt stub (λ/2 long at 6 GHz) and the low-pass filter (cutoff frequency 4 GHz). Package inductance and capacitance are therefore an integral part of the tuning structure at all three frequencies (f_0, $2f_0$, and $3f_0$).

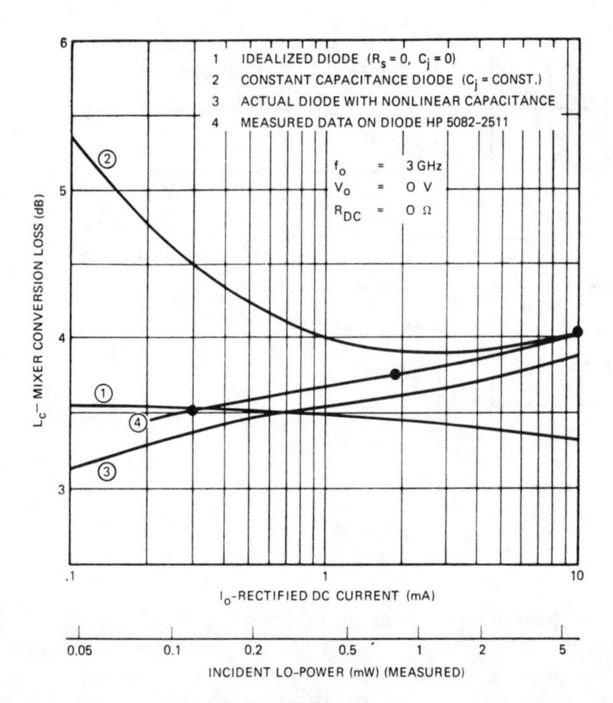

Fig. 3. Conversion loss L_c of mixer stage versus rectified dc current I_0. (At each LO power level, the mixer parameters are optimized for minimum receiver noise figure F_r.)

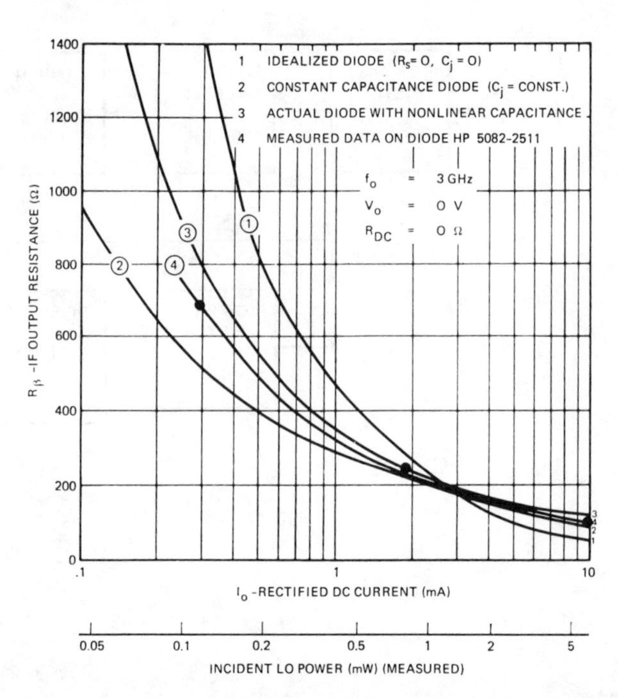

Fig. 5. IF output resistance R_β of mixer versus rectified dc current I_0. (At each LO power level, the mixer parameters are optimized for minimum receiver noise figure F_r.)

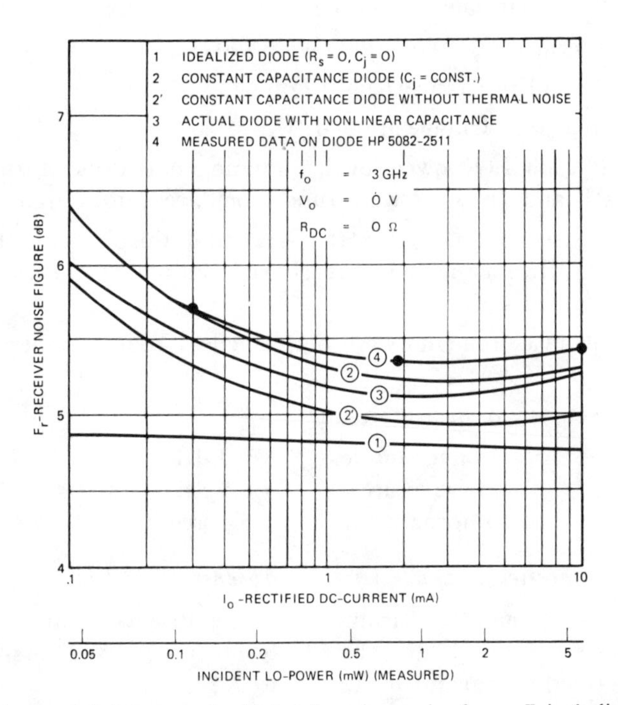

Fig. 4. Minimum single sideband receiver noise figure F_r including an IF amplifier noise figure of 1.5 dB versus rectified dc current I_0.

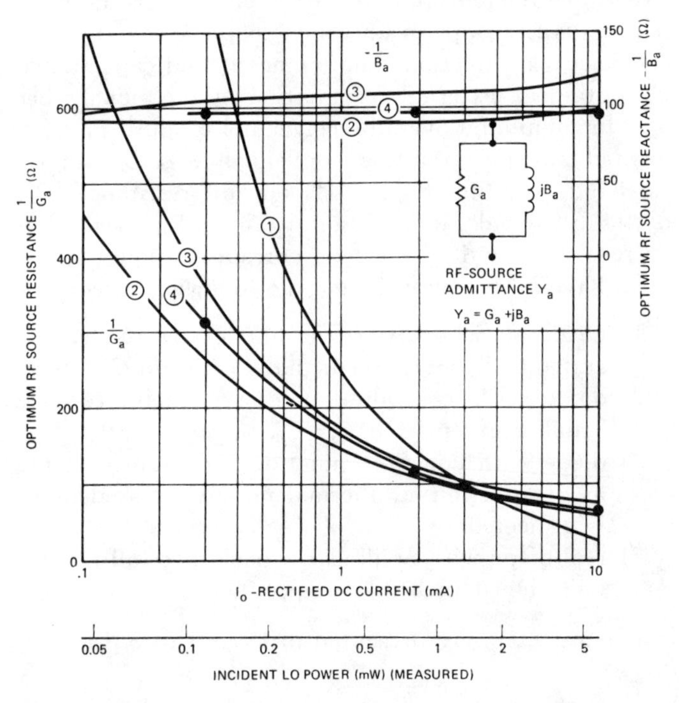

Fig. 6. Optimum signal and image (RF) source-admittance Y_a versus rectified dc current I_0. (At each LO power level, the mixer parameters are optimized for minimum receiver noise figure F_r.)

results from a change in the shape of the diode current waveform. The pulse width measured at half the peak current decreases, resulting in an increase of the conversion coefficient G_1/G_0 and G_2/G_0 and consequently in a decrease of the conversion loss as shown in Fig. 3. Barber [5] has presented mixer characteristics for idealized Schottky-barrier diodes as functions of an appropriately defined current–pulse duty ratio. In his approach, noise figure is computed under the assumption that the noise temperature ratio t_d of the pumped diode has a constant value of 1/2. Under these conditions, noise figure is a linear function of conversion loss and optimization can be achieved by using standard formulas for minimum conversion loss [1]. Recently, it has been proven [14] that for the case of a purely resistive diode, this approach is justified and that

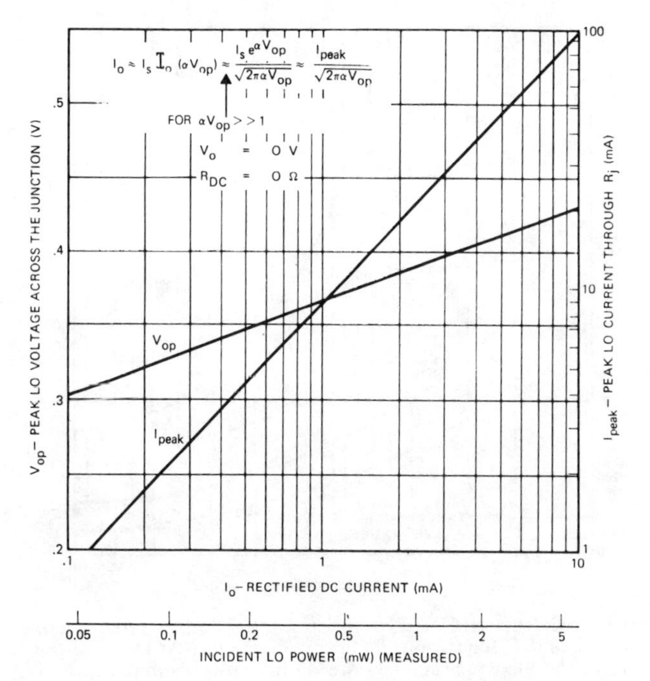

Fig. 7. Peak LO voltage V_{op} (sinusoidal) across the diode junction and peak LO current I_{peak} flowing through the junction resistance versus rectified dc current I_0.

it yields the same result as obtained from detailed analysis of the "amplitude modulated" shot-noise spectrum and its frequency conversion.[6] Consequently, the data presented in curve 1 of Figs. 3 to 6 (idealized diode) are identical to Barber's results. In comparing the data it is noticed that the current–pulse duty ratio changes only from 14 to 17 percent as the LO power is varied over the two decades plotted in the figures (50 μW to 5 mW). However, the peak current rises from 1 to 100 mA (Fig. 7).

In the case of the constant capacitance diode, the increase in conversion loss and noise figure at the low and high ends of the LO power range is due primarily to an increased loss in the R_s–C_j coupling network [2], [4], [12]. At low LO drive levels, the mean value of the junction conductance g_i becomes small and more current flows in the R_s–C_j path. Therefore, less power is coupled to g_j where conversion takes place. At high LO drive levels, g_j becomes comparable to $1/R_s$, and the ratio of signal power dissipated in R_s to the power absorbed by g_j increases. Similarly, IF-power lost in R_s increases with respect to the power coupled from the junction to the load. At still higher power levels, the junction voltage waveform becomes heavily distorted (clamped) owing to the nonlinear behavior of g_j in the voltage-divider consisting of R_s and $1/g_j$. Consequently, the conversion coefficients decrease, raising the conversion loss and noise figure.

As outlined above, the conversion loss of the constant capacitance diode increases rapidly as the LO power is reduced below 0.5 mW. In the LO power range considered, this is in contrast to the behavior of the actual nonlinear capacitance diode. Here, the conversion loss decreases as the LO power is reduced. This is because lower LO power reduces the shunting effect of the junction conductance. Below 0.3 mW, the nonlinear capacitance diode (curve 3) shows a conversion loss lower than the loss of an ideal Schottky-barrier diode with no series resistance or capacitance. In the experiment, this reduction in conversion loss at low LO power could be clearly demonstrated. The measured curve runs parallel to the predicted curve for the nonlinear capacitance diode. At 0.1 mW LO power, a conversion loss of 3.5 dB could be measured. To obtain this result, optimum equal termination of the signal and image ports ("broadband tuning"), the required match at the IF port, and short-circuited chip terminals at the LO, second, and third harmonics must be achieved simultaneously.

Basically, the nonlinear capacitance alone, without junction conductance and series resistance, is capable of yielding arbitrary gain in signal down conversion with positive signal input impedance. This is a direct consequence of the Manley–Rowe equations if power flow is allowed at the LO, IF, signal, and image frequencies [13] (four frequency nonlinear reactance circuit). The Manley–Rowe equations, however, do not state the necessary conditions to reach this goal such as reactive termination of other frequencies or the presence of LO harmonics. Only a detailed analysis of the circuit model reveals the necessary conditions. Torrey and Whitmer [1] have investigated the behavior of the IF output resistance as a function of signal source impedance for nonlinear admittance mixers with high signal frequency to IF ratio, with equally terminated signal and image ports, and with all frequencies above the upper sideband short circuited. They found that basically an infinite output resistance can be realized with a signal source impedance that has a positive real part. This indicates that the converter model could yield arbitrary down conversion gain because it can be shown that the conversion loss vanishes when the IF resistance approaches infinity [7]. For this purpose, the presence of a capacitance term C_2 varying with $2\omega_0$ is not necessary. Fig. 8 illustrates that under reverse bias condition net conversion gain is theoretically possible for a mixer incorporating the type of Schottky-barrier diode used (suitable tradeoff between high cutoff frequency and large dC/dV at the dc operating point).

The noise figure of the idealized diode, which acts as a shot-noise generator and frequency converter of external noise generated in the terminations, is plotted versus rectified dc current in curve 1 of Fig. 4. Curve 2′ shows the noise figure of the constant capacitance diode with "noiseless" series resistance R_s. The addition of R_s and C_0 to the idealized diode has increased the conversion loss more than the loss in coupling the shot-noise power from the junction to the IF termination. Therefore, the noise figure in curve 2′ is higher than in curve 1.

[6] As carried through in the derivation of formula 3 where no assumption for t_d has been made.

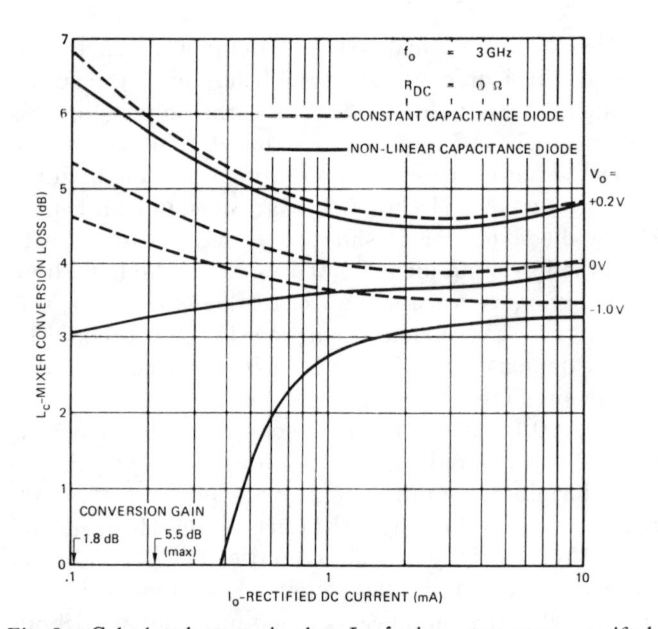

Fig. 8. Calculated conversion loss L_c of mixer stage versus rectified dc current I_0 for various dc bias voltages V_0. (At each LO power level and bias voltage, the mixer parameters are optimized for minimum receiver noise figure F_r.)

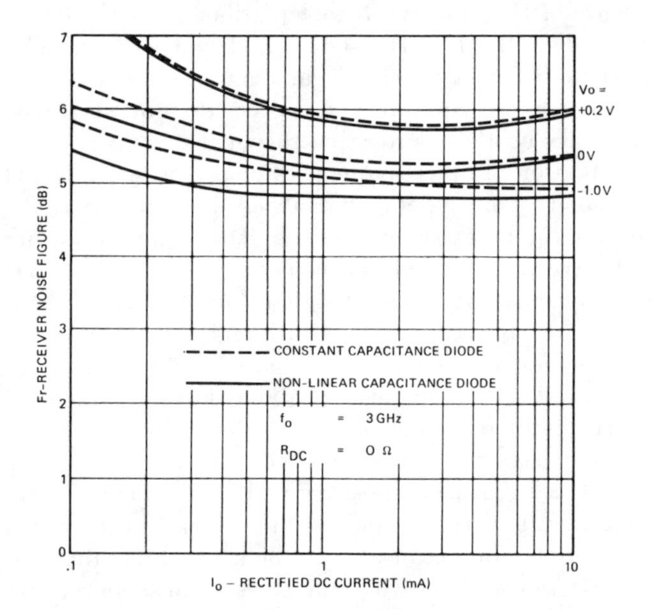

Fig. 9. Calculated minimum single sideband receiver noise figure F_r including an IF amplifier noise figure of 1.5 dB versus rectified dc current I_0 for various dc bias voltages V_0.

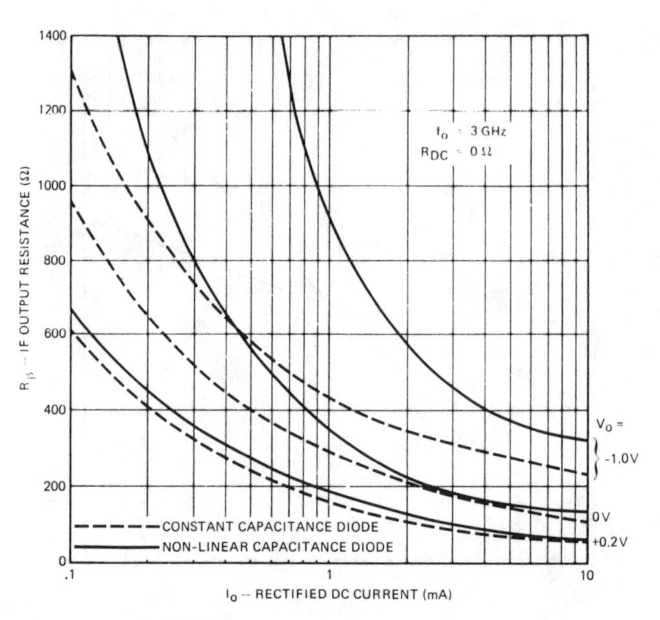

Fig. 10. Calculated IF output resistance R_β of mixer versus rectified dc current I_0 for various dc bias voltages V_0. (At each LO power level the mixer parameters are optimized for minimum receiver noise figure F_r.)

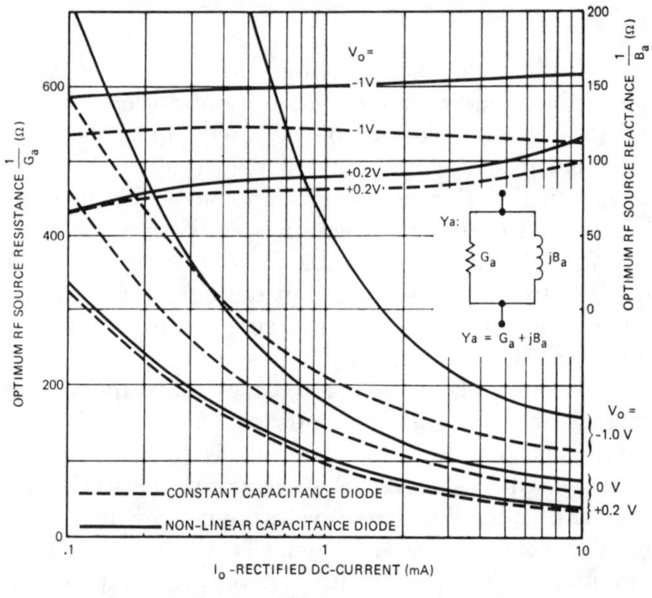

Fig. 11. Calculated optimum signal and image (RF) source-admittance Y_a versus rectified dc current I_0 for various dc bias voltages V_0. (At each LO power level the mixer parameters are optimized for minimum receiver noise figure F_r.)

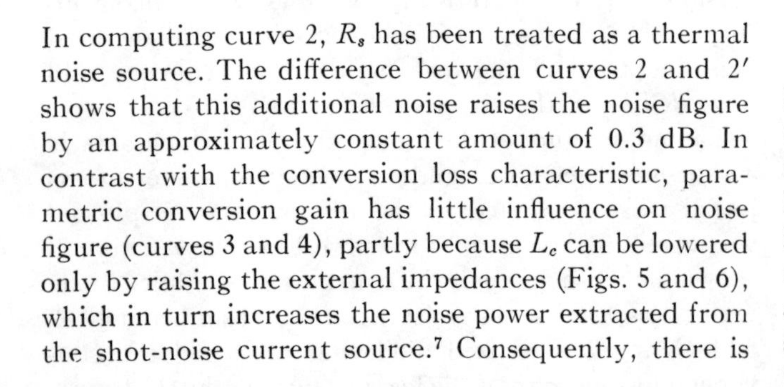

In computing curve 2, R_s has been treated as a thermal noise source. The difference between curves 2 and 2′ shows that this additional noise raises the noise figure by an approximately constant amount of 0.3 dB. In contrast with the conversion loss characteristic, parametric conversion gain has little influence on noise figure (curves 3 and 4), partly because L_c can be lowered only by raising the external impedances (Figs. 5 and 6), which in turn increases the noise power extracted from the shot-noise current source.[7] Consequently, there is little practical interest in low power mixer operation ($P_{LO} < 0.2$ mW) because the noise figure cannot be improved and because the transformation from input line characteristic impedance to the high RF input impedance results in very narrow-band matching. The measured noise figure on the diode (curve 4) is in good agreement with the theoretical prediction. The measured data are expected to be higher than calculated, because theory neglects noise components, generated in the diode above the signal frequency, which beat with higher harmonics of the local oscillator.

[7] This statement is only true for converters in which the signal and image ports are terminated by equal impedances.

C. The Influence of dc Bias on Converter Parameters

Converter parameters are plotted versus rectified dc current for various dc bias voltages in Figs. 8 to 11.

At +0.2 volts forward bias, conversion losses of the constant and nonlinear capacitance diodes approach equality (Fig. 8). As forward bias is removed and reverse bias is applied, the conversion loss of the constant capacitance diode decreases. This is caused primarily by the change in the shape (narrower pulse width) of the LO current waveform and the related increase in the conversion coefficients G_1/G_0 and G_2/G_0 and only secondarily by the reduction in junction capacitance.[8] However, the nonlinear capacitance diode even exhibits net conversion gain under reverse bias. In this operating region, though, the mixer must be matched to extremely high external impedances (Figs. 10 and 11), and the noise figure is only 0.4 dB less than the noise figure of the constant capacitance diode (Fig. 9).

A nonlinear resistance diode without series resistance ("idealized diode"), which is pumped by the LO, can be characterized by the time-invariant noise temperature ratio $t_d = 1/2$ [14]. If the diode operates in a down-converter where signal and image ports are equally terminated, then the IF output noise temperature ratio is always smaller than one; and it approaches unity when the conversion loss reaches the theoretical limit of 3 dB [3]. For a diode with voltage dependent junction capacitance, the assumption for $t_d = 1/2$ does not hold. Consequently, mixer output noise temperature ratios exceeding one occur, which is noticed by comparing conversion loss and noise figure in Figs. 8 and 9.

V. Conclusions

The properties of down-converters using Schottky-barrier diodes, where signal and image ports are equally terminated, were discussed. It was shown theoretically and experimentally that at low LO power levels (typically 0.1 mW), and with no dc bias, the presence of nonlinear capacitance reduces the conversion loss to 3.5 dB, which is 0.1 dB lower than the loss of a converter incorporating an ideal Schottky-barrier diode with no series resistance and no capacitance. With reverse bias theory predicts net conversion gain at certain LO power levels. However, in this operating condition the optimum external impedances become extremely high. Furthermore, the theoretical and experimental results demonstrate that in a resistive mixer with equal termination of signal and image ports the addition of a nonlinear capacitance lowers the conversion loss but does not reduce the noise figure appreciably. It was further shown that in absence of dc bias and in the LO power range considered (0.05 to 5 mW) the minimum noise figure of a receiver using an ideal Schottky-barrier diode with no series resistance and capacitance is approximately constant (4.8 dB)[9] that thermal noise contribution from R_s (3.4 ohms) raises the noise figure by a constant amount (0.3 dB), and that the frequency and LO power dependance of the noise figure for the constant capacitance diode is mainly due to the coupling loss in feeding the signal and IF power through the R_s–C_j two port.

[8] Constant capacitance diode: C_j = mean value of $C_j(t)$, averaged over one LO period.

[9] Including an IF amplifier noise figure of 1.5 dB.

Appendix I

Noise Factor Formula

The equivalent network of the frequency down-converter is shown in Fig. 12. The ideal filters are assumed to short circuit all frequencies but the one marked in each case. Therefore, the small-signal voltage v across the junction conductance g_j and capacitance C_j, pumped by the LO, will consist of components at signal (ω_α), image (ω_γ), and intermediate (ω_β) frequencies only. The basic equation for the small-signal current is

$$i = g_j(t)\cdot v + \frac{d}{dt}[C_j(t)\cdot v]. \tag{9}$$

Using a time-varying parameter analysis which assumes that the LO alone controls the variation of the nonlinear elements, the equations for the three voltage components v_α, v_β, and v_γ, defined in Fig. 12, may be written in matrix form as:

$$\begin{bmatrix} v_\alpha \\ v_\beta \\ v_\gamma^* \end{bmatrix} = \begin{bmatrix} Z_{\alpha\alpha} & Z_{\alpha\beta} & Z_{\alpha\gamma} \\ Z_{\beta\alpha} & R_{\beta\beta} & Z_{\beta\alpha}^* \\ Z_{\alpha\gamma}^* & Z_{\alpha\beta}^* & Z_{\alpha\alpha}^* \end{bmatrix} \cdot \begin{bmatrix} Y_a'\cdot E_a \\ 0 \\ 0 \end{bmatrix} \tag{10}$$

where the matrix elements $Z_{\beta\alpha}$ and $R_{\beta\beta}$ are given by (6) and (7) and the * denotes the complex conjugate operator. Y_a' is the admittance of Y_a in series with R_s (4), and E_a is a voltage source at the signal input. The matrix in (10) fully describes the small-signal conversion properties of the junction. It was derived under the assumption that the intermediate frequency is much smaller than the local oscillator frequency, that the susceptances at the IF can be neglected with respect to the conductances, that signal and image ports are equally terminated, and that the IF termination Z_b is matched to the output resistance R_β (5).

The mean-square shot-noise current $\overline{i_s^2}$ generated in the diode junction is proportional to the equivalent current I_{eq}

$$\overline{i_s^2} = 2qI_{eq}\Delta f \tag{11}$$

where Δf is the IF bandwidth, and I_{eq} is the net forward current flowing across g_j plus two times the magnitude of the saturation current I_s. Let us first assume I_{eq} to be a dc current; then the shot-noise spectrum at a certain time can be represented by an infinite sum of noise currents.

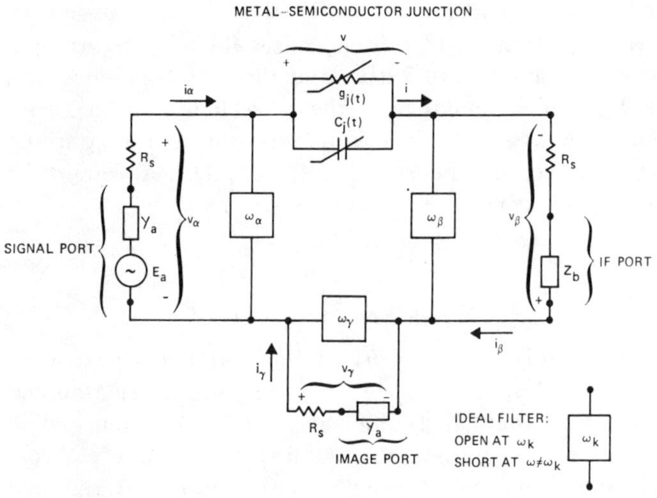

Fig. 12. Equivalent network of the down-converter.

$$i_s(t) = \sum_{l=1}^{\infty} I_{s0} \cos(\omega_l t + \theta_l). \tag{12}$$

In a mixer, however, I_{eq} is determined by the local-oscillator voltage and is therefore a periodic function of time.

$$i_{\text{eq}}(t) = I_{e0} + 2\sum_{k=1}^{\infty} I_{ek} \cos(k\omega_0 t). \tag{13}$$

Due to this variation of I_{eq}, the noise currents become "amplitude-modulated" and consequently, a noise component at angular frequency ω_l splits up into sidebands at $\omega = k\omega_0 \pm \omega_l$, $(k = 0, 1, \cdots)$

$$i_s(t) = \sum_{l=1}^{\infty}\Big[I_{s0}\cos(\omega_l t + \theta_l) + \sum_{k=1}^{\infty} I_{sk}\{\cos[(k\omega_0 - \omega_l)t - \theta_l] + \cos[(k\omega_0 + \omega_l)t + \theta_l]\}\Big]. \tag{14}$$

In deriving an expression for the available shot-noise power at the IF port, the following steps have to be carried out. First, all currents of this amplitude modulated noise spectrum which fall in the signal, image, and IF band (width Δf) have to be summed up. Second, frequency conversion of the currents in the three bands is considered and the resulting noise voltages across the IF terminals are calculated. Some of these noise components are phase-correlated because the original components have split into sidebands with related phase (14) and were then down converted to the IF. Finally, the available noise power is computed. This is done by calculating the mean-square of the sum of the noise voltages while taking into account the phase correlation between components. The detailed formulation of these steps have been published in [11]. Carrying through an analogous derivation, the following mean-square shot-noise voltage $\overline{v_{s\beta}^2}$ across the IF filter (Fig. 12) is obtained:

$$\overline{v_{s\beta}^2(t)} = 4kT\Delta f n[G_0(|Z_{\beta\alpha}|^2 + \tfrac{1}{2}R_{\beta\beta}^2) + 2G_1 R_{\beta\beta}\,\text{Real}\,(Z_{\beta\alpha}) + G_2\,\text{Real}\,(Z_{\beta\alpha}^2)]. \tag{15}$$

Here, k is the Boltzmann constant and T is the absolute junction temperature. From the series resistance R_s the mean-square thermal noise voltage $\overline{v_{t\beta}^2}$ is obtained across the IF filter:

$$\overline{v_{t\beta}^2} = 4kTR_s\Delta f\left[2|Z_{\beta\alpha}Y_a'|^2 + \frac{R_{\beta\beta}^2}{(R_s + R_\beta)^2}\right]. \tag{16}$$

Finally, the thermal noise generated by the signal termination Y_a, down-converted to the IF, yields

$$\overline{v_{T\beta}^2} = 4kT\,\text{Real}\left[\frac{1}{Y_a}\right]\Delta f|Z_{\beta\alpha}Y_a'|^2. \tag{17}$$

An equivalent contribution is obtained from the image termination. The single-sideband noise factor F_m of the converter is given by

$$F_m = 2 + \frac{\overline{v_{s\beta}^2} + \overline{v_{t\beta}^2}}{\overline{v_{T\beta}^2}}. \tag{18}$$

Substituting (15), (16), and (17) into (18) leads to expression (3).

Appendix II

Definition of Symbols

Symbol	Definition
C_j	junction capacitance
C_{j0}	junction capacitance at zero volt
C_0, C_1, C_2	Fourier coefficients of time-varying capacitance $C_j(t)$
D	determinant of mixer admittance matrix, defined in [1]
F_m	single-sideband noise factor of converter
F_r	single-sideband noise factor of receiver (converter+IF amplifier with noise figure of 1.5 dB)
f_0	LO frequency
g_j	small-signal junction conductance
G_0, G_1, G_2	Fourier coefficients of time-varying conductance $g_j(t)$
$G_{\beta\alpha}$, $G_{\beta\beta}$, $\cdots$	elements (real part) of mixer admittance matrix, defined in [1]
I_s	diode saturation current, defined in [12]
I_0	dc current flowing through junction where LO and bias voltage are applied
I_{peak}	peak diode current
L_c	conversion loss of mixer
n	diode ideality factor, defined in [12]
p	exponent in $C_j(v)$ relationship
R_β	IF output resistance
$R_{\beta\beta}$	matrix element in (10)
R_{dc}	load resistance in dc current loop
t_d	diode noise temperature ratio, defined in [14]

V_0	dc bias voltage
V_{df}, V_{dr}, V_r	voltage parameters in $C_j(v)$ approximation
Y_a	admittance terminating signal and image port
Y_a'	admittance of Y_a in series with R_s
$Y_{\alpha\alpha}$, $Y_{\alpha\beta}$, $Y_{\alpha\gamma}$	elements of mixer admittance matrix, defined in [1]
$Z_{\beta\alpha}$	matrix element in (10)

Acknowledgment

The author wishes to thank J. Botka for his technical assistance and his valuable experimental contributions. An acknowledgment is also due to G. Kaposhilin, Dr. M. Cowley, R. Zettler, and J. Lepoff for helpful suggestions and comments.

References

[1] H. C. Torrey and C. A. Whitmer, *Crystal Rectifiers*, Radiation Laboratory Series, vol. 15. New York: McGraw-Hill, 1948.
[2] P. D. Strum, "Some aspects of crystal mixer performance," *Proc. IRE*, vol. 41, pp. 875–889, July 1953.
[3] W. L. Pritchard, "Notes on a crystal mixer performance," *IRE Trans. Microwave Theory Tech.*, vol. MTT-3, pp. 37–39, January 1955.
[4] G. C. Messenger and C. T. McCoy, "Theory and operation of crystal diodes as mixers," *Proc. IRE*, vol. 45, pp. 1269–1283, September 1957.
[5] M. R. Barber, "Noise figure and conversion loss of the Schottky barrier mixer diode," *IEEE Trans. Microwave Theory Tech.*, vol. MTT-15, pp. 629–635, November 1967.
[6] C. F. Edwards, "Frequency conversion by means of a nonlinear admittance," *Bell Syst. Tech. J.*, vol. 35, pp. 1403–1416, November 1956.
[7] A. C. Macpherson, "An analysis of the diode mixer consisting of nonlinear capacitance and conductance and ohmic spreading resistance," *IRE Trans. Microwave Theory Tech.*, vol. MTT-5, pp. 43–52, January 1957.
[8] L. Becker and R. L. Ernst, "Nonlinear admittance mixers," *RCA Rev.*, vol. 25, pp. 662–691, December 1964.
[9] D. P. Howson, "A parametric converter using nonlinear resistance and reactance," *Proc. IEEE* (Correspondence), vol. 53, pp. 1228–1229, September 1965.
[10] M. J. O. Strutt, "Noise-figure reduction in mixer stages," *Proc. IRE*, vol. 34, pp. 942–950, December 1946.
[11] C. S. Kim, "Tunnel-diode converter analysis," *IRE Trans. Electron Devices*, vol. ED-8, pp. 394–405, September 1961.
[12] N. P. Cerniglia, R. C. Tonner, G. Berkovits, and A. H. Solomon, "Beam-lead Schottky-barrier diodes for low-noise integrated microwave mixers," *IEEE Trans. Electron Devices*, vol. ED-15, pp. 674–678, September 1968.
[13] D. K. Adams, "An analysis of four-frequency nonlinear reactance circuits," *IRE Trans. Microwave Theory Tech.*, vol. MTT-8, pp. 274–283, May 1960.
[14] C. Dragone, "Analysis of thermal and shot noise in pumped resistive diodes," *Bell Syst. Tech. J.*, vol. 47, pp. 1883–1902, November 1968.

Effects of the Diode Junction Capacitance on the Conversion Loss of Microwave Mixers

LUCIO MANIA AND GIOVANNI B. STRACCA, MEMBER, IEEE

Abstract—**The theory of the frequency down-converters, which use sinusoidally driven exponential junction diode, is reviewed here with the purpose of investigating the effects of the junction capacitance on the conversion loss. The analysis deals with single-ended mixers, both in the Z and Y configurations. Their performances are compared with those obtained from the same configurations of mixers using a purely resistive (PR) exponential diode, in cases of both short-circuited and open-circuited image frequency. This comparison shows that the junction capacitance modifies strongly the behavior of the single-ended mixer compared to the case of purely resistive diodes. It is pointed out that, when the junction capacitance may not be neglected, and the image frequency is either open- or short-circuited, good performance can still be achieved from a sinusoidally driven Y mixer. On the contrary, the sinusoidally driven Z mixer, which operates better than the Y mixer when the effect of the junction capacitance may be neglected, is strongly deteriorated, and it becomes worse than the Y mixer.**

I. FOREWORD

FREQUENCY converters are still used as the first stage of microwave receivers. The recent improvements of diodes and mixers allow one to obtain today low-noise figures with inexpensive equipment. Therefore the theory of frequency converters continues to receive considerable attention [1]–[7]. The theoretical case of frequency converters with purely resistive ideal diodes has recently been investigated, and the influences of the mixer configuration, imbedding network, diode characteristics, and pump waveforms, as well as various idler frequencies on the converter performance, have been pointed out [5]–[7].

Manuscript received October 19, 1973. This work was supported by the Consiglio Nazionale delle Ricerche of Italy.

The authors are with the Istituto di Elettrotecnica e di Elettronica, University of Trieste, Trieste, Italy.

At microwave frequencies, however, the presence of a junction capacitance may considerably affect the mixer performance. Some investigators [8]–[10] have performed computations concerning this problem considering the diode driven by a sinusoidal pump voltage, and supposing that the RF, IF, and image signal voltages are sinusoidal at the ends of the diode. With these assumptions, the RF, IF, and the image signal voltages across the diode junction may also be considered sinusoidal as a first approximation. This fact allows an approximate and simplified analysis of the behavior of the mixer, but it may not be applied to the case of diodes driven by a sinusoidal pump current. Furthermore, the validity of the approximate results obtained in [8]–[10] is strictly limited to the specific parameters of the diode considered in the computation, and to the value of the pump frequency.

In this paper the effects of the junction capacitance on the conversion loss of single-ended mixers are investigated in a general way, in order to obtain general results whose validity is not connected with a particular kind of diode or with a specific value of the pump frequency.

To this end, the analysis of [5]–[7] has been extended to a more sophisticated diode model, viz., a nonpurely resistive (NPR) diode instead of a purely resistive (PR) one. The adopted model, as discussed in detail in Section II, is different from that of [5]–[7] mainly because it takes into account the presence of a constant capacitance across the nonlinear component of the resistive model (Fig. 1), which is assumed to be that of an ideal junction diode. All the frequencies involved in the conversion process are in the form

$$\omega_{\pm n} = n\omega_p \pm \omega_0, \qquad n = 0,1,2 \cdots \tag{1}$$

Reprinted from *IEEE Trans. Commun.*, vol. COM-22, pp. 1428–1435, Sept. 1974.

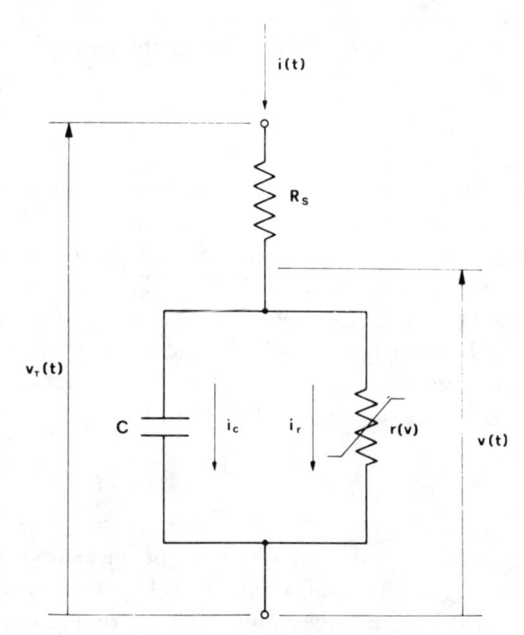

Fig. 1. Model of the NPR diode.

where ω_p is the pump frequency,[1] $\omega_1 = \omega_p + \omega_0$ is the input RF frequency, ω_0 is the output IF frequency, and all the other frequencies are called "idler frequencies." It is assumed that $\omega_0 \ll \omega_p$.

The available conversion loss L_0 (i.e., the reciprocal of the available conversion gain) has been calculated for two basic diode imbedding networks, shown in Fig. 2(a) and (b), where the network consisting of R_s, C, and $r(t)$ [or $g(t)$] represents the linearized equivalent circuit of the pumped and biased diode, as discussed in detail in Section II. Following Saleh's classification of mixers [6], these configurations are called Y mixer [Fig. 2(a)] and Z mixer [Fig. 2(b)], respectively.

Only two cases of idler frequencies terminations are considered here, viz., image frequency (ω_{-1}) port open-circuited or short-circuited. All the other idler frequencies are short-circuited in the Y mixer and open-circuited in the Z mixer. These terminations are provided by means of the ideal passband filters B_1, B_0, B_{-1}, and X_1, X_0, X_{-1}, respectively. A sinusoidal pump voltage is taken across the diode for the Y mixer, and a sinusoidal pump current for the Z mixer.

As pointed out in [5]–[7], in the case of the PR diode model the elements of the impedance (or admittance) matrix which describes the linearized small-signal current-voltage relations existing among the input, output, and image ports of the mixer are easily derived from the coefficients of the Fourier expansion of the periodically time-varying differential resistance $r(t)$ of the diode. In turn, the differential resistance of the diode is related to a sinusoidal pump waveform in a simple way. In the case of the NPR diode, the evaluation of these matrices becomes considerably involved due to the presence of the junction capacitance. In fact, the differential resistance $r(t)$ is no longer related in a simple way to the pump current (or voltage), and the presence of the junction capaci-

[1] The word "frequency" is occasionally used here to mean "angular frequency."

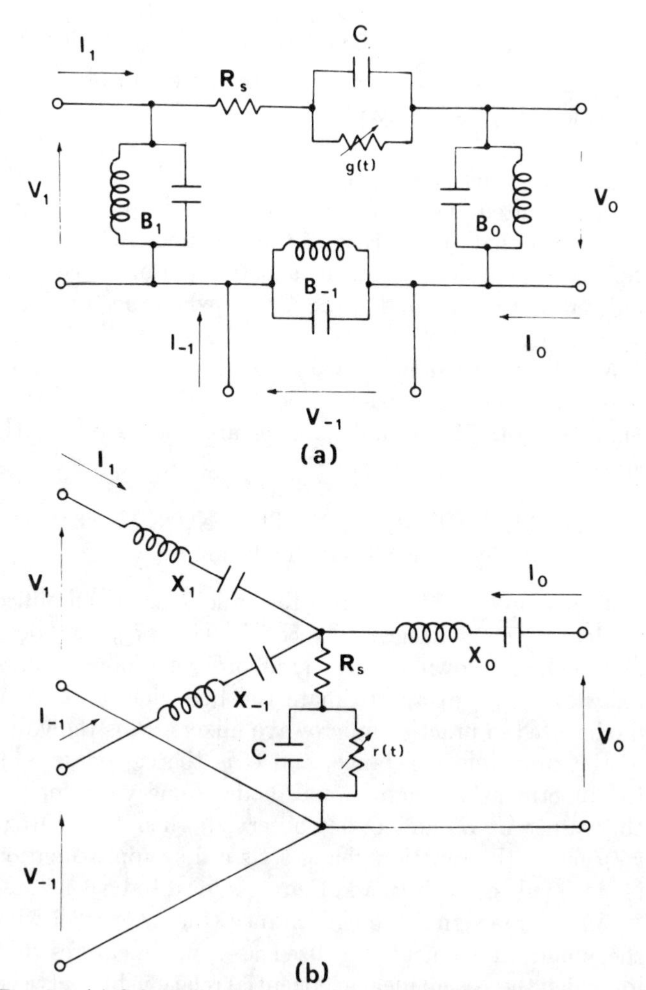

Fig. 2. (a) Schematic representation of the single-ended Y mixer. (b) Schematic representation of the single-ended Z mixer.

tance implies the solution of differential equations [involving $r(t)$]. However, when the matrix elements have been evaluated, the mixer can be again analyzed as a linear two-port network. The conversion losses for the Y and Z mixer have been computed for various values of the parameters on which the conversion process depends. The results of these computations are discussed in Sections III and IV. In Section V these results are compared with the results previously obtained with the PR model. The mathematical methods used in the analysis of the equivalent circuit of the mixers are outlined briefly in the Appendix.

The main results of the analysis can be summarized as follows.

1) The available conversion loss depends on many parameters. Nevertheless, considering that in the reverse region the differential resistance of the real diodes cannot grow without limit, and taking into account the dissipation limits of the practical devices, it is possible to define a minimum available conversion loss which essentially depends only on the quality factor Q of the diode (defined in Section II) both for Y and Z mixers, even if the presence of the signal at the image frequency is taken into account.

2) The presence of the junction capacitance C causes the value of the conversion loss to deteriorate compared to the case of the PR diode. However, this deterioration is

stronger for the Z mixer than for the Y mixer.

3) A significant comparison with the performance of a PR diode is possible by plotting the minimum conversion loss versus Q^{-2}.

4) The difference in the behavior of the mixers with short-circuited and open-circuited image frequency is less important than in the case of the PR diode. The kind of termination which results in a better performance coincides with that of the PR diode only when the Q value is very large.

5) Within the range of values of the parameters which determine the conversion process used in the present analysis, both the Y and Z mixer are always inherently stable.

II. THE MODEL FOR THE NONLINEAR ELEMENT OF THE MIXER

The simple nonlinear resistor, taken as a simplified model of the nonlinear element in [6] [7], has been modified, as shown in Fig. 1, in order to have a more realistic model to approximate the behavior of the real diodes used in practical microwave mixers. R_s is the value of the spreading resistance, and C is the capacitance of the junction. Although in real diodes some variations of the values of R_s and C are observed when the voltage $v(t)$ across the junction changes, as a first approximation in this analysis both R_s and C are assumed to be constant.

For a large signal the capacitance C is in parallel with the nonlinear element, idealized as a nonlinear resistor, for which there is an ideal exponential relationship between the applied voltage v and the current i_r:

$$i_r = I_s[\exp(v/Vc) - 1]. \tag{2}$$

In (2) I_s is the reverse saturation current, $V_c = kT/q$ is the electron-volt equivalent of the absolute temperature T of the junction (q is the electron charge and k is Boltzmann's constant). When the diode is used as the nonlinear element in frequency converters, small signals are superimposed on the large signal of the pump. For these small signals, the nonlinear resistive component of the diode is equivalent to a linear time-varying resistance $r(t)$ [or conductance $g(t) = 1/r(t)$], as shown in Fig. 2(a) and (b). From (2), $r(t)$ is

$$r(t) = r[v(t)] = r_{d0} \exp[-v(t)/V_c], \tag{3}$$

where $r_{d0} = V_c/I_s$ is the value of $r[v(t)]$ when $v = 0$. The current component i_c in the capacitance C is given by

$$i_c = C\frac{dv}{dt}. \tag{4}$$

The conversion loss of the mixer is also a function of the frequency because of the presence of the capacitance C. To normalize the results of the analysis it is useful to define a cut-off frequency ω_c of the diode,

$$\omega_c = \frac{1}{R_sC}, \tag{5}$$

and a quality factor Q of the diode at the pump frequency ω_p,

$$Q = \frac{\omega_c}{\omega_p}. \tag{6}$$

The quality factor Q may also be related to the ratio between two physical quantities, which may be easily measured: when the nonlinear resistor is switched from a value much smaller than both R_s and $1/\omega_pC$ to a value much larger than $1/\omega_pC$, then the real part G of the small-signal diode admittance at the frequency ω_p varies from a very large value $G_{\max} \cong 1/R_s$, to a very small value $G_{\min} \cong \omega_p^2C^2R_s$; the ratio between $G_{\min}$ and $G_{\max}$ is given by

$$\frac{G_{\min}}{G_{\max}} \cong \frac{1}{Q^2}. \tag{7}$$

Therefore, the value of Q^{-2} may be measured at the frequency ω_p as the small-signal VSWR presented by the diode biased with a large value of dc current provided it has been previously matched with a zero-bias dc current. Moreover, let us remark that, since in [5], [7] the ratio $k = g_{\min}/g_{\max}$ between the minimum and the maximum values of $g(t)$ was taken as the quality factor of a PR diode, the choice of Q^{-2} as a reference parameter for the NPR diode quality allows an interesting comparison between the conversion loss of the PR model and that of the NPR model in Fig. 1. In the case of the PR diode, $g_{\min}$ and $g_{\max}$ coincide with $G_{\min}$ and $G_{\max}$, respectively.

The effects of a series inductance have not been taken into account. One can show that they are negligible at the IF frequency ω_0, and may be neglected also at the frequencies ω_p, ω_1, and ω_{-1} if a careful design of the imbedding network of the mixer provides a suitable series circuit for tuning this inductance at ω_p, ω_1, and ω_{-1}. In any case, the inductance value may be reduced to a very small value by a careful choice of the diode package.

III. CONVERSION LOSS OF THE Y MIXER

All the filters of the Y mixer imbedding network of Fig. 2(a) are assumed to be ideal; only sinusoidal voltages v_p, v_1, v_0, and v_{-1} at the pump frequency ω_p, the RF frequency ω_1, the IF frequency ω_0, and the image frequency ω_{-1}, respectively, are allowed to exist across the diode. The pump signal is taken to be sufficiently larger than the RF, IF, and image signals in order to justify representing the pumped and biased diode as a periodic time-dependent equivalent circuit. A linear relationship may therefore be written among the RF, IF, and image sinusoidal components of the current flowing through the diode, and the corresponding components of the voltage across it.

$$\begin{aligned}
v_1 &= V_1 \exp(j\omega_1 t) + V_1^* \exp(-j\omega_1 t)\\
v_0 &= V_0 \exp(j\omega_0 t) + V_0^* \exp(-j\omega_0 t)\\
v_{-1} &= V_{-1} \exp(j\omega_{-1} t) + V_{-1}^* \exp(-j\omega_{-1} t)\\
i_1 &= I_1 \exp(j\omega_1 t) + I_1^* \exp(-j\omega_1 t)\\
i_0 &= I_0 \exp(j\omega_0 t) + I_0^* \exp(-j\omega_0 t)\\
i_{-1} &= I_{-1} \exp(j\omega_{-1} t) + I_{-1}^* \exp(-j\omega_{-1} t).
\end{aligned} \tag{8}$$

The linear relationship among these components may be expressed in terms of a 3×3 admittance matrix which characterizes the diode at the input, output, and image ports.

$$\begin{vmatrix} I_1 \\ I_0 \\ I_{-1}^* \end{vmatrix} = \{Y\} \begin{vmatrix} V_1 \\ V_0 \\ V_{-1}^* \end{vmatrix}. \tag{9}$$

In the Appendix the evaluation of matrix elements Y_{ij} is discussed. In both situations of open-circuited image ($I_{-1} = 0$), and short-circuited image ($V_{-1} = 0$), the matrix (9) can be easily transformed into a 2×2 matrix, which describes the mixer as a linear two-port network and which allows one to evaluate L_0 in a very simple way [11].

As shown in the Appendix, L_0 depends on four parameters:

1) the quality factor Q of the diode;

2) the normalized pump voltage amplitude $V_p' = V_p/V_c$;

3) the normalized dc bias voltage $V_{dc}' = V_{dc}/V_c$; and

4) the ratio R_s/r_{d0}.

Results of computations, performed over a wide range of these four parameters, are summarized in Figs. 3 and 4. These figures are computed for a typical ratio $\omega_0/\omega_1 = 0.01$ between the IF and RF frequencies. In practical situations, in fact, the ratio ω_0/ω_1 is very small, and the results of the computations for $\omega_0/\omega_1 = 0.01$ practically coincide with the limiting case $\omega_0/\omega_1 = 0$, i.e., when a PR diode is assumed at IF frequency.

Fig. 3 gives the behavior of L_0 versus V_p' for $V_{dc}' = 0$, for typical values of Q ranging from 6.25 to 100, and for $R_s/r_{d0} = 10^{-7}$ and 10^{-11}. Solid curves correspond to short-circuited image and dashed curves to open-circuited image. For the sake of comparison L_0 is drawn also for the limiting case $Q = \infty$ (i.e., $C = 0$). One can see the degradation of the mixer performance due to the presence of the junction capacitance. These curves give rise to the following interesting considerations.

1) For the NPR diode, unlike the case for the PR diode, the presence of a voltage at the image frequency across the diode affects the behavior of L_0 only when the diode is strongly pumped.

2) L_0 has a minimum value L_{0m} for each value of Q (including $Q = \infty$) for both situations of short- and open-circuited image frequency. The value of V_{dc}' has a small influence on L_{0m}, which is minimum for $V_{dc}' = 0$ when V_{dc}' is within the range $V_{dc}' = 0$ to $V_{dc}' = V_p'$.[2] Therefore, the values of L_{0m} of Fig. 3 represent the minimum values $L_{0\min}$ which can be achieved by L_0 as a function of Q when R_s/r_{d0} ranges from 10^{-7} to 10^{-11} and $V_{dc}' \geq 0$. They are also representative of the best performances obtainable in practical cases with the diode biased near zero V and ω_{-1} either open- or short-circuited. Also R_s/r_{d0} has a small influence on L_{0m}.

3) For values of Q smaller than approximately 25, the values of $L_{0\min}$ which correspond to the situation of short-circuited image frequency are smaller than those obtained with an open-circuited image frequency for equal values of Q. The contrary is true when the value of Q is larger than about 25. However, the difference is not so remarkable as in the PR diode.

The results given here show that it is possible to represent $L_{0\min}$ as a function only of one parameter Q^{-2}, as shown in Fig. 4, where $L_{0\min}$ is plotted versus Q^{-2} for $R_s/r_{d0} = 10^{-7}$ (curves a and b), and for $R_s/r_{d0} = 10^{-11}$ (curves c and d). Curves a and c correspond to the case of short-circuited image frequency, curves b and d to the case of open-circuited image frequency.

One can see that $L_{0\min}$ decreases very slowly with Q^{-2} when Q increases over values larger than 100. Therefore, it does not seem advantageous to use more expensive diodes having a quality factor better than this limit.

IV. CONVERSION LOSS OF THE Z MIXER

In the case of the Z mixer, the filters of the imbedding network allow only sinusoidal currents to flow through the diode at frequencies ω_p, ω_1, ω_0, and ω_{-1}. The same assumption made in the case of the Y mixer permits the writing of a linear relationship between the voltage sinusoidal components V_1, V_0, and V_{-1} across the diode at the frequencies ω_1, ω_0, and ω_{-1}, and the corresponding current components flowing through it.

$$\begin{vmatrix} V_1 \\ V_0 \\ V_{-1}^* \end{vmatrix} = \{Z\} \begin{vmatrix} I_1 \\ I_0 \\ I_{-1}^* \end{vmatrix}. \tag{10}$$

The evaluation of the matrix elements Z_{ij} is briefly discussed in the Appendix. The matrix (10) may be transformed into a 2×2 matrix and L_0 can be evaluated in the same way as for the Y mixer. Now L_0 depends on three parameters:

1) the quality factor Q;

2) the pump current amplitude I_p;

3) the ratio R_s/r_0, where r_0 is the value assumed by the differential resistance when only the dc bias current is present. This parameter takes into account the bias current.

Results of computations, performed for a wide range of values of the three quoted parameters, are summarized in Figs. 4, 5, and 6. In Fig. 5, L_0 (computed for $R_s/r_0 = 0.5$) is plotted for several values of Q versus the parameter $k = r_{\min}/r_{\max}$, where $r_{\min}$ and $r_{\max}$ are the minimum and the maximum values assumed by $r(t)$, and k is therefore representative of the pumping. For the sake of comparison in Fig. 5, L_0 is also drawn for the limiting case $Q = \infty$ (i.e., $C = 0$) for $R_s/r_0 = 0.5$ (curve 1) and $R_s/r_0 = 0$

[2] In the present analysis we have not considered negative values of V_{dc}'. In fact, when V_{dc}' is negative, the minimum value of L_0 (L_{0m}) is obtained with a pump voltage value which implies values of the differential resistance $r(t)$ that are too large and not achievable in the real diodes. Nevertheless, from a theoretical point of view, it can be seen that L_{0m} decreases further on when V_{dc}' is made negative (see also Footnote 3).

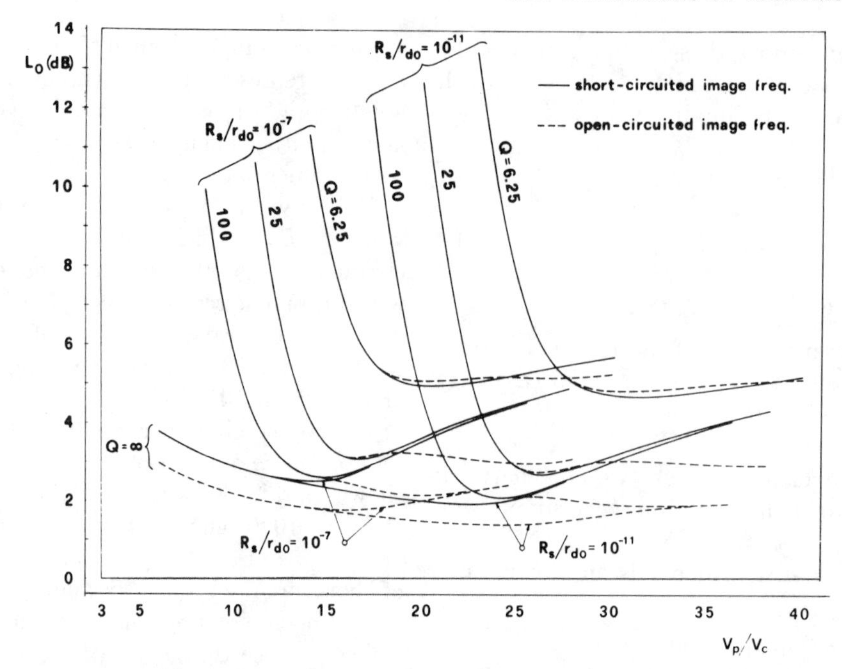

Fig. 3. Available conversion loss L_0 (in decibels) of the Y mixer versus V_p/V_c, for various values of Q ($V_{dc} = 0$, $R_s/r_{d0} = 10^{-7}$, and $R_s/r_{d0} = 10^{-11}$).

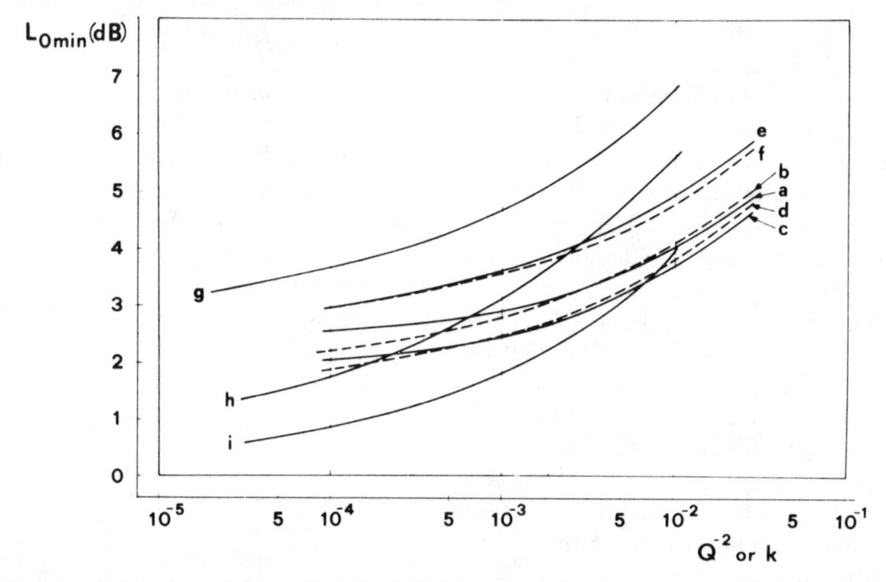

Fig. 4. Minimum available conversion loss L_{0min} (in decibels) versus Q^{-2} or $k = g_{min}/g_{max}$: a) Y mixer, short-circuited image frequency, $R_s/r_{d0} = 10^{-7}$; b) Y mixer, open-circuited image frequency, $R_s/r_{d0} = 10^{-7}$; c) Y mixer, short-circuited image frequency, $R_s/r_{d0} = 10^{-11}$; d) Y mixer, open-circuited image frequency, $R_s/r_{d0} = 10^{-11}$; e) Z mixer, open-circuited image frequency; f) Z mixer, short-circuited image frequency; g) Y mixer using an exponential PR diode with $R_s = 0$, driven sinusoidally (open-circuited image frequency); h) Z mixer using an exponential PR diode with $R_s = 0$, driven sinusoidally (open-circuited image frequency); i) Z and Y mixer using a PR diode driven by an optimum rectangular pump waveform.

(curve 2). All the curves of Fig. 5 refer to the situation of open-circuited image frequency. It can be observed that the presence of the junction capacitance again degrades the mixer performance compared to the case of the PR diode, and that the degradation of the Z mixer is greater than that of the Y mixer.

For the Z mixer, L_0 always decreases with k (i.e., when the pumping becomes stronger), even for $Q = \infty$. The decrease of L_0 with k is, however, very weak when k becomes very small. Furthermore, by using a more realistic model for the diode, the Z mixer also has a minimum for L_0 for very low values[3] of k. So it seems reasonable to take the value of L_0 which corresponds to $k = 10^{-11}$ as

[3] It must be pointed out that the assumption of an exponential law as in (2) for the nonlinear resistive part of the diode is not realistic for voltage values $v < 0$, since it presupposes a differential resistance which grows without limit when the voltage v is made more and more negative, while measurements on real diodes show that this resistance presents a maximum value which is not far from r_{d0}. Therefore, some calculations have also been performed with another diode model which involves the exponential law (2) for $v \geq 0$, and a differential resistance which has a maximum value slightly larger than r_{d0} for $v < 0$. The results of these calculations (not shown here) have demonstrated that L_0 also has a minimum value for the Z mixer.

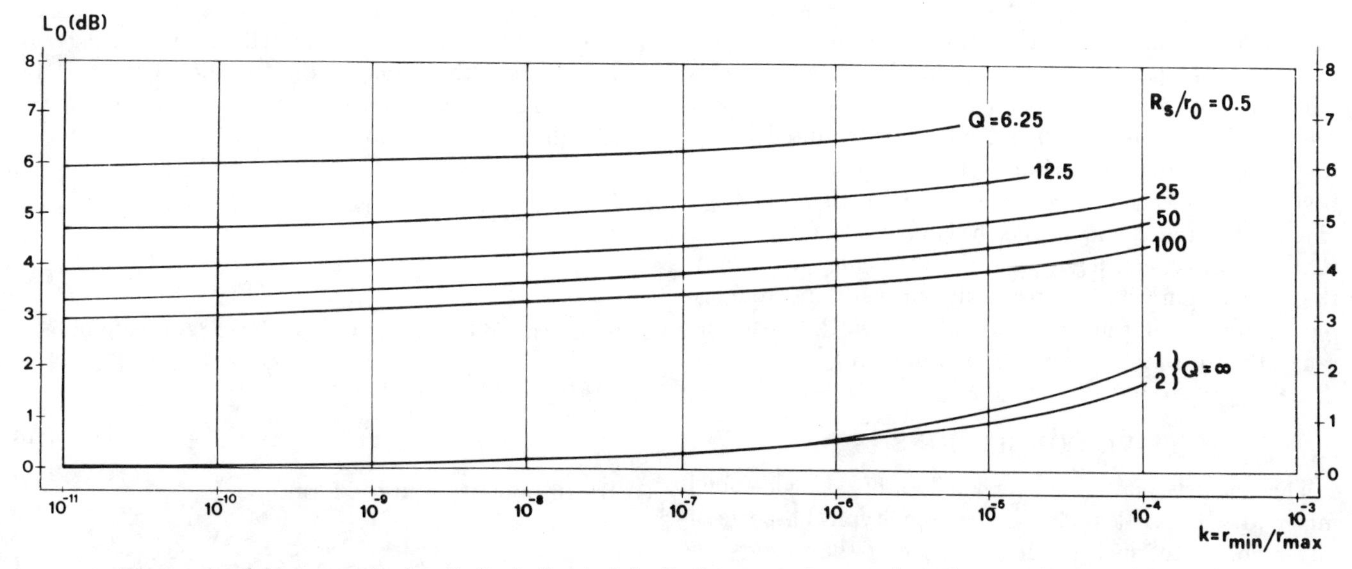

Fig. 5. Available conversion loss L_0 (in decibels) of the Z mixer versus $k = r_{\min}/r_{\max}$, for typical values of Q, and for $R_s/r_0 = 0.5$. Curves 1 and 2 represent the behavior of L_0 versus k, evaluated for $R_s/r_0 = 0.5$ and $R_s/r_0 = 0$, respectively, when $C = 0$.

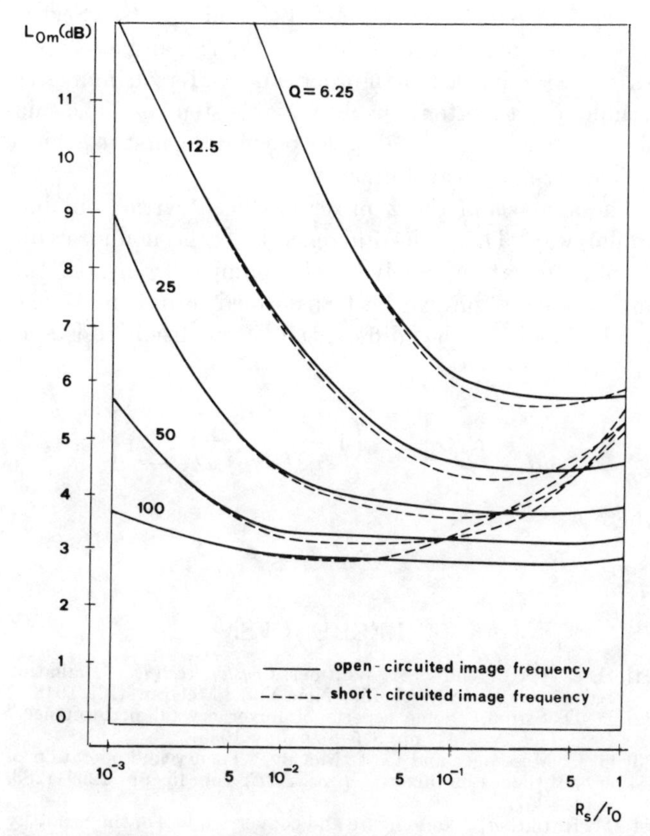

Fig. 6. Minimum available conversion loss L_{0m} (in decibels) of the Z mixer versus R_s/r_0, for typical values of Q.

representative of the minimum value L_{0m} achievable by L_0 in practical cases. Fig. 6 shows the dependence of L_{0m} on the parameter R_s/r_0 (i.e., on the bias current) for the cases of open-circuited (solid curves) and of short-circuited image frequency (dashed curves). In the range of Q values considered here (6.25–100) there exists an optimum value of R_s/r_0 which yields a minimum value $L_{0\min}$ of L_{0m}. In the case of open-circuited image frequency this minimum value is achieved when $R_s/r_0 \cong 0.5$, i.e., for the value used in the computation of Fig. 5. We note that, for high values of Q, L_{0m} becomes essentially independent of R_s/r_0 (for R_s/r_0 within the range 0.001 to 1). In the case of short-circuited image frequency the optimum value of R_s/r_0 depends on the Q value; lower values of optimum R_s/r_0 correspond to higher values of Q, as shown in Fig. 6 (dashed curves).

The curves e and f of Fig. 4 give the behavior of $L_{0\min}$ versus Q^{-2} for the situation of open- and short-circuited image frequency, respectively. A short-circuited image frequency port is slightly more favorable when the value of Q is small. The difference becomes negligible for increasing Q, and the situation is reversed for very large values of Q. In fact, when $Q = \infty$ the open-circuited image frequency case presents smaller losses than the short-circuited case [6], [7].

V. COMPARISON WITH THE PR DIODE

As pointed out in [5] and [7], a significant quality factor for the PR diodes is the ratio k between the minimum and the maximum values assumed by $r(t)$ [or $g(t)$] during the pumping. The relation between the parameters Q^{-2} and k, pointed out in Section II, suggests a possible comparison between an NPR diode, characterized by a given $G_{\min}/G_{\max}$, and an equivalent PR diode, i.e., a PR diode having $g_{\min}/g_{\max} = G_{\min}/G_{\max}$ (or, $k = Q^{-2}$).

Therefore, Fig. 4 also shows the behavior of $L_{0\min}$ (i.e., the minimum achievable conversion loss) versus k for open-circuited image frequency Y mixer (curve g), and open-circuited image frequency Z mixer (curve h), which use an ideal exponential PR diode with $R_s = 0$, driven sinusoidally. These two cases are the most favorable cases with a sinusoidal pump waveform (see [7]).

A comparison between the curves g and h shows that the behavior of the Z mixer with the PR diode is more favorable than that of the Y mixer. However, it should be remarked that the best performance with PR diodes is not obtained with a sinusoidal pump waveform, but when the diodes are driven by an optimum rectangular waveform [5], as shown by curve i. In this case the $L_{0\min}$ values for the Z mixer (open-circuited image frequency) are the same as for the Y mixer (short-circuited image frequency). The comparison between the losses of the best single-ended mixer configuration with an NPR diode (i.e., curves a, b, c, and d) and the curves g, h, and i, shows that the conversion losses of the equivalent PR diode in the Z configuration, sinusoidally driven, are roughly representative of the first ones when $k = Q^{-2}$ is in the range 10^{-4} to 10^{-3}, i.e., for typical values of Q of practical diodes ($Q = 30$–100).

VI. CONCLUSIONS

The analysis on the conversion loss of a single-ended mixer with an NPR diode clarifies some typical aspects of mixer operation, such as the influence of the mixer configuration and of the imbedding network. The introduction of the parameter Q as the quality factor of the diode has allowed a significant comparison between the PR diode and the NPR diode.

It is pointed out that, when the junction capacitance may not be neglected, and the image frequency is taken either open- or short-circuited, good performances can still be achieved from a sinusoidally driven Y mixer. On the contrary, the sinusoidally driven Y mixer, which operates better than the Y mixer when $C = 0$, is strongly deteriorated and becomes worse than the Y mixer when $C \neq 0$. Further investigations are being carried out to assess the influence of a reactive termination at ω_{-1} on L_0.

The analysis has also shown that it does not seem convenient to use diodes having quality factor values larger than 100.

APPENDIX

ANALYSIS OF THE Y AND Z MIXER

In the case of the Y mixer, in order to obtain the admittance matrix (9), it is necessary to evaluate the amplitude of the sinusoidal components at the frequencies ω_1, ω_0, and ω_{-1} of the current flowing into the biased and pumped diode [Fig. 2(a)] when a sinusoidal small-signal voltage is applied across it at the three frequencies of interest. So, the following system of linear differential equations must be resolved.

$$\frac{di_a}{dt} = \frac{1}{CR_s}\left[(\cos \omega_1 t - R_s i_a) g(t) - i_a - \omega_1 C \sin \omega_1 t\right]$$

$$\frac{di_b}{dt} = \frac{1}{CR_s}\left[(\cos \omega_0 t - R_s i_b) g(t) - i_b - \omega_0 C \sin \omega_0 t\right]$$

$$\frac{di_c}{dt} = \frac{1}{CR_s}\left[(\cos \omega_{-1} t - R_s i_c) g(t) - i_c - \omega_{-1} C \sin \omega_{-1} t\right] \tag{A1}$$

where i_a, i_b, and i_c are the nonsinusoidal signal junction currents which are generated by the sinusoidal voltages applied across the diode at the frequencies ω_1, ω_0, and ω_{-1}, respectively.

The conductance $g(t)$ is related to the voltage $v(t)$ existing across the junction due to the pump and bias voltages. The voltage $v(t)$ is obtained by solving the nonlinear differential equation

$$R_s\left(C\frac{dv}{dt} + I_s[\exp(v/V_c) - 1]\right) + v = V_{dc} + V_p \cos \omega_p t. \tag{A2}$$

Defining the following four normalized parameters:

$$\alpha = \omega_p t, \qquad R_{d0}' = R_s/r_{d0}, \qquad V_p' = V_p/V_c, \qquad V_{dc}' = V_{dc}/V_c, \tag{A3}$$

(A2) can be written in the form:

$$\frac{dv'}{d\alpha} = Q[V_{dc}' + V_p' \cos \alpha - R_{d0}'(e^{v'} - 1) - v'] \tag{A4}$$

where $v' = v/V_c$. The solution of (A4) depends, therefore, on the four parameters Q, R_{d0}', V_p', and V_{dc}' (see Section III). The simultaneous solution of (A1) and (A4) was performed using a fourth-order Kutta-Merson numerical technique with automatically variable step size. The solution had to evolve several cycles beyond the first to achieve the steady-state waveforms.

The analysis of the Z mixer has been carried out in a similar way. The main difference is in the nonlinear differential equation involving the pump current. In this case this equation can be transformed into a linear one, which involves the differential normalized resistance $r'(t) = r(t)/r_0$:

$$\frac{dr'}{d\alpha} = R_s'Q\left[1 - \left(1 - \frac{I_p \sin \alpha}{I_s - I_{dc}}\right) r'(\alpha)\right] \tag{A5}$$

where $R_s' = R_s/r_0$.

REFERENCES

[1] H. C. Torrey and C. A. Whitmer, *Crystal Rectifiers* (Radiation Laboratory Series), vol. 15. New York: McGraw-Hill, 1948.

[2] P. D. Strum, "Some aspects of mixer crystal performance," *Proc. IRE*, vol. 41, pp. 875–889, July 1953.

[3] G. C. Messenger and C. T. McCoy, "Theory and operation of crystal diodes as mixers," *Proc. IRE*, vol. 45, pp. 1269–1283, Sept. 1957.

[4] M. R. Barber, "Noise figure and conversion loss of the Schottky barrier mixer diode," *IEEE Trans. Microwave Theory Tech.*, vol. MTT-15, pp. 629–635, Nov. 1967.

[5] G. B. Stracca, "On frequency converters using nonlinear resistors," *Alta Freq.*, vol. 38, pp. 313–331, May 1969.

[6] A. A. M. Saleh, *Theory of Resistive Mixers* (Research Monograph), vol. 64. Cambridge, Mass.: M.I.T. Press, 1971.

[7] G. B. Stracca, "Noise in frequency mixers using nonlinear resistors," *Alta Freq.*, vol. 40, pp. 484–505, June 1971.

[8] L. Becker and R. L. Ernst, "Nonlinear-admittance mixers," *RCA Rev.*, pp. 662–691, Dec. 1964.

[9] C. A. Liechti, "Down-converters using Schottky-barrier diodes," *IEEE Trans. Electron Devices*, vol. ED-17, pp. 975–983, Nov. 1970.

[10] M. Katoh and Y. Akaiwa, "4-GHz integrated-circuit mixer," *IEEE Trans. Microwave Theory Tech.*, vol. MTT-19, pp. 634–637, July 1971.

[11] M. S. Ghausi, *Principles and Design of Linear Active Circuits.* New York: McGraw-Hill, 1965, pp. 62–73.

Conversion Loss Limitations on Schottky-Barrier Mixers

MALCOLM McCOLL

Abstract—**A new set of criteria involving diode area, material parameters, and temperature is introduced for the Schottky-barrier mixer diode that must be considered if its usage is to be extended to the submillimeter wavelength region or cryogenically cooled to reduce the noise contribution of the mixer. It has been well established that, in order to reduce the parasitic loss as the frequency is increased, it is necessary to reduce the area of the diode. What has not been analyzed heretofore is the effect that a reduction in diode area can have on the intrinsic conversion loss L_0 of the diode resulting from its nonlinear resistance. This analysis focuses on the competing requirements of impedance matching the diode to its imbedding circuit and the finite dynamic range of the nonlinear resistance. As a result, L_0 can increase rapidly as the area is reduced. Results are first expressed in terms of dimensionless parameters, and then some respresentative examples are investigated in detail. The following conclusions are drawn: a large Richardson constant extends the usefulness of the diode to smaller diameters, and hence, shorter wavelengths; cooling a thermionic emitting diode can have a very detrimental effect on L_0; impedance mismatching is found, in general, to be a necessity for minimum conversion loss; and large barrier heights are desirable for efficient tunnel emitter converters.**

I. Introduction

The metal–semiconductor contact, or Schottky-barrier diode, has a long history of utilization as a mixer element [1], [2]. Its use has progressed to higher and higher frequencies, with the highest frequency recently being demonstrated by Fetterman *et al.*, who observed mixing at 3 THz with a GaAs Schottky diode [3]. For efficient operation at submillimeter wavelengths, many previously accepted tenets applicable to the design of microwave Schottky diodes must be reexamined.

Mixer conversion loss L_c, defined as the ratio of available power from the RF source to the power absorbed in the IF load, can be expressed in the form

$$L_c = L_0 L_p. \tag{1}$$

The intrinsic conversion loss L_0 is the loss arising from the conversion process within the nonlinear resistance of the diode and includes the impedance mismatch losses at the RF and IF ports. The parasitic loss L_p is the loss associated with the parasitic elements of the diode, the junction capacitance, and spreading resistance. Defined as the ratio of total power absorbed by the impedance R_m of the nonlinear resistance at the signal frequency, L_p is given by [4]

$$L_p = 1 + R_s/R_m + \omega_1^2 C^2 R_m R_s \tag{2}$$

where ω_1 is the signal angular frequency and C is the junction capacitance. The spreading resistance R_s is the resistance resulting from constriction of current flow in the semiconductor near the contact and is in series with the parallel elements C and R_m. Since $C \propto d^2$ and $R_s \propto d^{-1}$, where d is the diameter of the junction, (2) indicates that d should be reduced as the frequency of interest is increased. With the development of electron beam fabrication techniques [5], [6], the ability to produce Schottky barriers with dimensions of the order of a few hundred angstroms is imminent. However, the effect of a reduction in area on the intrinsic conversion loss L_0 must also be evaluated to determine overall mixer performance. This consideration is the central topic of this short paper.[1]

The dependence of L_0 on area originates in the impedance requirements the circuit places on the device. In order for the diode, driven by a local oscillator (LO), to couple most efficiently to a circuit with a specified impedance, it must pass approximately the same current, independent of the junction size. Hence reducing the size of the diode increases the current density through the device and, as a consequence, the dc bias voltage V_0 must be increased. Increasing V_0 limits the useful amplitude of the LO voltage V_1 because the current–voltage (I–V) characteristic of the junction in the forward direction is only nonlinear for applied voltages less than the barrier height potential V_B of the metal–semiconductor interface. Since $V_0 + V_1 \leq V_B$, decreasing the area serves to limit V_1, and consequently may increase L_0.

Because of the inverse relationship between the RF impedance of the diode and the bias current, superior results should be obtained for small areas if the Richardson constant of the semiconductor and the impedance of the circuit are large. The much larger Richardson constant of silicon extends its usefulness to smaller diameters than gallium arsenide. Moreover, it is predicted that the diode should be operated in an impedance mismatched condition; cooling a thermionic emitting diode can have a very detrimental effect on L_0, and large values of barrier height are desirable for efficient tunnel emitter converters.

From the classical conversion loss equations developed in Section II, specific situations are analyzed in Section III. Optimum coupling between the diode and the circuit is first analyzed. This result is applied to both thermionic emitting n-GaAs and n-Si Schottky diodes operating at 290 and 77 K,

Manuscript received March 22, 1976; revised June 28, 1976.
The author is with the Electronics Research Laboratory, The Aerospace Corporation, El Segundo, CA 90245.

[1] For examples of L_p values with Schottky barriers on GaAs, Si, and Ge for wavelengths extending into the submillimeter, the reader is referred to [7].

Reprinted from *IEEE Trans. Microwave Theory Tech.*, vol. MTT-25, pp. 54–59, Jan. 1977.

TABLE I
LIST OF IMPORTANT SYMBOLS

L_0	Intrinsic conversion loss
L_{00}	Intrinsic conversion loss with optimum coupling
L_p	Parasitic loss
R_m	Barrier resistance at ω_1 with local oscillator
C	Junction capacitance
R_s	Spreading resistance
ω, ω_1, ω_2	Local oscillator, RF, and IF angular frequencies
i	Diode current
i_0	Preexponential current term
V	Voltage applied to the diode
V_0	DC bias voltage
V_1	Local oscillator voltage amplitude
g(t)	Diode conductance with applied local oscillator
g_0, g_1, g_2	Fourier series coefficients of g(t)
I_0, I_1, I_2	Modified Bessel functions of the first kind
R_1, R_2	RF source and IF load impedances
R_{10}, R_{20}	RF source and IF load impedances for optimum coupling
y_1, y_2	Impedance mismatch ratios
d	Junction diameter
d_m	Smallest junction diameter for which $y_1 = y_2 = 1$
J_{max}	Current density at flat-band voltage
V_b	Barrier height potential (flat-band voltage)
ε	Permittivity of semiconductor
N	Majority carrier concentration
A^*	Richardson constant
q	Electronic charge
k	Boltzmann's constant
T	Temperature (K)
$\hbar$	$(\frac{1}{2\pi})$ x Planck's constant

and to a tunnel emitting n-GaAs Schottky. The effect of impedance mismatching is then explored.

II. Classical Formulation

The I-V characteristic of a Schottky-barrier diode can be written in the form [8], [9]

$$i = i_0[\exp(SV) - 1], \qquad V < V_B. \tag{3}$$

Material parameters and the temperature of the junction determine the emission mechanism responsible for conduction and, consequently, determine i_0 and S in (3). The effects of these parameters on conversion are examined in subsequent sections, but first the basic mixing equations will be outlined. The symbols used are defined in Table I.

The broad-band Y-connected [10] circuit shown in Fig. 1 and the classical analysis found in Torrey and Whitmer [1] is employed. The crux of the analysis is the treatment of the conductance waveform of the nonlinear resistance. This conductance is a periodic function whose period is determined by

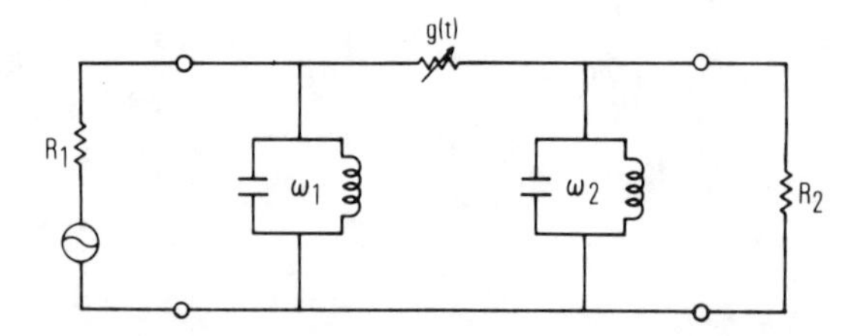

Fig. 1. A mixer in a Y-connected circuit with an RF source impedance R_1 and an IF load impedance R_2. The L-C parallel circuits are tuned to $\omega_{1,2}$ and are assumed to be open circuited at $\omega_{1,2}$ and short circuited at all other frequencies. In the broad-band mixer, the bandwidth of the input L-C circuit has sufficient width to include both the signal and image frequencies.

the large LO voltage. As such, the conductance $g(t)$ can be expanded into a Fourier series

$$g(t) = \tfrac{1}{2}g_0 + g_1 \cos \omega t + g_2 \cos 2\omega t + \cdots \tag{4}$$

where ω is the LO angular frequency.

Conventional small-signal analysis of the circuit then yields

$$L_0 = \frac{1}{2y_1y_2\eta}[y_1 + y_2 + (y_1y_2 + 1)\sqrt{1-\eta}]^2 \tag{5}$$

where

$$y_1 = R_{10}/R_1 \tag{6}$$

$$y_2 = R_{20}/R_2 \tag{7}$$

$$\eta = \frac{2g_1^2}{g_0(g_0 + g_2)} \tag{8}$$

$$R_{10} = \frac{2/g_0}{(1 + g_2/g_0)\sqrt{1-\eta}} \tag{9}$$

$$R_{20} = \frac{2/g_0}{\sqrt{1-\eta}}. \tag{10}$$

Minimization of L_0 with respect to the parameters y_1 and y_2 yields optimum conditions of $y_1 = y_2 = 1$. Therefore, the conversion loss with optimum coupling L_{00} becomes from (5)

$$L_{00} = \frac{2}{\eta}(1 + \sqrt{1-\eta})^2. \tag{11}$$

The significance of this equation is discussed further in Section III.

With a dc bias V_0 and a sinusoidal LO voltage $V_1 \cos \omega t$ applied to the junction, the conductance can be expanded into a series of modified Bessel functions of the first kind $[I_n(x) = j^{-n}J_n(jx)]$ yielding

$$g_0 = 2Si_0I_0(SV_1)\exp(SV_0) \tag{12}$$

$$g_1/g_0 = I_1(SV_1)/I_0(SV_1) \tag{13}$$

$$g_2/g_0 = I_2(SV_1)/I_0(SV_1). \tag{14}$$

With this brief review of mixing theory covered in much greater detail by Torrey and Whitmer [1], the conversion loss of Schottky-barrier diodes will now be explored in terms of their material parameters.

III. Schottky Diodes

A. General Formulation

The intrinsic conversion loss with optimum coupling of a Schottky diode is directly calculable given the area, temperature, and material parameters of the device and the impedance level of its imbedding circuit. The quantity i_0 for both thermionic emitting and tunnel emitting Schottky barriers can be expressed approximately in the following form [8], [9]:

$$i_0 = \tfrac{1}{4}\pi d^2 J_{\max} \exp(-SV_B). \tag{15}$$

With optimum coupling, the RF impedance R_1 can be expressed in terms of V_0, V_1, and the material parameters by setting R_1 equal to R_{10}. From (8), (9), (12)–(15)

$$R_1^{-1} = \tfrac{1}{4}\pi d^2 S J_{\max}(I_0 + I_2)\sqrt{1-\eta}\exp[-S(V_B - V_0)]. \tag{16}$$

Solving this equation for d yields

$$\frac{d}{d_m} = [(I_0 + I_2)\sqrt{1-\eta}]^{-1/2}\exp[\tfrac{1}{2}S(V_B - V_0)] \tag{17}$$

where

$$d_m = (\tfrac{1}{4}\pi R_1 S J_{\max})^{-1/2}. \tag{18}$$

The argument of the modified Bessel functions is SV_1. In (17), d represents the diameter required to yield optimum coupling. Since the expression on the right of (17) is ≥ 1 in the region of interest $V_0 + V_1 \leq V_B$, d_m represents the smallest possible diameter for which the optimum coupling condition can be met. Expressed in terms of operating conditions, d_m is the diameter required to achieve an optimum coupling with the diode biased at flat band with zero LO drive. The constraints on V_0 and V_1 and their relationship to (11) and (17) must now be analyzed.

Constraints on V_0 and V_1 arise from practical considerations as well as from the nature of the emission mechanism. With positive applied voltages, only voltages less than the flat-band value of V_B should be considered since the voltage dependence cannot be neglected as the flat-band ($V \to V_B$) condition is approached. That is, for a Schottky diode, the capacitance C is given by

$$C = \tfrac{1}{4}\pi d^2\left[\frac{q\epsilon N}{2(V_B - V)}\right]^{1/2}, \quad V < V_B \tag{19}$$

where V is the applied voltage. As the flat-band voltage is approached, the capacitance C approaches very high values leading to excessively long RC charging times.[2] Consequently, a mixer diode with the LO voltage equal to or exceeding V_B would be a low-frequency device.

Constraints on the magnitude of a negative applied voltage depend upon the emission mechanism responsible for conduction. With thermionic emission, the reverse current approaches the saturation value i_0 for small negative voltages as indicated in (3). As the reverse voltage is increased further, the diode goes into a breakdown condition. However, assuming the breakdown voltage is large enough to be ignored, the only constraint on V_1 for a thermionic emitter is from (19)

$$V_1 < V_B - V_0. \tag{20}$$

The substitution of the upper bound $V_1 = V_B - V_0$ of (20) into (17) yields a lower bound on the ratio d/d_m as a function of SV_1 for a thermionic emitting junction. This relationship is plotted in Fig. 2.

When the I–V behavior is dominated by tunnel emission, the junction conducts strongly for both positive and negative applied voltages. For efficient conversion, large negative LO

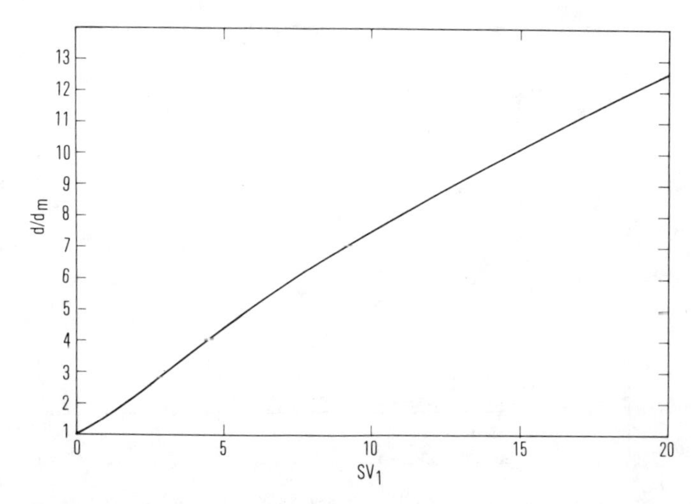

Fig. 2. Normalized diameter versus LO drive for a Schottky-barrier diode mixer under the constraints $V_1 = V_B - V_0$ and $y_1 = y_2 = 1$. This solution is applicable to a thermionic emitter for $V_0 < V_B$ and a tunnel emitter for $\frac{1}{2}V_B \leq V_0 < V_B$.

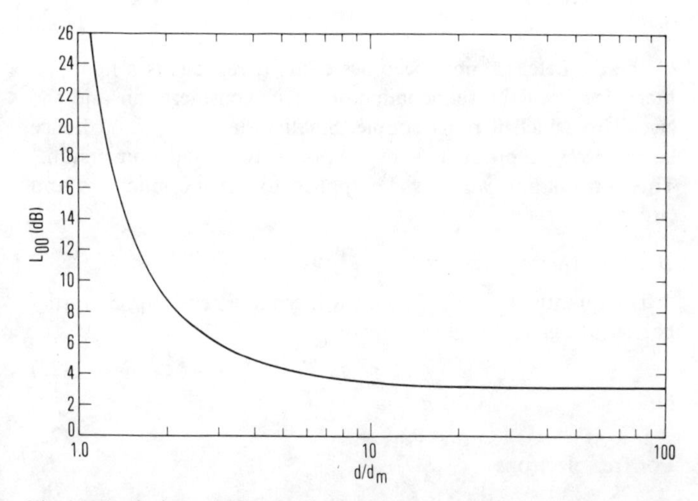

Fig. 3. Intrinsic conversion loss with optimum coupling versus normalized diameter for a Schottky-barrier diode mixer. This solution is applicable to a thermionic emitter for $V_0 < V_B$ and a tunnel emitter for $\frac{1}{2}V_B \leq V_0 < V_B$.

currents cannot be tolerated. Hence, to avoid voltage less than zero and greater than V_B, the following restrictions on V_1 are necessary for tunnel emitters:

$$V_1 < V_B - V_0, \quad \text{for } V_0 \geq \tfrac{1}{2}V_B \tag{21a}$$

$$V_1 < V_0, \quad \text{for } V_0 \leq \tfrac{1}{2}V_B. \tag{21b}$$

Therefore, the graph shown in Fig. 2 is also suitable as a lower limit on d/d_m for a tunnel emitter if the dc bias is constrained to $V_0 \geq \frac{1}{2}V_B$. For $V_0 \leq \frac{1}{2}V_B$, (17) and (21b) can be combined in similar fashion to yield a functional relationship between $(d/d_m)\exp(-\frac{1}{2}SV_B)$ and SV_1 which is applicable to larger diameters. As will be shown for a tunnel emitter, these constraints on V_0 and V_1 also lead to a minimum in L_{00} as a function of d.

From (11)–(14), (17), and (20) (or in graphical terms, combining Fig. 2 with the well-known plot of L_{00} versus SV_1 that results from (11)–(14) [1]), it is possible to establish a basic relationship between L_{00} and d/d_m for the two emission mechanisms. These relationships are plotted in Fig. 3.[3] The significance

[2] If the diode were being driven by a voltage source, the appropriate R would consist of the series and shunt resistances in parallel.

[3] From the standpoint of rigor, it should be remembered that $V_1 < V_B - V_0$, and hence the values of L_{00} in these and subsequent figures are lower limits on L_{00}.

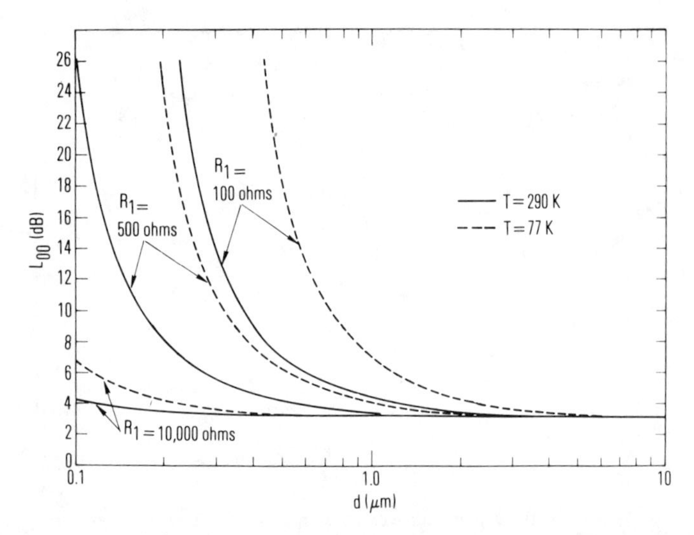

Fig. 4. Intrinsic conversion loss with optimum coupling versus diameter for a thermionic emitting n-type GaAs Schottky-barrier diode at 290 and 77 K with the source impedance as specified and $A^*/A = 0.076$[9]. With $R_1 = 100\ \Omega$ and $T = 290$ K, $d_m = 0.20\ \mu$m.

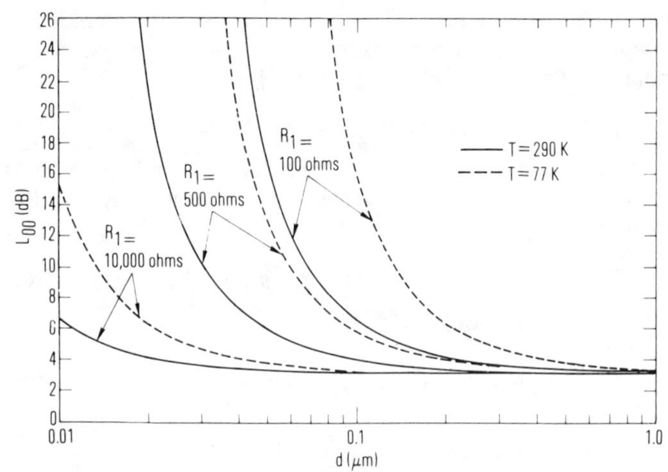

Fig. 5. Intrinsic conversion loss with optimum coupling versus diameter for a thermionic emitting n-type Si Schottky-barrier diode at 290 and 77 K with the source impedances as specified and $A^*/A = 2.2$ [11]. With $R_1 = 100\ \Omega$ and $T = 290$ K, $d_m = 0.038\ \mu$m.

of the diameter d_m now becomes clear. It represents a figure of merit for both the semiconductor under consideration and the circuit to which it must couple. Small values of d_m, and hence large R_1SJ_{max} products, are necessary for good conversion.[4] This formulation will now be applied to specific junctions and circuits.

B. *The Thermionic Emitter*

The quantities S and J_{max} for a thermionic emitting Schottky barrier are given by [9]

$$S = q/kT \tag{22}$$

and

$$J_{max} = A^*T^2. \tag{23}$$

For free electrons

$$A^* \equiv A = 120\ \text{A/cm}^2\text{K}^2 \tag{24}$$

where A is the Richardson constant for thermionic emission into a vacuum. The value A^* is determined by the band structure of the semiconductor. For n-type GaAs the ratio A^*/A is equal simply to the effective mass ratio m^*/m of the majority carriers [11]. Using the parameters stipulated in (22) and (23), (18) becomes

$$d_m = (\pi q R_1 A^* T/4k)^{-1/2}. \tag{25}$$

Equation (25) indicates that not only are large values of R_1 and A^* desirable, but that reducing the temperature of a thermionic emitting mixer diode will increase its conversion loss even though its nonlinear resistive behavior has greatly increased.

Plots of L_{00} versus d for n-type GaAs Schottky diodes are shown in Fig. 4 for room temperature and liquid nitrogen operation for specific values of RF circuit impedances. The critical nature of temperature, impedance level, and submicron dimensions is in evidence. The marked degradation in conversion with cooling is clearly shown. Although the cooled junction has greater nonlinearity, as given by a larger value of q/kT, the T^2 dependence contained in the term J_{max} overbalances this benefit and produces a larger value of d_m. The purpose of cooling this type of Schottky mixer is, of course, to reduce its noise temperature [12]–[14]. However, Fig. 4 suggests that the reduction in noise could be more than offset by the increase in intrinsic loss if impedances are low and dimensions are small.

Fig. 4 also illustrates the importance of high impedance circuits. An increase in impedance reduces the dc bias current and, hence, reduces the dc voltage which permits a larger level of LO excitation. Unfortunately, high impedance circuitry cannot be readily obtained at millimeter wavelengths, and, as such, an implementation of this approach will depend on future developments. On the other hand, low impedance circuitry is available and tempting to the designer in that it reduces L_p via the third term in (2). The difficulty it can inflict on the intrinsic loss is clearly evident in Fig. 4.

The A^* of n-type silicon, being much larger than that of n-GaAs, extends its usefulness to much smaller diameters. Fig. 5 shows the theoretical conversion loss of an n-type silicon thermionic emitter as a function of d. Its characteristic diameter d_m is smaller than that of n-GaAs by over a factor of 5. From this standpoint, n-Si has a better high-frequency capability than n-GaAs. Of course, in terms of the parasitic loss problem, the opposite situation is true with respect to these materials. However, the distinct advantage n-Si has with its greater current carrying ability suggests the merits of the two materials need further investigation. Multiple contacts [12], [15] can effectively reduce the resistivity of the semiconductor, and, with this technique, n-Si could surpass n-GaAs at very short wavelengths.

C. *The Tunnel Emitter*

Tunnel emitters have an advantage over thermionic emitters as high-frequency mixers in that they provide a small spreading resistance without having to resort to complex epitaxial structures [16] to reduce R_s. This advantage is somewhat offset by a larger junction capacitance and larger shot noise, which is not reducible with cooling [17].

Both J_{max} and S are only weakly dependent on voltage and for this analysis can be regarded as constants. Using [9, eq. (1)], an evaluation of L_{00} versus d for an n-type GaAs Schottky diode with an electron concentration of 5×10^{18} cm^{-3} is shown in Fig. 6. Since the conduction is a tunneling process, the results

[4] This statement is rigorously true for the conditions stipulated in Fig. 3. The statement breaks down for a tunnel emitter with $V_0 < \frac{1}{2}V_B$ whereupon large values of d_m are desirable. For example, an increase in d_m brought on by a decrease in R_1 would force an increase in V_0 to achieve an impedance match. From (21b) an increase in V_0 permits a corresponding increase in V_1 which would result in an improvement in conversion efficiency.

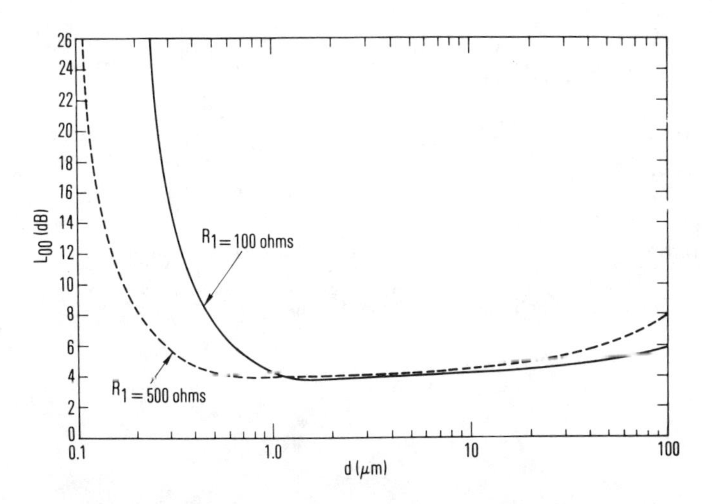

Fig. 6. Intrinsic conversion loss with optimum coupling versus diameter for a tunnel emitting n-type GaAs Schottky-barrier diode with $N = 5 \times 10^{18}$ cm^{-3}, $S = 18$ V^{-1}, $V_B = 0.90$ V, and the source impedances specified.

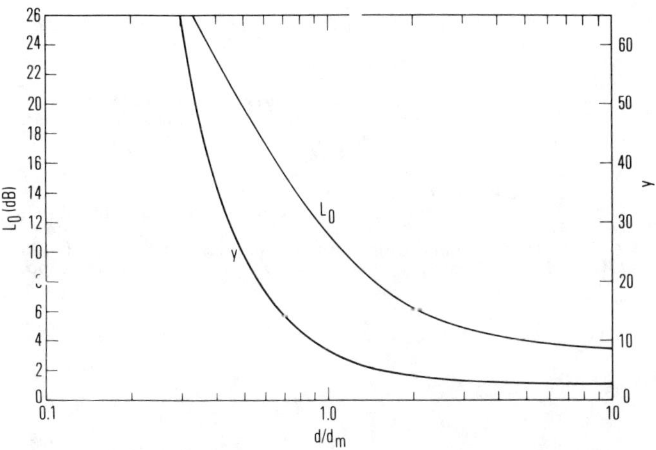

Fig. 7. Intrinsic conversion loss and impedance mismatch ratio versus normalized diameter for a Schottky-barrier diode mixer with $y \equiv y_1 = y_2 \neq 1$. This solution is applicable to a thermionic emitter for $V_0 < V_B$ and a tunnel emitter for $\frac{1}{2}V_B \leq V_0 < V_B$.

are independent of temperature to the degree that V_B and the band gap are temperature independent.

One interesting feature of Fig. 6 is the clear evidence of optimum diameters for conversion. These minima in L_{00}, occurring for the condition $V_1 = V_0 = \frac{1}{2}V_B$, can be reduced further by using larger values of V_B. Hence, as a general feature, large values of V_B are desirable for efficient tunnel emitter converters.

D. *Impedance Mismatching*

Equation (11) yields the minimum conversion loss for a mixer with given values of V_0 and V_1 having treated the circuit impedances as variable quantities. However, it is far more typical experimentally for V_0 and V_1 to be variables and the circuit impedances to be the constrained quantities. The situation can be explained more clearly in terms of the preceding analysis. Consider a junction with a very small diameter such that a dc bias almost equal to V_B is required to yield the optimum matching condition $y_1 = y_2 = 1$. The quantity SV_1 is so small and the L_{00} versus SV_1 curve so steep under these conditions that it would be far better to relax the $y_1 = y_2 = 1$ restriction by reducing the bias V_0 in order to allow an increase in V_1. As V_0 is reduced still further, the conversion loss would continue to decrease until the increase in the coupling loss due to the impedance mismatch begins to outweigh the benefit of larger V_1. By adjusting V_0 and V_1 in this manner, a minimum in L_0 would be established. A computer solution with $y_1 = y_2$ for this minimum in conversion loss and the resulting mismatch is shown in Fig. 7 as a function of diameter normalized to d_m. Notice that this mode of operation not only achieves smaller conversion losses for a given d but utilizes the region $d < d_m$. It does, however, raise the diode impedance considerably for small d.

At very short wavelengths this solution can lose some of its significance because of the parasitic loss problem. The high diode impedances that result from this mode of operation increase L_p via the third term in (2). Hence at very high frequencies the decrease in L_0 brought on by this solution can be more than offset by the increase in L_p. In general, at all frequencies a mismatch is advantageous but possibly not to the full extent illustrated in Fig. 7. The exact mismatch necessary to achieve a minimum in L_c would depend on the frequency, diode, and circuit in question.

IV. Conclusion

The size, material parameters, and temperature of a Schottky-barrier mixer diode place a lower fundamental limit on its conversion loss in a given circuit. The limit is a fundamental one in that the origin of the limit arises from the mixing process itself and takes on major significance in the design of millimeter and submillimeter wavelength heterodyne receivers. At these frequencies, in order to minimize the effects of the parasitic junction capacitance and series spreading resistance, the junctions must take on micron and submicron dimensions. The analysis presented shows that such dimensions can severely limit the performance of the device irrespective of the loss associated with the parasitic elements. In general, the intrinsic conversion loss is determined by the ratio d/d_m and the impedance matching conditions, as illustrated in Figs. 3 and 7. For thermionic emission, d_m is determined by the Richardson constant of the semiconductor, the temperature of the junction, and the impedance level of the circuit. The larger the product of these three quantities, the more efficient will be the converter.

Note Added in Proof: The Schottky diode has recently been extended to wavelengths of 70 μm as a mixer and 42 μm as a video detector.

Acknowledgment

The author wishes to thank Dr. M. F. Millea and Dr. J. R. Tucker for helpful suggestions and comments.

References

[1] H. C. Torrey and C. A. Whitmer, *Crystal Rectifiers* (M.I.T. Radiation Lab. Ser., vol. 15), New York: McGraw-Hill, 1948.
[2] F. L. Warner, in *Millimetre and Submillimetre Waves*, edited by F. A. Benson (Iliffe Books Ltd., London, 1969), pp. 457–476.
[3] H. R. Fetterman, B. J. Clifton, P. E. Tannenwald, and C. D. Parker, "Submillimeter detection and mixing using Schottky diodes," *Appl. Phys. Lett.*, vol. 24, pp. 70–72, Jan. 15, 1974.
[4] G. C. Messenger and C. T. McCoy, "Theory and operation of crystal diodes as mixers," *Proc. IRE*, vol. 45, pp. 1269–1283, Sept. 1957.
[5] M. McColl, W. A. Garber, and M. F. Millea, "Electron beam fabrication of submicrometer diameter mixer diodes for millimeter and submillimeter wavelenghts," *Proc. IEEE* (Corresp.), vol. 60, pp. 1446–1447, Nov. 1972.
[6] E. D. Wolf, F. S. Ozdemir, W. E. Perkins, and P. J. Coane, "Response of the positive electron resist, Elvacite 2041, to kilovolt electron beam exposure." presented at the Eleventh Symp. Electron, Ion, and Laser Beam Technology, Boulder, CO, May 1971.
[7] T. H. Oxley and J. G. Summers, "Metal-gallium arsenide diodes as mixers," in *1966 Proc. Int. Symp. on GaAs*, I.P.P.S. Conf. Series No. 3, pp. 138–150.
[8] F. A. Padovani and R. Stratton, "Field and thermionic-field emission in Schottky barriers," *Solid State-Electron*, vol. 9, pp. 695–707, July 1966.

[9] M. F. Millea, M. McColl, and C. A. Mead, "Schottky barriers on GaAs," *Phys. Rev.*, vol. 177, pp. 1164–1172, Jan. 1969.
[10] A. A. M. Saleh, "Theory of resistive mixers," Ph.D. dissertation, M.I.T., Cambridge, MA, 1970.
[11] S. M. Sze, *Physics of Semiconductor Devices*, New York: Wiley, 1969.
[12] M. McColl, R. J. Pedersen, M. F. Bottjer, M. F. Millea, A. H. Silver, and F. L. Vernon, Jr., "The super-Schottky diode microwave mixer," *Appl. Phys. Lett*, vol. 28, pp. 159–162, Feb. 1, 1976.
[13] S. Weinreb and A. R. Kerr, "Cryogenic cooling of mixers for millimeter and centimeter wavelengths," *IEEE J. Solid-State Circuits* (*Special Issue on Microwave Integrated Circuits*), vol. SC-8, pp. 58–63, Feb. 1973.
[14] A. R. Kerr, "Low-noise temperature and cryogenic mixers for 80–120 GHz," *IEEE Trans. Microwave Theory Tech.*, vol. MTT-23, pp. 781–787, Oct. 1975.
[15] H. M. Day, A. C. Macpherson, and E. F. Bradshaw, "Multiple contact Schottky barrier microwave diode," *Proc. IEEE* (Corresp.), vol. 54, pp. 1955–1956, Dec. 1966.
[16] M. McColl and M. F. Millea, "Advantages of Mott barrier mixer diodes," *Proc. IEEE* (Corresp.), vol. 61, pp. 499–500, Apr. 1973.
[17] T. J. Viola, Jr., and R. J. Mattauch, "Unified theory of high-frequency noise in Schottky barriers," *J. Appl. Phys.*, vol. 44, pp. 2805–2808, June 1973.
[18] D. T. Hodges and M. McColl, "Extension of the Schottky barrier detector to 70 μm (4.3 THz) using submicron-dimensional contacts," *Appl. Phys. Lett.*, vol. 30, pp. 5–7, Jan. 1, 1977.
[19] M. McColl, D. T. Hodges, and W. A. Garber, "Submillimeter-wave detection with submicron size Schottky barrier diodes," presented at the 2nd Int. Conf. and Winter School on Submillimeter Waves and Their Applications, San Juan, Puerto Rico, Dec. 6–11, 1976.

Influence of mixer diode reactances on noise figure and conversion loss

F. Günes, D.P. Howson and K.J. Glover

Indexing terms *Microwave devices, Schottky-barrier diodes*

Abstract: An analysis of Y and Z microwave mixers is presented, incorporating a Schottky-diode model with nonlinear resistance and reactance. The analysis suggests that Y mixers have satisfactory conversion loss and noise figure up to higher frequencies than Z mixers using the same type of diodes. A comparison of four mixer configurations is made.

1 Introduction

A recent paper[1] has summarised the basic principles of microwave mixer analysis and presented a theory based on a Schottky-diode model including both linear and nonlinear reactances. A numerical nonlinear analysis is used to establish the Fourier coefficients of the diode conductance and capacitance for particular local oscillator conditions, followed by a small-signal analysis using these coefficients in an admittance matrix for the diode, to calculate conversion loss and noise figure.

This form of analysis was also developed, independently, at Bradford[2] and this paper shows the results of our work when applied to the problem of deciding which configuration of mixer circuit is most affected by the inevitable parasitic diode reactances.

Our model for the local-oscillator circuit and diode has been given previously[2] (Fig. 1). Note that the local-oscillator source impedance is assumed to be a pure resistance. The mixer circuits examined are, in the notation of Saleh,[3]

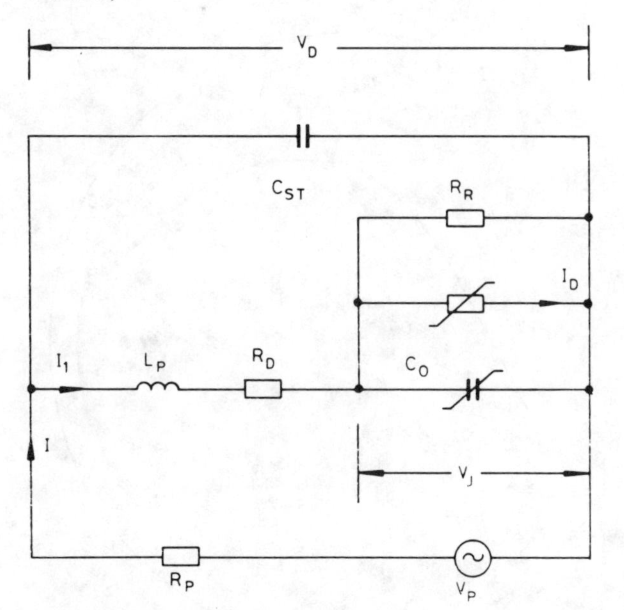

Fig. 1 *Diode model*

Paper T287 M, first received 26th June and in revised form 23rd October 1978

The authors are with the Postgraduate School of Electrical & Electronic Engineering, University of Bradford, Bradford BD7 1DP, West Yorkshire, England

(*a*) the Y mixer with short-circuited image termination (Y1)

(*b*) the Y mixer with open-circuited image termination (Y2)

(*c*) the Z mixer with open-circuited image termination (Z1)

(*d*) the Z mixer with short-circuited image termination (Z2)

The diode parameters chosen were (HP2817) $C_0 = 0{\cdot}8\,\text{pF}$, $C_{ST} = 0{\cdot}17\,\text{pF}$, $L_p = 2{\cdot}3\,\text{nH}$, $R_D = 9\Omega$, $I_{so} = 7{\cdot}5 \times 10^{-10}\,\text{A}$, $q/nKT = 38{\cdot}0228/\text{V}$, $V_{diff} = 0{\cdot}9\,\text{V}$, $\gamma = 0{\cdot}5$, and the junction capacitance law is $C_0/\{1-(V_J/V_{diff})\}^{\gamma}$. The local oscillator was considered to have 50Ω output impedance, 10 mW available power, and the d.c. bias voltages corresponding to the minimum loss were assumed to bias the diode in each configuration. The i.f. frequency was chosen as 70 MHz but this could have been varied over a wide range without upsetting the results, and the i.f. amplifier was considered to have a 1·2 dB noise figure.

The large-signal nonlinear analysis and the small-signal linear analysis were included in a compact computer program[5] in which small-signal voltage- and current-conversion coefficients are calculated automatically in the frequency domain, the idler frequencies can be terminated with any impedance including reactances, and any diode capacitance law can be assumed.

The thermal noise generated in the spreading resistance, the reverse limit resistance and the shot noise of the nonlinear junction resistance were taken into account[6] in the noise-figure analysis which is compatible with the formulation of the small-signal mixer analysis.

The assumed loading conditions are conjugate matching at signal and i.f. ports, separately adjusted at each frequency, the variations of which with respect to r.f. being given by Figs. 2*b*, 3*b*, 4*b* and 5*b*. It should be noted that the input and output termination requirements at the optimum biasing point are easier to realise for the Z mixer than the Y mixer. This is because the shunt configuration gives input and output impedances approaching 50Ω, whilst the series configuration gives terminations remote from 50Ω. with nonzero reactive terms. The graphs shown in Figs. 2*a*, 3*a*, 4*a* and 5*a* give calculated system performance including the i.f. amplifier.

The results show clearly that for a given diode, satisfactory operation up to a higher frequency is possible with a Y mixer as compared with a Z mixer. Furthermore, the model suggests that the simpler Y1 mixer gives as good a

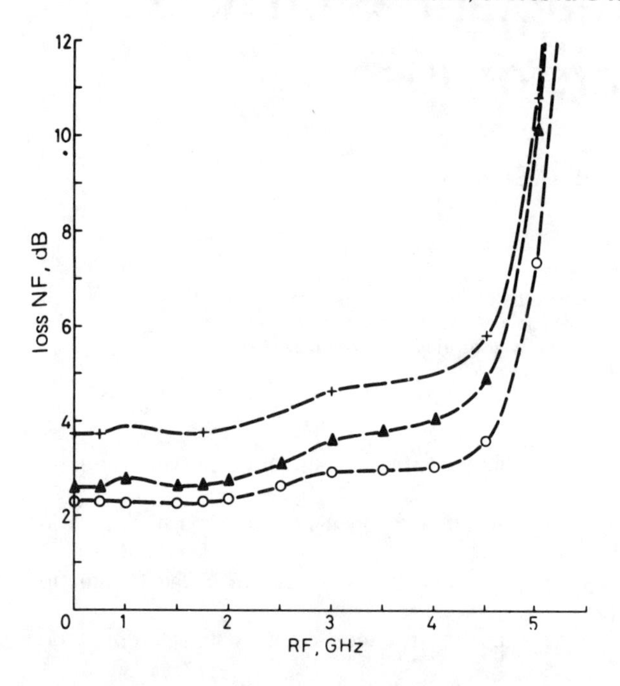

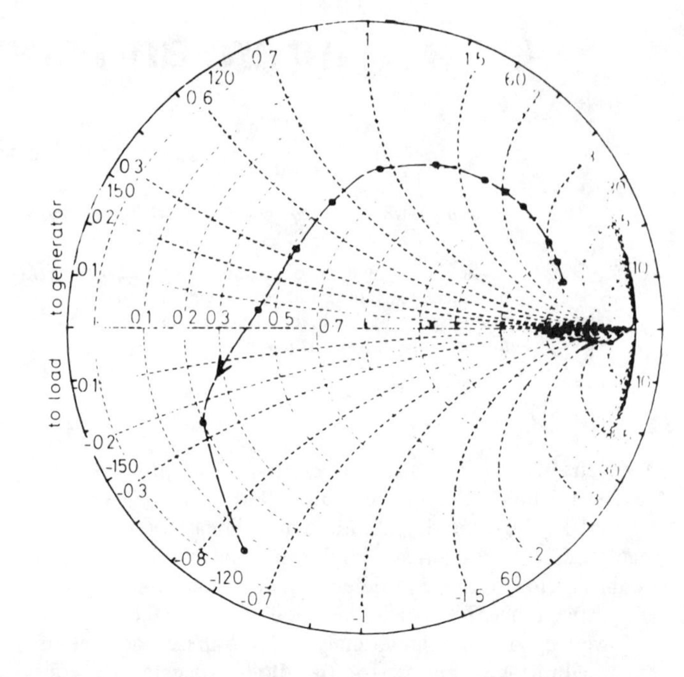

Fig. 2 *Y1 mixer*

a Loss and noise figure
Bias voltage = − 1·40 V
—⊖— mixer conversion loss
—▲— mixer noise figure
—+— signal noise fiture

b R.F. and i.f. terminations
Direct voltage/Pump amplitude = − 0·70
—●— norm ZS
—▲— norm Z1
Intermediate frequency = 0·07 GHz

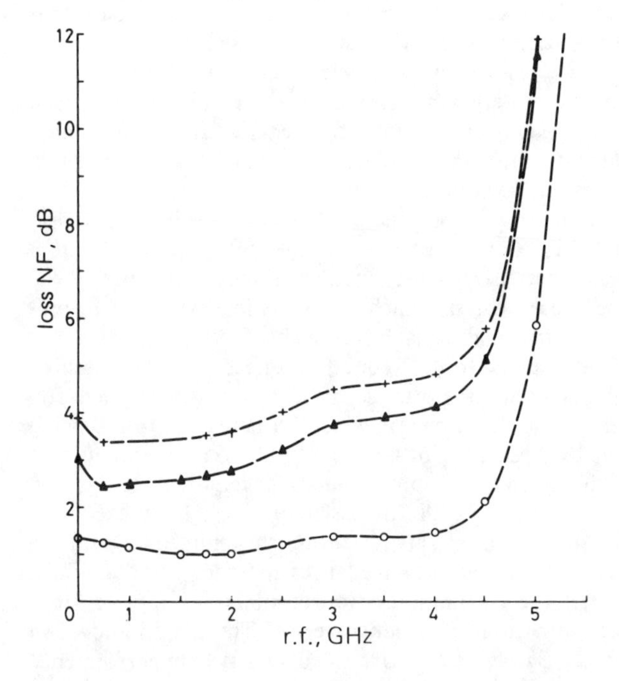

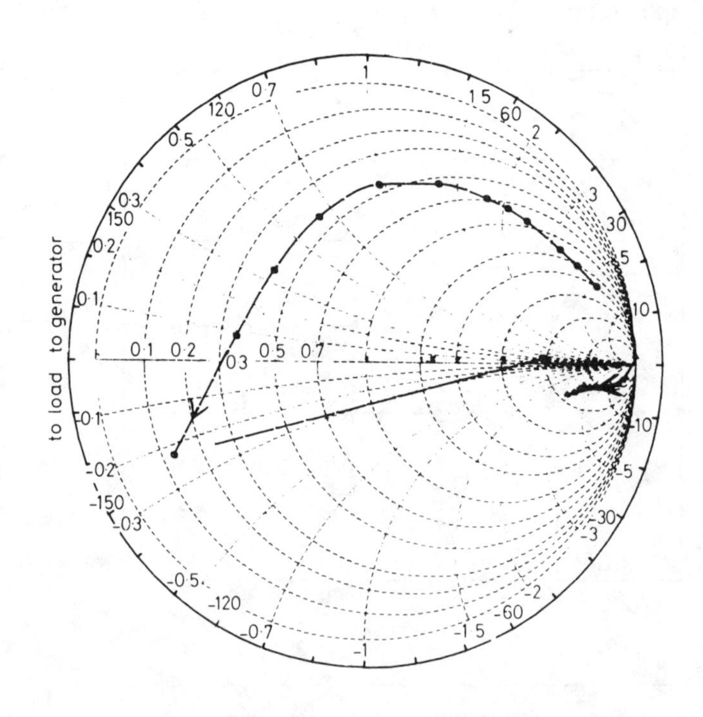

Fig. 3 *Y2 mixer*

a Loss and noise figure
Bias voltage = − 1·40 V
—⊖— mixer conversion loss
—▲— mixer noise figure
—+— system noise figure

b R.F. and i.f. terminations
Direct voltage/pump amplitude = − 0·70
—●— norm ZS
—▲— norm ZI
Intermediate frequency = 0·07 GHz

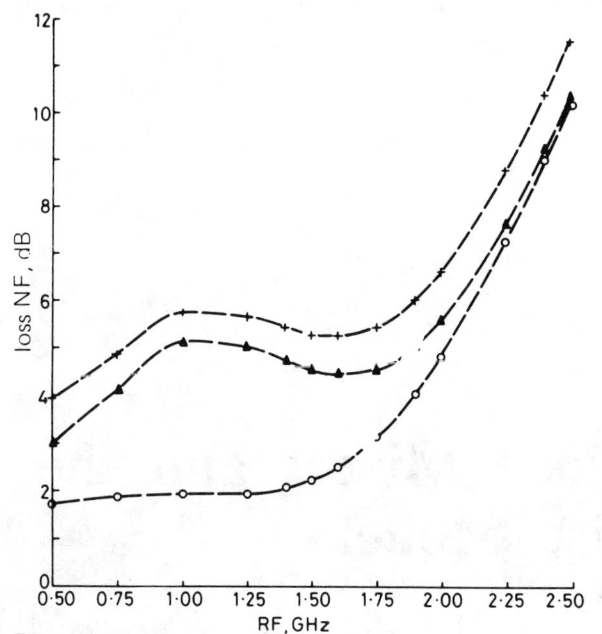

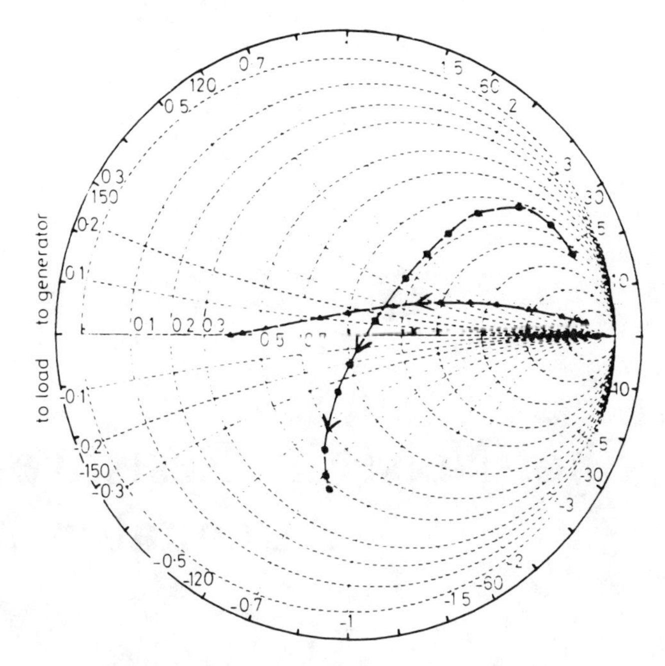

Fig. 4 *Z1 mixer*

a Loss and noise figure
Bias voltage = + 2·1 V
— ⊖ — mixer conversion loss
— ▲ — mixer noise figure
— + — system noise figure

b R.F. and i.f. terminations
Direct voltage/pump amplitude = + 1·05
—●— norm ZS
—▲— norm ZI
Intermediate frequency = 0·07 GHz

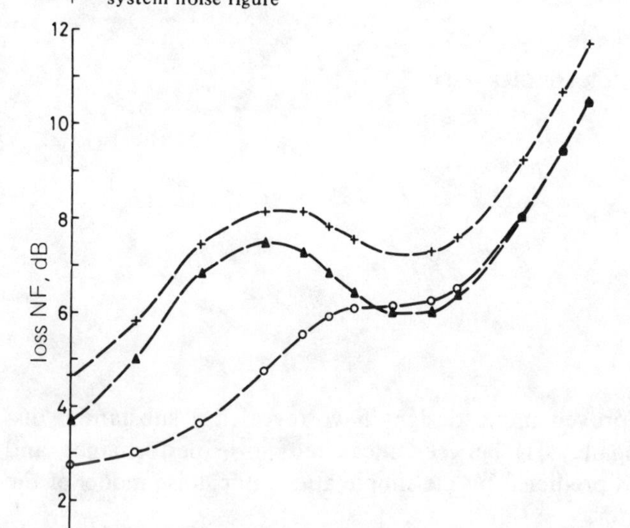

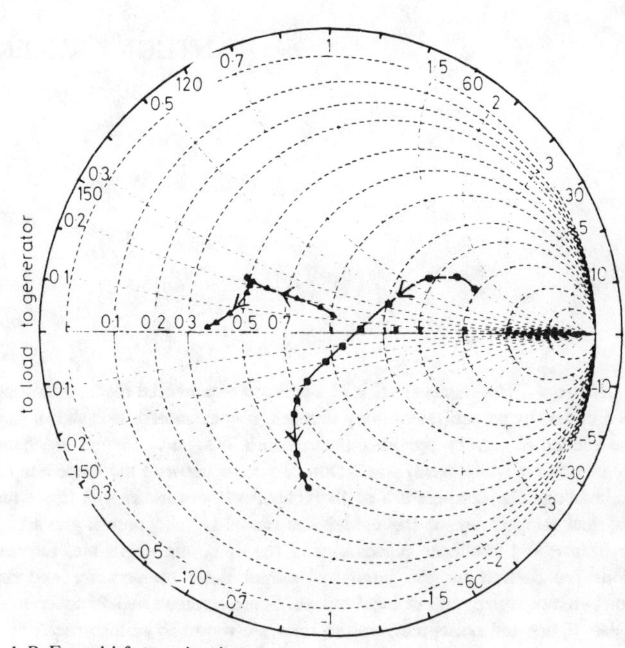

Fig. 5 *Z2 mixer*

a Loss and noise figure
Bias voltage = + 2·1 V
— ⊖ — mixer conversion loss
— ▲ — mixer noise figure
— + — system noise figure

b R.F. and i.f. terminations
Direct voltage/pump amplitude = + 1·05
—●— norm ZS
—▲— norm ZI
Intermediate frequency = 0·07 GHz

system noise figure as the more complex Y2 mixer, although the conversion loss is inferior.

The results tend to confirm the suggestions of earlier workers[4] who used less accurate device models.

2 References

1 HELD, D.N., and KERR, A.R.: 'Conversion loss and noise of microwave and millimeter wave mixers. Pts. 1 and 2', *IEEE Trans.*, 1978, **MTT-26**, pp. 49–61

2 GLOVER, K.J., GARDINER, J.G., and HOWSON, D.P.: 'Microwave mixer performance computation incorporating nonlinear diode capacitance'. Presented at the international conference on electronic circuits, Prague, 1973

3 SALEH, A.A.M.: 'Theory of resistive mixers' (MIT Press, 1971)

4 MANIA, L., and STRACCA, G.B.: 'Effects of the diode junction capacitance on the conversion loss of microwave mixers', *IEEE Trans.*, 1974, **COM-22**, pp. 1428–1435

5 HOWSON, D.P., GLOVER, K.J., and GÜNES, F.: 'Computer-aided design of Schottky diode mixers', SSCT 77, Prague, 1977

6 HOWSON, D.P., GLOVER, K.J., and GÜNES, F.: 'Noise figure analysis for Y and Z microwave Mixers'. Proceedings of the 6th colloquium on microwave communications, Budapest, Aug. 1978

Shot-Noise in Resistive-Diode Mixers and the Attenuator Noise Model

ANTHONY R. KERR, SENIOR MEMBER, IEEE

Abstract—**The representation of a pumped exponential diode, operating as a mixer, by an equivalent lossy network, is reexamined. It is shown that the model is correct provided the network has ports for all sideband frequencies at which (real) power flow can occur between the diode and its embedding. The temperature of the equivalent network is $\eta/2$ times the physical temperature of the diode. The model is valid only if the series resistance and nonlinear capacitance of the diode are negligible. Expressions are derived for the input and output noise temperature and the noise–temperature ratio of ideal mixers. Some common beliefs concerning noise–figure and noise–temperature ratio are shown to be incorrect.**

I. Introduction

IN RECENT YEARS, the need for low-noise mixers, especially in the field of millimeter-wave radio astronomy, has stimulated a considerable amount of research into the theory and design of mixers and mixer diodes. Improved mixer designs have revealed a substantial discrepancy [1] between measured noise performance and that predicted by the simple attenuator noise model of the mixer.

In the attenuator noise model, the mixer is represented as a lossy network whose port-to-port power loss is equal to the mixer conversion loss, and whose physical temperature T_A accounts for the mixer noise. Uncertainty has existed concerning the value of T_A. One widely held belief is that the output noise–temperature ratio[1] t_M of the mixer should be close to unity, from which it follows that T_A is equal to the physical temperature T of the mixer, and that the noise figure of a room-temperature mixer is equal to its conversion loss. An alternative view is that t_M is equal

Manuscript received February 15, 1978; revised May 18, 1978.

The author is with the National Aeronautics and Space Administration, Goddard Space Flight Center, Institute for Space Studies, New York, NY 10025.

[1]The noise–temperature ratio t_M of a mixer is defined as [2] (the available IF output noise power in bandwidth Δf)$\div(kT\Delta f)$, when the mixer and all its input terminations are maintained at ambient temperature T.

Reprinted from *IEEE Trans. Microwave Theory Tech.*, vol. MTT-27, pp. 135–140, Feb. 1979.

to the time average of the noise–temperature ratio[2] t_D of the dc-biased diode, usually close to 0.5. Neither of these agrees well with measured mixer noise, particularly at the shorter microwave and millimeter wavelengths.

The theory of shot-noise in mixers was first examined by Strutt [4] and more recently by van der Ziel and Watters [5], van der Ziel [6], and Kim [7], all of whom considered two- or three-frequency mixer models. Dragone [8] extended this work, deriving a general expression for the correlation between various frequency components of the shot-noise in a pumped diode, and showed that an ideal pumped diode is equivalent to a lossy multiport network at a constant temperature. Saleh [9] used this to examine the noise behavior of mixers with exponential diodes and reactive terminations at all sideband frequencies above the signal and image.

The observed discrepancies between measured mixer noise and that predicted by the three-port attenuator model have recently been studied by Held and Kerr, [10], [11] and shown to have three main causes: 1) the time-varying diode admittance, generally complex, acting on correlated components of the time-varying shot-noise of the diode, 2) lossy terminations at higher sidebands, and 3) appreciable series resistance in the diode. Scattering effects in the semiconductor material were also shown to cause a slight increase in the noise of room-temperature mixers.[3]

In this paper, the general mixer theory of Held and Kerr is used to investigate the noise behavior of ideal mixers and the validity of some widely held ideas about mixer noise. An ideal exponential diode is assumed, having full shot-noise but negligible series resistance and nonlinear junction capacitance.[4] It is shown that, in agreement with Dragone [8] and Saleh [9], such an ideal diode, operated as a mixer, is equivalent to a lossy network at a temperature T_A. The equivalent network has ports for all sideband frequencies at which (real) power flow can occur between the diode and its embedding. The temperature T_A is shown to be $\eta T/2$, where η is the ideality factor of the diode and T is its physical temperature. Such a noise model has practical applications at lower microwave frequencies where Schottky diodes are available with low series resistance and small capacitance and where the embedding impedance is well defined up to several harmonics of the LO. At very low frequencies, below ~1 MHz, $1/f$-noise will become significant, thereby invalidating the simple model. At frequencies where the nonlinear diode capacitance has a significant susceptance, parametric effects will be present, and a more general analysis is required.

[2]The noise–temperature ratio t_D of a dc-biased diode is defined as [3] (the available IF output noise power in bandwidth Δf)$\div(kT\Delta f)$, when the diode is at physical temperature T.

[3]Other investigations of noise in millimeter-wave mixers have been made recently by Fei and Mattauch [12] and Keen [13].

[4]The diode may have finite *static* capacitance, which can be regarded as part of the embedding circuit for purposes of analysis.

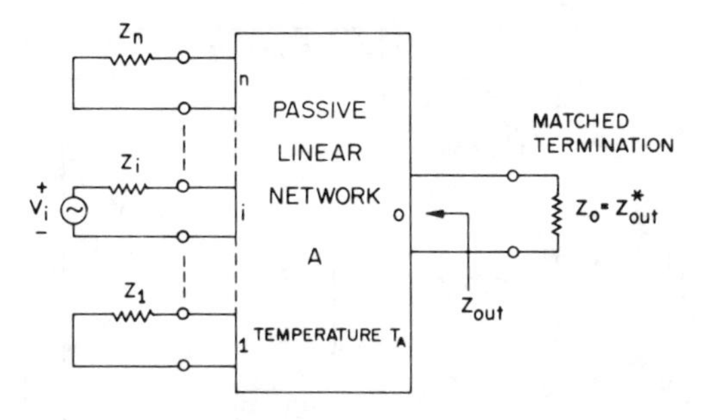

Fig. 1. Passive linear network at temperature T_A.

II. The Attenuator Noise Model

In this section, the noise properties of a passive linear multiport network will be examined. It will be shown in Section IV that the noise behavior of a certain class of pumped mixer diodes can be described by such a multiport network or attenuator.

Consider the passive linear multiport A, shown in Fig. 1, to be terminated at ports 1 to n by impedances Z_i. The output port 0 is conjugate-matched with impedence $Z_0 = Z^*_{out}$. The loss L_i from any port i to the output is defined as

$$L_i = \frac{\text{power available from source } (V_i \text{ and } Z_i)}{\text{power from } V_i \text{ delivered to load } Z_0}. \tag{1}$$

This is analogous to the conversion loss of a mixer with a matched output termination.

Let the network A and all its terminations be maintained at a temperature T_A. Then, for thermal equilibrium, the available noise power in a narrow band Δf at the output port is

$$P_{0_{\text{avail}}} = kT_A \Delta f. \tag{2}$$

This is the sum of thermal noise contributions from each of the terminations Z_i $(i = 1, \cdots, n)$ and a contribution P' from the network A itself:

$$P_{0_{\text{avail}}} = P' + \sum_{i=1}^{n} \frac{kT_A \Delta f}{L_i}. \tag{3}$$

With (2), this gives

$$P' = kT_A \left[1 - \sum_{i=1}^{n} \frac{1}{L_i}\right] \Delta f. \tag{4}$$

P' is the available noise power at the output port 0 of network A when all the input terminations Z_i $(i = 1, \cdots, n)$ are held at absolute zero temperature.

For the more general case, with the network A at temperature T_A and the terminations Z_i at temperature T_E, the available output power at port 0 is

$$P_{0_{\text{avail}}} = k\Delta f \left\{ T_A \left[1 - \sum_{i=1}^{n} \frac{1}{L_i}\right] + T_E \sum_{i=1}^{n} \frac{1}{L_i} \right\}. \tag{5}$$

It will be shown below that, for a certain class of mixer diode, there exists a noise-equivalent passive linear network whose output noise is described by an equation similar to (5), with T_A determined only by the physical temperature and ideality factor of the diode.

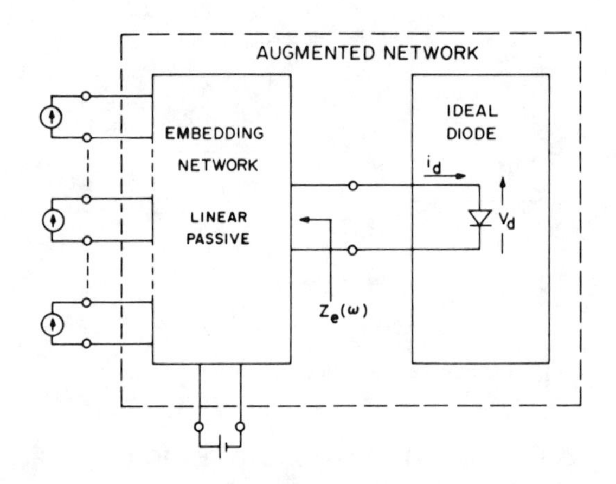

Fig. 2. The ideal diode connected to its linear passive embedding network, which together comprise the augmented network (broken line). All sources are external to the augmented network.

III. Outline of Mixer Theory

The mixer theory of Held and Kerr [10], [11] is used here to study the noise power flow at the terminals of an ideal exponential diode embedded in a linear passive network and pumped by an arbitrary local oscillator waveform. The ideal diode has zero nonlinear capacitance and series resistance and exhibits full shot-noise.

The diode current and voltage are related by

$$i_d = i_0[\exp(\alpha v_d) - 1] \tag{6}$$

where

$$\alpha = q/\eta kT. \tag{7}$$

The incremental conductance of the diode

$$g(i) = \frac{di_d}{dv_d} \cong \alpha i. \tag{8}$$

When the diode is biased at a dc current i, the shot-noise is given by the standard shot-noise equation:

$$\langle \delta i_n^2 \rangle = 2qi\Delta f. \tag{9}$$

The diode is connected to the passive linear embedding network which includes all external source and load impedances, as shown in Fig. 2. Local oscillator power applied to the mixer produces a periodic current waveform at the diode

$$i_d(t) = \sum_{m=-\infty}^{\infty} I_m \exp jm\omega_p t, \qquad I_{-m} = I_m^*. \tag{10}$$

The corresponding diode conductance waveform is

$$g(t) = \sum_{m=-\infty}^{\infty} G_m \exp jm\omega_p t, \qquad G_{-m} = G_m^*. \tag{11}$$

It follows from (8) that

$$I_m = G_m/\alpha. \tag{12}$$

In analyzing the small-signal and noise behavior of the mixer diode, the concise subscript notation of Saleh [9] is used to denote the various sideband frequencies: if ω_p and ω_0 are the local oscillator and intermediate (angular) frequencies, then ω_m denotes sideband frequency $\omega_0 + m\omega_p$. Thus ω_1, ω_{-1}, and ω_2 are the upper sideband, lower sideband, and sum frequencies. The small-signal behavior of the diode is described by a conversion admittance matrix $\boldsymbol{Y}$ relating currents and voltages δI_m and δV_m at all the sideband frequencies ω_m. Thus

$$\boldsymbol{\delta I} = \boldsymbol{Y \delta V} \tag{13}$$

where

$$\boldsymbol{\delta V} = \{\cdots, \delta V_m, \cdots, \delta V_1, \delta V_0, \delta V_{-1}, \cdots\}^T$$

$$\boldsymbol{\delta I} = \{\cdots, \delta I_m, \cdots, \delta I_1, \delta I_0, \delta I_{-1}, \cdots\}^T$$

and

$$\boldsymbol{Y} = \begin{array}{c} \text{row \#} \\ \vdots \\ 1 \\ 0 \\ -1 \\ \vdots \\ \end{array} \begin{bmatrix} & \vdots & \vdots & \vdots & \\ \cdots & Y_{11} & Y_{10} & Y_{1-1} & \cdots \\ \cdots & Y_{01} & Y_{00} & Y_{0-1} & \cdots \\ \cdots & Y_{-11} & Y_{-10} & Y_{-1-1} & \cdots \\ & \vdots & \vdots & \vdots & \end{bmatrix}$$
$$\begin{array}{ccccc} \cdots & 1 & 0 & -1 & \cdots \end{array}$$
$$\text{column \#}$$

where [3]

$$Y_{mn} = G_{m-n}. \tag{14}$$

It follows from the Y matrix that the pumped diode can be represented as a multiport network with one port for each sideband frequency, as shown in Fig. 3. Each port is connected to its embedding impedance $Z_{e_m} = Z_e(\omega_0 + m\omega_p)$.

The embedded diode is represented by the augmented network outlined by the broken line in Figs. 2 and 3. In normal mixer operation, all ports of the augmented network are open-circuited or connected to signal or noise current sources. For the augmented network

$$\boldsymbol{\delta I'} = \boldsymbol{Y' \delta V} \tag{15}$$

where

$$\boldsymbol{Y'} = \boldsymbol{Y} + \mathbf{diag}\,\{\cdots, 1/Z_{e_m}, \cdots\}. \tag{16}$$

Inverting (15) gives

$$\boldsymbol{\delta V} = \boldsymbol{Z' \delta I'} \tag{17}$$

where

$$\boldsymbol{Z'} = (\boldsymbol{Y'})^{-1}. \tag{18}$$

The small-signal properties of the mixer can be deduced from (17).

If the conversion loss from any sideband ω_m to the IF ω_0 is defined as L_m = (signal power at sideband ω_m available to the diode from the embedding network)/(down-converted signal power available at ω_0 to the embedding network from the diode), then it follows from (17) and Fig. 2 that

$$L_m = \frac{1}{4|Z'_{0m}|^2} \frac{|Z_{e_m}|^2}{\mathrm{Re}\,[Z_{e_m}]} \frac{|Z_{e_0}|^2}{\mathrm{Re}\,[Z_{e_0}]}. \tag{19}$$

The mean-square output noise voltage due to shot-noise

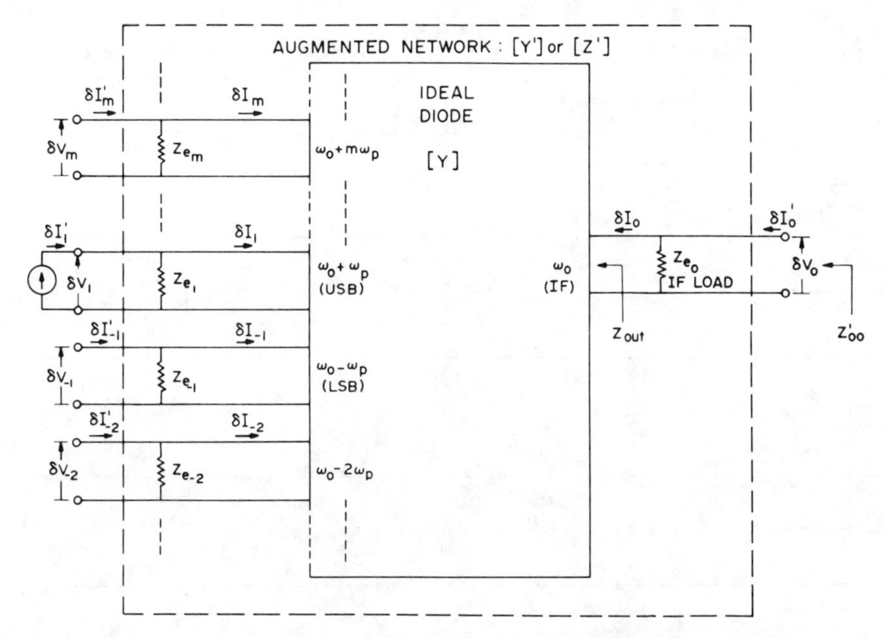

Fig. 3. Frequency-domain representation of the mixer. The diode is represented as the inner multiport network, characterized by conversion admittance matrix $\boldsymbol{Y}$. The complete mixer is represented by the augmented network (broken line), which includes all embedding impedances Z_{e_m}, but excludes all sources, and is described by the augmented matrices $\boldsymbol{Y}'$ or $\boldsymbol{Z}'$.

in the diode is given by[5]

$$\langle \delta V_{n_0}^2 \rangle = \boldsymbol{Z}_0' \boldsymbol{C} \boldsymbol{Z}_0'^{\dagger} \tag{20}$$

where $\boldsymbol{Z}_0'$ is the center row of matrix $\boldsymbol{Z}'$ and $\boldsymbol{C}$ is the shot-noise correlation matrix given by [8]

$$C_{mn} = 2qI_{m-n}\Delta f \tag{21}$$

where I_{m-n} is the $(m-n)$th Fourier coefficient of the LO current in the diode as defined in (10). Note that (20) does not include the effects of thermal noise in the embedding network.

If the output of the mixer is conjugate-matched, $Z_{\text{out}} = Z_{e_0}^*$. Then the central element of the matrix $\boldsymbol{Z}'$ is (see Fig. 3)

$$Z_{00}' = \left[\frac{1}{Z_{e_0}} + \frac{1}{Z_{\text{out}}} \right]^{-1} = \frac{|Z_{e_0}|^2}{2\,\text{Re}\left[Z_{e_0}\right]}. \tag{22}$$

Equations (18)–(22) are the basis for the following noise analysis.

IV. Shot-Noise in the Resistive-Diode Mixer

In this section, it will be shown that the shot-noise equation (20) for a pumped exponential diode implies an equivalent lossy network with the same noise and conversion properties as the diode. The equivalent temperature T_L of this network depends only on the physical temperature T of the diode, and its ideality factor η.

From (7), (8), (16), and (21), we have

$$\boldsymbol{C} = 2\eta k T \Delta f \boldsymbol{Y} \tag{23}$$

$$= 2\eta k T \Delta f [\boldsymbol{Y}' - \boldsymbol{D}] \tag{24}$$

where

$$\boldsymbol{D} = \textbf{diag}\left\{ \cdots, 1/Z_{e_m}, \cdots \right\}. \tag{25}$$

Now form the product $\boldsymbol{Z}'\boldsymbol{C}\boldsymbol{Z}'^{\dagger}$ in two stages. First, using (18) and (24)[6],

$$\boldsymbol{Z}'\boldsymbol{C} = 2\eta k T \Delta f [\boldsymbol{Z}'\boldsymbol{Y}' - \boldsymbol{Z}'\boldsymbol{D}] \tag{26}$$

$$= 2\eta k T \Delta f [\boldsymbol{I} - \boldsymbol{Z}'\boldsymbol{D}]. \tag{27}$$

Then post-multiply by $\boldsymbol{Z}'^{\dagger}$ to obtain the square matrix

$$\boldsymbol{Z}'\boldsymbol{C}\boldsymbol{Z}'^{\dagger} = 2\eta k T \Delta f [\boldsymbol{Z}'^{\dagger} - \boldsymbol{Z}'\boldsymbol{D}\boldsymbol{Z}'^{\dagger}]. \tag{28}$$

The center element of this matrix is, from (20), $\boldsymbol{Z}_0'\boldsymbol{C}\boldsymbol{Z}_0'^{\dagger} = \langle \delta V_{n_0}^2 \rangle$.

Therefore,

$$\langle \delta V_{n_0}^2 \rangle = 2\eta k T \Delta f \left\{ Z_{00}'^{*} - \sum_m |Z_{0m}'|^2 \frac{1}{Z_{e_m}} \right\}. \tag{29}$$

Using (22),

$$\langle \delta V_{n_0}^2 \rangle = \eta k T \Delta f \frac{|Z_{e_0}|^2}{\text{Re}\left[Z_{e_0}\right]} \cdot \left\{ 1 - 2\sum_m \frac{|Z_{0m}'|^2}{Z_{e_m}} \frac{\text{Re}\left[Z_{e_0}\right]}{|Z_{e_0}|^2} \right\}. \tag{30}$$

[5]The symbol † denotes the conjugate transpose of a matrix.

[6]The identity matrix, **diag** $\{\cdots, 1, \cdots\}$, is denoted by $\boldsymbol{I}$.

TABLE I
NOISE PARAMETERS FOR IDEAL DIODE IN IDEAL SINGLE- AND DOUBLE-SIDEBAND EMBEDDING CIRCUITS

	SSB MIXER	DSB MIXER
Input noise temp. (referred to ω_1)	$T_M = \frac{\eta T}{2}[L_1 - 1]$	$T_M = \frac{\eta T}{2}[L_1 - 1 - \frac{L_1}{L_{-1}}]$
Input noise temp. (double sideband)	———	$T_{M_{DSB}} = \frac{\eta T}{2}[1 - \frac{1}{L_1} - \frac{1}{L_{-1}}] \div [\frac{1}{L_1} + \frac{1}{L_{-1}}]$
Mixer contribution to output noise temp.	$T'_M = \frac{\eta T}{2}[1 - \frac{1}{L_1}]$	$T'_M = \frac{\eta T}{2}[1 - \frac{1}{L_1} - \frac{1}{L_{-1}}]$
Mixer noise-temp.-ratio	$t_M = \frac{\eta}{2}[1 - \frac{1}{L_1}] + \frac{1}{L_1}$	$t_M = \frac{\eta}{2}[1 - \frac{1}{L_1} - \frac{1}{L_{-1}}] + \frac{1}{L_1} + \frac{1}{L_{-1}}$

The right side of (30) must be real,[7] so it is necessary only to consider the real parts in the summation. Hence,

$$\langle \delta V_{n_0}^2 \rangle = \eta k T \Delta f \frac{|Z_{e_0}|^2}{\operatorname{Re}[Z_{e_0}]} \cdot \left\{ 1 - 2\sum_m |Z'_{0m}|^2 \frac{\operatorname{Re}[Z_{e_m}]}{|Z_{e_m}|^2} \frac{\operatorname{Re}[Z_{e_0}]}{|Z_{e_0}|^2} \right\}. \quad (31)$$

From (22), the $m=0$ element of the summation is equal to $1/4$. Using (19), (31) can be written as

$$\langle \delta V_{n_0}^2 \rangle = \eta k T \Delta f \frac{1}{2} \frac{|Z_{e_0}|^2}{\operatorname{Re}[Z_{e_0}]} \left\{ 1 - \sum_{m \neq 0} \frac{1}{L_m} \right\}. \quad (32)$$

The shot-noise power delivered to the matched load Z_{e_0} is

$$P' = \langle \delta V_{n_0}^2 \rangle \frac{\operatorname{Re}[Z_{e_0}]}{|Z_{e_0}|^2} = \frac{\eta k T \Delta f}{2} \left\{ 1 - \sum_{m \neq 0} \frac{1}{L_m} \right\} \quad (33)$$

c.f., (4).

If the embedding network is at temperature T_E, thermal noise will be down-converted from each sideband to the (matched) IF load Z_{e_0}. The available noise power from the mixer diode at the IF port is then

$$P_{0_{\text{avail}}} = k\Delta f \left\{ \frac{\eta T}{2} \left[1 - \sum_{m \neq 0} \frac{1}{L_m} \right] + T_E \sum_{m \neq 0} \frac{1}{L_m} \right\}. \quad (34)$$

By comparison with (5), it is clear that the available IF noise power from the mixer diode is equal to that at the output of a linear passive network having the following properties.

1) The physical temperature T_A of the network is related to that of the diode T by

$$T_A = \eta T/2. \quad (35)$$

2) The loss L_m between port m of the network and the output is equal to the corresponding conversion loss L_m of the embedded mixer diode for all m.

A. Noise Characterization of the Mixer Diode in Ideal Embedding Circuits

From (34), it is possible to derive the commonly used mixer noise parameters for the ideal mixer diode. However, it is clear that $P_{0_{\text{avail}}}$ depends on the embedding impedance (and its temperature) at all sideband frequencies ω_m. For simplicity, only two embedding configurations will be examined here.

1) A single-sideband (SSB) mixer with reactive embedding impedances at all sidebands other than the upper sideband (ω_1) and IF (ω_0). It follows from (19) that $L_m \to \infty$ for $m \neq 1$.

2) A double-sideband (DSB) mixer with reactive embedding impedances at all sidebands other than the upper sideband (ω_1), lower sideband (ω_{-1}), and IF (ω_0). In this case, $L_m \to \infty$ for $m \neq \pm 1$.

Commonly used noise parameters are: 1) the equivalent input noise temperature T_M of the mixer, referred to one sideband, 2) the available output noise temperature T'_M of the mixer, evaluated with the input terminations (Z_e) at absolute zero temperature, and 3) the mixer noise–temperature ratio $t_M = P_{0_{\text{avail}}}/kT\Delta f$ with the input terminations and the diode at $T_0 = 290$ K. Expressions for these quantities are given in Table I.

For radiometric applications, in which the ideal DSB mixer sees equal temperatures in the upper and lower sidebands, it is common to use a DSB equivalent input noise temperature $T_{M_{\text{DSB}}}$, and a DSB conversion loss L_{DSB}.

[7]That $\langle \delta V_{n_0}^2 \rangle$ must be real is evident from physical considerations. This can also be deduced mathematically from (20) using the fact that C is Hermitian: then for any vector V, $VCV^\dagger$ is real.

For the ideal DSB mixer,

$$L_{\mathrm{DSB}} = \left[\frac{1}{L_1} + \frac{1}{L_{-1}} \right]^{-1} \tag{36}$$

and

$$\begin{aligned} T_{M_{\mathrm{DSB}}} &= T'_M L_{\mathrm{DSB}} \\ &= \frac{\eta T}{2} \left[1 - \frac{1}{L_1} - \frac{1}{L_{-1}} \right] \left[\frac{1}{L_1} + \frac{1}{L_{-1}} \right]^{-1}. \end{aligned} \tag{37}$$

B. Comments on the Noise–Temperature Ratio t and Noise Figure

The noise–temperature ratio t_M of a mixer, as defined above, is an easily measured parameter, and, consequently, it is often quoted when describing mixer performance. Some confusion has existed between t_M for the mixer, and the noise–temperature ratio t_D of a dc-biased diode. It can be shown from (7)–(9) that, under dc bias, the available noise power from a diode at temperature T is $\eta k T \Delta f / 2$, whence $t_D = \eta / 2$. It is clear from Table I that, for the ideal mixers considered, $t_M \neq t_D$.

For the ideal SSB and DSB mixers considered here, the minimum possible conversion losses are [9], [14] 1 (0 dB) and 2 (~3 dB). It follows from Table I that, in both cases, when operating with minimum conversion loss, $t_M \rightarrow 1$.

It is widely believed that the noise figure of a mixer is equal to its conversion loss. That this is not strictly true, even for the idealized two-and three-frequency mixers described above, can be shown easily using the equations in Table I. For many practical receivers, especially those in which the overall receiver noise is dominated by IF amplifier noise, the error in equating conversion loss and noise figure may be tolerable, but, in general, this rule should not be expected to give accurate results.

V. Conclusion

The ideal exponential diode, operating as a mixer, has been investigated using the analysis of Held and Kerr [10], [11]. The diode is assumed to exhibit full shot-noise, and is pumped by an arbitrary LO waveform. It has been shown that the diode is equivalent to a passive lossy multiport network having ports at all sideband frequencies at which (real) power flow can occur between the diode and its embedding. The physical temperature T_A of the network is related to that (T) of the diode by $T_A = \eta T / 2$, which is in agreement with Dragone [8] and Saleh [9]. It follows that an SSB (DSB) mixer is accurately represented by a two-port (three-port) attenuator only if the diode is reactively terminated at all sidebands other than the signal and IF (and image).

Two common beliefs concerning the noise performance of diode mixers are examined and shown to be *generally incorrect*. These are: 1) that the noise–temperature ratio t_M of a pumped mixer is equal to the noise–temperature ratio t_D of the dc-biased diode, and 2) that the noise figure of a mixer is equal to its conversion loss. For the case of ideal SSB and DSB mixers, it is shown that, as the conversion loss approaches the theoretical minimum value (0 or 3 dB), the mixer noise–temperature ratio t_M approaches unity.

The attenuator noise model has application to Schottky diode mixers operating in the region from ~1 MHz to several GHz, provided the diode has small series resistance and nonlinear capacitance, and the embedding (mount) impedance is well defined at frequencies up to several times the LO.

For mixer diodes with appreciable nonlinear capacitance, a more general analysis [10], [11] must be used to include the (parametric) effects of the time-varying capacitance acting on correlated components of the time-varying shot-noise. The more general analysis must also be used for diodes in which the relationship between the current-dependent noise and the diode current differs from the simple shot-noise equation (9).

References

[1] A. R. Kerr, "Anomalous noise in Schottky-diode mixers at millimeter wavelengths," *IEEE MTT-S Int. Microwave Symp. Dig. Tech. Papers*, pp. 318–320, May 1975.

[2] G. C. Messenger and C. T. McCoy, "Theory and operation of crystal diodes as mixers," *Proc. IRE*, vol. 45, p. 1269–1283, 1957.

[3] H. C. Torrey and C. A. Whitmer, *Crystal Rectifiers* (M.I.T. Radiation Lab. Series, vol. 15). New York: McGraw-Hill, 1948.

[4] M. J. O. Strutt, "Noise figure reduction in mixer stages," *Proc. IRE*, vol. 34, no. 12, pp. 942–50, Dec. 1946.

[5] A. van der Ziel and R. L. Watters, "Noise in mixer tubes," *Proc. IRE*, vol. 46, pp. 1426–1427, 1958.

[6] A. van der Ziel, *Noise: Sources, Characterization, and Measurement*. Englewood Cliffs, NJ: Prentice-Hall, 1970.

[7] C. S. Kim, "Tunnel diode converter analysis," *IRE Trans. Electron Devices*, vol. ED-8, pp. 394–405, Sept. 1961.

[8] C. Dragone, "Analysis of thermal shot noise in pumped resistive diodes," *Bell Syst. Tech. J.*, vol. 47, pp. 1883–1902, 1968.

[9] A. A. M. Saleh, *Theory of Resistive Mixers*. Cambridge, MA: M.I.T. Press, 1971.

[10] D. N. Held, "Analysis of room temperature millimeter-wave mixers using GaAs Schottky barrier diodes," Sc. D. dissertation, Dep. Elec. Eng., Columbia Univ., New York, NY, 1976.

[11] D. N. Held and A. R. Kerr, "Conversion loss and noise of microwave and millimeter-wave mixers: Part 1—Theory," and "Part 2—Experiment," *IEEE Trans. Microwave Theory Tech.*, vol. MTT-26, Feb. 1978.

[12] R. J. Mattauch and F. S. Fei, "Local-oscillator-induced noise in GaAs Schottky mixer diodes," *Electron. Lett.*, vol. 13, no. 1, pp. 22–23, Jan. 6, 1977.

[13] N. J. Keen, "Evidence for coherent noise in pumped Schottky diode mixers," *Electron. Lett.*, vol. 13, pp. 282–284, 1977.

[14] A. J. Kelly, "Fundamental limits on conversion loss of double sideband resistive mixers," *IEEE Trans. Microwave Theory Tech.*, vol. MTT-25, pp. 867–869, Nov. 1977.

Fundamental Limits on Conversion Loss of Double Sideband Resistive Mixers

ALEXANDER J. KELLY, SENIOR MEMBER, IEEE

Abstract—**Although the resistive mixer has been the subject of numerous studies [1]–[3], these have all dealt with specific cases for terminations at the higher order mixing products (idlers). This paper deals with the general case of the double sideband mixer, and demonstrates that when no energy is dissipated at the idler frequencies the fundamental limit on conversion loss is 3 dB, with the lost energy being equally divided between conversion to the image and reflection loss at the signal port. Also treated is the case where matched loads are presented to each idler. It is shown that, in this case, the theoretical limit on conversion loss is 3.92 dB (20 log $\pi/2$), independent of the mixer configuration.**

Introduction

THE RESISTIVE MIXER comprises one or more diodes pumped by a local oscillator (LO). The first-order analysis assumes that the signal level is significantly lower than that of the LO, and does not, therefore, perturb the LO-pumped diode conductance waveform. Under this assumption, no harmonics of the signal are generated, and the mixing products are defined by

$$v_{\text{OUT}} = v_{\text{SIG}} \cdot \sum_{n=1}^{\infty} [k_n \cos (n\omega_{\text{LO}} - \omega_{\text{SIG}})t + k_n \cos (n\omega_{\text{LO}} + \omega_{\text{SIG}})t]. \quad (1)$$

Fig. 1 shows the spectral distribution for a resistive mixer. The key frequencies are

$$\omega_{\text{IF}} = \omega_{\text{LO}} - \omega_{\text{SIG}} \quad (2)$$

$$\omega_{\text{IMAGE}} = 2\omega_{\text{LO}} - \omega_{\text{SIG}}. \quad (3)$$

The remaining mixing products (idlers) are grouped in pairs about the harmonics of the LO frequency.

Double Sideband Mixers

The double sideband mixer is one in which an IF output is generated for a signal above or below the LO. Each is the "image" of the other. Double sideband mixers are utilized in radiometers. They are also utilized as a basic building block in single sideband receivers.

Fig. 2 is a schematic representation of a double sideband mixer. Although a single physical port serves as the signal and image terminals, they are shown as two separate ports for mathematical analysis. The signal port is designated as port 1; the image port, port 2; the IF port, port 3.

Manuscript received March 29, 1976; revised November 2, 1976.

The author was with LNR Communications, Inc., Hauppauge, NY. He is now with the Hazeltine Corporation, Research Laboratories, Greenlawn, NY 11740.

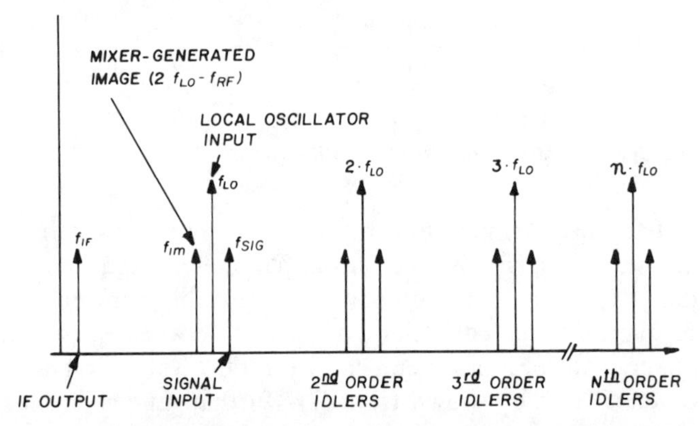

Fig. 1. Frequency spectrum of a resistive mixer excited by a local oscillator and signal.

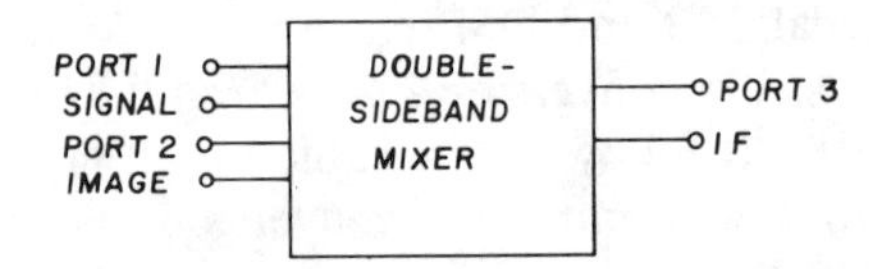

Fig. 2. Three-port equivalent circuit of a double-sideband mixer.

Fundamental Limit on Conversion Loss

The most efficient double sideband mixer is one in which real power flow is allowed to take place only at the signal, image, and intermediate frequencies. For this to occur, the idlers must be reactively terminated. To maintain generality, the nature of these terminations is left unspecified in this analysis.

To establish the fundamental limit on conversion loss, a "perfect" diode is assumed. By "perfect," it is meant that the diode has no parasitics and is driven between two states—perfect open circuit and perfect short circuit. With these assumptions, and the fact that the idlers are reactively terminated, the mixer reduces to a lossless three port, where port 1 is the signal, port 2 is the image, and port 3 is the IF:

$$S = \begin{bmatrix} S_{11} & S_{12} & S_{13} \\ S_{21} & S_{22} & S_{23} \\ S_{31} & S_{32} & S_{33} \end{bmatrix}. \quad (4)$$

Since the mixer and image ports are physically one and the same, the matrix can be rewritten as

$$S = \begin{bmatrix} S_{11} & S_{12} & S_{13} \\ S_{21} & S_{11} & S_{13} \\ S_{31} & S_{31} & S_{33} \end{bmatrix}. \quad (5)$$

Since the mixer is lossless, the following matrix equation holds (unnormalized s parameters are used):

$$\tilde{S}^* Y_0 S = Y_0 \quad (6)$$

Reprinted from *IEEE Trans. Microwave Theory Tech.*, vol. MTT-25, pp. 867–869, Nov. 1977.

where

$$Y_0 = \begin{bmatrix} Y_{01} & 0 & 0 \\ 0 & Y_{01} & 0 \\ 0 & 0 & Y_{03} \end{bmatrix}. \tag{7}$$

Performing the matrix multiplication shown in (6) yields four independent equations:

$$Y_{01}|S_{11}|^2 + Y_{01}|S_{21}|^2 + Y_{03}|S_{31}|^2 = Y_{01} \tag{8}$$

$$Y_{01}S_{11}^*S_{12} + Y_{01}S_{21}^*S_{11} + Y_{03}|S_{31}|^2 = 0 \tag{9}$$

$$Y_{01}S_{11}^*S_{13} + Y_{01}S_{21}^*S_{13} + Y_{03}S_{31}^*S_{33} = 0 \tag{10}$$

$$2Y_{01}|S_{13}|^2 + Y_{03}|S_{33}|^2 = Y_{03}. \tag{11}$$

The transducer gain is given by

$$G = |S_{31}|^2 \frac{Y_{03}}{Y_{01}}. \tag{12}$$

However, from reciprocity,

$$Y_{01} \cdot S_{13} = Y_{03} \cdot S_{31}. \tag{13}$$

Substituting (13) into (12),

$$G = |S_{13}|^2 \frac{Y_{01}}{Y_{03}}. \tag{14}$$

From (11)

$$G = \tfrac{1}{2}(1 - |S_{33}|^2). \tag{15}$$

By inspection of (15), it is seen that the maximum conversion gain is obtained when the mixer circuit constants are chosen such that $|S_{33}|$ is zero. Under this condition: *the theoretical limit of conversion loss for a double sideband mixer is 3 dB.*

Since $S_{33} \equiv 0$, (10) reduces to

$$Y_{01} \cdot S_{13} \cdot (S_{11}^* + S_{21}^*) = 0. \tag{16}$$

Since $S_{13} \neq 0$

$$S_{11}^*|_{\text{opt}} = -S_{21}^*|_{\text{opt}}. \tag{17}$$

Equation (9) then reduces to

$$-2|S_{11}|_{\text{opt}}^2 + \frac{Y_{03}}{Y_{01}}|S_{31}|_{\text{opt}}^2 = 0. \tag{18}$$

But, since the second term is G_{opt}, which is one half,

$$|S_{11}|_{\text{opt}} = \tfrac{1}{2} = |S_{21}|_{\text{opt}}.$$

This proves that, in the limit, the lowest conversion loss that can be achieved in a double sideband mixer is 3 dB. The power that is lost is equally divided between conversion to the image and reflection at the signal port (3:1VSWR).

These results have been derived without specifying either the nature of the reactive idler terminations or the ratio of the time the diode is driven into the short-circuit state, to the period of the LO (pulse duty ratio).

This result, therefore, establishes the fundamental limit on performance of a double sideband mixer.

Ultra-Broadband Double Sideband Mixer

In an ultra-broadband double sideband mixer (greater than an octave), the signal and idler spectra overlap. It is not possible to reactively terminate all of the idlers. A meaningful limit on conversion loss can be established by assuming matched loads at all of the idlers. This is most easily done by using an s-matrix model of the diode:

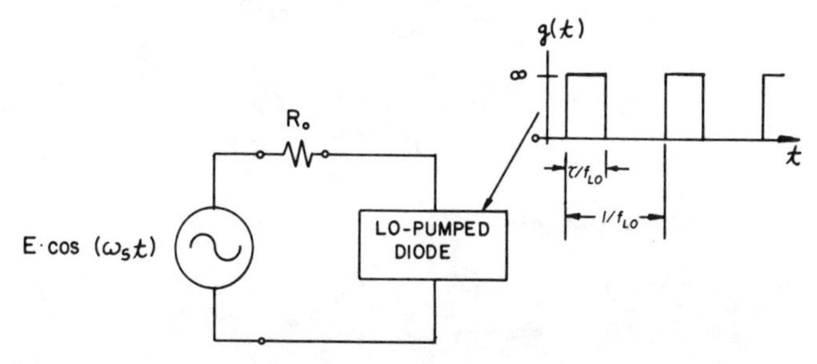

Fig. 3. Circuit used to derive the s parameters of a mixer diode.

$$[b] = [S] \cdot [a]. \tag{19}$$

Truncating the matrix assigns $a_i = 0$ for all ports excluded from the matrix expansion. Therefore, writing a three-port matrix automatically assigns matched loads to the idlers.

To determine the s parameters, the circuit of Fig. 3 is used. A small-signal voltage source with a source impedance R_0 is connected to a time-varying conductance. This conductance is driven between an open-circuit state and a short-circuit state by a local oscillator. The ratio of the time the diode is a short circuit relative to the LO period is denoted as τ.

The voltage and current across the diode are given by

$$v = E \cos(\omega_s t) \cdot [1 - f(t)] \tag{20}$$

$$i = \frac{E \cos(\omega_s t)}{R_0} \cdot f(t) \tag{21}$$

where $f(t)$ is a train of rectangular pulses with unity magnitude, the period is $1/f_{\text{LO}}$, and the width is τ/f_{LO}.

The Fourier expansion of $f(t)$ is

$$F\{f(t)\} = \tau + \frac{2}{\pi} \cdot \sum_{n=1}^{\infty} (-1)^n \frac{\sin(n\pi\tau)}{n} \cdot \cos(n\omega_{\text{LO}}t). \tag{22}$$

Substituting (22) into (20) and (21),

$$v = (E \cdot (1 - \tau)) \cdot \cos(\omega_s t) - \frac{2E}{\pi} \cdot \sum_{n=1}^{\infty} (-1)^n \frac{\sin(n\pi\tau)}{n} \cdot \cos(n\omega_{\text{LO}}t) \cdot \cos(\omega_s t) \tag{23}$$

$$i = \frac{E \cos(\omega_s t)}{R_0} \cdot \tau + \frac{2E}{\pi R_0} \cdot \sum_{n=1}^{\infty} (-1)^n \frac{\sin(n\pi\tau)}{n} \cdot \cos(n\omega_{\text{LO}}t) \cdot \cos(\omega_s t). \tag{24}$$

If one postulates a multiplexer at the diode junction, with each output terminated in R_0, the currents and voltages can be combined to determine scattering parameters of an infinite-port network, where each port corresponds to a different frequency.

Scattering parameters are defined as follows:

$$a_1 = \tfrac{1}{2}(V_1/\sqrt{R_0} + I_1\sqrt{R_0}) \tag{25}$$

$$b_1 = \tfrac{1}{2}(V_1/\sqrt{R_0} - I_1\sqrt{R_0}) \quad (26)$$

$$a_N = \tfrac{1}{2}(V_N/\sqrt{R_0} + I_N\sqrt{R_0}) \quad (27)$$

$$b_N = \tfrac{1}{2}(V_N/\sqrt{R_0} - I_N\sqrt{R_0}). \quad (28)$$

Substituting (23) and (24),

$$a_1 = \frac{1}{2}\cdot\frac{E\cos(\omega_s t)}{\sqrt{R_0}} \quad (29)$$

$$a_N = 0 \quad (30)$$

$$b_1 = \frac{1}{2}\cdot\frac{E\cos(\omega_s t)}{\sqrt{R_0}}\cdot(1-2\tau) \quad (31)$$

$$b_N = -\frac{E}{2\sqrt{R_0}}\cdot\left[\frac{2}{\pi}(-1)^n\frac{\sin(n\pi\tau)}{n}\cos(n\omega_{LO}\pm\omega_s)t\right]. \quad (32)$$

Since the scattering matrix represents complex wave amplitudes, the frequency terms are not carried, but are understood:

$$S_{11} = \frac{b_1}{a_1} = 1 - 2\tau \quad (33)$$

$$S_{N1} = \frac{b_N}{a_1} = -\frac{2}{\pi}\cdot(-1)^n\cdot\frac{\sin(n\pi\tau)}{n}. \quad (34)$$

A similar analysis can be performed for the excitation at each remaining port. The resultant matrix, carried for the three ports, signal, IF, and image, is given by

$$S = \begin{bmatrix} 1-2\tau & \frac{2}{\pi}\sin(\pi\tau) & -\frac{1}{\pi}\sin(2\pi\tau) \\ \frac{2}{\pi}\sin(\pi\tau) & 1-2\tau & \frac{2}{\pi}\sin(\pi\tau) \\ -\frac{1}{\pi}\sin(2\pi\tau) & \frac{2}{\pi}\sin(\pi\tau) & 1-2\tau \end{bmatrix}. \quad (35)$$

Since this mixer is ultra broadband, and the idlers have been assigned matched loads, R_0, it is assumed that the image is also matched and that the signal source impedance is R_0. The degree of freedom is the "pulse duty ratio" [2], τ.

S_{21} is a maximum for $\tau = 0.5$. At this point, the conversion loss is 3.92 dB. This is the value normally assigned as the theoretical limit for double-balanced mixers [4]. As can be seen, at $\tau = 0.5$, the signal and IF ports self-match. This present result shows that this is a general limit for any resistive mixer where all of the higher order mixing products are presented with matched loads.

Conclusions

This paper has examined the fundamental limits on double sideband mixer conversion loss, under the assumption of a "perfect" diode—one with no parasitics, infinite forward conductance, and zero reverse conductance. As would be expected, results have been obtained which are independent of the absolute values of signal source impedance or IF load impedance, since the perfect mixer diode acts as a transformation network.

The first result demonstrated that a double sideband mixer with reactive idler terminations will exhibit, in the limit, a 3-dB conversion loss, with half of the power lost being converted to the image, and the other half lost in a 3 : 1 mismatch at the signal port. Note that any matching network at the signal port also affects the image, so that the mismatch can only be improved at the expense of increased overall conversion loss. This result was determined strictly on the basis of the network properties of the mixer.

A second limitation was obtained for the case where all of the idlers are matched—a good approximation for a multi-octave mixer. There, s parameters were employed, and the pumped conductance waveform was used. This result showed a limit of 3.92 dB, independent of mixer configuration (single-ended, balanced, or double-balanced). This optimum occurs for the diode driven into the forward region for 50 percent of the LO period.

Acknowledgment

The author wishes to thank Dr. H. Okean and S. Okwit for their support and encouragement, and to acknowledge the many stimulating discussions with his colleague, S. Foti. He also thanks the reviewers for many important comments on this paper.

References

[1] H. C. Torrey and C. A. Whitmer, "Crystal rectifiers," *Radiation Laboratory Series*, vol. 15. New York: McGraw-Hill, 1948.

[2] M. P. Barber, "Noise figure and conversion loss of the Schottky barrier mixer diode," *IEEE Trans. Microwave Theory Tech.*, vol. MTT-15, pp. 629–635, Nov. 1967.

[3] A. A. M. Saleh, *Theory of Resistive Mixer*. Boston, MA: M.I.T. Press, 1971.

[4] R. B. Mouw and S. M. Fukuchi, "Broadband double balanced mixer/modulators, Part I," *Microwave J.*, pp. 131–134, Mar. 1969.

Effect of source resistance in microwave broadband balanced mixers

E. Korolkiewicz and B.L.J. Kulesza

Indexing terms: *Mixers (circuits), Solid-state microwave circuits*

Abstract: A broadband mixer may be designed to obtain matched conditions at the r.f. port, or the r.f. port may be mismatched to produce a minimum conversion power loss. Using a practical diode law this paper compares the performance of a broadband balanced mixer designed by the above two methods and shows that, under certain conditions, the conversion loss can be made to be independent of the diode series resistance by mismatching the r.f. port.

1 Introduction

Mixers are classified according to the kind of termination that is 'seen' by the image frequency component. Three special cases called broadband, narrowband-image open circuit and narrowband-image short circuit mixers are usually considered. For narrowband (-image open circuit or -image short circuit) mixers the same results are obtained whether the source and load impedances are chosen to produce matched conditions at the r.f. and i.f. ports, a condition recognised in microwave circuits, or the expression for the conversion power loss is optimised and then the load and source impedances determined.[1-3] In the case of broadband mixers the two methods produce different results, although for single-diode series and shunt mixers the difference in the conversion power loss is small (less than 0·1 dB) over a wide range of bias conditions around the optimum.[4]

What is generally not recognised is that for broadband balanced mixers the two methods of design lead to a considerable difference in performance. This type of mixer is normally used in practice as it has the advantage over a single-diode mixer of producing a lower number of unwanted harmonic products and a lower noise contribution from the local oscillator. Balanced mixers are also increasingly applied as harmonically pumped down-convertors.[5,6] This paper analytically compares the performance of a balanced mixer designed to produce a minimum conversion loss with one designed to produce matched conditions at the input and output ports. A new approach is presented in that a practical diode law is assumed to include the effect of diode series resistance in the performance of the broadband balanced mixer.

2 Signal analysis of a broadband balanced mixer

A balanced H mixer (Reference 7), shown in Fig. 1*a*, may be treated as a passive linear network having three conceptual ports for signal, image and i.f. frequencies and may be described by the following matrix equation:

$$\begin{bmatrix} v_0 \\ i_{-1} \\ v_{-2} \end{bmatrix} = \begin{bmatrix} h_{11} & h_{12} & h_{13} \\ -h_{12} & h_{22} & -h_{12} \\ h_{13} & h_{12} & h_{11} \end{bmatrix} \begin{bmatrix} i_0 \\ v_{-1} \\ i_{-2} \end{bmatrix} \tag{1}$$

Paper T266 M, received 8th Sepember 1978

Mr. Korolkiewicz is with the Department of Electrical Engineering & Physical Electronics, Newcastle upon Tyne Polytechnic, Ellison Building, Newcastle upon Tyne NE1 8ST, England. Dr. Kulesza is with the Department of Applied Physics and Electronics, University of Durham, South Road, Durham, England

In a broadband mixer the image and signal real R_s components 'see' the same termination (R_s) and, therefore, using the relationship

$$v_{-2} = -i_{-2}R_s$$

eqn. 1 may be expressed in the form

$$\begin{bmatrix} v_0 \\ i_{-1} \end{bmatrix} = \begin{bmatrix} H_{11} & H_{12} \\ -H_{12} & H_{22} \end{bmatrix} \begin{bmatrix} i_0 \\ v_{-1} \end{bmatrix} \tag{2a}$$

where

$$H_{11} = h_{11}(1 - a^2/R) \tag{2b}$$

$$H_{12} = h_{12}(1 - a/R) \tag{2c}$$

$$H_{22} = h_{22}(1 + K_0/R) \tag{2d}$$

$$K_0 = h_{12}/h_{11}h_{22} \tag{2e}$$

$$a = h_{13}/h_{11} \tag{2f}$$

$$R = 1 + R_s/h_{11} \tag{2g}$$

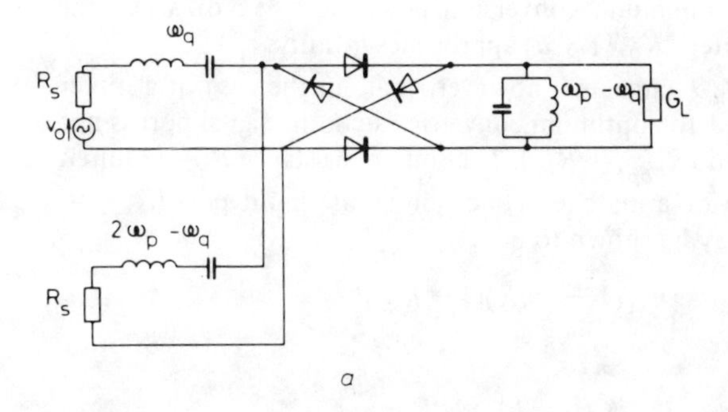

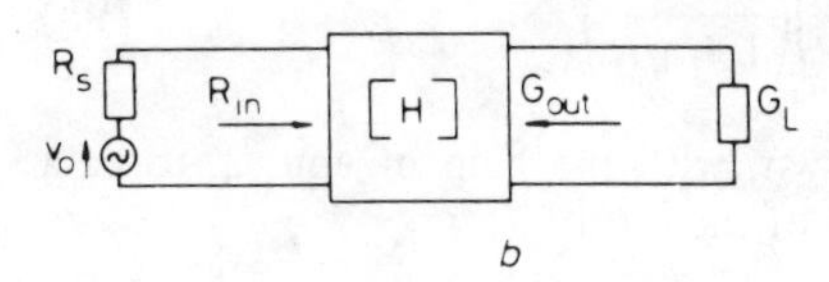

Fig. 1 *Broadband balanced H mixer*

The network described by eqn. 2*a* is shown in Fig. 1*b*. The conversion power loss (c.p.l.) is defined as

$$\text{c.p.l.} = 10\log_{10}\left(\frac{\text{available input power}}{\text{output power}}\right) = 10\log_{10}L \tag{3}$$

Reprinted with permission from *IEE J. Microwave Opt. Acoust.*, vol. 2, pp. 183–187, Nov. 1978.

Using eqn. 2, L can be expressed in the form

$$L = \frac{\{(1+y)(R+a) + 2K_0\}^2}{4K_0(R-1)y} \tag{4a}$$

where

$$y = G_L/h_{22} \tag{4b}$$

The minimum conversion loss is obtained by setting the derivatives with respect to R and y to zero.

Setting $\partial L/\partial y = 0$ results in the expression for optimum y and G_L as follows:

$$y_{opt} = 1 + 2K_0/(R+a) \tag{5a}$$

and

$$G_{L_{opt}} = h_{22}\{1 + 2K_0/(R+a)\} \tag{5b}$$

Eqn. 5b corresponds to the condition of the i.f. port being matched, i.e. $G_{out} = G_{L_{opt}}$.

Substituting eqn. 5a into eqn. 4a for y_{opt} and setting $\partial L/\partial R = 0$ leads to the following equation in R:

$$R^2 - 2R - (1+a)(2K_0 + a) = 0 \tag{6}$$

The optimum source resistance is obtained by using eqn. 2g and the positive root of eqn. 6 and is given by

$$R_{s_{opt}} = h_{11}(1+a)\left\{1 + \frac{2K_0}{(1+a)}\right\}^{1/2} \tag{7}$$

Finally, an expression for L_{opt} is obtained by substituting eqn. 7 into eqn. 4a,

$$L_{opt} = 2\left[\frac{1 + \{1 + 2K_0/(1+a)\}^{1/2}}{-1 + \{1 + 2K_0/(1+a)\}^{1/2}}\right] \tag{8}$$

This result is in agreement with Torrey and Whitmer[1] and gives a minimum conversion power loss of 3 dB when the parameter $2K_0/(1+a)$ approaches infinity.

Eqn. 7 indicates, however, that, in the case of a mixer designed for optimum conversion loss, the signal port is not matched ($R_{s_{opt}} \neq R_{in}$). The source resistance R_{s_m} required to provide a matched condition at the input port ($R_{s_m} = R_{in}$) may be shown to be

$$R_{s_m} = h_{11}(1 - a^2/R)(1 + K_m)^{1/2} \tag{9a}$$

where

$$K_m = \left(\frac{K_0}{1 + K_0/R}\right)\left(\frac{1 - a/R}{1 + a/R}\right) \tag{9b}$$

Eqn. 9a may be shown, with the help of eqn. 2, to be a quartic equation in R of the form

$$R^4 + R^3(K_0 - 2) - 3K_0R + Ra\{2 + K_0(2+a)\} + a^2\{K_0(1-2a) - a^2\} = 0 \tag{10}$$

An estimate of the largest real root (β) of eqn. 10 can be made using Tillots[8] criterion. The largest real root is limited by the following inequality:

$$\beta < 1 + \frac{|3K_0|}{K_0 - 2} \tag{11}$$

For most diodes the parameter K_0 (as discussed in Section 3) is much greater than 2 and, hence, the root $\beta < 4$. By approximation, therefore, eqn. 10 can be reduced to a cubic:

$$R^3 - 3R^2 + Ra(2+a) + a^2(1-2a) = 0 \tag{12}$$

The roots of eqn. 12 are found using Cardan's formula,[5] where the largest root leading to a realisable source resistance is

$$R = 2\left(\frac{(a+3)(a-1)}{3}\right)^{1/2}\cos(\theta/3) + 1 \tag{12a}$$

where

$$\cos\theta = \frac{27(1+a^2)(a-1)}{|(a+3)(a-1)|^{3/2}} \tag{12b}$$

The load conductance necessary to obtain a match at the output port is readily shown, using eqn. 2 to be

$$G_{L_m} = h_{22}(1 + K_0/R)(1 + K_m)^{1/2} \tag{13}$$

Finally, the corresponding expression for L, when both ports of the mixer are matched, is shown to be[1]

$$L_m = \frac{1 + (1+K_m)^{1/2}}{-1 + (1+K_m)^{1/2}} \tag{14}$$

By means of the relationships derived in this section, it is possible to compare the performance of a broadband mixer designed for minimum conversion loss with that of a similar mixer having matched conditions at input and output ports.

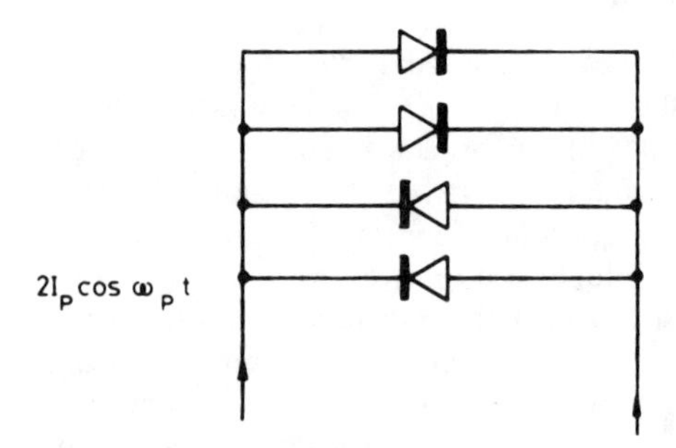

Fig. 2 *Diode circuit seen by the local oscillator drive*

3 Large-signal analysis of a broadband balanced mixer

The diode circuit seen by the local-oscillator current drive is shown in Fig. 2. The voltage developed across the diode's terminals may be expressed, using the diode practical v/i relationship, in the form

$$V = Ir_s + \frac{1}{\alpha}\log_e(1 + I/I_s) \tag{15}$$

where I is the current through each diode and α and I_s are the diode parameters. The time-varying diode incremental resistance is given, using eqn. 15, by

$$r(t) = \frac{1}{\alpha I_s}\left(\frac{1}{1 + I/I_s}\right) \tag{16}$$

The current through each diode may be conveniently written in the form

$$I = \tfrac{1}{2}I_p\cos\omega_p t\{1 + S(t)\} \tag{17}$$

where I_p is the peak amplitude of the local oscillator

current drive in each diode and $S(t)$ is a switching function defined as

$$S(t) = 1 \qquad -\pi/2 \leqslant \omega_p t \leqslant \pi/2$$
$$S(t) = -1 \qquad \pi/2 \leqslant \omega_p t \leqslant 3\pi/2 \tag{18}$$

Substituting eqn. 17 into eqn. 16, the general expression for the time-varying incremental resistance becomes

$$r(t) = r_s + \frac{r_b}{1 + X(t) + X(t)S(t)} \tag{19}$$

where $X(t) = X \cos \omega_p t$, $X = I_p/2I_s$ and r_b is the incremental diode resistance at the origin.

The balanced mixer shown in Fig. 1*a* may, taking into account the 180° phase difference of the local oscillator between two pairs of diodes, be described by the following relationship:

$$\begin{bmatrix} v_s \\ \\ i_L \end{bmatrix} = \begin{bmatrix} 2r_s + \dfrac{r_b}{1 + X(t)S(t)} & \dfrac{X(t)}{1 + X(t)S(t)} \\ \\ \dfrac{-X(t)}{1 + X(t)S(t)} & \dfrac{1}{r_b}\left[\dfrac{1 + X(t)S(t)}{1 + X(t)S(t)}\right] \end{bmatrix} \begin{bmatrix} i_s \\ \\ v_L \end{bmatrix} \tag{20}$$

provided $r_b \gg r_s$.

Frequency-selective circuits are normally used in a mixer and only the following signals are present:

$$i_s = i_0 \cos \omega_q t + i_{-2} \cos (2\omega_p - \omega_q) t \tag{21a}$$

and

$$v_L = v_{-1} \cos (\omega_p - \omega_q) t \tag{21b}$$

Substituting eqns. 21*a* and 21*b* into eqn. 20 and performing a frequency-balance operation results in the matrix equation (eqn. 1), where the Fourier coefficients are given by the following relationships:

$$h_{11} = 2r_s + \frac{r_b}{2\pi} \int_{-\pi/2}^{3\pi/2} \frac{d(\omega_p t)}{1 + X(t)S(t)} \tag{22a}$$

$$h_{12} = \frac{1}{2\pi} \int_{-\pi/2}^{3\pi/2} \frac{\cos^2 (\omega_p t)}{1 + X(t)S(t)} d(\omega_p t) \tag{22b}$$

$$h_{13} = \frac{r_b}{2\pi} \int_{-\pi/2}^{3\pi/2} \frac{\cos (2\omega_p t)}{1 + X(t)S(t)} d(\omega_p t) \tag{22c}$$

$$h_{22} = \frac{1}{2r_b \pi} \int_{-\pi/2}^{3\pi/2} \frac{1 + 2X(t)S(t)}{1 + X(t)S(t)} d(\omega_p t) \tag{22d}$$

These integrals are evaluated[9] resulting in the following expressions for the h coefficients of eqn. 1:

$$h_{11} = 2\left(r_s + \frac{r_b}{\pi X} \log_e 2X\right) \tag{23a}$$

$$h_{12} = 2/\pi \tag{23b}$$

$$h_{13} = \frac{2r_b}{\pi X}(2 - \log_e 2X) \tag{23c}$$

$$h_{22} = 2/r_b \tag{23d}$$

4 Results

Practical microwave Schottky-barrier diode (HP 2833) parameters were measured and were found to be $r_b = 2{\cdot}8 \times 10^7\ \Omega$ and $\alpha = 40\ \mathrm{V}^{-1}$.

Common parameters influencing mixer performance are K_0 and a and are defined by eqns. 2*e* and 2*f*, respectively. The parameter K_0 may be expressed in terms of local oscillator drive X and diode parameters using eqn. 23, and is of the form

$$K_0 = \frac{r_b}{\pi^2}\left(r_s + \frac{r_b}{\pi X} \log_e 2X\right)^{-1} \tag{24}$$

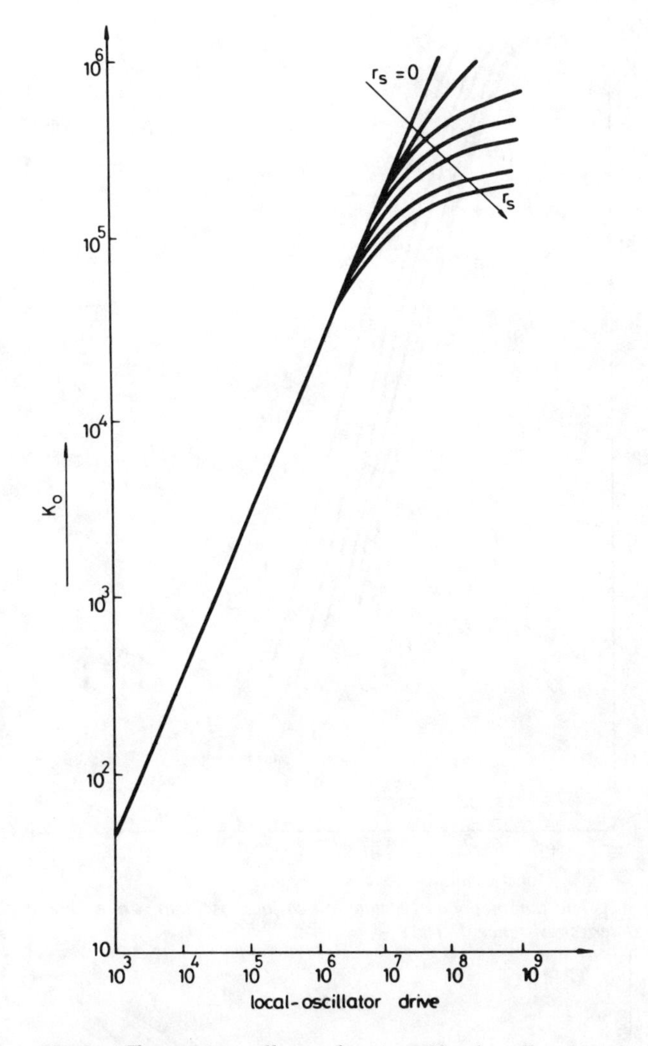

Fig. 3 *The parameter K_0 as a function of local-oscillator drive for diode series resistance 0–18 Ω*

Fig. 3 shows how the parameter K_0 behaves as a function of X. Examining eqn. 24 the parameter K_0 tends to a limiting value of $r_b/\pi^2 r_s$ at high local-oscillator drive.

The parameter a may also be expressed in terms of local-oscillator drive X and diode parameters, and using eqns. 2*f* and 23, is given by

$$a = \frac{r_b}{\pi X}\left[\frac{2 - \log_e 2X}{r_s + \dfrac{r_b}{\pi X} \log_e 2X}\right] \tag{25a}$$

Fig. 4 shows the relationship between the parameter a and the local oscillator drive X. It is seen that a always lies between the limits 0 and −1. For finite diode series resistance r_s and high oscillator drive eqn. 25a tends to the following form:

$$a \simeq \frac{r_b}{r_s \pi X}\,[2 - \log_e 2X] \qquad (25b)$$

Figs. 5–7 show the effect of diode series resistance and local-oscillator drive on the conversion loss, source resistance and load conductance for both methods of design of broadband balanced mixers.

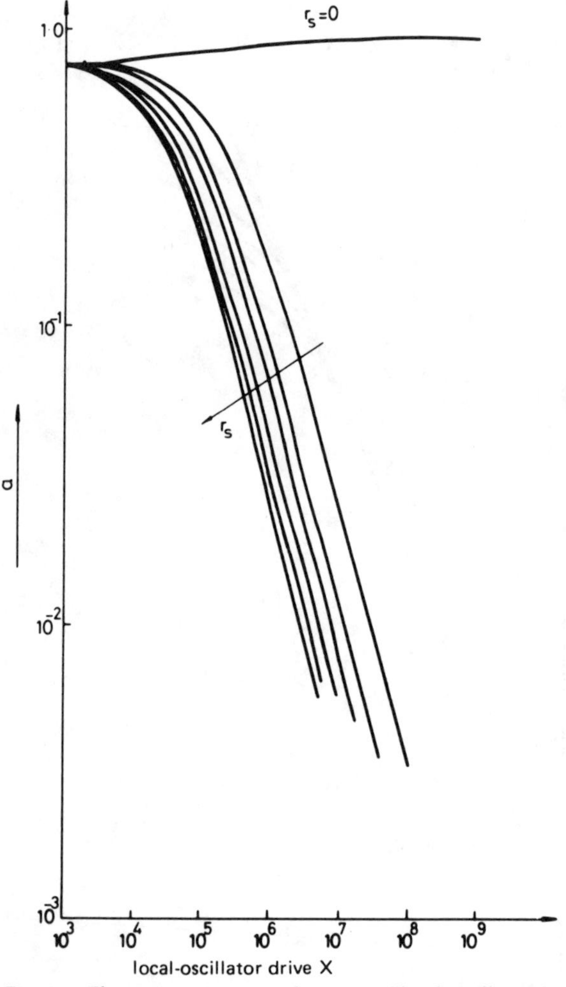

Fig. 4 *The parameter a as a function of local-oscillator drive for diode series resistance 0–18 Ω*

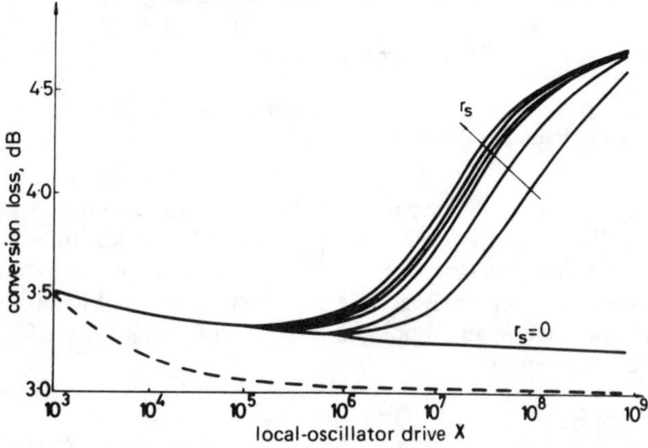

Fig. 5 *Computed conversion loss for an optimum designed and matched broadband balanced mixer for diode series resistance r_s 0–18 Ω*

---- optimum
—— matched

For a mixer designed to have a minimum conversion loss, the parameter $2K_0/(1+a)$ tends, at high oscillator drive, towards a limiting value of $2r_b/\pi^2 r_s$ and, provided this limiting value is much greater than one, the conversion loss defined by eqn. 8 becomes independent of the diode series resistance (see Fig. 5). The conversion loss for a matched mixer is dependent on the diode-series resistance, and Fig. 5 shows that it passes a minimum value before reaching a limiting value at high oscillator drive. Examining eqn. 12 and Fig. 4 it is seen that, as the parameter a approaches zero, the root of eqn. 12 approaches three. Computer analysis indicates that this root remains approximately equal to three for a wide range of local-oscillator drive. Using this value for R in eqn. 9b, the local-oscillator drive X necessary to obtain minimum conversion loss may be shown to be governed by the following transcendental equation:

$$\frac{\pi X}{2}\,[\log_e 2X - 3] = \frac{r_b}{r_s} \qquad (26)$$

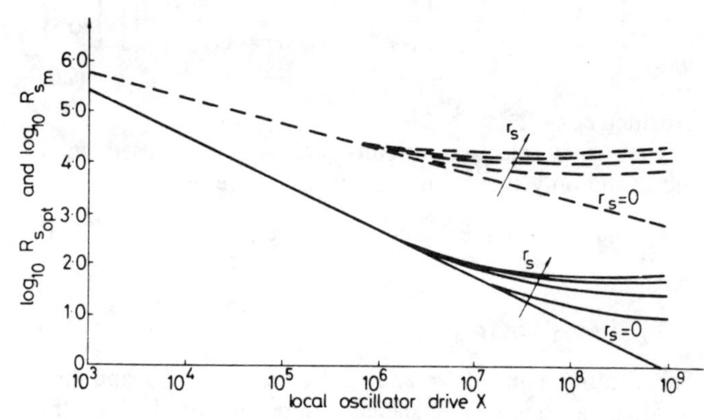

Fig. 6 *Computed source resistance for an optimum designed and matched broadband balanced mixer for diode series resistance r_s 0–18 Ω*

---- optimum
—— matched

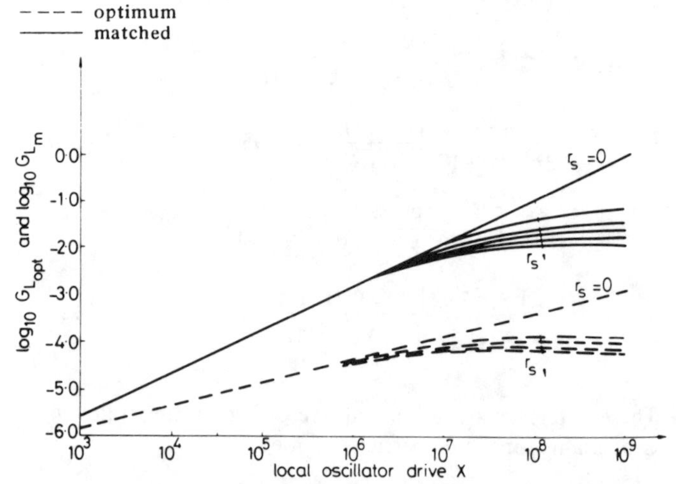

Fig. 7 *Computed load conductance for an optimum designed and matched broadband balanced mixer for diode series resistance 0–18 Ω*

---- optimum
—— matched

The limiting value of the conversion at high oscillator drive is, using eqn. 14, 4·7 dB.

Figs. 6 and 7 show the effect of diode series resistance and local-oscillator drive X on the source resistance and load conductance for both methods of design of broadband mixers. In both cases the source resistance and load conductance tend to limiting values at high local-oscillator drive.

For a mixer designed for minimum conversion loss the limiting values of source resistance and load conductance are

$$R_{s_{opt}} \rightarrow \frac{2}{\pi} [2r_b r_s]^{1/2} \tag{27a}$$

$$G_{L_{opt}} \rightarrow \frac{2}{\pi} \left[\frac{2}{r_b r_s}\right]^{1/2} \tag{27b}$$

For a broadband mixer designed to obtain a match at the input and output ports the limiting values are

$$R_{s_m} \rightarrow 4r_s \tag{28a}$$

$$G_{L_m} \rightarrow 4/(3\pi^2 r_s) \tag{28b}$$

5 Conclusion

The performance of broadband balanced mixers is influenced by the choice of the source impedance at the r.f. port. An improvement in the conversion loss is obtained by mismatching the r.f. port and, as shown in Fig. 5, a theoretical limit of 3 dB can be approached for a large oscillator drive. The diode-series resistance does not significantly affect the conversion loss provided the diode incremental resistance at the origin (r_b) is large, a condition normally satisfied by Schottky-barrier diodes. The basic problem of achieving the theoretical limit of 3 dB at large local-oscillator drive is that large values of source impedance are required.

If the source is chosen to produce matched conditions at the r.f. port, the conversion loss is not only increased but is also influenced by the diode-series resistance (r_s) and the diode incremental resistance at the origin. The local-oscillator drive necessary to obtain a minimum conversion loss is governed by the transcendental equation (eqn. 26). Matching at the r.f. port has an advantage of requiring a lower source resistance as compared with a mixer designed to produce minimum conversion loss. It is interesting to note that the increase of the conversion loss at high local-oscillator drive for a matched mixer is not due to the diode series resistance, as has been assumed, but is actually due to the deviations of source resistance from its value to give minimum conversion loss.

6 References

1 TORREY, H.C., and WHITMER, C.A.: 'Crystal rectifiers' (MIT Press, 1948)
2 STRUM, P.D.: 'Some aspects of mixer crystal performance', *Proc. IRE*, 1953, **41**, pp. 875–884
3 BURKLEY, C.J., and O'BRIEN, R.S.: 'Optimization of an 11 GHz mixer circuit using image recovery', *Int. J. Electron.*, 1975, **38**, pp. 777–787
4 HOWSON, D.P.: 'Minimum conversion loss and input match conditions in the broad-band mixer', *Radio & Electron. Eng.*, 1972, **42**, pp. 237–242
5 SCHNEIDER, M.V., and SNELL, W.W.: 'Harmonically pumped stripline down-converter', *IEEE Trans.*, 1975, **MIT-23**, pp. 271–275
6 BUCHS, J.D., and BEGEMANN, G.: 'Frequency conversion using harmonic mixers with resistive diodes', *IEE J. Microwaves, Opt. & Acoust.*, 1978, **2**, pp. 71–76
7 SALEH, A.A.M.: 'Theory of resistive mixers' (MIT Press, 1971)
8 'Survey of applicable mathematics' (Iliffe, 1968)
9 BOIS, G.P.: 'Tables of indefinite integrals' (Dover, 1961)

INFLUENCE OF DIODE CAPACITANCE ON PERFORMANCE OF BALANCED MICROWAVE MIXERS

Indexing terms: Microwave circuits, Mixers

The effect of the diode capacitance parasitics on the performance of microwave double balanced lattice mixers, driven by a local oscillator having a large internal resistance and a current sinusoidal waveform, is examined. It is found that the current present in the diodes is modified by the diode capacitance and this in turn influences the performance of the mixers. To reduce the effect of the diode capacitance on the performance of the mixer it is necessary to reduce the internal resistance of the local oscillator. There is therefore a compromise between the need of a low noise figure, and hence a requirement of having a large internal resistance of the local oscillator, and the undesired effect on the diode current, and hence on the performance of the mixer.

Introduction: In the analysis of mixers the effect of the diode parasitics on the waveform of the local oscillator has usually been ignored. Rustom and Howson,[1] using a resistive diode model, have confirmed the conclusion reached by Stracca[2] that H and G mixers[3] driven by a local oscillator with a large internal resistance and having a current waveform are the most promising mixer circuits for low system noise figure. This letter analytically examines the performance of lattice H mixers driven by a local oscillator having a current waveform and a large internal resistance when the effect of the diode capacitance parasitics is included.

Large signal analysis of a lattice mixer with capacitance: The effective circuit of a lattice configuration of the four diodes as 'seen' by the local oscillator is shown in Fig. 1*a*. Reducing the circuit to that shown in Fig. 1*b*, it can be shown that the current present in each diode to a first approximation is a truncated half-wave rectified sinewave shown in Fig. 1*c*, where the angle of truncation is given by[4]

$$\sin^2 \theta_{c/2} = \frac{\varepsilon}{2} \log_e[4X^2(2\varepsilon/\pi)^{1/2}]$$

where $\varepsilon = X_c r_b/4X$, X_c is the reactance of the capacitance C_e, r_b is the incremental diode resistance at the origin and X is the normalised current drive $(I_p/2I_s)$. The resultant current waveforms in the 'on' diode may be expressed in the form

$$i_1 = \tfrac{1}{2}[1 + S_1(t)]I_p \cos \omega_p t \tag{2a}$$

and in the 'off' diode in the form

$$i_2 = \tfrac{1}{2}[1 + S_2(t)]I_p \cos \omega_p t \tag{2b}$$

The two switching functions $S_1(t)$ and $S_2(t)$ are defined as

$$S_1(t) = \begin{cases} -1 \text{ for } -\pi/2 \le \omega_p t \le -\pi/2 + \theta_c \\ 1 \text{ for } -\pi/2 + \theta_c \le \omega_p t \le \pi/2 \\ -1 \text{ for } \pi/2 \le \omega_p t \le 3\pi/2 + \theta_c \end{cases} \tag{3a}$$

$$S_2(t) = \begin{cases} 1 \text{ for } -\pi/2 \le \omega_p t \le \pi/2 + \theta_c \\ -1 \text{ for } \pi/2 + \theta_c \le \omega_p t \le 3\pi/2 \end{cases} \tag{3b}$$

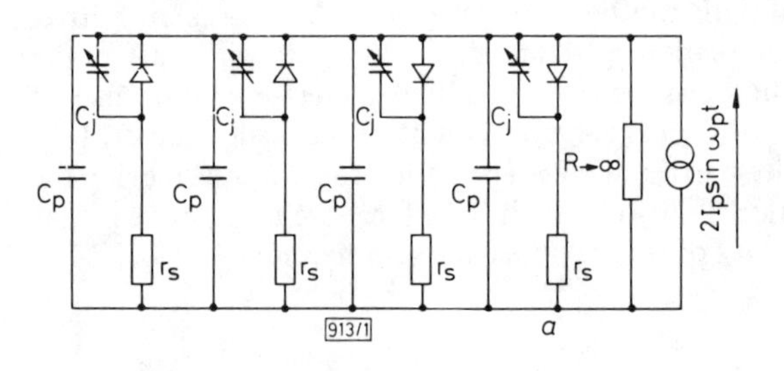

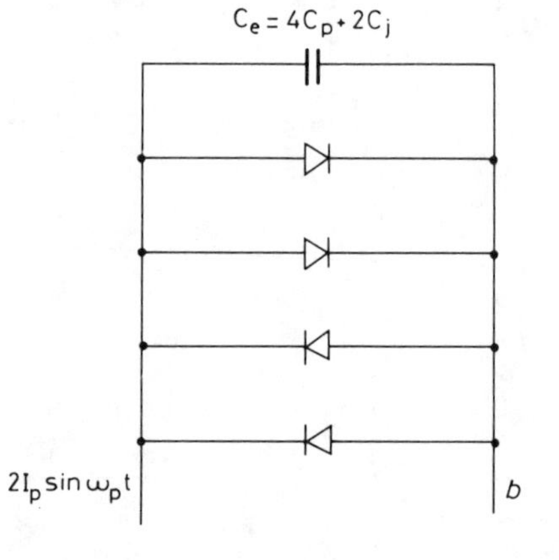

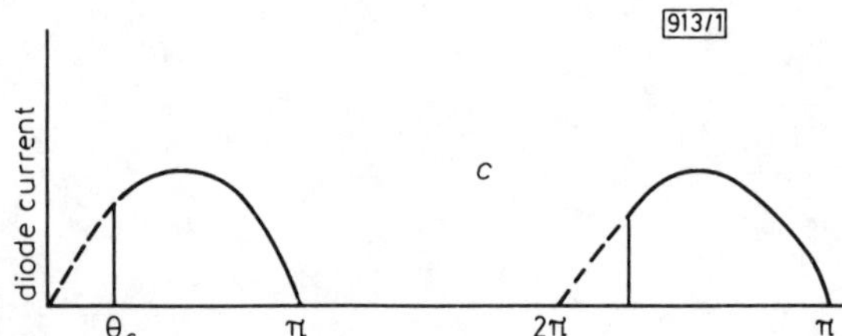

Fig. 1

Reprinted with permission from *Electron. Lett.*, vol. 17, pp. 144–146, Feb. 5, 1981.

The time varying resistances of the 'on' and 'off' diodes, taking into account the 180° phase difference of the local oscillator current at the two pairs of the diodes are

$$r_+(t) = r_s + \frac{r_b}{1 + X(t) + X(t)S_1(t)} \quad (4a)$$

$$r_-(t) = r_s + \frac{r_b}{1 - X(t) + X(t)S_2(t)} \quad (4b)$$

where $X(t) = X \cos \omega_p t$.

Small signal analysis of a lattice H mixer: Using a bisection theorem, the equivalent circuit of a lattice mixer is found as shown in Fig. 2. In the case of an H mixer, the effective capacitance at the IF port can be incorporated in the parallel tuned circuit. The overall circuit matrix $[a]$ for the circuit in Fig. 2 is given by

$$[a] = \begin{bmatrix} A + Cr_s & B + r_s[A + D + r_s C] \\ j(\omega/\omega_c)\frac{A}{r_s} + C[1 + j(\omega/\omega_c)] & j(\omega/\omega_c)\frac{B}{r_s} + D[1 + j(\omega/\omega_c)] \end{bmatrix} \quad (5)$$

where $\omega_c = 1/C_e r_s$. Provided $(\omega/\omega_c) \ll 1$, eqn. 5 reduces to

$$[a] = \begin{vmatrix} A + Cr_s & r_s[A + D + r_s C] \\ C & D[1 + r_s] \end{vmatrix} \quad (6)$$

and the effect of the diode capacitance C_e on the small signal analysis can be ignored. Thus a general matrix equation describing an H mixer is

$$\begin{vmatrix} V_1 \\ I_2 \end{vmatrix} = \begin{vmatrix} \Sigma a_n \cos n\omega_p t & \Sigma b_n \cos n\omega_p t \\ -\Sigma b_n \cos n\omega_p t & \Sigma c_n \cos n\omega_p t \end{vmatrix} \begin{vmatrix} I_1 \\ V_2 \end{vmatrix} \quad (7)$$

By evaluating the Fourier coefficients of eqn. 7, it may be shown that the coefficients of the H matrix for a small but finite angle of truncation θ_c are given by

$$h_{11} \simeq 2r_s + \frac{r_b\theta_c}{\pi} \simeq \frac{r_b\theta_c}{\pi}; \quad h_{12} \simeq 2/\pi$$

$$h_{13} \simeq \frac{-r_b}{2\pi} \sin 2\theta_c; \quad h_{22} \simeq 2/r_b \quad (8)$$

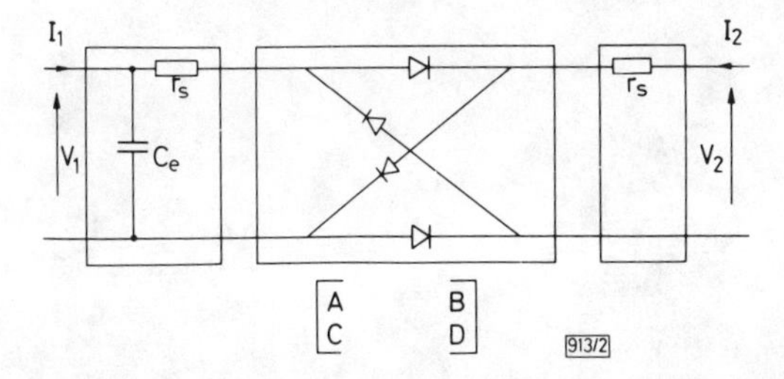

Fig. 2

Performance of lattice H mixers: Using the general expressions derived by Saleh[3] the performance of the lattice H mixers having different terminations at the image port can be readily determined. For a lattice H mixer with image open circuit the optimum terminations at the RF and IF ports and the resulting conversion loss are

$$R_{S_{opt}} \simeq \frac{r_b\theta_c}{\pi}[1 + (2/\pi\theta_c)]^{1/2}$$

$$G_{I_{opt}} \simeq \frac{2}{r_b}[1 + (2/\pi\theta_c)]^{1/2} \quad (9)$$

$$L_{opt} \simeq \frac{1 + [1 + (2/\pi\theta_c)]^{1/2}}{-1 + [1 + (2/\pi\theta_c)]^{1/2}}$$

For a lattice H mixer with image short circuit the corresponding equations are

$$R_{S_{opt}} \simeq \frac{4r_b\theta_c^3}{3\pi}[1 + (3/\theta_c^2)]^{1/2}$$

$$G_{I_{opt}} \simeq \frac{2}{r_b}[1 + (3/\theta_c^2)]^{1/2} \quad (10)$$

$$L_{opt} = \frac{1 + [1 + (3/\theta_c^2)]^{1/2}}{-1 + [1 + (3/\theta_c^2)]^{1/2}}$$

Finally, for the broadband H mixer, the equations for optimum terminations and conversion loss are

$$R_{S_{opt}} \simeq \frac{2r_b\theta_c^3}{3}[1 + (6/\pi\theta_c^3)]^{1/2}$$

$$G_{I_{opt}} \simeq \frac{2}{r_b}[1 + (6/\pi\theta_c^3)]^{1/2} \quad (11)$$

$$L_{opt} \simeq \frac{1 + [1 + (6/\pi\theta_c^3)]^{1/2}}{-1 + [1 + (6/\pi\theta_c^3)]^{1/2}}$$

Conclusion: The main effect of the diode effective capacitance is to considerably increase the conversion loss for low local oscillator drive and increase in magnitude the required terminations for the three types of lattice H mixers.

To minimise the effect of the diode capacitive parasitics on the performance of a lattice mixer, it is necessary to reduce the parameter ε. This can be done by lowering the frequency of the local oscillator and/or increase the local oscillator current drive X. There are, however, practical limitations to both these solutions. An alternative method is to allow the effective diode capacitance to discharge during the switching action of the four diodes by reducing the output resistance of the local oscillator. This effect was practically verified using an analogue model of a high frequency lattice mixer.

A compromise must be made in practice between the need to have a large output impedance of the local oscillator to reduce the noise figure of the mixer, as suggested by Howson and Stracca, and the need to lower the output impedance of the local oscillator so that the angle of truncation of the diode current can be minimised.

E. KOROLKIEWICZ

School of Electronic Engineering
Faculty of Engineering
Newcastle upon Tyne Polytechnic, England

B. L. J. KULESZA

Department of Applied Physics and Electronics
University of Durham, Durham, England

22nd December 1980

References

1 RUSTOM, S., and HOWSON, D. P.: 'Mixer noise figure using an improved resistive model', *Int. J. Electronics*, 1976, **41**
2 STRACCA, G. B.: 'Noise in frequency mixers using non-linear resistors', *Alta Freq.*, 1971, 6
3 SALEH, A. A. M.: 'Theory of resistive mixers' (MIT Press, 1971)
4 ARMSTRONG, R., KOROLKIEWICZ, E., and KULESZA, B. L. J.: 'Large signal waveforms in microwave balanced mixers with capacitance', *Proc. IEE*, 1978, **125**, (8), pp. 728–729

Predicted conversion gain in superconductor-insulator-superconductor quasiparticle mixers

J. R. Tucker
The Aerospace Corporation, P.O. Box 92957, Los Angeles, California 90009

(Received 8 October 1979; accepted for publication 27 December 1979)

A computer simulation is constructed to describe the performance of superconductor-insulator-superconductor (SIS) quasiparticle heterodyne receivers in terms of photon-assisted tunneling theory. The results predict that conversion efficiencies greater than unity can be realized at millimeter wave frequencies, with noise temperatures approaching the fundamental quantum limit. The nonlinearity responsible for the mixing is the extremely rapid onset of quasiparticle tunneling current in SIS junctions near the full gap voltage $2\Delta/e$, and is independent of the Josephson pair currents. Substantial gain is expected under a variety of conditions and for a wide range of source and load impedances.

PACS numbers: 73.40.Gk, 73.40.Ei, 74.50. + r

Attention has recently been focused on the use of the quasiparticle nonlinearity in superconductor-insulator-superconductor (SIS) tunnel junctions for mixing and detection at microwave and millimeter wave frequencies.[1-3] Richards, Shen, Harris, and Lloyd[1] reported single sideband heterodyne conversion efficiencies of 0.16 and a mixer noise temperature, $T_M < 14\,^{\circ}\mathrm{K} = 8\hbar\omega/k$, within an order of magnitude of the quantum limit at 36 GHz. Simultaneously, Dolan, Phillips, and Woody[2] reported conversion efficiencies of roughly 0.10 with an upper bound to the noise temperature of $T_M < 100\,^{\circ}\mathrm{K}$ at 115 GHz. Both groups used Pb-alloy junctions whose quasiparticle *I-V* characteristics were quite rounded compared to the step discontinuity of an ideal junction at the gap voltage, and substantial leakage currents were also present. In both sets of experiments, hysteretic Josephson mixing was observed at low dc bias which was unstable and very noisy.[4] At bias voltages approaching $2\Delta/e$, however, a much quieter mode was observed that can be attributed to mixing on the quasiparticle nonlinearity. The behavior in this region was substantially unaffected by the application of a dc magnetic field, implying that the Josephson pair currents play no significant role. SIS quasiparticle receiver performance obtained to date is roughly consistent with classical microwave mixer analysis, given the extreme nonlinearity of the *I-V* characteristic in the neighborhood of the gap voltage, although distinct features characteristic of photon-assisted tunneling are observed.

The theory of photon-assisted tunneling has been extended to construct a quantum generalization of classical microwave mixer analysis, including noise, for nonlinear single-particle tunneling devices.[5,6] This work predicts that quasiparticle heterodyne receivers are intrinsically capable of approaching the ultimate photon shot noise limit for sensitivity in the detection of electromagnetic radiation. The theory may also be used to model the performance expected for a particular mixer diode. Selected results from a simulation developed on this basis for an ideal SIS quasiparticle heterodyne receiver will be presented here. The properties of the junction are characterized by a complex response function $j(\omega)$. The imaginary part of this function is simply the dc current-voltage relation $\mathrm{Im}[j(\omega) = I_{\mathrm{dc}}(\hbar\omega/e)]$. The real part $\mathrm{Re}[j(\omega)]$ represents the reactive response, and is related to the dc *I-V* characteristic through a Kramers-Kronig transform. Closed form expressions for these quasiparticle response functions in an ideal SIS tunnel junction at low temperatures $kT \ll \Delta$ were obtained by Werthamer[7] and analyzed by Harris.[8,9] The results for a symmetric junction are reproduced in Fig. 1.

The equivalent circuit employed in the mixer simulation consists of the basic 3-port heterodyne receiver model described in Appendix C of Ref. 6. A sinusoidal local oscillator drive $V(t) = V_0 + V_{\mathrm{LO}} \cos\omega t$ applied to the nonlinear junction couples the output frequency ω_0 with all sidebands $\omega_m = m\omega + \omega_0$. Here only the signal, output, and image frequencies, $\omega_1, \omega_0, \omega_{-1}$ are included. All higher harmonics and their sidebands are assumed shorted through the junction capacitance. The output frequency is also assumed low in the sense $\omega_0 \ll \omega$, so that the approximation $\omega_1 \approx \omega \approx -\omega_{-1}$ may be used in computing the elements of the small signal admittance matrix for the pumped diode. The conductance components $G_{mm'}$ are then given for this model by Eq. (C4) of Ref. 6. In addition, the complete admittance matrix also contains intrinsically quantum-mechanical reactive components which may be obtained from Eq. (7.22) of Ref. 6 for $\omega_0 \ll \omega$ in the form

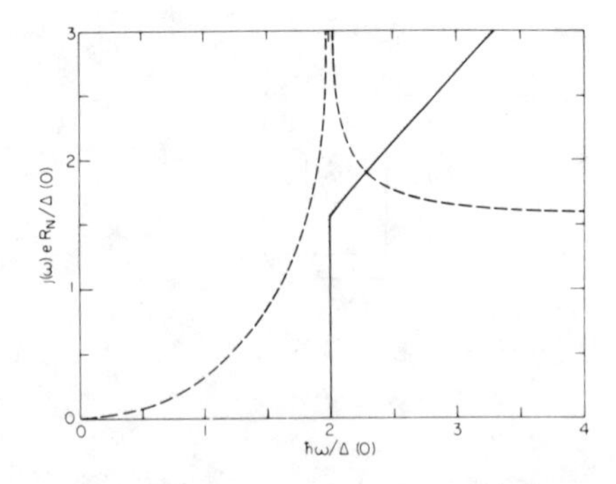

FIG. 1. Quasiparticle current response function for an ideal SIS junction between identical superconductors in the limit $kT \ll \Delta$. The dashed curve represents $\mathrm{Re}[j(\omega)]$, and $\mathrm{Im}[j(\omega)]$ is indicated by the solid line.

$$B_{00} = \frac{\hbar\omega_0}{2e} \sum_{n=-\infty}^{\infty} J_n^2(\alpha) \times \frac{d^2}{dV_0^2} \mathrm{Re}[\, j(n\omega + eV_0/\hbar)] \,,$$

$$B_{10} = -B_{-10}$$

$$= \frac{1}{2} \sum_{n=-\infty}^{\infty} J_n(\alpha)[J_{n-1}(\alpha) - J_{n+1}(\alpha)] \times \frac{d}{dV_0} \mathrm{Re}[\, j(n\omega + eV_0/\hbar)] \,,$$

$$B_{01} = B_{0\text{-}1} = (\omega_0/\omega) B_{10} \,,$$

$$B_{11} = -B_{\text{-}1\text{-}1}$$

$$= \frac{e}{2\hbar\omega} \sum_{n=-\infty}^{\infty} J_n^2(\alpha) \, \mathrm{Re}\{\, j[(n+1)\omega + eV_0/\hbar] - 2j(n\omega + eV_0/\hbar) + j[(n-1)\omega + eV_0/\hbar]\},$$

$$B_{1\text{-}1} = -B_{\text{-}11}$$

$$= \frac{e}{2\hbar\omega} \sum_{n=-\infty}^{\infty} J_{n-1}(\alpha) J_{n+1}(\alpha) \times \mathrm{Re}\{\, j[(n+1)\omega + eV_0/\hbar] - 2j(n\omega + eV_0/\hbar) + j[(n-1)\omega + eV_0/\hbar]\} \,, \tag{1}$$

where $\alpha = eV_{\mathrm{LO}}/\hbar\omega$. Inclusion of these reactive terms is required in order to obtain a physically consistent simulation at high frequencies. In the limit $\omega_0 \ll \omega$, however, the $B_{01} = B_{0\text{-}1}$ terms are seen to become vanishingly small and are therefore neglected. It is further assumed that the diagonal reactive components $B_{11} = -B_{\text{-}1\text{-}1}$ and B_{00} are tuned out, along with the ordinary junction capacitance, so that the effective signal and load terminations $Y_1 = Y_{-1} = G_S$ and $Y_0 = G_L$ may be represented as purely resistive.

The calculated conversion loss obtained for this simple heterodyne receiver model may be conveniently expressed in the form

$$L_c = L_0 \frac{1}{4\eta y_S y_L} \frac{[(\xi + y_S)(1 + y_S) - \gamma^2]^2}{[(\xi + y_S)^2 + \gamma^2]} (y_L + y_L^0)^2 \,, \tag{2}$$

where

$$y_L^0 = 1 - \eta \frac{[(\xi + y_S) - \beta\gamma]}{[(\xi + y_S)(1 + y_S) - \gamma^2]} \,, \tag{3}$$

and

$$L_0 = 2G_{10}/G_{01}, \quad \eta = 2G_{01}G_{10}/[G_{00}(G_{11} + G_{1\text{-}1})] \,,$$
$$y_S = G_S/(G_{11} + G_{1\text{-}1}) \,, \quad y_L = G_L/G_{00} \,, \tag{4}$$
$$\xi = (G_{11} - G_{1\text{-}1})/(G_{11} + G_{1\text{-}1}) \,, \quad \beta = B_{10}/G_{10} \,,$$
$$\gamma = B_{1\text{-}1}/(G_{11} + G_{1\text{-}1}) \,.$$

For a particular choice of effective source impedance $R_S = G_S^{-1}$, the conversion loss in Eq. (2) is minimized for $y_L = y_L^0$, corresponding to a load impedance $R_L^0 = (G_{00} y_L^0)^{-1}$, so long as this quantity remains positive. The dependence of L_c on the load impedance is seen, in fact, to be characterized by a simple linear mismatch relationship,

$$L_c = L_c^0 \frac{(R_L + R_L^0)^2}{4R_L^0 R_L} \,, \tag{5}$$

where L_c^0 represents the conversion loss expression of Eq. (2) evaluated for $y_L = y_L^0$. Within the context of this 3-port heterodyne receiver model, therefore, conversion gain becomes possible if practical values of dc bias V_0, local oscillator amplitude V_{LO}, and effective source impedance R_S can be found for which $L_c^0 < 1$. If the quantity y_L^0 of Eq. (3) should become negative, moreover, the conversion loss is seen to vanish—implying arbitrarily large conversion gain—as the load impedance approaches $R_L = |R_L^0|$. Neither of these conditions is possible in a classical resistive mixer. The quantum-mechanical reactive terms of Eq. (1) have been shown[6] to vanish in the limit where the photon energy becomes small relative to the voltage scale of the dc nonlinearity. The real components of the admittance matrix $G_{mm'}$ reduce to the Fourier coefficients $G[(m - m')\omega]$ of the modulated conductance in this same limit. The resulting classical 3-port mixer model has been extensively analyzed, and Torrey and Whitmer[10] proved that gain cannot be achieved in this context without a region of negative resistance. Both of the above conditions for conversion gain are,

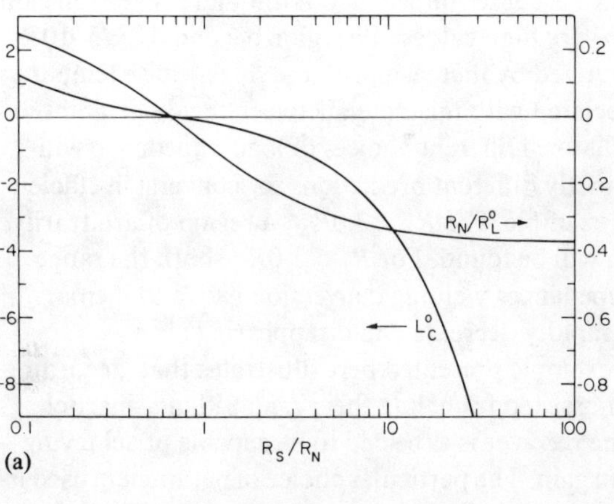

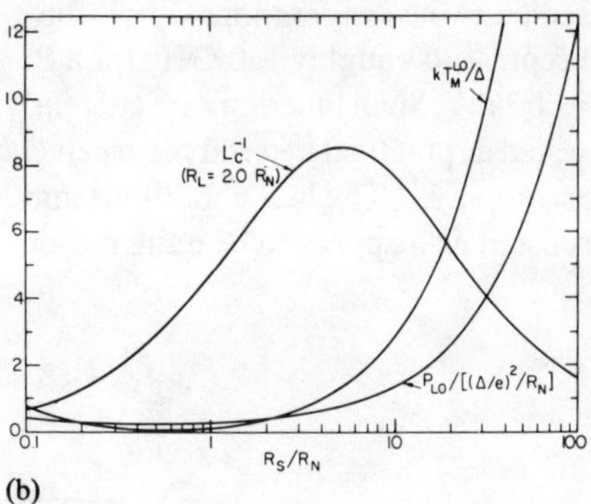

FIG. 2. Computed results of the SIS heterodyne receiver simulation in the limit $kT \ll \Delta$ for $\hbar\omega = 0.4\Delta$, $eV_0 = 1.8\Delta$, and $eV_{\mathrm{LO}} = 0.5\Delta$. (a) The parameters L_c^0 and R_L^0 appearing in Eq. (5), and (b) the conversion efficiency L_c^{-1} for $R_L = 2.0R_N$, local oscillator contribution T_M^{LO} to the mixer noise temperature, and incident power P_{LO} as functions of effective source impedance.

however, predicted to occur in the SIS quasiparticle mixer simulation based on photon-assisted tunneling theory.

Figure 2 summarizes the results of this SIS heterodyne receiver simulation under the following set of conditions. The signal frequency is given in terms of the gap parameter by $\hbar\omega = 0.40\Delta$. The dc bias voltage and the amplitude of the local oscillator waveform are taken to be $V_0 = 1.80\Delta/e$ and $V_{\rm LO} = 0.50\Delta/e$, respectively. For this particular example, the calculated values obtained for the parameters L_c^0 and $1/R_L^0$ of Eq. (5) are plotted in Fig. 2(a) as functions of effective source impedance R_S. Both source and load impedances are referenced to the normal resistance R_N of the junction. In Fig. 2(a), L_c^0 is seen to decrease below unity and to become negative with increasing source impedance. The minimum conversion loss in Eq. (5) is achieved for $R_L = |R_L^0|$. When R_L^0 becomes negative, the conversion loss vanishes as this point is approached. Arbitrarily large conversion gain is then predicted within the context of this linearized mixer theory.

Figure 2(b) illustrates the computed conversion efficiency L_c^{-1} in this example for a particular choice of load impedance $R_L = 2.0R_N$. Also shown are the contribution $T_M^{\rm LO}$ of the local oscillator shot noise to the overall mixer noise temperature, computed using Eqs. (C5), (7.13), and (7.17) of Ref. 6, together with the level of incident power $P_{\rm LO}$ associated with the waveform amplitude $V_{\rm LO} = 0.50\Delta/e$. Substantial gain is seen to be available for $R_L = 2.0R_N$ over a wide range of source impedance. Although conversion gain persists to very high values, the region beyond $R_S \gtrsim 5$–$10R_N$ is characterized by increasingly large mixer noise temperatures associated with mismatch between the diode and the local oscillator. Different choices of load impedance will yield distinctly different predictions for conversion efficiency in this example. For $R_L \gtrsim 3.0R_N$, a region of arbitrarily large gain will be found. For $R_L \lesssim 1.0R_N$, both the range of source impedances yielding conversion gain and its magnitude will rapidly decrease and disappear.

The example presented here illustrates that, according to photon-assisted tunneling theory, an SIS quasiparticle heterodyne receiver is expected to be capable of achieving conversion gain. The particular choice of parameters used in this illustration is based on two considerations. First, the frequency considered represents roughly 120 GHz for a Pb-alloy junction with $\Delta \approx 1.3$ mV. Such junctions are currently being developed and tested as practical heterodyne receivers for millimeter-wave astronomy.[11] The bias conditions and impedance ranges discussed also appear to be quite reasonable, and it should eventually prove feasible to approximate them in an operating receiver. The second criterion for selecting this example involves the singularities in both the dissipative and reactive components of the response function at the full gap voltage for the ideal SIS junction shown in Fig. 1. The performance of the diode within the context of photon-assisted tunneling theory is determined by the values for these response functions at integral multiples $V_0 + n\hbar\omega/e$ of the photon energy above and below the dc bias point. At a frequency $\hbar\omega = 0.40\Delta$, the gap edge in Fig. 1 lies half-way between $V_0 = 1.80\Delta/e$ and $V_0 + \hbar\omega/e = 2.20\Delta/e$; the separation here is sufficiently large that the results should be relatively insensitive to a small rounding of the *I-V* characteristic in the immediate neighborhood of the gap voltage. Although some refinements in both diode fabrication and mixer construction may be required, the conditions described here lie well within the potential of available technology. Experimental confirmation of conversion gain in this type of SIS quasiparticle mixer, coupled with their already demonstrated low noise properties, would then open the door to construction of practical millimeter wave receivers operating near the ultimate quantum limit for sensitivity.

This work was performed at the Department of Physics, University of California, Berkeley, on a short-term study assignment sponsored by The Aerospace Corporation. I am most grateful to Professor P.L. Richards and his group for their generous hospitality during this period. I am also pleased to acknowledge stimulating discussions with P.L. Richards, T.M. Shen, R.Y. Chiao, and C.H. Townes on the substance of the present work. A portion of this effort was supported by the U.S. Office of Naval Research.

[1]P.L. Richards, T.M. Shen, R.E. Harris, and F.L. Lloyd, Appl. Phys. Lett. **34**, 345 (1979).
[2]G.J. Dolan, T.G. Phillips, and D.P. Woody, Appl. Phys. Lett. **34**, 347 (1979).
[3]S. Rudner and T. Claeson, Appl. Phys. Lett. **34**, 711 (1979).
[4]Y. Taur, J.H. Claassen, and P.L. Richards, Appl. Phys. Lett. **24**, 101 (1974).
[5]J.R. Tucker and M.F. Millea, Appl. Phys. Lett. **33**, 611 (1978).
[6]J.R. Tucker, IEEE J. Quantum Electron. **QE-15**, 1234 (1979).
[7]N.R. Werthamer, Phys. Rev. **147**, 255 (1966).
[8]R.E. Harris, Phys. Rev. B **10**, 84 (1974).
[9]R.E. Harris, Phys. Rev. B **11**, 3329 (1975).
[10]H.C. Torrey and C.A. Whitmer, *Crystal Rectifiers*, MIT Radiation Lab. Series, Vol. 15 (McGraw-Hill, New York, 1948). See Sections 5.7 and 13.5.
[11]D.P. Woody (private communication).

Part II
Accurate Modeling and Computer-Aided Analysis of Schottky Diode Mixers

IN this part methods of more detailed analysis are presented. The difficult problem of solving the nonlinear equations describing the voltage and current waveform of the pumped diode has been attacked by various authors. In the first two papers Egami and Kerr, respectively, discuss methods for this purpose. Kerr's technique has subsequently been used by several authors and is probably preferred if high accuracy is required. In the now classic papers by Held and Kerr (papers 3 and 4), the authors present a complete model for single-ended mixers which accurately describes the conversion properties as well as the noise behavior.

Faber and Gwarek in their paper have developed a technique for analyzing a mixer with any number of diodes. As an example they investigate properties of the two-diode crossbar balanced mixer.

In the next paper Hicks and Khan show a somewhat different approach of how to calculate properties of subharmonically pumped mixers. The nonlinear part of the analysis is an extension of a technique developed by the same authors for single diode analysis [1]. Balanced as well as two-diode subharmonically pumped mixers have also been extensively developed by Kerr [2].

In the final paper of this part Hines investigates some single diode mixer properties using a time domain analysis. From an analytical viewpoint, this method provides some new insights into mixer behavior and fundamental limits which are not so easily realized by the frequency domain methods used by most other authors.

References

[1] R. G. Hicks and P. J. Khan, "Numerical analysis of nonlinear solid-state device excitation in microwave circuits," *IEEE Trans. Microwave Theory Tech.*, vol. MTT-30, pp. 251–259, Mar. 1982.

[2] A. R. Kerr, "Noise and loss in balanced and subharmonically pumped mixers, Parts I and II, theory and application," *IEEE Trans. Microwave Theory Tech.*, vol. MTT-27, pp. 938–950, Dec. 1979.

Nonlinear, Linear Analysis and Computer-Aided Design of Resistive Mixers

SHUNICHIRO EGAMI, MEMBER, IEEE

Abstract—Nonlinear large-signal analysis of local current shape and linear small-signal analysis of small-signal products are made for resistive mixers. Iteration adapted from Newton's method was used in the nonlinear analysis. The conjugate match method was used in the linear analysis to find the minimum conversion loss. These theories were applied to a new type spurious suppressed mixer and a reliable computer simulation was made.

I. Introduction

WITH THE ADVENT of the Schottky barrier diode, performance of the mixer was substantially improved. At the same time, mathematical treatment of the mixer was also made by many investigators [1]–[5]. In spite of the existence of detailed mixer theories, actual mixer circuit design still involves a laborious trial and error method, and realistic realizable minimum conversion loss cannot be found theoretically, even if diode characteristics are correctly given. The difficulty in applying existing mixer theory to the actual circuit design arises from the fact that the amplitude and phase of the local and its harmonic current, which is a fundamental determinant of mixer performance, cannot be obtained for an actual mixer in general, and that the necessary order of harmonics to make a realistic mixer analysis is not clearly given. The latter difficulty arises because, although the power of the nth local harmonic decreases by $1/n^2$ [6], amplitude of the nth harmonic current has no such restriction. Therefore, there is no sure reason to consider the first two harmonics only, as was always done in previous papers.

The aim of this paper is to provide a more realistic theoretical estimate of the mixer performance, by which actual circuit design can be made. The local and its harmonic current are obtained, when load admittances to each harmonic are given, which was not considered in previous papers. The conversion matrix, which is determined by the harmonic current, is used to get an optimum mixer performance. These nonlinear and linear analyses are combined into a single computer program, in which the order of the harmonics considered can be increased as desired. These analyses are applied to a new type mixer in which leakage of unnecessary harmonic and small-signal product are suppressed, thus making a low conversion loss possible.

Manuscript received June 25, 1973; revised September 24, 1973.
The author is with the Electrical Communication Laboratories, Nippon Telegraph and Telephone Public Corporation, Yokosuka-shi, 238-03, Japan.

II. Nonlinear Large-Signal Analysis to Find the Elements of the Conversion Matrix

When local oscillator power with frequency ω_p is fed into the diode, currents with frequency $n\omega_p$ $(n = 0,1,2,\cdots)$ flow across the junction of the diode. These currents determine the small-signal admittance of the diode. Therefore, the conversion matrix which determines the relation among small-signals is determined by these currents. Since finding these large-signal currents requires a nonlinear analysis, it is impossible to get a general analytical solution. However, using the iteration method, a solution can be found, if the load admittances to each harmonic are prescribed.

At the junction of the diode, voltage V and current I satisfy the following relation:

$$I = I_s \exp(\beta V) \tag{1}$$

where I_s and β are determined by the diode used.

Actually, the diode has a series resistance R_s, junction capacitance C_j, and lead inductance L_s, but these elements can be included in the load admittances which represent the outer circuit seen from the junction. Differentiating (1) by V, one obtains the small-signal admittance G_j:

$$G_j = \beta I. \tag{2}$$

This equation indicates that the small-signal admittance is determined by the current of the diode.

When local oscillator power is fed into the diode, harmonic $(n\omega_p,\ n = 0,1,2,\cdots)$ currents and voltages arise in the diode. Therefore, V and I can be represented as follows, if the first N harmonics are taken into consideration:

$$V = \sum_{n=-N}^{N} V_n \exp(jnx) \tag{3}$$

$$I = \sum_{n=-N}^{N} I_n \exp(jnx) \tag{4}$$

where

$$x = \omega_p t \qquad V_n = V_{-n}^* \qquad I_n = I_{-n}^*.$$

The equivalent circuit for this large-signal nonlinear analysis is shown in Fig. 1. Y_n is the load admittance to the nth harmonic $(n\omega_p)$ generated in the diode, and E_p is the voltage of the local oscillator. Since V and I of (3) and (4) satisfy (1), the mth harmonic current I_m, for example, can be determined by V_n $(n = 0,1,2,\cdots N)$, from the following equation:

Reprinted from *IEEE Trans. Microwave Theory Tech.*, vol. MTT-22, pp. 270–275, Mar. 1973.

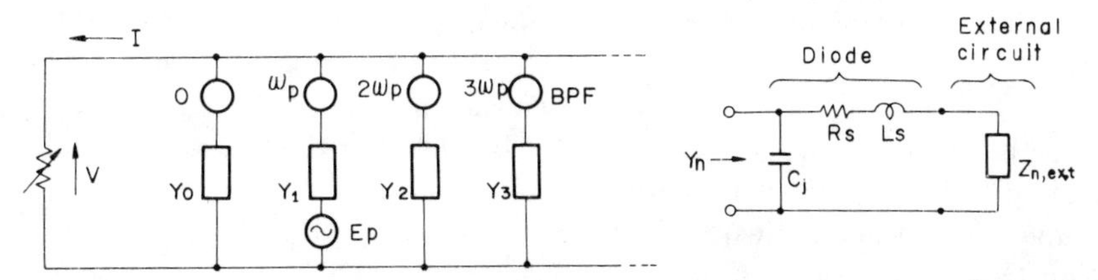

Fig. 1. Equivalent circuit for nonlinear analysis, which determines the shape of the local current. Y_n: load admittance to the nth harmonic, $Z_{n,\text{ext}}$: External circuit impedance at the nth harmonic.

$$I_m = \frac{1}{2\pi}\int_{-\pi}^{\pi} I_s \exp\left\{\beta \sum_{n=-N}^{N} V_n \exp(jnx) - jmx\right\} dx. \quad (5)$$

Besides, each harmonic satisfies the following equations, as understood by Fig. 1:

$$I_k = -Y_k\{V_k - E_p\cdot\delta(k-1)\}, \qquad k = 0,1,2,\cdots,N \quad (6)$$

where δ means the delta function, $\delta(0) = 1$, otherwise $= 0$.

Here, these harmonic voltages and currents and the load admittances are represented in matrix form as

$$\boldsymbol{V} = \begin{pmatrix} V_0 \\ V_1 \\ \vdots \\ V_N \end{pmatrix}, \quad \boldsymbol{E}_p = \begin{pmatrix} 0 \\ E_p \\ 0 \\ \vdots \\ 0 \end{pmatrix}, \quad \boldsymbol{I} = \begin{pmatrix} I_0 \\ I_1 \\ \vdots \\ I_N \end{pmatrix},$$

$$Y = \begin{pmatrix} Y_0 & & & 0 \\ & Y_1 & & \\ & & \ddots & \\ 0 & & & Y_N \end{pmatrix} \quad (7)$$

Using these relations, (6) can be expressed as follows:

$$\boldsymbol{I} + Y\cdot(\boldsymbol{V} - \boldsymbol{E}_p) = 0. \quad (8)$$

Since the current vector $\boldsymbol{I}$ is a function of the voltage vector, $\boldsymbol{V}$ as understood from (5), the above equation can be considered as a nonlinear equation for $\boldsymbol{V}$. This nonlinear equation is solved by an iteration method adapted from Newton's method [10].

We represent the left-hand side of (8) by $\boldsymbol{F}(\boldsymbol{V})$, as

$$\boldsymbol{F}(\boldsymbol{V}) = \boldsymbol{I}(\boldsymbol{V}) + Y\cdot(\boldsymbol{V} - \boldsymbol{E}_p) \quad (9)$$

where the kth element of this vector is given as follows:

$$F_k = \frac{1}{2\pi}\int_{-\pi}^{\pi} I_s \exp\left\{\beta \sum_{n=-N}^{N} V_n \exp(jnx) - jkx\right\} dx + Y_k\{V_k - E_p\cdot\delta(k-1)\}. \quad (10)$$

Then (8) is written as

$$\boldsymbol{F}(\boldsymbol{V}) = 0. \quad (11)$$

If we represent the 0th order approximate solution of (11) by $\boldsymbol{V}^{(0)}$ (this is the initial value of the iteration), correction vector $\delta\boldsymbol{V}$ is determined by the following equation:

$$\boldsymbol{F}(\boldsymbol{V}^{(0)}) + D\cdot\delta\boldsymbol{V} = 0 \quad (12)$$

where vectors $\boldsymbol{F}$ and $\delta\boldsymbol{V}$ and matrix D are defined as follows:

$$\boldsymbol{F} = \begin{pmatrix} F_0 \\ F_1 \\ \vdots \\ F_N \end{pmatrix}, \quad \delta\boldsymbol{V} = \begin{pmatrix} \delta V_0 \\ \delta V_1 \\ \vdots \\ \delta V_N \end{pmatrix}$$

$$D = \begin{pmatrix} \frac{\partial}{\partial V_0}F_0 & \frac{\partial}{\partial V_1}F_0 & \cdots & \frac{\partial}{\partial V_N}F_0 \\ \cdots & \cdots & \cdots & \cdots \\ \frac{\partial}{\partial V_0}F_N & \frac{\partial}{\partial V_1}F_N & \cdots & \frac{\partial}{\partial V_N}F_N \end{pmatrix}. \quad (13)$$

The element of matrix D, for example, $(\partial/\partial V_m)F_k$ is given by the differentiation of (10) as

$$\frac{\partial}{\partial V_m}F_k = \frac{\beta}{2\pi}\int_{-\pi}^{\pi} I_s \exp\left\{\beta \sum_{n=-N}^{N} V_n \exp(jnx) - j(k-m)x\right\} dx + Y_k\cdot\delta(k-m)$$
$$= \beta I_{k-m} + Y_k\cdot\delta(k-m). \quad (14)$$

Therefore, matrix D can be written as

$$D = \beta\begin{pmatrix} I_0 & I_{-1} & \cdots & I_{-N} \\ I_1 & I_0 & \cdots & I_{-N+1} \\ \cdots & \cdots & \cdots & \cdots \\ I_N & I_{N-1} & & I_0 \end{pmatrix} + \begin{pmatrix} Y_0 & & & 0 \\ & Y_1 & & \\ & & \ddots & \\ 0 & & & Y_N \end{pmatrix}. \quad (15)$$

Since the elements of this matrix I_n are given by (5) for $\boldsymbol{V} = \boldsymbol{V}^{(0)}$, correction vector $\delta\boldsymbol{V}$ can be calculated by the following equation:

$$\delta\boldsymbol{V} = -D^{-1}\cdot\boldsymbol{F}(\boldsymbol{V}^{(0)}). \quad (16)$$

Using this correction vector, the first-order approximate solution $\boldsymbol{V}^{(1)}$ is given as

$$\boldsymbol{V}^{(1)} = \boldsymbol{V}^{(0)} + \delta\boldsymbol{V}. \quad (17)$$

Repeating this procedure, if $\boldsymbol{V}^{(n)}$ converges, the converged vector can be considered as the solution of (11). Convergence of this iteration depends on the initial vector

$\boldsymbol{V}^{(0)}$. Although the author could not specify the boundary of convergence for $\boldsymbol{V}^{(0)}$, intuitive selection of $V^{(0)}$ as

$$V^{(0)} = \boldsymbol{E}_p \tag{18}$$

was always successful. This is credible, because, when all harmonics are short circuited, solution vector $\boldsymbol{V}$ is equal to $\boldsymbol{E}_p$. Application of this method to the practical problem will be described in Section IV.

III. Linear Analysis to Find Minimum Conversion Loss

Frequency conversion of the input signal is carried out by the modulated admittance of the diode. Since input signal power is very small compared with the local oscillator power, the relation among small-signal products generated by the input signal can be considered linear. Admittance of the diode, which is modulated by the local oscillator power, can be expressed in accordance with the former section as

$$G_j = \sum_{n=-N}^{N} g_n \exp(jn\omega_p t) \tag{19}$$

where

$$g_n = g_{-n}^*.$$

g_n can be related to the harmonic current of the diode using (2) and (4) as

$$g_n = \beta I_n, \qquad n = 0,1,2,\cdots N. \tag{20}$$

When an input signal is fed into this modulated admittance, small-signal products are generated. Here these small-signal voltages and currents are represented as [9]

$$v = \sum_{n=-N}^{N} v_n \exp[j(n\omega_p + \omega_{\mathrm{IF}})t] \tag{21}$$

$$i = \sum_{n=-N}^{N} i_n \exp[j(n\omega_p + \omega_{\mathrm{IF}})t] \tag{22}$$

where ω_{IF} is the output IF frequency. These small-signal voltages and currents and the conversion matrix are conveniently represented in matrix form as follows:

$$\boldsymbol{v} = \begin{pmatrix} v_{-N} \\ \vdots \\ v_0 \\ \vdots \\ v_N \end{pmatrix}, \quad \mathring{\boldsymbol{e}} = \begin{pmatrix} i_{-N} \\ \vdots \\ i_0 \\ \vdots \\ i_N \end{pmatrix}$$

$$G = \begin{pmatrix} g_0 & g_{-1} & \cdots & g_{-N} \\ g_1 & g_0 & \cdots & g_{-N+1} \\ \cdots & \cdots & \cdots & \cdots \\ g_N & g_{N-1} & \cdots & g_0 \end{pmatrix}. \tag{23}$$

Using these representations, the following relation is obtained between the small-signal voltages and currents:

$$\mathring{\boldsymbol{e}} = G \cdot \boldsymbol{v}. \tag{24}$$

This equation represents the relation across the nonlinear resistance of the diode. The parasitic elements of the diode R_s, C_j, L_s can be included in the equivalent circuit as shown in Fig. 2.

By defining the matrices P and Q as

$$P = \begin{pmatrix} j(-N\omega_p + \omega_{\mathrm{IF}})C_j & & & & & & 0 \\ & \ddots & & & & & \\ & & & j\omega_{\mathrm{IF}}C_j & & & \\ & & & & \ddots & & \\ 0 & & & & & & j(N\omega_p + \omega_{\mathrm{IF}})C_j \end{pmatrix}$$

$$Q = \begin{pmatrix} R_s + j(-N\omega_p + \omega_{\mathrm{IF}})L_s & & & & 0 \\ & \ddots & & & \\ & & R_s + j\omega_{\mathrm{IF}}L_s & & \\ & & & \ddots & \\ 0 & & & & R_s + j(N\omega_p + \omega_{\mathrm{IF}})L_s \end{pmatrix} \tag{25}$$

the modified conversion matrix G', which includes the effects of R_s, C_j, L_s is given as follows:

$$G' = (1 + PQ + GQ)^{-1} \cdot (P + G). \tag{26}$$

If load admittances to the small-signal products other than input signal and output IF are given, from the conversion matrix G', one can deduce a two-port matrix which relates the input signal to output IF. If the load

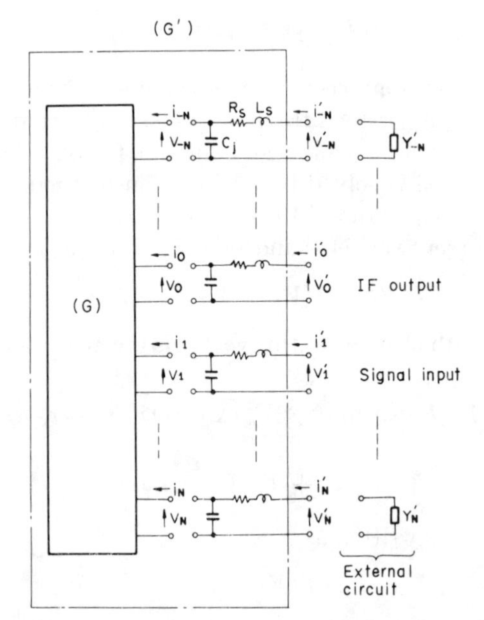

Fig. 2. Equivalent circuit for the small-signal products. Matrix G represents the conversion at the junction. Matrix G' includes the effects of R_s, C_j, and L_s.

admittance to the small-signal product $k\omega_p + \omega_{IF}$ is represented by Y_k' as shown in Fig. 2, and matrix Y' is defined as

$$Y' = \begin{pmatrix} Y'_{-N} & & & & & & & & \\ & \cdot & & & & & 0 & & \\ & & \cdot & & & & & & \\ & & & Y'_{-1} & & & & & \\ & & & & 0 & & & & \\ & & & & & 0 & & & \\ & & & & & & Y'_2 & & \\ & & & & & & & \cdot & \\ 0 & & & & & & & & Y'_N \end{pmatrix} \begin{matrix} \\ \\ \\ \cdots \mathrm{IF} \\ \cdots \mathrm{signal} \\ \\ \\ \\ \end{matrix} \tag{27}$$

then, conversion to a two-port matrix is made by matrix inversion as

$$Z = (G' + Y')^{-1}. \tag{28}$$

Taking out the elements of this matrix, a two-port matrix can be made as follows:

$$\begin{pmatrix} v_0' \\ v_1' \end{pmatrix} = \begin{pmatrix} Z_{00} & Z_{01} \\ Z_{10} & Z_{11} \end{pmatrix} \begin{pmatrix} i_0' \\ i_1' \end{pmatrix}. \tag{29}$$

This impedance matrix can be transformed to a transmission matrix [11] by the following equation:

$$A = Z_{11}/Z_{01} \quad B = -Z_{10} + Z_{11} \cdot Z_{00}/Z_{01}$$
$$C = 1/Z_{01} \quad D = Z_{00}/Z_{01}. \tag{30}$$

If these parameters are real, signal-IF conversion loss takes a minimum value when the signal and IF port are matched by the image impedances. This can be extended to the network with complex A,B,C,D by the conjugate matching method [5]. Conjugate match means that impedances of the network, seen from outside, are equal to the complex conjugate of the load impedances, and the conjugate matched signal and IF impedances can be considered as the optimum impedances which give the minimum conversion loss. Thus optimum impedance to the signal input is given as

$$Z_{\mathrm{signal}} = -\alpha_1 + (\alpha_1^2 + \alpha_2)^{1/2} \tag{31}$$

where

$$\alpha_1 = j\frac{\mathrm{Im}\,(B^*C + A^*D)}{2\,\mathrm{Re}\,(CD^*)} \qquad \alpha_2 = \frac{\mathrm{Re}\,(AB^*)}{\mathrm{Re}\,(CD^*)} \tag{32}$$

and optimum impedance to the IF output is given as

$$Z_{\mathrm{IF}} = -\gamma_1 + (\gamma_1^2 + \gamma_2)^{1/2} \tag{33}$$

where

$$\gamma_1 = j\frac{\mathrm{Im}\,(AD^* + B^*C)}{2\,\mathrm{Re}\,(AC^*)} \qquad \gamma_2 = \frac{\mathrm{Re}\,(BD^*)}{\mathrm{Re}\,(AC^*)} \tag{34}$$

and conjugate matched conversion loss L_c is given as

$$L_c = \frac{\mathrm{Re}\,Z_{\mathrm{signal}}}{\mathrm{Re}\,Z_{\mathrm{IF}}} \cdot \frac{|AZ_{\mathrm{IF}}^* + B|^2}{|Z_{\mathrm{signal}}|^2}. \tag{35}$$

These are the inherent values of the two-port matrix. Thus we can find the minimum conversion loss and optimum signal IF impedances if the conversion matrix is reduced to a two-port matrix.

IV. Spurious Suppressed Mixer—A Model of Computer Simulation

By the modulated admittance of the diode, the input signal is converted to many frequencies. In these frequencies, only the IF frequency power is derived as the output. Though other frequencies are not necessary to the mixer operation, termination impedances at these frequencies considerably affect the mixer performance. Since these unnecessary frequency products have a higher frequency, they can transmit in the waveguide by a higher order mode. Therefore, it is difficult to suppress the leakage of these powers by a signal bandpass filter (BPF). Leakage of these small-signal powers to the outer circuit degrades the conversion loss and the delay time characteristics of the mixer. To prevent this, usually a waffle-iron type lowpass filter (LPF), which has no spurious response over several multiples of its passband frequency, was used [8]. The characteristics of the waffle-iron type LPF can be realized by the coaxial LPF, which approximates the lumped constant LPF. Use of the latter type LPF, for rejection of unwanted harmonics and small-signal products, is preferable for the following reasons. 1) Unwanted small-signals and harmonic are reflected near the diode, and leakage to IF and signal ports are prevented simultaneously. Also, ambiguity to the terminations at these frequencies is greatly reduced. 2) It is easy to design, especially in the upper microwave and millimeter-wave band.

Fig. 3(a) shows the 20-GHz mixer using the latter type coaxial LPF. LPF 1 is the 20-GHz LPF which pass $\omega_{\mathrm{IF}}, \omega_p \pm \omega_{\mathrm{IF}}$ and reject $2\omega_p, 2\omega_p \pm \omega_{\mathrm{IF}} \cdots$. A lumped constant LPF, which is designed by an image impedance method, was approximated by the coaxial line (capacitance by low impedance line, inductance by high impedance line). Fig. 3(b) shows the computer simulated transmission characteristic of the designed coaxial LPF. Thus it can be understood that by this type of coaxial LPF, rejection is possible to over several multiples of the passband frequency.

Hereafter the characteristics of this mixer are computer simulated by the method described in Sections II and III. Local frequency and IF frequency are chosen as 20 and 0.5 GHz, respectively. The typical diode used in this frequency band is a $10\mu\phi$ GaAs planar Schottky diode, which has the series resistance $R_s = 3\ \Omega$, lead inductance $L_s = 0.3$ nH, junction capacitance $C_j = 0.15$ pF, saturation current $I_s = 10^{-13}$ A (in Fig. 6, $I_s = 10^{-9}$ A is also considered), and $\beta = 30\ \mathrm{V}^{-1}$.

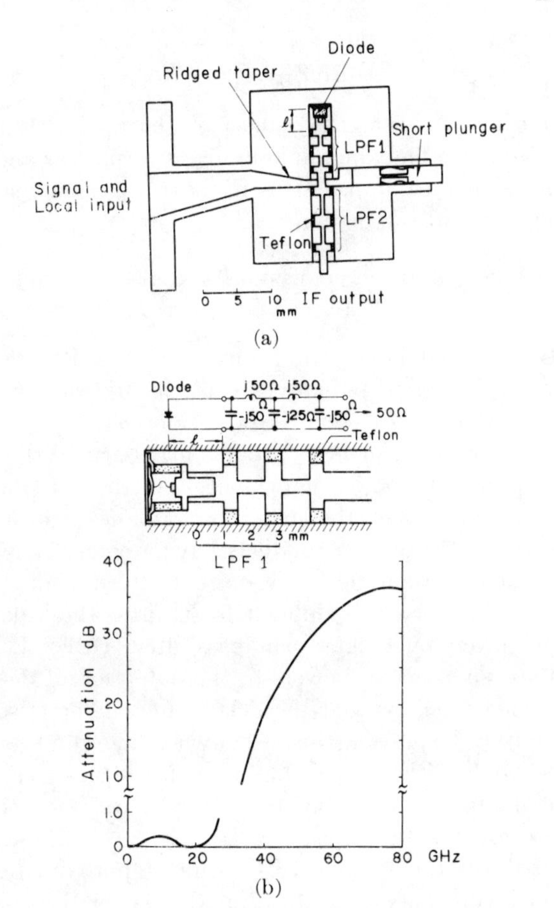

Fig. 3. (a) Spurious suppressed 20-GHz mixer which is used as a model of the computer simulation. (b) Design and characteristics of LPF 1.

Load admittances to the harmonic and small-signal products are changed by the length l (shown in Fig. 3) between diode and LPF 1. On the other hand, image frequency power can pass LPF 1, and is reflected at the signal BPF. Therefore, load admittance to the image frequency does not depend on l, and can be determined independently, changing the line length to the signal BPF. Fig. 4 shows the local and its harmonic current, determined by the method stated in Section II as a function of line length l. In the computation, up to the third local harmonic (which means $N = 3$) was considered. Iteration was converged by $30 \sim 50$ steps for reasonable local power. More steps were necessary for comparatively large local power input. Conjugate matched conversion loss and signal IF impedances, which correspond to the complex local current were computed and are also shown in the figure. Image termination was optimized independently to give a lowest conversion loss.

Fig. 5 shows the effect of image termination. Image termination can be changed by line length L between diode and signal BPF. $L/\lambda_{\text{image}} = 0$ and 0.5 correspond to image short, and $L/\lambda_{\text{image}} = 0.25$ to image open. Optimum image termination is neither short nor open, if the actual mixer as shown here is considered. Fig. 6 shows the effect of the local oscillator power to conjugate matched conversion loss and optimum signal IF impedances. Saturation current of 10^{-13} A corresponds to the GaAs diode, and of 10^{-9} A corresponds to the silicon diode. Lower values of saturation current give a lower conversion loss, as expected. The result of the preliminary experiment is shown in Fig. 7. Line length l between diode and LPF 1, and image termi-

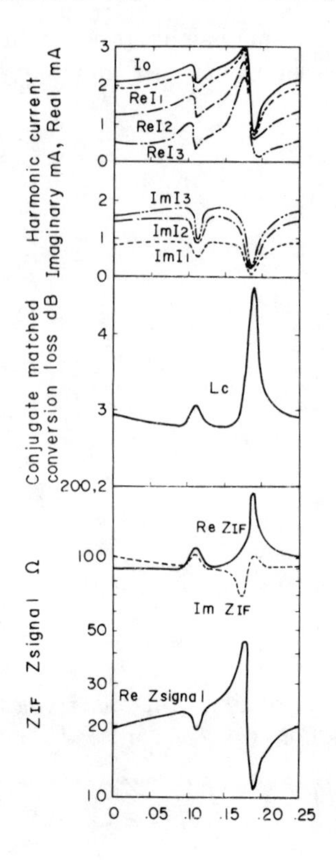

Fig. 4. Local current, conjugate matched conversion loss and signal, IF impedances are obtained as functions of l/λ_p. λ_p: wavelength of local input. Im Z_{signal} is lower than 10 Ω. Image termination is optimized at any point of l. $E_p = 1.1$ V.

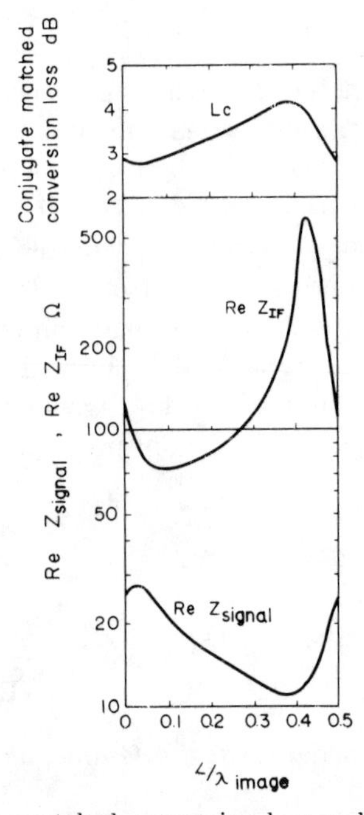

Fig. 5. Conjugate matched conversion loss and signal IF impedances as functions of image termination. L: line length between diode and signal BPF. λ_{image}, image frequency wavelength. $l/\lambda_p = 0.125$; $E_p = 1.1$ V.

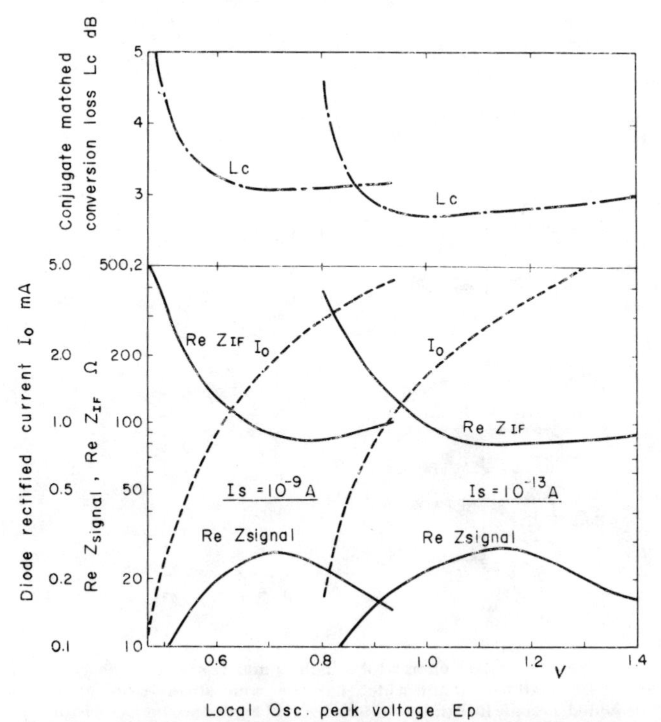

Fig. 6. Conjugate matched conversion loss and signal IF impedances as functions of local oscillator peak voltage. Image termination and l are optimized.

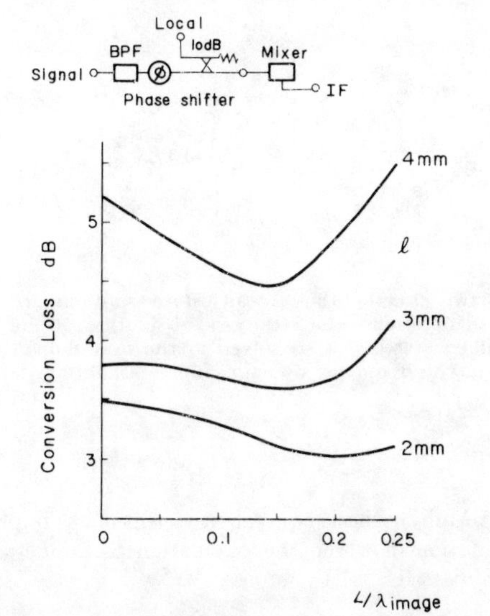

Fig. 7. Experimental conversion loss of the 20-GHz mixer shown in Fig. 2. Image termination and l are changed.

nation are changed for a constant rectified current of 5 mA. This experimental conversion loss may include the mismatch loss in the signal or IF port. Therefore, this result cannot be directly compared with the computer-simulated conversion loss of Fig. 5. However, it shows the advantage of the mixer shown here, compared with the conventional mixer.

V. Conclusion

Large-signal nonlinear analysis of the local current and small-signal linear analysis of the small-signal products were made. The first analysis, which has been rather bypassed heretofore, is solved using the iteration method. The correct solution of the first analysis made the solution of the second analysis reliable.

Since the conversion matrix was complex, the conjugate matching method was used to get a minimum conversion loss. These analyses were applied to the spurious suppressed mixer. Results of the simulation showed the feasibility of the method.

References

[1] K. Garbrecht and W. Heinlein, "Theorie Des Empfangsmischers Mit Gestuertem Wirkleitwert," *Frequenz*, vol. 19, pp. 377–385, Nov. 1965.

[2] M. R. Barber, "Noise figure and conversion loss of the Schottky barrier mixer diode," *IEEE Trans. Microwave Theory Tech.*, vol. MTT-15, pp. 629–635, Nov. 1967.

[3] G. B. Stracca, "On frequency converters using nonlinear diode," *Alta Freq.*, vol. 38, pp. 318–331, May 1969.

[4] C. A. Liechti, "Down-converters using Schottky-barrier diodes," *IEEE Trans. Electron Devices*, vol. ED-17, pp. 975–983, Nov. 1970.

[5] C. F. Edwards, "Frequency conversion by means of a nonlinear admittance," *Bell Syst. Tech. J.*, vol. 35, pp. 1403–1416, Nov. 1956.

[6] C. H. Page, "Frequency conversion with positive nonlinear resistors," *J. Nat. Bur. Stand.*, vol. 56, pp. 179–182, Apr. 1956.

[7] C. Dragone, "Performance and stability of Schottky barrier mixers," *Bell Syst. Tech. J.*, vol. 51, pp. 2169–2195, Dec. 1972.

[8] T. A. Abele *et al.*, "Schottky barrier receiver modulator," *Bell Syst. Tech. J.*, vol. 47, pp. 1257–1287, Sept. 1968.

[9] H. C. Torry and C. H. Whitmer, *Crystal Rectifiers* (M.I.T. Radiation Laboratories Series, vol. 15). New York: McGraw-Hill, 1948.

[10] P. Henrici, *Elements of Numerical Analysis.* New York: Wiley, 1964.

[11] F. Kuo, *Network Analysis and Synthesis.* New York: Wiley, 1962.

[12] D. A. Fleri and L. D. Cohen, "Nonlinear analysis of the Schottky-barrier mixer diode," *IEEE Trans. Microwave Theory Tech.*, vol. MTT-21, pp. 39–43, Jan. 1973.

[13] A. N. Willson, Jr., "On the solutions of equations for nonlinear resistive networks," *Bell Syst. Tech. J.*, vol. 47, pp. 1755–1773, Oct. 1968.

A Technique for Determining the Local Oscillator Waveforms in a Microwave Mixer

A. R. KERR, ASSOCIATE MEMBER, IEEE

Abstract—**A technique is described which enables the large-signal current and voltage waveforms to be determined for a mixer diode. This technique is applicable to any configuration where the impedance seen by the diode at the local oscillator (LO) frequency and its harmonics is known.**

I. INTRODUCTION

The performance of a diode mixer is largerly determined by the current and voltage waveforms produced at the diode by the local oscillator (LO). These waveforms depend on the diode itself, and on the impedance of its embedding network at the LO frequency and its harmonics. This short paper describes a method for computing the diode current and voltage waveforms for any mixer in which the impedance seen by the diode at the LO frequency and its harmonics is known.

In the past there have been various approaches to this problem. Torrey and Whitmer [1] and others have assumed a sinusoidal driving voltage at the diode, all the harmonics of the LO being assumed to be short-circuited. Fleri and Cohen [2] used both digital and analog computers to solve the nonlinear problem, assuming simple lumped-element embedding networks.

Egami [3] and Gwarek [4] have used a harmonic balance approach in the frequency domain. However, convergence has been found difficult to achieve for some circuits when many harmonics are considered, and especially at large LO drive levels; and the initial guess has a strong effect on the rate of convergence.

A recent approach by Gwarek [4] uses a time-domain analysis to determine the diode waveforms in an embedding network consisting of a simple lumped-element network in series with a string of voltage sources, one at each harmonic of the LO. The voltage sources are input-voltage dependent so that the embedding network is able to simulate any complex network as it appears at the LO frequency and its harmonics. It is reported that this method is convergent, and more economical of computer time and memory than the harmonic balance technique.

In the approach described here, the circuit of Fig. 1 (a) is modified by the insertion of a transmission line, Fig. 1 (b), which, by virtue of its electrical length at the LO frequency and its harmonics, has no effect on the steady-state solution of the problem. It will be shown that this enables the problem to be solved iteratively by alternately solving the simpler circuit problems shown in Fig. 2 (a) and (b).

Hypothesis: The steady-state i_d and v_d waveforms for the two circuits of Fig. 1 are the same. Certainly the solution for Fig. 1 (b) is a valid solution for Fig. 1 (a), but there exists the possibility of more than one steady-state solution for Fig. 1 (a), the one which is finally reached being determined by the particular path taken to reach the steady state.[1] An hypothesis equivalent to this one is implicit in any method of solution in which the calculation of the actual turn-on transient is bypassed.

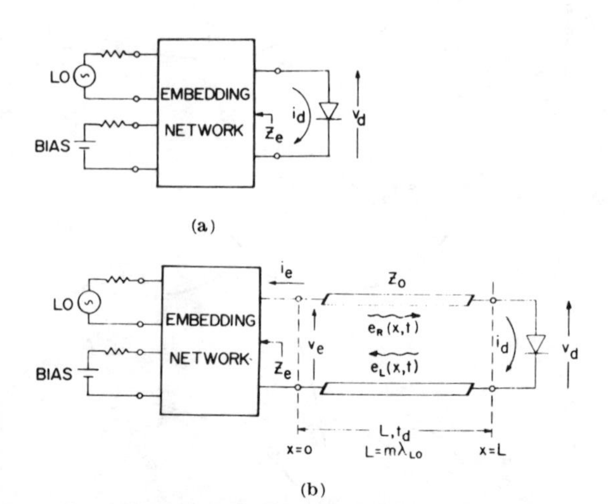

Fig. 1. (a) The mixer circuit for which v_d and i_d are to be determined. (b) The modified circuit which has the same steady-state v_d and i_d provided L is an integral number of wavelengths at the LO frequency. The right- and left-propagating waves on the transmission line are denoted by e_R and e_L.

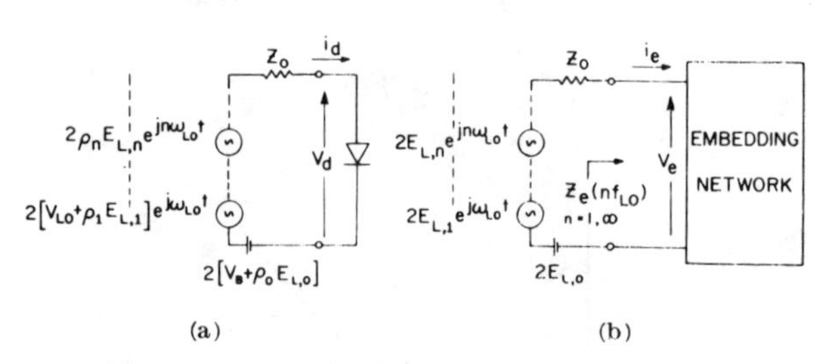

Fig. 2. The two circuits whose steady-state solutions are alternately computed to determine the steady-state solution of Fig. 1 (b). The simple nonlinear circuit (a) is solved in the time domain, the linear circuit (b) in the frequency domain. Source amplitudes are given by (9)–(11).

II. METHOD

Consider the circuit of Fig. 1 (b) with the diode initially disconnected. When the diode is connected transient reflections occur alternately at the two ends of the transmission line until eventually the steady-state condition is approached. In the steady state, waves of constant amplitude, containing many LO harmonics generated by the diode, propagate in each direction. The approach taken here is to let the transmission line become so long that in the periods between transient reflections a steady-state condition is reached

Manuscript received October 4, 1974; revised May 2, 1975. This work was supported by Associated Universities, Inc., under contract with the National Science Foundation.

The author was with the National Radio Astronomy Observatory, Charlottesville, Va. 22901. He is now with the NASA Goddard Institute for Space Studies, New York, N. Y. 10025.

[1] As an example of this, it has been observed that for some mixers, as the LO power is increased from zero, parametric oscillation will occur. At higher power levels the oscillation ceases, and it cannot be made to reappear unless the LO power is first reduced below some threshold value and then increased again.

Reprinted from *IEEE Trans. Microwave Theory Tech.*, vol. MTT-23, pp. 828–831, Oct. 1975.

between the transmission line and the diode on the one hand, and between the transmission line and the embedding network on the other. The problem then reduces to that of alternately determining the steady-state solutions for the two simple circuits of Fig. 2, each time changing the voltage sources in a predetermined way, until the terminal currents and voltages are the same for the two circuits.

For clarity, a simple exponential diode will be assumed in describing the method: the inclusion of diode capacitance and series resistance will be described later. For the simple diode

$$i_d = i_0[\exp(\alpha v_d) - 1]. \tag{1}$$

The right- and left-propagating waves on the transmission line are denoted by $e_R(x,t)$ and $e_L(x,t)$ as shown in Fig. 1(b). These waves must conform to the boundary condition imposed by (1), with

$$e_R(L,t) + e_L(L,t) = v_d(t) \tag{2a}$$

and

$$[e_R(L,t) - e_L(L,t)]/Z_0 = i_d(t). \tag{2b}$$

At the left-hand end of the transmission line the voltage and current are

$$v_e(t) = e_L(0,t) + e_R(0,t) \tag{3a}$$

and

$$i_e(t) = [e_L(0,t) - e_R(0,t)]/Z_0. \tag{3b}$$

It is convenient to express v_e and i_e as Fourier series

$$v_e(t) = \sum_{n=0}^{\infty} V_{e,n} \exp(jn\omega_{LO}t) \tag{4a}$$

and

$$i_e(t) = \sum_{n=0}^{\infty} I_{e,n} \exp(jn\omega_{LO}t) \tag{4b}$$

where the Fourier coefficients $V_{e,n}$ and $I_{e,n}$ are, in general, complex. The steady-state boundary condition defined by the embedding network is

$$\frac{V_{e,n}}{I_{e,n}} = Z_e(jn\omega_{LO}), \qquad n > 1 \tag{5a}$$

$$\frac{V_{e,1} - V_{LO}}{I_{e,1}} = Z_e(j\omega_{LO}) \tag{5b}$$

and

$$\frac{V_{e,0} - V_B}{I_{e,0}} = Z_e(0) \tag{5c}$$

where V_{LO} and V_B are the open-circuit LO and dc voltages of the embedding network.

The solution is commenced at time $t = 0$ when the diode is first connected to the circuit of Fig. 1(b). The equivalent circuit of Fig. 2(a) applies, but with only two voltage sources, V_B at dc, and V_{LO} at the LO frequency. For the simple diode assumed in this example $v_d(t)$ and $i_d(t)$ are easily computed. During the time interval $0 < t < 2t_d$

$$e_R(L,t) = V_B + V_{LO} \exp(j\omega_{LO}t) \tag{6a}$$

and

$$e_L(L,t) = v_d(t) - i_d(t)Z_0 \tag{6b}$$

where t_d is the propagation delay on the transmission line. At the left-hand end of the transmission line the left-propagating wave is

$$e_L(0,t) = e_L(L,t - t_d). \tag{7}$$

Since e_L is composed only of the LO frequency and its harmonics for which the transmission line is an integral number of wavelengths long

$$e_L(L,t - t_d) = e_L(L,t). \tag{8}$$

Therefore, during the time interval $t_d < t < 3t_d$

$$e_L(0,t) = v_d(t) - i_d(t)Z_0. \tag{9}$$

Expressing (9) as a Fourier series gives

$$e_L(0,t) = \sum_{n=0}^{\infty} E_{L,n} \exp(jn\omega_{LO}t). \tag{10}$$

The embedding network reaches steady state after some time δ so that in the interval $t_d + \delta < t < 3t_d$ a new steady-state right-propagating wave $e_R(0,t)$ exists. This may be determined knowing the complex reflection coefficient of the embedding network at each harmonic of the LO

$$\rho_n = \frac{Z_e(jn\omega_{LO}) - Z_0}{Z_e(jn\omega_{LO}) + Z_0}. \tag{11}$$

Then

$$e_R(0,t) = V_B + V_{LO} \exp(j\omega_{LO}t) + \sum_{n=0}^{\infty} \rho_n E_{L,n} \exp(jn\omega_{LO}t). \tag{12}$$

Again e_R contains only components at the LO frequency and its harmonics, and so after the appropriate delay time t_d we have

$$e_R(L,t) = e_R(0,t). \tag{13}$$

This applies during the time interval $2t_d + \delta < t < 4t_d$. The diode voltage and current $v_d(t)$ and $i_d(t)$ are again calculated, giving the next value of $e_L(L,t)$ from (6b), which commences the next cycle of iteration.

Summary of Iteration Cycle

The typical cycle of iteration may be summarized as follows.

Step 1): From the diode voltage and current, $v_d(t)$ and $i_d(t)$, compute the left-propagating wave $e_L(L,t)$ using (6b).

Step 2): After a time t_d the left-propagating wave reaches the embedding network causing a response which reaches steady state after a further time δ. Since the transmission line can be made as long as required, it is always possible to ensure that $\delta \ll t_d$. The new steady-state right-propagating wave at the embedding network $e_R(0,t)$ is calculated using (11) and (12).

Step 3): After a further propagation delay t_d this wave becomes the new right-propagating wave at the diode, $e_R(L,t)$ from which new values of diode voltage and current are computed using the equivalent circuit of Fig. 2(a).

Convergence

Under steady-state operation the voltages and currents at the two ends of the transmission line are equal—that is $v_d(t) = v_e(t)$ and $i_d(t) = -i_e(t)$. In the frequency domain this is equivalent to

$$V_{d,n} = V_{e,n} \tag{14a}$$

and

$$I_{d,n} = -I_{e,n} \tag{14b}$$

where $V_{d,n}$, $I_{d,n}$, $V_{e,n}$, and $I_{e,n}$ are the nth Fourier coefficients of v_d, i_d, v_e, and i_e, respectively. It follows that

$$\frac{V_{d,n}}{I_{d,n}} = -\frac{V_{e,n}}{I_{e,n}}. \tag{15}$$

It is convenient to define a *diode impedance*

$$Z_d(jn\omega_{LO}) = V_{d,n}/I_{d,n}. \tag{16}$$

Then with (5a) and (15)

$$Z_d(jn\omega_{LO}) = -Z_e(jn\omega_{LO}), \qquad \text{for } n > 1. \tag{17}$$

We define a convergence parameter

$$|Z_d(jn\omega_{LO})|/|Z_e(jn\omega_{LO})|, \qquad n > 1$$

which must be equal to unity for a completely converged solution. For the example given in the following (Fig. 3), the convergence parameter is shown in Fig. 4 as a function of n.

The choice of the hypothetical transmission-line characteristic impedance Z_0 has some effect on the rate of convergence of the solution. A value of 50 Ω has been used in the example; the effects of varying it have not been investigated in detail.

Finite Number of Harmonics

In a practical computer solution of the mixer problem it is possible to consider only a finite number N of harmonics of the LO. This is electrically equivalent to defining the embedding impedance for the higher LO harmonics equal to the transmission-line characteristic impedance

$$Z_e(jn\omega_{LO}) = Z_0, \qquad n > N. \tag{18}$$

To show this equivalence, consider the circuit of Fig. 1(b), with Z_e as defined in (18), so that $\rho_n = 0$ for $n > N$ (11). From (12) and (13) we see that the right-propagating wave $e_R(L,t)$ contains no frequency components above the Nth LO harmonic. It follows that there are no voltage sources above the Nth harmonic in the circuit of Fig. 2(a), and that the computed v_d and i_d waveforms are the same (up to the Nth harmonic) as would be computed for the mixer if (18) were not valid and if all harmonics above the Nth were neglected in the computation.

Diode Capacitance and Series Resistance

In explaining the method of solution of the mixer problem it was convenient to assume a simple diode with no parasitic elements, as described by (1). We now apply the method to two more realistic diode models. The first, in which the diode capacitance C and series resistance R_s are both constant, requires no modification of the method. The elements C and R_s are simply considered as part of the embedding network and the problem is solved as before.

For the second diode model the capacitance and series resistance are functions of the diode terminal voltage. In this case the simple circuit of Fig. 2(a) is modified to include $C(v_d)$ and $R_s(v_d)$, and numerical integration of the circuit equations is necessary in order to determine the steady-state solutions for $v_d(t)$ and $i_d(t)$ each time Step 3 of the cycle is performed. The right-propagating wave $e_R(L,t)$ now produces an initial transient response at the diode which must be allowed to die away before the steady-state values of v_d, i_d, and $e_L(L,t)$ are calculated. This situation is the same as occurs at the embedding network when a new wave $e_L(0,t)$ arrives at it.

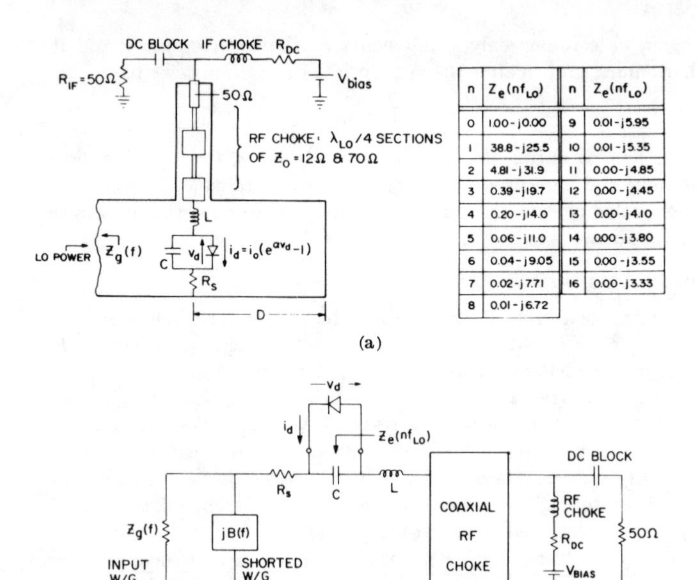

n	$Z_e(nf_{LO})$	n	$Z_e(nf_{LO})$
0	1.00-j0.00	9	0.01-j5.95
1	38.8-j25.5	10	0.01-j5.35
2	4.81-j31.9	11	0.00-j4.85
3	0.39-j19.7	12	0.00-j4.45
4	0.20-j14.0	13	0.00-j4.10
5	0.06-j11.0	14	0.00-j3.80
6	0.04-j9.05	15	0.00-j3.55
7	0.02-j7.71	16	0.00-j3.33
8	0.01-j6.72		

Fig. 3. (a) The waveguide mixer used in the example, and (b) its equivalent circuit. The embedding impedance Z_e seen by the diode is tabulated as a function of harmonic number n. Guide impedance Z_g is given by (19). Parameter values are: f_{LO} = 15 GHz, i_0 = 5 nA, α = 40 V^{-1}, V_{bias} = 0, R_{dc} = 1.0 Ω, C = 0.2 pF, L = 0.72 nH, R_s = 5 Ω, D = 0.377 in, and the waveguide is 0.070 in high × 0.662 in wide.

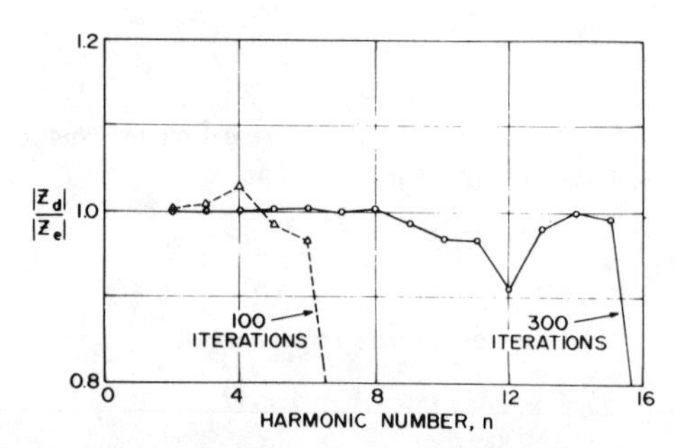

Fig. 4. Convergence parameter $|Z_d(nf_{LO})| / |Z_e(nf_{LO})|$ as a function of harmonic number n after 100 iterations (dashed curve) and 300 iterations (solid curve). For 500 iterations (not shown) all harmonics except the 16th were within 0.5 percent of unity.

III. EXAMPLE

To test the method a computer program was written which determines the diode waveforms i_d and v_d in the waveguide mixer of Fig. 3. The diode is mounted across the middle of a reduced-height waveguide, and connected through a coaxial RF choke to the bias and IF circuits. To one side of the diode there is a waveguide short circuit at a distance D, and on the other side the diode sees the guide impedance

$$Z_g(f) = 2\left(\frac{\mu}{\epsilon}\right)^{1/2} \frac{b}{a} \frac{1}{[1-(f_c/f)^2]^{1/2}}. \tag{19}$$

The impedance of the embedding network, including the RF choke and the shorted waveguide section, is calculated at each LO harmonic. This characterization of the mixer is not an accurate representation at the harmonics of the LO frequency [5], [6], but it

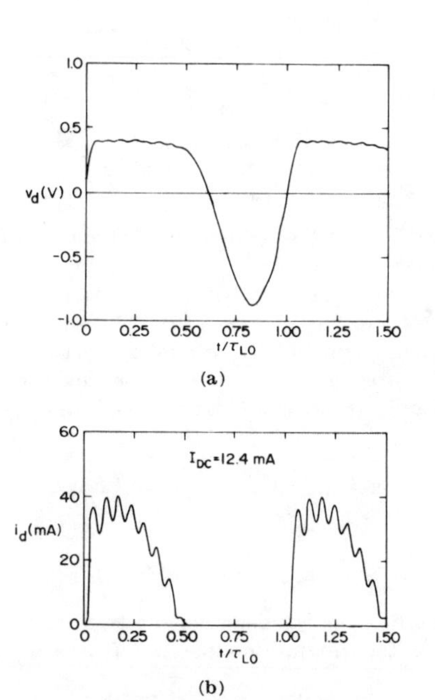

Fig. 5. (a) Diode voltage. (b) Diode current for the mixer shown in Fig. 3, after 500 iterations, considering 16 harmonics of the LO frequency. Time t is normalized to the LO period τ_{LO}.

provides a useful approximation for the purposes of this example. The embedding impedance Z_e seen by the diode at the LO frequency and its harmonics is tabulated in Fig. 3.

Fig. 4 shows the convergence parameter $|Z_d(jn\omega_{LO})|/|Z_e(jn\omega_{LO})|$ as a function of harmonic number n, when 16 harmonics are considered. For 300 iterations, convergence is reasonably complete up to the 11th harmonic. For 500 iterations $|Z_d|$ is within 0.5 percent of $|Z_e|$ up to the 15th harmonic.

The diode current and voltage waveforms are shown in Fig. 5. The computation time per cycle of iteration, when 16 harmonics are considered, is 3 ms on an IBM 360/95, and 2 s on an IBM 360/50.

REFERENCES

[1] H. C. Torrey and C. A. Whitmer, *Crystal Rectifiers* (M. I. T. Radiation Lab. Ser., vol. 15). New York: McGraw-Hill, 1948.

[2] D. A. Fleri and L. D. Cohen, "Nonlinear analysis of the Schottky-barrier mixer diode," *IEEE Trans. Microwave Theory Tech.*, vol. MTT-21, pp. 39–43, Jan. 1973.

[3] S. Egami, "Nonlinear, linear analysis and computer-aided design of resistive mixers," *IEEE Trans. Microwave Theory Tech.* (*Special Issue on Computer-Oriented Microwave Practices*), vol. MTT-22, pp. 270–275, Mar. 1974.

[4] W. K. Gwarek, "Nonlinear analysis of microwave mixers," M. S. thesis, Mass. Inst. Technol., Cambridge, Sept. 1974.

[5] A. R. Kerr, "Low-noise room-temperature and cryogenic mixers for 80–120 GHz," this issue, pp. 781–787.

[6] R. L. Eisenhart and P. J. Khan, "Theoretical and experimental analysis of a waveguide mounting structure," *IEEE Trans. Microwave Theory Tech.*, vol. MTT-19, pp. 706–719, Aug. 1971.

Conversion Loss and Noise of Microwave and Millimeter-Wave Mixers: Part 1—Theory

DANIEL N. HELD, MEMBER, IEEE, AND ANTHONY R. KERR, ASSOCIATE MEMBER, IEEE

***Abstract*—An analysis is presented for the conversion loss and noise of microwave and millimeter-wave mixers. The analysis includes the effects of nonlinear capacitance, arbitrary embedding impedances, nonideality of microwave diodes, and shot, thermal, and scattering noise generated in the diode. Correlation of down-converted components of the time-varying shot noise is shown to explain the "anomalous" noise observed in millimeter-wave mixers. Part 1 of the paper presents the theoretical basis for predicting mixer performance, while Part 2 compares theoretical and experimental results for mixers operating at 87 and 115 GHz.**

I. Introduction

THE BASIC principles of frequency conversion using crystal diodes were first studied in depth by Torrey and Whitmer [1] in 1948. Since then attempts at a more accurate analysis of microwave mixer performance have been limited by the complexity of the nonlinear problem and by a lack of understanding of the noise properties of pumped diodes. Practical developments in the design of mixers and mixer diodes have resulted in a number of commonly used designs whose conversion loss and noise figure are often within a few decibels of the theoretical best values predicted for idealized switching mixers. This paper and its companion paper (Part 2) attempt to close the gap between theory and practice.

Following Torrey and Whitmer's original work, the assumption of a sinusoidal local oscillator (LO) voltage across an exponential resistive diode was often used, with its implicit assumption that the diode was short-circuited at all harmonics of the LO. The effects of the parasitic series resistance and capacitance (assumed constant) of the mixer diode were investigated using approximate methods by Sharpless [2], Messenger and McCoy [3], Mania and Stracca [4], and Kerr [5]; and a new and intuitive approach to mixer analysis was taken by Barber [6], who approximated the diode by a switch whose pulse-duty ratio determined the conversion properties of the mixer. However, all these approaches required simplifying assumptions about the termination of higher order sidebands and ignored the effects of the nonlinear diode capacitance. Agreement between theory and experiment was at best within a few decibels in conversion loss and noise figure.

An important work by Saleh [7] in 1971 studied the effects of local oscillator waveforms and higher order sideband terminations on resistive mixers, and demonstrated that these characteristics must be considered if an accurate analysis is to be made. The problem of a more exact analysis of a microwave mixer has been tackled recently by Egami [8], who performed a numerical nonlinear analysis for a known diode and embedding impedance, followed by a small-signal conversion loss analysis. Egami assumed a constant junction capacitance and used a harmonic balance technique to solve the nonlinear problem, considering three harmonics of the local oscillator.

In recent years considerable misunderstanding has arisen on the subject of mixer noise. The correlation properties of shot noise in vacuum tube mixers were understood by Strutt [9] in 1946 and since then van der Ziel and Waters [10], [11], Kim [12], and Dragone [13] have further developed this theory. The misunderstanding seems to have arisen with the assumption, by Messenger and McCoy [3], that a mixer and a passive attenuator have similar noise properties and that the temperature of this equivalent attenuator is "the time average of the static noise characteristic" of the diode. This led to a widely held belief that for a Schottky diode mixer, the noise–temperature ratio [3] $t \simeq 1$, and to the subsequent observation of an "anomalous" component of noise in millimeter-wave mixers [14].

In the present paper the large-signal nonlinear problem and the small-signal linear problem are presented in forms suitable for computer solution. Arbitrary embedding impedances are allowed at the harmonics of the LO and at all the sideband frequencies, and any diode capacitance law can be assumed. The shot-noise theories of Strutt [9], van der Ziel [11], Kim [12], and Dragone [13] are extended to include the effects of arbitrary sideband terminations, as well as thermal (Johnson) noise and scattering noise, resulting in a noise correlation matrix which is compatible with the formulation of the small-signal mixer analysis. The microwave properties of small Schottky-barrier diodes are discussed, and it is shown that skin effect, thermal time constants, and depletion layer fringing effects can all be significant in determining the mixing properties of the diode.

The theory presented in this paper is experimentally verified in a companion paper (Part 2) in which theoretical loss and noise predictions for millimeter-wave mixers are shown to be in good agreement with experiment.

II. Nonlinear Analysis

The small-signal loss and noise properties of a mixer are governed by the large-signal current and voltage waveforms

Manuscript received February 18, 1977; revised May 12, 1977.

D. N. Held was with the Department of Electrical Engineering, Columbia University, New York, NY 10027. He is now with the Jet Propulsion Laboratory, California Institute of Technology, Pasadena, CA 91103.

A. R. Kerr is with the NASA Institute for Space Studies, Goddard Space Flight Center, New York, NY 10025.

Reprinted from *IEEE Trans. Microwave Theory Tech.*, vol. MTT-26, pp. 49–55, Feb. 1978.

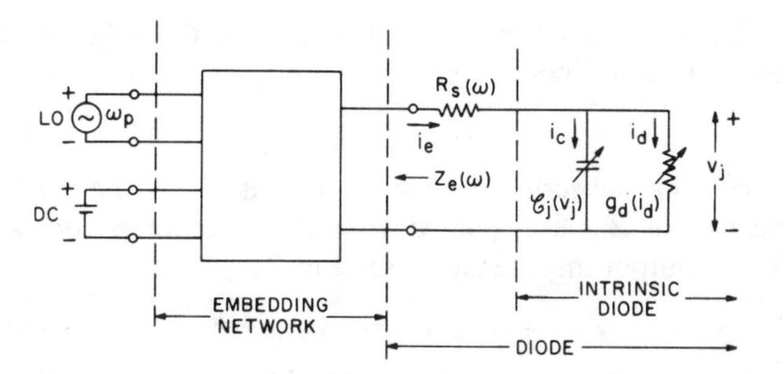

Fig. 1. Equivalent circuit of the mixer. The intrinsic diode $\mathscr{C}_j$ and g_d is nonlinear and is characterized in the time domain, while the diode series resistance R_s and the embedding impedance Z_e are linear and are best represented in the frequency domain.

produced at the diode by the LO. The steady-state large-signal response of the mixer circuit of Fig. 1 can be described in terms of the Fourier coefficients of the voltage and current v_j and i_e. Thus

$$v_j(t) = \sum_{k=-\infty}^{\infty} V_k e^{jk\omega_p t}, \qquad V_k = V^*_{-k} \tag{1}$$

$$i_e(t) = i_c(t) + i_d(t) = \sum_{k=-\infty}^{\infty} I_{e_k} e^{jk\omega_p t}, \qquad I_{e_k} = I^*_{e-k} \tag{2}$$

where ω_p is the LO frequency. Two sets of boundary conditions must be satisfied simultaneously by these quantities. The first, imposed by the diode, is most easily expressed in the time domain, while the second set, imposed by the embedding network, is for our purposes more conveniently considered in the frequency domain. At the diode

$$i_d = i_0[\exp(\alpha v_j) - 1] \tag{3}$$

and

$$i_c = \mathscr{C}_j \frac{dv_j}{dt} \tag{4}$$

where

$$\mathscr{C}_j = \mathscr{C}_{j0}\left(1 - \frac{v_j}{\phi}\right)^{-\gamma} \tag{5}$$

and

$$\alpha = q/\eta kT. \tag{6}$$

From (3) the incremental conductance of the diode

$$g_d = \frac{di_d}{dv_j} = \alpha(i_d + i_0) \simeq \alpha i_d. \tag{7}$$

The embedding network requires that

$$V_k = -I_{e_k}[Z_e(k\omega_p) + R_s(k\omega_p)], \qquad k = \pm 2, \pm 3, \cdots, \pm\infty \tag{8a}$$

$$V_{\pm 1} = V_p - I_{e\pm 1}[Z_e(\pm\omega_p) + R_s(\pm\omega_p)] \tag{8b}$$

$$V_0 = V_{dc} - I_{e0}[Z_e(0) + R_s(0)] \tag{8c}$$

where V_p and V_{dc} are the LO and dc-bias voltages, respectively. The frequency dependence of R_s is due to skin effect and is discussed in Section V.

Methods for solving the nonlinear mixer problem have been described by several authors [15]–[17]. However, only that of Gwarek [17] appears capable of handling the case in which the diode capacitance is a function of junction voltage and in which a large number (six or seven) of harmonics of the LO are considered. Using an IBM 360/95 computer, we have found that with Gwarek's method convergence is achieved in about two seconds. The degree of convergence of the solution is determined using the convergence parameter described by Kerr [16]. It is assumed that there is one unique steady-state solution; the possibility of multiple solutions, mentioned in [16] is beyond the scope of this work.

Having determined the LO waveforms at the diode, (3)–(7) give $i_d(t)$, $g_d(t)$, and $\mathscr{C}_j(t)$. These may be expressed as Fourier series:

$$i_d(t) = \sum_{k=-\infty}^{\infty} I_k \exp(jk\omega_p t), \qquad I_k = I^*_{-k} \tag{9}$$

$$g_d(t) = \sum_{k=-\infty}^{\infty} G_k \exp(jk\omega_p t), \qquad G_k = G^*_{-k} \tag{10}$$

and

$$\mathscr{C}_j(t) = \sum_{k=-\infty}^{\infty} C_k \exp(jk\omega_p t), \qquad C_k = C^*_{-k}. \tag{11}$$

These quantities, together with the embedding impedance $Z_e(\omega)$, determine the small-signal properties of the mixer.

III. Small-Signal Analysis

Knowing the Fourier coefficients of the diode capacitance and conductance, it is possible to construct the small-signal conversion matrix for the diode [1], [17]. This matrix interrelates the various sideband frequency components of the small-signal current and voltage δI_m and δV_m shown in Fig. 2. The subscript notation for the sideband quantities follows that of Saleh [7]; subscript m indicates frequency $\omega_0 + m\omega_p$, where ω_p and ω_0 are the LO and intermediate frequencies. Thus $Z_{e_m} = Z_e(\omega_m) = Z_e(\omega_0 + m\omega_p)$. The conversion matrix $\boldsymbol{Y}$ is a square matrix defined by

$$\boldsymbol{\delta I} = \boldsymbol{Y\delta V}$$

where

$$\boldsymbol{\delta I} = [\cdots, \delta I_1, \delta I_0, \delta I_{-1}, \cdots]^T$$

and

$$\boldsymbol{\delta V} = [\cdots, \delta V_1, \delta V_0, \delta V_{-1}, \cdots]^T. \tag{12}$$

For convenience, the row and column numbering of all matrices and vectors in this paper will correspond with the sideband numbering. For example,

$$\boldsymbol{Y} \equiv \begin{array}{c} \text{row \#} \\ \vdots \\ 1 \\ 0 \\ -1 \\ \vdots \\ \\ \end{array} \begin{bmatrix} & \vdots & \vdots & \vdots & \\ \cdots & Y_{11} & Y_{10} & Y_{1-1} & \cdots \\ \cdots & Y_{01} & Y_{00} & Y_{0-1} & \cdots \\ \cdots & Y_{-11} & Y_{-10} & Y_{-1-1} & \cdots \\ & \vdots & \vdots & \vdots & \end{bmatrix} . \\ \begin{array}{ccccc} \cdots & 1 & 0 & -1 & \cdots \end{array} \\ \text{column \#}$$

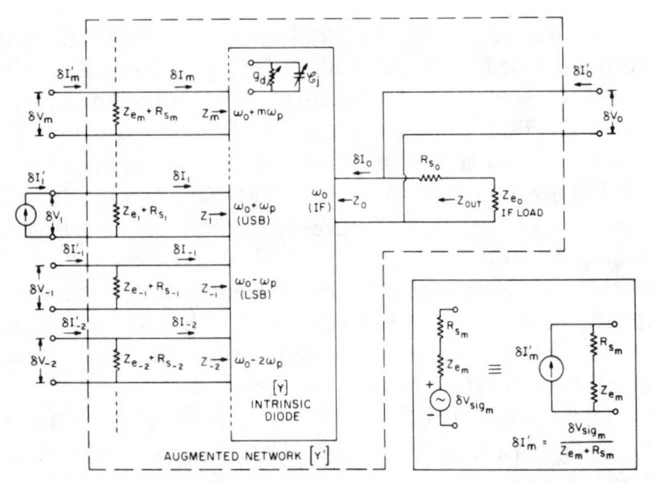

Fig. 2. Small-signal representation of the mixer as a multifrequency linear multiport network. The voltage and current δV_m and δI_m at any port m are the small-signal components at frequency $(\omega_0 + m\omega_p)$ appearing at the intrinsic diode; each port represents one sideband frequency. The conversion matrix Y is the admittance matrix of the intrinsic diode. The augmented network (broken line) includes all the sideband embedding impedances Z_{e_m} and is characterized by the augmented admittance matrix Y'. During normal mixer operation the equivalent signal current generator $\delta I'$ is connected at port 1 of the augmented network, the other ports being open-circuited. In the noise analysis, equivalent noise current sources $\delta I'_{S_m}$ and $\delta I'_{T_m}$ are connected to all ports. The inset shows the relation between the signal source δV_{sig_m} at the mth sideband and its equivalent current source $\delta I'_m$.

Using this notation the elements of $\boldsymbol{Y}$ are given by [1], [17]

$$Y_{mn} = G_{m-n} + j(\omega_0 + m\omega_p)C_{m-n} \tag{13}$$

where G_k and C_k are the Fourier coefficients of the diode conductance and capacitance as defined in (10) and (11).

It is convenient to form an augmented Y matrix, $\boldsymbol{Y}'$, which is the admittance matrix of the multiport network outlined by the broken line in Fig. 2. This augmented network contains the whole mixer, including all its external terminating impedances Z_{e_m}, but does not contain signal sources associated with these terminations. Signal sources are replaced by equivalent current sources $\delta I'_m$ connected across $(Z_{e_m} + R_{s_m})$ as shown in Fig. 2 (inset). The ports of the augmented network are all normally open-circuited. For the augmented network

$$\boldsymbol{\delta I}' = \boldsymbol{Y}'\boldsymbol{\delta V} \tag{14}$$

where the elements of $\boldsymbol{\delta I}'$ are defined in Fig. 2, and

$$\boldsymbol{Y}' = \boldsymbol{Y} + \mathbf{diag}\left[\frac{1}{Z_{e_m} + R_{s_m}}\right]. \tag{15}$$

Inverting (14) gives

$$\boldsymbol{\delta V} = \boldsymbol{Z}'\boldsymbol{\delta I}' \tag{16}$$

where

$$\boldsymbol{Z}' = (\boldsymbol{Y}')^{-1}. \tag{17}$$

A. Mixer Port Impedances

To determine the port impedance Z_m, defined in Fig. 2, the corresponding embedding impedance Z_{e_m} is open-circuited, enabling Z_m to be measured at port m of the augmented network. It follows that

$$Z_m = Z'_{mm,\infty} \tag{18}$$

where the subscript ∞ indicates that Z'_{mm} (the mmth element of $\boldsymbol{Z}'$) is evaluated with $Z_{e_m} = \infty$. In particular the IF output impedance is (see Fig. 2)

$$Z_{\text{out}} = Z_0 + R_{s0} = Z'_{00,\infty} + R_{s0}. \tag{18a}$$

For a microwave mixer the embedding impedance at all frequencies other than the IF is defined by the mixer geometry. However, the IF load impedance Z_{e0} can usually be adjusted for optimum performance using a matching circuit. *Throughout the rest of this paper it will be assumed that the IF port is conjugate-matched,*[1] and IF-matched $\boldsymbol{Y}'$ and $\boldsymbol{Z}'$ matrices will be implied.

B. Conversion Loss

The conversion loss of a mixer

$$L = \frac{\text{Power available from source } Z_{e_1}}{\text{Power delivered to load } Z_{e0}}. \tag{19}$$

For the purposes of analysis L may be expressed as the product of three separate loss components:

$$L = K_0 L' K_1 \tag{20}$$

where

$$K_0 = \frac{\text{Power delivered to } (Z_{e0} + R_{s0})}{\text{Power delivered to } Z_{e0}} = \frac{\text{Re}\,[Z_{e0} + R_{s0}]}{\text{Re}\,[Z_{e0}]} \tag{21}$$

$$L' = \frac{\text{Power available from } (Z_{e_1} + R_{s_1})}{\text{Power delivered to } (Z_{e0} + R_{s0})} \tag{22}$$

and

$$K_1 = \frac{\text{Power available from } Z_{e_1}}{\text{Power available from } (Z_{e_1} + R_{s_1})} = \frac{\text{Re}\,[Z_{e_1} + R_{s_1}]}{\text{Re}\,[Z_{e_1}]}. \tag{23}$$

K_0 and K_1 account for loss in the series resistance at the IF and signal frequencies, while L' is the conversion loss of the intrinsic mixer with no series resistance. Using (22) and (16),

$$L' = \frac{1}{\text{Re}\,[\delta V_0 \delta I_0^*]} \frac{|\delta I'_1|^2 |Z_{e_1} + R_{s_1}|^2}{4\,\text{Re}\,[Z_{e_1} + R_{s_1}]} = \frac{1}{4|Z'_{01}|^2} \frac{|Z_{e0} + R_{s0}|^2}{\text{Re}\,[Z_{e0} + R_{s0}]} \frac{|Z_{e_1} + R_{s_1}|^2}{\text{Re}\,[Z_{e_1} + R_{s_1}]}. \tag{24}$$

From (20), (21), (23), and (24), the conversion loss

$$L = \frac{1}{4|Z'_{01}|^2} \frac{|Z_{e0} + R_{s0}|^2}{\text{Re}\,[Z_{e0}]} \frac{|Z_{e_1} + R_{s_1}|^2}{\text{Re}\,[Z_{e_1}]}. \tag{25}$$

[1] Note that although a conjugate-match at the IF port results in maximum power transfer, the signal port should not in general be conjugate-matched for minimum conversion loss [7].

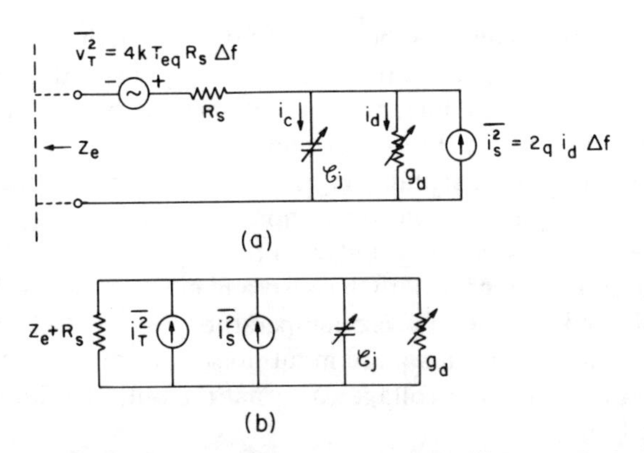

Fig. 3. (a) Noise equivalent circuit of the mixer. (b) The equivalent circuit with the thermal noise source $\overline{v_T^2}$ transformed to a current source $\overline{i_T^2}$ given by equation (27). The sideband components of the noise sources $\overline{i_T^2}$ and $\overline{i_S^2}$ can then be treated in the same way as the equivalent external small-signal currents $\delta I'_m$ in Fig. 2.

Equation (25) may be generalized to give the conversion loss from any sideband j to any other sideband i.

$$L_{ij} = \frac{1}{4|Z'_{ij}|^2} \frac{|Z_{e_i} + R_{s_i}|^2}{\mathrm{Re}\,[Z_{e_i}]} \frac{|Z_{e_j} + R_{s_j}|^2}{\mathrm{Re}\,[Z_{e_j}]}. \tag{26}$$

IV. Mixer Noise

The sources of noise in a Schottky diode are 1) thermal noise in the series resistance, 2) shot noise generated by current flow across the barrier, and 3) noise due to phonon scattering and, in gallium arsenide, intervalley scattering [22]. In room-temperature mixers shot and thermal noise predominate, with scattering noise contributing typically only 5–10 percent to the overall mixer noise, as is shown in Part 2 of this paper. We have not attempted an exact analysis of scattering noise but assume that it can be approximated by an increase in the noise temperature of the series resistance.

The equivalent circuit of the diode including noise sources is shown connected to the embedding network in Fig. 3(a). Fig. 3(b) shows the same circuit transformed to a configuration more compatible with the mixer equivalent circuit of Fig. 2; the thermal noise source $\overline{v_T^2}$ has been replaced by an equivalent current source $\overline{i_T^2}$. As seen by the intrinsic diode (g_d and $\mathscr{C}_j$),

$$\overline{i_T^2} = \frac{4kT_{eq}R_s\Delta f}{|Z_e + R_s|^2} \tag{27a}$$

but, as seen by the IF load impedance Z_{eo},

$$\overline{i_T^2} = \frac{4kT_{eq}R_s\Delta f}{|Z_0|^2} \tag{27b}$$

where Z_0 is the output impedance of the intrinsic diode as defined in Fig. 2. The shot-noise current source

$$\overline{i_S^2} = 2qi_d\Delta f. \tag{28}$$

A. Shot Noise

The diode shot noise given by (28) is a function of the instantaneous flow of current i_d across the barrier, produced by the LO. Since $i_d(t)$ is periodic, the shot-noise current can be regarded as a stationary white noise current modulated at the LO frequency ω_p. The properties of this modulated shot noise have been studied by Strutt [9], Kim [12], van der Ziel [11], and Dragone [13], and the present analysis is based on their work.

First, we shall consider the *unmodulated* shot-noise current to be composed of pseudosinusoidal currents [18], [19] $\delta\mathscr{I}_{S_m} \exp\,[j(\omega_0 + m\omega_p)t + j\phi_m]$ at each of the sideband frequencies $\omega_m = \omega_0 + m\omega_p$, $m = 0, \pm 1, \cdots, \pm\infty$. All other frequency components of the noise can be disregarded since they cannot contribute to the IF output of the mixer. Next, consider each of the currents $\delta\mathscr{I}_{S_m}$ to be modulated by the local oscillator (ω_p), generating modulation products at all the frequencies $\omega_m + n\omega_p$, $n = 0, \pm 1, \pm 2, \cdots, \pm\infty$, which are the sideband frequencies ω_{m+n}. The modulation products are present at the terminals of the intrinsic diode and will be converted to the output frequency as described by (16). Clearly their individual outputs will be correlated. The random phase variable ϕ_m is preserved in these output components but is eliminated when we finally consider a finite bandwidth and take the ensemble average of a set of pseudosinusoidal currents in a narrow band about each of the frequencies ω_m.

Let $\delta I'_{S_m}$ denote the complex amplitude of the pseudosinusoidal noise current at sideband frequency ω_m. $\delta I'_{S_m}$ contains components due to the modulation products of all the currents $\delta\mathscr{I}_S$ and is therefore the shot-noise input current at port m of the mixer, which corresponds to a signal input current $\delta I'_m$ in Fig. 2. The vector $\boldsymbol{\delta I'_S}$ is defined as $[\cdots, \delta I'_{S_1}, \delta I'_{S_0}, \delta I'_{S_{-1}}, \cdots]^T$. Equation (16) now enables the IF output noise voltage due to shot noise to be determined:

$$\delta V_{S_0} = \boldsymbol{Z_0}\,\boldsymbol{\delta I'_S} \tag{29}$$

where $\boldsymbol{Z_0}$ is the zeroth row of the square matrix $\boldsymbol{Z'}$. Taking the ensemble average[2] of $|\delta V_{S_0}|^2$ (which corresponds considering a nonzero bandwidth),

$$\langle|\delta V_{S_0}|^2\rangle = \boldsymbol{Z'_0}\langle\boldsymbol{\delta I'_S}\,\boldsymbol{\delta I'^{\dagger}_S}\rangle\boldsymbol{Z'^{\dagger}_0}. \tag{30}$$

The square matrix $\langle\boldsymbol{\delta I'_S}\,\boldsymbol{\delta I'^{\dagger}_S}\rangle$ is known as the *correlation matrix* for the mixer, since the (m,n) element describes the correlation between the components of shot noise at sideband frequencies ω_m and ω_n. Dragone [13] has evaluated this matrix, obtaining

$$\langle\delta I'_{S_m}\,\delta I'^{*}_{S_n}\rangle = 2qI_{m-n}\Delta f \tag{31}$$

where I_{m-n} is one of the Fourier coefficients of the local oscillator current $i_d(t)$ as defined in (9).

A different approach leading to the same result has been taken by van der Ziel [11] in his analysis of shot noise in mixers. The shot noise, caused by electrons crossing the depletion layer of the diode, is considered as a series of impulsive deviations from the noiseless diode current. The

[2] The symbol $\langle\cdots\rangle$ denotes statistical (or ensemble) average as opposed to time average which is denoted by an overbar. Superscripts T and $\dagger$ denote, respectively, the transpose and the complex conjugate transpose of a matrix or vector.

spectrum of each current impulse is flat and the phase is linear with frequency, its slope depending on the time of occurrence of the impulse. The different frequency components ω_m are down-converted according to (16) and have phases related to the time of occurrence of the impulse and to the LO phase. Summing the separate effects of all such impulses gives the output voltage due to shot noise, including the effects of correlation between the components. Van der Ziel gives the example of a simple two-port mixer as an illustration of the effect of shot-noise correlation.

B. Thermal Noise

It was shown above ((27) and Fig. 3) that the thermal noise generated in the series resistance of the diode can be regarded as a noise current source $\overline{i_T^2}$ across the terminals of the intrinsic diode.

Since R_s is assumed to be time-invariant, down-converted thermal noise will have no correlated components; quasisinusoidal components $\delta I'_{T_m}$ at the sideband frequencies ω_m give rise to an output voltage (using (16))

$$\delta V_{T_0} = \boldsymbol{Z}_0 \boldsymbol{\delta I'}_T \tag{32}$$

where $\boldsymbol{\delta I'}_T = [\cdots, \delta I'_{T_1}, \delta I'_{T_0}, \delta I'_{T_{-1}}, \cdots]^T$, and $\boldsymbol{Z}_0$ is a row of the matrix $\boldsymbol{Z}'$.

Taking the ensemble average of $|\delta V_{T_0}|^2$ gives

$$\langle \delta V_{T_0}^2 \rangle = \boldsymbol{Z}'_0 \langle \boldsymbol{\delta I'}_T \, \boldsymbol{\delta I'}^\dagger_T \rangle \boldsymbol{Z}'^\dagger_0. \tag{33}$$

The square matrix $\langle \boldsymbol{\delta I'}_T \, \boldsymbol{\delta I'}^\dagger_T \rangle$ is again the correlation matrix (cf. (30)), but since $\delta I'_{T_m}$ and $\delta I'_{T_n}$ are uncorrelated for $m \neq n$, it is a diagonal matrix with the elements

$$\langle \delta I'_{T_m} \delta I'^*_{T_m} \rangle = \frac{4kT_{eq} R_{s_m} \Delta f}{|Z_{e_m} + R_{s_m}|^2}, \qquad m \neq 0 \tag{34a}$$

$$= \frac{4kT_{eq} R_{s0} \Delta f}{|Z_0|^2}, \qquad m = 0. \tag{34b}$$

C. Total Mixer Noise

Combining the shot and thermal noise output voltages given by (30) and (33) gives the total noise output voltage appearing across the series combination of the IF load and the diode series resistance (see Fig. 2):

$$\langle \delta V_{N_0}^2 \rangle = \boldsymbol{Z}'_0 \{ \langle \boldsymbol{\delta I}_S \, \boldsymbol{\delta I}^\dagger_S \rangle + \langle \boldsymbol{\delta I'}_T \, \boldsymbol{\delta I'}^\dagger_T \rangle \} \boldsymbol{Z}'^\dagger_0 \tag{35}$$

where the shot and thermal noise correlation matrices $\langle \boldsymbol{\delta I}_S \, \boldsymbol{\delta I}^\dagger_S \rangle$ and $\langle \boldsymbol{\delta I'}_T \, \boldsymbol{\delta I'}^\dagger_T \rangle$ are given by (31) and (34). It is assumed here that the effects of scattering noise are fairly small and are equivalent to a small increase in the value of the equivalent noise temperature of the diode series resistance T_{eq} in (27) and (34). The magnitude of this increase can be estimated from the current waveform of the pumped diode and from measurements on the dc biased diode, as described in Part 2 of this paper.

The equivalent input noise temperature T_M of a two-frequency mixer (no external image termination) may be defined in terms of a noiseless but otherwise identical mixer: T_M is the temperature to which the source conductance of the noiseless mixer must be heated in order to deliver to the IF load the same noise power as the noisy mixer delivers when its source conductance is maintained at absolute zero temperature. For a three-frequency mixer (having external terminations at the signal, image, and intermediate frequencies), the single-sideband mixer noise temperature $T_{M_{SSB}}$ is defined in the same way, but with the stipulation that the image conductance of both the noisy and noiseless mixers be maintained at absolute zero temperature.[3]

The effective mean-square input noise current necessary to produce the noise voltage $\langle \delta V_{N_0} \rangle$ at the output, is found using (16):

$$\langle \delta I'^2_{N_1} \rangle = \langle \delta V_{N_0}^2 \rangle / |Z'_{01}|^2. \tag{36}$$

This corresponds to input current source $\delta I'_1$ in Fig. 2 and is associated with the impedance $(Z_{e_1} + R_{s_1})$. The corresponding mean-square noise voltage associated with the source impedance Z_{e_1} is $\langle \delta I'^2_{N_1} \rangle |Z_{e_1} + R_{s_1}|^2$. The effective input noise temperature of the mixer T_M is the temperature to which $\mathrm{Re}\,[Z_{e_1}]$ must be heated to generate an equal noise voltage. Hence, using (36),

$$T_M = \frac{\langle \delta V_{N_0}^2 \rangle}{4k\Delta f} \frac{|Z_{e_1} + R_{s_1}|^2}{|Z'_{01}|^2 \, \mathrm{Re}\,[Z_{e_1}]}. \tag{37}$$

$\langle \delta V_{N_0}^2 \rangle$ is given by (35) in terms of the internal shot and thermal noise sources of the diode and Z'_{01} is an element of the matrix $\boldsymbol{Z}'$ defined in (17).

The single sideband noise figure of the mixer is defined as

$$F_{SSB} = 1 + \frac{T_M}{T_0} \tag{38}$$

where $T_0 = 290$ K by convention.

V. The Microwave Schottky Diode

The Schottky diode is conventionally represented as shown in Fig. 1. The series resistance is assumed equal to its dc value and independent of frequency, while the capacitance exponent γ (5) is assumed independent of voltage. For a diode of the kind used for microwave and millimeter-wave mixers these assumptions are not generally valid, and it is important to include the departures from ideality in the nonlinear and small-signal analyses if accurate results are to be obtained.

A. Series Resistance

There are three effects which cause the series resistance R_s of a diode to differ from the value determined from the dc log I–V curve. These are 1) thermal time-constants in the diode, 2) RF skin effect in the diode material, and 3) the voltage dependence of the depletion-layer width.

It has been shown by Decker and Weinreb [20] that the incremental junction resistance of a diode, measured at a

[3] For the three-frequency mixer a double-sideband noise temperature $T_{M_{DSB}}$ is frequently used. For cases in which the signal and image conversion loss L_{01} and L_{0-1} are equal $T_{M_{DSB}} = \frac{1}{2} T_{M_{SSB}}$. If $L_{01} \neq L_{0-1}$, which is particularly probable when a high IF is used, then $T_{M_{DSB}} = T_{M_{SSB}}/(1 + L_{01}/L_{0-1})$.

low frequency (audio or dc), contains a negative component caused by heating of the diode by the test signal. Thus

$$r_{lf} \triangleq \frac{dv_j}{di_d} = \left.\frac{\partial v_j}{\partial i_d}\right|_T + \left.\frac{\partial v_j}{\partial T}\right|_{i_d} \cdot \frac{dT}{di_d} \tag{39}$$

$$= r_{hf} + \left.\frac{\partial v_j}{\partial T}\right|_{i_d} \cdot \frac{dT}{di_d}. \tag{40}$$

The second term on the RHS of (39) is the (negative) component of the low-frequency junction resistance which results from the diode temperature varying in phase with the test signal. At higher frequencies, above about 10 MHz for our 2.5-μm GaAs diodes, thermal time constants in the diode prevent its temperature from varying appreciably and only the term r_{hf} is measured. It follows that the value of the series resistance R_s deduced from the dc log I–V curve is lower than the true constant temperature value which would be observed at microwave frequencies. For our millimeter-wave diodes the error in R_s is 1–3 Ω [20], [23].

The RF skin effect contributes an additional resistance R_{skin} in series with the diode. R_{skin} is proportional to $\sqrt{f}$ and can be estimated if the diode geometry and resistivity are known [24]. The diode contact wire is also likely to have a significant skin resistance, which should be included in R_{skin}. For our 2.5-μm GaAs diodes R_{skin} is 2–3 Ω at 100 GHz.

The width of the depletion layer in a diode depends on the junction voltage and, therefore, the component of series resistance contributed by the undepleted epitaxial material will also be voltage-dependent. In the present paper we have assumed that this voltage dependence has negligible effect on the performance of the mixer.

B. Junction Capacitance

The simple diode capacitance law, (5), is derived in many texts [21] for planar diodes. If the doping profile at the barrier (or junction) is uniform or linear with distance, the exponent γ has the value $\frac{1}{2}$ or $\frac{1}{3}$, respectively. However, for the very small diameter diodes used at millimeter wavelengths the assumption of a planar device is no longer valid: fringing effects at the edge of the depletion layer are significant and the shape of the depletion layer is voltage-dependent [21]. The simple capacitance law, (5), can still be used provided the exponent γ is allowed to be voltage dependent, i.e., $\gamma \rightarrow \gamma(v_j)$.

VI. Application to Practical Mixers

The mixer analysis presented in the preceeding sections assumes a knowledge of the diode and embedding impedance at all harmonics of the LO and at all the sideband frequencies $\omega_0 + k\omega_p$, $k = 0, \pm 1, \pm 2, \cdots, \pm \infty$. In practice the number of frequencies which can be considered is limited by the ability to make meaningful measurements or calculations of the embedding impedance beyond some frequency limit, and by the size of complex matrix which can be inverted by available computers. The computer time required to solve the nonlinear problem increases rapidly with the number of LO harmonics. It is necessary therefore to work with a finite number N of LO harmonics and a corresponding number $(N + 1)$ of sideband frequencies. The infinite series and matrices of Sections II–IV can then be replaced by finite ones. The electrical implication of such a truncation is that all higher harmonics or sidebands are either open- or short-circuited. In particular, truncation of the Y matrix in (12) implies that all higher sidebands are short-circuited. This is likely to be a good approximation for millimeter-wave mixers because the mean junction capacitance approaches a short circuit for very high frequencies. A similar situation results from ignoring frequencies above $N\omega_p$ in the nonlinear analysis. Using Gwarek's method it is tacitly assumed that the junction voltage v_j has no components above frequency $N\omega_p$. Again this should be a good approximation provided N is high enough.

VII. Conclusion

The nonlinear, small-signal, and noise analysis given in Sections II–V provides the means for accurately determining the performance of microwave and millimeter-wave mixers. In the companion paper, Part 2, the analysis is applied to mixers operating at 87 and 115 GHz. Good agreement with measured results is obtained, and the "anomalous" mixer noise [14] is shown to be primarily shot noise for which the phase relations of down-converted components must be taken into account.

References

[1] H. C. Torrey and C. A. Whitmer, *Crystal Rectifiers*, (MIT Radiation Lab. Series, vol. 15). New York: McGraw-Hill, 1948.

[2] W. M. Sharpless, "Wafer-type millimeter-wave rectifiers," *Bell Syst. Tech. J.*, vol. 35, pp. 1385–1402, Nov. 1956.

[3] G. C. Messenger and C. T. McCoy, "Theory and operation of crystal diodes as mixers," *Proc. IRE*, vol. 45, pp. 1269–1283, 1957.

[4] L. Mania and G. B. Stracca, "Effects of the diode junction capacitance on the conversion loss of microwave mixers," *IEEE Trans. Communications*, vol. COM-22, pp. 1428–1435, Sept. 1974.

[5] A. R. Kerr, "Low-noise room-temperature and cryogenic mixers for 80–120 GHz," *IEEE Trans. Microwave Theory Tech.*, vol. MTT-23, pp. 781–787, Oct. 1975.

[6] M. R. Barber, "Noise figure and conversion loss of the Schottky barrier mixer diode," *IEEE Trans. Microwave Theory Tech.*, vol. MTT-15, pp. 629–635, Nov. 1967.

[7] A. A. M. Saleh, *Theory of Resistive Mixers*. Cambridge, MA: MIT Press, 1971.

[8] S. Egami, "Nonlinear, linear analysis and computer aided design of resistive mixers," *IEEE Trans. Microwave Theory Tech.*, vol. MTT-22, pp. 270–275, Mar. 1974.

[9] M. J. O. Strutt, "Noise figure reduction in mixer stages," *Proc. IRE*, vol. 34, no. 12, pp. 942–950, Dec. 1946.

[10] A. van der Ziel and R. L. Waters, "Noise in mixer tubes," *Proc. IRE*, vol. 46, pp. 1426–1427, 1958.

[11] A. van der Ziel, *Noise: Sources, Characterization, and Measurement*. Englewood Cliffs, NJ: Prentice-Hall, 1970.

[12] C. S. Kim, "Tunnel diode converter analysis," *IRE Trans. Electron Devices*, vol. ED-8, no. 5, pp. 394–405, Sept. 1961.

[13] C. Dragone, "Analysis of thermal shot noise in pumped resistive diodes," *Bell Syst. Tech. J.*, vol. 47, pp. 1883–1902, 1968.

[14] A. R. Kerr, "Anomalous noise in Schottky-diode mixers at millimeter wavelengths," *IEEE MTT-S Int. Microwave Symp., Digest of Technical Papers*, pp. 318–320, May 1975.

[15] D. A. Fleri and L. D. Cohen, "Nonlinear analysis of the Schottky-barrier mixer diode," *IEEE Trans. Microwave Theory Tech.*, vol. MTT-21, pp. 39–43, Jan. 1973.

[16] A. R. Kerr, "A technique for determining the local oscillator waveforms in a microwave mixer," *IEEE Trans. Microwave Theory Tech.*,

vol. MTT-23, pp. 828–831, Oct. 1975.

[17] W. K. Gwarek, "Nonlinear analysis of microwave mixers," M.S. thesis, MIT, Cambridge, Sept. 1974.

[18] S. O. Rice, "Mathematical analysis of random noise," *Bell Syst. Tech. J.*, vol. 23, no. 3, pp. 282–332, July 1944.

[19] V. R. Bennet, *Electrical Noise.* New York: McGraw-Hill, 1960.

[20] D. R. Decker and S. Weinreb, private communication.

[21] S. M. Sze, *Physics of Semiconductor Devices.* New York: Wiley, 1969 (see p. 90 *et seq.*).

[22] W. Baechtold, "Noise behavior of GaAs field-effect transistors with short gates," *IEEE Trans. Electron Devices*, vol. ED-19, pp. 674–680, 1972.

[23] D. N. Held, "Analysis of room temperature millimeter-wave mixers using GaAs Schottky barrier diodes," Sc.D. dissertation, Department of Electrical Engineering, Columbia University, New York, 1976.

[24] J. A. Calviello, J. L. Wallace, and P. R. Bie, "High performance GaAs quasi-planar varactors for millimeter waves," *IEEE Trans. Electron Devices*, vol. ED-21, pp. 624–630, Oct. 1974.

Conversion Loss and Noise of Microwave and Millimeter-Wave Mixers: Part 2—Experiment

DANIEL N. HELD, MEMBER, IEEE, AND ANTHONY R. KERR, ASSOCIATE MEMBER, IEEE

***Abstract*—The theory of noise and conversion loss in millimeter-wave mixers, developed in a companion paper, is applied to an 80–120-GHz mixer. Good agreement is obtained between theoretical and experimental results, and the source of the recently reported "anomalous noise" is explained. Experimental methods are described for measuring the embedding impedance and diode equivalent circuit, needed for the computer analysis.**

I. Introduction

In Part 1 of this paper [1] the theory of microwave and millimeter-wave mixers was presented in a form suitable for analysis by digital computer. The present paper gives the results of such an analysis, comparing computed and measured conversion loss, noise, and output impedance for an 80–120-GHz mixer under various operating conditions.

The significance of this work in relation to earlier work is discussed in the introduction to Part 1, the salient points being that this method of analysis gives unprecedented agreement between theory and experiment and that the "anomalous noise" [2] reported in millimeter-wave mixers is entirely accounted for.

The analysis requires a knowledge of the embedding impedance seen by the diode at a finite number of frequencies, and of the equivalent circuit of the diode, including noise sources. Experimental methods for determining these input quantities are described in Sections II and III. The mixer analysis is described in Section IV and typical sets of theoretical and measured results are shown to be in good agreement. Section V discusses the sources of mixer noise and loss, and small-signal power flow in the mixer.

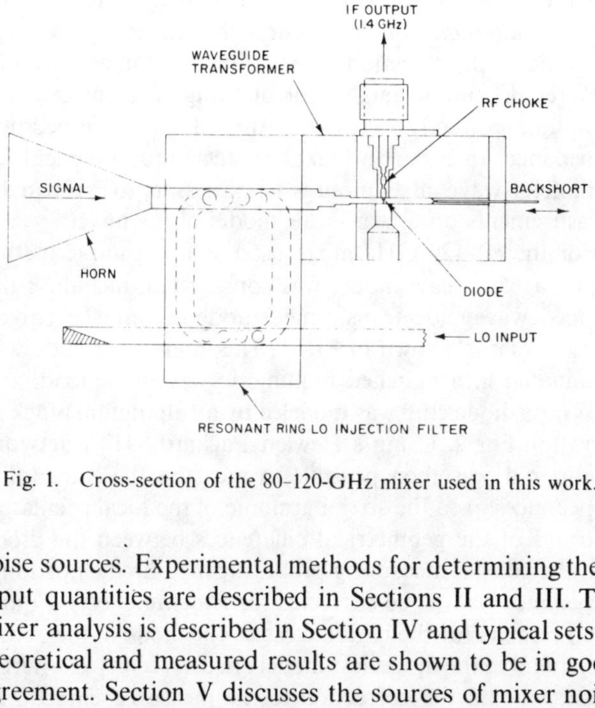

The mixer used for these experiments is the room-temperature 80–120-GHz mixer described in [3] and is shown here in Fig. 1. It uses a 2.5-μm-diameter GaAs Schottky diode in a quarter-height waveguide mount. The diode was made by Professor R. J. Mattauch at the University of Virginia.

Fig. 1. Cross-section of the 80–120-GHz mixer used in this work.

Manuscript received February 18, 1977; revised May 12, 1977.

D. N. Held was with the Department of Electrical Engineering, Columbia University, New York, NY 10027. He is now with the Jet Propulsion Laboratory, California Institute of Technology, Pasadena, CA 91103.

A. R. Kerr is with the NASA Institute for Space Studies, Goddard Space Flight Center, New York, NY 10025.

Reprinted from *IEEE Trans. Microwave Theory Tech.*, vol. MTT-26, pp. 55–61, Feb. 1978.

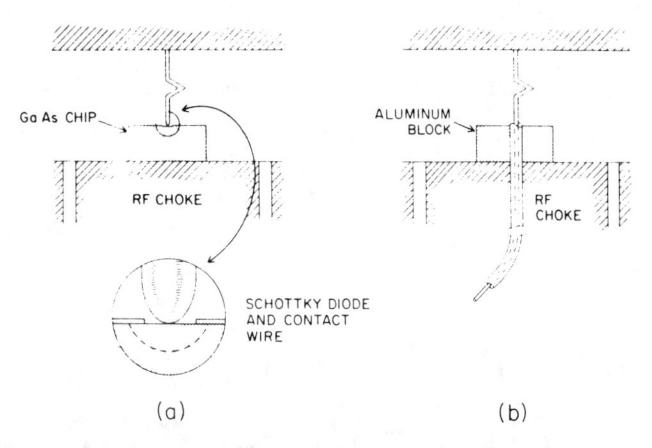

Fig. 2. Technique for measuring the embedding impedance seen by the diode. The diode in the real mixer (a) is replaced in the scale model (b) by the end of a small coaxial cable through which the embedding impedance can be measured directly.

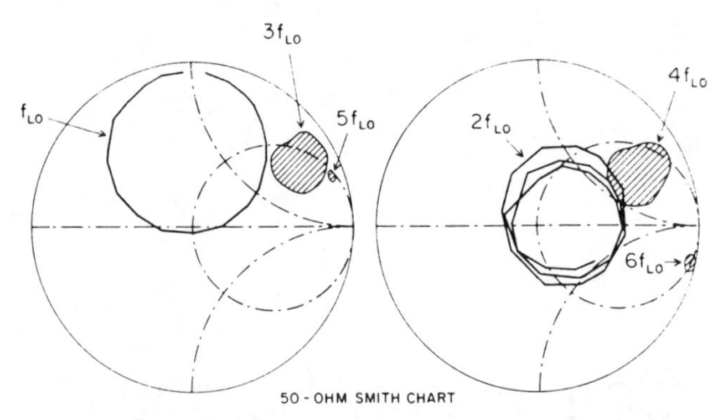

Fig. 3. The embedding impedance at harmonics of the LO frequency, for $f_{LO} = 87$ GHz. The impedances are shown as functions of backshort position.

II. The Diode Embedding Impedance

Eisenhart and Kahn [4], [5] have described a technique whereby it is possible to measure the embedding impedance seen by a waveguide-mounted diode. The diode is removed from its mount and replaced by a small coaxial cable, so the impedance seen by the end of the cable is essentially equal to the embedding impedance normally seen by the diode. The other end of the cable is connected to a slotted line or network analyzer which measures this impedance. Except for some small corrections due to differences in geometry in the region immediately surrounding the diode, the technique directly provides the desired embedding impedance. In order to make this technique practical for millimeter-wave mixer mounts, it is necessary to perform the measurements on a large-scale model of the mixer.

For the 80–120-GHz mixer used as an example in this paper, a 65 × scale model was constructed, including the stepped waveguide transformer, diode mount, RF choke, and backshort, shown in Fig. 1. The input waveguide was terminated in a matched multimode waveguide load, and the GaAs diode chip was modeled by an aluminum block as shown in Fig. 2. Using a Hewlett-Packard 8410A network analyzer it was then possible to measure the embedding impedance up to the sixth harmonic of the local oscillator. Because of the geometrical differences between the diode and the coaxial measuring probe, the measured embedding impedance had to be corrected for fringing effects. Equivalent circuits for the region close to the diode in the actual mixer and in the diode-less model are derived in [6], where it is shown that the correction to the measured embedding impedance is equivalent to adding a 0.63 fF (0.00063 pF) capacitor in shunt with the measured impedance (cf. the zero-bias diode capacitance $\mathscr{C}_{j0} \simeq 7.0$ fF).

Typical embedding impedances measured at the scale-model equivalent of 87 GHz are illustrated in Fig. 3. The experimental results illustrated in this figure are presented as a function of the mixer backshort position, since in subsequent sections of this paper mixer performance will be evaluated as a function of backshort position. It is clear from Fig. 3 that the embedding impedances at the harmonics of the local oscillator cannot be considered as open-circuits, short-circuits, or even simple reactive terminations, as has sometimes been assumed in the literature. In particular, the embedding impedance at the second harmonic has a substantial and widely varying real part.

III. The Diode Equivalent Circuit

In Section V of Part 1 the characteristics of very small Schottky diodes were discussed. It was pointed out that the RF skin effect could contribute substantially to the series resistance, while nonuniform epilayer doping and fringing effects at the periphery of the depletion layer might cause the capacitance-law exponent γ to be voltage-dependent. It was further pointed out that scattering, observed in Schottky diodes at high-current levels, would effect mixer noise, and that thermal time constants must be taken into account when measuring the series resistance of a diode.

The equivalent circuit of a Schottky barrier diode, including noise sources, is shown in Fig. 4. In the following paragraphs procedures are described for determining the element values of this equivalent circuit, which are appropriate at millimeter wavelengths.

A. *Junction Capacitance*

The capacitance of the junction as a function of voltage is given by [7]

$$\mathscr{C}_j(v_j) = \frac{dQ}{dv_j} = \mathscr{C}_{j0}\left(1 - \frac{v_j}{\phi}\right)^{-\gamma(v_j)} \tag{1}$$

where $\mathscr{C}_{j0}$ is the zero-voltage capacitance, ϕ is the barrier potential, and $\gamma(v_j)$ is related to the geometry of the junction and the semiconductor doping profile. If the junction can be considered planar, with negligible fringing effects, and if the epilayer doping is uniform, then $\gamma = 0.5$. However, when the diameter of the diode is not large compared with the thickness of the depletion layer, fringing effects at the edges of the diode may be significant, and will be most pronounced under reverse bias when the depletion layer is widest.

The junction capacitance can be measured at 1 MHz

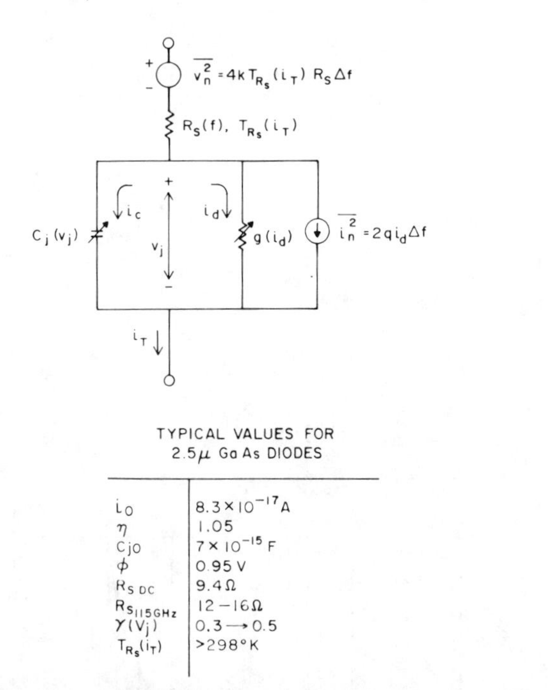

Fig. 4. The equivalent circuit of a Schottky diode, including frequency dependent series resistance and current dependent resistor noise temperature.

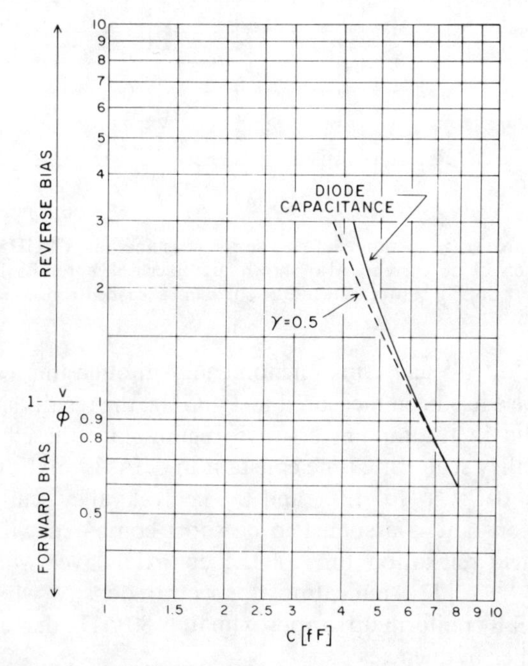

Fig. 5. Experimental $\mathscr{C}_j$ versus $(1 - v_j/\phi)$ for a typical 2.5-μm diode, illustrating the voltage dependence of γ.

using a capacitance bridge[1] [6], [8]. By plotting log $\mathscr{C}_j$ versus log $(1 - v_j/\phi)$, a line whose slope is equal to $-\gamma$ is obtained. Experimental results indicate that the value of γ for some diodes exhibits a pronounced voltage dependence, a typical result being shown in Fig. 5.

The question arises as to whether the capacitance measured at 1 MHz is meaningful at 100 GHz and above. There is, however, no evidence to suggest any frequency

[1] A Boonton model 75D was used for this work.

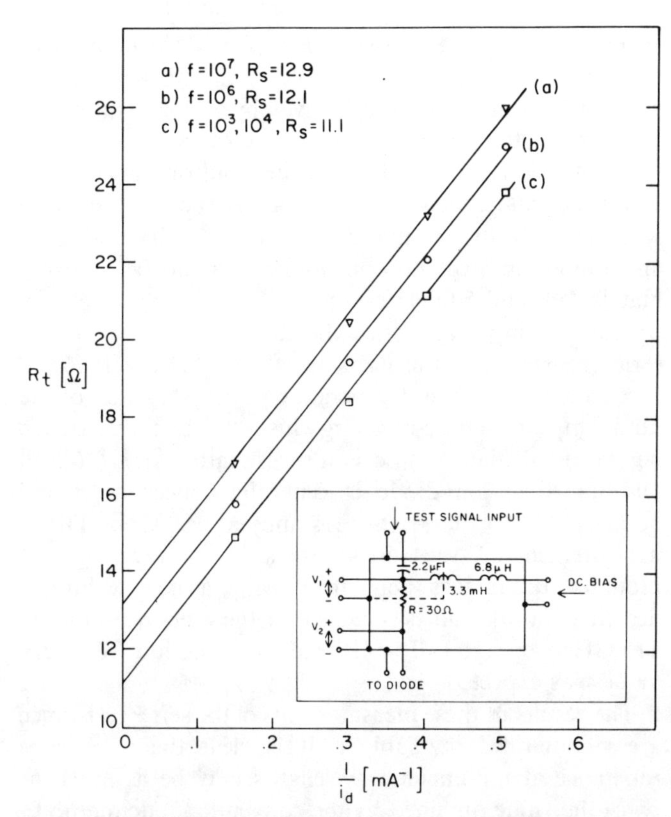

Fig. 6. Measured incremental resistance of a 2.5-μm gallium arsenide diode as a function of reciprocal dc bias current, showing apparent frequency dependence of the series resistance (the y-axis intercept). This is due to heating effects. Inset: Circuit of device for measuring the incremental resistance R_t at frequencies up to 10 MHz.

dependence of the junction capacitance over this range of frequencies. There have been some papers [10]–[12] dealing with the effect of deep-level donors on frequency dependence of the transition capacitance of p-n junctions. However, the effect is only significant when the number of deep traps is comparable to the number of donors, which should not be the case for the heavily doped diodes under study. A second factor which can cause a frequency dependence of the junction capacitance is the scattering limited velocity of the electrons in the semiconductor. This limits the rate at which the depletion layer can be depleted or refilled, and is expected to have negligible effects below ~ 100 GHz [13].

B. The Barrier Resistance and the Series Resistance

It has been shown by Weinreb and Decker [14] that the conventional method of determining the series resistance R_s from the dc log I–V curve contains an inherent error due to heating at the junction. To determine the true value of R_s measurements must be performed at frequencies greater than the reciprocal of the shortest thermal time constant of the diode and its mount. Measurements of these thermal time constants [14] indicate that errors should be negligible above ~ 20 MHz.

A simple circuit, shown in Fig. 6, enables the incremental resistance R_t of a forward-biased diode to be measured in the frequency range 1 kHz–10 MHz. Measurements on a 2.5 μm GaAs diode indicate that the resistance starts to increase above ~30 kHz and levels off at ~10 MHz. These results are shown in Fig. 6 where R_t is plotted as a function of reciprocal bias current for several different frequencies; the value of R_s at each frequency is given by the y-axis intercept. It is seen that the resistance increases by $\sim$2–3 Ω between 10 kHz and 10 MHz. Further confirmation of these measurements comes from Weinreb and Decker [14] who measured the thermal time constants of the diode chip and mounting structure and found them to be approximately 1 μs and 50 ms, respectively.

Another important phenomenon effecting the diode's series resistance is the RF skin effect. This effect occurs predominantly in the degenerately doped substrate of the diode chip and in the contacting whisker. The skin resistance has been calculated by several investigators [15], [16], [6] and for a 2.5-μm diode is generally conceded to add about 2–3 Ω to the series resistance at 100 GHz. Direct measurement of the series resistance of a waveguide-mounted diode was performed using a new technique described in [6]. Values obtained for the series resistance at 115 GHz were typically 2 Ω larger than the low frequency values[2] as expected.

The results of these measurements of the series resistance are summarized in Table I. It is clear that the series resistance at millimeter wavelengths may be as much as twice the value obtained by the conventional dc method.

C. Noise Temperature of the Series Resistance

It is well known that at electric field strengths in the vicinity of 3×10^3 V/cm^3, bulk GaAs displays a nonlinear I–V characteristic which is due primarily to central valley and intervalley phonon scattering.[3] It will be shown in Section IV that under normal mixer operating conditions the instantaneous value of the current through the series resistance of a diode can reach levels $\sim$ 10 mA. For the diodes used in this work it can be shown that the peak value of the electric field in the undepleted epitaxial material may then be as large as 1.9×10^3 V/cm.[4] Although this value is comfortably below the critical field required for the onset of the Gunn effect at low frequencies, the field strength is sufficient to increase substantially the effective temperature of the electron gas in the GaAs and hence the noise temperature of the series resistance.

The noise temperature of the series resistance, measured at a frequency of 1.4 GHz on a dc biased diode, is shown in Fig. 7 as a function of the dc current. The noise temperature of the series resistance was derived by measuring the diode output temperature and correcting for the effect of the shot noise produced in the junction [6], [17]. Also shown in Fig. 7 are results reported by Baechtold [18] who measured the noise temperature of a bulk sample of GaAs. It is clear, then, that at high currents excess noise is generated in GaAs mixer

TABLE I
SERIES RESISTANCE VALUES

R_s measured at DC (including thermal effects)	6.8±0.1 ohm
R_s measured at 10 MHz	9.4±0.2 ohm
Thermal time-constant error	-2.6±0.3 ohm
True low frequency R_s (measured at 10 MHz)	9.4±0.2 ohm
Calculated skin-effect at 115 GHz	2.5±0.5 ohm
Predicted value of R_s at 115 GHz	11.9±0.7 ohm
Measured value of R_s at 115 GHz	11.2±2.0 ohm

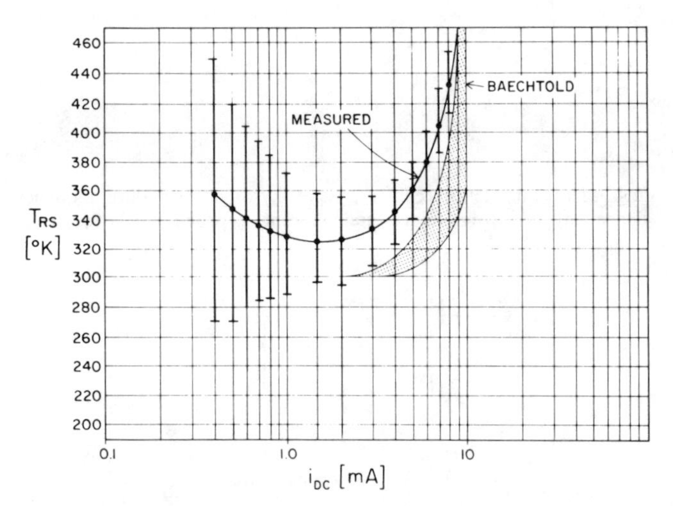

Fig. 7. Noise temperature of the series resistance at 1.4 GHz as a function of dc current. Also shown are Baechtold's results for an epilayer doping between 2 and 3×10^{17} cm^{-3} (shaded region).

diodes due to scattering phenomena. Although the scattering noise has been measured at 1.4 GHz, the results should be valid in the millimeter-wave region since the effective intervalley scattering time constant in GaAs is on the order of 2×10^{-12} s [19]–[21], and the central valley scattering time constant is assumed to be short compared with the dielectric relaxation time of the epitaxial layer which is $\sim 10^{-14}$ s [22], indicating a spectral density which is essentially uniform up to approximately 80 GHz, decreasing at higher frequencies.

When the diode is operating as a mixer the noise temperature of the series resistance will be a function of time, and the down-converted components of this noise will be partially correlated [1]. For a room-temperature mixer this correlation is expected to have a small effect, and it will be assumed here that scattering noise can be approximated by an increased average noise temperature of the series resistance,[5] an assumption supported by the good agreement obtained between theory and experiment in Section

[2] Low-frequency values corrected for thermal time-constant effects.

[3] The critical field for highly doped GaAs is approximately 5×10^3 V/cm [9], [18].

[4] This figure is based on the assumption of uniform current flow through a 2.5-μm diameter diode, with a resistivity of 0.096 Ω-cm in the epilayer, corresponding to an impurity concentration of 2.5×10^{17}/cm^3.

[5] The average noise temperature is derived by integrating the measured noise temperature (Fig. 7) at the total diode current, i_T (Fig. 8), over the LO cycle. Typical average noise temperatures are approximately 60 K–100 K above ambient at room temperature.

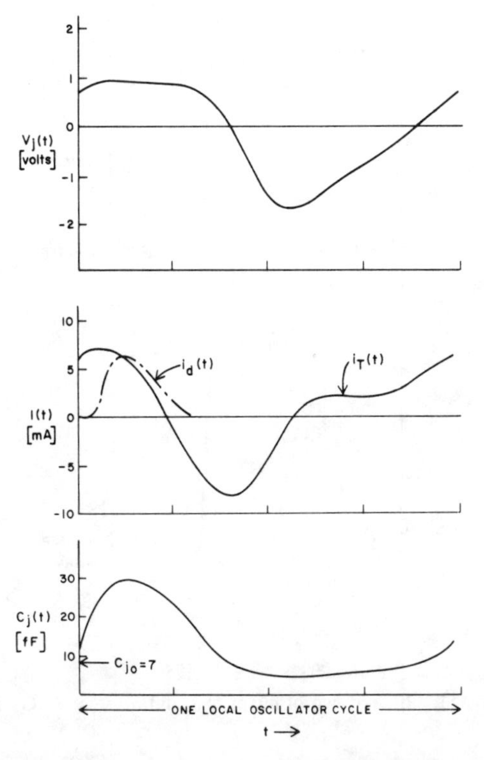

Fig. 8. Typical diode waveforms at 87 GHz. The junction voltage $v_j(t)$, total current in R_s, $i_T(t)$, current in the diode conductance, $i_d(t)$, and junction capacitance $\mathscr{C}_j(t)$ are illustrated for a single local oscillator cycle. The dc bias conditions are: $I_{dc} = 2.0$ mA, $V_{dc} = 0.5$ V.

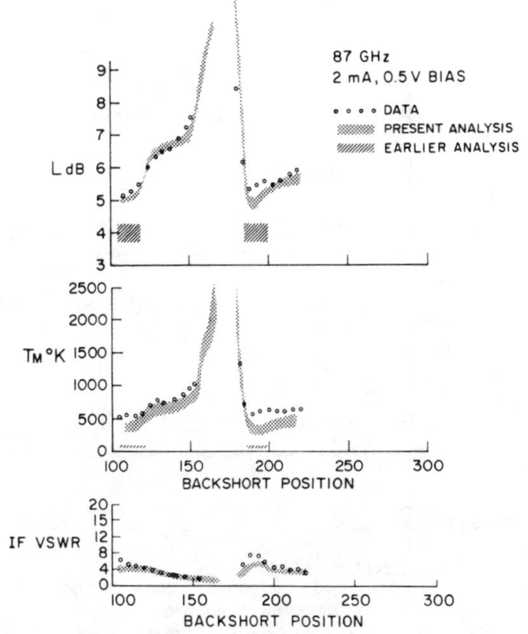

Fig. 9. Comparison of theoretical and experimental results at 87 GHz. The SSB conversion loss L, equivalent SSB input noise temperature T_M of the mixer (excluding IF amplifier noise), and IF VSWR with respect to 50 Ω, are shown as functions of backshort position. The theoretical results are computed for a broadband mixer (equal signal and image response) with the capacitance exponent $\gamma = 0.3$–0.5 and the series resistance $R_S(100\text{ GHz}) = 12$–16 Ω. Also shown are the results of an earlier analysis [2], [3] which led to the observation of anomalous mixer noise. T_M is related to the noise temperature ratio t of a broad-band mixer by $T_M = T_a(Lt - 2)$, where T_a is the ambient temperature of the mixer and its input termination.

IV. In cryogenic mixers however, scattering noise and its partially correlated down-converted components are expected to be very significant and this simplifying assumption may be quite invalid.

IV. Mixer Performance Analysis

It was shown in Part 1 [1] that if the embedding impedance and the equivalent circuit of the diode are known, it is possible to solve the large-signal nonlinear problem to determine the LO waveforms at the diode, and then to determine the small-signal conversion loss and noise figure of the mixer. In this section the results of such an analysis are given for the 80–120 GHz mixer shown in Fig. 1 [3]. The embedding impedance was measured on a model, as described in Section II of the present paper, and the large-signal LO waveforms were computed as described in Section II of Part 1. Typical waveforms are shown in Fig. 8.

Knowing the LO waveforms, the conversion loss and noise temperature of the mixer can be determined as described in Sections II and III of Part 1. Since the mixer backshort setting has the strongest effect of all the external variables on mixer performance (other variables are dc bias and LO power), the measured and computed results given here are plotted as functions of backshort position. This serves not only to verify the accuracy of the computed solution for a wide variety of embedding impedances, but also provides useful insight into the behavior of the mixer. Typical sets of results are shown in Figs. 9 and 10 for 87 and 115 GHz. Computed results are shown for $\gamma = 0.3$–0.5 and $R_s(100\text{ GHz}) = 12$–16 Ω; these correspond to the likely range of values of these quantities allowing for skin effect and fringing effects as discussed in Section III. It has been assumed in this work that $\gamma(v_j)$ can be approximated by a voltage-independent mean value and that R_s exhibits the normal skin-effect type of frequency dependence. Also shown in Fig. 9 are the results of an earlier analysis [2], [3] which led to the observation of an anomalous component of mixer noise, thereby stimulating the present work.

The main sources of error in the computed results are expected to be due to high-Q effects and to loss in the backshort. It was assumed for simplicity that the embedding impedance at a LO harmonic was equal to the impedance at the adjacent sideband frequencies, i.e., $Z_e(n\omega_p \pm \omega_{if}) = Z_e(n\omega_p)$. Clearly this assumption will lead to errors if there are any sharp features on the loss or noise-temperature curves of Figs. 9 and 10. The conversion-loss minimum in Fig. 10 is an example of this and it can be seen that the measured and computed results differ most in this region. The backshort used in the experimental mixer was of the contacting spring-finger type, generally assumed to have low loss. Measurement of the loss of this particular backshort revealed a surprisingly large loss, equivalent to a resistance of ~ 8 Ω at the plane of the short-circuit. At certain backshort positions this can contribute substantially to the measured conversion loss. Experimental repeatability was somewhat degraded by this backshort, presumably due to poor electrical contact with the waveguide. Experience during these measurements suggests experimental uncer-

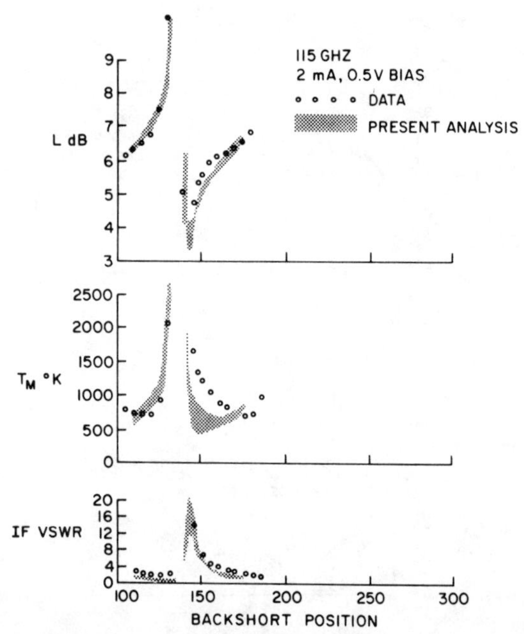

Fig. 10. Comparison of theoretical and experimental mixer performance at 115 GHz. See Fig. 9 caption for details.

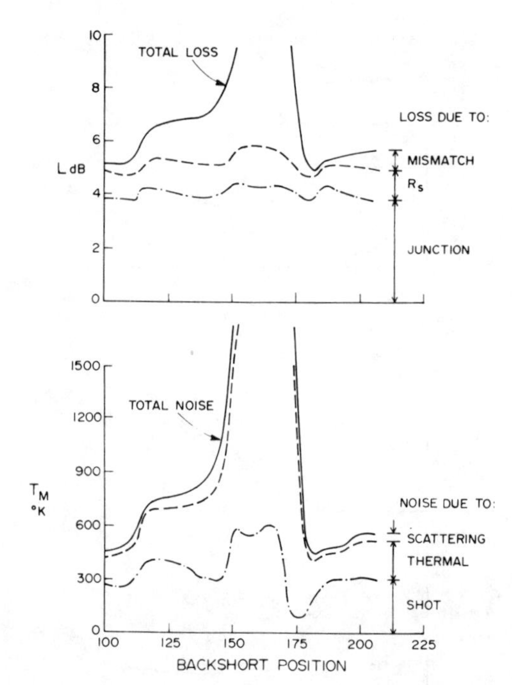

Fig. 11. Breakdown of mixer loss and noise at 87 GHz.

tainties of ± 0.3 dB in L and $\pm 130°$ in T_M for most backshort settings, doubling in the vicinity of the sharp conversion-loss minimum. All measurements were performed using the 1.4-GHz IF radiometer/reflectometer apparatus described by Wienreb and Kerr [23].

V. Discussion

Having established that the performance of a mixer can be accurately predicted by the theory given in the companion paper [1], it is useful to evaluate the various contributions to the conversion loss and the mixer noise.

A. Sources of Mixer Loss and Noise

In Fig. 11 T_M and L are shown broken down into their constituant parts. The three major components of the conversion loss are 1) RF input mismatch loss, 2) dissipation in the series resistance at the signal and intermediate frequencies, and 3) loss in the junction, which includes power dissipated in the junction resistance at the signal and intermediate frequencies and also power lost by conversion to other frequencies. It should be noted that the mismatch loss can not necessarily be eliminated by RF tuning, since for broadband mixers the minimum overall conversion loss usually occurs for a mismatched input [24].

Noise in this room-temperature mixer is composed almost entirely of shot and thermal noise; the effect of the scattering mechanisms is fairly small. In the vicinity of the minimum of T_M the contributions from the shot and thermal sources are comparable.

B. Power Flow in the Mixer

From the conversion loss analysis given in Part 1 it is possible to calculate the conversion loss between any two frequencies and to analyze the power flow in the mixer. A typical result is shown in Fig. 12 for two backshort positions at 115 GHz. It is interesting to observe from Fig. 12(a)

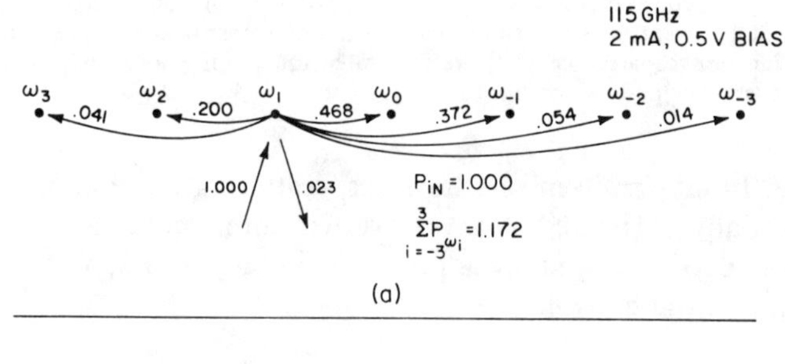

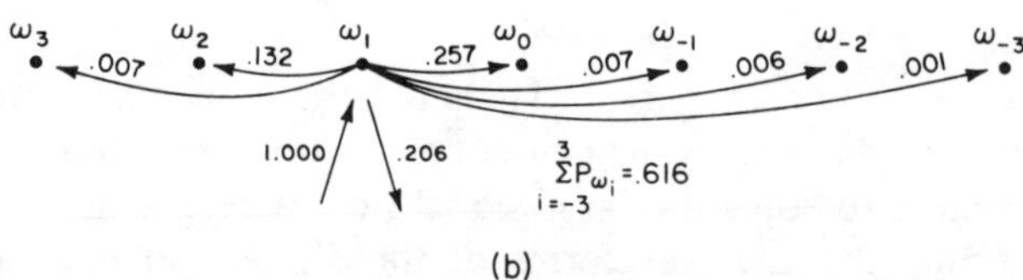

Fig. 12. Power flow at 115 GHz for two backshort positions. (a) Backshort position corresponding to minimum conversion loss. (b) Backshort position corresponding to minimum T_M.

that the mixer can behave as an active device, more small signal power being generated at the sideband frequencies than is incident at the signal frequency. This is a consequence of the time variation of the junction capacitance which produces "parametric" effects. It is also interesting to note that a substantial amount of power is dissipated at the sum frequency $\omega_2 = 2\omega_p + \omega_0$. Thus the embedding impedance at that frequency may be as important in mixer design as is the termination of the image frequency.

VI. Conclusion and Summary

The theory given in Part 1 has been used to predict the conversion loss and noise of an 80–120-GHz mixer. Close agreement was obtained between measured and computed results for a variety of operating conditions. The analysis has demonstrated the following.

1) For a room temperature mixer the "anomalous" noise can be explained almost entirely by shot and thermal noise. The contribution due to scattering is typically only 10 percent.

2) The shot noise has components which, when down converted via the action of the mixer, are correlated.

3) The assumptions that the series resistance is time invariant and that scattering noise can be approximated by an increase in the noise temperature of the series resistance, are reasonable for this room-temperature mixer. This is not expected to be the case for cryogenic mixers, however, as scattering noise may be a substantial part of the overall noise.

4) The value of the series resistance in the vicinity of 100 GHz may be as much as twice its apparent dc value. This is due to the RF skin effect, and to an error, caused by thermal effects, inherent in the conventional method of determining the low-frequency series resistance from the dc log I–V characteristic of the diode. The difference between the apparent dc series resistance and the actual RF value accounts for approximately 1 dB of passive loss in the mixer used in this work.

Acknowledgment

The authors wish to thank the following for their substantial contributions to the work reported here: R. J. Mattauch and G. Green of the University of Virginia; S. Weinreb and D. R. Decker of the National Radio Astronomy Observatory; S. P. Schlessinger and E. Yang of Columbia University; and P. Thaddeus, J. Grange, I. Silverberg, and H. Miller of the Goddard Institute for Space Studies.

References

[1] D. N. Held and A. R. Kerr, "Conversion loss and noise of microwave and millimeter-wave mixers: Part 1—Theory," *IEEE Trans. Microwave Theory Tech.*, vol. MTT-26, Feb. 1978.

[2] A. R. Kerr, "Anomalous noise in Schottky diode mixers at millimeter wavelengths, *IEEE MTT-S Int. Microwave Symp., Digest of Technical Papers*, 1975.

[3] ——, "Low-noise room-temperature and cryogenic mixers for 80–120 GHz," *IEEE Trans. Microwave Theory Tech.*, vol. MTT-23, pp. 781–787, 1975.

[4] R. L. Eisenhart, "Impedance characterization of a waveguide microwave circuit," U.S. Army Electronics Command, Fort Monmouth, NJ, Technical Report 208, 1971.

[5] R. L. Eisenhart and P. J. Kahn, "Theoretical and experimental analysis of a waveguide mounting structure," *IEEE Trans. Microwave Theory Tech.*, vol. MTT-19, pp. 706–719, 1971.

[6] D. N. Held, "Analysis of room temperature millimeter-wave mixers using GaAs Schottky barrier diodes," Sc.D. dissertation, Columbia University, 1976.

[7] S. M. Sze, *Physics of Semiconductor Devices*. New York: Wiley, 1969.

[8] A. R. Kerr, R. J. Mattauch, and J. Grange, "A new mixer design for 140–220 GHz," *IEEE Trans. Microwave Theory Tech.*, vol. MTT-25, pp. 399–401, May 1977.

[9] R. J. Mattauch, Private communication.

[10] E. Schibli and A. G. Milnes, "Effects on deep impurities on n + p junction reverse-biased small-signal capacitance," *Solid State Electron.*, vol. 11, pp. 323–334, 1968.

[11] W. Schultz, "A theoretical expression for the impedance of reverse-biased P-N junctions with deep traps," *Solid State Electron.*, vol. 14, pp. 227–231, 1971.

[12] K. Hesse and H. Strack, "On the frequency dependence of GaAs Schottky barrier capacitances," *Solid State Electron.*, vol. 15, pp. 767–774, 1972.

[13] J. C. Irvin, T. P. Lee, and D. R. Decker, "Varactor diodes," ch. in *Microwave Semiconductor Devices and their Circuit Application*, H. A. Watson, ed. New York: McGraw-Hill, 1969.

[14] S. Weinreb and D. R. Decker, Private communication.

[15] J. A. Calviello, J. L. Wallace, and P. R. Bie, "High performance GaAs quasi-planar varactors for millimeter waves," *IEEE Trans. Electron Devices*, vol. ED-21, pp. 624–630, Oct. 1974.

[16] M. Schneider, Private communication.

[17] R. J. Mattauch and T. J. Viola, "High frequency noise in Schottky barrier diodes," *Research Laboratories for the Engineering Sciences*, Univ. of Virginia, Report No. EE-4734-101-73U, 1973.

[18] W. Baechtold, "Noise behavior of GaAs field-effect transistors with short gates," *IEEE Trans. Electron Devices*, vol. ED-19, pp. 674–680, 1972.

[19] B. Culshaw and P. A. Blakey, "Intervalley scattering in gallium arsenide avalanche diodes," *Proc. IEEE (Lett.)*, vol. 64, pp. 569–571, 1976.

[20] T. Ohmi, "A limitation on the frequency of Gunn effect due to the intervalley scattering time," *Proc. IEEE (Lett.)*, vol. 55, pp. 1739–1740, 1967.

[21] V. Szekely and K. Tarnay, "Intervalley scattering model of the Gunn domain," *Electronics Letters*, vol. 4, pp. 592–594, 1968.

[22] D. E. McCumber and A. G. Chynoweth, "Theory of negative conductance amplification and of Gunn instabilities in 'two-valley' semiconductors," *IEEE Trans. Electron Devices*, vol. ED-13, pp. 4–21, 1966.

[23] S. Weinreb and A. R. Kerr, "Cryogenic cooling of mixers for millimeter and centimeter wavelengths," *IEEE Journal Solid State Circuits*, vol. SC-8, pp. 58–63, 1973.

[24] A. A. M. Saleh, *Theory of Resistive Mixers*, Res. Monograph 64. Cambridge, MA: MIT Press, 1971.

Nonlinear-Linear Analysis of Microwave Mixer with Any Number of Diodes

MAREK T. FABER AND WOJCIECH K. GWAREK

Abstract—**A theory is presented for analyzing mixers with any number of diodes. Both the nonlinear and linear steps of the analysis are included. The diodes are characterized by both nonlinear conductance and nonlinear capacitance. Any linear embedding network is allowed. It is assumed that both the parameters of the linear part of a mixer circuit and the parameters of the diodes are known. This general approach to microwave circuits with diodes, which is a qualitatively new problem in circuits analysis, allows to investigate any diode mixer with deep insight into its operation. A computer program has been developed to perform the analysis and all computations. The program has been utilized to analyze a crossbar mixer configuration which exhibits extremely encouraging performance. Some computed results are presented herein.**

I. Introduction

MICROWAVE mixer analysis has been an important problem for more than thirty years, and is still an evolving art. Starting from the fundamental work of Torrey and Whitmer [1] and up to the late sixties [2], [3] very simplified mixer models were analyzed. The computer era has opened the possibility of more accurate computer-aided analyses based on less simplified mixer models. It was the first basic problem to analyze a single-diode mixer without the assumptions of sinusoidal drive, linearity of the diode capacitance, and short- or open-circuit terminals seen by the diode at pump (LO) harmonics and various mixing products. Egami [4], Gwarek [5], [6], Kerr [7], and Held [8] have contributed to the problems solution and satisfactory practical results have been reported by Held and Kerr [9].

No theory has been still available for accurate analysis of multidiode mixers and such mixers have been analyzed in very simplified ways and their development has mainly been empirical. However, it is possible to generalize the aforementioned one-diode mixer analyses to the multidiode case. A generalization to subharmonically pumped and balanced two-diode mixers has been recently reported [10].

In this paper a method for analyzing a mixer with any number of diodes is presented. The diodes and the circuit models used in the analysis are similar to those used in the one-diode mixer analyses. No extra simplifying assumptions are taken.

Let us consider a mixer with M diodes. Its model is shown on Fig. 1. A linear time-invariant network (LTIN) possesses $M+3$ ports.[1] Each of the ports $1,2,\cdots,M$ is loaded with parallel connection of a nonlinear conductance and a nonlinear capacitance representing the junction of the respective diode. All other elements of the equivalent circuits of the diodes (including series resistances, package capacitances, whiskers inductances, and so on) are included in the LTIN. Port numbered $M+1$ is connected to the pump (LO) source. Ports $M+2$ and $M+3$ are signal input and output, respectively.

In practical circuits the level of the pumping signal is much higher than that of the input signal. This allows to divide the analysis into two steps [1], [2], [3]. In the first step, called *nonlinear analysis*, the signal source is not taken into account and the goal of this step is to find the waveforms of voltages $u_{jm}(t)$, $m=1,2,\cdots,M$, due to the driving from the pump source $E_p e^{j\omega_p t}$, where ω_p is the pump frequency. The functions $g_m(u_{jm})$ and $c_m(u_{jm})$ are known, thus determination of the waveforms u_{jm} allows to find the dependences $g_m(t)$ and $c_m(t)$. From this point on in the analysis, the pump source may be ignored, and the circuit can be treated as a circuit with parametric elements $g_m(t)$, $c_m(t)$ periodically varying with time. Such a circuit can be described by a set of linear equations, and the mixer performance can be determined. This is the task of the *linear* step of the mixer *analysis*.

II. Nonlinear Analysis

The method of nonlinear analysis presented here arises from previously published methods of one-diode mixer analysis, namely correction sources method [5] and reflecting waves method [7]. It was found [6] that in their advanced forms both lead to the same computer algorithm. The latter method is superior by the clearity of physical interpretation of the employed iteration process.

The aim of the nonlinear analysis is to find the waveforms $u_{jm}(t)$ on the diodes' junctions. Since the input signal has a much smaller level than the pumping signal, the mixer circuit is, for the purpose of the nonlinear analysis, reduced to that of Fig. 2. A linear time-invariant network LTIN2 has been formed from the LTIN of Fig. 1 by loading the ports $M+2$ and $M+3$ with the impedance

Manuscript received March 28, 1980; revised July 24, 1980.

M. T. Faber is with the Institute of Electronics Fundamentals, Technical University of Warsaw, Nowowiejska 15/19, 00-665 Warszawa, Poland.

W. K. Gwarek is with the Institute of Radioelectronics, Technical University of Warsaw, Nowowiejska 15/19, 00-665 Warszawa, Poland.

[1] If the diodes are dc biased, one or more extra ports have to be added for connecting the biasing sources. Since this is not a typical case for microwave mixer, the dc sources will not be considered here. The discussion presented in the paper can be easily extended to the case when the dc bias is present.

Reprinted from *IEEE Trans. Microwave Theory Tech.*, vol. MTT-28, pp. 1174–1181, Nov. 1980.

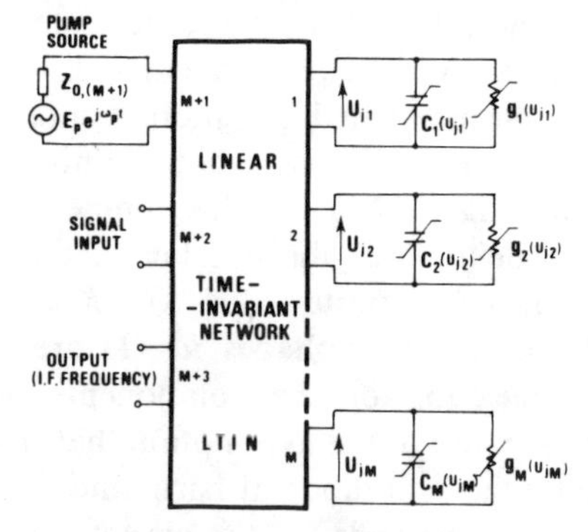

Fig. 1. General equivalent circuit of a mixer with M diodes.

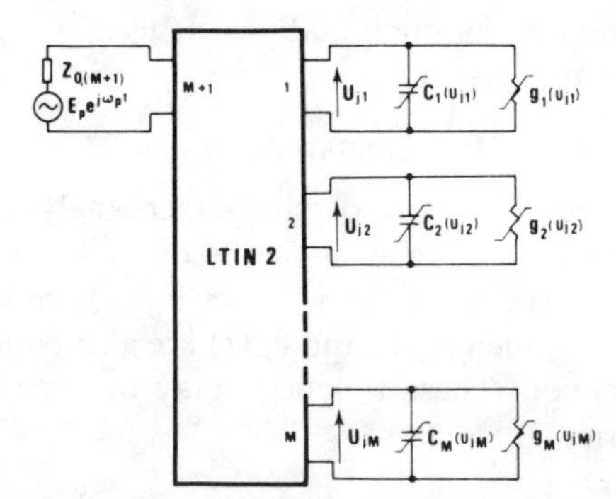

Fig. 2. Mixer model used in the nonlinear analysis.

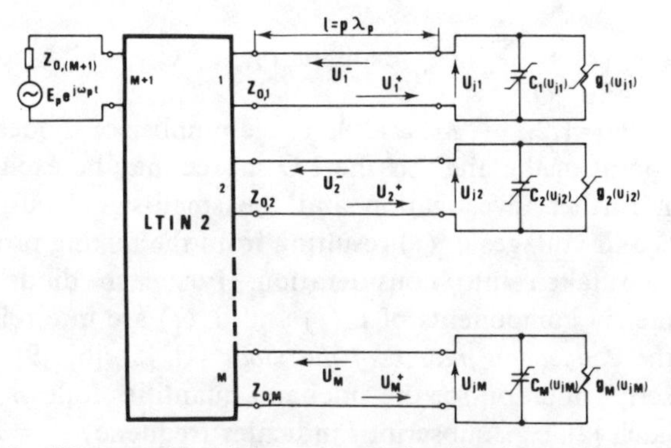

Fig. 3. Mixer model with long transmission lines inserted between the diodes' junctions and the rest of the circuit.

of a signal generator and the input impedance of an intermediate frequency amplifier, respectively. In the method it is generally assumed that the parameters of LTIN2 are known for relevant frequencies. Let us assume that the circuit will be described by the elements of the S matrix defined with the loading impedances Z_{om} at the respective ports.

The idea of the method is to change the circuit of Fig. 2 in a way that the new circuit has the same steady-state but in which this steady-state can be determined easier.[2] Satisfactory results are obtained by changing the circuit of Fig. 2 to that of Fig. 3. Dispersionless lossless transmis-

[2] It is assumed that there is only one steady-state in the circuit. In the circuit of Fig. 2 containing nonlinear reactances there is a theoretical possibility of existing more than one steady-state for a particular pumping. However operation with two (or more) possible steady-states, if achieved, would cause instability of the mixer parameters. That kind of operation should not occur in a properly designed mixer and is not considered here.

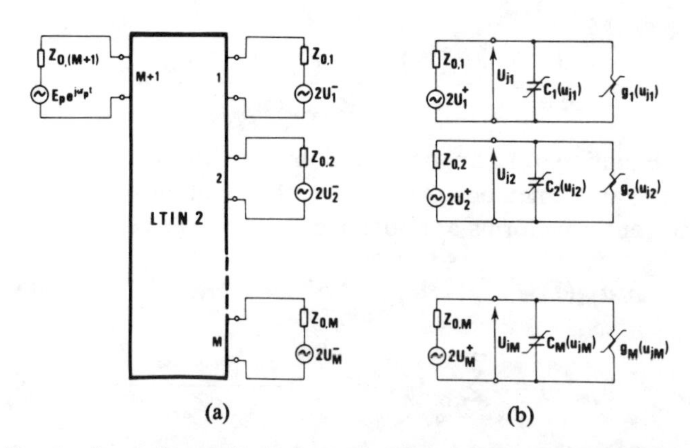

Fig. 4. Separation of the diodes from the rest of the circuit employing the concept of incident and reflected waves.

sion lines inserted between the junctions' nodes and respective ports of LTIN2 have the lengths

$$l=p\cdot\lambda_p$$

where λ_p is the wavelength for the pump angular frequency ω_p, and p is the large integer. For the steady-state those transmission lines are "transparent" for the pump frequency, any of its harmonics and for dc signals.

Let us investigate the circuit of Fig. 3 in the time following connection of the source E_p to the port number $M+1$ of LTIN2. If the transmission lines in the circuit are sufficiently long (p is large) the circuit LTIN2 will reach the local steady-state with the right-hand side propagating waves u_m^+

$$u_m^+(t)=\sum_{k=-\infty}^{\infty} U_{mk}^+ e^{jk\omega_p t} \tag{1a}$$

where $m=1,2,\cdots,M$.

Before the reflected waves return from the diodes, the LTIN2 circuit sees, on ports $1,2,\cdots,M$, only the characteristic impedances of the lines. Since the circuit is driven sinusoidally initially only the terms of the frequency ω_p are present in the sum of (1). Thus

$$U_{mk}^+=\frac{1}{2}E_pS_{m,M+1}(k\omega_p)\sqrt{\frac{Z_{om}}{Z_{o,M+1}}}, \quad \text{for } |k|=1$$

$$U_{mk}^+=0, \quad \text{for } |k|\neq 1$$

and therefore

$$u_m^+(t)=\frac{1}{2}\,\mathrm{Re}\left[E_pS_{m,M+1}(\omega_p)\sqrt{\frac{Z_{om}}{Z_{o,M+1}}}\,e^{j\omega_p t}\right]. \tag{1b}$$

When the propagating waves reach the ends of the lines loaded with the diodes they become sources of the electromotive forces

$$e_m^+(t)=2u_m^+(t)$$

with the internal impedances Z_{om}. At that moment this causes the waveforms of voltage u_{jm} at the mth diode junction to be calculated from the differential equation (2) describing the relations in the diodes' circuits shown in Fig. 4(b)

$$\frac{du_{jm}(t)}{dt}=\frac{2u_m^+(t)-u_{jm}(t)[1+g_m(u_{jm})Z_{om}]}{Z_{om}c_m(u_{jm})}. \quad (2)$$

Equation (2) can be solved by one of the numerical integration methods. As a result, the junction voltage and current waveforms are obtained

$$u_{jm}(t)=\sum_{k=-\infty}^{\infty} U_{jmk}e^{jk\omega_p t} \quad (3)$$

$$i_{jm}(t)=\sum_{k=-\infty}^{\infty} I_{jmk}e^{jk\omega_p t}=\frac{2u_m^+(t)-u_{jm}(t)}{Z_{om}}. \quad (4)$$

The voltages and currents calculated above represent the local steady-state at the diodes junctions reached after arrival of the waves u_m^+. When that state is reached, the left-hand side waves u_m^- are propagating towards the LTIN2 circuit. The waves u_m^- are those reflected from the diodes upon incidence of the waves u_m^+ and can be expressed as

$$u_m^-(t)=\sum_{k=-\infty}^{\infty} U_{mk}^-e^{jk\omega_p t}=\sum_{k=-\infty}^{\infty} U_{mk}^+\Gamma_{mk}e^{jk\omega_p t}$$
$$=\sum_{k=-\infty}^{\infty} U_{mk}^+\frac{Z_{jmk}-Z_{om}}{Z_{jmk}+Z_{om}}e^{jk\omega_p t} \quad (5)$$

where $Z_{jmk}=U_{jmk}/I_{jmk}$.

Formula (5) needs to be modified for practical purposes since in a nonlinear circuit it may happen that $U_{mk}^+=0$ and $\Gamma_{mk}=\infty$[3] which is unacceptable for a computer. Let us notice that in the circuit of Fig. 4(b)

$$u_m^+(t)=\frac{1}{2}[u_{jm}(t)+Z_{om}i_{jm}(t)]. \quad (6)$$

Thus (5) can be transformed to the form

$$u_m^-(t)=\frac{1}{2}\sum_{k=-\infty}^{\infty}(Z_{jmk}-Z_{om})I_{jmk}e^{jk\omega_p t}. \quad (7)$$

After some amount of time the waves u_m^- reach the ports of LTIN2. The wave u_m^- incident on port m is interpreted as a source of electromotive force $e_m^-=2u_m^-$ and internal impedance Z_{om} as shown on Fig. 4(a). The wave is partially reflected from the port of entry and partially passes through the LTIN2 to the other ports. The new local steady-state in LTIN2 is achieved and new right hand side waves described by formula (8) start travelling towards the diodes

$$u_m^{+\prime}(t)=\sum_{k=-\infty}^{\infty}\left[\sum_{p=1}^{M}U_{pk}^-S_{mp}(k\omega_p)\sqrt{\frac{Z_{om}}{Z_{op}}}\right]e^{jk\omega_p t}. \quad (8)$$

The process of calculating sequential values of u_m^+ resulting from reflections in the transmission lines can be treated as an iteration process convergent to the value describing the steady-state of the all circuit. Formula (1) gives the initial value of u_m^+ in this process; formulas (2), (7), (8) show how to calculate the next step of iteration when the former step has been calculated.

[3]It actually happens for the first reflection from the diodes because $U_{mk}^+=0$ for $|k|\neq 1$ while in general $U_{mk}^-\neq 0$.

Several of the presented equations contain sums with index k varying from minus to plus infinity which corresponds to investigation of infinite number of harmonics. In practical calculations, the number of these harmonics has to be reduced to finite value $K(-K\leqslant k\leqslant K)$. Eventual waveforms of the voltages $u_{jm}(t)$ are obtained by solving (2). Deletion of the components of $u_m^+(t)$ for $|k|>K$ is equivalent to the assumption that the impedance Z_{om} is seen by the mth diode at harmonics higher than K. The choice of K depends on the circuit under investigation, required accuracy of the calculations, and restrictions on the time of computations. It usually varies from three to seven.

III. Linear Analysis

In the nonlinear step of the mixer analysis the LO waveforms $u_{jm}(t)$ at each of the M diode junctions have been determined. Since the voltages $u_{jm}(t)$ are time periodic functions, then $g_m(t)$ and $c_m(t)$ are also periodic with the same period. These functions may be expanded into Fourier series

$$g_m(t)=\sum_{l=-\infty}^{\infty}G_{m,l}\exp(jl\omega_p t), \qquad G_{m,l}=G_{m,-l}^* \quad (9)$$

$$c_m(t)=\sum_{l=-\infty}^{\infty}C_{m,l}\exp(jl\omega_p t), \qquad C_{m,l}=C_{m,-l}^* \quad (10)$$

where $m=1,2,\cdots,M$, and M is the number of diodes. At this point of the analysis the LO source may be excluded from further investigation and the small-signal currents $i_m(t)$ and voltages $u_m(t)$ resulting from the mixing process can be taken into consideration. For each diode the frequency components of $i_m(t)$ and $u_m(t)$ are interrelated by the *conversion matrix of the diode* [1], [5], [6], [9]. The subscript notation for the sideband quantities follows that of Saleh [2], e.g., subscript l indicates frequency $\omega_l=\omega_0+l\omega_p$. For the mth diode the conversion matrix $\boldsymbol{Y}_m$ is a square matrix with the elements given by

$$Y_{m,l,n}=G_{m,l-n}+j(\omega_0+l\omega_p)C_{m,l-n} \quad (11)$$

where the subscript m indicates the diode while the subscripts l and n indicate the row and the column in the matrix $\boldsymbol{Y}_m$. $G_{m,k}$ and $C_{m,k}$ are the Fourier coefficients of the mth diode conductance and capacitance at the frequency $\omega_k=k\omega_p$ as defined in (9) and (10). If only L components of the small-signal currents and voltages are considered each of the conversion matrices $\boldsymbol{Y}_m$ has the size $(2L+1)\times(2L+1)$.

Having determined all the $\boldsymbol{Y}_m$ matrices of all the M diodes, the *admittance conversion matrix* $\boldsymbol{Y}_C$ *of the whole mixer* can be composed. The mixer may be represented as a multifrequency linear multiport network. Since the time-invariant part of the mixer circuit is linear, the diodes and external load admittances are interconnected at each frequency ω_l by $M+1$ port networks. Load admittances at frequencies other than signal, image, and IF may be included into respective multiport networks. For example,

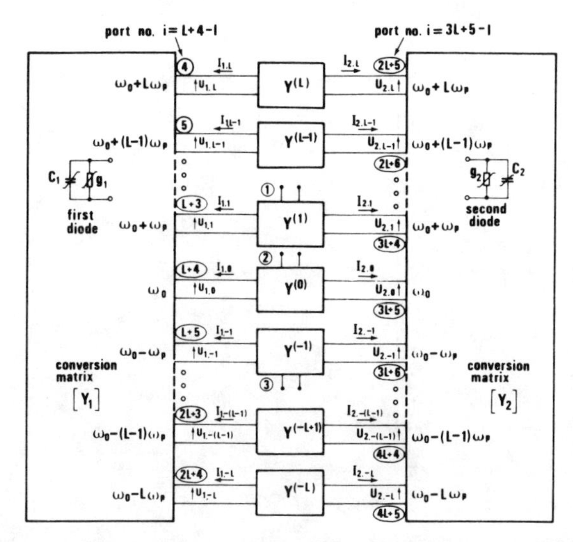

Fig. 5. Small-signal representation of a two-diode mixer as a multiport network. Frequency ω_l at diodes' ports $i=L+4-l$ and $i=3L+5-l$, i.e. $I_{L+4-l}=I_{1,l}$; $U_{L+4-l}=U_{1,l}$; $I_{3L+5-l}=I_{2,l}$; $U_{3L+5-l}=U_{2,l}$. Usually input at frequency ω_1 or ω_{-1} (port 1 or 3), output at intermediate frequency ω_0 (port 2).

if we consider a mixer having input, output, and image signals at frequencies ω_1, ω_0, ω_{-1}, respectively, the $M+1$ port networks at frequencies ω_l, $l\neq -1,0,1$, may be reduced to M port networks. This is pictured in Fig. 5 for the case $M=2$.

It is necessary to determine $Y^{(l)}$ matrices describing the time-invariant part of the mixer at frequencies $\omega_l=\omega_0+l\omega_p$ $(-L\leqslant l\leqslant L)$. Since $Y_{i,j}(-\omega)=Y^*_{i,j}(\omega)$, the Y parameters may be determined at frequencies ω_0, $l\omega_p+\omega_0$ and $l\omega_p-\omega_0$, $l=1,2,\cdots,L$ (e.g., computed or by measurements of the S parameters) and then the $Y^{(l)}$ matrices can be computed.

Let us introduce the quantities

$$\begin{aligned} i_0&=2\\ i_1&=L+4\\ i_2&=3L+5\\ i_3&=5L+6\\ &\vdots\\ i_m&=L(2m-1)+m+3\\ &\vdots\\ i_M&=L(2M-1)+M+3 \end{aligned} \tag{12}$$

and let us number the ports in the way as it was done in Fig. 5 for the two-diode mixer, i.e.,

$i=i_1-l$, for the first diode

$i=i_2-l$, for the second diode

$\vdots$

$i=i_m-l$, for the mth diode

$\vdots$

$i=i_M-l$, for the last Mth diode,

where $-L\leqslant l\leqslant L$

and

$i=i_0-l$, for the mixer input and output, where $l=-1,0,1$.

It is convenient to express the conversion matrix of the mixer $\boldsymbol{Y_C}$ by means of matrices $\boldsymbol{Y_D}$ and $\boldsymbol{Y_S}$

$$\boldsymbol{Y_C}=\boldsymbol{Y_D}+\boldsymbol{Y_S}.$$

The matrix $\boldsymbol{Y_D}$ is composed of the matrices $\boldsymbol{Y_m}$ of the diodes and its nonzero elements are

$$Y_{D_{i_m-l,i_m-n}}=Y_{m,l,n} \tag{13}$$

where $m=1,2,\cdots,M$; $l=-L,\cdots,-1,0,1,\cdots,L$; $n=-L,\cdots,-1,0,1,\cdots,L$ and $Y_{m,l,n}$ and i_m are given by (11) and (12), respectively.

The matrix $\boldsymbol{Y_S}$ is composed of the matrices $\boldsymbol{Y^{(l)}}$ of the time-invariant part of the mixer circuit. Its nonzero elements are

$$Y_{S_{i_p-l,i_r-l}}=Y^{(l)}_{p,r}, \qquad \text{for } l=-L,\cdots,-2,2,\cdots,L$$

$$Y_{S_{i_t-l,i_w-l}}=Y^{(l)}_{t+1,w+1}, \qquad \text{for } l=-1,0,1 \tag{14}$$

where $p=1,2,\cdots,m,\cdots,M$; $r=1,2,\cdots,m,\cdots,M$; $t=0,1,2,\cdots,m,\cdots,M$; $w=0,1,2,\cdots,m,\cdots,M$; and $i_0,i_1,i_2,\cdots,i_m,\cdots,i_M$ are defined in (12). The matrices $\boldsymbol{Y_D}$ and $\boldsymbol{Y_S}$ are pictured on Figs. 6 and 7 for the case of two-diode mixer $(M=2)$.

The conversion matrix $\boldsymbol{Y_C}$ of the mixer is a *sparse square matrix* having the size $(2LM+M+3)\times(2LM+M+3)$. Its *density factor d* depends on L and M

$$d=\frac{M+1}{2LM+M+3}+\frac{2M(2L^2+1)}{(2LM+M+3)^2}. \tag{15}$$

When the matrix $\boldsymbol{Y_C}$ is determined, the *impedance matrix of the mixer* can be found

$$\boldsymbol{Z_C}=(\boldsymbol{Y_C})^{-1}$$

and the three-frequency mixer model having ports associated with the signal, image, and intermediate frequencies may be used to determine all the small-signal properties of the mixer [11]. The stability problems and matching conditions can be investigated.

As far as the noise properties of the mixer are concerned, it should be noted that *the noise produced by one diode is not correlated with the noise of any other diode.* Thus for each of the diodes, the *shot noise* correlation matrices can be found in the way described by Held and Kerr [9] for a one-diode mixer. *Thermal noise* generated in the time-invariant part of the mixer can be described by the correlation matrix as it has been done by Twiss [12]. The noise correlation matrix of the whole mixer can be composed in the way described above in this paper. Then the noise properties of the mixer (the equivalent input noise temperature or the noise figure) can be determined. Noise analysis, however, will not be dealt with in this paper.[4]

[4]This will be done in a paper to be published shortly.

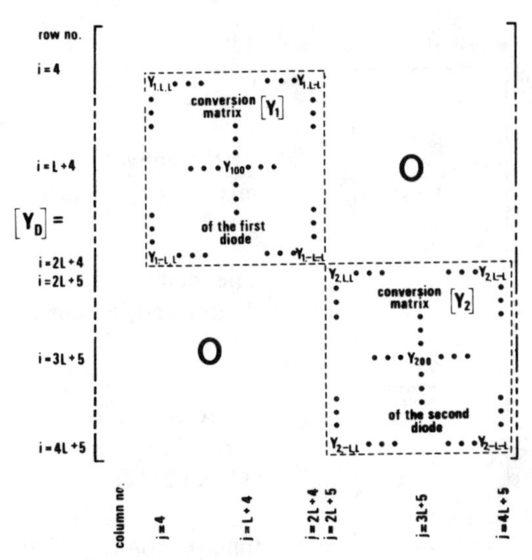

Fig. 6. Matrix Y_D in the case of a two-diode mixer ($M=2$). The elements of the conversion matrix Y_1 of the first diode are in rows from $i=4$ to $i=2L+4$ and columns from $j=4$ to $j=2L+4$. The elements of the conversion matrix Y_2 of the second diode are in rows from $i=2L+5$ to $i=4L+5$ and columns from $j=2L+5$ to $j=4L+5$. The nonzero elements of the matrix Y_D are given in (13) if $M=2$.

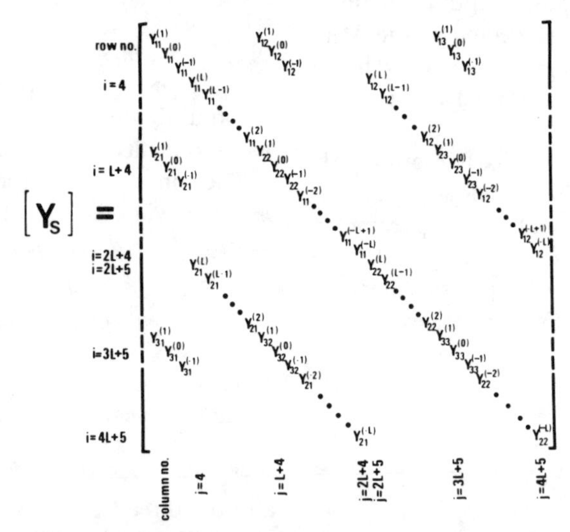

Fig. 7. Matrix Y_S in the case of a two-diode mixer ($M=2$). The matrix Y_S is composed of the elements of the $Y^{(l)}$ matrices. The nonzero elements of the matrix Y_S are given in (14) if $M=2$.

IV. Computer Implementation of the Theory

A computer has to be employed for practical application of the above theory. Therefore, a computer program written in Fortran has been developed to perform the analysis and all the computations. The program allows the study of the influence of the diode parameters on mixer performance as well as effects of loads at the pump frequency, its harmonics, signal, image, and idle frequencies. Frequency and thermal dependences of mixer parameters can be investigated. Each diode may have different parameters, thus unbalance effects due to the diodes and the embedding network can be studied. The program has been utilized to analyze microwave mixers with two Schottky diodes.

We have found that in most practical cases, it is sufficient to consider 4 to 6 harmonics in the nonlinear step and 6 to 8 harmonics in the linear step of the mixer analysis. It should be noted that in the nonlinear analysis, deletion of the higher harmonics has the physical interpretation equivalent to the assumption that the characteristic impedances Z_{om} of the transmission lines are seen by the diodes at the deleted harmonics. In the linear analysis, such a physical interpretation does not exist. One run of the program takes approximately 1 min on CDC CYBER 73 computer, if standard procedures for numerical integration and inversing of complex matrices are called from the computer library. It can be speeded up by utilizing procedures written specially for these particular applications. It looks profitable to utilize a sparse matrix technique when analyzing mixers with more than two diodes.

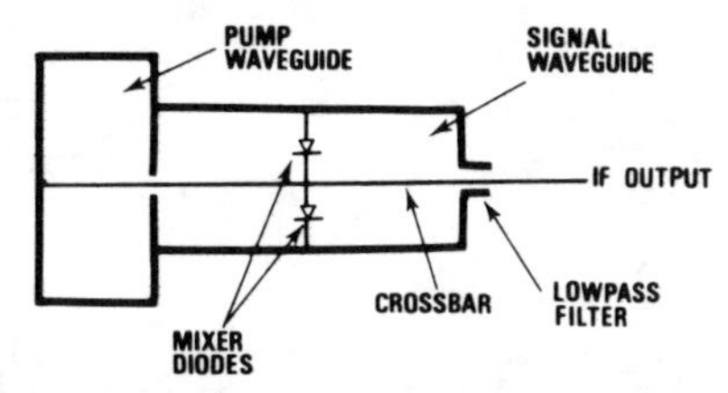

Fig. 8. Schematic diagram of the crossbar mixer (transverse cross section).

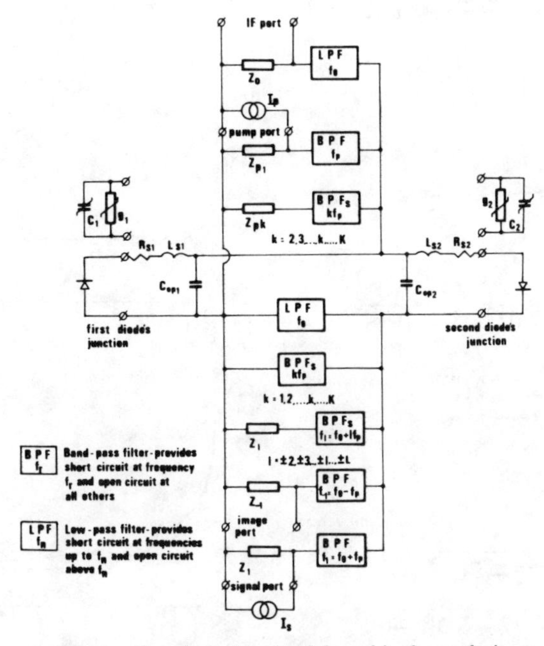

Fig. 9. Crossbar mixer model used in the analysis.

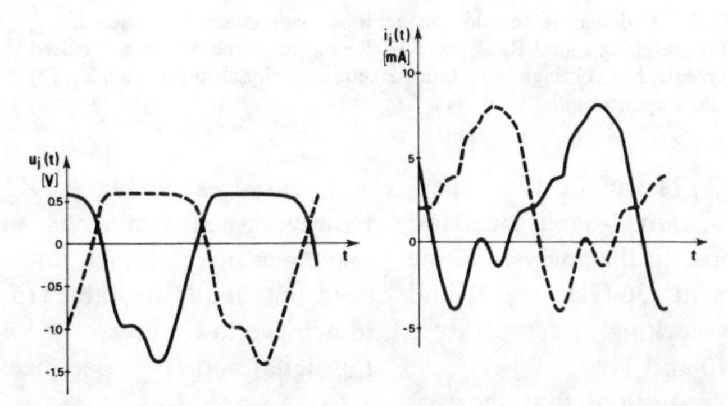

Fig. 10. Crossbar mixer: voltage and current waveforms at diodes' junctions ($I_o = 2.4$ mA).

V. Crossbar Mixer Example

A schematic diagram of the crossbar mixer is shown in Fig. 8. In this mixer configuration a pair of diodes in connected in series across the broadwalls of the signal waveguide. A metal crossbar serves as a mechanical support for the diodes and as a transmission line for the incoming LO pumping signal and the IF output signal. The diodes are electrically in parallel with respect to the transmission line formed by the crossbar and in series with respect to the waveguide. The diodes therefore can simultaneously match the relatively high impedance of the waveguide and the relatively low impedance of the transmission line without impedance transformers. The diodes are driven out of phase by the pump which leads to balanced mixer properties. Encouraging performance of the crossbar mixer configuration has been confirmed [13], [14], and this type of mixer is commercially available [15].

The mixer model used in the analysis is given in Fig. 9. The Schottky-barrier diodes are characterized by

$$i_g = I_s[\exp(qu_j/\eta kT) - 1]$$

$$c = C_o(1 - u_j/\Phi)^{-\gamma}$$

and the series resistances R_s. In the example the GaAs diodes' parameters are $R_s = 5\ \Omega$, $C_o = 9.5$ fF, $\eta = 1.1$, $\Phi = 0.9$ V, $\gamma = 0.5$, $I_s = 12.7 \cdot 10^{-12}$ A. Whiskers' inductances

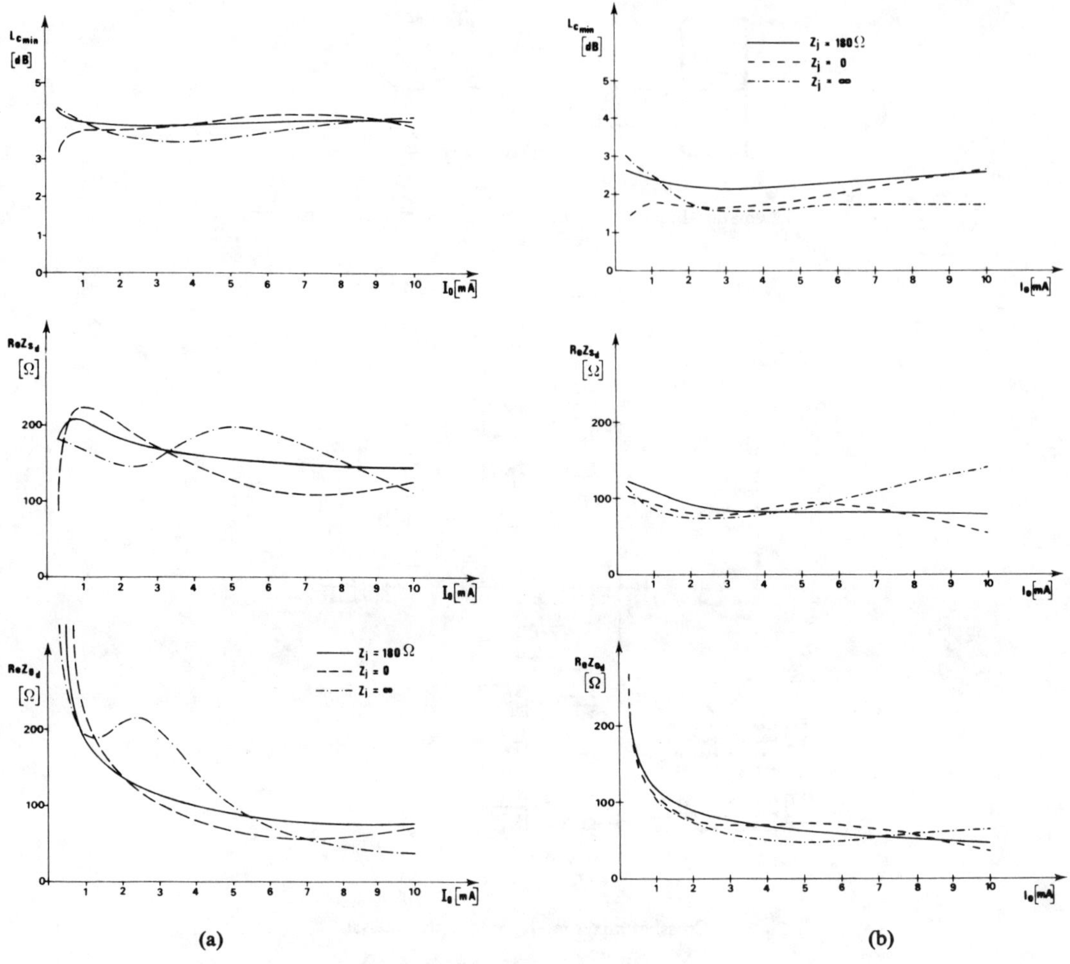

Fig. 11. Crossbar mixer: lowest achieveable conversion losses $L_{C\min}$ and matching signal $\mathrm{Re}\,Z_{s_d}$ and IF $\mathrm{Re}\,Z_{0_d}$ resistances versus rectified currents I_o. (a) Image impedance Z_i equal to signal impedance Z_s. (b) Image opencircuited $Z_i=\infty$.

and parasitic capacitances are 0.2 nH and 7.5 fF, respectively. Six pump harmonics and six small-signal sideband components ($L=6$) were considered in the analysis. Some of the results computed for Q-band (36-GHz signal and 750-MHz IF frequencies) mixer working at temperature $T=298$ K are presented in Figs. 10 and 11.

Comparing the results it can be noticed that short or open circuiting of the idle frequencies do not lead to distinct improvements in conversion losses if the image impedance Z_i is equal to the signal impedance Z_s. In this case 4-dB conversion losses are achieveable providing that $\mathrm{Re}\,Z_s=180\ \Omega$ and $\mathrm{Re}\,Z_0=120\ \Omega$. In image recovery mixers much better practical results can be obtained with opencircuited image rather than with shortcircuited. Open circuited image ($Z_i=\infty$) leads to an unconditionally stable low sensitive mixer with 2.2 dB lowest achieveable losses. Conversion losses can be further decreased by means of reactive loads at the idle frequencies. It is more profitable to open circuit the idle frequencies. This gives $L_{C\min}=1.6$ dB together with easier realizeable Z_s and Z_0. Opencircuits at the idle frequencies are better also in the case of short circuited image ($Z_i=0$). $L_{C\min}=2.4$ dB are possible with easy realizeable $\mathrm{Re}\,Z_s=210\ \Omega$ and $\mathrm{Re}\,Z_0=50\ \Omega$. However such a mixer is sensitive to the pump and the reactance of the image impedance (Fig. 11(d)). It can be even potentially unstable. In this mixer case it is difficult to achieve low conversion losses due to high sensitivity to the signal and IF impedances.

VI. Summary and Conclusions

The theory given in the paper permits both the nonlinear and linear analyses of mixers with any number of diodes. In this theory, diodes are assumed to have both nonlinear conductance and capacitance. No restrictions are placed on the linear embedding network. It is only assumed that properties of the time-invariant part of the circuit are known at respective frequencies. It is not necessary to determine the equivalent circuit of the network. No simplifying assumptions other than those of most advanced one-diode mixer analyses are taken.

A computer program written in Fortran has been developed for practical application of this theory. The program has been thoroughly tested and its value in two-diode

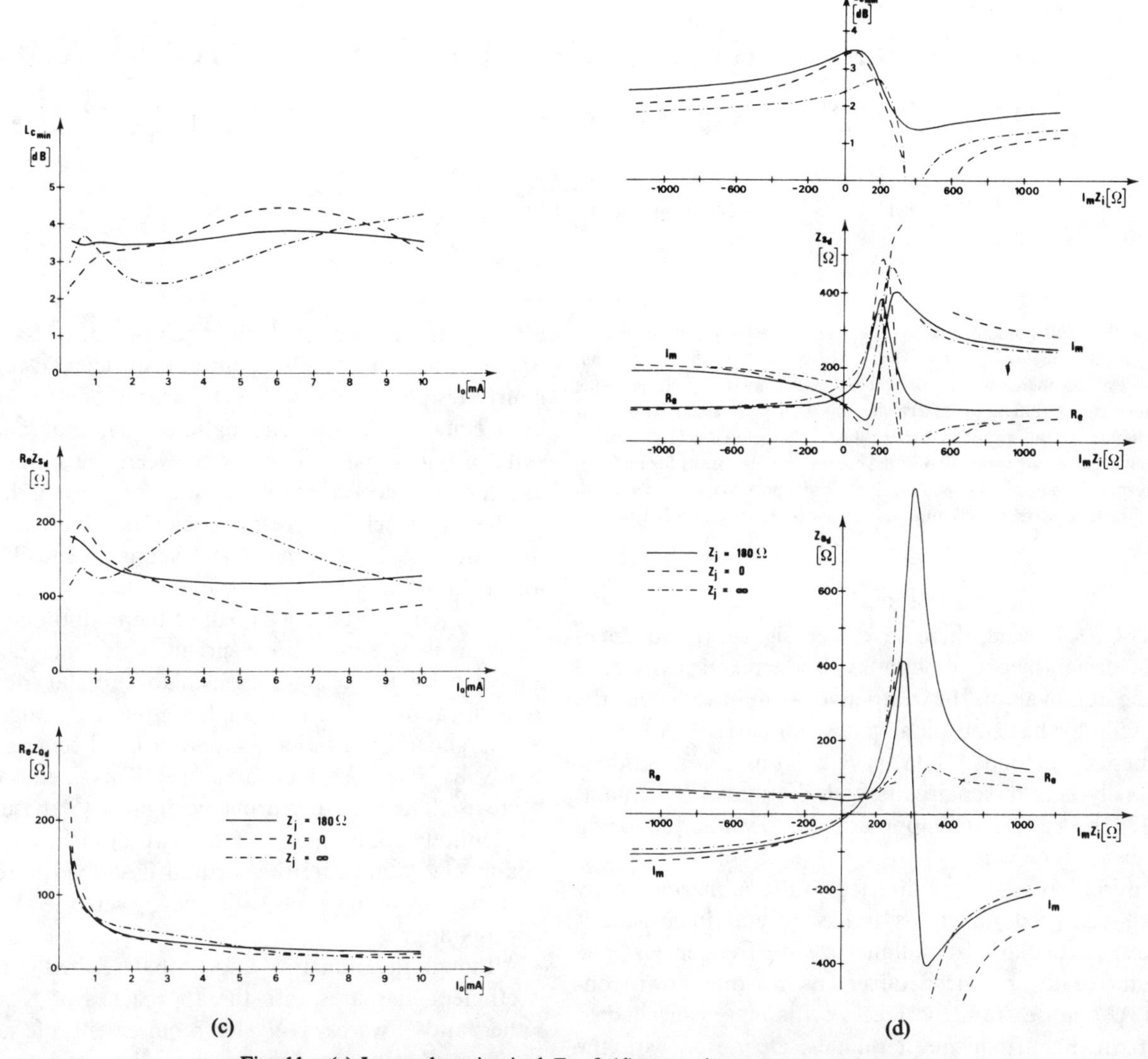

Fig. 11. (c) Image shortcircuited $Z_i=0$ (d) versus image reactance Im Z_i for Re $Z_i=0$ and $I_o=2.4$ mA. Z_j are external impedances at idle frequencies.

mixers analyses has been proved. In the paper, the cross-bar mixer example and some practical results of its analysis are provided to illustrate the theory.

The authors believe that the theory introduced here for mixer analyses may also be applied in analyzing other microwave circuits with diodes.

References

[1] H. C. Torrey and C. A. Whitmer, *Crystal Rectifiers*, (M.I.T. Radiation Lab. Series, vol. 15). New York: McGraw-Hill, 1948.

[2] A. A. M. Saleh, *Theory of Resistive Mixers*. Cambridge, MA: M.I.T. Press, 1971.

[3] M. R. Barber, "Noise figure and conversion loss of the Schottky barrier mixer diode," *IEEE Trans. Microwave Theory Tech.*, vol. MTT-15, pp. 629–635, Nov. 1967.

[4] S. Egami, "Nonlinear, linear analysis and computer aided design of resistive mixers," *IEEE Trans. Microwave Theory Tech.*, vol. MTT-22, pp. 270–275, Mar. 1974.

[5] W. K. Gwarek, "Nonlinear analysis of microwave mixers," M. S. thesis, M.I.T., Cambridge, MA, Sept. 1974.

[6] W. K. Gwarek, "Computer aided analysis of a one-diode microwave mixer," Ph.D. dissertation, Dep. Electron. Tech. Univ. Warsaw, Warsaw, Poland, 1977.

[7] A. R. Kerr, "A technique for determining the local oscillator waveforms in a microwave mixer," *IEEE Trans. Microwave Theory Tech.*, vol. MTT-23, pp. 828–831, Oct. 1975.

[8] D. N. Held, "Analysis of room temperature millimeter-wave mixers using GaAs Schottky barrier diodes," Sc.D. dissertation, *Dep. Elec. Eng., Columbia Univ.*, New York, NY., 1976.

[9] D. N. Held and A. R. Kerr, "Conversion loss and noise of microwave and millimeter-wave mixers: Part 1—Theory," and "Part 2—Experiment," *IEEE Trans. Microwave Theory Tech.*, vol. MTT-26, pp. 49–61, Feb. 1978.

[10] A. R. Kerr, "Noise and loss in balanced and subharmonically pumped mixers: Part I—Theory" and "Part II—Application," *IEEE Trans. Microwave Theory Tech.*, vol. MTT-27, pp. 938–950, Dec. 1979.

[11] M. T. Faber, "Linear analysis of a multidiode mixer" (in Polish), in *Proc. III National Conf. Circuit Theory Electronic Circuits*, (Stawiska, Poland), 1979, pp. 369–378.

[12] R. Q. Twiss, "Nyquist's and Thevenin's theorems generalized for nonreciprocal linear networks," *J. Appl. Phys.*, vol. 26, no. 5, pp. 599–602, May 1955.

[13] G. B. Stracca, F. Aspesi, and T. D'Arcangelo, "Low-noise microwave down-converter with optimum matching at idle frequencies," *IEEE Trans. Microwave Theory Tech.*, vol. MTT-21, pp. 544–547, August 1973.

[14] L. T. Yuan, "Design and performance analysis of an octave bandwidth waveguide mixer," *IEEE Trans. Microwave Theory Tech.*, vol. MTT-25, pp. 1048–1054, Dec. 1977.

[15] Data Sheet, SpaceKom Mixers (Strip-Guide Mixers).

Numerical Analysis of Subharmonic Mixers Using Accurate and Approximate Models

ROSS G. HICKS, STUDENT MEMBER, AND PETER J. KHAN, SENIOR MEMBER, IEEE

Abstract —**A full nonlinear numerical analysis technique is applied to subharmonically pumped mixer circuits where the two diodes are not identical. Results indicate that a slight imbalance in the diode parasitic parameters can significantly affect the mixer performance. A bilinear approximation of the Schottky-barrier diode characteristic is described, permitting accurate determination of the conversion loss peaks for millimeter-wave subharmonically pumped mixers. This approximation provides an analysis which requires significantly less computer time than a full nonlinear analysis.**

I. Introduction

In RECENT years, there have been significant advances in the design theory of Schottky-diode mixers, associated with the removal of the restrictive assumption that the local oscillator has sinusoidal voltage or current. A variety of numerical methods [1]–[6] have been used for the nonlinear analysis of mixer circuits, with the study of Held and Kerr [7], based on the approach of Gwarek [2], being particularly significant.

Attention has also been directed to the subharmonically pumped balanced mixer, as it has several intrinsic advantages, particularly at millimeter-wave frequencies. The symmetry of the balanced mixer ensures that down-converted AM noise from the local oscillator is cancelled at the intermediate-frequency terminals. Operation with the local oscillator near half the signal frequency is cost efficient at millimeter wavelengths, where the cost of pump power increases rapidly with frequency. Kerr [8] has presented a detailed analysis of these mixers for the case where the two diodes are assumed to be identical. In practice, it has not been possible to reproduce the theoretical results readily under experimental conditions, although some encouraging experimental results have been reported [9]–[12]. Researchers have found that both slight changes in diode mounting and replacement of diodes have had a pronounced effect on performance [13].

This paper reports on the development of an accurate numerical method for the nonlinear analysis of balanced subharmonically pumped mixers, for the important practical case where the two diodes differ, either in device characteristics or in mounting configurations. In addition, the paper presents an approximate approach based on a bilinear diode model, which requires much less computational effort and yields results of sufficient accuracy for many design purposes. The bilinear model is applied to both equal-diode and unequal-diode circuits, and the results of a comparative study between the bilinear model and the nonlinear analysis method are presented.

The approach of Kerr [8] can be applied but with difficulty to the unsymmetrical subharmonically pumped mixer diode case, since in its form described in [8], it relies on the circuit symmetry to reduce the multidiode circuit to an equivalent single-diode circuit which may be readily analyzed by the techniques available in the literature. A generalization of the Kerr multiple reflection algorithm [4] to the general multidiode situation has been reported recently by Faber and Gwarek [14]. This method has not been used here as performance figures [5] based on the single-diode counterparts of the two-diode analysis methods available indicate the method described here has significant advantages of efficiency over the Faber and Gwarek approach.

Although the numerical approach described in this paper is efficient, it shares with the approaches of Kerr [8] and Faber and Gwarek [14] the requirement of substantial amounts of computer time for the following reasons: 1) the large number of calculations involved in the numerical integration of the nonlinear differential equations representing the junction varactor capacitance and the Schottky-barrier junction; and 2) the number of iterations required to ensure the linear embedding circuit constraints at the pump frequency and its harmonics are satisfied.

Because of the computational effort required for this nonlinear analysis, attention has also been directed to the accuracy attainable using simplified bilinear diode models. Barber [15] first proposed such a model for the single-diode mixer, with the mixer properties being chiefly determined by the pulse duty ratio of the switch. Bordonskiy *et al.* [16] expanded on Barber's switch model by extending the number of sidebands under consideration together with making an allowance for broad-band nonzero terminating impedances at both the image and sum frequencies. Both Barber and Bordonskiy commented on the diode cutoff frequency as being a parameter of fundamental importance in that it sets the upper frequency limit of performance for the parallel RC junction device. Zabyshnyi *et al.* [17] extended the work of Bordonskiy to the case of subharmonic mixers, using the assumption of identical diodes. However, the

Manuscript received March 16, 1982; revised June 9, 1982. This work was supported in part by United States ARO Grant DAA G29-76-G-0279, the Australian Research Grants Committee Grant F76/15147, and the Radio Research Board.

The authors are with the Department of Electrical Engineering, University of Queensland, St. Lucia, Queensland, 4067, Australia.

Reprinted from *IEEE Trans. Microwave Theory Tech.*, vol. MTT-30, pp. 2113–2119, Dec. 1982.

approach of Zabyshnyi fails to account for the crucial effect of the parasitic lead inductance, which at resonance induces multiple conductions in each voltage waveform [8]. The analysis, however, did emphasize the importance of the diode cutoff frequency in determining mixer performance.

II. Multidiode Update Nonlinear Analysis Approach

The approach [18] described here is based upon an extension of a single-diode nonlinear analysis technique previously published [5], [6] by the present authors. It is based upon subdivision of the circuit into three subnetworks, comprising two nonlinear one-ports and a connecting linear two-port containing the embedding network within which the pump source is located, as shown in Fig. 1.

The two-port network will be described at each harmonic of the pump frequency, using Z parameters, as follows:

$$V_1(\omega) = -Z_{11}(\omega)I_1(\omega) + Z_{12}(\omega)I_2(\omega) + V_1^{oc}(\omega) \quad (1)$$

$$-V_2(\omega) = -Z_{21}(\omega)I_1(\omega) + Z_{22}(\omega)I_2(\omega) - V_2^{oc}(\omega) \quad (2)$$

where $V_1^{oc}(\omega)$ = voltage at port 1 with both ports 1 and 2 open-circuited and represents the component due to the exciting source; $V_2^{oc}(\omega)$ = voltage at port 2 with both ports 1 and 2 open-circuited and again represents the component attributable to the exciting source. For a symmetrical reciprocal embedding network

$$Z_{12}(\omega) = Z_{21}(\omega) \quad (3)$$

$$Z_{11}(\omega) = Z_{22}(\omega) \quad (4)$$

$$V_1^{oc}(\omega) = -V_2^{oc}(\omega). \quad (5)$$

The procedure then goes as follows.

1) To commence the algorithm, sinusoidal waveforms are assumed for $I_1^o(t)$ and $V_2^o(t)$, where the superscript indicates this is the initial iteration. These values are taken at the excitation frequency, without harmonics, and may be found by an approximate calculation. Each iteration will seek to improve the values of $I_1(t)$ and $V_2(t)$.

2) The periodic voltage response $V_1^o(t)$ that diode 1 produces in response to the input $I_1^o(t)$ may be determined by successively integrating (using the classical Runge–Kutta algorithm) the nonlinear Schottky-diode equation until the transients decay, i.e.,

$$\frac{dV_1^o(t)}{dt} = \frac{I_1^o(t) - i_1[\exp(qV_1^0(t)/\eta kT) - 1]}{C_1(V_1^o(t))} \quad (6)$$

where the capacitance term is calculated using the normal varactor equation, i.e.,

$$C_1(V_1(t)) = C_{01}(1 - V_1(t)/\phi)^{-\gamma} \quad (7)$$

where

C_{01} zero bias capacitance for diode 1
γ doping profile index
ϕ contact potential
i_1 diode 1 saturation current, and
η diode ideality factor.

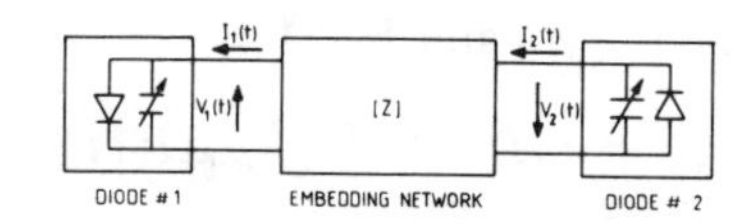

Fig. 1. Subdivision of the subharmonic mixer circuit into the two nonlinear diodes and the linear embedding network. The pump source is contained within the embedding network.

Typically, 64 points are considered in the integration computation, allowing consideration of harmonics in $V_1^o(t)$ up to the eighth order with a minimal truncation error in the Runge–Kutta algorithm.

3) Using the assumed waveforms for $I_1^o(t)$ and $V_2^o(t)$, together with the calculated waveform of $V_1^o(t)$, as given by (6), enables the fourth variable of the two-port network to be calculated. This variable, $I_2^o(t)$, the current flowing into the linear network, may be calculated using the known Z parameters of the two-port as follows:

$$I_2(\omega) = \frac{V_2(\omega) - Z_{21}(\omega)I_1(\omega) - V_2^{oc}(\omega)}{-Z_{22}(\omega)}. \quad (8)$$

The above linear embedding network calculations are most efficiently done in the frequency domain. The fast Fourier transform is used to convert between the frequency and the time domain.

4) With $I_2^o(t)$ available, a revised voltage $V_2^*(t)$ at port 2 may be calculated in the same manner as for diode 1 (revised voltages are indicated by a superscript *)

$$\frac{dV_2^*(t)}{dt} = \frac{I_2(t) - i_2[\exp(qV_2^*(t)/\eta kT) - 1]}{C_2(V^*(t))} \quad (9)$$

where

$$C_2(V_2^*(t)) = C_{02}(1 - V_2^*(t)/\phi)^{-\gamma} \quad (10)$$

where

C_{02} zero bias capacitance for diode 2
γ doping profile index
ϕ contact potential
i_2 diode 2 saturation current, and
η diode ideality factor.

5) Similarly, a revised $I_1^*(\omega)$ may be calculated in the frequency domain since $V_2^*(\omega)$, $I_2^o(\omega)$, and $V_1^o(\omega)$ are known quantities as is the embedding network information

$$I_1(\omega) = \frac{V_1(\omega) - Z_{12}(\omega)I_2(\omega) - V_1^{oc}(\omega)}{-Z_{11}(\omega)}. \quad (11)$$

6) It remains now to specify how the new estimates of voltage $V_2^1(t)$ and $I_1^1(t)$ are determined at the beginning of the next iteration and thereafter for each successive iteration. Two convergence parameters p_1 and p_2, one for each diode, are introduced. The range of values of these parameters is restricted to $0 < p \leqslant 1$. Values of p may be permitted to become complex [6] but experience has shown there is little advantage over using real p values. Determination of their values is based on criteria derived from a detailed convergence analysis to be given in the Appendix. These parameters are used to provide the next iteration of $I_1(t)$

and $V_2(t)$, namely $I_1^1(t)$ and $V_2^1(t)$, i.e.,

$$I_1^1(t) = p_1 I_1^*(t) + (1 - p_1) I_1^o(t) \tag{12}$$

$$V_2^1(t) = p_2 V_2^*(t) + (1 - p_2) V_2^o(t). \tag{13}$$

7) One iteration of the loop has now been completed, giving revised values of the periodic waveforms $I_1(t)$ and $V_2(t)$ for the next iteration cycle which begins at step 2. Iterations proceed until stationary solutions are achieved for these waveforms. The resulting computer program required 10K words of 32-bit VAX 11/780 memory.

The Appendix details the convergence mechanism of the above iteration algorithm. None of the available two-diode methods previously reported in the literature have yet provided a detailed convergence assessment to assist the user by providing prior information on the likelihood of convergence. It is clear that for the purposes of efficient automated nonlinear computation, the analysis algorithm and convergence mechanism should be thoroughly understood. The analysis in the Appendix produces a matrix for which convergence is assured if each and every element of that matrix is small. Control over the size of these matrix elements is provided by the use of the convergence parameters p_1 and p_2.

It is clear that this technique may be readily extended to the case of more than two diodes.

III. Small Signal and Noise Analysis

The small signal analysis used follows that of Kerr [8], with the mixing elements being a parallel combination of the diode varactor capacitance and Schottky-barrier conductance. The Fourier coefficients of the capacitance and conductance waveforms permit the construction of the small-signal conversion matrix for each diode. An overall mixer admittance matrix may then be formed using the two diode conversion matrices plus the diagonal matrix representing the embedding admittance network and external load admittances. This combined overall system matrix permits the calculation of the output IF impedance and the conversion loss. As eight harmonics of the local oscillator were determined by the nonlinear analysis, this permits four upper and lower sidebands to be considered in addition to the intermediate or output frequency.

The noise analysis comprises two components: the thermal noise emanating from both the diode series resistance, and the embedding network is determined using the theory of Twiss [19]; the shot noise contribution of the two diodes is calculated using Dragone's [20] noise correlation theory. Using these two components together with the superposition principle, the input noise temperature is calculated directly in an identical manner to that done by Kerr [8].

IV. Application of the Dual Diode Update Analysis

Studies were carried out on the subharmonically pumped mixer circuit examined by Kerr, given as example 1 in his paper [8]. This circuit presents a short-circuit coupling between the two diodes at frequencies above the signal frequency. At other frequencies, the load in parallel with the diodes is 50 Ω. In the unperturbed (or equal diode case), the diode parameters used were: $R_s = 10\ \Omega$, $C_0 = 7.0$ fF, $L_s = 0.4$ nH, $\eta = 1.12$, $\phi = 0.95$ V, $\gamma = 0.5$, and $i_0 = 8 \times 10^{-17}$ A. The signal, pump, and IF are at 103, 50, and 3 GHz. The convergence parameters p_1 and p_2 were both set at 0.5 for these studies. Subsequently, both the lead inductance and zero-bias capacitance values of diode 2 were allowed to vary, yielding the effects shown in Figs. 2 and 3. In all cases, the LO power was adjusted to give a constant rectified current of 2 mA in diode 1. When the capacitance was varied, the series resistance of diode 2 was modified such that the $C_0 R_s$ product remained constant. In both Figs. 2 and 3, the rectified current of diode 2 is plotted in

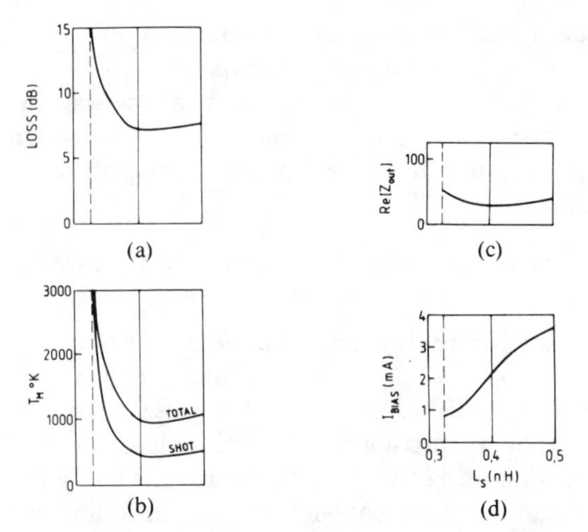

Fig. 2. The variation of mixer conversion properties with diode 2 lead inductance: (a) conversion loss, (b) input noise temperature, (c) real part of the IF-output impedance, (d) diode 2 bias current. Diode 1 parameters are: $R_s = 10\ \Omega$, $C_0 = 7.0$ fF, $L_s = 0.4$ nH, $\eta = 1.12$, $\phi = 0.95$ V, $\gamma = 0.5$, $i_0 = 8 \times 10^{-17}$ A, $i_{\text{bias}} = 2$ mA. Diode 2 parameters are identical to those of diode 1 but with the lead inductance and bias current allowed to vary.

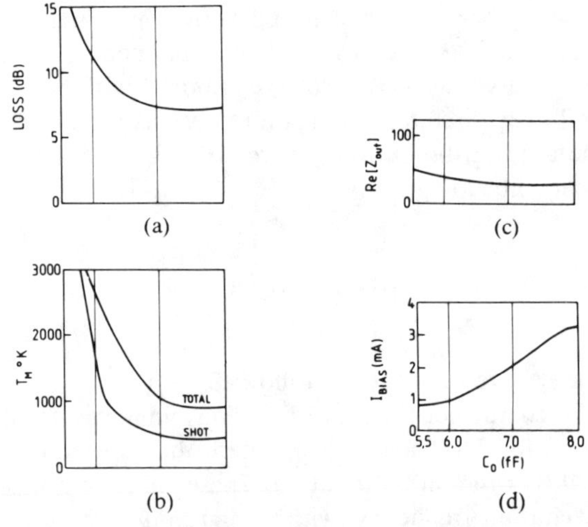

Fig. 3. The variation of mixer conversion properties with diode 2 zero-bias capacitance: (a) conversion loss, (b) input noise temperature, (c) real part of the IF-output impedance, (d) diode 2 bias current. Diode 1 parameters are identical to those given in Fig. 2. Diode 2 parameters are identical to those of diode 1 but with the zero-bias capacitance and bias current allowed to vary.

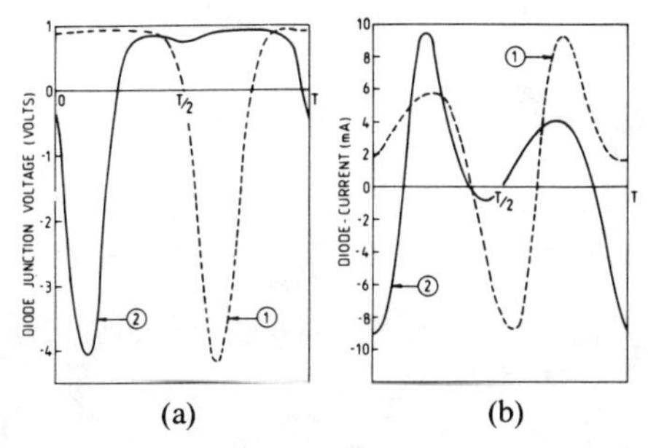

Fig. 4. Waveforms for the two diodes: (a) junction voltage, (b) diode current. Diode 2 parameters are $L_s = 0.325$ nH, $C_0 = 7.0$ fF. Diode 2 is in the resonant condition. Diode 1 parameters are $L_s = 0.4$ nH, $C_0 = 7.0$ fF.

addition to the mixer performance figures of conversion loss, input noise temperature, and IF output impedance. Typical current and voltage waveforms showing the phenomenon of double conduction [8] in diode 2 are given in Fig. 4.

It is clear from an examination of Figs. 2 and 3 that as in the identical diode situation analyzed by Kerr [8], variations in the lead inductance and zero bias capacitance in only one of the two diodes have significant effects on the overall performance of the mixer. A resonance between the lead inductance and junction capacitance of diode 2 is responsible for the poor conversion losses depicted in Figs. 2 and 3, with the resonant frequency being in the vicinity of the signal frequency. An examination of the current waveforms in Fig. 4 illustrates this point further. The resonant conditions of diode 2 are evident in the second negative current excursion; it is the increased second harmonic content which reduces the coupling of the IF current to the external IF load.

Compensation for the poor conversion losses may be achieved by addition of a separate dc bias supply for diode 2, as shown in Fig. 5. In this case, the bias voltage is adjusted until the rectified currents of both diodes (2 mA) are equal. The resulting conversion-loss diagram shows the absence of the resonant peak. In this case, the shift in the bias point on the varactor capacitance curve to a larger value of capacitance with the addition of the dc voltage tends to offset the fall in the inductance, and the resonant frequency is kept below the signal frequency.

Alternatively, when the diodes are mounted in waveguide, compensation for the variation in inductance and capacitance may be obtained readily by adjustment of the gap height. The coupled high-order TE and TM modes provide a variable reactive shunt across the gap terminals [21]. Adjustment of the gap height gives a range of compensatory reactance values for use in adjusting the diode current values for equality.

An interesting feature in both Figs. 2 and 3 is the sensitivity of the rectified current in diode 2 to the diode parameter imbalance. This has also been found experimentally, particularly in the millimeter-wave region [22]. This sharp variation in current is due to the proximity of the lead inductance-junction capacitance resonance to the fundamental pump frequency. In this situation, an increase in lead inductance tunes out a portion of the junction capacitance seen at the pump frequency and thereby results in an increase in the dc-rectified current.

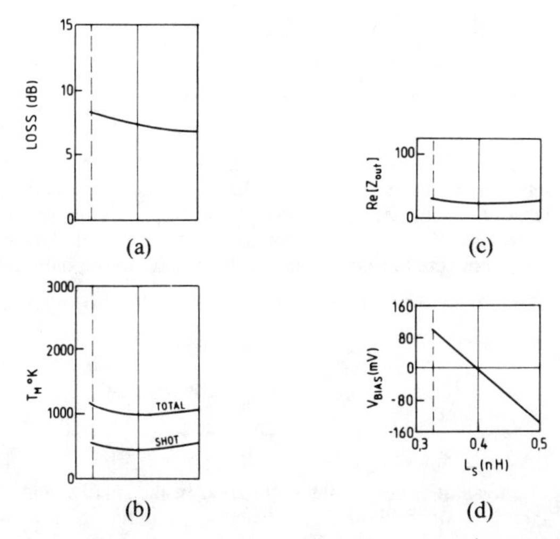

Fig. 5. The variation of mixer conversion properties with diode 2 lead inductance: (a) conversion loss, (b) input noise temperature, (c) real part of the IF output impedance, (d) diode 2 bias voltage. Diode 1 parameters are identical to those given in Fig. 2. Diode 2 parameters are identical to those of diode 1 but with lead inductance and bias voltage allowed to vary, thereby fixing the bias current of diode 2 at 2 mA.

V. Bilinear Model

The analysis of a millimeter-wave subharmonic mixer, using a bilinear model which incorporates all the important parasitic elements, involves two steps, a large-signal analysis followed by small-signal calculations [23].

The large signal analysis is carried out using the simplified model, shown in Fig. 6, for each diode in the anti-parallel pair. In this model, C_j is taken to be the zero-bias capacitance value, and the turn-on voltage is taken to be equal to the dc bias voltage which gives rise to the required dc bias current. The pump source is assumed to be ideal and have zero internal source impedance. Two linear analyses are required for each of the two states of the switch. The switch is closed for the period during which the diode voltage across C_j is greater than or equal to $V_{\text{turn-on}}$. Clearly, at each change of the state of the bipolar switch, the two initial conditions comprising the capacitance charge together with the lead inductance flux must be determined. It is possible due to resonances and ringing to obtain multiple conductions per cycle. In this case, care must be taken to ensure this phenomenon is correctly characterized in the calculations. For each case, the conduction angle and pump voltage amplitude are required following specification of the dc bias current; in practice, a rapid iteration gives the required pump voltage value to provide the specified bias current.

The small-signal analysis involves determination of the conductance and capacitance waveforms, from which harmonic components are readily found by Fourier analysis. This analysis is carried out with the diode model shown in Fig. 7, which differs from that of Fig. 6 in provision of elements C_{j2} and R_j when diode conduction is occurring and the switch is closed. R_j is the diode conductance and the sum of C_{j1} and C_{j2} is the diode capacitance for the

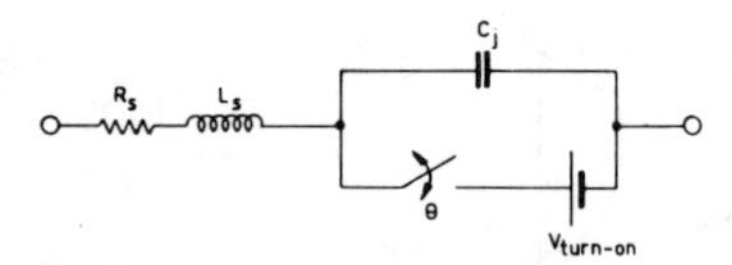

Fig. 6. Bilinear model of the mixer diode to be used in the approximate large signal nonlinear analysis. R_s is the series resistance, L_s is the lead inductance, C_j is the zero-bias capacitance, $V_{\text{turn-on}}$ is the forward bias turn-on voltage, θ is the conduction angle of the switch. Two of these diodes are connected in antiparallel to form a subharmonic mixer.

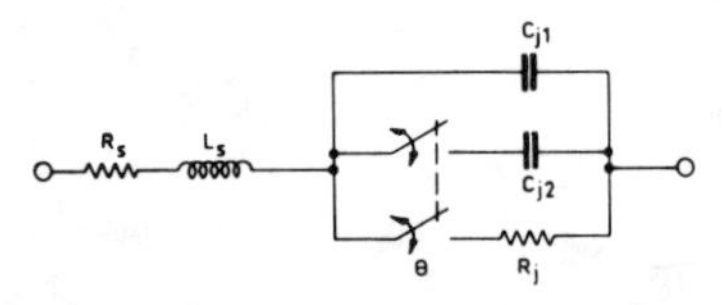

Fig. 7. Equivalent circuit of the diode used in the small signal analysis. This circuit features both a switched capacitance and a switched conductance. C_{j1} is the zero-bias capacitance, R_j is the diode conductance, and $C_{j1}+C_{j2}$ the diode capacitance for the specified dc-bias current.

specified dc bias current value; both are found using the exact diode $I-V$ and $C-V$ relations. C_{j1} now denotes the zero-bias capacitance. Using this model, together with the switch duty ratio found from the large signal bilinear model analysis, conductance and capacitance waveforms are readily found and a small-signal conversion matrix constructed for each diode.

An overall mixer admittance matrix may then be formed using the two diode conversion matrices plus the diagonal matrix representing the embedding admittance network and external load admittances. This combined overall system matrix enables calculation of the output IF impedance, conversion loss, and the input signal impedance. As in the analysis in Section III, four upper and lower sidebands were considered.

The noise analysis proceeds in two steps. Firstly, the thermal noise emanating from the diode series resistances and the embedding network is determined using the theory of Twiss [19]. The shot-noise contribution of the two diodes, each represented as an ideal switch, may be determined using the following theorem presented by Kerr [8]. A two-diode mixer, *using ideal exponential diodes* mounted in a lossless circuit, has essentially the same output noise as a lossy multiport network maintained at a temperature $\eta T/2$, where T is the physical temperature of the diodes and η is their ideality factor. From the total noise power delivered to the IF load admittance, the previously calculated conversion loss enables the input SSB noise temperature to be determined. For the switched bilinear model presented in this paper, it is assumed that the switch in the model contributes an amount of shot noise equivalent to that of the ideal exponential diode analyzed by Kerr in his theorem.

The resulting computer program required 7K words of 32-bit VAX 11/780 memory.

VI. Application of the Bilinear Model

Comparative studies were carried out on the subharmonically pumped mixer of Kerr, the circuit analyzed in Section IV using the dual-diode update nonlinear analysis.

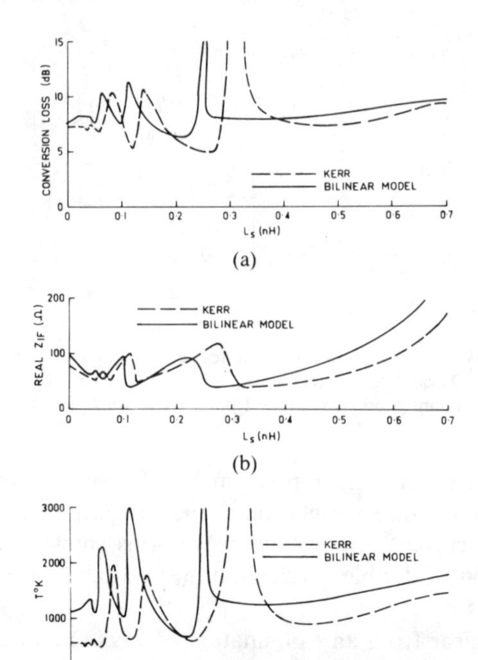

Fig. 8. Comparison of: (a) calculated conversion loss, (b) IF output impedance and (c) input noise temperature values given by the full nonlinear analysis reported by Kerr [8] and the bilinear model described in this paper. Identical diodes were used. Pump frequency = 50 GHz, signal frequency = 103 GHz, IF frequency = 3 GHz, bias current = 2 mA, $R_s = 10\ \Omega$, $C_j = 7.0$ fF.

Fig. 8 shows the variation of mixer performance with lead inductance, the diode lead inductances being constrained to be equal. The exact analysis results [8] are shown for comparison.

It is clear that the peaks given by the analysis of Kerr [8] are also predicted by the simplified bilinear model. This applies to all three properties of the mixer, namely conversion loss, IF-output impedance, and the input-signal noise temperature. However, it is equally apparent that there is a systematic horizontal displacement of the conversion peaks. Elementary LC resonance calculations using the zero-bias capacitance (based on resonance at the second pump harmonic) predict the largest conversion loss peak to occur at 0.36 nH in Fig. 8. As the average pumped capacitance is higher than the zero bias capacitance, it is to be expected that the full nonlinear analysis of Kerr [8] will shift the peak to a smaller value of inductance than 0.36 nH. The bilinear model analysis will naturally shift the peaks to a still further smaller value of inductance as the forward conduction region is effectively modelled as an infinite capacitance (short circuit), thus further increasing the average capacitance.

As may be expected, the bilinear model predicts sharper resonances than that given by Kerr [8], who used a full nonlinear analysis. This is a consequence of the absence of any diode-junction resistance in the perfect switch which was used in the diode bilinear model. On the other hand, the exponential Schottky-barrier equation inherently adds resistance to the circuit of the more comprehensive model, thereby providing damping to the LC resonant behavior of the circuit.

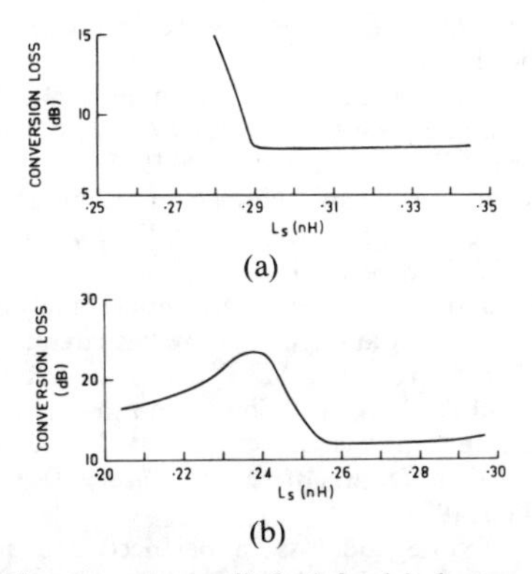

Fig. 9. Conversion loss versus diode 2 lead inductance for 2 different unequal diode situations: (a) diode 1 lead inductance = 0.30 nH, (b) diode 1 lead inductance = 0.25 nH. Other diode parameters (for both diodes) are: $R_s = 10\ \Omega$, $C_0 = 7.0$ fF. Diode 1 bias current = 2.0 mA. Note in (b) the conversion loss is always high as diode 1 is resonant at the signal frequency.

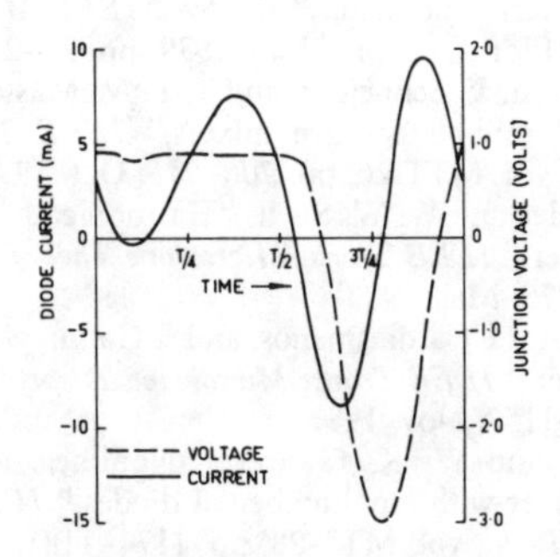

Fig. 10. Typical voltage and current waveforms calculated by the bilinear model where the phenomenon of multiple conduction is occurring. Bias current = 2 mA, $R_s = 10\ \Omega$, $C_j = 7.0$ fF, $L_s = 0.25$ nH. The two diodes are identical in this case.

The unequal diode results given in Fig. 9 reinforce the observations made in Section IV where the full nonlinear analyses were used. Diode 2 in Fig. 9(a) is in a nonresonant condition, and the subharmonic mixer performance deteriorates only when the lead inductance of diode 1 approaches 0.27 nH and the resonance at the signal frequency. On the other hand, the lead inductance of diode 2 in Fig. 9(b) is such that it produces an LC resonance at the signal frequency and thus the subharmonic mixer provides a poor response (12-dB loss) which deteriorates still further as diode 1 approaches resonant conditions. It is thus clear that, should one diode be at a resonant condition, the total mixer performance will suffer, irrespective of the condition of the other diode.

Finally, Fig. 10 depicts typical calculated voltage and current waveforms (using the bilinear model) for the case of double conduction leading to a conversion loss peak. The shape of these waveforms is virtually identical to those given in Fig. 4, which are calculated by the full update nonlinear analysis approach.

VII. Conclusions

This paper has presented a full dual-diode nonlinear analysis iteration technique whose convergence properties have been investigated. The dual-diode nonlinear analysis method requires a significant amount of computer time but enables the diodes to be fully characterized by their Schottky-barrier equations together with their nonlinear varactor capacitances. This technique has been applied to a study of a subharmonically pumped mixer with a view to establishing its performance with nonidentical diodes present. The study demonstrated the importance of the lead inductance–junction capacitance resonance which, if present in either one or both diodes, will cause a large conversion loss. This work complements the "identical diode" subharmonic mixer work reported by Kerr [8].

More rapid subharmonic mixer performance calculations may be achieved by using the bilinear model approach described in this paper. This model retains only the most significant features of the more comprehensive equivalent circuit used in the full analysis, yet still enables reasonably accurate mixer performance calculations to be made.

Appendix
Convergence Analysis

The convergence mechanism of the approach described above is investigated through a determination of the rate of error decrease per iteration cycle. Because of the two nonlinearities present, two error rates need to be monitored. Convergence to a stable solution imposes the constraint that both error magnitudes approach zero with increasing iteration number.

Let the nth harmonics of the correct solutions at the two interfaces be denoted by I_{1n}^{∞} and V_{2n}^{∞}. After m iterations

$$I_{1n}^{m} = I_{1n}^{\infty} + \delta_{1n}^{m} \quad \text{(A1)}$$

$$V_{2n}^{m} = V_{2n}^{\infty} + Z_{2n}^{NL}\delta_{2n}^{m} \quad \text{(A2)}$$

where δ_{1n}^{m} and δ_{2n}^{m} are the respective error terms, and Z_{2n}^{NL} is the impedance of diode 2 presented at the nth harmonic pump frequency.

The next iteration cycle would proceed as follows:

$$V_{1n}^{m} = Z_{1n}^{NL} I_{1n}^{m} \quad \text{(A3)}$$

where Z_{1n}^{NL} is the impedance of diode 1 presented at the nth harmonic pump frequency.

Using the embedding network information

$$I_{2n}^{m} = \left| \frac{V_{2n}^{m} - Z_{21} I_{1n}^{m} - V_{2}^{oc}}{-Z_{22}} \right|. \quad \text{(A4)}$$

The nonlinearity due to diode 2 then requires

$$V_{2n}^{*} = Z_{2n}^{NL} I_{2n}^{m}.$$

I_{1n}^{*} may then be calculated as follows:

$$I_{1n}^{*} = \left| \frac{V_{1n}^{m} - Z_{12} I_{2n}^{m} - V_{1}^{oc}}{-Z_{11}} \right|. \quad \text{(A5)}$$

The succeeding iterates, denoted with the superscript $(m+1)$, are given by

$$I_{1n}^{m+1} = p_1 I_{1n}^{*} + (1-p_1) I_{1n}^{m}$$
$$= p_1 \left| \frac{Z_{1n}^{NL} I_{1n}^{\infty} + Z_{1n}^{NL}\delta_{1n}^{m} - Z_{12} I_{2n}^{m} - V_{1}^{oc}}{-Z_{11}} \right|$$
$$+ (1-p_1)(I_{1n}^{\infty} + \delta_{1n}^{m}) \quad \text{(A6)}$$

where

$$I_{2n}^{m} = \left| \frac{V_{2n} + Z_{2n}^{NL}\delta_{2n}^{m} - Z_{21}I_{1n}^{\infty} - Z_{21}I_{1n}^{\infty} - Z_{21}\delta_{1n}^{m} - V_{2}^{oc}}{-Z_{22}} \right|. \tag{A7}$$

Using the definitions of V_{2n}^{∞} and I_{1n}^{∞}, the following simplifications can be made:

$$I_{1n}^{m+1} = I_{1n}^{\infty} + \left| p_1 \left(\frac{-Z_{1n}^{NL}Z_{22} + Z_{12}Z_{21}}{Z_{11}Z_{22}} \right) + (1 - p_1) \right| \delta_{1n}^{m} + p_1 \left(\frac{-Z_{12}Z_{2n}^{NL}}{Z_{11}Z_{22}} \right) \delta_{2n}^{m} \tag{A8}$$

$$= I_{1n}^{\infty} + \delta_{1n}^{m+1} \tag{A9}$$

where δ_{1n}^{m+1} is defined by (A8) and (A9). Similarly

$$V_{2n}^{m+1} = V_{2n}^{\infty} + Z_{2n}^{NL}\delta_{2n}^{m+1} \tag{A10}$$

where

$$\delta_{2n}^{m+1} = p_2 \frac{Z_{21}}{Z_{22}} \delta_{1n}^{m} + \left| -p_2 \frac{Z_{2n}^{NL}}{Z_{22}} + (1 - p_2) \right| \delta_{2n}^{m}. \tag{A11}$$

The cross coupling of the errors given by (A8)–(A11) gives rise to a matrix formulation, i.e.,

$$\boldsymbol{\delta}_{n}^{m+1} = M_n \boldsymbol{\delta}_{n}^{m} \tag{A12}$$

where

$$\text{the error vector } \boldsymbol{\delta}_{n}^{m} = \begin{pmatrix} \delta_{1n}^{m} \\ \delta_{2n}^{m} \end{pmatrix} \tag{A13}$$

and the four elements of the matrix M_n are defined by (A8)–(A11).

For convergence, a norm of M_n, denoted $\|M_n\|$, must be less than unity [24]. The norm of a matrix quantifies its magnifying power when used in vector multiplication; a small matrix norm is guaranteed by having small matrix elements. For the mixer circuits analyzed here, the lead inductances of the diodes ensure large values for Z_{11} and Z_{22}, which in turn keeps the size of the matrix elements down. In addition, the convergence parameters provide some flexibility.

References

[1] D. A. Fleri and L. D. Cohen, "Nonlinear analysis of the Schottky-barrier mixer diode," *IEEE Trans. Microwave Theory Tech.*, vol. MTT-21, pp. 39–43, Jan. 1973.

[2] W. K. Gwarek, "Nonlinear analysis of microwave mixers," M.S. thesis, Massachusetts Institute of Technology, Cambridge, MA., Sept. 1974.

[3] S. Egami, "Nonlinear, linear analysis, and computer-aided design of resistive mixer," *IEEE Trans. Microwave Theory Tech.*, vol. MTT-22, pp. 270–275, Mar. 1974.

[4] A. R. Kerr, "A technique for determining the local oscillator waveforms in a microwave mixer," *IEEE Trans. Microwave Theory Tech.*, vol. MTT-23, pp. 828–831, Oct. 1975.

[5] R. G. Hicks and P. J. Khan, "Numerical technique for determining pumped nonlinear device waveforms," *Electron. Lett.*, vol. 16, no. 10, pp. 375–376, May 8, 1980.

[6] R. G. Hicks and P. J. Khan, "Numerical analysis of nonlinear solid-state device excitation in microwave circuits," *IEEE Trans. Microwave Theory Tech.*, vol. MTT-30, pp. 251–259, Mar. 1982.

[7] D. N. Held and A. R. Kerr, "Conversion loss and noise of microwave and millimeter-wave mixers: Parts 1 and 2: Theory and experiment," *IEEE Trans. Microwave Theory Tech.*, vol. MTT-26, pp. 49–61, Feb. 1978.

[8] A. R. Kerr, "Noise and loss in balanced and subharmonically pumped mixers: Parts I and II: Theory and application," *IEEE Trans. Microwave Theory Tech.*, vol. MTT-27, pp. 938–950, Dec. 1979.

[9] T. F. McMaster, M. V. Schneider, and W. W. Snell, "Millimeter-wave receivers with subharmonic pump," *IEEE Trans. Microwave Theory Tech.*, vol. MTT-24, pp. 948–952, Dec. 1976.

[10] R. E. Forsythe, V. T. Brady, and G. T. Wrixon, "Development of a 183-GHz subharmonic mixer," in *Proc. IEEE MTT-S Int. Microwave Symp.*, (Florida), April–May 1979, pp. 20–21.

[11] E. R. Carlson, M. V. Schneider, and T. F. McMaster, "Subharmonically pumped millimeter-wave mixers," *IEEE Trans. Microwave Theory Tech.*, vol. MTT-26, pp. 706–715, Oct. 1978.

[12] M. V. Schneider and W. W. Snell, "Harmonically pumped stripline down-converter," *IEEE Trans. Microwave Theory Tech.*, vol. MTT-23, pp. 271–275, Mar. 1975.

[13] P. T. Parrish, A. G. Cardiasmenos, and I. Galin, "94-GHz beam-lead balanced mixer," *IEEE Trans. Microwave Theory Tech.*, vol. MTT-29, pp. 1150–1157, Nov. 1981.

[14] M. T. Faber and W. K. Gwarek, "Nonlinear-linear analysis of microwave mixer with any number of diodes," *IEEE Trans. Microwave Theory Tech.*, vol. MTT-28, pp. 1174–1181, Nov. 1980.

[15] M. R. Barber, "Noise figure and conversion loss of the Schottky barrier mixer diode," *IEEE Trans. Microwave Theory Tech.*, vol. MTT-15, pp. 629–635, Nov. 1967.

[16] G. S. Bordonskiy *et al.*, "Frequency converter at wavelength of 2.5 mm," *Radio Eng. Electron. Phys.*, vol. 21, no. 3, pp. 88–93, Mar. 1976.

[17] A. I. Zabyshnyi *et al.*, "Millimetre-range mixers with subharmonic pumping," *Radio Eng. Electron. Phys.*, vol. 25, no. 4, pp. 287–290, Apr. 1980.

[18] R. G. Hicks and P. J. Khan, "Analysis of balanced subharmonically pumped mixers with unsymmetrical diodes," in *Proc. IEEE MTT-S Int. Microwave Symp.*, (Los Angeles, CA), June 1981, pp. 457–459.

[19] R. Q. Twiss, "Nyquist's and Thevenin's theorems generalized for nonreciprocal linear networks," *J. Appl. Phys.*, vol. 26, no. 5, pp. 599–602, May 1955.

[20] C. Dragone, "Analysis of thermal and shot noise in pumped resistive diodes," *Bell Syst. Tech. J.*, vol. 47, pp. 1883–1902, Nov. 1968.

[21] R. L. Eisenhart, "Understanding the waveguide diode mount," in *Proc. IEEE MTT-S Int. Microwave Symp.*, (Arlington Heights), May 1972, pp. 154–156.

[22] M. Cohn, J. E. Degenford, and B. A. Newman, "Harmonic mixing with an antiparallel diode pair," *IEEE Trans. Microwave Theory Tech.*, vol. MTT-23, pp. 667–673, Aug. 1975.

[23] R. G. Hicks and P. J. Khan, "Numerical analysis of subharmonic mixers using a bilinear diode model," in *Proc. IEEE MTT-S Int. Microwave Symp.*, (Dallas, TX), June 1982, pp. 382–384.

[24] L. W. Johnson and R. D. Riess, *Numerical Analysis.* Reading: Addison–Wesley, 1977, pp. 50–61.

Inherent Signal Losses in Resistive-Diode Mixers

MARION E. HINES, FELLOW, IEEE

Abstract—**A new time-domain method is presented for the characterization and analysis of resistive-diode mixers. The method has been found to be helpful in evaluating experimental models of new mixer designs. From an analytical viewpoint, this method has provided some new insights into mixer behavior and the fundamental limits to mixer performance. In analyzing equivalent mixer models, the method has been found to be in agreement with the classical frequency-domain approach.**

In using the time-domain method to determine the minimum available loss, theoretical studies were made using an "ideal-diode" model which is presumed to have zero forward-bias resistance and infinite reverse-bias impedance. Significant signal losses were found to occur, even in this "lossless" condition, when reactive filtering was used to suppress unwanted frequency responses. The lost signal power was not reflected and it did not appear at other signal-related frequencies. This result has also been found in a frequency-domain analysis using a new formulation suggested to the author in private correspondence.

This loss is explained in two different ways, depending upon the model used and the method of analysis. In one example, using the time-domain approach, it was found that signal energy is converted into dc in the rectified current. In the frequency domain analysis, the loss is explained as the result of frequency conversions into a large number of high-order modulation products. The paper includes some newly formulated conjectures concerning the ultimate limits on conversion loss in single-diode mixers.

I. Introduction

IN THIS PAPER, a new time-domain scattering-matrix method is presented for the characterization and analysis of diode mixers of the class commonly used for RF and microwave receivers. The characterization method has been found useful in the experimental evaluation of new designs. Using this method, the frequency conversion characteristics and signal impedences can be deduced from data taken at the LO frequency only. The theory has been useful in providing new insights into the physical principles involved.

In Section II, the basic equations for the time-domain method are derived. This method is applicable to single-diode mixers in which the intermediate frequency is a small fraction of the signal and local oscillator (LO) frequencies, wherein the network impedance is essentially constant over the band which includes the signal, LO, and image frequencies. The results provide a three-frequency (signal, IF, image) scattering matrix which describe the mixer's performance as though it were a three-port network. The method is then extended to include the effects of suppressing the image response by the insertion of a band-rejection filter into the input RF line. An analysis shows the degree of improvement obtainable in this way.

In Section III, comparisons are made between this time-domain and the classical frequency-domain approaches to mixer analysis [4]–[6]. No discrepancy has been found between them. When properly carried out, using an adequately high-order admittance matrix in the frequency-domain approach, the results have been found to agree when analyzing "practical" models of mixers using realistic circuits and exponential type diodes. It is shown that significant errors can occur in the frequency-domain analysis of a low-loss mixer if the matrix is truncated to a low order.

In order to obtain an expression for the lowest possible mixer loss, the time-domain approach was used to obtain the scattering matrix of some simple mixer networks using an "ideal diode." This hypothetical device has zero resistance in forward bias and infinite resistance in reverse. Some of the results were reported in oral papers in [1]–[3]. In the case of a tuned mixer in which all higher responses were suppressed ($nf_p \pm f_0$, $n>1$) it was found that the minimum conversion loss in the double-sideband case was 3 dB, and that when the LO, signal, and IF frequencies were impedance-matched, *no image response appeared.* Half of the signal power was *lost* and did not appear at any signal-related frequency. This was a surprising result[1] which seemed paradoxical in that it was not immediately clear how a "lossless" network could absorb energy. The paradox has been resolved and some of the explanations are given in [1]–[3]. The analysis of this ideal case is carried out in the Appendix, using both the time-domain and frequency-domain approaches. The image-suppression case is also analyzed there in the frequency domain. The results are in full agreement.

In the time-domain analysis of a tuned ideal-diode mixer, it is found that the lost power can be explained by a second-order effect in which energy is converted from the signal into the dc rectified current. In the frequency-domain analysis the diode is modeled as a periodically operated switch with a vanishingly small resistance. In this case, it is explained that the lost power is converted into an indefinitely large number of high-order modulation products which are dissipated in this infinitesimal resistance.

[1]This author is not aware of such a result in the literature. There is a widespread misconception that all of the input signal energy could be converted to other signal-related frequencies if the diode's resistance could be eliminated and the input impedances matched.

Manuscript received April 8, 1980; revised December 1, 1980.
The author is with Microwave Associates, Inc., Burlington, MA 01803.

Reprinted from *IEEE Trans. Microwave Theory Tech.*, vol. MTT-29, pp. 281–292, Apr. 1981.

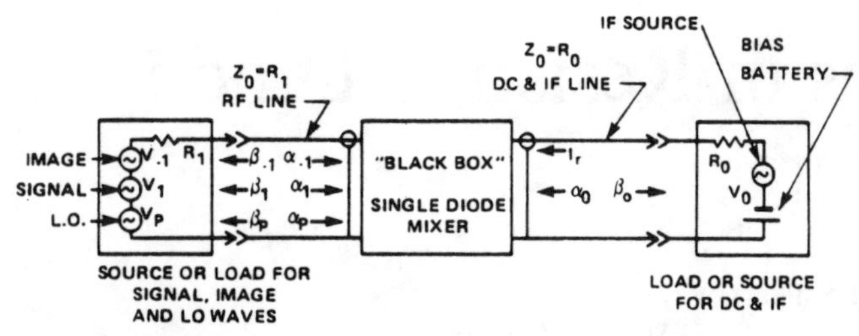

Fig. 1. A "black-box" double-sideband mixer circuit showing the external connections and the interpretation of the incident and emerging wave quantities α and β. The black box includes circuitry to suppress dc and IF from the RF line and RF from the dc–IF line. Fig. 4 shows a highly simplified tuned mixer of this class using and "ideal diode."

II. Time-Domain Analysis of Single-Diode Mixers

A. The "Black-Box" Model

The time-domain approach used here is an extension of the "incremental method" of mixer characterization described by Torrey and Whitmer [4]. The method applies only to mixers in which the IF is a small fraction of the LO and signal frequencies. The double sideband case is treated in Sections II-B, II-C, and II-D which follow. This is extended to the "image-rejection" case in Section II-E.

Fig. 1 shows a simple "black box" diode rectifier circuit which serves as a mixer. Typically, two such circuits are combined to form a balanced mixer. Combining at RF is usually done through a four-port hybrid circuit such as a 180° "magic tee" or a 90° 3-dB directional coupler, not shown here. The RF signal is applied at one of the hybrid ports, a local oscillator at another, and the two diode mounts at the other two ports. The feed line to each mount carries both the signal and the LO waves, each at half strength. This input line is treated as a single Thevenin-type resistive generator with signal, LO, and image sources as shown in Fig. 1. The two mounts are combined at IF and the pair act in parallel. We need analyze only one mount inasmuch as the net gain is the same for either one as for the combination.

Except for the diode, the network is assumed to be linear and reciprocal. It contains reactive frequency-separation networks to exclude dc and IF from the RF port and RF from the IF port. Further filtering is often used to suppress unwanted high-order responses. The IF port includes the dc rectification current as well as the IF signal. The signal is assumed to be small so that $V_1 \ll V_p$, and the IF current is small compared with the dc rectified current.

B. Theoretical Basis

The analysis will involve first-order perturbation methods. First, we express the two input RF waves as a single modulated wave where the LO is the "carrier" whose amplitude and phase both vary slowly with time at the "low" IF frequency. Let α_p and α_1 represent the input complex wave quantities, and $\hat{\alpha}_p$ and $\hat{\alpha}_1$ represent their magnitudes, equal to the square root of the power involved. Using complex vector addition and the law of cosines, the sum of these waves can be expressed by an identity, valid for large or small signals

$$\hat{\alpha}_p \cos \omega_p t + \hat{\alpha}_1 \cos \omega_1 t \equiv \sqrt{\hat{\alpha}_p^2 + \hat{\alpha}_1^2 + 2\hat{\alpha}_p \hat{\alpha}_1 \cos \omega_0 t} \cdot \cos\left\{\omega_p t + \tan^{-1} \frac{\hat{\alpha}_1 \sin \omega_0 t}{\hat{\alpha}_p + \hat{\alpha}_1 \cos \omega_0 t}\right\} \quad (1)$$

where ω_0 is the IF, ω_p is the LO or pump, and ω_1 is the signal, and

$$\omega_0 = \omega_1 - \omega_p.$$

We see that the wave magnitude, expressed by the radical, varies with $\cos \omega_0 t$; and the phase, expressed in the $\tan^{-1}$ function, varies with $\sin \omega_0 t$. If $\hat{\alpha}_1 \ll \hat{\alpha}_p$, then we can use the first-order approximation

$$\hat{\alpha}_p \cos \omega_p t + \hat{\alpha}_1 \cos \omega_1 t \simeq (\hat{\alpha}_p + \hat{\alpha}_1 \cos \omega_0 t) \cos\left(\omega_p t + \frac{\hat{\alpha}_1}{\hat{\alpha}_p} \sin \omega_0 t\right). \quad (2)$$

If $\omega_0 \ll \omega_p$, and if the network contains no sharply tuned resonances or other long-time-constant circuitry, $\sin \omega_0 t$ and $\cos \omega_0 t$ are slowly varying functions and the response of the mixer mount at any instant will be indistinguishable from the steady-state response for a single sinusoidal wave whose amplitude and phase are given by (2). As the input wave amplitude slowly varies, the "dc" current will vary, the "dc" bias will vary, and the input RF reflection coefficient also. However, at any instant, these quantities would be the same as for single-frequency LO excitation at the level indicated by the radical of (1). If we characterize the diode mount by a full set of *single-frequency* measurements (or a single-frequency analysis) over a range of power levels at the LO frequency, we can determine the full response for a composite wave as in (1) or (2) under these assumptions.

To do so, we presume that the two functions of (3) and (4) are known for the mount, obtained either through analysis or by measurement. Each is a function of two independent variables, the input LO wave amplitude[2] $\hat{\alpha}$ and the dc bias voltage V_b. The first is the complex RF reflection coefficient Γ and the second is for the rectified current i_r

$$\Gamma = \Gamma_a(\hat{\alpha}, V_b) \quad (3)$$

$$i_r = I_r(\hat{\alpha}, V_b). \quad (4)$$

These functions are sketched graphically in Figs. 2 and 3. Fig. 2 is a plot of the dc rectified current, drawn as a family of curves for different input RF wave amplitudes, plotted versus the bias voltage V_b.[3] Fig. 3 is a *complex* polar plot of the reflection coefficient (as in the Smith Chart) for various values of $\hat{\alpha}$ and V_b. The complex value of Γ is the vector from the center to any point specified by choosing values for $\hat{\alpha}$ and V_b. Fig. 2 is a theoretical chart, obtained by computer analysis of a "practical" circuit which included an exponential diode with several parasitic

[2] $\hat{\alpha}$ is a real quantity, normalized as the square root of the available input LO power, in watts. Later, α_p, α_1, α_0, and α_{-1} will represent complex input waves at three frequencies.

[3] This chart is similar to one presented by Gerst [8].

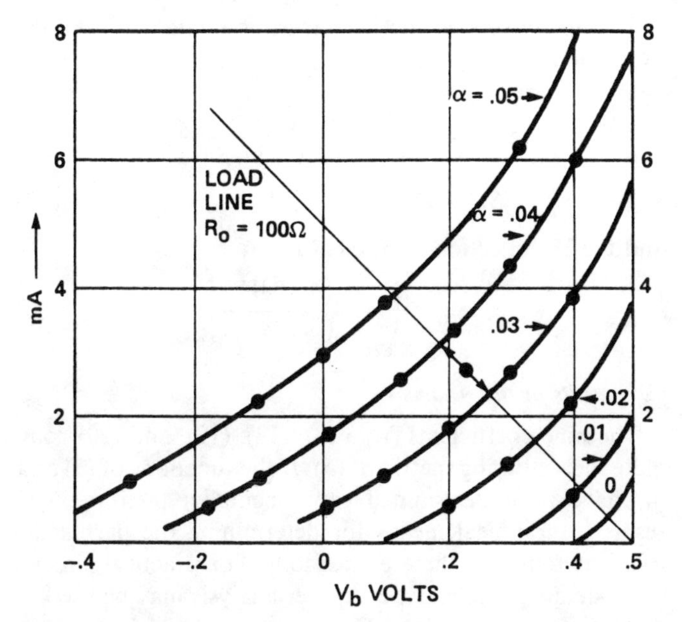

Fig. 2. Rectification chart for a mixer, illustrating the behavior of the function $I_r(\alpha, V_b)$. The points plotted were obtained by computer analysis of a "practical" mixer network using an exponential diode with several tuning and parasitic elements included.

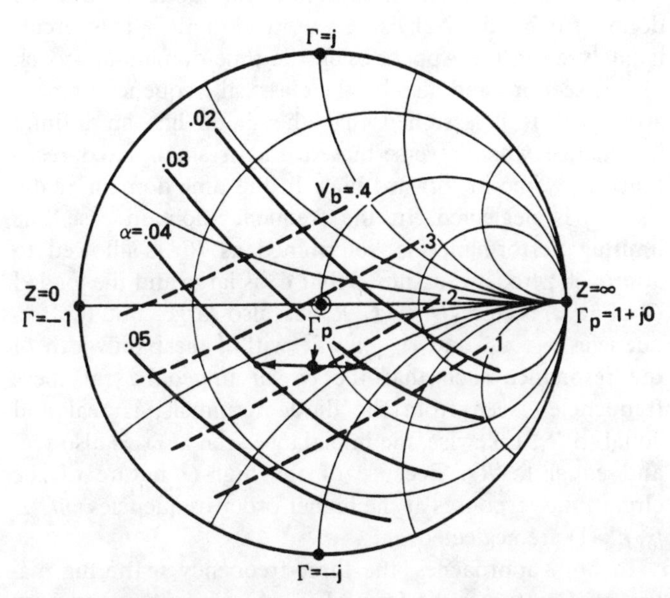

Fig. 3. A sketch illustrating the reflection coefficient function $\Gamma_a(V_b, \alpha)$. In actual charts of this type the two families of curves tend to coalesce into a narrow zone across the chart.

elements included in the circuit model. Fig. 3 is a sketch intended only to illustrate the principles involved in this theory. In actual charts of this type, the curves tend to coalesce into a single line across the unit circle.

An average operating point is chosen by setting the LO level and the average or dc bias. One such point is shown on Figs. 2 and 3. Fig. 2 shows an IF "load line" which represents the variation of bias with rectified current *as determined by the bias battery and the load resistance* R_0. Consider, now, the effect of slowly varying the input wave amplitude as implied by (1). On Fig. 2, this moves the operating point back and forth along the load line from one α curve to another, with the current i_r, and the bias V_b, varying together. As a function of time, the operating point moves sinusoidally along this load line as between the arrowheads shown. Meanwhile, on Fig. 3, the operating point will also move back and forth sinusoidally causing a variation in the RF reflection coefficient Γ.

Using these two charts, a graphical analysis can be made which yields a complete three-frequency scattering matrix for this mixer. The "slow" variation of the "dc" represents the IF wave. The reflected wave varies, in amplitude and phase, and is interpreted as a wave which is modulated differently from that of the input wave. It is then resolved into its three sinusoidal frequency components, which are interpreted as reflected "signal", LO, and "image" waves.

It is convenient to express the characteristics of a microwave mixer in scattering matrix form. Corresponding to the Y-matrix formulation of Peterson and Llewellyn [5], a three-frequency S-matrix has the form of (5). It will be noted that α_{-1} and β_{-1} are to be entered here in complex-conjugate form (*) to account for the spectral inversion which occurs in the image response

$$\begin{array}{c|ccc} & \alpha^*_{-1} & \alpha_0 & \alpha_1 \\ \hline \beta^*_{-1} & S_{-1-1} & S_{-10} & S_{-11} \\ \beta_0 & S_{0-1} & S_{00} & S_{01} \\ \beta_1 & S_{1-1} & S_{10} & S_{11} \end{array}. \tag{5}$$

Here, α_0 is the input wave quantity at the IF frequency, ω_0, α_1 is that for an upper-sideband "signal" at the frequency $\omega_p + \omega_0$, and α_{-1} for the lower sideband "image" at $\omega_p - \omega_0$. Similarly β_0, β_1, and β_{-1} represent the emerging waves. These are complex quantities and are normalized such that their magnitudes are expressed as the square roots of the powers involved. The diagonal terms represent the reflection coefficients. The off-diagonal terms are the complex frequency conversion transmission coefficients. From Fig. 1, $\hat{\alpha}_1 = V_1/\sqrt{8R_1}$, etc.

The mathematical derivations for determining these scattering parameters from the functions of (3) and (4) are straightforward, although somewhat tedious and detailed. These will be summarized briefly. We will require four partial derivatives from (3) and (4), $\partial i_r/\partial\hat{\alpha}$, $\partial i_r/\partial V_b$, $\partial\Gamma/\partial\hat{\alpha}$, $\partial\Gamma/\partial V_b$, taken at the operating point. These can be obtained graphically from the charts of Figs. 2 and 3, by direct measurements or by analytic means. Using a Taylor expansion with (4), to a first approximation

$$\Delta i_r = \frac{\partial i_r}{\partial\hat{\alpha}}\Delta\hat{\alpha} + \frac{\partial i_r}{\partial V_b}\Delta V_b. \tag{6}$$

Noting that $\Delta\hat{\alpha} = \hat{\alpha}_1\cos\omega_0 t$, and $\Delta V_b = -\Delta i_r R_0$, we obtain, from (2)

$$\Delta i_r = \frac{\partial i_r/\partial\hat{\alpha}}{1 + R_0(\partial i_r/\partial V_b)}\hat{\alpha}_1\cos\omega_0 t. \tag{7}$$

With normalization, we obtain the two down-conversion S-coefficients.

$$S_{0-1} = S_{01} = \frac{(\partial i_r/\partial\hat{\alpha})\sqrt{R_0/2}}{1 + R_0(\partial i_r/\partial V_b)} = S_d. \tag{8}$$

This is maximized if R_0 is chosen to match the impedance of the mixer, as

$$R_0=\left(\frac{\partial i_r}{\partial V_b}\right)^{-1} \tag{9}$$

giving the *maximum* down-conversion gain for the operating point at which the derivatives are determined:

$$(G_d)_{\max}=\frac{(\partial i_r/\partial \hat{\alpha})^2}{8(\partial i_r/\partial V_b)}. \tag{10}$$

Although expressed differently, this is equivalent to the conversion loss formulas for the "incremental method" of [4, pp. 213–218].

Similarly, the coefficients S_{11}, S_{-1-1}, S_{1-1}, and S_{-11} can be determined by the use of partial derivatives obtained from Fig. 3. Noting that from (3)

$$\Delta\Gamma=\frac{\partial\Gamma}{\partial\hat{\alpha}}\Delta\hat{\alpha}+\frac{\partial\Gamma}{\partial V_b}\Delta V \tag{11}$$

and from (7) we obtain

$$\Delta\Gamma=\left[\frac{\partial\Gamma}{\partial\hat{\alpha}}-\frac{(\partial\Gamma/\partial V_b)(\partial i_r/\partial\hat{\alpha})R_0}{1+R_0\dfrac{\partial i_r}{\partial V_b}}\right]\hat{\alpha}_1\cos\omega_0 t. \tag{12}$$

If Γ_p represents the RF reflection coefficient at the chosen operating point, then

$$\Gamma_{\text{sum}}=\Gamma_p+\Delta\Gamma \tag{13}$$

and the composite reflected wave is expressed as

$$\beta_{\text{sum}}=\alpha_{\text{sum}}\Gamma_{\text{sum}} \tag{14}$$

where α_{sum} is expressed from (2) in complex form

$$\alpha_{\text{sum}}=\hat{\alpha}_p\left(1+\frac{\hat{\alpha}_1}{\hat{\alpha}_p}\cos\omega_0 t\right)e^{j(\hat{\alpha}_1/\hat{\alpha}_p)\sin\omega_0 t}. \tag{15}$$

Using standard trigonometric identities and the correspondence between the complex exponential and sinusoids, the modulated-wave expression for β_{sum} can be resolved into three frequency components, with ω_p as the carrier and ω_1 and ω_{-1} as sidebands.

To simplify the expressions, let

$$\Gamma_d=\hat{\alpha}_p\left[\frac{\partial\Gamma}{\partial\hat{\alpha}}-\frac{R_0(\partial\Gamma/\partial V_b)(\partial i_r/\partial\hat{\alpha})}{1+R_0(\partial i_r/\partial V_b)}\right]. \tag{16}$$

Detailed algebraic manipulations yield three sinusoids

$$\beta_{\text{sum}}e^{j\omega_p t}=\Gamma_p\alpha_p e^{j\omega_p t}+\alpha_1\left(\frac{\Gamma_d}{2}+\Gamma_p\right)e^{j\omega_1 t}+\alpha_1^*\frac{\Gamma_d}{2}e^{j\omega_{-1}t}. \tag{17}$$

For excitation at ω_{-1} instead of ω_1 the results are similar. These relations provide four more of the scattering coefficients

$$S_{11}=\frac{\Gamma_d}{2}+\Gamma_p$$
$$S_{-11}=\frac{\Gamma_d^*}{2}$$
$$S_{-1-1}=\frac{\Gamma_d^*}{2}+\Gamma_p^*$$
$$S_{1-1}=\frac{\Gamma_d}{2}. \tag{18}$$

The up-conversion coefficients can be obtained from the expression for $\partial\Gamma/\partial V_b$

$$S_{-10}=\left(\frac{\partial\Gamma}{\partial V_b}\right)^*\frac{\hat{\alpha}_p\sqrt{2R_0}}{1+R_0(\partial i_r/\partial V_b)}$$
$$S_{10}=S_{-10}^* \tag{19}$$

and the IF reflection coefficient is

$$S_{00}=\frac{1-R_0(\partial i_r/\partial V_b)}{1+R_0(\partial i_r/\partial V_b)}. \tag{20}$$

C. Results of the Analysis

The nine coefficients from (8), (18), (19), and (20) complete the scattering matrix of (5). If the functions of (3) and (4) have been determined experimentally, then we may resort to graphical means for determining the partial derivatives found in these expressions. For practical circuits of moderate complexity, computer analysis may be used to determine these functions. However, entirely analytic means have also been used in the case of three highly simplified circuits in which an "ideal diode" was assumed. These results are reported in [1]–[3].

Fig. 4 shows a highly simplified circuit model for a tuned double-sideband mixer using an "ideal diode". This circuit is analyzed in the Appendix by the time-domain approach of this section; and also by the classical frequency-domain approach. It is assumed that the diode has an infinite impedance when reverse-biased and a small fixed resistance R_d when in forward bias. In the time domain analysis, R_d is neglected. In the frequency-domain case, the limiting performance is determined as R_d is allowed to approach zero. It is assumed that C_2 is large and the loaded Q of the resistor ($Q=\omega_p C_1 R_1$) is also large, but the frequencies ω_p, ω_1, and ω_{-1} all lie within the bandwidth of the resonance such that the *circuit* impedance at these frequencies, seen from the diode terminals, is real and equal to R_1. Likewise, the impedance seen at ω_0 is also real and equal to R_0. Because of the high-Q nature of the circuit, the responses at the higher order frequencies ($nf_p \pm f_0$, $n>1$) are neglected.

In both approaches, the three-frequency scattering matrix of (5) assumes the form of (21-A), *in the limit*, as these assumed conditions are approached

$$\bar{S}=\begin{vmatrix} \dfrac{V_b}{V_p}-\dfrac{2R_1}{2R_1+R_0} & \dfrac{2\sqrt{R_0R_1}}{2R_1+R_0} & \dfrac{R_0}{2R_1+R_0}-\dfrac{V_b}{V_p} \\ \dfrac{2\sqrt{R_0R_1}}{2R_1+R_0} & \dfrac{2R_1-R_0}{2R_1+R_0} & \dfrac{2\sqrt{R_0R_1}}{2R_1+R_0} \\ \dfrac{R_0}{2R_1+R_0}-\dfrac{V_b}{V_p} & \dfrac{2\sqrt{R_0R_1}}{2R_1+R_0} & \dfrac{V_b}{V_p}-\dfrac{2R_1}{2R_1+R_0} \end{vmatrix}. \tag{21A}$$

The signal-to-IF conversion loss is minimized at 50 percent if $R_0=2R_1$, independent of the applied bias level. This also provides an impedance match at IF. The signal and image reflection coefficients and the signal-to-image

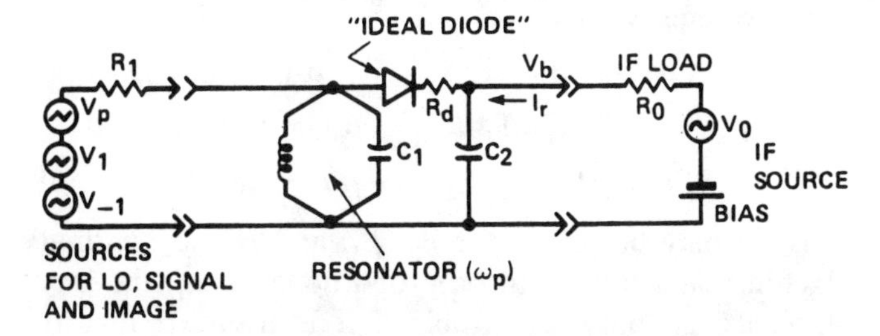

Fig. 4. A simple circuit for a tuned double-sideband mixer to determine the minimum conversion loss of a single-diode device using an "ideal diode". The resistance R_d is neglected in the time-domain analysis, and is assumed to become vanishingly small in the frequency-domain case.

conversion coefficients depend upon the dc bias at the diode. If $V_b = V_p/2$, then the LO wave is impedance-matched (see A-3 and A-5 in the Appendix), the signal and image waves are also matched and there is no conversion between signal and image. This condition occurs in this circuit if the bias battery is eliminated. For these conditions the S-matrix of (21-A) reduces to (21-B).

$$\bar{S} = \begin{vmatrix} 0 & \sqrt{2}/2 & 0 \\ \sqrt{2}/2 & 0 & \sqrt{2}/2 \\ 0 & \sqrt{2}/2 & 0 \end{vmatrix}. \tag{21B}$$

Both (21-A) and (21-B) describe lossy circuits. In down-conversion in (21-B), 50 percent of an input signal is absorbed, but in up-conversion, 50 percent of the input power appears in each sideband. For (21-A), varying the bias will not change the conversion loss but will affect the amount of signal power absorbed by the circuit. In the limit, if $R_0 = 2R_1$, and if $V_b \to 0$ or $V_b \to V_p$, the coefficients S_{11}, S_{-1-1}, S_{1-1}, and S_{-11} each approach 0.5 in magnitude, such that 25 percent of the signal power can be converted to the image and 25 percent can be reflected. In this way, the lossless assumption made by Kelly [9] can be approached. This results in a severe mismatch of the LO wave.

In [1], two other "ideal-diode" circuits were treated by the time-domain approach, giving conversion losses greater than 3 dB. In each of these cases, under "optimum" conditions, the signal impedances were matched and no signal-to-image conversion was found. As described in Section III, computer analyses have also been made for a number of more realistic mixer models using lossy exponential type diodes in more complex circuits. In each case, when near-optimum conditions were found by trial-and-error, the input signal impedances were nearly matched and image conversion was negligibly small. The S-matrix in these cases resembled (21-B), but with conversion coefficients somewhat smaller than $\sqrt{2}/2$.

In further study of this class of mixer, it was found that the addition of a small signal wave to a strong LO wave results in a second-order increase in the dc rectified current. Second-order changes may also occur in the reflected LO waves and/or in its harmonics. Such a second-order change, when added to an already finite direct current or ac wave, results in a *power* change which is proportional to the applied perturbing *power*. In this way, signal energy can be transferred into dc or into other waves not related in frequency to the signal.

As an example, consider again the idealized mixer circuit of Fig. 4. This device will now be treated as a *detector* with the bias battery eliminated. Consider the case where $R_0 = 2R_1$ so that the device will be impedance-matched at both input frequencies. The rectified or "dc" output voltage follows the peak value of the sum of the two input waves, with a "slow" variation as given by (1). If we substitute into (1) the relations $V_p = \alpha_p\sqrt{8R_1}$ and $V_1 = \alpha_1\sqrt{8R_1}$, the rectified output voltage will be one-half the sum value, given by

$$V_{\text{out}} \simeq \frac{1}{2}\sqrt{V_p^2 + V_1^2 + 2V_1V_p\cos\omega_0 t}\,. \tag{22}$$

The radical may be expanded as a power series using the binomial theorem, giving, to second order in V_1.

$$V_{\text{out}} \simeq \frac{V_p}{2}\left[1 + \frac{V_1^2}{4V_p^2} + \frac{V_1}{V_p}\cos\omega_0 t - \frac{V_1^2}{4V_p^2}\cos 2\omega_0 t \cdots\right]. \tag{22A}$$

The dc output power, to second order in V_1, is

$$P_{\text{dc}} \simeq \frac{V_p^2}{8R_1} + \frac{V_1^2}{16R_1} + \cdots \tag{23}$$

and the IF power is

$$P_{\text{if}} \simeq \frac{V_1^2}{16R_1} + \cdots. \tag{24}$$

The total available power from the generator is

$$P_{\text{av}} = \frac{V_1^2 + V_p^2}{8R_1} \tag{25}$$

so that one-half of the available signal power appears at IF, and one-half at dc. All of the LO power is theoretically rectified to dc. Thus, we should not be surprised that one-half of the RF signal power is lost in an ideal tuned mixer of the type shown in Fig. 4.

D. An Equivalent Circuit for an Optimized Double-Sided Mixer

A scattering matrix of the form of (21) is similar to that for a waveguide "magic tee" hybrid circuit in which the fourth arm has been internally loaded with an appropriate resistor. This is illustrated in Fig. 5. The signal and image waves of (21) correspond to the symmetrically opposed ports of the tee. The series (E-plane) arm corresponds to the IF, and the loaded fourth arm corresponds to the internal sink which represents losses to other frequencies, including dc. The normalized impedance of this internal load resistor depends upon the relative impedance levels of the signal source and that presented to the LO. A "conjugator" is shown in the image arm to account for the spectrum inversion between ω_{-1} and ω_0 and between ω_{-1} and ω_1. Additional attenuators are shown to account for additional losses which appear in practice where S_{10}, S_{01}, S_{-10}, S_{0-1} are smaller than 0.7071.

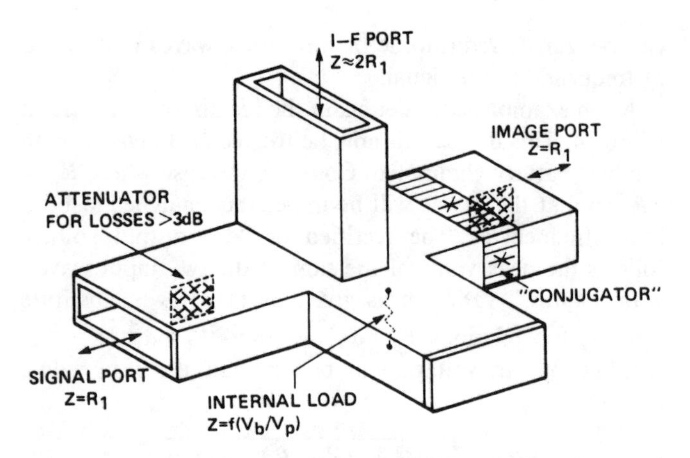

Fig. 5. An equivalent circuit for a double-sideband mixer, shown as a waveguide magic-tee.

E. Application of the Scattering Matrix to an "Image-Rejection" Mixer

One of the known methods of improving the conversion gain of a mixer is to insert a narrow bandstop filter into the input line, tuned to reflect the image frequency only. This "image-rejection" method has received considerable attention in recent years. It has been commonly assumed that the improvement results from an avoidance of signal power loss at the image frequency, the image wave being reflected back into the mixer for further conversion to the IF. Although it has been shown that little or no image signal is generated in a double-sideband mixer which is fully impedance-matched, some improvement in conversion gain can be obtained in this way if the circuit is reoptimized by modifying the external IF and RF impedances, or by changing the LO impedance which is a sensitive function of the dc bias applied.

This theory applies when an image filter is added to a double-sideband mixer whose scattering parameters are known. It is assumed that the IF load impedance can be assigned any desired value different from R_0, the three-frequency optimum. Looking *outward* into the microwave line, at ω_{-1}, the image reflection coefficient Γ_i is assumed to have a magnitude of 1.0, but may have any desired phase.

Looking outward toward the IF load, the reflection coefficient Γ_{if} will have a magnitude less than 1.0, and may have any desired phase angle. (For a real IF impedance R_{if} and Γ_{if} will be real.) For an arbitrary complex IF impedance Z_{if}, Γ_{if} is given by

$$\Gamma_{if} = \frac{Z_{if} - R_0}{Z_{if} + R_0}. \tag{26}$$

Because of these reflections, the mixer will see input waves at all three frequencies, given by

$$\begin{aligned} \alpha_1 &= \text{applied value} \\ \alpha_{-1} &= \Gamma_i \beta_{-1} \\ \alpha_0 &= \Gamma_{if} \beta_0. \end{aligned} \tag{27}$$

Substituting these values into the three equations which are represented by the scattering matrix of (5), we obtain three modified equations

$$\begin{aligned} S_{-11}\alpha_1 &= \beta^*_{-1}(1 - \Gamma_i^* S_{-1-1}) + \beta_0(-S_{-10}\Gamma_{if}) \\ S_{01}\alpha_1 &= \beta^*_{-1}(-\Gamma_i^* S_{0-1}) + \beta_0(1 - \Gamma_{if} S_{00}) \\ S_{11}\alpha_1 &= \beta^*_{-1}(-\Gamma_i^* S_{1-1}) + \beta_0(-\Gamma_{if} S_{10}). \end{aligned} \tag{28}$$

This set may be solved for β_0, β_1, and β_{-1} by ordinary algebra. Each result will be proportional to α_1, the input wave. To simplify the problem, it is assumed that the scattering matrix is similar to (21-B), but more lossy, such that the image conversion is zero, and all inputs are impedance-matched to the reference line impedances:

	α^*_{-1}	α_0	α_1
β^*_{-1}	0	S_u^*	0
β_0	S_d	0	S_d
β_1	0	S_u	0

(29)

Typically, S_d will be real. S_u will be complex with a phase angle which depends upon the placement of the reference plane in the microwave line. Typically, also, S_u and S_d will have the same magnitudes, with a maximum possible value of $\sqrt{2}/2$. Substituting the values of (29) into (28) and solving for β_0 and β_1 gives the results

$$\begin{aligned} \beta_0 &= \alpha_1 \frac{S_d}{1 - S_u^* S_d \Gamma_i^* \Gamma_{if}} \\ \beta_1 &= \alpha_1 \frac{S_u S_d \Gamma_{if}}{1 - S_u^* S_d \Gamma_i^* \Gamma_{if}}. \end{aligned} \tag{30}$$

The net power delivered to the IF load is

$$P_{if} = \hat{\beta}_0^2 (1 - \hat{\Gamma}_{if}^2).$$

The *net* input power, subtracting that which is reflected, is

$$P_{sig} = \hat{\alpha}_1^2 \left(1 - \frac{\hat{\beta}_1^2}{\hat{\alpha}_1^2} \right). \tag{31}$$

The operating power gain, if the input is left in a mismatched condition, is

$$G_{do} = \frac{\hat{S}_s^2 (1 - \hat{\Gamma}_{if}^2)}{|1 - S_u^* S_d \Gamma_i^* \Gamma_{if}|^2} \tag{32}$$

and the *net* power gain, if the signal input is impedance-matched with a narrow-band transformer[4] is

$$G_{dm} = \frac{\hat{S}_d^2 (1 - \hat{\Gamma}_{if}^2)}{|1 - S_u^* S_d \Gamma_i^* \Gamma_{if}|^2 - |S_u S_d \Gamma_{if}|^2}. \tag{33}$$

It is evident that (32) and (33) will both be maximized if the phase angle of Γ_i has a value such that the product $S_u^* S_d \Gamma_i^* \Gamma_{if}$ is real and positive. If Γ_{if} is restricted to be a real quantity, it may be positive or negative, depending upon whether R_{if} is greater or less than R_0, respectively. By choosing the proper position of the filter at a distance from the mixer, Γ_i can be given any desired phase angle. One position will be correct if $R_{if} > R_0$ and another will be

[4]Note that the use of an impedance transformer which affects other frequencies will modify the effective scattering coefficients in ways which are difficult to predict.

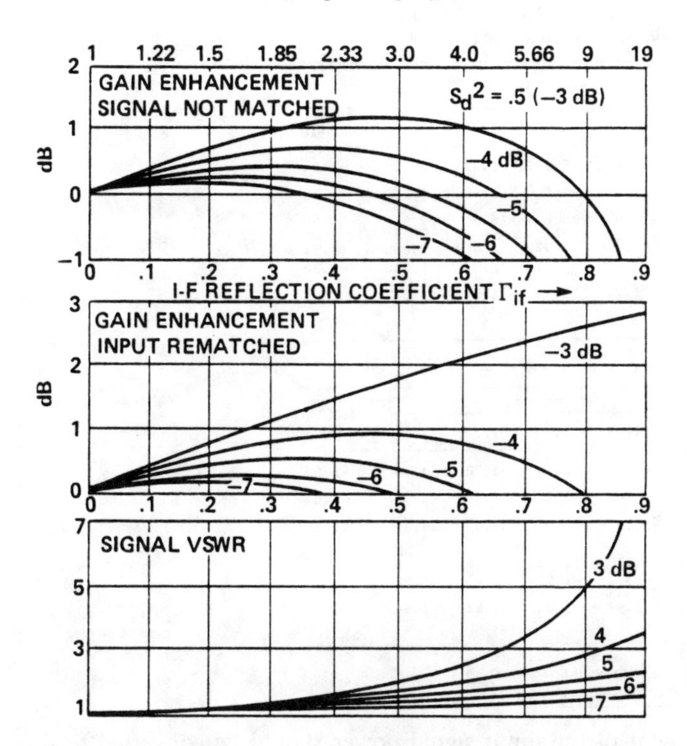

Fig. 6. Effects of adding an image-rejection filter for mixers with double-sideband conversion losses in the range 3–7 dB. It is assumed that the original S-matrix has the form of equation (29). S_d^2 and R_0 apply to the unmodified mixer. Γ_{if} applies to the IF load, and is to be taken with respect to R_0 as the reference impedance. The VSWR values are those which must be nullified by RF transformation to achieve the "input-rematched" condition.

correct if $R_{if} < R_0$. For either case, the maximum gains are

$$(G_{do})_{max} = \frac{\hat{S}_d^2(1-\hat{\Gamma}_{if}^2)}{(1-\hat{S}_u\hat{S}_d\hat{\Gamma}_{if})^2}$$

$$(G_{dm})_{max} = \frac{\hat{S}_d^2(1-\Gamma_{if}^2)}{1-2\hat{S}_u\hat{S}_d\hat{\Gamma}_{if}} \tag{34}$$

where the ($\hat{}$) indicates the magnitude of the complex quantity.

Fig. 6 shows plots of the available gain enhancement by this method, plotted as a function of Γ_{if}. At the top, corresponding values are shown for VSWR, R_{if}/R_0 or R_0/R_{if}. In this case it is assumed that the coefficients S_d and S_u have equal magnitudes. The different curves apply for different values of conversion loss for the mixer when optimized without the image filter. Fig. 6 also shows the effect on the input impedance of the variation in Γ_{if}.

On Fig. 6, the 3-dB loss curves apply to the ideal-diode tuned mixer which is discussed in earlier parts of this section and is analyzed in the Appendix. As in earlier work, it is predicted that the loss can be made to approach 0 dB as an ultimate limit, but it is shown here that this requires that the two signal impedances be much greater or much less than the impedance presented to the LO wave,[5] depending upon whether the image is open-circuited or short-circuited. In cases where the double-sideband losses exceed 3 dB, we see that limited improvements are available by modifying these impedances by finite amounts.

[5]In practice, the impedance presented to the LO wave can be easily changed by changing the dc bias level. If large changes are needed, this may require a substantial increase in the available LO power.

In Part II of the Appendix, the *ideal diode* case is treated also by the frequency-domain approach for both double-sideband and image-rejection mixers. The results are in agreement.

III. Comparisons with Frequency-Domain Analysis

The classical frequency-domain approach to mixer analysis was presented by Peterson and Llewellyn in 1945 [5]. Their approach has usually been followed in various texts such as [4],[6] and in innumerable papers in the interim. Little purpose would be served by reviewing it here. Some discussion, however, is in order. In that approach, a diode or other nonlinear resistor, which is driven by a strong local oscillator wave, is modeled as a periodically time-variable, but *linear*, conductance. This model may be thought of as having the behavior of a motor-driven rheostat. In the low-loss limit it can be modeled as a periodically operated switch. The conductance (or resistance) variation is allowed to assume a highly nonsinusoidal time variation, with the time-function being described by a Fourier time-series, with an infinite number of harmonic components. These form the coefficients of an *infinite-order* matrix which relates the diode currents, at each one of the infinite set of interrelated signal frequencies, to all of the voltages which may exist at these frequencies. These frequencies are $nf_p \pm f_0$ where nf_p is the nth harmonic of the local oscillator and f_0 is the intermediate frequency.

We cannot manipulate an infinite-order matrix. In most analyses, the matrix has been truncated to a low order, 3×3 for a double-sideband mixer or 2×2 for a single-sideband. This is justifiable only if the Fourier coefficients for the conductance (or impedance) are very small for the higher harmonics and/or the circuit impedances are very small (or very large) for the neglected response frequencies. In low-loss practical mixers, such drastic truncation can lead to significant errors. It should be noted that the use of reactive filtering to suppress external responses does not imply that the circuit impedance is either zero or infinite for such frequencies.

In background studies which led to this paper, computer analyses were carried out to simulate the behavior of "practical" mixers which utilized diodes with exponential I–V curves, imbedded in networks which included several tuning elements and parasitic reactances and resistances. In one program, the diode current and voltage waveforms were determined with some precision. Following the time-domain procedure of Section II, this program then determined the coefficients of the three-frequency scattering matrix of (5). In a second program, using the same network and diode, the waveform at the diode junction already determined was entered as a Fourier series to the 20th harmonic. These Fourier components formed the coefficients of a Peterson and Llewellyn [5] frequency-domain Y-matrix which was then inverted. The final results were

TABLE I
COMPARISON OF TIME-DOMAIN AND FREQUENCY-DOMAIN RESULTS FOR THE CIRCUIT OF FIG. 4 USING AN EXPONENTIAL DIODE

	MATRIX ORDER USED WITH FREQUENCY DOMAIN							
	3	5	7	9	11	17	21	TIME DOMAIN
Down-conversion (dB)	-3.5	-3.5	-3.5	-3.5	-3.3	-3.23	-3.25	-3.25
Up-conversion (dB)	-3.5	-3.5	-3.6	-3.6	-3.5	-3.25	-3.28	-3.46
Image conversion (dB)	-9.2	-9.8	-12.0	-18.8	-45.6	-28.0	-21.5	-38.0
RF input impedance/R_1	.412 +j.0	.436 -j.14	.515 -j.277	.700 -j1.40	.929 -j.47	.899 -j.344	.809 -j.338	.93 -j.38
IF input impedance/R_1	1.73	1.75	1.75	1.74	1.75	1.84	1.84	1.81 (opt)

A. Low-loss case with strong L.O. drive

	3	5	7	9	11	17	21	TIME DOMAIN
Down-conversion (dB)	-3.52	-3.56	-3.58	-3.58	-3.59	-3.58	-3.58	-3.59
Up-conversion (dB)	-3.52	-3.55	-3.58	-3.61	-3.63	-3.64	-3.64	-3.65
Image conversion (dB)	-28.5	-29.1	-30.0	-30.9	-30.8	-29.4	-29.3	-29.1
RF input impedance/R_1	1.06 -j.12	1.06 -j.12	1.08 -j.19	1.10 -j.24	1.12 -j.27	1.16 -j.28	1.16 -j.28	1.16 -j.29
IF input impedance/R_1	1.94	1.94	1.94	1.94	1.94	1.94	1.94	1.94

B. Moderately lossy case with weaker L.O. drive

presented in the scattering matrix format of (5). In this way, the time-domain and frequency-domain approaches could be accurately compared. It was found in all cases that the two scattering matrixes were in close agreement when a high-order matrix was used in the frequency domain case, but significant errors appeared when a truncated matrix was used when analyzing low-loss tuned mixers.

One set of results are presented here for the simple tuned mixer network of Fig. 4, in which an exponential diode was used instead of an "ideal" one. The normalized parameters used in the program corresponded to the following network conditions: $C_1=3.18$ pF; $L_1=0.0795$ nH; $C_2 \gg C_1$; $f_p=10.0$ GHz; $f_0 \ll f_p/\omega C_1 R_1$; I diode$=2\times10^{-15}\exp(40.0\ V)$; $P_{LO}=8.8$ mW or 1.13 W; $Q_L=\omega_p C_1 R_1=10$; diode conduction angle$\simeq 80°$; $V_{bias}=-2$ V or -3.4 V; $R_1=50\ \Omega$; $R_0=90.5\ \Omega$; and circuit impedance seen at the diode at the second harmonic$=0.124-j2.5\ \Omega$.

Two different LO drive levels were simulated, one at 8.8 mW and one at 1.13 W. Although relatively strong compared with common practice, the low-drive case is within the range of practical circuitry. The strongly driven case represents an attempt to approach switch-like behavior with the diode alternating between very high and very low conductance during each LO cycle. Table I shows the comparison of the results obtained. The time-domain analysis is shown in the right-hand column. Various orders of the frequency conversion matrix were used for the other results shown. We see close agreement between the time-domain and frequency-domain results when the highest order matrix was used. However, when the matrix was truncated to 3×3, significant errors appeared, particularly in the strongly drive lower-loss case. The most significant errors are in the predictions for image conversion and input impedance. These 3×3 results indicate that 12 percent of the input signal power should be converted to the image and 17 percent should be reflected at the signal frequency for these conditions. However, the correct result is that image conversion and signal reflections are quite insignificant for the chosen operating point.

The final results of Table I are similar to those of the matrix (21-B). Exact agreement is not to be expected because 1) the diode is still slightly lossy even though very strongly driven, and 2) the loaded Q is only 10 so that weak frequency conversions will be found at $2f_p \pm f_0$, $3f_p \pm f_0$, etc. As expected from the theory, more than 50 percent of an input signal at ω, or ω_{-1} is *lost* in that it is neither reflected nor converted into other modulation products.

In the Appendix, the simple high-Q tuned mixer of Fig. 4 is treated for the case of an "ideal diode". Both the time-domain and frequency-domain methods were used, giving identical results. The image-rejection mixer is also analyzed there in the frequency domain, giving results which agree with those of Section II. The lossy S-matrix of (21-B) was obtained by both methods for the case in which the impedance presented to the LO wave is equal to that of the RF signal source.

IV. Discussion of Loss Mechanisms in Ideal-Diode Mixers

In Section II-C, it was shown that the circuit of Fig. 4, in the ideal low-loss limit, acts as a peak detector with an overall rectification and conversion efficiency approaching 100 percent. It was shown, in one case, that the application of a weak signal wave to a strong LO wave results in 50-percent conversion to IF and 50-percent conversion to dc.

However, in the frequency-domain analysis in the Appendix, where the diode was modeled as a periodically operated switch, there is no mechanism for signal energy conversion to dc or to harmonics of the LO. The two *models* are conceptually different.

Imagine the circuit of Fig. 4 with the LO source eliminated, and with the diode replaced by a *periodically* operated switch with a small residual resistance. This is the model used in the frequency-domain analysis in the Appendix. Whenever the switch closes, there will be a voltage difference between the two capacitors, and a "spike" of current will flow to equalize these voltages. Some energy will be lost each time. The amount lost will depend only upon the voltages and capacitances involved, *and this loss will occur no matter how small the residual resistance is in the switch*. In unpublished work, this author has analyzed this situation in some depth and has concluded that this is the preferred explanation for the signal energy losses *when the mixer is modeled in this way*.[6] The sequence of current spikes, when resolved by Fourier analysis, will be found to consist of the same set of converted frequencies as those which appear in the infinite order Y-matrix of [5]. Thus, in the frequency-domain analysis of this model in the Appendix, the lost energy is absorbed after being converted into a large number of the higher order conversion products which have been omitted from the matrix. When the diode resistance is assumed to be vanishingly small, the power dissipated at each frequency is also vanishingly small but the number of such products is indefinitely large, resulting in a finite total loss.

It should be pointed out, however, that an "ideal diode" changes states only when its voltage is zero. Thus, the "spikes" of current described above cannot occur. In a *diode* mixer, when a signal is added to the LO wave, the *times* of "switching" of the ideal diode are variable, to first order, so that the switching action is not a truly periodic time function. This allows other energy conversion effects to appear.

Kerr and Taur [10] have pointed out that if second-order terms are included in the Taylor expansion of the diode conductance function in a more complex small-signal frequency-domain analysis, it will be found that signal energy will be converted to dc and/or to various harmonics of the LO, in a manner analogous to that found in Section II-C where the "ideal mixer" was analyzed as a detector.

The paradox has been resolved in that the limiting losses found in the time-domain analysis are also found in an equivalent frequency domain analysis, and in answer to the question "Where does the lost energy go?" different answers are obtained depending upon the model and method of analysis. For a realistic diode mixer, a second-order small signal theory is necessary to answer that question properly, but that is beyond the scope of this paper.

V. Conclusions

A new method of mixer characterization and analysis has been presented which is based upon a time-domain approach. The method is particularly useful in evaluating experimental models of double-sideband devices. Mixer performance can be predicted from a set of single-frequency measurements under LO drive at various bias levels.

Comparisons have been made with the classical frequency-domain method of analysis and in all cases, the results of the two approaches have been found to agree. When the time-domain method was applied analytically to some idealized low-loss mixer circuits, some surprising results were obtained that had not been evident in previous publications based upon frequency-domain analysis. However, a new frequency-domain analysis for the same conditions has been found to provide the same results.

Analyses have been presented of some particular single-diode mixer circuits in which lossless circuitry was assumed and the diode was idealized to have, or to approach, ideal switch-like behavior under LO drive. The author was led to present the following somewhat generalized conjectures concerning the ultimate low-loss limits in such cases.

1) In single-diode mixers which use reactive filtering to suppress unwanted responses, internal signal power losses occur in the general case, even with ideally lossless diodes and circuitry.
2) In double-sideband single-diode mixers in which all responses are reactively suppressed except for the signal, IF, and image, the minimum signal loss in any one frequency conversion is 50 percent. If all impedances are matched, *including that presented to the LO*, conversion between the signal and image frequencies is normally suppressed, and the unconverted signal energy is absorbed. The scattering matrix of such a mixer is analogous to that of a hybrid junction with its fourth port internally loaded. The normalized impedance of this internal load depends upon the impedance which the diode presents to the LO, and this can be changed by varying the dc bias level, or by other means.
3) Lower losses are achievable in two-frequency single-diode mixers in which all other responses are reactively suppressed. To approach the zero-loss ultimate limit, the impedance presented to the LO wave must be much greater than that of the two signal waves or much less. This ratio is most easily changed by variation of the dc bias applied, but other means are possible.

Appendix
Scattering Matrix Coefficients for a Tuned Ideal-Diode Mixer

I. Time-Domain Approach

In this section, the coefficients of the three-frequency scattering matrix are derived for the circuit of Fig. 4, as presented in (21). It is assumed that the mixer network is lossless, and the diode is an "ideal" one with zero forward and infinite reverse impedance. The resonator L_1C_1 is assumed to have a *large* loaded Q, that is $\omega_p C_1 R_1 \gg 1$. The capacitor C_2 is also large, comparable to C_1. In the steady-state under LO drive, at the resonant frequency of L_1C_1, the diode will conduct only during short intervals at the

[6] Periodically switched capacitors are now being used as substitutes for resistors in some types of monolithic integrated circuits [7].

RF voltage peaks. The RF voltage wave will be very nearly sinusoidal, and the voltage at R_0 will be very nearly constant. Also, it is evident that the peak RF voltage at the input V_{in} will be very nearly equal to V_b. The current drawn by the diode assumes the character of a sequence of δ-functions. This current, expressed as a Fourier series is

$$i_d = i_0 + \sum_{n=1}^{\infty} i_n \cos(n\omega_p t). \tag{A-1}$$

For the lower harmonics, it is easily shown that

$$i_n \simeq 2i_0. \tag{A-2}$$

Here a_0 is the dc or average current. The coefficient a_1 is the component at the LO frequency and its peak value is nearly twice the dc rectified current. The resonator is tuned at ω_p and draws no current, therefore, the peak current at ω_p from the generator is also $i_1 \simeq 2i_0$.

The following equation specifies the RF voltage in terms of the RF impedance R_p, seen at the LO frequency:

$$\frac{V_{in}}{V_p} = \frac{V_b}{V_p} = \frac{R_p}{R_p + R_1}. \tag{A-3}$$

The reflection coefficient Γ is specified by

$$\Gamma = \frac{R_p - R_1}{R_p + R_1}. \tag{A-4}$$

Combining these equations gives

$$\Gamma = 2\frac{V_b}{V_p} - 1. \tag{A-5}$$

Using the relation

$$\alpha_p = \frac{V_p}{\sqrt{8R_1}}. \tag{A-6}$$

We obtain the function $\Gamma_a(\alpha, V_b)$ applicable to (4)

$$\Gamma_a(\alpha, V_b) = \frac{V_b}{\alpha\sqrt{2R_1}} - 1. \tag{A-7}$$

Noting that the peak value of the LO wave current is given by

$$i_1 = 2i_0 = \frac{\alpha}{\sqrt{\frac{R_1}{2}}}(1 - \Gamma) \tag{A-8}$$

the function $I_r(\alpha, V_b)$ applicable to (3) to

$$I_r(\alpha, V_b) = \frac{V_b}{2R_1} - \frac{\alpha}{\sqrt{R_1/2}}. \tag{A-9}$$

The four partial derivatives applicable to (3) and (4) are

$$\frac{\partial \Gamma_a}{\partial \alpha} = -\frac{V_b}{\alpha^2\sqrt{2R_1}}$$

$$\frac{\partial \Gamma_a}{\partial V_b} = \frac{1}{\alpha\sqrt{2R_1}}$$

$$\frac{\partial I_r}{\partial \alpha} = \frac{-1}{\sqrt{\frac{R_1}{2}}}$$

$$\frac{\partial I_r}{\partial V_b} = \frac{1}{2R_1}. \tag{A-10}$$

Substituting (A-5), (A-6), and (A-10) into (8), (18), (19), and (20) yields the nine S coefficients

$$S_{11} = S_{-1-1} = \frac{V_b}{V_p} - \frac{2R_1}{2R_1 + R_0}$$

$$S_{00} = \frac{2R_1 - R_0}{2R_1 + R_0}$$

$$S_{01} = S_{1-0} = S_{0-1} = S_{-10} = \frac{2\sqrt{R_0 R_1}}{2R_1 + R_0}$$

$$S_{1-1} = S_{-11} = \frac{R_0}{2R_1 + R_0} - \frac{V_b}{V_p}. \tag{A-11}$$

II. Frequency-Domain Approach

A. Three-Frequency Case[7]

In analyzing a single-diode mixer by the frequency-domain method of [5], the diode is modeled as a periodically operated switch with a small series resistance R_d. It is assumed to be in the conduction state during a small fraction of each LO cycle between the phase angles $-\theta$ and θ, in an even function of time. The conductance is expressed in a Fourier time series as

$$G(t) = \sum_{n=-\infty}^{\infty} g_n e^{jn\omega_p t}$$

where

$$g_0 = \frac{\theta}{\pi R_d}$$

$$g_{-1} = g_1 = \frac{\sin\theta}{\pi R_d}$$

$$g_{-2} = g_2 = \frac{\sin\theta\cos\theta}{\pi R_d}. \tag{A-12}$$

If the responses at other frequencies are to be suppressed, the loaded Q of the resonator must be large ($Q_L = \omega_p C_1 R_1$). If the resistance R_d is finite, and it is assumed that the LO wave is sinusoidal, then the current during the conduction interval will be given by

$$i(t) = \frac{V_{in}\cos\omega_p t - V_b}{R_d} \tag{A-13}$$

where V_{in} is the peak voltage of the LO wave, and V_b is the bias voltage, equal to $V_{in}\cos\theta$. Integrating over the conduction period and averaging over a full cycle, the dc current will be

$$i_{dc} = \frac{V_{in}}{\pi R_d}(\sin\theta - \theta\cos\theta). \tag{A-14}$$

The diode current wave will have the form of a sequence of short truncated cosine wave sections. Expressing this as

[7]This method of analysis was suggested to the author in private correspondence from two independent sources, A. A. M. Saleh of the Bell Telephone Laboratories, and A. R. Kerr and Y. Taur of the Goddard Institute for Space Studies.

a Fourier series, as in (A-1) and (A-2), the fundamental component reduces to

$$i_1 \simeq \frac{2V_{\text{in}}\theta^3}{3\pi R_d}\cos\omega_p t \tag{A-15}$$

when higher order terms in θ are ignored.

The impedance at the LO frequency, presented by the diode, is designated as R_p, equal to V_{in}/i_1. In the general case, the LO need not be impedance matched to the input line. Substituting into (A-14) we obtain the relation

$$\theta^3 = \frac{3\pi R_d}{2R_p}. \tag{A-15}$$

It follows that we can express g_0, g_1, and g_2 as functions of θ and R_p only

$$g_0 = \frac{3}{2\theta^3 R_p}\theta$$

$$g_1 = \frac{3}{2\theta^3 R_p}\sin\theta$$

$$g_2 = \frac{3}{2\theta^3 R_p}\sin\theta\cos\theta. \tag{A-16}$$

Because of the high-Q assumption, the circuit impedances can be neglected for the higher order conversion products, allowing matrix truncation to 3×3. Including the circuit admittances Y_1, Y_{-1}, and Y_0, the complete Y-matrix becomes

$$\bar{Y} = \begin{vmatrix} Y_{-1}+g_0 & g_1 & g_2 \\ g_1 & Y_0+g_0 & g_1 \\ g_2 & g_1 & Y_1+g_0 \end{vmatrix}. \tag{A-17}$$

When inverted, we obtain an impedance matrix

$$\bar{Z} = \begin{vmatrix} Z_{-1-1} & Z_{-10} & Z_{-11} \\ Z_{0-1} & Z_0 & Z_{01} \\ Z_{1-1} & Z_{10} & Z_{11} \end{vmatrix}. \tag{A-18}$$

This Z-matrix can be directly converted to a scattering matrix. For the diagonal terms

$$S_{nn} = \frac{2Z_{nn}}{R_n} - 1 \tag{A-19}$$

and for the off-diagonal terms

$$S_{nm} = \frac{2Z_{nm}}{\sqrt{R_n R_m}} \tag{A-20}$$

where the R_n terms represent the various source impedances.

In the Y-matrix the sinusoidal functions were entered as power series in θ. It was assumed that $Y_{-1} = Y_1 \neq Y_0$ and that each was assumed to be real, with impedances R_1, R_1, and R_0. After some tedious algebra, the Z-matrix coefficients were obtained, each having the form of a ratio of two power series in θ. Each series began with a constant term, which was a function of R_p, R_1, and R_0. In the low-loss limit, all the θ^n terms were assumed to approach zero as R_d and θ approach zero.

The Z-coefficients reduced to the following expressions:

$$Z_{-1-1} = Z_{11} = R_p \frac{1+(2R_p/R_1)+(2R_p/R_0)}{(4R_p/R_0)+(2R_p/R_1)+(4R_p^2/R_1R_0)+(2R_p^2/R_1^2)}$$

$$Z_{-11} = Z_{1-1} = R_p \frac{-2(R_p/R_0)+1}{(4R_p/R_0)+(2R_p/R_1)+(4R_p^2/R_1R_0)+(2R_p^2/R_1^2)}$$

$$Z_{00} = Z_{10} = Z_{01} = Z_{0-1} = Z_{-10} = \frac{R_0R_1}{2R_1+R_0}. \tag{A-21}$$

The scattering coefficients become

$$S_{10} = S_{01} = S_{-10} = S_{0-1} = \frac{2\sqrt{R_0R_1}}{2R_1+R_0}$$

$$S_{1-1} = S_{-11} = \frac{R_0}{2R_1+R_0} - \frac{R_p}{R_1+R_p}$$

$$S_{00} = \frac{2R_1-R_0}{2R_1+R_0}$$

$$S_{11} = S_{-1-1} = \frac{R_p}{R_1+R_p} - \frac{2R_1}{R_0+2R_1}. \tag{A-22}$$

Using (A-3), it will be found that all of the S-coefficients agree with those of (A-11).

B. Two-Frequency Image-Rejection Case

If the image impedance is zero at the diode, the Y-matrix of (A-18) includes only the signal and IF frequencies,

$$\bar{Y} = \begin{vmatrix} \frac{1}{R_0}+g_0 & g_1 \\ g_1 & \frac{1}{R_1}+g_0 \end{vmatrix}. \tag{A-23}$$

Following the same procedure, the Z-matrix obtained by inversion has the coefficients

$$Z_{00} = Z_{11} = Z_{10} = Z_{01} = \frac{2R_p}{1+2(R_p/R_0)+2(R_p/R_1)} \tag{A-24}$$

and the scattering coefficients are

$$S_{00} = \frac{2(R_p/R_0)-2(R_p/R_1)-1}{2(R_p/R_0)+2(R_p/R_1)+1} \tag{A-25}$$

$$S_{01} = S_{10} = \frac{4R_p}{\sqrt{R_0R_1}\,(2(R_p/R_0)+2(R_p/R_1)+1)} \tag{A-26}$$

$$S_{11}=\frac{2(R_p/R_1)-2(R_p/R_0)-1}{2(R_p/R_0)+2(R_p/R_1)+1}. \tag{A-27}$$

Consider two special cases.

1) Let $R_0=2R_1$, the optimum condition for double-sideband operation. Then

if $R_p \neq R_1$:

$$S_{00}=-\frac{R_p+R_1}{3R_p+R_1}$$

$$S_{01}=S_{10}=\frac{2\sqrt{2}\,R_p}{3R_p+R_1}$$

$$S_{11}=\frac{R_p-R_1}{3R_p+R_1}$$

if $R_p=R_1$:

$$S_{00}=-0.5$$

$$S_{01}=S_{10}=0.707$$

$$S_{11}=0. \tag{A-28}$$

If $R_p=R_1$, we see that the input impedance is matched for ω_1 but not for ω_0, and the conversion loss is 3 dB as before. This is to be expected inasmuch as no image current will flow for an input at ω_1, but it will flow for an input at ω_0. Shorting the image in a matched double-sideband mixer as in (21-B) will not affect the input impedance at ω_1, nor will it affect the conversion loss. However, the input impedance at ω_0 will be reduced.

2) Let $R_0=R_1$. Then

if $R_p \neq R_0$:

$$S_{00}=S_{11}=\frac{-1}{1+(4R_p/R_0)}$$

$$S_{10}=S_{01}=\frac{1}{1+(R_0/4R_p)}$$

if $R_p=R_0$:

$$S_{00}=S_{11}=-0.2$$

(4-percent reflection)

$$S_{10}=S_{01}=0.8$$

$(\simeq 2\text{-dB loss})$. (A-29)

It is evident that the conversion loss can approach 0 dB and the impedances approach a matched condition, but only if $R_0 \ll 4R_p$. Further analysis will show that it is possible to impedance-match at either ω_0 or ω_1 but not at both frequencies simultaneously for finite values of R_p, R_1, and R_0.

Acknowledgment

The author is indebted to Dr. A. A. M. Saleh, Dr. M. Gupta, Dr. A. R. Kerr, Dr. Y. Taur, and those anonymous reviewers who took the time and trouble to read this paper carefully, and to provide constructive criticisms, helpful suggestions, and even alternative methods of analysis which have been utilized in this final version.

References

[1] M. E. Hines, "Image conversion effects in diode mixers," in *1977 IEEE MTT-5 Int. Symp. Dig.*, (San Diego, CA), pp. 487–490.

[2] M. E. Hines, "Failure of the classical circuit model in the analysis of low-loss band-limiter mixers," in *1978 IEEE MTT-5 Int. Microwave Symp. Dig.*, (Ottawa, Canada), pp. 402–404.

[3] M. E. Hines, "Failure of the mathematical model in the analysis of band-limited mixers,"in *Proc. 8th European Microwave Conf.*, 1978, pp. 691–695.

[4] C. T. Torrey and C. A. Whitmer, *Crystal Rectifiers* (M.I.T. Rad. Lab. Series VI4). New York: McGraw Hill, 1948.

[5] L. C. Peterson and F. B. Llewellyn, "The performance and measurement of mixers in terms of linear-network theory," *Proc. IRE*, vol. 33, pp. 458–476, July 1945.

[6] A. A. M. Saleh, *Theory of Resistive Mixers* (Res. Monograph #64). Cambridge, MA: M.I.T. Press, 1971.

[7] B. J. Hostica, R. W. Broderson, and P. R. Gray, "MOS sampled-data recursive filters using switched-capacitor integrators," *IEEE J. Solid-State Circuits*, vol. SC-12, pp. 600–608, Dec. 1977. (Also, see other papers in that issue and in vol. SC-14, Dec. 1979.)

[8] C. W. Gerst, Jr., "New mixer designs boost D/F performance," *Microwaves*, vol. 12, Oct. 1973, p. 60.

[9] A. J. Kelly, "Fundamental limitations on the conversion loss of double-sideband resistive mixers," *IEEE Trans. Microwave Theory Tech.*, vol. MTT-25, pp. 867–869, Nov. 1977.

[10] A. R. Kerr and Y. Taur, Goddard Institute for Space Studies, New York, NY, private communication.

[11] A. A. M. Saleh, Bell Laboratories, Inc., Crawford Hill Laboratory, Holmdel, NJ, private communication.

Part III
Practical Implementations of Image-Enhanced Mixers

THE choice of embedding impedance of the mixer is essential for the mixer performance. Besides optimum input and output impedances, it is worthwhile to reactively terminate the image port and the harmonic sidebands ($nf_{LO} \pm f_{IF}$) to prevent power loss at those frequencies. By proper reactances, in particular at the image frequency, further enhanced performance can be achieved. In this part we will describe ways of making image-enhanced single-ended and balanced mixers.

In his often quoted paper, the first one of this part, Johnson describes an elegant approach for designing an integrated circuit balanced mixer with a reactively terminated image. Experimentally, he clearly shows that there is an optimum image reactance for the image termination that yields a minimum in the noise figure of the mixer including the IF amplifier. The paper by Katoh and Akaiwa presents a similar discussion for a single diode mixer, while Dickens and Maki introduce a circuit for sum enhancement (the sum frequency is the LO frequency plus the signal frequency) in a two-diode balanced mixer. In all of these first three papers short theoretical discussions are included.

A somewhat more elaborate discussion is presented in the fourth paper, by Burkley and O'Brien. They study an 11 GHz balanced image recovery mixer in which the relationship between complex terminations at the signal, image, and IF ports and the image power generated are considered.

In the next three papers wide-band image rejection mixers are described, based on a pair of balanced mixers and using appropriate phasing techniques for coupling them together. The three papers describe various techniques for realizing such mixers. The last of these papers, by Oxley, gives a short discussion concerning advantages and disadvantages with different coupling circuits.

X-Band Integrated Circuit Mixer with Reactively Terminated Image

KENNETH M. JOHNSON, MEMBER, IEEE

Abstract—An X-band mixer using GaAs Schottky barrier diodes with a thin-film 500-MHz IF preamplifier was developed using hybrid microwave integrated circuit techniques. The balanced mixer had filters to provide a short circuit at the image frequency.

The entire mixer preamplifier occupied an area of only 0.38 square inches and had a noise figure of 6.7 dB which corresponded quite closely to the theoretical noise figure considering all losses. The thin-film IF amplifier alone had a 2.2-dB noise figure and the mixer IF amplifier coupling network had a loss of 0.4 dB.

I. Introduction

THE DEVELOPMENT of the "true" Schottky barrier diode along with microwave integrated circuit techniques makes possible miniature high-performance microwave detectors. This paper describes a microwave integrated circuit mixer which used Schottky barrier diodes and reactive image termination with a 500-MHz thin-film IF amplifier to achieve low-noise figure. The entire mixer–IF amplifier occupies an area of only 0.38 square inch and has a noise figure of 6.7 dB.

The basic theory of mixer operation has been known for a number of years.[1]–[3] A rather complete analysis is given by Torrey and Whitmer[1] including various conditions of image termination. They showed that a lower conversion loss is obtained if the image is reactively terminated in an open or short circuit rather than being terminated in the signal port resistance. Strum[3] extended the mixer theory by providing a method for calculating the conversion conductances of the diode. Van Der Ziel,[4] Guggenbuhl and Strutt,[5] and Nicoll[6] as well as others developed the noise theory for mixer diodes. More recently, since the development of the Schottky barrier diode, papers by a number of authors have appeared extending previous theory to the Schottky barrier diode characteristic. With previous point contact diodes such as the 1N23B, the improvement to be gained by reactively terminating the image was relatively quite small. This was because the point contact diode's series resistance and noise ratio was so large. Schottky barrier diodes recently developed have series resistance an order of magnitude less than the 1N23B and noise ratios of approximately unity.

There are, however, some uncertainties with respect to using the Schottky barrier diode in a circuit. As will be described in the following, the terminations of the local oscillator harmonics and other out-of-band frequencies, especially the sum of the signal and local oscillator frequencies, are quite important in determining the conversion loss of the diode and the type of image termination to use, whether short or open circuit. The effect of the series resistance is also quite important, especially at higher frequencies. The series resistance is the primary loss contributor and also affects the conversion loss mechanism of the mixer.

II. Theory

A. Conversion Loss

The equivalent circuit of a mixer has been developed by a number of authors previously mentioned and therefore will not be given here. They also derived the equations for conversion loss, input conductance, and IF con-

Manuscript received January 25, 1968; revised April 4, 1968. This work was supported by the U. S. Air Force under Contract AF33 (615)-67-1639.

This special issue on Microwave Integrated Circuits is published jointly with IEEE Transactions on Electron Devices, July 1968, and the IEEE Journal of Solid-State Circuits, June 1968.

The author is with Texas Instruments Incorporated, Dallas, Tex. 75222

Reprinted from *IEEE Trans. Microwave Theory Tech.*, vol. MTT-16, pp. 388–397, July 1968.

ductance as a function of the diode's pumped characteristic. That is, the diode is assumed to have the following conductance variation due to the local oscillator voltage:

$$g = g_0 + \sum_{n=1}^{\infty} g_n \cos n\omega t. \tag{1}$$

From this can be defined conductance ratios,

$$\gamma_n = \frac{g_n}{g_0}. \tag{2}$$

The conversion loss, input conductance, and IF conductance for various conditions of image termination are as follows.

Conversion Loss:

Matched image

$$L_2 = 2\frac{1+\sqrt{1-\epsilon_2}}{1-\sqrt{1-\epsilon_2}} \qquad \epsilon_2 = \frac{2\gamma_1^2}{1+\gamma_2}. \tag{3}$$

Short-circuited image

$$L_1 = \frac{1+\sqrt{1-\epsilon_1}}{1-\sqrt{1-\epsilon_1}}, \qquad \epsilon_1 = \gamma_1^2. \tag{4}$$

Open-circuited image

$$L_3 = \frac{1+\sqrt{1-\epsilon_3}}{1-\sqrt{1-\epsilon_3}}, \qquad \epsilon_3 = \frac{\gamma_1^2}{1-\gamma_1^2}\cdot\frac{1-\gamma_2}{1+\gamma_2}. \tag{5}$$

Signal Input "Image" Conductance:

Matched image

$$\frac{g_{a2}}{g_0} = \sqrt{(1+\gamma_2)(1+\gamma_2-2\gamma_1^2)}. \tag{6}$$

Short-circuited image

$$\frac{g_{a1}}{g_0} = \sqrt{1-\gamma_1^2}. \tag{7}$$

Open-circuited image

$$\frac{g_{a3}}{g_0} = \sqrt{1-\gamma_2^2}\sqrt{\frac{(1-\gamma_2)(1+\gamma_2-2\gamma_1^2)}{1-\gamma_1^2}}. \tag{8}$$

IF Conductance:

Matched image

$$\frac{g_{b2}}{g_0} = \sqrt{\frac{1+\gamma_2-2\gamma_1^2}{1+\gamma_2}}. \tag{9}$$

Short-circuited image

$$\frac{g_{b1}}{g_0} = \sqrt{1-\gamma_1^2}. \tag{10}$$

Open-circuited image

$$\frac{g_{b3}}{g_0} = \sqrt{(1-\gamma_2)^2\left(\frac{1+\gamma_2-2\gamma_1^2}{1-\gamma_1^2}\right)}. \tag{11}$$

The preceding equations are valid if the signal generator conductance equals the signal "image" conductance. Usually the IF conductance is made equal to the optimum source conductance for the IF preamplifier. This is, in general, not equal to the IF conductance g_b. There is, therefore, a mismatch "reflected" to the signal port of the mixer due to the IF mismatch for optimum noise figure. This is a fairly small amount for the mixer–IF amplifiers used in this study, but amounts to approximately 0.3-dB loss when matching the local oscillator to the diode.

In order to calculate the conversion loss and matching conditions for the Schottky barrier diode, the conductance (or resistance) coefficients of (1) must be determined. This is essentially what was done by Barber and Ryder.[7] Simply stated, the current through a Schottky diode is

$$i = i_0[\exp(\alpha V) - 1]. \tag{12}$$

Assuming that the applied voltage is of the form $V = V_0 + V_1 \cos \omega t$, the small signal conductance $g = \partial i/\partial V$ is

$$\begin{aligned} g &= \alpha i_0[\exp(\alpha V_0)][\exp(\alpha V_1 \cos \omega t)] \\ &= \alpha i_0 \exp(\alpha V_0)\left[I_0(\alpha V_1) + 2\sum_{n=1}^{\infty} I_n(\alpha V_1) \cos n\omega t\right] \end{aligned} \tag{13}$$

where I_n is the nth order modified Bessel function.

Using the value of $\gamma_n = g_n/g_0$ obtained from (13) and typical value of α and i_0, the conversion loss, signal image conductance, and IF conductance for matched, short- and open-circuited images are plotted in Figs. 1, 2, and 3, respectively. The bias voltage V_0 is assumed equal to zero. Fig. 4 shows the conductance and current as a function of V_1.

These curves show that the open-circuited image provides the best conversion loss, being about 0.5 dB better than the short-circuit case. However, this conclusion arises from an oversimplification, since when it was assumed that the local oscillator voltage is a simple sinusoid, this means that all harmonics of the local oscillator are short-circuited. This will be, in general, not the case especially in an *X*-band mixer for a number of reasons.

First, the presence of diode series resistance loss, line loss, and filter loss results in imperfect lossy terminations. This is particularly true in microstrip where line losses are large. Second, in a balanced mixer, a perfect constant voltage source is not possible since each diode sees a local oscillator signal which has been shaped by the other diode. The microstrip mixer described here had a 90° hybrid to separate signal and local oscillator signals. (This means that the reversing switch model described by Rafuse[8] does not apply since his balanced mixer model requires a 180° hybrid.) Third, usually little or no effort is made to terminate the harmonics, which often cannot be made without introducing more loss than is gained by using the proper termination.

Furthermore, it is necessary to consider the termination of other out-of-band components,[8] especially the sum frequency (the sum of the signal and local oscillator frequencies, $\omega_s+\omega_L$) as well as the diode's series resistance in calculating conversion loss. These factors are quite important in determining the type of image terminator to use, whether short circuit or open circuit. Calculation on the effect of these factors reveal that the open-circuited image is much more strongly affected by series resistance, out-of-band, and harmonic terminations than the short-circuited image.

This can be demonstrated by considering the case in which all the harmonics are open-circuited. This requires a local oscillator which is a constant current source, i.e.,

$$i = I_0 + I_1 \cos \omega t. \tag{14}$$

In this case, the resistance coefficients are computed from

$$v = \frac{1}{\alpha} \ln (i/i_0).$$

Thus,

$$r = \frac{\partial v}{\partial i} = \frac{1}{\alpha i} = \frac{1}{\alpha I_0 \left(1 + \dfrac{I_1}{I_0} \cos \omega t\right)} \tag{15}$$

$$= \frac{1}{\alpha I_0 (1 - q^2)^{1/2}} \left[1 + 2 \sum_{n=1}^{\infty} a_n \cos n\omega t\right] \tag{16}$$

$$a_n = \left(\frac{-q}{1 + \sqrt{1 - q^2}}\right)^n \tag{17}$$

where

$$q = I_1/I_0 \tag{18}$$

and the parameter

$$\gamma_n = a_n. \tag{19}$$

Using these values for γ_n and assuming similar values for α and i_0, the conversion loss, signal "image" conductance, and IF conductance for matched short- and open-circuited images are plotted in Figs. 5, 6, and 7, respectively. Also shown is the zero-order resistance r_0 on the function of the current in Fig. 8.

These curves show that the short-circuited image is considerably better than the open-circuited. Comparing the curves of conversion loss for the constant voltage (Fig. 1) and constant current (Fig. 5), the constant

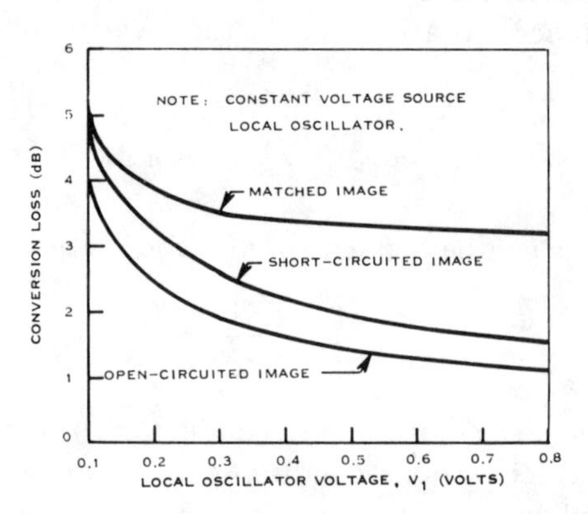

Fig. 1. Conversion loss versus local oscillator voltage for matched, open-, and short-circuited images. Constant voltage source local oscillator.

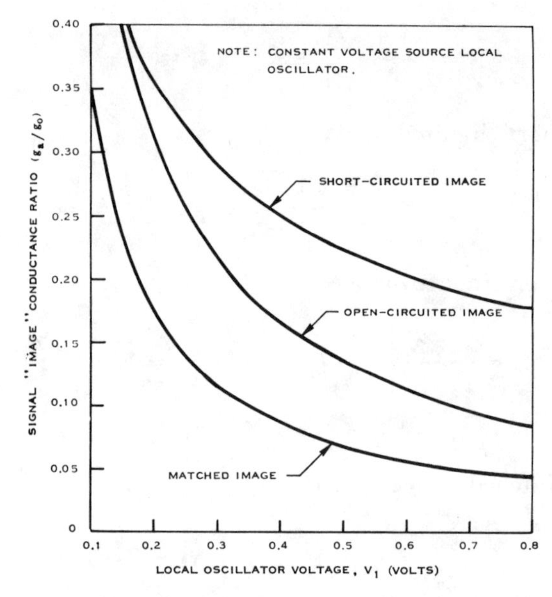

Fig. 2. Signal "image" conductance ratio g_a/g_0 versus local oscillator voltage for matched, open-, and short-circuited images. Constant voltage source local oscillator.

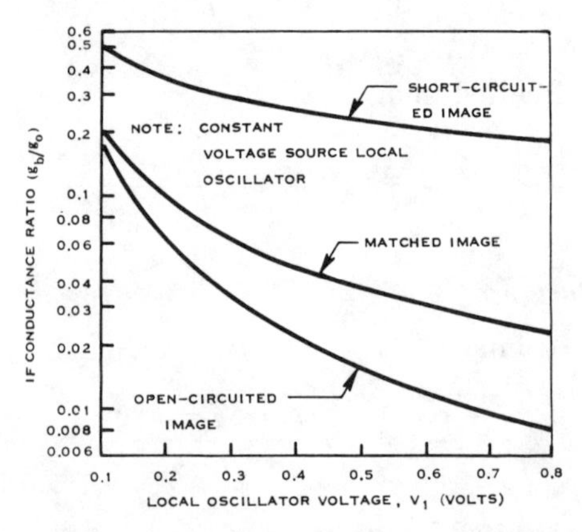

Fig. 3. IF conductance ratio g_b/g_0 versus local oscillator voltage for matched, open-, and short-circuited images. Constant voltage source local oscillator.

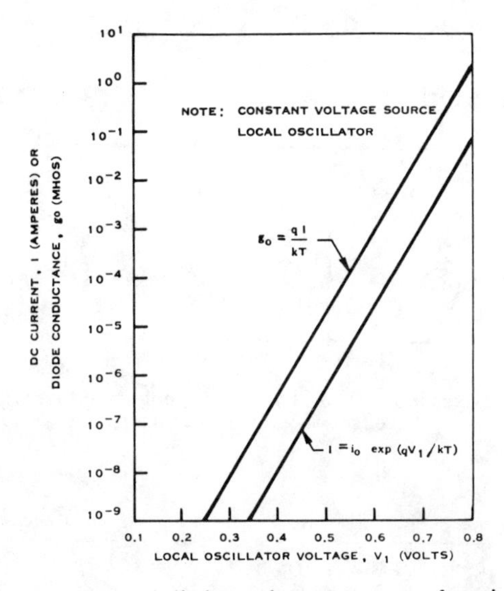

Fig. 4. Dc current and diode conductance g_0 as a function of local oscillator voltage for matched, open-, and short-circuited images. Constant voltage source local oscillator.

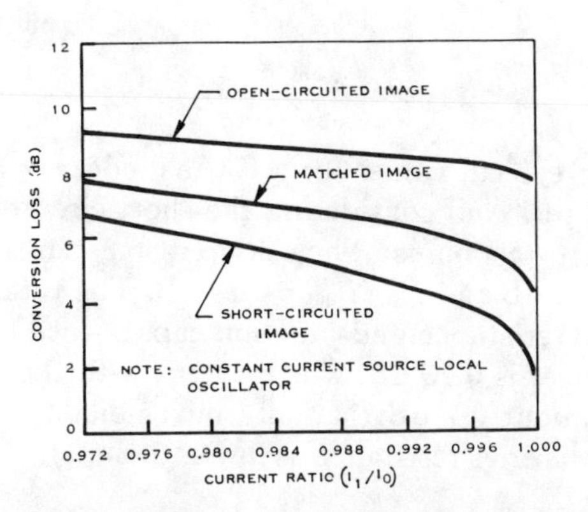

Fig. 5. Conversion loss versus current ratio I_1/I_0 for matched, open-, and short-circuited images. Constant current source local oscillator.

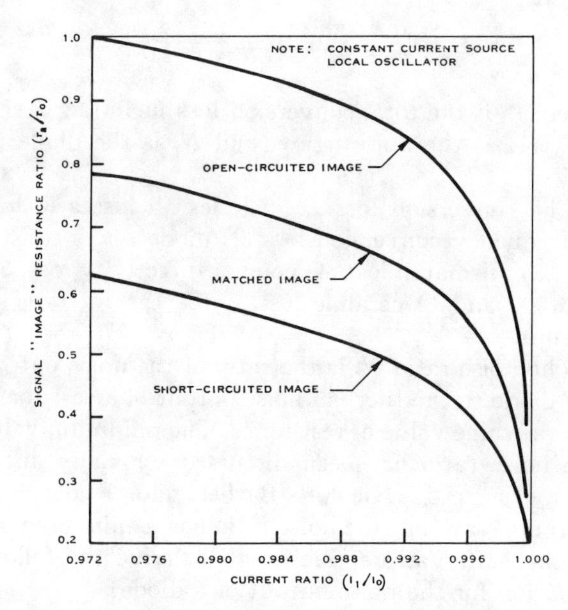

Fig. 6. Signal "image" resistance ratio r_a/r_0 versus current ratio I_1/I_0 for matched, open-, and short-circuited images. Constant current source local oscillator.

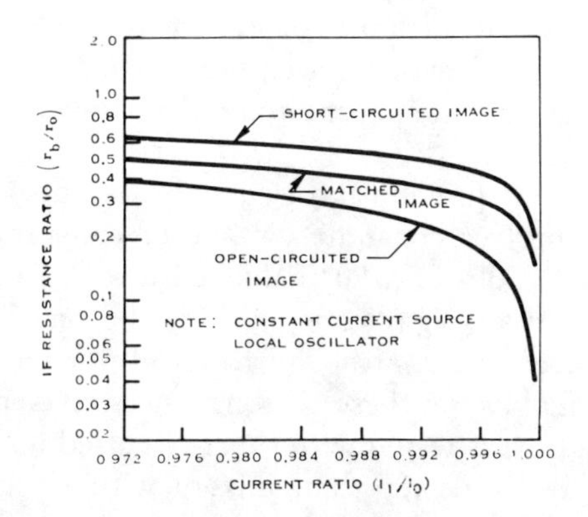

Fig. 7. IF resistance ratio r_b/r_0 versus current ratio I_1/I_0 for matched, open-, and short-circuited images. Constant current source local oscillator.

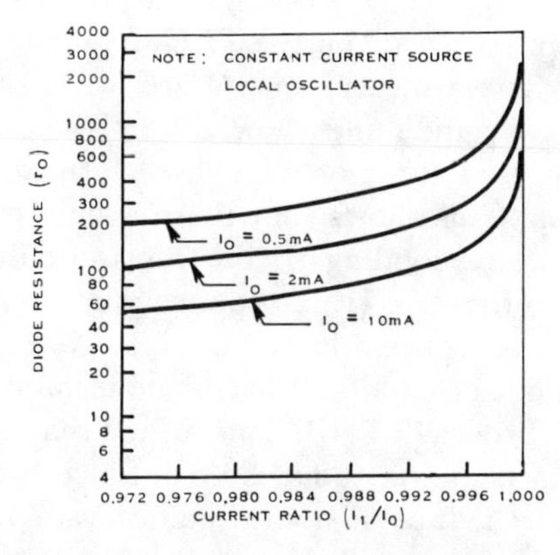

Fig. 8. Diode resistance r_0 for various currents as a function of current ratio I_1/I_0. Constant current source local oscillator.

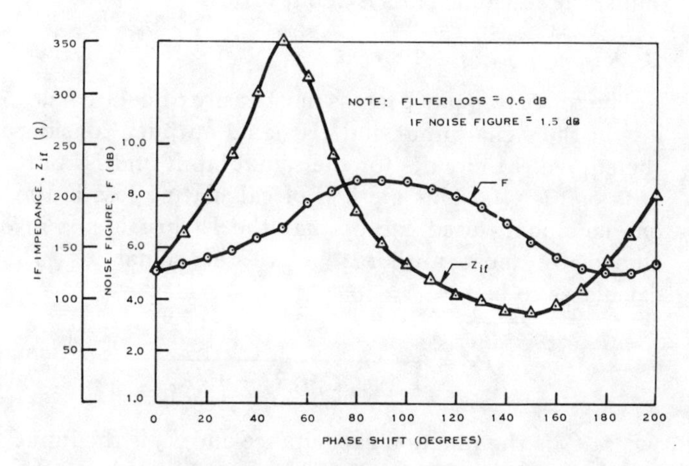

Fig. 9. Noise figure and IF impedance of TI GaAs diode as a function of phase shift of image short circuit from diode. Filter loss 0.6 dB and IF noise figure 1.5 dB.

voltage source for the local oscillator is superior to the constant current source. It also appears that as the local oscillator deviates from a constant voltage to a constant current source, the open-circuited image conversion loss degrades, while the short-circuited image changes only slightly.

For reasons previously discussed it is not possible to obtain a perfect constant voltage or constant current source especially in a 90° balanced mixer. In addition, perhaps a more significant effect is the termination of the sum frequency, although this will not be discussed here. Probably for these reasons, the single-ended and balanced mixers tested with short-circuited images were at least 1 dB better than mixers with images open-circuited. However, this does not mean to imply that the open-circuited image is necessarily worse than the short-circuited but only that in practical X-band microstrip mixers it is more difficult to achieve low-noise figure with the open-circuited than with the short-circuited image. The experimental results on mixer termination measured on the TI GaAs diode are shown in Fig. 9. These measurements are for a single-ended mixer in waveguide and show the noise figure and IF impedance as a function of phase shift of the image termination. This shows that there is only one distinct minimum corresponding to the short-circuited image. Notice, too, that the IF impedance minimum is shifted from the noise figure minimum. This again is probably due to the out-of-band and harmonic terminations. The IF noise figure was 1.5 dB. This means that the conversion loss measured is about 2.9 dB. Since these are the best measured data in a waveguide system, the integrated circuit mixer will be somewhat worse than this. The 2.9 dB includes the loss due to the diode's series resistance as well as waveguide and reflection loss. These losses, however, are probably not more than 0.4 dB. Thus, the resistance loss is about 0.7 dB.

B. Resistance Loss

The series resistance loss is most easily calculated as a loss in the signal input and IF signal output. Consider the equivalent circuits for the signal input and IF outputs of Fig. 10. This method of calculating loss is the one commonly used except that the IF loss is not included by most authors.[9] Here the signal loss P_s calculates to be

$$P_s = \frac{P_{\text{in}}}{1 + \omega_s^2 C_j^2 r_s R_g + r_s/R_g} \tag{20}$$

where C_j is the junction capacitance and R_g is the input resistance of the mixer. The IF loss similarly is[1]

[1] Strictly speaking, (21) is correct only if the IF amplifier input impedance is matched to the mixer IF output impedance. With a very low noise IF amplifier, this is not usually done, and, as a result, the IF loss may be less. However, for the IF amplifier described in this paper, (21) is a good approximation.

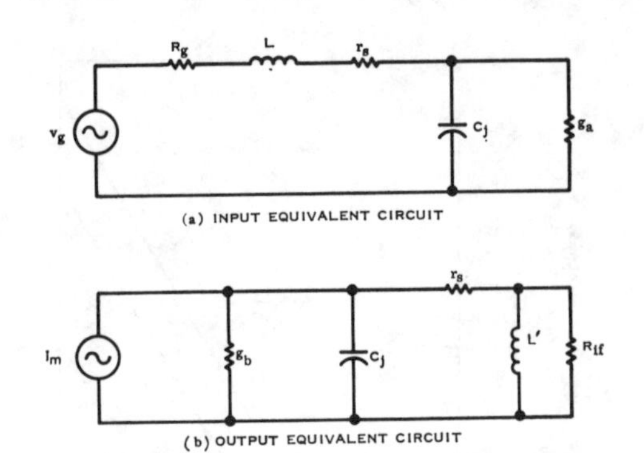

Fig. 10. Mixer input (a) and output (b) equivalent circuits showing effect of series resistance r_s.

$$P_{IF} = \frac{P_{\text{out}}}{1 + \omega_{IF}^2 C_d^2 r_s R_{IF} + r_s/R_{IF}} \approx \frac{P_{\text{out}}}{1 + r_s/R_{IF}} \tag{21}$$

where R_{IF} is the IF output resistance of the mixer. The optimum R_g, by the way, can be shown to be

$$R_{g\text{ opt}} = \frac{1}{\omega C_j} \cdot \tag{22}$$

Using typical values for a GaAs diode, $r_s = 3.0$ ohms, $C_j = .20$ pF. And considering the short-circuited image, $R_{\text{in}} = R_{\text{out}} = 50$ ohms. Then $R_{g\text{ opt}} = 56.8$ ohms and $P_s/P_{\text{in}} = 0.44$ dB and $P_{\text{in}}/P_{\text{out}} = 0.29$ dB. The total loss due to series resistance and junction capacitance for the X-band mixer is 0.73 dB, which agrees with the measured loss of about 0.7 dB (actually mixer diodes have been seen with series resistance as low as 2 ohms).

C. Noise Figure

The noise figure of the mixer is given by the well-known equation.

$$F = L_c(N_r + F_{\text{IF}} - 1). \tag{23}$$

where L_c is the total conversion loss including resistive loss, F_{IF} is the noise figure, and N_r is the diode noise ratio.

The conversion loss L_c includes all losses including 1) frequency conversion loss, 2) diode series resistance loss, 3) mismatch loss, 4) coupler directivity loss, 5) filter loss, and 6) ceramic loss in the case of integrated circuits.

The noise ratio N_r is the ratio of the noise output of the diode to the thermal noise output of a resistor having the same value of resistance. The minimum value of the noise ratio has been discussed by many authors. Many such as Nicol,[6] Barber and Ryder,[7] and Barber[1] say that the noise ratio has a minimum value of one half. Van der Ziel *et al.*[4] derive the following equation for the noise output of a diode:

$$\overline{i_N^2} = 4kT\Delta f g - 2q\, i_{10}\Delta f. \tag{24}$$

Now, for $i=i_0 \exp[qV/kT]$ and $g=\partial i/\partial V=qi/kT$,

$$\overline{i_N^2} = 2kT\Delta fg, \tag{25}$$

so that the noise output of the diode, exclusive of the series resistance, is one half the thermal noise output of an equivalent resistance. However, when the diode is operated in a mixer, the noise output is no longer given simply by (24). There is, first of all, the "attenuation effect" of the conversion loss. This has been derived by Pritchard[11] and others and the mathematics will not be repeated here. This has the effect of altering the noise ratio as

$$N_{tr} = \frac{1}{L_c}[N_r(L_c - 1) + 1] \quad \text{for open- or short-circuit image} \tag{26}$$

$$N_{tr} = \frac{2}{L_c}\left[N_r\left(\frac{L_c}{2} - 1\right) + 1\right] \quad \text{for matched image.}$$

The second thing that will alter the diode's noise output when operated as a mixer is the fact that there will be shot-noise components from many of the harmonic mixing products of the local oscillator and signal frequencies which have different frequencies at the IF frequency. Most existing analyses neglect mixing products of the harmonics. For this reason the intrinsic noise ratio of the diode when operated in a mixer will probably be not less than unity. Analytical proof of this is beyond the scope of this paper and will not be given here. It will be assumed for the rest of this discussion, however, that the noise ratio $N_r=1$ and therefore $N_{tr}=1$. This assumption, it should be added, seems to be borne out by experimental measurement.

III. Mixer Development

A. Filter

There are a number of different filter types which were considered for terminating the image of the mixer. These include Chebychev high-pass, cavity-type bandpass, and cavity-type band reject filters. Most of these were rejected because of size and passband loss. The effect of having a signal filter attenuation and of not having a perfect image termination is to degrade the conversion loss as

$$L_s' = L_s\left(1 + \frac{L_s}{L_i}\right)L_f \tag{27}$$

with

L_s = signal conversion loss
L_i = image conversion loss
L_f = filter loss at signal frequency.

Thus, for a filter with a passband loss of 0.5 dB and an image to signal conversion loss ratio of 18 dB, the conversion loss is degraded by 0.57 dB.

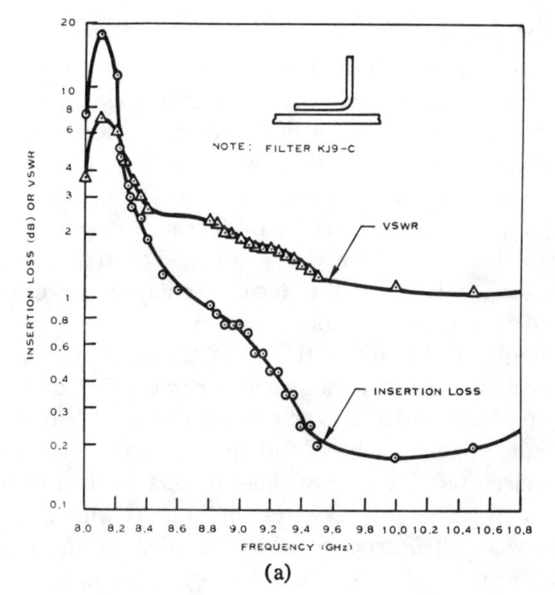

(a)

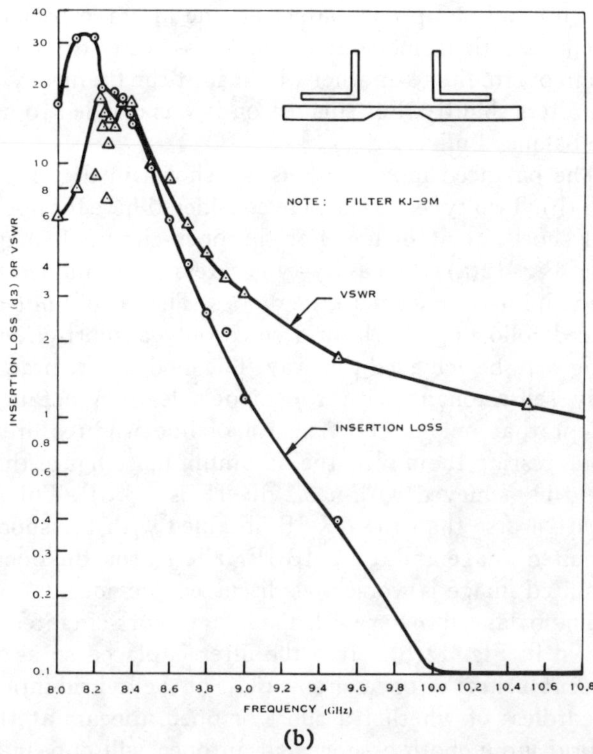

(b)

Fig. 11. (a) Cavity-type band rejection filter insertion loss and VSWR as function of frequency. (b) Cascaded cavity-type rejection filter insertion loss and VSWR as function of frequency.

Two types of filters were tested having the desired characteristics. These are shown in Fig. 11(a) and (b) along with their bandpass characteristics. They are made up of a half-wavelength line coupled to the main transmission line over a quarter wavelength and provide an attenuation of 18 dB at the image frequency and only 0.5 dB at the signal frequency. It was decided to use the smaller of the two since size was of more importance in this study than bandwidth. The use of this type of filter in an integrated circuit mixer was first reported by Tatsuguchi.[12]

B. Mixer

There were two basic mixer types which were considered, the single-ended mixer and the balanced mixer. The space question was not important here since the single-ended mixer required two additional filters, while the balanced mixer required a hybrid. There were, however, a couple of reasons for preferring the single-ended mixer. One is that only one diode was required and, therefore, only half the local oscillator power. The second was that bias could more readily be applied to the single-ended mixer than the balanced mixer. For this reason the single-ended mixer was tried first.

Two single-ended mixer types were tested, one with the image open-circuited and the other with the image short-circuited. Neither of these mixers worked particularly well. The primary reason is that the rejection filters were all narrow band and subject to interaction with the other filters and circuitry. It was extremely difficult to line up adequately all the filters, especially when variations in ceramic thickness, etc., from one group of circuits to another would shift the frequency of the filters slightly. For this reason it was decided to use the balanced mixer.

The balanced mixer designs are shown in Fig. 12(a) and (b). Two types again were considered here for open- and short-circuit image. For the open-circuited image case, Fig. 12(a), the cavity-type filters were coupled to the two lines adjacent to the diodes. They could not be placed following the hybrid since only a short-circuit image can be achieved this way. The diodes were practically self-resonant with the diode's lead inductance, except that only a short length of line was required. Upon testing the mixer, the minimum noise figure that could be achieved with a 2.2-dB IF is 8.3 dB. This is 2.0 dB worse than the 6.3 dB obtained with the short-circuited image and a 2.2-dB IF. The reason the open-circuited image is worse was discussed previously.

The balanced mixer with the image short-circuited is shown in Fig. 12(b). Here the filter is placed so as to present a short or open circuit right at the hybrid input. Regardless of whether a short or open appears at the hybrid input, both a short and an open will appear at the two output terminals due to the two phase shifts which are 90° and 180°. The result is that the short circuit is seen by the image. Thus, with the diodes close to the hybrid the image is short-circuited.

With the mixer of Fig. 12(b) the minimum noise figure measured was 6.3 dB with a 2.2-dB IF. Fig. 13 shows a plot of the measured noise figure as a function of frequency. The noise figure increases off-center frequency because the filter no longer completely rejects the image. There is some fluctuation in the noise figure due to the filter reactance.

The conversion loss versus frequency characteristics are shown in Fig. 14. Here the effect of the filter can be seen. It shows that there is a 16-dB difference between

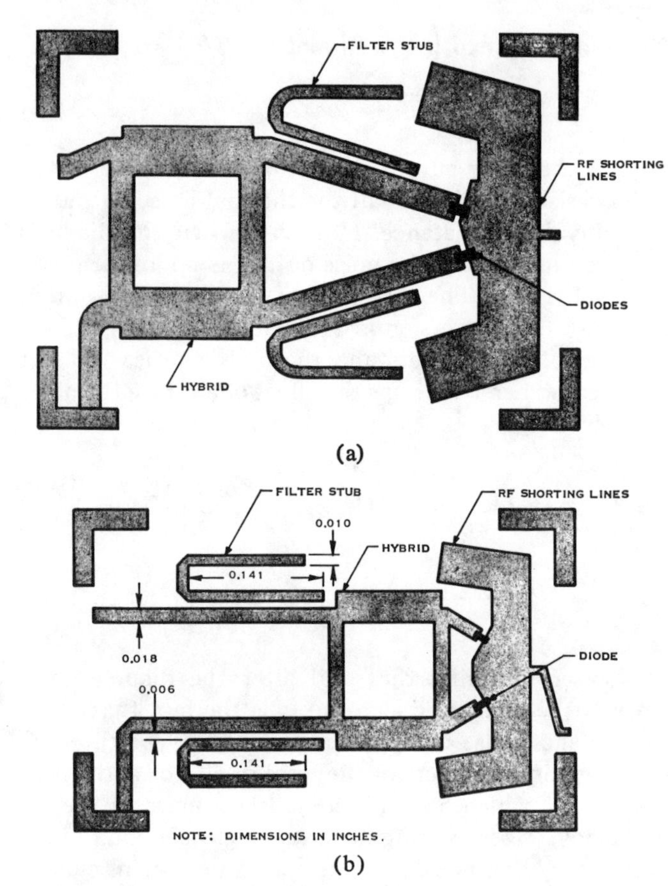

Fig. 12. (a) Balanced image rejection mixer with open-circuited image rejection filters. (b) Balanced image rejection mixer with short-circuited image rejection filters.

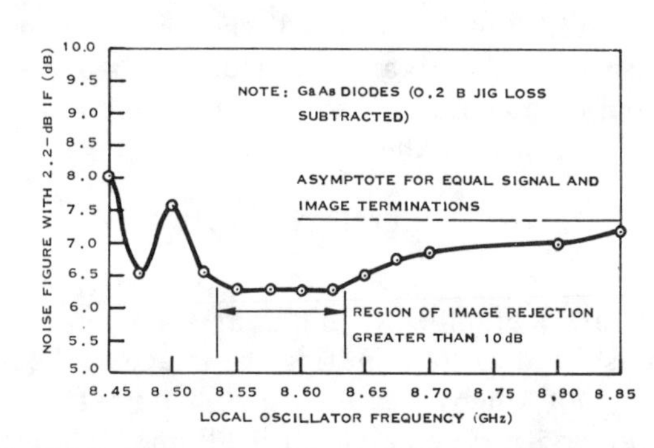

Fig. 13. Noise figure versus local oscillator frequency for mixer of Fig. 12(b).

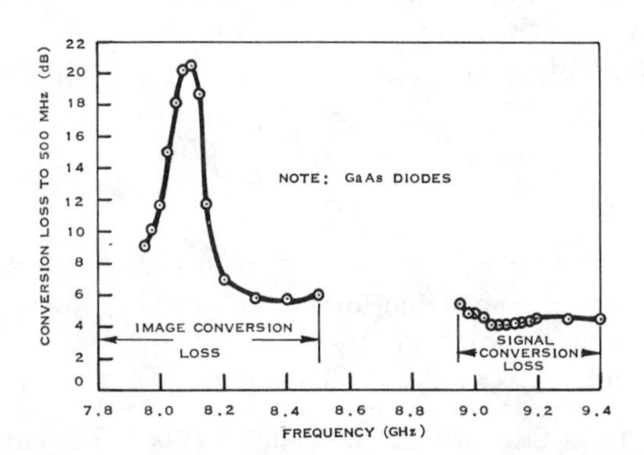

Fig. 14. Conversion loss versus frequency for mixer of Fig. 12(b) showing conversion loss at image and signal frequency bands.

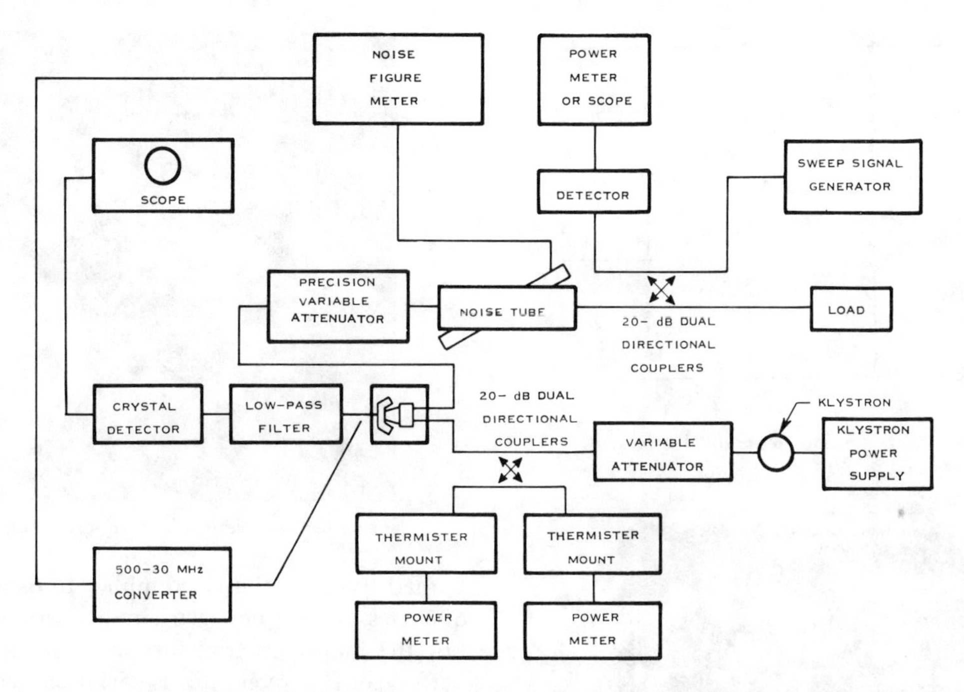

Fig. 15. Conversion loss and noise figure measuring setup.

TABLE I

Conversion Loss	1.80 dB
Diode Resistance Loss	0.73 dB
Filter Loss	0.57 dB
Hybrid Loss	0.30 dB
Other Ceramic Loss (not including filter and hybrid loss)	0.20 dB
Mismatch Loss	0.20 dB
Output Filter Loss	0.20 dB
	4.00 dB

the signal conversion and image conversion. The setup for measuring the frequency conversion is shown in Fig. 15. Feeding the local oscillator arm is a local oscillator source followed by an attenuator and directional couplers with detectors for monitoring incident and reflected powers. The signal arm is fed by a signal generator followed by a directional coupler with a thermistor for monitoring incident power level. Following this is a precision attenuator to accurately determine conversion loss differences as a function of frequency. Following the mixer of mixer–IF preamplifier or mixer–IF preamplifier plus converter is a low-pass filter and a crystal detector. The detected output is used either as a reference or, in conjunction with a swept signal generator, to display the conversion loss versus frequency characteristic.

The 6.3-dB measured noise figure is quite close to the noise figure predicted if all losses are considered. These losses are tabulated in Table I. This gives a calculated noise figure of 6.2 dB with a 2.2-dB IF which agrees quite closely with the measured noise figure of 6.3 dB.

IV. Integrated Mixer–IF Preamplifier

A. 500-MHz Preamplifier

The 500-MHz IF preamplifier circuit used is a thin-film circuit with three transistor stages designed to achieve a wide bandwidth with a low-noise figure. Fig. 16 shows a schematic of the preamplifier circuit, and Fig. 17 shows one of the circuits with the transistor devices mounted. The circuits for the image terminated mixer used selected low-noise transistors in the first stage in order to obtain the low-noise figure required.

With the new low-noise selected transistors the following data were measured on one of the IF preamplifiers:

Center frequency	500 MHz
3-dB bandwidth	400 MHz
Power gain	24.5 dB at 500 MHz
Optimum noise figure	2.2 dB at 500 MHz

B. Coupling Network

With 4.1-dB conversion loss in the mixer and 2.2-dB noise figure in the 500-MHz IF preamplifier, the remaining problem is to combine these two circuits in integrated form with the minimum coupling loss. Since the image is short-circuited and two diodes appear in parallel, the mixer–IF output impedance is low, on the order of 30 ohms. In addition, the mixer has some capacitance in shunt with the 30 ohms due to the RF quarter-wavelength shorting stubs.

The IF amplifier also looks capacitive on its input with a capacitive reactance of about 100 ohms. Its optimum source impedance is high also, about 100-ohms

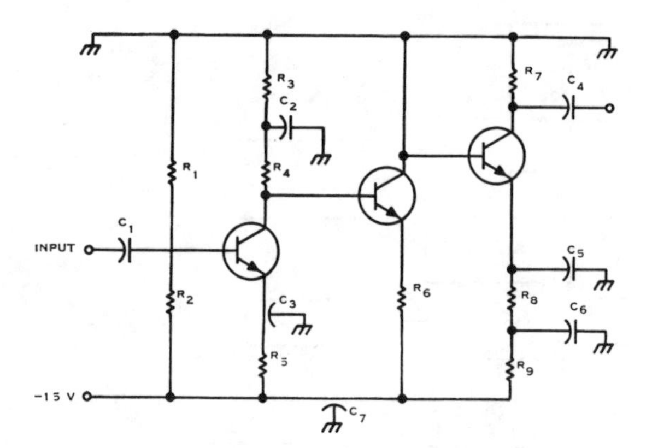

Fig. 16. 500-MHz preamplifier equivalent circuit.

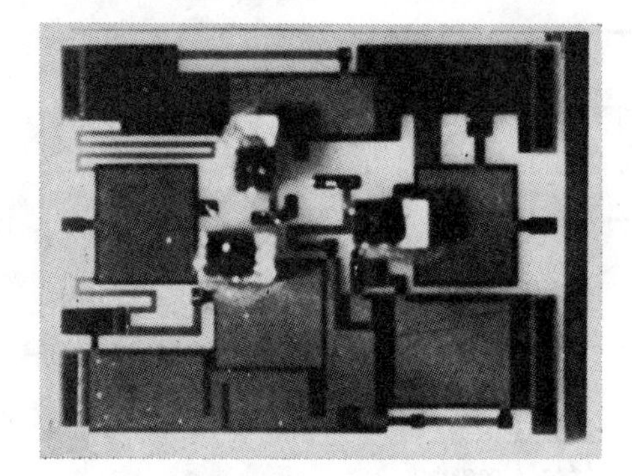

Fig. 17. 500-MHz preamplifier circuit.

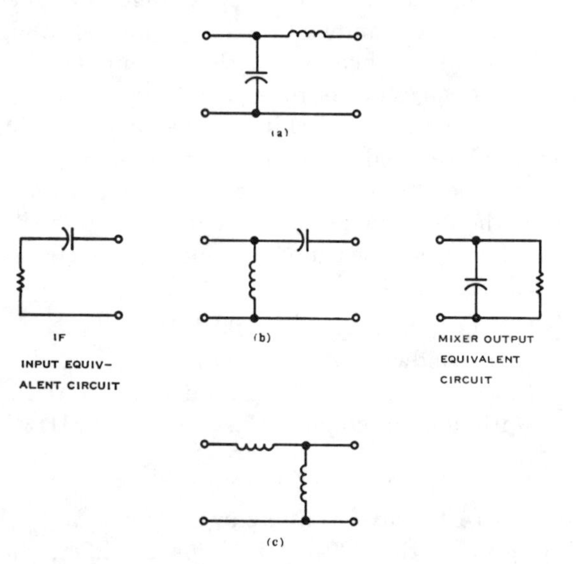

Fig. 18. Mixer–IF amplifier matching networks.

resistive. Thus, some inductance is required to resonate the circuits and the 30-ohms impedance of the mixer must be transformed to about 100 ohms. This impedance match can be accomplished in a number of ways, three of which are illustrated in Fig. 18, which shows a low-pass transforming filter, a high-pass transforming filter, and a simple tapped inductor. The low-pass transforming filter did not work too well since it

Fig. 19. Complete mixer–IF preamplifier receiver.

tended to cause the IF amplifier to oscillate at low frequencies. The other two circuits worked satisfactorily but the high-pass transforming filter was chosen since better control over mixer variations could be had by simply changing series capacitors. The loss through the coupling network was somewhat high, about 0.4 dB.

The final-mixer preamplifier design is shown in Fig. 19. Here may be seen the OSM connectors for bringing the signal and local oscillator in, the mixer, the IF coupling network, and the IF preamplifier.

With the 4.1-dB mixer conversion loss, the 0.4-dB mixer–IF amplifier coupling network loss, and a 2.2-dB IF preamplifier, the total receiver noise figure was 6.7 dB.

V. Conclusion

The analysis of the image terminated mixer revealed that out-of-band and harmonic terminations have less effect on conversion loss with the image short-circuited than with the image open-circuited. The conversion loss for the short-circuited image case is about 1.8 dB. Conversion losses lower than 1.3 dB are possible, but only at lower frequencies where terminations of the out-of-band and local oscillator harmonics can be controlled and where the series resistance is lower.

The series resistance losses were calculated for the GaAs diode and found to be about 0.7 dB. The other circuit losses amounted to 1.5 dB. This gave a mixer noise figure of 6.3 dB with a 2.2-dB IF amplifier noise figure.

A number of filters were tested for terminating the image. Most of these could not be used because of size or loss considerations. The final filter was a half-wavelength line coupled over a quarter wavelength to the main line which provided 18-dB image rejection with only 0.50-dB signal loss.

By selection of transistors and slight circuit adjustment a 500-MHz preamplifier was developed which has a 2.2-dB noise figure. The coupling loss to the amplifier from the mixer added, however, an additional 0.4-dB loss. This gave an overall noise figure of 6.7 dB.

REFERENCES

[1] H. C. Torrey and A. C. Whitmer, *Crystal Rectifiers*, M.I.T. Rad. Lab. Ser., vol. 15. New York: McGraw-Hill, 1948, pp. 111–178.

[2] E. W. Herold, R. R. Bush, and W. R. Ferris, "Conversion loss of diode mixers having image—frequency impedance," *Proc. IRE*, vol. 33, pp. 603–609, September 1945.

[3] P. D. Strum, "Some aspects of mixer crystal performance," *Proc. IRE*, vol. 41, pp. 875–889, July 1953.

[4] A. Van der Ziel and A. G. T. Becking, "Theory of junction diode and junction transistor noise," *Proc. IRE*, vol. 45, pp. 589–594, March 1957.

[5] M. J. O. Strutt, "Noise-figure reduction in mixer states," *Proc. IRE*, vol. 34, pp. 942–950, December 1946.

[6] G. R. Nicoll, "Noise in silicon microwave diodes," *Proc. IEE* (London), vol. 101, pp. 317–324, 1954.

[7] M. R. Barber and R. M. Ryder, "Ultimate noise figure and conversion loss of the Schottky barrier mixer diode," presented at the 1966 Electron Devices Conf.

[8] R. Rafuse, "Low noise and high dynamic range using symmetric circuits at VHF and microwaves," presented at the 1967 Conf. on High Frequency Generation and Amplification, Cornell University, Ithaca, N. Y.

[9] "The hot carrier diode theory, design and application," Hewlett Packard Assoc., Application Note 907, May 1967.

[10] M. R. Barber, "Noise figure and conversion loss of the Schottky barrier diode mixer," *IEEE Trans. Microwave Theory and Techniques*, vol. MTT-15, pp. 629–635, November 1967.

[11] W. L. Pritchard, "Notes on a crystal mixer performance," *IRE Trans. Microwave Theory and Techniques*, vol. MTT-3, pp. 37–39, January 1955.

[12] I. Tatsuguchi, "Integrated 4 GHz balanced mixer assembly," 1967 *ISSCC Digest of Tech. Papers*, pp. 60–61.

4-GHz Integrated-Circuit Mixer

MOTONOBU KATOH AND YOSHIHIKO AKAIWA

Abstract—**We have calculated the conversion loss for microwave diode mixers taking into account the effects of series resistance and barrier capacitance in the diode and the internal resistance of the local oscillator. The relations between the conversion loss and the parameters, which are important for the design of the diode mixer, are clarified. A 4-GHz integrated-circuit low-noise mixer is developed. The minimum overall noise figure obtained is 4.1 dB with a short-circuited image-frequency termination.**

I. Introduction

The calculation of conversion loss for microwave diode mixers for image-frequency short-circuited termination and image-frequency open-circuited termination has been carried out by several authors under simplified assumptions [1]–[5]. The results obtained so far are not necessarily sufficient for designing a mixer with a reactively terminated image. Recently a detailed study has been performed on a mixer diode in the millimeter-wave range; however, the results are not sufficient for applying in the microwave range [6].

We have calculated the conversion loss of a mixer with a reactively terminated image and studied the effects of parameters on the conversion loss in order to obtain a low-noise mixer in the 4-GHz region. We have made a 4-GHz integrated-circuit mixer and investigated the effect of the image rejection filter on the conversion loss experimentally. A 4-GHz integrated-circuit low-noise mixer using a GaAs Schottky diode has been developed by referencing the calculated results and experimental information.

The noise figure of a mixer with a short-circuited image is 4.1-dB with the IF preamplifier having a 1.7-dB noise figure.

II. Calculation of Conversion Loss

For the diode model we take into account the series resistance (R_s) and the constant barrier capacitance (C_j).

In an ideal Schottky diode the current i is given by

$$i = i_0(\exp(\alpha v) - 1)$$

where v is the voltage applied to the barrier and i_0 is the saturation current. α is given by

$$\alpha = e/nkT$$

where e is the basic electron charge, k is Boltzmann's constant, T is the absolute temperature, and n is a constant which is 1.0 for the ideal diode. The constants of the microwave GaAs mixer diode which we prepared have the following values:

$$R_s = 1\ \Omega; \quad C_j = 0.2\ \text{pF}; \quad i_0 = 3.8 \times 10^{-14}\ \text{A}; \quad n = 1.1.$$

In the following calculation we have used the value $\alpha = 34.7$. The following differential equation should hold:

$$(R_g + R_s)\left(i_0(\exp(\alpha v) - 1) + C_j \frac{dv}{dt}\right) + v = V_1 \sin(\omega t) + V_0 + E', \qquad E' = R_g I_{dc}$$

where R_g is the internal resistance of the local oscillator, V_1 and ω are the amplitude and the angular frequency of the local oscillator, respectively, I_{dc} is the rectified current, and V_0 is the dc bias voltage. We have only taken the signal, IF, and image frequencies into consideration and have assumed that the other frequency components are short-circuited.

We have calculated the conversion loss for the case in which a mixer is conjugate-matched at the ports of input and output. We have assumed two cases in which the image-frequency port is both short-circuited and open-circuited and have chosen signal, local, and IF frequencies of 4070 MHz, 4000 MHz, and 70 MHz, respectively.

In the ideal diode ($R_g = R_s = C_j = 0$), the conversion loss decreases monotonically and approaches 0 dB with the increase of local voltage. The conversion loss is independent of the dc bias. The input and output matching conductance can take any values by means of dc bias without changing the conversion loss. In the actual diode the signal loss due to R_s appears in addition to conversion loss.

Neglecting the signal loss due to R_s, we considered the effect of voltage waveform deformation of the local oscillator on conversion loss. The results are given in Fig. 1 with dashed lines. The voltage waveform deformation is caused by the nonlinear voltage drop at R_g and R_s (Fig. 2). As the local oscillator amplitude becomes large, the deformation becomes noticeable and the conversion loss begins to increase beyond a minimum.

Calculation was carried out by taking the signal loss due to R_s and C_j into consideration. The results are also given in Fig. 1 with solid lines. The signal loss for a small local oscillator amplitude occurs because the signal is short-circuited by the parallel barrier capacitance, when the diode impedance is very high and C_j cannot be canceled due to R_s.

Manuscript received November 10, 1970; revised February 1, 1971.

The authors are with the Central Research Laboratory, Nippon Electric Company, Kawasaki, Japan.

Reprinted from *IEEE Trans. Microwave Theory Tech.*, vol. MTT-19, pp. 634–637, July 1971.

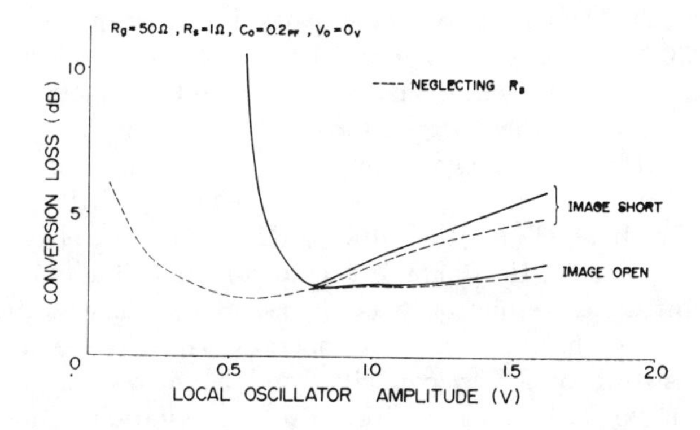

Fig. 1. Conversion loss versus local oscillator amplitude.

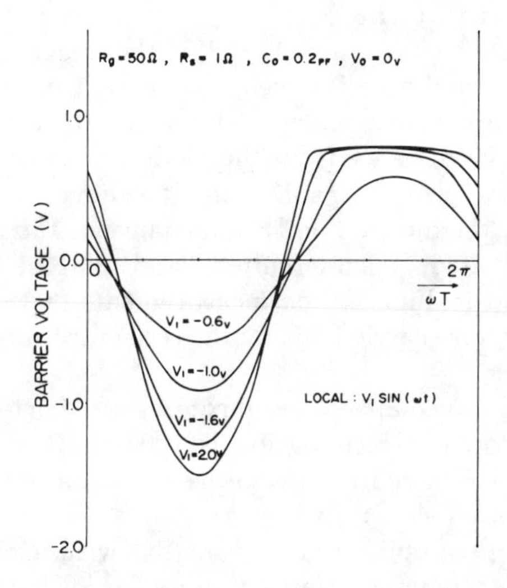

Fig. 2. Barrier voltage waveform.

The signal loss for the large local oscillator amplitude occurs by the series resistance R_s since the diode impedance becomes low in this case.

We see that there is a minimum in the conversion-loss versus local-oscillator-amplitude curve. We should make the minimum an operating point as far as the conversion loss is concerned. However, we must take into consideration that the diode should be matched at the ports of signal input and IF output. In Fig. 3 the IF-matching-resistance (reciprocal of matching conductance) versus local-oscillator-amplitude relation is given. Considering the fact that the input impedance of the IF amplifier is of the order of 100 Ω, the IF matching resistance is too high at the minimum point of the conversion loss. Let us apply the dc forward bias. The result is also given in Fig. 3 for the case in which the dc forward 0.2 V is applied. It is shown that the matching resistance can be diminished. However, the conversion loss increases by 1.5 dB for a short-circuited image frequency and by 0.3 dB for an open-circuited image frequency at the local oscillator amplitude of 0.8 V. The conversion loss decreases as the reverse dc bias becomes deep.

We have used an R_g of 50 Ω in the calculations, since the impedance of the microwave GaAs mixer diode was

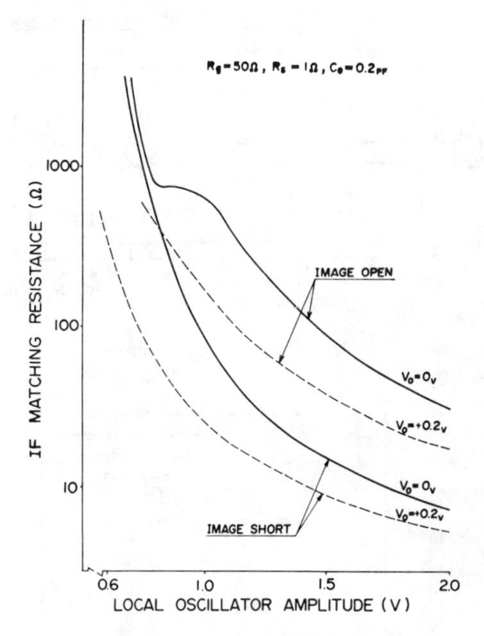

Fig. 3. IF matching resistance.

determined to be about 50 Ω under the usual conditions. We have found by calculation that small conversion loss is not obtained by making R_g smaller.

We studied the effects of R_s and C_j on the conversion loss by varying their values. The signal loss was sensitive to the R_s value at larger local oscillator amplitudes. The conversion loss has been calculated for several C_j (0.2 pF–1.2 pF). The minimum conversion loss is nearly constant for the different barrier capacitance values; however, a larger local oscillator amplitude is needed for a larger barrier capacitance.

III. 4-GHz Integrated-Circuit Mixer

From the results of the calculation described above, it follows that the image should be open-circuited in order to obtain a lower noise figure in a mixer, although the signal and IF impedances become high. When the circuit of a mixer has to be made using a stripline, or when an IF preamplifier has a low input impedance, the image open circuit is not suitable because the circuit becomes a complicated and narrow band due to inevitable matching circuits. However, signal and IF impedances with a short-circuited image attain suitable values which are convenient for the integrated circuit such as stripline. The wide-band characteristics can be expected.

The 4-GHz integrated-circuit mixer developed, therefore, has a filter to provide a short-circuited image. Investigations were carried out using a mixer like that shown in Fig. 4. The circuit of the mixer was manufactured by carrying out evaporation and electroplating on a 1-mm ceramic substrate (Al_2O_3) placed at the center of two ground conductors 4 mm apart. The IF was 70 MHz. A GaAs Schottky barrier diode V 584 developed at the Nippon Electric Company was used.

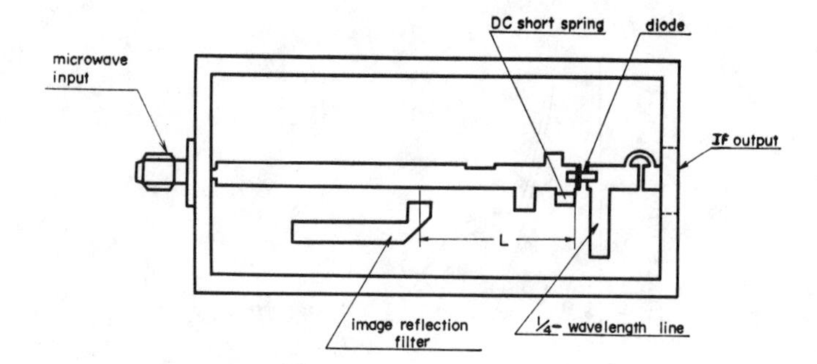

Fig. 4. Configuration of mixer.

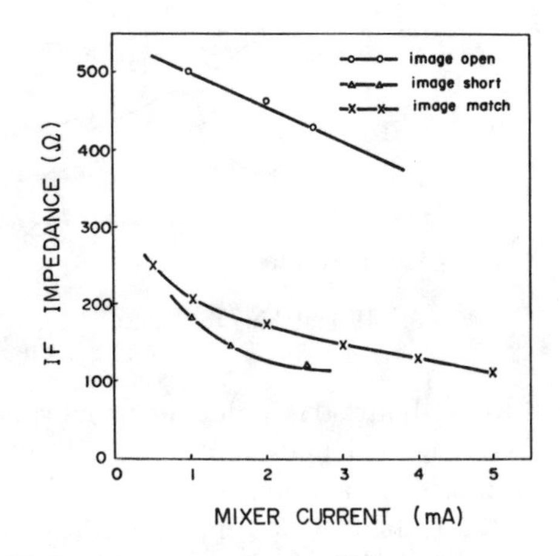

Fig. 5. IF impedance of the mixer for short-circuited, open-circuited, and matched image-frequency termination.

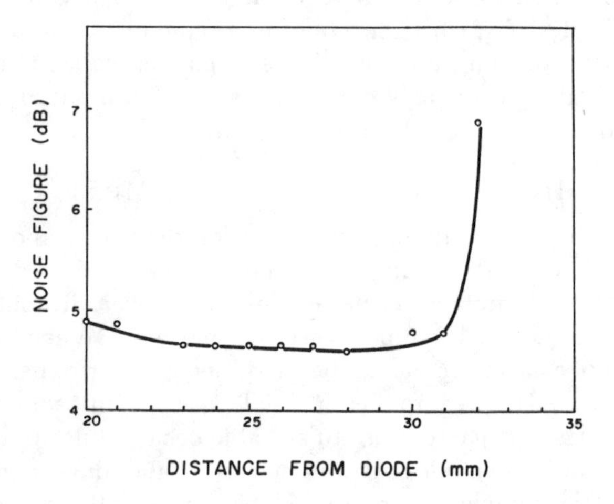

Fig. 6. Noise figure as function of distance of image filter from diode.

The image filter consisting of a stripline a bit shorter than a half wavelength is loosely coupled through a gap to the main transmission line of 50 Ω on which a mixer diode is mounted. The transmission line is short-circuited on the spot where the rejection filter is located, due to the series resonance of the distributed constant inductance with the coupling lumped constant capacitance.

At the local frequency of 4000 MHz and the signal frequency of 4070 MHz, the rejection characteristics over 20 MHz is the vicinity of the image frequency of 3930 MHz was greater than 10 dB. The insertion loss of the image filter was 0.4 dB at the signal frequency. The frequency of the image filter can be tuned from 3.7 to 4.2 GHz only by changing the insertion length of the tuning screw set up at the edge of the image filter.

The band elimination filter of RF grounding is placed at the back of the diode; it is made of a distributed constant series resonator making use of a quarter-wavelength stub and a lumped constant parallel resonator consisting of a gap capacitor and a narrow stripline inductance. By means of this filter, attenuation of more than 20 dB over a 20-percent bandwidth of the center signal frequency was obtained. The insertion loss at 70 MHz was 0.1 dB.

In the mixer shown in Fig. 4, if the image-frequency filter is located a half-wavelength apart from the diode ($l=27$ mm) it behaves as a short circuit at the diode, while if the distance from the diode is around a three-quarter wavelength ($l=33$ mm) it behaves as an open circuit at the diode. The IF impedance of the mixer for short-circuited, open-circuited, and matched images are shown in Fig. 5. The measurements of the IF impedances were carried out by the resistance substitution method.

The IF impedance of the mixer with a short-circuited image-frequency termination is 100–200 Ω, while that with an open-circuited image frequency termination is 400–500 Ω as shown in Fig. 5.

An IF preamplifier in conjunction with a mixer is a two-stage common emitter amplifier using 2SC823 transistors. The measured input impedance of the preamplifier was 119 Ω. From Fig. 5 we see that the IF impedance of the mixer with a short-circuited image-frequency termination at a mixer current of 2 mA is 130 Ω. The mixer can therefore be connected to the IF preamplifier without a matching circuit.

The noise figure measured by the Y-parameter method as a function of the distance of the image filter from the diode is shown in Fig. 6. A minimum noise figure of 4.5 dB was obtained with a short-circuited image-frequency termination at a distance of $l=27$ mm. The noise figure increases rapidly when the distance l exceeds 30 mm, since the image frequency impedance approaches an open-circuit value and the IF impedance becomes high.

A mixer with an image filter placed right in front of the diode was fabricated in order to improve the characteristics dependency on the electrical length and make the delay distortion independent of image frequency [7]. The circuit of the mixer was formed on a 0.5-mm ceramic substrate.

The mixer is shown in Fig. 7. The IF preamplifier is the same as that described. The minimum overall noise figure at 3.77 GHz was 4.1 dB with a mixer current 2–3 mA (3.9-dBm local power) and an IF preamplifier

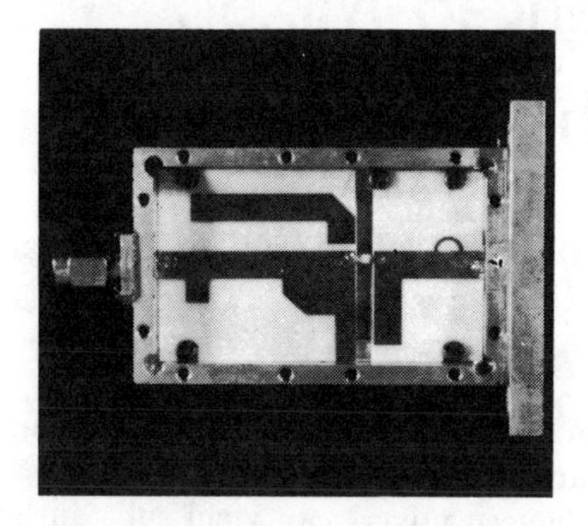

Fig. 7. Improved mixer.

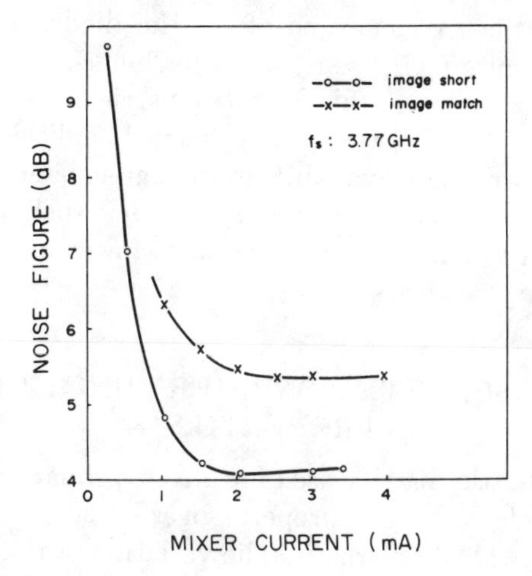

Fig. 8. Minimum noise figure obtained by improved mixer.

having a 1.7-dB noise figure (Fig. 8). The accuracy of the measurement is within 0.5 dB. The amplitude variation is less than ±0.05 dB over a 20-MHz band centered at 70 MHz. The differential gain is less than 0.2 percent. This mixer can be used in a multichannel FM repeater system. Noise-figure variation in the 3.7- to 4.2-GHz band was within ±0.1 dB when only the frequency of the image filter was changed.

IV. Conclusion

We have investigated the conversion loss and noise figure of a mixer with a reactively terminated image by means of calculation and experiment in order to obtain a low-noise mixer in the 4-GHz band. The conversion loss with an open-circuited image-frequency termination is theoretically found to be smaller than that with a short-circuited image-frequency termination. The IF impedance for the former, however, becomes so high compared with the impedance for the latter that the image open circuit is not suitable due to inevitable matching circuits between a mixer and a preamplifier.

We fabricated a 4-GHz integrated-circuit mixer with an image short circuit using a GaAs Schottky barrier diode. The minimum overall noise figure obtained was 4.1 dB with an IF preamplifier having a 1.7-dB noise figure. This mixer has good characteristics of conversion gain and differential gain sufficient to be used in a multichannel FM repeater system.

Acknowledgment

The authors wish to thank Dr. M. Uenohara and Dr. H. Murakami for their encouragement and guidance and Dr. K. Ayaki for his valuable advice throughout the study.

References

[1] E. W. Herold, R. R. Bush, and W. R. Ferris, "Conversion loss of diode mixers having image-frequency impedance," *Proc. IRE*, vol. 33, Sept. 1945, pp. 603–609.
[2] H. C. Torrey and C. A. Whitmer, *Crystal Rectifiers* (Radiation Laboratory Series), vol. 15. New York: McGraw-Hill, 1948.
[3] P. D. Strum, "Some aspects of crystal mixer performance," *Proc. IRE*, vol. 41, July 1953, pp. 875–889.
[4] M. R. Barber, "Noise figure and conversion loss of the Schottky barrier mixer diode," *IEEE Trans. Microwave Theory Tech.*, vol. MTT-15, Nov. 1967, pp. 629–635.
[5] K. M. Johnson, "*X*-band integrated circuit mixer with reactively terminated image," *IEEE Trans. Microwave Theory Tech.*, vol. MTT-16, July 1968, pp. 388–397.
[6] M. Akaike and S. Okamura, "Semiconductor diode mixer for millimeter wave region," *Trans. Inst. Electron. Commun. Eng. Jap.*, vol. 52-B, Oct. 1969, pp. 601–609.
[7] T. Kawahashi and T. Uchida, "Delay distortion in crystal mixers," *IRE Trans. Microwave Theory Tech.*, vol. MTT-7, Apr. 1959, pp. 247–256.

An Integrated-Circuit Balanced Mixer, Image and Sum Enhanced

LAWRENCE E. DICKENS, SENIOR MEMBER, IEEE, AND DOUGLAS W. MAKI, MEMBER, IEEE

Abstract—GaAs Schottky-barrier diodes with a zero-bias cutoff frequency of 800 GHz have been used in an integrated-circuit balanced diode mixer operating with a signal frequency centered at 9.3 GHz and a local-oscillator (LO) frequency at 7.8 GHz. For an instantaneous bandwidth of 1.0 GHz, the conversion loss (including all circuits and connector losses) was under 3.15 dB. Over the center 0.5 GHz of the band, the conversion loss was less than or equal to 2.8 dB. The conversion loss at the image-band edges was greater than 25 dB; the loss at the center of the image band was greater than 35 dB.

I. INTRODUCTION

THIS PAPER reports generally on a microwave mixer for operation at X band, and more specifically on an integrated mixer circuit having very low conversion loss. Microwave diode mixers have been used for many years to obtain conversion of a signal at microwave frequencies to one at a much lower frequency. Such mixers have been the subject of much study and development. However, there is a continuing need for improvements which can result in better electrical performance, higher reliability, improved reproducibility, and lower production costs.

Well known [1]–[4] are the techniques for the enhancement of mixer operation by the proper control of the impedances at each of the mixer terminals and at each of the frequencies of importance. The frequencies of importance are the modulation products which exist according to the heterodyne principle by which the mixer operates. The received signal (RF), together with a higher level signal from a local oscillator (LO), are applied to a nonlinear element. The signal is mixed with the LO producing the sum frequency LO + RF, the difference (or intermediate) frequency (IF), LO − RF, and the image frequency 2LO − RF.

It has been known for some time that this loss in converting an RF signal to an IF signal can be minimized by properly terminating the sum and image frequencies. However, the realization of the proper termination can represent a severe problem. Prior integrated-circuit forms [3], [5] of image-enhanced mixers have generally been single-ended (unbalanced) mixers as opposed to balanced mixers, and have suffered the limitation of narrow-band operation imposed by the use of a narrow-band filter for image termination control.

Image-enhanced mixers can yield substantial improvement in performance only if high-performance diodes are available. The measure of potential performance is indicated by the frequency cutoff of the diode. Very high frequency cutoff (f_{co}) is required for low conversion loss. Until the advent of the Schottky-barrier diode, and in particular, the GaAs Schottky barrier, sufficiently high f_{co} diodes were not available and image-enhanced mixers were only an academic curiosity. Now such Schottky barriers are readily available and low-conversion-loss mixers are a reality.

II. SCHOTTKY-BARRIER JUNCTION PROPERTIES

The Schottky barrier used for mixers primarily requires the variable resistance property of a junction, thus it is commonly called a varistor. The variation of resistance of a varistor as a function of applied voltage is dramatic. The reverse-bias resistance is on the order of many megohms. The resistance decreases rapidly with increasing forward bias until the forward-bias series resistance R_s dominates over the effect of the junction resistance.

The junction resistance is in parallel with a junction capacitance C_j which is also a voltage variable component. The varistor must be designed such that the junction capacitance is minimized for a given series-limiting resistance. To compare varistors of differing R_s and C_j values, it is useful to define a cutoff frequency, $f_{co} = (2\pi C_j R_s)^{-1}$. For this comparison the zero-bias value for f_{co} is useful. It has been found that his value correlates well with measured results.

Figs. 1 and 2 show the f_{co} as computed for silicon and GaAs Schottky barriers. The value for R_s is made up of two parts. The first is the resistance of the epitaxial region X_e, and the second is the parasitic resistance due to the spreading of the current from the epitaxial layer into a substrate of finite conductivity ρ_s. Two values of junction diameter D_j were assumed. Three values of X_e were assumed. The first curve shows the limiting value of f_{co} due to the junction alone (no substrate resistance), and the epi-layer thickness is taken to be $X_e = W_B$. The remaining curves assumed a substrate spreading resistance. The second curve was calculated assuming the epitaxial-layer thickness X_e to be just enough to accommodate the space charge region at breakdown W_B. The third and

Manuscript received April 22, 1974; revised September 27, 1974. This work was supported in part by the United States Army Electronics Command, Fort Monmouth, N. J., under Contract DAAB07-72-C-0221.

The authors are with the Systems Development Division, Westinghouse DESC, Baltimore, Md. 21203.

Reprinted from *IEEE Trans. Microwave Theory Tech.*, vol. MTT-23, pp. 276–281, Mar. 1975.

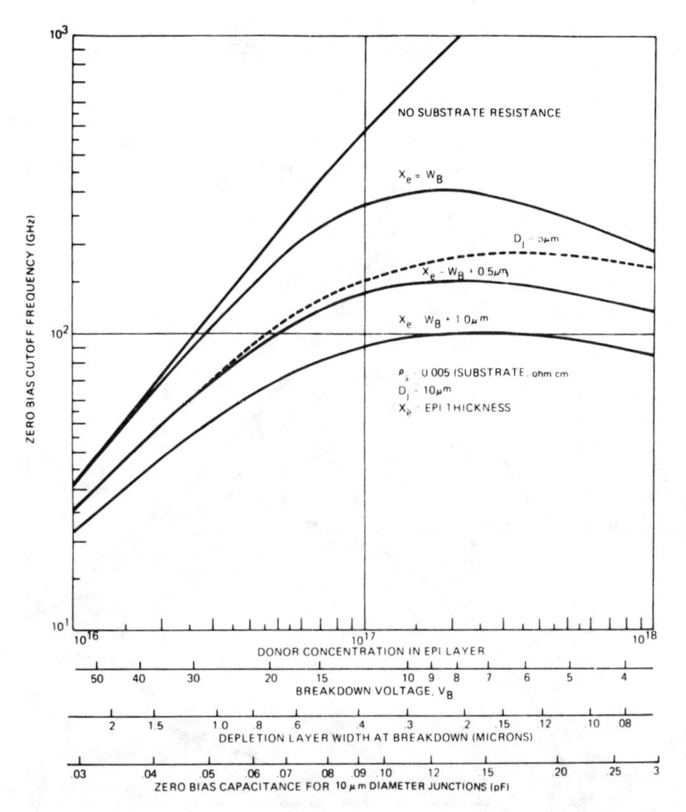

Fig. 1. Theoretical parameters of epitaxial silicon Schottky barriers.

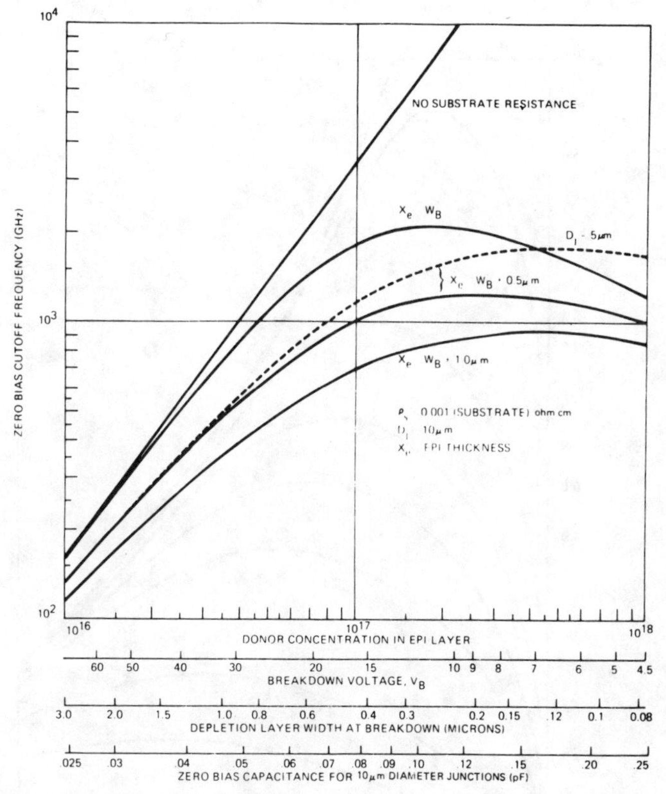

Fig. 2. Theoretical parameters of epitaxial GaAs Schottky barriers.

fourth curves make an allowance in the epi-layer thickness of 0.5 and 1.0 μm, respectively.

The effect of the parasitic substrate resistance is clearly apparent in the figures, both in the drastically reduced f_{co} and in the occurrence of a maximum in the curves [3]. The breakdown-voltage–impurity-density relationship of Sze and Gibbons [6] for an abrupt junction has been plotted also in the figures. These breakdown voltages represent bulk breakdown characteristics. Note that the f_{co} calculation has been made assuming at least an epi-thickness sufficient to be fully depleted at breakdown. This allows maximum breakdown voltage for a given doping level. If a much reduced epi-thickness is used, then an improvement in f_{co} can be obtained. Assume a GaAs epi-layer doping of 10^{16} and a thickness of 0.5 μm. Then one can calculate an f_{co} (for the 10-μm-diam junction on GaAs) of 2283 GHz, a breakdown voltage of 22 V, and a punch-through voltage of 2.0 V. The zero-bias capacitance is 0.024 pF. Now suppose that an excess of 0.5 μm were left on this epi-layer (due to processing variables). Now the value for f_{co} drops to 604 GHz, a drastic change in the value for R_s. Thus the epi-layer thickness for this diode becomes an extremely critical control parameter. However, such a thin-layer diode is attractive, especially for high-burnout diodes, because, for a given application, a junction diameter of 20 μm on a 0.5-μm epi-layer of 10^{16} will give about the same impedance level and f_{co} as a junction dimaeter of 10 μm on a 0.5-μm epi-layer of 2×10^{17}. Such a diode would have a two-to-one thermal impedance improvement over the smaller diode. The case for which the barrier depletion layer at zero bias extends through, or nearly through, the entire lightly doped epi-layer represents a special form of metal-semiconductor barrier known as a Mott barrier [3].

Note in Figs. 1 and 2 that there is an order of magnitude difference in the f_{co} scales between Si and GaAs. Realistic substrate resistances have been assumed. It can be seen that if one assumes a nominal 0.5-μm excess of epi-layer material for both Si and GaAs, that for 10-μm junctions the Si devices will yield an f_{co} of no more than 150 GHz; the GaAs devices can be expected with f_{co} on the order of 1000 GHz. These numbers are realistic and have been readily approximated in practice. Thus GaAs is the natural choice for very low-conversion-loss mixers.

III. DESIGN CONSIDERATIONS

Barber [4] has presented an analysis of microwave mixers and has shown that the pulse-duty ratio of the Schottky-diode current waveform is the most fundamental parameter for defining mixer operation because the diode current pulse retains its typical (switched) shape even when the voltage waveform becomes highly nonsinusoidal.

It can be shown that most microwave mixer diodes (adjusted for lowest conversion loss) behave as though the barrier itself were switched on and off at the LO rate, and that the resistance in the ON state is just that of the limiting series resistance (R_s), and the impedance in the OFF state is just that expected of the series resistance R_s in series with the barrier capacitance C_j. Of course the barrier capacitance is a function of voltage and time, but good correlation with measured results are obtained if the zero-bias capacitance value is used. Thus the frequency cutoff is $f_{co} = (2\pi R_s C_j)^{-1}$.

Using these considerations, an extension of Barber's analysis [7] has allowed the calculation of the conversion loss as a function of the pulse-duty ratio and as limited by the operating frequency to cutoff-frequency ratio (f/f_{co}). Fig. 3 shows the expected mixer conversion loss that would be obtained for the broad-band case (wherein the image termination equals the signal termination). Fig. 4 shows the computed mixer conversion loss for the case wherein the image is short circuited.

Figs. 5, 6, and 7 show the computed values of mixer terminal impedances plotted as functions of the pulse-duty ratio (t). In each case the RF-signal impedance R_{RF} has been chosen to minimize the mixer noise figure.

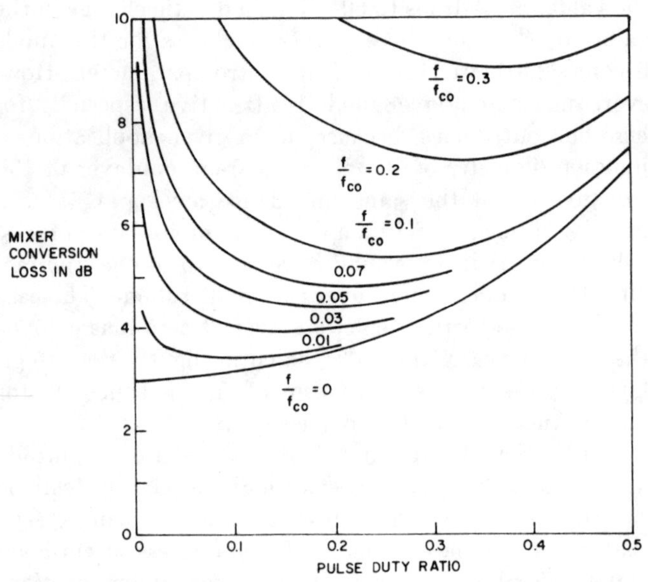

Fig. 3. Computed mixer conversion loss for the broad-band case. (Image termination equals signal termination.)

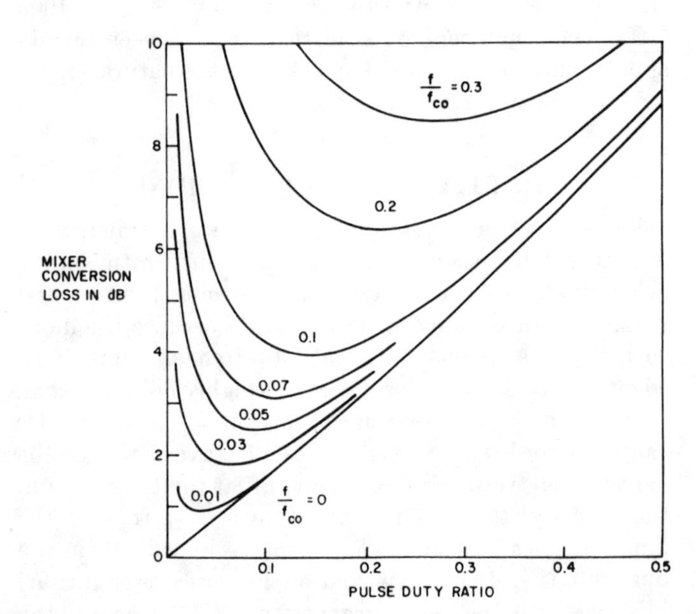

Fig. 4. Computed mixer conversion loss for the short-circuited image case.

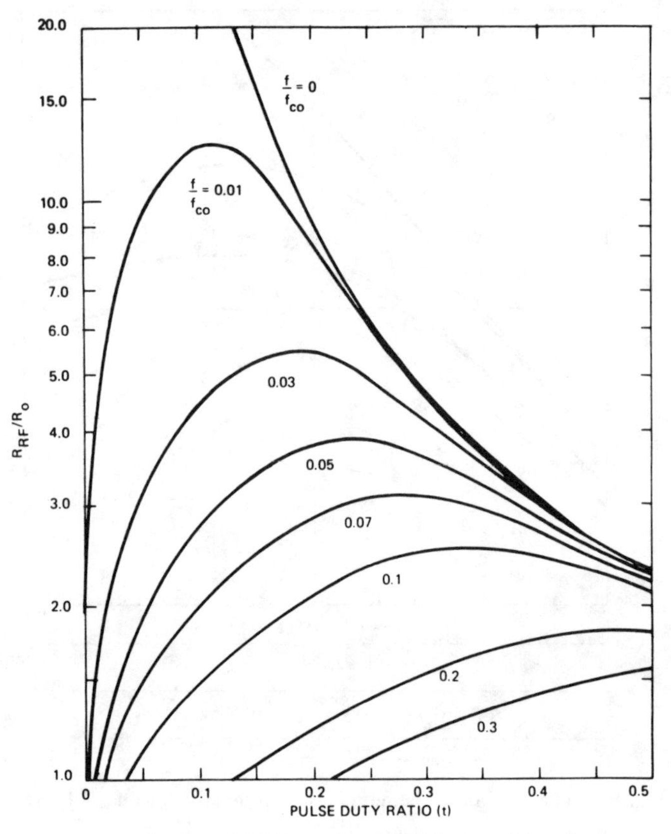

Fig. 5. Computed RF impedance for the broad-band case.

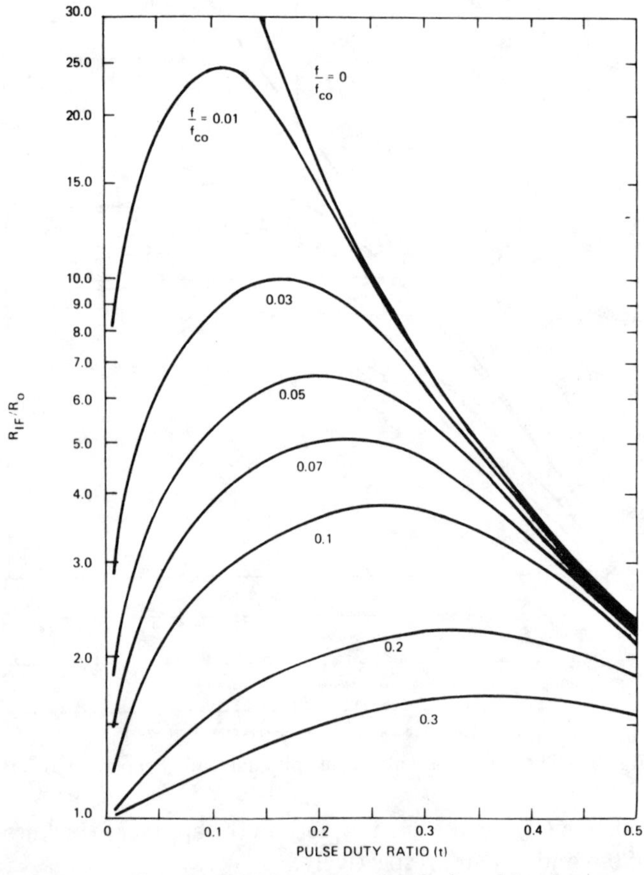

Fig. 6. Computed IF impedance for the broad-band case.

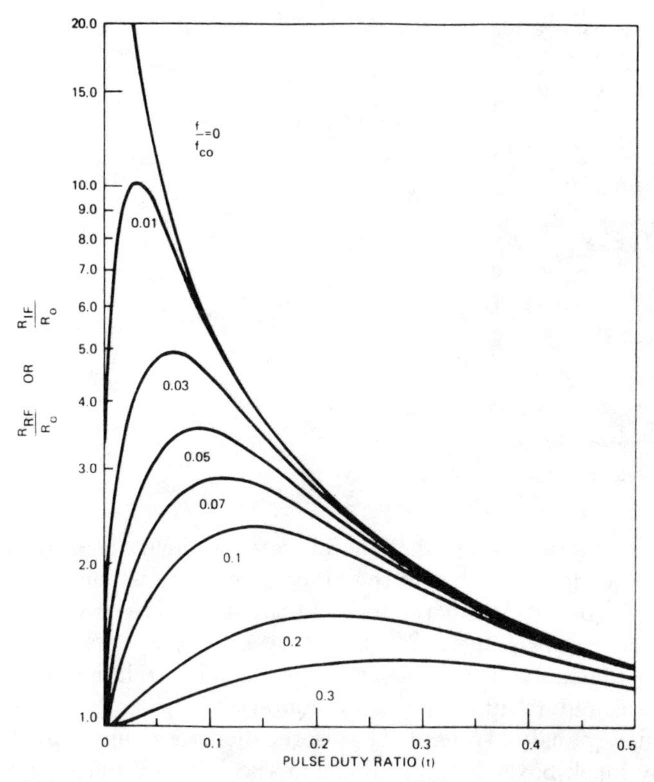

Fig. 7. Computed terminal impedance for the short-circuited image case.

This condition also results in an input impedance match. The RF and IF impedances have been computed as quantities normalized to R_0. R_0 is the average diode impedance and is well approximated by the simple expression

$$R_0 = R_s/t. \tag{1}$$

This is the time-averaged diode impedance and thus is the impedance presented to the LO.

The rectified dc current is a useful quantity for checking mixer operation and can be calculated by using (2).

$$I_{dc} \cong \left(\frac{Pt}{R_s}\right)^{1/2} \left[\frac{0.9 \sin(\pi t) - \sqrt{2} t \cos(\pi t)}{1 + t}\right] \tag{2}$$

where P is the LO power, t is the pulse-duty ratio, and R_s is the diode-limiting resistance.

The LO power can be estimated by (3).

$$P = \frac{t(V_0 - V_b)^2}{R_s[1 + \cos(2\pi t)]} \tag{3}$$

where V_0 is the forward potential drop of the Schottky barrier and V_b is the bias voltage. Typical values for V_0 are 0.75 V (GaAs), 0.5 V (Si-Schottky barrier), and 0.15 V (Si-point contact).

IV. RESULTS

Fig. 8 shows the implemented design which allows complete realization of the desired image-enchanced balanced mixer using GaAs Schottky-barrier diodes. This is a plan view of the microwave-integrated-circuit (MIC) mixer as viewed from the ground-plane side of the alumina substrate. The RF signal of frequency 9.3 GHz enters the substrate on the right edge via the microstrip (signal-input) port. The RF is coupled to the diodes via a broadband microstrip-to-slot-line transition and through a bandpass image-reject impedance-matching filter consisting of microstrip lines coupled to the slot line. The pair of mixer diodes terminate this filter.

The LO at a frequency of 7.8 GHz is injected via the LO-input microstrip terminal. The LO power then passes through the directional filter and to the mixer diodes by way of a microstrip-to-coplanar-line transition (pin through the substrate). Slot-line stubs at the end of the

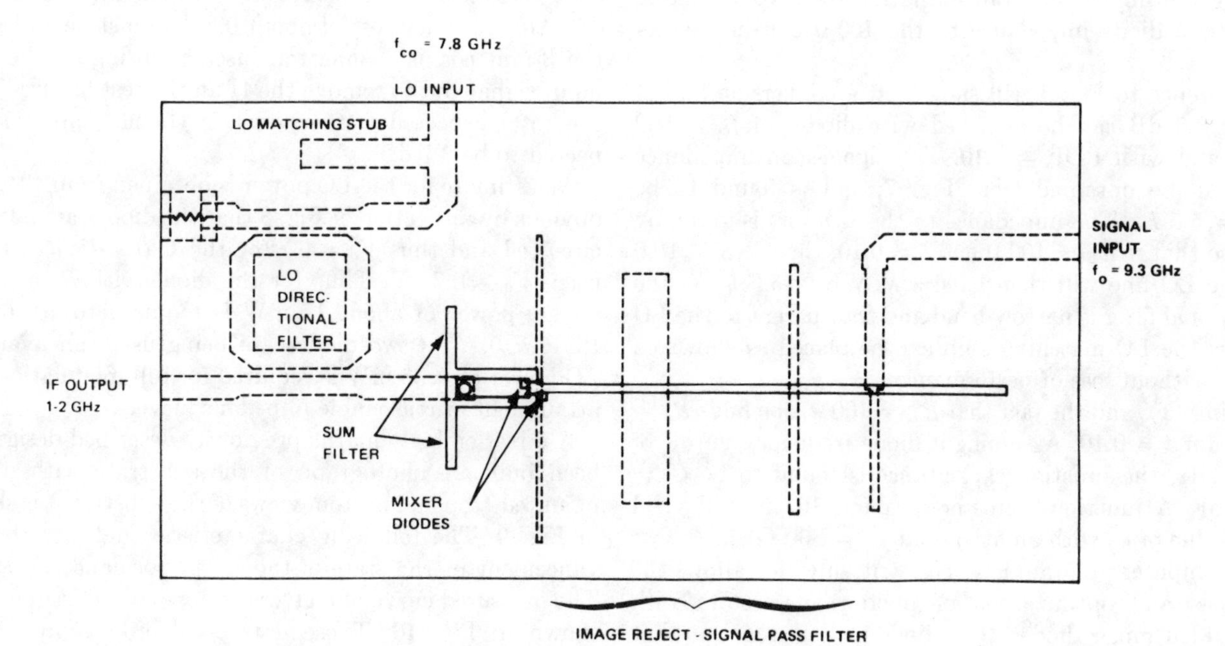

Fig. 8. *X*-band image-enhanced balanced mixer.

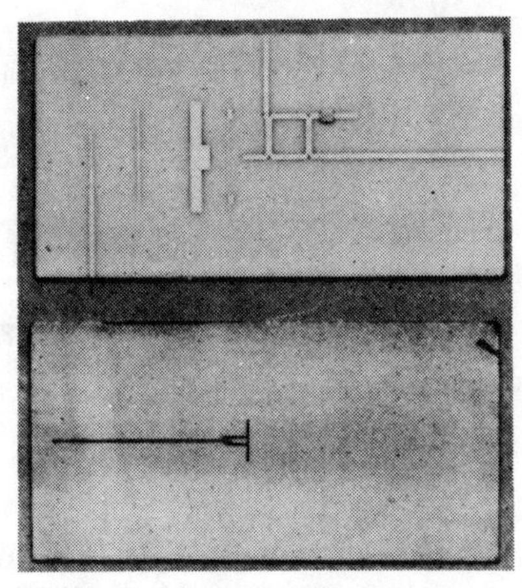

Fig. 9. Photograph of the X-band balanced mixer.

coplanar-line section present a short circuit to the mixer diodes at the sum frequency. The diodes are arranged such that the IF-output line is common with the LO input. The wide frequency separation of the IF and LO, and the directional- and frequency-selective properties of the directional filter allow very simple diplexing of the two signals with essentially zero bandwidth limiting of the IF port.

The two diodes are in parallel to the IF and LO ports, but in series to the signal port. As the conventional IF amplifier input impedance is nominally 50 Ω, the microstrip-to-coplanar-line characteristic impedance is set to 50 Ω. In a short-circuited image mixer the signal impedance is equal to the IF impedance. Thus the diode impedance must be 100 Ω, the two in parallel then matching the 50-Ω line. But the two in series will present 200 Ω to the signal slot line. The slot line can readily be made with a characteristic impedance of nominally 100 Ω. The 50-Ω microstrip to 100-Ω slot-line transition has well in excess of 10-percent bandwidth. The signal filter then is designed to supply the impedance transformation required to match the 200-Ω diode impedance to the 100-Ω slot-line impedance.

Reference to Fig. 4 will show that a conversion loss of under 2.0 dB can be achieved with diodes of $f_{co} \cong 800$ GHz and with PDR $\cong 0.10$. The signal-load impedance R_{RF} can be obtained from Fig. 7, and is found to be $R_{\mathrm{RF}} = 5.5\ R_0$. The impedance to the LO port is given by (1), so that if $R_{\mathrm{L}} = 100\ \Omega$ and $t = 0.10$, then $R_{\mathrm{LO}} = 18.0$ Ω. The LO line will then have a VSWR $\cong$ 5.5:1. As the directional filter is narrow-band and transparent to the LO power, the LO matching stub can be placed as shown in Fig. 8 without loss of performance.

Using (1), and the fact that $R_{\mathrm{L}} = 100\ \Omega$, one finds $R_s = 1.8\ \Omega$ for $t = 0.10$. Assuming a diode frequency cutoff of 800 GHz, the junction capacitance is found to be $C_j = 0.11$ pF. A junction diameter of about 10 μm will yield this value of C_j with an attendant $f_{co} = 800$ GHz.

A computer program has been written which allows the analysis and optimization of microwave circuits with embedded mixer diodes. The diode model used was that which assumes that the Schottky diode can be represented as a junction switched at the LO rate. In the ON condition the resistance is that of the limiting series resistance; and in the OFF state it is the series resistance in series with the barrier capacitance. The zero-bias capacitance was used for the analysis and appears to give good correlation with measured results. In this computer routine a three-frequency analysis is used. That is, the diodes are represented by black boxes with terminals at the RF, IF, and image frequencies with couplings set by the 3×3 conductance matrix. All other frequencies generated are assumed to be short circuited. A nodal analysis routine is used to model the external circuitry at the three frequencies of interest and the diodes are added in parallel. An optimization program is then used to vary circuit values until the desired results are obtained. The computer results tracked the measured values within 0.4 dB over the band.

The signal filter loss was estimated to be $\approx$0.85 dB. The conversion loss was $L_c \cong 1.6$ dB. The microstrip-to-slot-line transition is no more than 0.05 dB so that an overall conversion loss for the complete mixer should be about 2.5 dB. An additional loss of about 0.2 dB must be added for the 3-mm coaxial connectors used to bring in the RF-input signal and to remove the IF in the test fixture. Thus the total expected conversion loss (band center) is expected to be 2.7 dB.

An estimate for the LO power is obtained from (3). It is obvious by inspection of Fig. 8 that the diodes are dc short circuited and thus $V_b = 0$. For the GaAs devices being used, $V_0 \cong 0.75$ V. Thus for the diode with $R_s = 1.8\ \Omega$, an LO power of about 17 mW is required to attain the PDR = 0.1. As two diodes are being used, an available LO power of $\approx$35 mW is required for full modulation and attainment of reasonable impedance levels.

A mixer embodying the previously described design has been built. A photograph of the substrate with diodes mounted (top and bottom views of the substrate) is shown in Fig. 9. The following characteristics indicate the advancement of the state-of-the-art performance obtained. The measured curve of conversion loss versus frequency is shown in Fig. 10. These data (Table I) represent all circuit losses including connector losses.

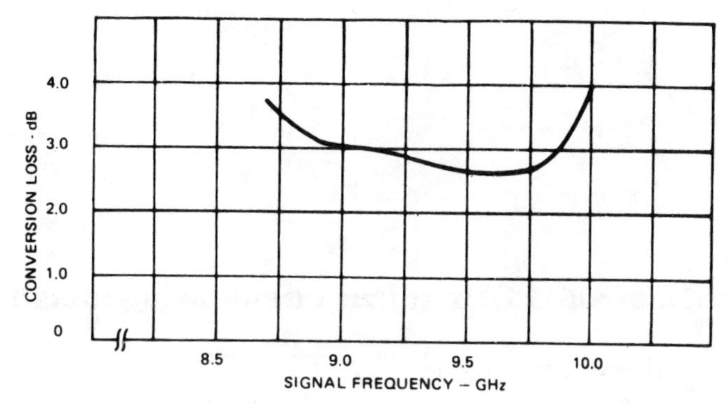

Fig. 10. Curve of conversion loss versus frequency.

TABLE I
BALANCED MIXER PERFORMANCE

Signal frequency (center)	9.4 GHz
LO frequency	7.8 GHz
Signal bandwidth	1.0 GHz
Conversion loss (1.0 - GHz band)	≤3.15 dB
Conversion loss (0.5 - GHz band)	≤ 2.8 dB
Conversion loss (best point)	2.6 dB
LO power	≈ 40 mW
Dynamic range (input signal for 1.0 - dB compression)	+ 13 dBm
Image band isolation	> 25 dB
VSWR (across signal band)	< 1.4:1

REFERENCES

[1] E. W. Herold *et al.*, "Conversion loss of diode mixers having image frequency impedance," *Proc. IRE*, vol. 33, pp. 603–609, Sept. 1945.

[2] H. C. Torrey and C. A. Whitmer, *Crystal Rectifiers* (M.I.T. Radiation Lab. Series), vol. 15. New York: McGraw-Hill, 1948.

[3] H. A. Watson, *Microwave Semiconductor Devices and Their Circuit Applications*. New York: McGraw-Hill, 1969.

[4] M. R. Barber, "Noise figure and conversion loss of the Schottky barrier mixer diode," *IEEE Trans. Microwave Theory Tech.*, vol. MTT-15, pp. 629–635, Nov. 1967.

[5] J. B. Cahalan *et al.*, "An integrated, *X*-band, image and sum frequency enhanced mixer with 1 GHz IF," in *1971 G-MTT Symp. Digest*, pp. 16–17, May 1971.

[6] S. M. Sze and G. Gibbons, "Avalanche breakdown voltages of abrupt and linearly graded p-n junctions in Ge, Si, GaAs, and GaP," *Appl. Phys. Lett.*, vol. 8, pp. 111–113, Mar. 1966.

[7] L. E. Dickens, "Low conversion loss millimeter wave mixers," in *1973 G-MTT Symp. Digest*, pp. 66–68, June 1973.

Optimization of an 11 GHz mixer circuit using image recovery

C. J. BURKLEY†‡ and R. S. O'BRIEN†

A detailed theoretical study of an 11 GHz balanced image recovery mixer is presented in which the relationships between complex terminations at the signal, image and i.f. ports and the image power generated are considered. The low level of image power implies that the concept of image recovery is considerably more complex than hitherto realized. Furthermore, the i.f. port impedance is complex in general and very dependent on the phase of the image termination but this dependence can be reduced by changing the l.o. drive or by applying a suitable d.c. bias to the mixer diodes.

1. Introduction

The sensitivity of a microwave receiver is limited by the conversion loss of the mixer diodes. Dissipation of power in the unwanted harmonics which are generated during the mixing process becomes the most important contribution to the conversion loss when the high quality diodes that are currently available are considered (Barber 1967, Johnson 1968, Stracca *et al.* 1973). The image frequency is considered to be the most important harmonic and thus the principal terms, which will influence the behaviour of the mixer, are the signal, image and i.f. frequencies (Katoh and Akaiwa 1971, Oxley *et al.* 1971). Since the generation of image power results in a loss in the available signal power, it is desirable not only to control image generation, but to recover the power if possible. Accordingly, the variation of image power with i.f., signal and image terminations has been studied with a view to optimizing conditions for its recovery. Previously the image power has been calculated, assuming that the various mixer ports were terminated in pure conductances (O'Neill 1965) but the present paper considers *complex* terminations, thereby enabling a more general model to be developed. The i.f. output impedance is found to be very dependent on the phase of the image reflection and hence the possibility of reducing this dependence by changing the d.c. operating conditions of the diode has also been investigated.

The effect of the series resistance and non-linear junction capacitance of the diode, which have relatively little effect below a l.o. voltage of 0·5 V (Fleri and Cohen 1973) and the effect of barrier capacitance which increases the conversion loss by a constant factor (Torrey and Whitmer 1948, Liechti 1970) are excluded from the present study as only the relative decreases in conversion loss are considered here.

2. Theory

The mixer is considered as a passive linear network having complex terminations at the signal, image and i.f. ports (Fig. 1) and can be represented by an

Received 14 January 1975.

† Department of Electrical Engineering, University College, Cork, Ireland.

‡ Present address: European Space Research and Technology Centre, Noordwijk, Holland.

Reprinted with permission from *Int. J. Electron.*, vol. 38, pp. 777–787, June 1975.

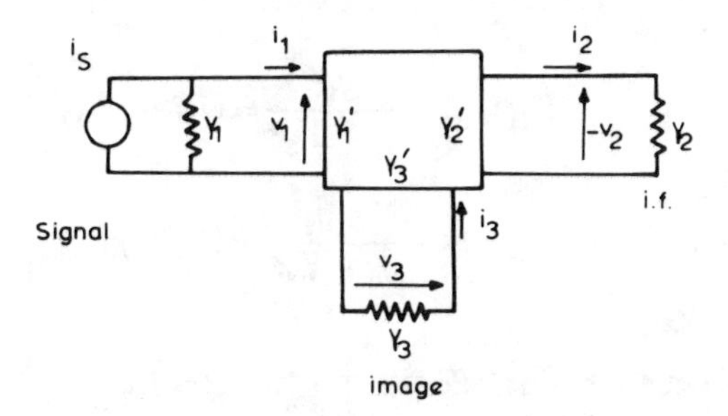

Figure 1. Linear network representation of mixer.

admittance matrix equation :

$$\begin{vmatrix} i_1 \\ i_2 \\ i_3^* \end{vmatrix} = \begin{vmatrix} y_0 & y_1 & y_2 \\ y_1 & y_0 & y_1^* \\ y_2 & y_1^* & y_0^* \end{vmatrix} \begin{vmatrix} v_1 \\ v_2 \\ v_3^* \end{vmatrix} \tag{1}$$

where y_0, y_1 and y_2 are incremental admittances. If the diode follows the well-known exponential characteristic

$$i = I_S\,[\exp(\alpha V) - 1]$$

where I_S is the reverse saturation current and α is the exponential coefficient, then the incremental admittances are given by

$$y_n = \alpha I_S \exp(\alpha V_{d.c.}) I_n(\alpha V_0), \quad n = 0, 1, 2 \tag{2}$$

where I_n is a modified Bessel function of order n, V_0 is the local oscillator voltage amplitude and $V_{d.c.}$ is the direct bias voltage. Thus y_0, y_1 and y_2 are real and can be replaced by the incremental conductances g_0, g_1 and g_2 respectively.

For image recovery, the load admittance at the image terminal $Y_3(=G_3+jB_3)$ will be different from the signal source admittance $Y_1(=G_1+jB_1)$ and if the load admittance at the i.f. terminal is $Y_2(=G_2+jB_2)$ then, from Fig. 1 :

$$\left.\begin{aligned} i_1 &= i_S - v_1 Y_1 \\ i_2 &= -v_2 Y_2 \\ i_3^* &= -v_3^* Y_3^* \end{aligned}\right\} \tag{3}$$

Y_1 is kept equal to the complex conjugate of Y_1', thus ensuring maximum power transfer between the signal source and the mixer. In practice the l.o. matched condition, which gives a slightly higher conversion loss (Anand 1971), is sometimes preferred, due to the simplicity with which it may be set up.

To calculate the mixer conversion loss L_c, a direct relationship between the signal and i.f. ports is necessary and eliminating i_3 and v_3 from eqns. (1) and (3) gives

$$\begin{vmatrix} i_1 \\ i_2 \end{vmatrix} = \begin{vmatrix} Y_{11} & Y_{12} \\ Y_{21} & Y_{22} \end{vmatrix} \begin{vmatrix} v_1 \\ v_2 \end{vmatrix} \tag{4}$$

where

$$\left.\begin{aligned} Y_{11}=g_0-\frac{g_2^2}{g_0+Y_3^*}=G_{11}+jB_{11} \\ Y_{12}=Y_{21}=g_1-\frac{g_1g_2}{g_0+Y_3^*}=G_{12}+jB_{12} \\ Y_{22}=g_0-\frac{g_1^2}{g_0+Y_3^*}=G_{22}+jB_{22} \end{aligned}\right\} \tag{5}$$

The signal input and i.f. output powers and hence L_c can now be calculated. L_c is then minimized with respect to the signal load admittance Y_1 (i.e. matched signal input conditions). This gives

$$L_c=\frac{1+\sqrt{(1-\epsilon)}}{1-\sqrt{(1-\epsilon)}} \tag{6}$$

where

$$\epsilon=\frac{G_{12}^2+B_{12}^2}{B_{12}^2+G_{11}G_{22}} \tag{7}$$

Equations (6) and (7) express the variation of conversion loss with the image termination indirectly. A direct relationship is obtained by rearranging eqn. (5) and substituting into eqn. (7) to give

$$\epsilon=\frac{g_1^2\{(g_0+G_3)^2+B_3^2\}\{(g_0+G_3-g_2)^2+B_3^2\}}{g_0^2\left\{(g_0+G_3)\left(g_0+G_3-\frac{g_2^2}{g_0}\right)+B_3^2\right\}\times\left\{(g_0+G_3)\left(g_0+G_3-\frac{g_1^2}{g_0}\right)+B_3^2\right\}+(g_1g_2B_3)^2} \tag{8}$$

The i.f. output admittance of the mixer circuit can be calculated from eqns. (3) and (4) giving

$$Y_2'=Y_{22}-\frac{Y_{12}}{Y_{11}+Y_1}$$

and substituting the value of Y_1 corresponding to minimum conversion loss gives

$$Y_2'=G_{22}\left[1-\frac{G_{12}^2-B_{12}^2}{G_{11}G_{22}}-\left(\frac{G_{12}B_{12}}{G_{11}G_{22}}\right)^2\right]^{1/2}+j\left[B_{22}-\frac{G_{12}B_{12}}{G_{11}}\right] \tag{9}$$

Thus, expressions have been derived for conversion loss, signal load admittance and i.f. output admittance as functions of image termination for all cases where the signal and i.f. ports are properly matched. These parameters can now be calculated when the diode parameters (and hence from eqn. (2) the incremental conductances) are known.

2.1. *Image power*

The image power becomes an important parameter when an image recovery circuit is envisaged. By obtaining a direct relationship between the signal and image ports an expression for the available image power, relative to the signal input power, can be obtained. Eliminating i_2 and v_2 from eqns. (1) and (3) gives :

$$\begin{vmatrix} i_1 \\ i_3^* \end{vmatrix} = \begin{vmatrix} Y_{13} & Y_{33} \\ Y_{33} & Y_{13} \end{vmatrix} \begin{vmatrix} v_1 \\ v_3^* \end{vmatrix} \tag{10}$$

where

$$\left. \begin{aligned} Y_{13} &= g_0 - \frac{g_1^2}{g_0 + Y_2} = G_{13} + jB_{13} \\ Y_{33} &= g_2 - \frac{g_1^2}{g_0 + Y_2} = G_{33} + jB_{33} \end{aligned} \right\} \tag{11}$$

Y_3 can now be evaluated in the same manner as was Y_2 and the ratio of the image to signal powers is found to be

$$P_{\text{im}} = \frac{P_3}{P_1} = \frac{G_1 |Y_{33}|^2}{G_{13}|Y_1 + Y_{13}|^2 - (G_{33}^2 - B_{33}^2)(G_1 + G_{13}) - 2G_{33}B_{33}(B_1 + B_{13})} \tag{12}$$

3. Circuit characteristics

The theoretical behaviour of an experimental 11 GHz balanced image recovery mixer, with an i.f. of 70 MHz, has been analysed. MA492EMR diodes are incorporated into the mixer and from their d.c. characteristics the parameters α and I_S were evaluated as 29·3 V^{-1} and 0·375 μA respectively. Image recovery is effected by locating an image rejection filter in the signal input circuit. A three-cavity elliptic function filter, which had a stopband rejection of greater than 30 dB and a pass-band ripple of less than 0·1 dB was designed. Since its performance adequately met the needs of an image rejection filter its characteristics were incorporated into the theoretical analysis of the mixer.

4. Theoretical results

The relationships between the conversion loss, image power generation and signal load and i.f. output admittances as functions of the image termination have been calculated. The image termination is changed by varying the electrical distance between the mixer and the image rejection filter.

(i) Conversion loss. L_c can be calculated from eqns. (6) and (8) for any arbitrary image termination and is plotted for the case of a lossless rejection filter in Fig. 2 (*a*) and for a lossy filter in Fig. 3 (*a*)—the mixer is assumed to be correctly matched at the signal and i.f. ports for both cases. To give an indication of the improvements available, the broadband conversion loss is also plotted. With the lossless filter an improvement of the order of 1·6 dB in L_c is obtained—the exact amount of the improvement will depend on the quality of the diodes. When a filter having a reflection loss of 0·94 dB at the image frequency is considered the improvement in conversion loss is about 0·1 dB less. However, the transmission loss of the filter will further degrade the overall conversion loss by adding directly to it, so assuming a transmission loss of 0·1 dB the overall degradation in conversion loss would be 0·2 dB.

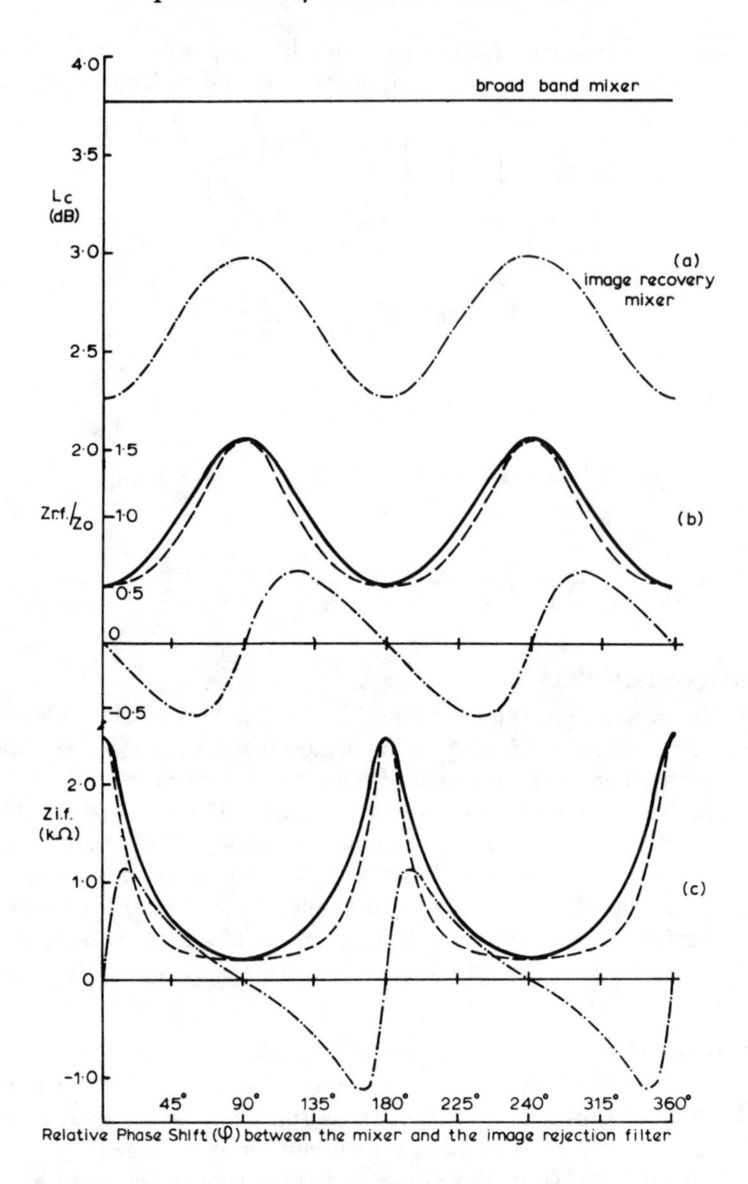

Figure 2. Variation of (*a*) conversion loss, L_c, (*b*) input impedance, $Z_{r.f.}$, relative to local oscillator characteristic impedance, Z_0, and (*c*) output impedance, $Z_{i.f.}$ of mixer circuit with phase of image termination ψ when the image frequency is terminated in a lossless filter. ——— : Impedance : - - - - : resistance ; — · — · — : reactance.

(ii) Input and output impedances. Introduction of image recovery also affects the input and output impedances and knowledge of the changes in these is necessary to match the mixer circuit correctly. These complex impedances are calculated for the same range of image termination (plotted for convenience as the relative phase shift between the mixer and the image rejection filter) as the conversion loss and are plotted together with their resistive and reactive components in Figs. 2 (*b*) and 2 (*c*) respectively for a lossless filter and in

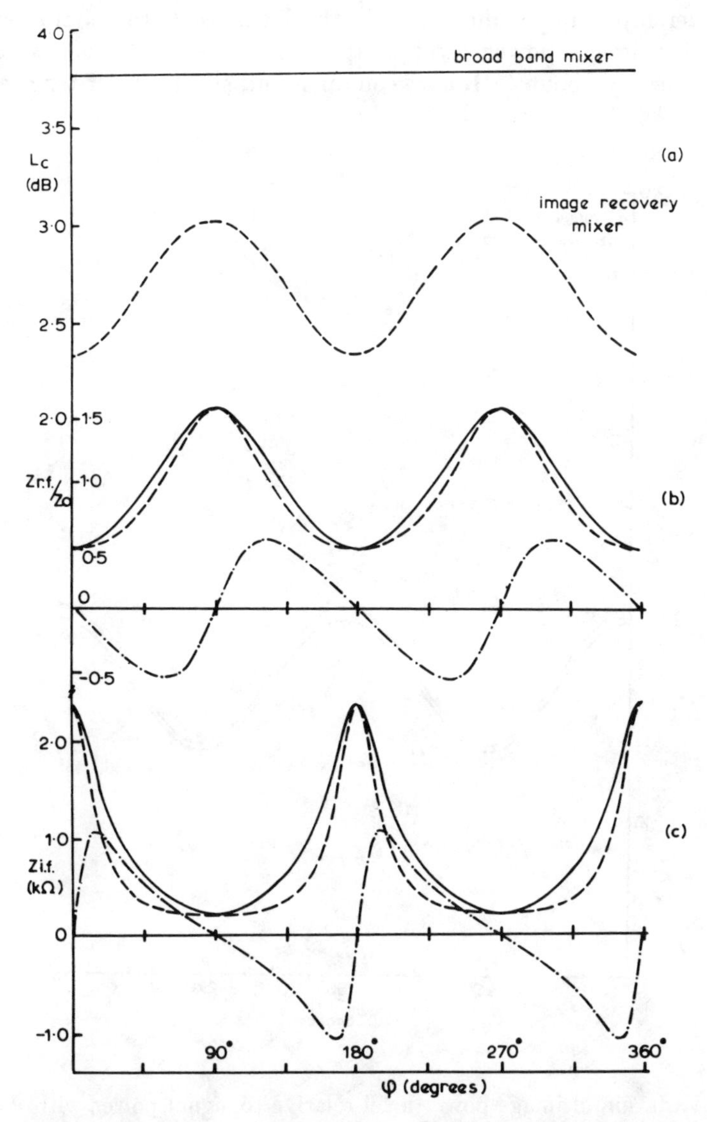

Figure 3. Variation of (*a*) L_c, (*b*) $Z_{r.f.}/Z_0$ and (*c*) $Z_{i.f.}$ with ψ, when the image frequency is terminated in a filter with a reflection loss of 0·94 dB.

Figs. 3 (*b*) and 3 (*c*) for the lossy filter case. The results show that the output i.f. impedance varies between 0·2 and 2·5 kΩ and is a pure resistance only for the cases of open-circuited and short-circuited image termination : at intermediate points a reactive term is also present. The r.f. impedance also varies over quite a considerable range, both in phase and magnitude. If the image rejection filter has a reflection loss of 0·94 dB then neither the i.f. nor the r.f. impedances are changed very appreciably.

(iii) Image power. The calculated image power level as a function of image termination is plotted in Fig. 4, assuming that matched output conditions are maintained, for the cases where the signal impedance is matched to (*a*) the mixer input impedance and (*b*) the local oscillator characteristic impedance. The image power level is approximately 12 dB below the signal level and varies by about 4 dB between open and short circuit terminations.

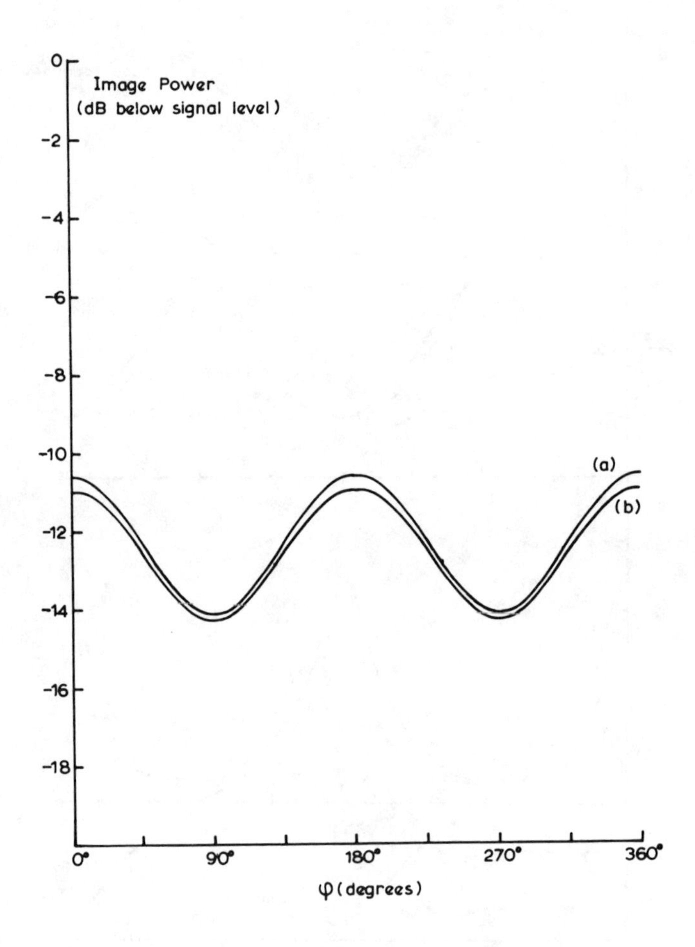

Figure 4. Variation of image power level relative to signal power with ψ when the signal impedance is matched to (*a*) the mixer input impedance $Z_{r.f.}$ and (*b*) the local oscillator characteristic impedance Z_0.

5. Discussion of results

The basic model of a broadband mixer assumes it to be a lossless linear detector with the incident signal power dividing equally in the mixing process—half being converted to i.f. power and half to the image frequency (O'Neill 1965, Oxley *et al.* 1971). The image power is subsequently dissipated uselessly in the broadband termination at the signal port and the amount of useful i.f. power then available will depend on the quality of the diodes and the match at the mixer i.f. output port. Thus for an ideal diode and a perfect match the conversion loss would be 3 dB. In an image recovery circuit the image signal is reactively terminated and reflected back to the mixer for conversion to i.f.

The conversion efficiency will then be improved or degraded depending on the phase of the reflected signal (Anand and Moroney 1971, Dorsi 1971). However, this model does not explain why the image is rejected initially by the mixer if, subsequently on reflection from the rejection filter, it is converted to i.f. Also if in fact half the signal input power is reflected as image then by reciprocity half the image power on reflection back to the mixer should be converted to the original signal frequency and subsequently rejected by the mixer to be uselessly dissipated in the signal load admittance. Thus the maximum improvement in the ideal case using image recovery techniques should be only 1·5 dB and not 3 dB as predicted by the full image recovery model.

The results of the present analysis are in sharp disagreement with the simple model of image recovery. Firstly, it is seen that when image recovery techniques are employed, an improvement in L_c is obtained regardless of the phase of the image reflection and secondly the level of image power is very low and even its complete recovery would give a maximum improvement in L_c of only 0·4–0·5 dB. Thus the concept of image recovery is quite complex and not simply a question of reconverting the normally wasted image power into useful i.f. power. The image and i.f. terminations are also important parameters and optimizing them results in an improvement in L_c better than that obtained by recovering the image alone. Reactively terminating the image frequency may affect the power generated and recovered at other higher order harmonics which in fact may not be negligible relative to the image, thus accounting for the greater reduction in L_c. (Note that recovery of the power at a particular frequency has the same effect on the overall conversion efficiency as if that particular frequency was never in fact generated.) Stracca (1969) has investigated the effects of terminating the various harmonics and concludes that they do in fact contribute to the overall L_c, but no details regarding the power levels of the various harmonics and their variation with the termination loads is yet available. Fekete (1971), in an examination of an image recovery circuit, concluded that there is a special relationship between the signals at the input and image frequencies and that voltage addition rather than power addition occurs at the i.f. Thus recovery of the image frequency leads to a greater reduction in L_c than would normally be expected. Fekete used a two mixer arrangement and injected the image frequency from one into the input port of the other. However, in view of the dependence of the image generation on the image and i.f. output impedances, this approach may alter the internal operation of the diodes and hence the level of image generation. In fact the mode of operation of this circuit was subsequently disputed (Howson and Gardiner 1972) and it was suggested that the action of the two mixers produced image cancellation rather than image recovery. Thus voltage rather than power addition at the i.f. may be only an apparent effect, introduced by the interaction between the two mixers.

Optimum image termination appears to be an open circuit but the choice of termination will also be influenced by the input and output impedances. For an open-circuit image termination the signal input impedance and i.f. output impedance are both maximum and purely resistive. The high i.f. output impedance could in practical circuits lead to serious mismatch losses due to stray capacitances and therefore it may be preferable to terminate the image port in some other impedance which would give a lower and more definitely defined impedance in return for a slightly higher conversion loss. The i.f. impedance will now have a reactive as well as a resistive component and a complex conjugate match will be necessary.

The input and output impedances of a mixer are also affected by the diode operating conditions. Hence an alternative method of providing an i.f. output impedance suitable for direct matching into the i.f. amplifier is to vary the diode operating conditions. There are two parameters which can be varied readily—the l.o. power and the d.c. bias on the diode. .

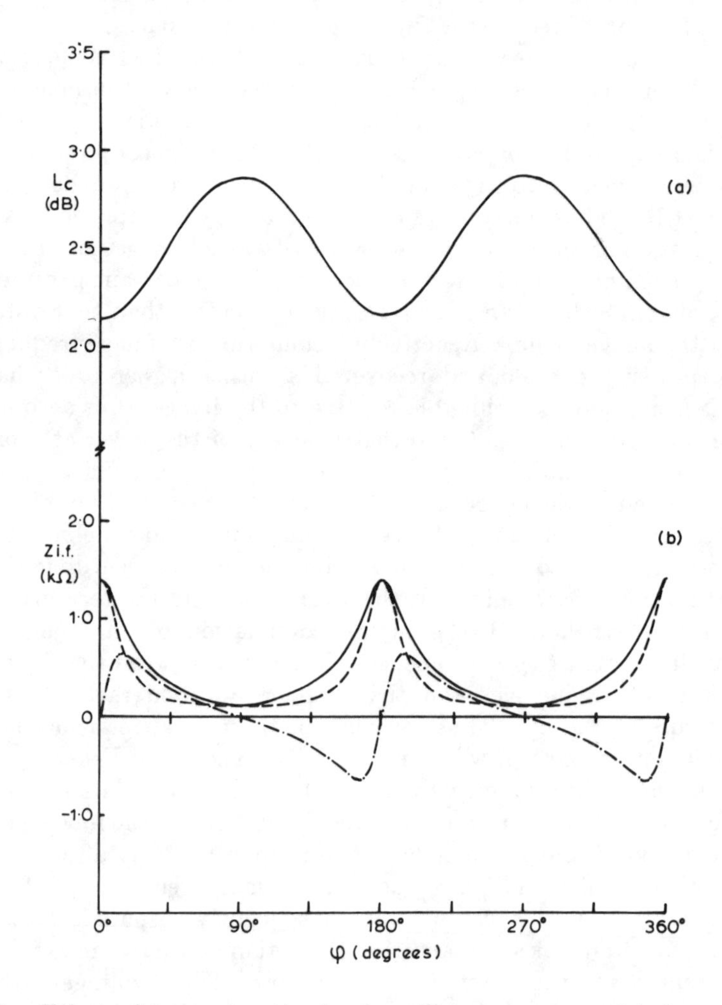

Figure 5. Effect of increasing the local oscillator power on (*a*) L_c and (*b*) $Z_{i.f.}$

Figure 5 shows that the effect of increasing the local oscillator power which leads to the rectified current increasing from 0·5 to 1·0 mA, is to decrease the conversion loss slightly and at the same time the i.f. impedance is almost halved. Since in practice L_c reaches a minimum over a small range of l.o. power the rectified current must preferably be chosen within this range of values thereby limiting the range of output impedances that can be achieved by this means.

The effect of applying a d.c. bias voltage to the mixer diodes is shown in Fig. 6. L_c and the i.f. output impedance for the mixer with l.o. drive of 0·5 mA and a forward bias of 0·05 V are presented. This forward bias voltage reduces the i.f. output impedance from a maximum of 2·5 kΩ to a maximum of 0·5 kΩ. This maximum impedance still coincides with the open-circuit image port but its variation about this value is much more gradual, so that small variations in

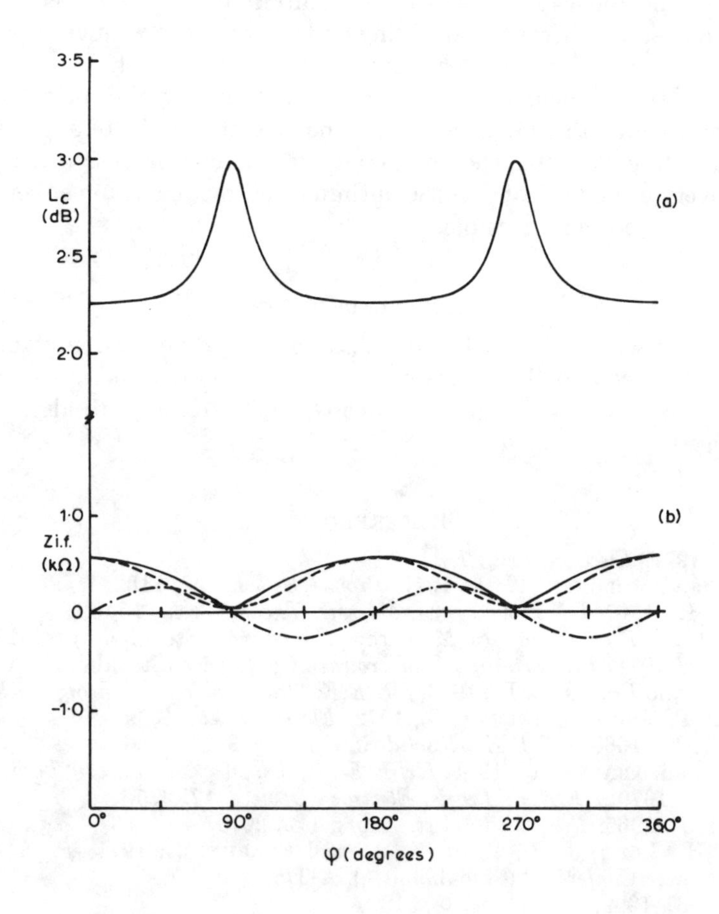

Figure 6. **Effect of applying a d.c. bias to mixer diodes on (*a*) L_c and (*b*) $Z_{i.f.}$**

the phase of the image termination will not lead to severe mismatches at the output. Also the conversion loss will now remain close to a minimum value over most of the range of image terminations. Thus the application of a small d.c. bias voltage to the diode in the forward direction can considerably reduce the matching problems associated with the mixer without significantly degrading the conversion loss, especially for a mixer whose image port is terminated in an open-circuit.

6. Conclusions

The study clearly shows that the concept and implementation of image recovery techniques are considerably more complex than was initially envisaged. This is principally due to the complex interdependence of the various parameters involved. The level of the image power itself is found to be less than that necessary to give the improvement obtained by apparently recovering it, indicating that reactively terminating the image significantly affects the operation of the diodes themselves. The different output impedances vary sharply and over a wide range depending on the phase of the image termination so that precise matching is necessary to avoid mismatch losses. The i.f. output impedance which is critical from a matching point of view can be modified by varying the d.c. bias on the diodes or the applied l.o. power. The analysis therefore indicates the complexity of the circuit design necessary in image recovery circuits to obtain the optimum improvement in conversion loss, hitherto considered quite feasible.

Acknowledgments

The authors wish to thank Dr. M. C. Sexton for advice and guidance. The financial assistance and the award of a post-doctoral fellowship to one of the authors (C.J.B.) by the National Science Council, Dublin, are also gratefully acknowledged.

References

Anand, Y., 1971, *Electron. Lett.*, **7,** 11.
Anand, Y., and Moroney, W. J., 1971, *Proc. I.E.E.E.*, **59,** 1182.
Barber, M. R., 1967, *I.E.E.E. Trans. Microw. Theory Tech.*, **15,** 629.
Dorsi, D., 1971, *Proc. European Microwave Conference* (Stockholm), p. A11/4.
Fekete, J. P., 1971, *Proc. European Microwave Conference* (Stockholm), p. A12/1.
Fleri, D. A., and Cohen, L. D., 1973, *I.E.E.E. Trans. Microw. Theory Tech.*, **21,** 39.
Howson, D. P., and Gardiner, J. G., 1972, *Electron. Lett.*, **8,** 352.
Johnson, K. M., 1968, *I.E.E.E. Jl Solid-St. Circuits*, **3,** 50.
Katoh, M., and Akaiwa, Y., 1971, *I.E.E.E. Trans. Microw. Theory Tech.*, **19,** 634.
Liechti, C. I., 1970, *I.E.E.E. Trans. Electron. Devices*, **17,** 975.
O'Neill, H. J., 1965, *Proc. Instn elect. Engrs*, **112,** 2019.
Oxley, T. H., Lord, J. F., Ming, K. J., and Clarke, J., 1971, *Proc. European Microwave Conference* (Stockholm), p. A11/5.
Stracca, G. B., 1969, *Alta Freq.*, **38,** 318.
Stracca, G. B., Aspesi, F., and D'Arcangelo, T., 1973, *I.E.E.E. Trans. Microw. Theory Tech.*, **21,** 544.
Torrey, H. C., and Whitmer, C. A., 1948, *Crystal Rectifiers* (New York : McGraw-Hill).

Wide-Band X-Band Microstrip Image Rejection Balanced Mixer

***Abstract*—A compact image rejection balanced mixer developed in microstrip medium is described. This integrated mixer gives over 20-dB image rejection and 10.6 ±0.6-dB noise figure (including 2.5-dB IF contribution) over 8- to 12-GHz range; it was designed on a 40-ohm impedance basis and is scalable to any 40-percent fractional bandwidth in the 1- to 12-GHz range.**

This correspondence describes a wide-band *X*-band image rejection mixer constructed in microstrip using a thin-glazed alumina substrate. This mixer has a 10- to 11.3-dB noise figure and over 20-dB image rejection over an 8- to 12-GHz frequency band. Image rejection is achieved by using a pair of balanced mixers with signal and LO voltages fed in quadrature and in phase, respectively; the two IF outputs are combined in a quadrature hybrid and channeled into the sum-output port, while the image is dissipated in the internal resistor at the difference-output port. Such circuits have been previously constructed in waveguide and conventional stip transmission line; the microstrip mixer described herein achieves comparable electrical performance in a much smaller volume and weight than *X*-band designs developed in these other transmission media.

All of the mixer components, except the IF power combining circuit, are located on a 1.960- by 1.375-inch alumina substrate as shown in Fig. 1. Fig. 2 shows the same components in a block diagram form. The remaining IF components are contained within a 1.000- by 1.375-inch area of another substrate as shown in Fig. 3. Both substrates are glazed 0.023-inch-thick alumina, covered with vacuum-deposited copper over a chromium layer on either side.

It is interesting to note that the RF portion of the mixer was designed on a 40-ohm impedance basis. (This is a deviation from the usual 50-ohm basis.) This resulted in lower impedance levels for all the coupler branches in the 3-dB 90° hybrids. Consequently, the two outer branches could be widened from less than 0.001 to 0.004 inch, which is much more compatible with the presently available thin-film etching techniques. The 50- to 40-ohm quarter-wavelength transformers were built at the signal-input port. All three 3-dB 3-branch 90° hybrids exhibited over 15-dB minimum (about 30-dB peak) isolation; the split-power outputs were 3.5 ±0.4 dB below the input signal level. Dissipation losses were estimated at 0.3 to 0.6 dB per wavelength (subject to the line impedance level) based on data collected during evaluation tests. Microstrip-to-OSM connector transitions (maximum VSWR 1.15:1) were developed on this project since commercially available units were found to be inadequate at that time.

The 3-dB power divider used for LO insertion also performed a 50- to 40-ohm impedance transformation [1] by having the internal resistor changed from 100 to 80 ohms and the quarter-wavelength segment impedances reduced from 70.7 to 63.1 ohms.

The RF rejection filters are made of two open-circuited low-impedance stubs, each filter providing 50-dB rejection at its resonance point. The dc blocking capacitors and the IF rejection filters (self-resonant at 200 MHz) are commercially available miniature units. The beam-lead Schottky barrier diodes D5818 are made by Sylvania. The dc ground return loop was completed in the LO input line by a grounded quarter-wavelength gold ribbon.

Manuscript received June 18, 1970; revised July 22, 1970. This work was sponsored by the Rome Air Development Center under Contract AF 30(602)-4164 and the Air Force Avionics Laboratory under Contract F33615-69-C-1859. This paper was presented at the 1970 International Microwave Symposium, Newport Beach, Calif., May 11–14.

Reprinted from *IEEE Trans. Microwave Theory Tech.*, vol. MTT-18, pp. 1181–1182, Dec. 1970.

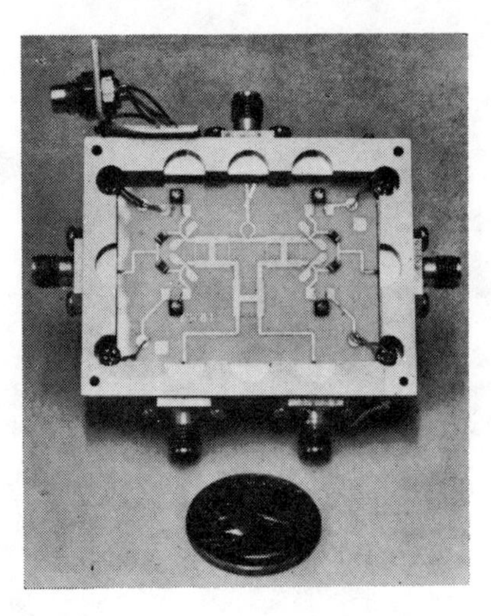

Fig. 1. Wide-band X-band image rejection balanced mixer.

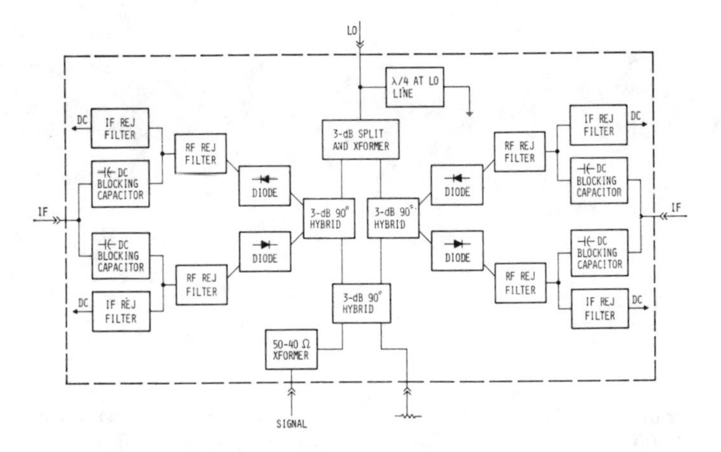

Fig. 2. Block diagram of image rejection balanced mixer shown in Fig. 1.

The lines in the IF power combiner and the 90° delay line (equivalent of a quadrature hybrid) were tightly folded on the substrate surface to conserve space. This resulted in some stray coupling effectively shortening the electrical lengths of the lines. A nearly 30-percent length adjustment (above the calculated nominal values) was needed to retain quarter wavelength at 200 MHz. Isolation between the two division ports at that frequency was measured at 31 dB.

The beam-lead diodes were tested singly on a network analyzer to adjust RF impedance matching. Embedding the diodes in the 50-ohm line, without any transformers, gave sufficient (if not the optimum) broad-band matching from 1 to 12 GHz. The impedance display over the 8- to 12-GHz range showed 2:1 maximum VSWR, and 1.5:1 or less over most of the band.

When the balanced mixer units were measured, it was noted that the noise figure was improved slightly when the 40-ohm impedance basis was used (with the diodes situated directly in the 40-ohm line). The best noise figure achieved with the balanced mixer unit over the 8- to 12-GHz frequency range was 10.0±0.4 dB (single-sideband noise figure, including the 2.5-dB IF noise figure) at 3 to 4-mW LO power input and 0.1-volt dc forward bias. The LO leakage into the signal port was a minimum of −16 dB for 2-mW LO power and 0.1-volt dc bias.

The initial tests of the image rejection balanced mixer indicated that optimum performance was achieved at 0.1-volt forward dc bias and 2 to 4-mW LO power input levels (0.5 to 1.0-mW per diode). Image rejection, as shown in Fig. 4, does not deteriorate when LO power is reduced to 1 mW, but noise figure suffers under those conditions by about 0.5 dB. At 2-mW LO power input, image rejection averages 23 dB over the 8- to 12.4-GHz range.

Fig. 3. IF power combiner with 90° delay line.

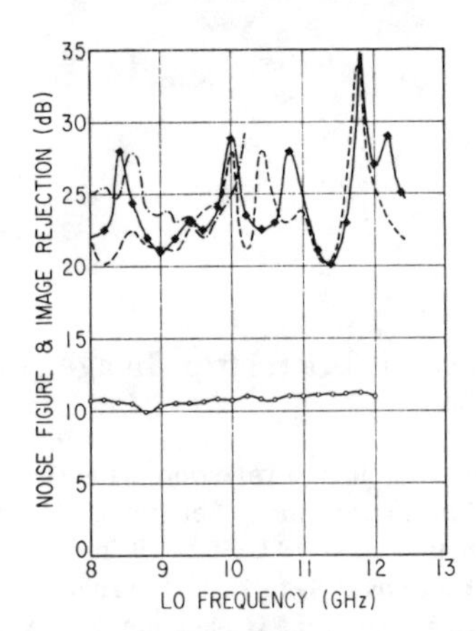

Fig. 4. Image rejection and noise figure versus frequency. IR LO power: solid line, 2mW; dashed line, 4 mW; dashed-and-dotted line, 1 mW. NF LO power: dashed-and-circled line, 2 mW. dc forward bias 0.1 volt.

Noise figure data for 0.1-volt dc forward bias and 2-mW LO power input is also given in Fig. 4. It ranges from 10 to 11.3 dB, including the 2.5-dB IF noise figure; it can be slightly improved with increasing LO power levels, but that eventually decreases image rejection.

The RF impedance data on the beam-lead Schottky barrier type diode used in the mixer indicates that the diodes can be well matched over the 1- to 12-GHz range. Therefore, we believe that this X-band mixer design is easily scalable to any 40-percent fractional bandwidth within the 1- to 12-GHz range.

ACKNOWLEDGMENT

The authors thank K. Mair for his assistance in making the necessary measurements, and P. Meier for providing the special X-band microstrip launcher design.

G. P. KURPIS
J. J. TAUB
AIL Div.
Cutler-Hammer, Inc.
Melville, N.Y. 11749

REFERENCES

[1] J. J. Taub and G. P. Kurpis, "A more general N-way hybrid power divider," *IEEE Trans. Microwave Theory Tech.*, vol. 17, pp. 406–408, July 1969.

A NEW "PHASED-TYPE" IMAGE ENHANCED MIXER

L. E. Dickens and D. W. Maki
Westinghouse D&ESC, Systems Development Division
Box 1521, MS 3717, Baltimore, Maryland 21203

Abstract

GaAs Schottky Barrier diodes with a zero bias cutoff frequency of 800 GHz have been used in an integrated circuit balanced mixer operating with a signal frequency centered at 12 GHz and an IF of 70 MHz. L_c at center-band was 2.2 dB. For an LO tunable bandwidth of over .65 GHz, the conversion loss was under 3.0 dB. No image filter is used in this mixer. The conversion loss across the image band was greater than 25 dB.

Introduction

This paper reports generally on a microwave mixer for operation at X-band, and more specifically on an integrated mixer circuit having very low conversion loss. Microwave diode mixers have been used for many years to obtain conversion of a signal at microwave frequencies to one at a much lower frequency. Such mixers have been the subject of much study and development. However, there is a continuing need for improvements which can result in better electrical performance, higher reliability, improved reproducibility, and lower production costs.

Briefly, the present paper describes an improved integrated circuit mixer. The mixer includes an integrated circuit dielectric substrate on which is defined a pair of balanced mixers of the coplanar line-slot line type. The pair of balanced mixers are so configured that one of the pair is the mirror image of the other. The RF inputs to each are then directly connected together and the LO inputs and IF outputs are coupled by 90 degree (quadrature) hybrid 3 dB couplers. This arrangement causes the image signals appearing at the RF input line and generated by each of the balanced mixers to be out of phase with each other, resulting in maximum image current or equivalently the short-circuited image condition, but now without the use of an image filter.

Theory of Image Enhancement

Well known are the fundamental techniques for the enhancement of mixer operation by the proper control of the impedances at each of the mixer terminals and at each of the frequencies of importance. The frequencies of importance are the modulation products which exist according to the heterodyne principle by which the mixer operates. The received signal (RF), together with a higher level signal from a local oscillator (LO), are applied to a nonlinear element. If the RF signal is sufficiently small, then the resulting frequencies can be given as $f_n = nf_p + f_o$; $n = -\infty, \ldots, +\infty$; where $f_o = |f_s - f_p|$ is the output (IF) frequency, f_s and f_p being the RF and LO frequencies, respectively. Note also that for the present application, $-f_{-1}$ corresponds to the signal frequency, and f_{+1} to the conventionally designated image frequency. For most mixer applications $f_o \ll f_p$, thus this notation has the advantage of that $|f_n| \approx f_{+n} \approx nf_p$; $n = 1, \ldots + \infty$; and the magnitude of the frequency is readily identifiable by its subscript. Further, for a particular group about nf_p, the plus (+) subscript always refers to the upper sideband and the negative (-) subscript always refers to the lower sideband. The three frequencies at the nth level are sometime referred to as the nth order idler frequencies.

The loss in converting an RF signal to an IF signal depends not only on properly matching the RF and IF impedances, but also upon properly terminating the sum and image frequencies as well as various other idler frequencies.[1] Equally as important as proper signal termination is the attainment of the proper form of mixer modulation by the LO.

It has been shown that, in a mixer without limiting parasitics, the lowest theoretically attainable conversion loss is obtained with a symmetrical rectangular LO drive waveform and dual terminations at the even and odd idler frequencies (that is, open circuits for the even and short circuits for the odd idlers or vice versa). These are the G- and H-type mixers.

In practical cases, mixer diodes are not purely resistive. The diode parasitics play an important role in the determination of the type of mixer to be used and the type of analysis that applies. The diode junction capacitance and series resistance represent the known diode parasitics. In the literature are found many authors who have dealt with this problem in Y-mixers. These authors assumed that the junction capacitance short circuits across the variable resistance all the higher order out-of-band frequencies. Thus, only the signal, IF, and image frequency voltages were considered. This assumption reduced the Y-mixer equations to a complex 3 x 3 Y-matrix which could be easily handled.

Barber[2] has presented such an analysis of microwave Y-type mixers. An extension[3,4] of Barber's analysis has allowed the calculation of the conversion loss of the three frequency Y-type mixer as a function of the pulse duty ratio and as limited by the operating frequency to cutoff frequency ratio (f/f_{co}). As noted by Saleh, this type of analysis can be extended to treat all mixer circuits by considering more than three frequencies. If one takes all the out-of-band frequencies above the m-order idlers to be short-circuited across the variable resistance, the mixer can be treated as a Y-mixer with 2 m + 1 frequencies. Such a (2 m + 1) frequency Y-mixer can be provided with the appropriate external idlers according to the type of mixer being analyzed. The idler termination would include the junction capacitance and series resistance as well as any external termination.

In an earlier paragraph it was pointed out that for the mixer without limiting parasitics, the minimum conversion loss was attained with a symmetrical square wave modulation of the resistance. This is an equivalent pulse duty ratio of 0.5. Figure 1 shows the computed minimum conversion loss for the Y-mixers (curve A) and the H- and G-mixers (curve B). Note that the mixers of curve A require a variable pulse duty ratio (up to 0.5) while the mixers of curve B need only the fixed pulse duty ratio of 0.5.

In a practical circuit we are limited by the parasitics of the diode and the fact that in a broadband circuit the open-/short-circuited idler requirements of the G-type and H-type mixers cannot be readily

Reprinted from *IEEE MTT-S Int. Microwave Symp. Dig.*, 1975, pp. 149–151.

attained. Thus we apply the extended analysis, as suggested previously, to a mixer with more than three frequencies so that a reasonable number of idlers can be controlled. Such control is then expected to yield a conversion loss between the two curves shown in Figure 1, and for an equivalent pulse duty ratio of less than 0.5 but greater than the ≤ 0.1 as required by the Y-type mixer for low conversion loss; the low value of pulse duty ratio of < 0.1 being very difficult to realize in actual practice. The impedance which will be generated across any reasonable frequency band to the idlers, in general, will not be the theoretically desirable open/short circuits but will usually be restricted to a range of large mismatch in the form of a dominately reactive termination. The magnitude of the reactance will be large or small relative to the signal (and IF) impedances when approximating an open-or short-circuited idler.

The control of the idlers is important in obtaining the low conversion loss, but the dominant frequency to be controlled is the image. If the image cannot be well shorted or opened across the full band, then control of the other idlers will do no good. Whatever scheme is used (there are an infinite number of possibilities), they all begin with means to either open-or short-circuit the image. Traditionally, image enhanced mixers have been single-ended (unbalanced) mixers[5,6], as opposed to balanced mixers, and have suffered the limitation of narrow band operation imposed by the use of a narrow band filter for image termination control. Single-ended mixers do not affect even and odd idler separations and so are usually constrained to Y-type mixer operation. Balanced mixers do affect idler separation as the odd idlers appear at the input and all the even idlers appear at the output. This more easily allows the imposition of the constraints of the G- and H-type mixers.

Mixer Configuration

The idler control features of the balanced mixer are desirable and so the balanced mixer was selected for the present design. Because of the desired large bandwidth and low IF, image control cannot be obtained by any form of bandwidth limiting filter. Therefore, two separate balanced mixers are used, the RF inputs of which are directly connected together and the LO inputs and IF outputs of which are coupled by 90-degree (quadrature) hybrids. See Figure 2. This configuration causes the image signals appearing at the RF input line and generated by each of the balanced mixers to be out of phase with each other, resulting in maximum image current or equivalently the short-circuited image condition. This short-circuit condition is not bandwidth limited because the electrical length between the diodes of the two balanced mixers is essentially zero.

The complete circuit is a truly integrated circuit comprising: microstrip transmission lines, coplanar strip transmission lines, slot transmission lines, a coplanar line-slot line hybrid junction for balanced semiconductor diode modulation, a broadband transition from slot line to microstrip line, and a broadband microstrip to coplanar line transition. See Figure 3 and 4.

Figure 3 is a sketch showing the layout of the two balanced mixers and the LO hybrid (Lange type) for LO injection. The IF hybrid is not shown. Figure 4 shows the details of the coplanar line-slot line-microstrip coupler circuits as well as the four diode mounting details.

The mixer configuration is designed to short-circuit the image, open-circuit the second idlers, and short-circuit all higher idlers. The parameters of the circuit, when adjusted for maximum bandwidth, will closely approximate these constraints. The circuit consists of a 180-degree hybrid ("T" junction) at the signal input port which feeds two balanced mixers. Local oscillator (LO) power supplied to the balanced mixers has a constant phase difference of 90 degrees at the output ports of the LO quadrature hybrid (Lange coupler). IF outputs from the mixers are combined in an IF quadrature hybrid. Due to their difference in phase, the image and signal components at IF due to image and signal components received at RF are summed in separate (and isolated) arms of the IF hybrid.

A photograph of the mixer is shown in Figure 5. The circuit is made up of a single substrate on which is mounted the RF circuitry. The IF hybrid is a 4-port quadrature hybrid of lumped element design and mounted in a low profile flatpack package.

Results

The following characteristics indicate the advancement of the state-of-the-art performance obtained. The measured curve of conversion loss versus frequency is shown in Figure 6. Two curves are shown. The one is for the complete mixer including all connector losses IF hybrid losses and track losses. The lower curve is a constant 0.4 dB less than the first and represents the conversion process without the 0.4 dB loss of the IF hybrid. Ultimately this mixer will be connected (without IF hybrid) to a pair of balanced, low-noise IF amplifiers. The IF will then be combined by the hybrid and after preamplification. Thus the lower curve is used for L_c in determining the ultimate noise figure of the overall mixer. At 70 MHz, a 1.0 dB noise figure preamp is to be used; thus, an overall noise figure of ≤ 3.0 dB is anticipated for this combination over a bandwidth of 250 MHz. The instantaneous bandwidth is 40 MHz and is limited by the IF hybrid used. The band shown in Figure 6 is the tunable bandwidth. This does not mean circuit tuning. Only the frequency of the LO is tuned across the band to realize the results in Figure 6.

Acknowledgement

The authors gratefully acknowledge the effort of Mr. F. G. Trageser in providing the GaAs Schottky diodes, and Mr. W. F. Stortz for assembling the circuits and performing the measurements.

References

1. A. A. M. Saleh, Theory of Resistive Mixer, Boston, Mass: M.I.T. Press, 1971.

2. M. R. Barber, "Noise Figure and Conversion Loss of the Schottky Barrier Mixer Diode," I.E.E.E. Trans. on MTT, Vol. MTT-15, No. 11, pp 629 - 635, November 1967.

3. L. E. Dickens, "Low Conversion Loss Millimeter Wave Mixers," 1973 G-MTT Symposium Digest, pp 66-68, June 1973.

4. L. E. Dickens and D. W. Maki, "An Integrated Circuit Balanced Mixer, Image and Sum Enhanced," I.E.E.E. Trans. on MTT, Vol. 23, No. 3, March 1975.

5. H. A. Watson, Microwave Semiconductor Devices and Their Circuit Applications, New York: McGraw-Hill, 1969.

6. J. B. Cahalan, et al., "An Integrated X-Band, Image and Sum Frequency Enhanced Mixer with 1 GHz IF," 1971 G-MTT Symposium Digest, pp 16 - 17, May 1971.

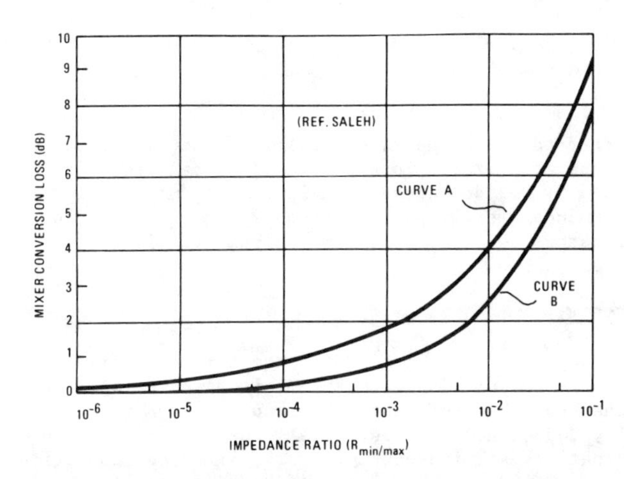

Figure 1 - Computed Minimum Conversion Loss for: Curve A Z-Mixer or Y-Mixer; Curve B, H-Mixer or G-Mixer.

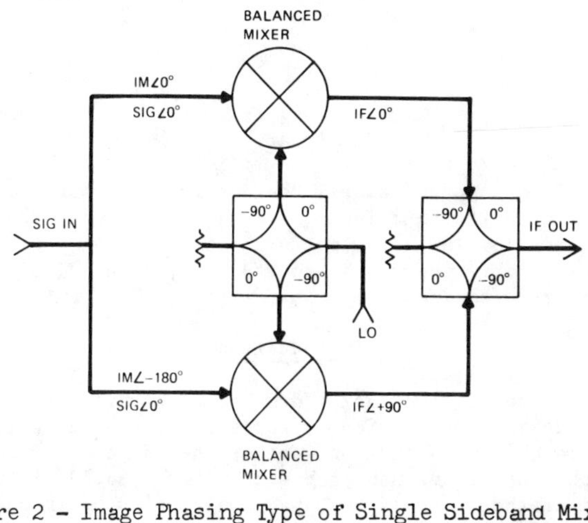

Figure 2 - Image Phasing Type of Single Sideband Mixer.

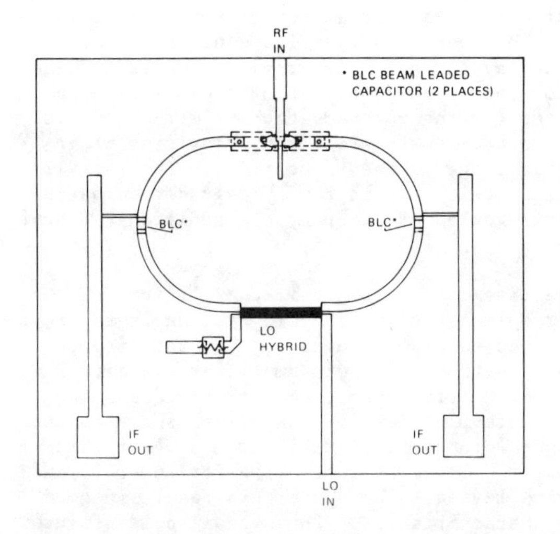

Figure 3 - Complete MIC Mixer Circuit.

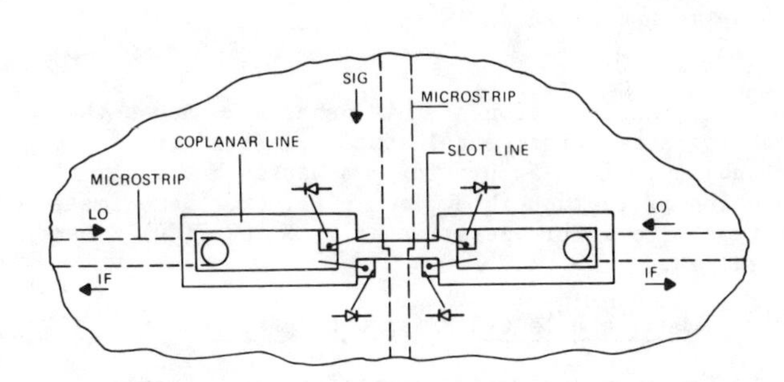

Figure 4 - Blow-up Mixer Plan View.

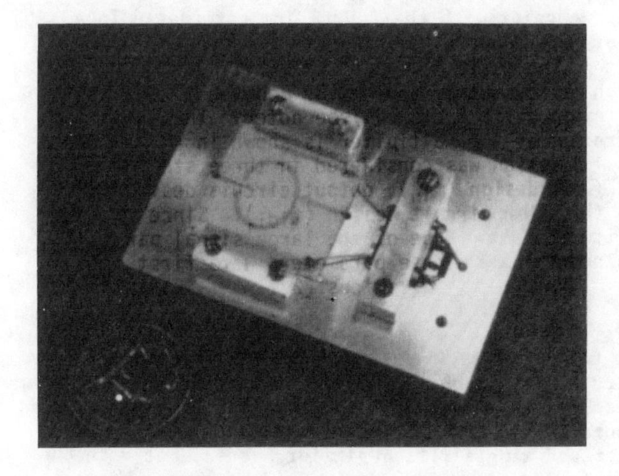

Figure 5 - Photograph Complete Mixer.

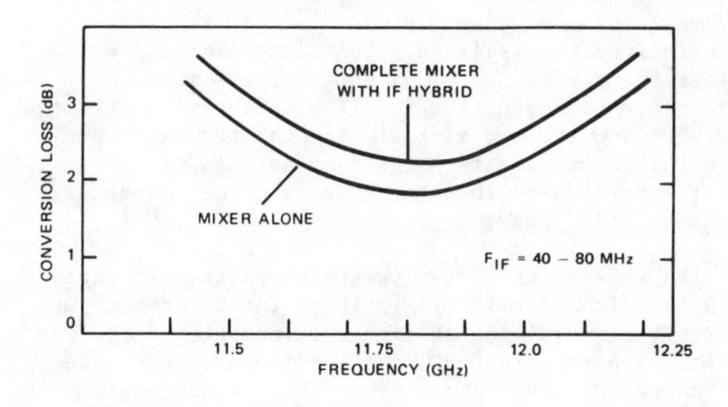

Figure 6 - Curve of Measured Conversion Loss versus Frequency.

PHASING TYPE IMAGE RECOVERY MIXERS

T.H. Oxley
AEI Semiconductors Ltd
Lincoln, England

Summary

This paper discusses the design of double balanced mixer circuits to effect image rejection, with the potential of low noise performance by recovery of image power. Performance of circuits at S, X and J(Ku) - bands is described.

Introduction

There are two basic forms of mixer circuits to provide image rejection (suppression) properties for single channel operation, either the use of a filter or by phasing techniques provided by two coupled mixers[1]. Both of these should be capable of achieving improved noise figure performance (compared with image matched) if reactive image termination is provided; this leads to the so called image enhancement process. Experience of many workers with phasing mixers, however, has indicated that critical operating conditions and severe performance reproducibility difficulties may be encountered, and in many cases any conversion loss improvement resulting from the image recovery process may not necessarily be reflected in the overall noise figure performance.

It is the intention of this paper to discuss the results of some experimental studies carried out on this subject, aimed at providing a design basis for the production of reproducible, low noise, broadband image rejection mixers for operation at S, X and J(Ku) - band frequencies.

Double Balanced Mixer Configuration

The experimental results obtained with X-band mixers of the MIC type shown in Figure 1 are briefly discussed in this section. The circuit configurations utilise two balanced mixers which employ 3 dB branch arm couplers and commercially available LID mounted GaAs Schottky barrier mixer diodes (i.e. AEI Type No. DC 1301).

Several circuit variants are considered. Briefly these include the application of two matched i.f. amplifiers, one following each mixer before i.f. power combining in the 90° hybrid, and the application of a single i.f. amplifier positioned after the 90° hybrid i.f. power combiner. In each case the effects of different forms of signal power divider are examined: these include the isolated Y junction Wilkinson to isolate the mixers and provide near image matched operation, the T junction to allow interaction of image power between the two mixers[1], and the rat race coupler (with the fourth arm terminated by an open or short circuit) to reflect the generated frequency products, including the image[3].

The experimental results **are summarised** in Table 1. A single balanced mixer is included for reference purposes. The overall noise figure maintaining image rejection properties >20 dB is the main parameter used for the performance comparison. The conversion loss is in general calculated from the measured i.f. amplifier contribution for a noise temperate ratio of 1.0: evaluation of the conversion loss has confirmed that this assumption may be used with confidence for all circuit configurations except the double amplifier and T-power splitter combination. The diode conversion loss (i.e. referred to the diode terminals) is an estimated value from assumed circuit losses. The r.f. bandwidth is defined by 0.5 dB degradation in overall noise figure performance (for image rejection >20 dB), and the local oscillator power level is for optimum overall noise figure.

Comments

The performance tabled for the twin amplifier system in conjunction with the T signal power splitter is the best observed, but the circuit design is critically dependent on achieving all the correct impedance terminations simultaneously, (r.f., i.f. and modulation products), for accurate image o/c or s/c conditions over the required r.f. bandwidth. Amplitude tracking of the twin i.f. amplifiers is also essential for acceptable image rejection properties. This system can thus result in severe problems of reproducibiltiy and critical operating conditions. In addition, circuits optimised for conversion loss may not necessarily reflect this parameter in the overall noise figure performance. The twin amplifier system can however yield consistent and useful performance results by use of the Y Wilkinson power splitter (or the rat race coupler with a 50 ohm termination on the fourth arm), thus providing mixer operation near image matched conditions. The overall noise figure in this case is simply degraded compared with the single balanced mixer by the additional circuit losses.

An objective of the single i.f. amplifier - T signal power splitter circuit configuration is to provide 50 ohm impedance levels throughout, thus reducing the undesirable effects of the twin amplifier system. The 50 ohm 90° hybrid i.f. power combiner provides a low impedance i.f. termination for the mixers and a constant load impedance for the i.f. amplifier. However, this does imply a high level of l.o. drive for low r.f. and i.f. impedance levels. The system can result in good performance characteristics. The overall noise figure is comparable with the single balanced mixer but with the additional properties of image rejection. Interaction of image power between the two mixers results in a lowering of the diode conversion loss by about 1 dB compared with the image matched situation. This level of image enhancement can also be effected by the signal rat race coupler with the fourth arm terminated in a short or open circuit. However the symmetry of the circuit is no longer ensured by the rat race coupler approach and this can result in inferior image rejection characteristics. A typical noise figure/image rejection frequency response of the form of circuit shown in Figure 1 provides an overall noise figure of about 6.0 dB (60 MHz F_{if} = 2.3 dB) and image suppression >20 dB for 10% r.f. bandwidth. The overall bandwidth is limited by the 3 dB couplers used in the l.o. feed and balanced mixers.

Double Ring Mixer Configuration

A further form of phasing type MIC image rejection/recovery mixer structure is shown in Figure 2. Following the results outlined in the previous section, the circuit configuration is based on the combined signal T power splitter and single i.f. amplifier system, but with the potential advantage of wider bandwidths and shorter transmission line-lengths provided by the compact ring mixer structure.

Reprinted from *IEEE MTT-S Int. Microwave Symp. Dig.*, 1980, pp. 270–273.

The mixers consist of a quad diode with r.f. connections (signal and l.o.) via balanced line feeds. The diode mounting structure is appropriate to the inclusion of LID, beam lead or chip devices, diode pairs being mounted on both sides of the MIC substrate. The circuit example of Figure 2 illustrates the use of a broadband Lange quadrature coupler for a l.o. feed to take advantage of the broadband properties of the ring mixers, and in this case shows the mixers positioned physically close together.

Experimental results of this form of mixer are summarised in Table 2. A single ring mixer is included for reference purposes. The separation 'd' between the mixers is varied between 0 (i.e. as close together as physically possible) and 3/8λ, in 1/8λ steps. The overall results are obtained from swept frequency measurements covering the range of 7 to 12 GHz.

The performance characteristics which may be obtained are illustrated in Figure 3. In this case 'd' = 0 and the diode quad consists of commercially available LID mounted GaAs Schottky barrier diode (AEI Type No. DC 1301). This example indicates an overall noise figure of about 5.0 dB (the spread observed has been 4.8 to 5.2 dB) including an i.f. contibution of 2.0 dB (i.e. 1.7 dB for the 60 MHz amplifier and 0.3 dB loss for the i.f. hybrid). A noise figure performance better than 5.5 dB is maintained for about 20% r.f. bandwidth, with image rejection greater than 20 dB, and i.f. impedance about 50 ohms.

Comments

This type of circuit configuration can result in good and repeatable performance characteristics and behaves favourably, particularly as regards bandwidth and noise figure, compared with its equivalent using the 3 dB coupler form of balanced mixers. Interaction of image power between the two ring mixers results in a lowering of the diode conversion loss approaching 2 dB, compared with the image matched situation. This results in an improvement in overall noise figure of about 1 dB compared with the single ring mixer, after allowing for circuit losses.

The high l.o. power requirement for 'd' = 0 suggests that this diode spacing results in circuit operation near image o/c conditions, with the conclusion that the electrical lengths associated with the quad diode structure prevent image s/c operation even when the mixers are positioned as close together as physically possible. This is confirmed by the 'd' = 1/4λ situation, which reduces the l.o. requirement to satisfy operation near image s/c conditions.

Adjusting 'd' to 1/8λ and 3/8λ suggests that the frequency range of 7 to 12 GHz encompasses both image s/c and o/c conditions near the band edges, thus providing two operating frequencies with their associated low and high l.o. power requirements for minimum noise figure. The observed degradation in noise figure compared with the 'd' = 0 and 1/4λ situation is the result of the image s/c and o/c responses falling outside the overall circuit bandwidth, which is centred about 9.5 GHz. Examination of the mixer i.f. impedance - frequency response clearly identifies the frequencies for image s/c and o/c, and is illustrated in Figure 4 for 'd' = 3/8λ. It is of considerable interest to note the reduction in the i.f. impedance levels with reduced dependency on image termination, by increasing the l.o. power; this has particular relevance to the operation of the double ring mixer configurations near image o/c conditions.

J(Ku) - Band Performance

Similar studies to those carried out at X-band with the double ring mixer configuration are also being carried out in the 12 to 18 GHz frequency range. In this case the quad diode mixers have been constructed using commercially available beam lead GaAs Schottky barrier diodes AEI Type No. DC 1306. The present results are tending to substantiate the experimental evidence obtained at X-band, particularly as regards the effect of varying the spacing between the diode quads. For example an overall noise figure of about 6.0 dB (including an i.f. noise figure contribution of 2.0 dB), for a 10% r.f. bandwidth centred about 17 GHz, has been obtained for a diode spacing of 1/4λ with 40 mW l.o. drive. Comparing this performance with a single ring mixer implies a diode conversion loss of about 3.0 dB (reactive image) and 4.5 dB (matched image) for the double and single ring mixer configurations respectively, when referred to the diode terminals.

System Application

The repeatability and acceptability for application to military environments of the forms of circuits discussed, have been demonstrated by employing these types of image rejection/recovery mixers in production MIC units for military radars in both the S and X frequency bands. For example an S-band version using the double ring mixer configuration utilising silicon LID mounted Schottky barrier diodes (AEI Type No. DC 1506) has provided a 5.0 dB overall noise figure and >20 dB image rejection for a 15% r.f. bandwidth. The l.o. power requirement is 5 mW. At these frequencies it is found that positioning the ring mixers as physically close together as possible represents near image s/c reactive conditions.

Conclusions

The phasing type of mixer configuration described in this paper can be designed to effect improvement in the diode conversion loss, presumably due to the image recovery process resulting from interaction of image power between the two mixers.

The application of twin amplifiers to image recovery designs presents too many parameter variables to optimise simultaneously, resulting in reproducibility problems and critical operating conditions. However, the application of the single amplifier system can realise practical circuits in which the improvement in conversion loss is reflected in the overall performance. The conversion loss and overall noise figure performance enhancement is more pronounced in the case of the quad diode ring mixer configuration than in the 3 dB coupler balanced mixer configuration; this is probably due to the better harmonic and intermodulation product suppression characteristics of the ring mixer. The reduction in conversion loss, referred to the diode terminals, is approaching about 2 dB for the double ring mixer configuration, compared with its single mixer counterpart. This is approaching the theoretically predicted improvement for an image reactive terminated mixer compared with an image matched mixer[2], and would thus suggest that the theoretically predicted improvement in diode conversion loss by reactive image termination may be achieved by the phasing double ring mixer circuit configuration. Further overall noise figure improvement with this type of circuit can only be accomplished by attention to circuit losses and the i.f. amplifier performance. The preferred circuit design of image s/c operation with minimal spacing between the two mixers is difficult to achieve above S-band frequencies due to the electrical lengths associated with the mixer structure. However, the circuits can be operated under image s/c or o/c conditions provided that the low impedance levels (particularly i.f.) are obtained by adjusting the l.o. power.

Finally, both forms of phasing type mixers can be designed and constructed to meet the reproducibility and environmental requirements for production military systems.

Acknowledgements

Part of this work has been carried out with the support of Procurement Executive, Ministry of Defence, sponsored by DCVD.

These studies are the result of many contributions, particularly Mr. P.L. Lowbridge, Mr. N.D.R. Shepherd, Mr. R.E. Scarman of AEI Semiconductors Ltd., and Mr. K.J. Ming, Dr. J.E. Curran of GEC Hirst Research Centre.

References

1. Oxley T.H., et al, Image Recovery Mixers, European Microwave Conference, Sweden, 1971

2. Torrey H.C., Whitmer C.A., Crystal Rectifiers, MIT Radiation Laboratory Series (McGraw-Hill, New York, 1948)

3. Hallford B.R., 2 dB Conversion Loss Mixer at 11 GHz Using a PRM Circuit, European Microwave Conference, Rome, 1976

TABLE 1

3 dB Coupler Mixer Configurations Performance Summary

Mixer System	Signal Power Divider	F_0 (dB)	F_{IF} (dB)	L_C (dB)	Diode L_C (dB)	P_{LO} mW	r.f. Bandwidth %
Single Bal.	—	6.0	1.5	4.5	4.1	4	10
Double Bal.-Two i.f.	T	5.3	1.7	3.6	2.8	2	2
"	Y	7.2	2.0	5.2	4.4	20	10
Double Bal.-One i.f.	Y	7.5	2.3	5.2	4.4	20	10
"	T	6.0	2.3	3.7	3.0	20	10
"	Rat Race o/c and s/c	6.0	2.3	3.7	3.0	20	5
"	Rat Race Matched	7.5	2.3	5.2	4.4	20	10

Input Level for 1 dB L_C Compression

≃ +5 dBm for P_{LO} = 20 mW

TABLE 2

Quad Diode Ring Mixer Configurations Performance Summary

Mixer System	'd'	F_0 (dB)	F_{IF} (dB)	L_C (dB)	Diode L_C (dB)	P_{LO} (mW)	F_C (GHz)	r.f. Bandwidth %
Single Ring	—	6.1	1.6	4.5	4.2	12	9.5	50
Double Ring-One i.f.	0	5.0	2.1	2.9	2.5	50	9.0	20
"	1/2λ	5.5	2.0	3.5	3.0	20	9.5	20
"	1/8λ	6.3	2.0	4.2	3.8	16	8.0	10
		6.3	2.0	4.2	3.8	50	11.0	—
"	3/8λ	6.5	2.0	4.4	3.8	16	8.0	10
		6.5	2.0	4.4	3.8	50	11.0	—

Input Signal Level for 1 dB L_C Compression

≃ +8 dBm for P_{LO} = 20 mW

≃ +15 dBm for P_{LO} = 50 mW

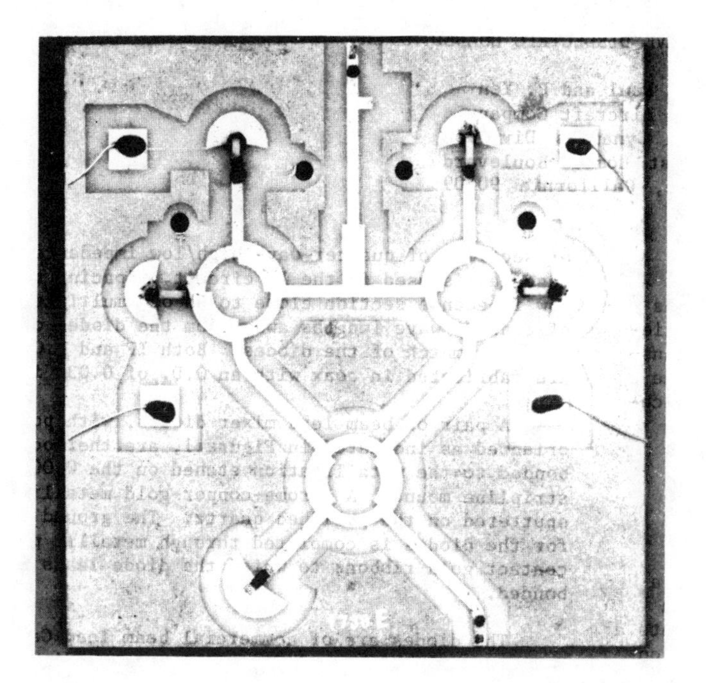

Fig. 1 X-band MIC double balanced image rejection mixer

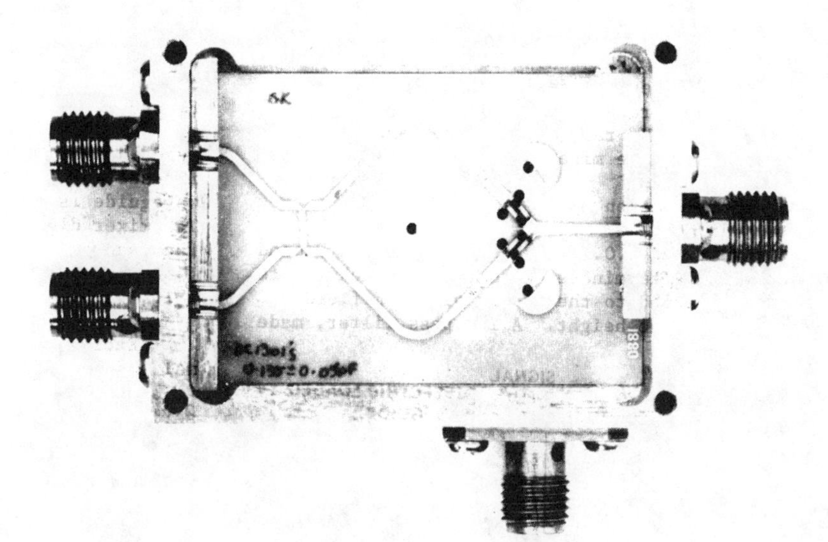

Fig. 2 X-band double ring image rejection mixer

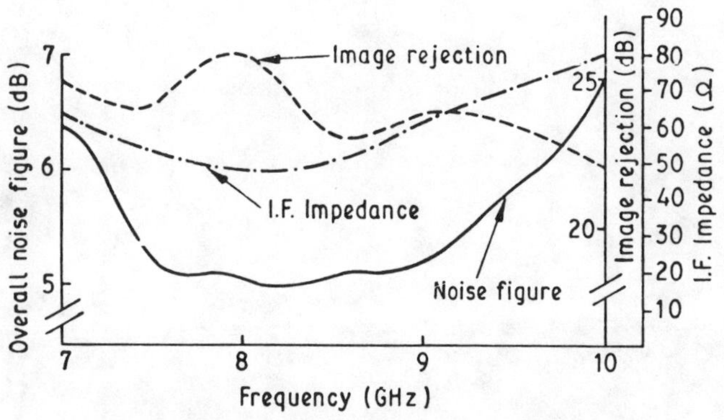

Fig. 3 X-band double ring phasing mixer

Overall noise figure, image rejection, i.f. impedance as a function of frequency

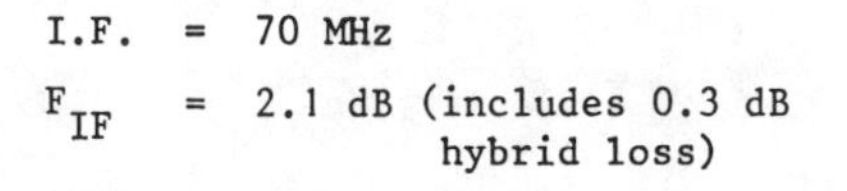

I.F. = 70 MHz

F_{IF} = 2.1 dB (includes 0.3 dB hybrid loss)

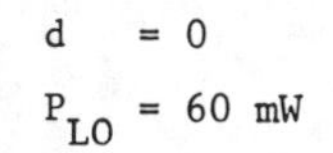

d = 0

P_{LO} = 60 mW

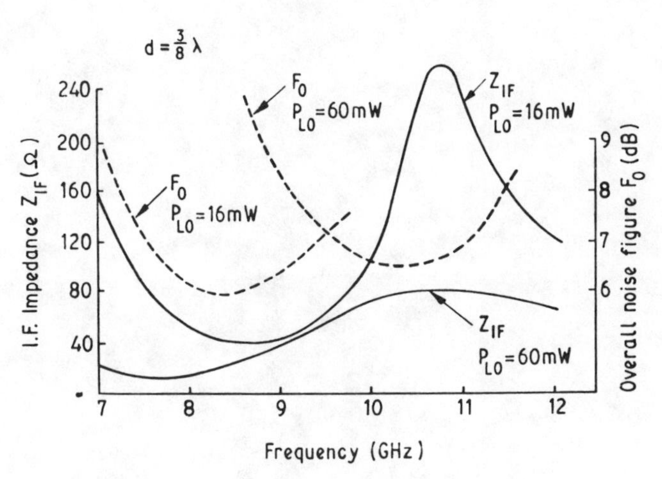

Fig. 4 X-band double ring phasing mixer

I.F. impedance, overall noise figure as a function of frequency

Part IV
Other Practical Implementations of Microwave Mixers

IN this part we have selected a few papers out of an overwhelming number one can find in the literature describing all kinds of mixer circuits. The fact that all the selected papers but one (the fourth paper) describe broad-band balanced and sometimes double-balanced mixers indicates that there are numerous applications where a broad or a very broad bandwidth is important. The aim of this part is certainly not to give a complete survey of possible mixer circuits, but hopefully to offer an overview which can serve as a source of design ideas for the microwave engineer.

In the first paper Mouw gives a detailed discussion on one class of hybrid junctions and applies his results to a four-diode star mixer/modulator. This mixer can typically be used over a wide bandwidth (octave) and has inherently high isolation between the LO and signal port. In the next paper Hunton and Takeuchi discuss four different approaches using slot-line and coplanar line techniques and two or four diodes for broad-band balanced mixer designs. In the third paper Pflieger describes a balanced mixer using slot-line techniques combined with microstrip techniques and wires. The next paper, by Konish *et al.*, describes an elegant way of making a single diode 12 GHz narrow-band mixer using a metal sheet with proper patterns in the *E*-plane of a waveguide. A Gunn-diode oscillator and an image rejection filter as well as the mixer itself are integrated on the same metal sheet.

Stracca *et al.* show the importance of taking into account higher order sideband frequencies and using a proper termination at the image frequency for achieving the ultimate performance of a waveguide crossbar balanced mixer. In the following paper, by Yuan, a modification of the crossbar mixer is shown to yield excellent performance over an octave bandwidth.

Papers 7–13 all show further approaches for designing broad-band balanced mixers. Using the symmetry properties of fin-lines (slot-lines), coplanar lines, and microstrip lines in various combinations, some quite interesting designs are described in papers 7, 8, 12, and 13 (in addition to the already mentioned papers 2 and 3). Begemann's paper shows an elegant and fairly straightforward approach for an ordinary type balanced mixer. The double-balanced mixer described by Ogawa *et al.* in the eighth paper is necessarily more complicated but has some interesting potentials.

The paper by Meier shows how a 94 GHz balanced mixer can be built on a piece of substrate mounted in the *E*-plane of standard waveguide. Hallford and Culberson in papers 10 and 11, respectively, discuss how balun-coupled mixers with broad bandwidth and single sideband response can be constructed.

For further references we recommend that the reader consult the reference lists of the papers included in this part.

A Broad-Band Hybrid Junction and Application to the Star Modulator

ROBERT B. MOUW, MEMBER, IEEE

Abstract—**A class I hybrid junction, $S_{13}=S_{23}$; $S_{14}=-S_{24}$ [1], [2] is described consisting of two separate pairs of parallel transmission lines or transformers connected to conjugate ports 1 and 2 meeting at two central terminal pairs which are conjuate ports 3 and 4. An analysis is made in terms of the admittance and scattering parameters which reveals the "magic tee" matrix. Conjugate port isolation is infinite and equality of coupling is perfect in principle for all frequencies. The potentials at the central terminals of the hybrid junction are suitable for driving elements connected as a four-branch star. The *four-diode star mixer/modulator* is described and realizations in lumped elements, coaxial line, stripline, and waveguide are discussed. Data are reported for coaxial line models covering the frequency range of 1 to 8 GHz in octave bandwidths. Other applications are discussed.**

Introduction

HYBRID junctions may be divided into two broad categories with respect to application: 1) signal separation, in which the ports of the hybrid are connected to transmission lines or system components; 2) applications where the ports of the hybrid are connected to devices such as diodes or transistors. In the first application it is usually necessary to have the ports unbalanced and sufficiently separated to enable connectors to be attached. The available hybrid junction configurations have generally satisfied this requirement.

The present status of semiconductor fabrication allows matched pairs of chips and doubly balanced quads to be installed as pretested integrated assemblies rather than individually. In this application, a hybrid junction with one set of conjugate terminal pairs inextricable to external unbalanced connectors is not impractical if convenient connection can be made to semiconductor circuits or other elements.

A hybrid junction of the latter type was evolved[1] which is capable of broad bandwidth and is suitable as a coupling network to a four-branch star of semiconductors or other elements. Application may also be found in antenna excitation, instrumentation, or broad-band impedance transformation. A novel characteristic is the symmetrical and central location of the balanced conjugate terminal pairs 33′ and 44′ shown in Fig. 1. A four-branch star diode network connected to these terminals (Fig. 8) is able to perform at frequencies through the microwave region all of the well-known functions of the four-diode ring modulator. An advantage of the four-diode star network over the ring network is the central node (junction of the four diodes) which allows a direct connection to the IF/modulation terminals. The hybrid junction, used with the four-diode star network, is a practical doubly balanced mixer/modulator and up-converter.[1] The configuration lends itself to a compact construction with sizes ranging from a one-inch (2.5-cm) cube in L-band to a half-inch (1.25-cm) cube in X-band.

Manuscript received December 18, 1967; revised March 27, 1968, and July 1, 1968. The work reported here was sponsored by a development program at Aertech, Sunnyvale, Calif.

The author was with Aertech, Sunnyvale, Calif. He is now with DeMornay–Bonardi, Division of Systron-Donner, Pasadena, Calif.

[1] Patent pending.

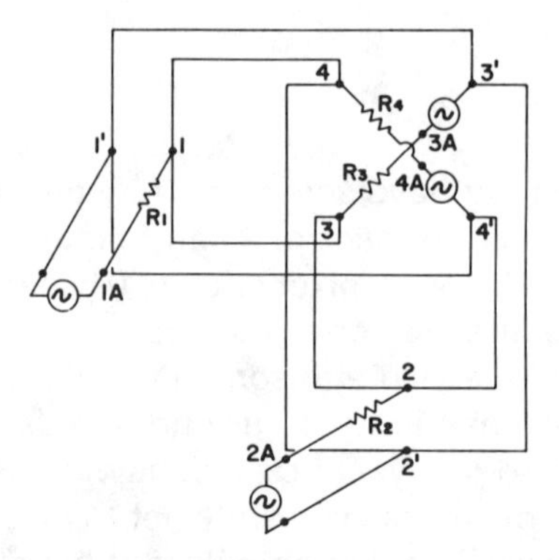

Fig. 1. Two-wire line hybrid junction augmented with resistors in series at each port. All lines are of length θ and characteristic impedance Z_{01}.

Hybrid Junction Description and Analysis

Conjugate Port Isolation

The hybrid junction is shown in Fig. 1 as a two-wire line network which is convenient for analysis. Pairs of transmission lines, all of equal length and characteristic impedance Z_{01}, are connected to terminal pairs 11′ and 22′ and meet at terminal pairs 33′ and 44′ in such a way that port 1 is conjugate to port 2, and port 3 is conjugate to port 4.

If $R_3=R_4$, terminals 34 and 3′4′ are equipotential in the presence of excitation at port 1:

$$V_{43} = V_{3'4'} = 0, \tag{1}$$

and port 2 is isolated from port 1 independent of frequency. When excitation is at port 3, terminals 44′ are equipotential for $R_1=R_2$:

$$V_{44'} = 0, \tag{2}$$

and port 4 is isolated from port 3 independent of frequency.

Equality of Coupling

If $R_1=R_2$ and $R_3=R_4$, it will be shown that the following holds.

1) With an input at port 1, coupling is equal and cophasal to ports 3 and 4.
2) With an input at port 2, coupling is equal and contraphasal to ports 3 and 4.

Reprinted from *IEEE Trans. Microwave Theory Tech.*, vol. MTT-16, pp. 911–918, Nov. 1968.

3) With an input at port 3, coupling is equal and cophasal to ports 1 and 2.
4) With an input at port 4, coupling is equal and contraphasal to ports 1 and 2.

In Fig. 2, the hybrid junction is redrawn with R_3 and R_4 as series connections of equal resistors. In this symmetrical configuration, it is apparent that the wires crossing at terminal 0 are equipotential for excitation at any port and may be short-circuited without affecting the operation of the hybrid junction. Under these conditions, each line connected to port 1 or port 2 is identically terminated and

$$V_{43'} = V_{34'}. \tag{3}$$

Since (1) applies,

$$V_{33'} = V_{44'} \tag{4}$$

and port 1 couples equally to ports 3 and 4. The argument is similar for an input at port 2, except that it follows from the choice of terminals in Fig. 2 that

$$V_{33'} = - V_{44'}. \tag{5}$$

Thus port 2 couples equally in amplitude and contraphasally to ports 3 and 4. Referring to Fig. 1, for excitation at port 3, terminals 44′ may be short circuited from (2) and by symmetry

$$V_{43'} = V_{34'} = V_{34} = V_{4'3'}. \tag{6}$$

Therefore, port 3 couples equally to ports 1 and 2. Finally, for an input to port 4, terminals 33′ may be short-circuited, whereby

$$V_{43'} = V_{34'} = - V_{34} = - V_{4'3'} \tag{7}$$

and port 3 couples equally in amplitude and contraphasally to ports 1 and 2. Thus, conjugate port isolation and equality of coupling are perfect in principle for all frequencies, independent of the hybrid junction input impedance and port-to-port transmission loss.

Input Impedance

When the input impedance and transmission loss are considered, it is necessary to include the reactive contribution of lines which effectively shunt ports 3 and 4. The hybrid junction is redrawn again in Fig. 3(a) and (b) to represent the equivalent circuits for the input impedance calculation at port 1. It is assumed that $R_1 = R_2 = r$, $R_3 = R_4 = R$, and that the lines are all of equal length θ and characteristic impedance Z_{01}. The short-circuited stubs of characteristic impedance Z_{02} and electrical length ϕ represent the two-wire transmission lines formed by the wires 42′–3′2′ and 32–4′2. Since from (1) $V_{43} = V_{3'4'}$, there is no propagation along the pairs 42′–32 or 3′2′–4′2 and port 2 may be deleted. The input impedance to port 1 is then that of two parallel transmission lines of characteristic impedance Z_{01} and electrical length θ terminated in resistances $\frac{1}{2}R_3 + \frac{1}{2}R_4 = R_3 = R_4$, shunted by short-circuited stubs of characteristic impedance Z_{02} and electrical length ϕ. The input impedance (derived in the

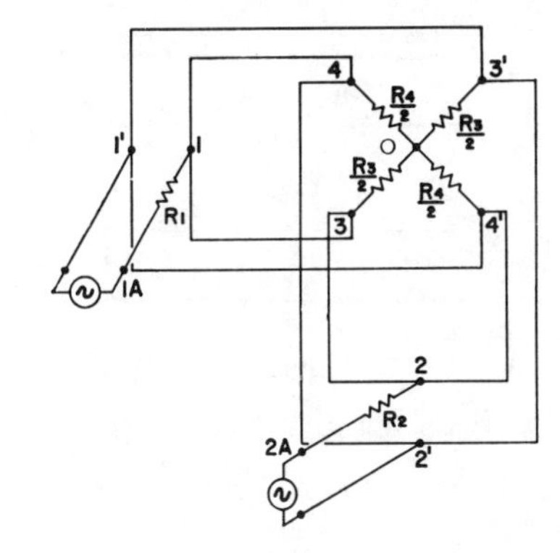

Fig. 2. Equivalent circuit for excitation at port 1 or port 2, $R_3 = R_4$.

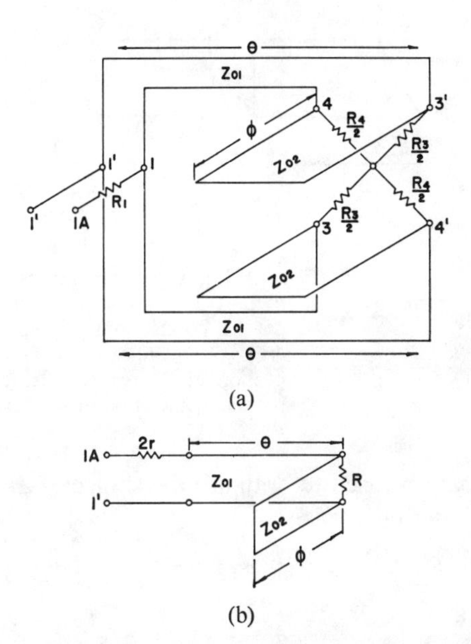

Fig. 3. Equivalent circuits as viewed from port 1. (a) Showing the effect of lines connected to port 2. (b) Equivalent half-circuit.

Appendix) is

$$Z_{i1} = Z_{i2}$$

$$= \frac{Z_{01}}{2} \frac{(Z_{01}/R) - j[\cot\theta + (Z_{01}/Z_{02})\cot\phi]}{1 - (Z_{01}/Z_{02})\cot\theta\cot\phi - j(Z_{01}/R)\cot\theta}. \tag{8}$$

The choice of Z_{01} for maximum bandwidth will be indicated by the prescribed value of R/r. Z_{02} will be fixed by the type of transmission line, Z_{01}, and the size of the enclosure. The value of Z_{02} is normally quite high, usually above $4R$, so that its effect is not great in applications whose bandwidth is less than one octave.

When the effect of the parasitic shunt stubs is neglected, (8) reduces to

$$Z_{i1} = Z_{i2} = \frac{Z_{01}}{2} \frac{(Z_{01}/R) - j\cot\theta}{1 - j(Z_{01}/R)\cot\theta}. \tag{9}$$

Maximum bandwidth in this case is obtained when

$$2r = R = Z_{01}. \tag{10}$$

When other ratios of the terminating resistances are required,

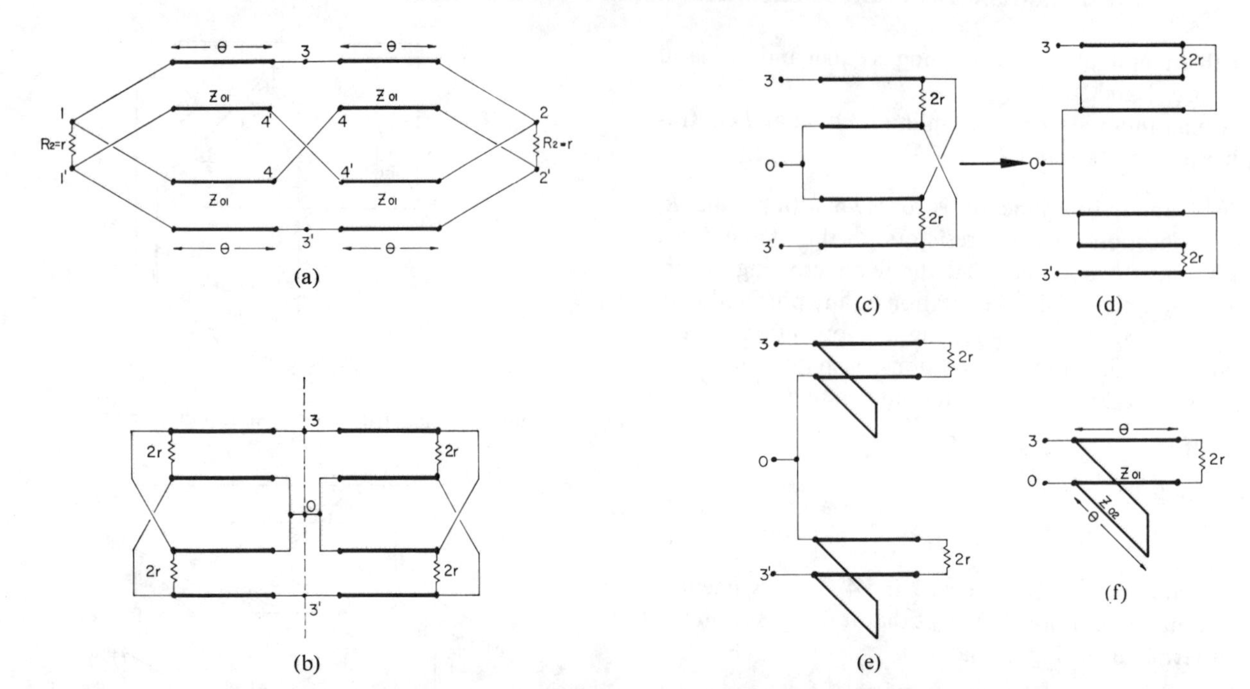

Fig. 4. Equivalent circuits as viewed from port 3. (a) Drawn symmetrically with respect to port 3. (b) Port 4 short-circuited with terminations at ports 1 and 2 divided into parallel connections of two resistors. (c) Right half of the symmetrical network in Fig. 4(b). (d) Partial extraction of the transmission lines which are an effect of the interconnections at the resistors $2r$. (e) Equivalent half-circuit. (f) Single element of the series-parallel connection of four elements.

(8) or (9) may be used to optimize the choice of Z_{01}. A practical first-order choice is

$$Z_{01} = \sqrt{2Rr}$$
$$\theta = \frac{\pi}{4} \tag{11}$$
$$\phi = \theta.$$

For input impedance calculations at port 3, the hybrid junction is redrawn to illustrate the symmetry of ports 1 and 2 with respect to ports 3 in Fig. 4(a). In Fig. 4(b), $R_1=r$ and $R_2=r$ are shown as parallel connections of resistors $2R_1=2r$ and $2R_2=2r$, and terminals 44′ are short-circuited to a single node 0, justified by (2). The following development will show that Fig. 4(b) is a series-parallel connection of four networks of the type shown in Fig. 4(f). The network in Fig. 4(b) may be bisected along the dashed line into identical left and right halves due to the symmetry with respect to terminals 33′. The right half shown in Fig. 4(c) is not a simple series connection of two terminated transmission lines because of the interconnections at the resistors $2r$. The effect of the interconnections is an additional pair of transmission lines shunting terminals 30 and 03′. The lines are extracted by the steps in Fig. 4(d) and (e), and what is now a true series connection of the network is shown in Fig. 4(f). The characteristic impedance of the shunt transmission lines is Z_{02} and the electrical length is ϕ since they are formed by the same elements 42′–3′2′ and 32–4′2 described in the equivalent circuits of Fig. 2. The input impedance to port 3 is then that of a series-parallel connection of four transmission lines of length θ and characteristic impedance Z_{01}, each terminated in $2R_1=2R_2=2r$ and shunted by short-circuited stubs of characteristic impedance Z_{02} and length ϕ. The input impedance (derived in the Appendix) is

$$Z_{i3} = Z_{i4} = Z_{01} \frac{(Z_{01}/2r) - j\cot\theta}{1 - (Z_{01}/Z_{02})\cot\theta\cot\phi - j(Z_{01}/2r)\cot\theta + (Z_{01}/Z_{02})\cot\phi}. \tag{12}$$

When the effect of the shunt stubs is neglected, (12) reduces to

$$Z_{i3} = Z_{i4} = Z_{01} \frac{(Z_{01}/2r) - j\cot\theta}{1 - j(Z_{01}/2r)\cot\theta}. \tag{13}$$

When design equations (11) are used, the input impedance to each of the ports at band center is

$$\left.\begin{aligned} Z_{i1} = Z_{i2} &= \frac{Z_{01}^2}{2R} = r \\ Z_{i3} = Z_{i4} &= \frac{Z_{01}^2}{2r} = R \end{aligned}\right\} \tag{14}$$

where $R_1=R_2=r$ and $R_3=R_4=R$. VSWR is plotted for the hybrid junction (including the effect of the parasitic lines) in Fig. 5(a) with $Z_{01}=R=2r$ and in Fig. 5(b) with $Z_{01}=\sqrt{2Rr}$,

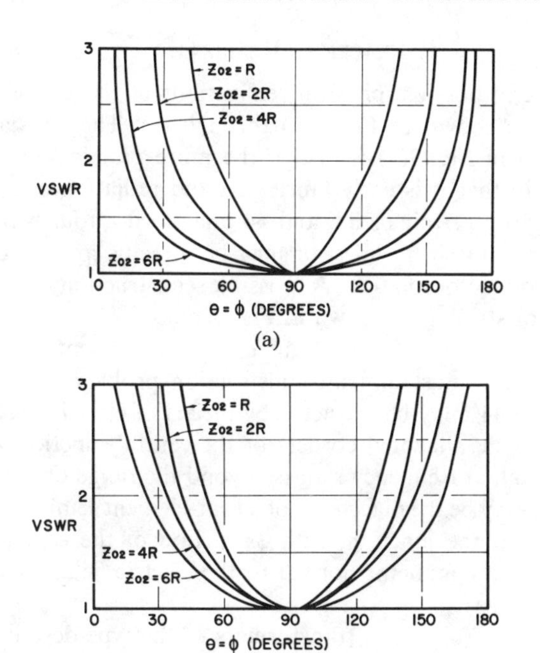

Fig. 5. Computed VSWR of the hybrid junction shown in Fig. 1. (a) With $Z_{01}=R=2r$. (b) With $Z_{01}=\sqrt{2Rr}$, $R=r$. In each case $\theta=\phi$.

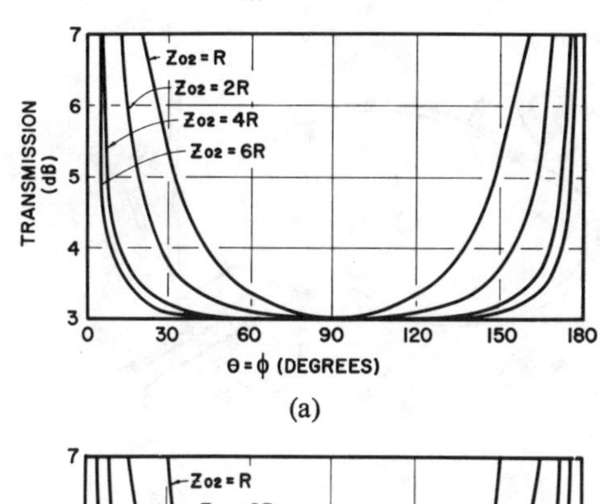

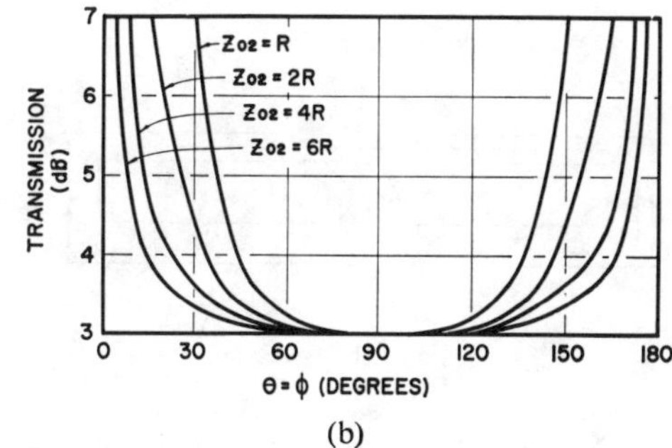

Fig. 6. Computed transmission loss of the hybrid junction of Fig. 1. (a) With $Z_{01}=R=2r$. (b) With $Z_{01}=\sqrt{2Rr}$, $R=r$. In each case $\theta=\phi$.

$R=r$. Electrical lengths θ and ϕ are made equal and curves are given for assumed values of Z_{02}.

A significant improvement in bandwidth may be obtained by using compensating series-matching stubs at ports 1 and 2 as described previously [3]. Using this procedure, a 1.23:1 VSWR bandwidth of 13:1 is possible for $Z_{02}=6R$, $Z_{01}=R=2r$.

Transmission Loss

The analysis of the hybrid junction is completed in the Appendix and is obtained by first computing the admittance matrix and from this the scattering matrix. The magnitude of the transfer coefficient is plotted in Fig. 6 as insertion loss versus electrical length $\theta=\phi$ with Z_{02} as a parameter.

When the effect of the parasitic lines shunting ports 3 and 4 is neglected, the scattering coefficients (42) in the Appendix simplify considerably and are tabulated as follows:

$$S_{11}=S_{22}=\left[\frac{\left(\dfrac{Z_{01}}{2r}-\dfrac{R}{Z_{01}}\right)-j\left(\dfrac{R}{2r}-1\right)\cot\theta}{\left(\dfrac{Z_{01}}{2r}+\dfrac{R}{Z_{01}}\right)-j\left(\dfrac{R}{2r}+1\right)\cot\theta}\right] \tag{15}$$

$$S_{12}=S_{21}=S_{34}=S_{43}=0 \tag{16}$$

$$\left.\begin{aligned} S_{31}&=S_{13}=\\ S_{41}&=S_{14}=\\ S_{32}&=S_{23}=\\ -S_{42}&=-S_{24}=\end{aligned}\right\}\frac{1}{\sqrt{2}}\left[\frac{\sqrt{\dfrac{R}{2r}}}{\dfrac{1}{2}\left[\left(1+\dfrac{R}{2r}\right)\cos\theta+j\left(\dfrac{R}{Z_{01}}+\dfrac{Z_{01}}{2r}\right)\sin\theta\right]}\right] \tag{17}$$

$$S_{33}=S_{44}=\left[\frac{\left(\dfrac{Z_{01}}{R}-\dfrac{2r}{Z_{01}}\right)-j\left(\dfrac{2r}{R}-1\right)\cot\theta}{\left(\dfrac{Z_{01}}{R}+\dfrac{2r}{Z_{01}}\right)-j\left(\dfrac{2r}{R}+1\right)\cot\theta}\right]. \tag{18}$$

When $R=2r=Z_{01}$, the scattering coefficients (15) through (18) become those of the "magic tee."

$$S=\frac{1}{\sqrt{2}}\begin{bmatrix}0 & 0 & e^{-j\theta} & e^{-j\theta}\\ 0 & 0 & e^{-j\theta} & -e^{-j\theta}\\ e^{-j\theta} & e^{-j\theta} & 0 & 0\\ e^{-j\theta} & -e^{-j\theta} & 0 & 0\end{bmatrix}. \tag{19}$$

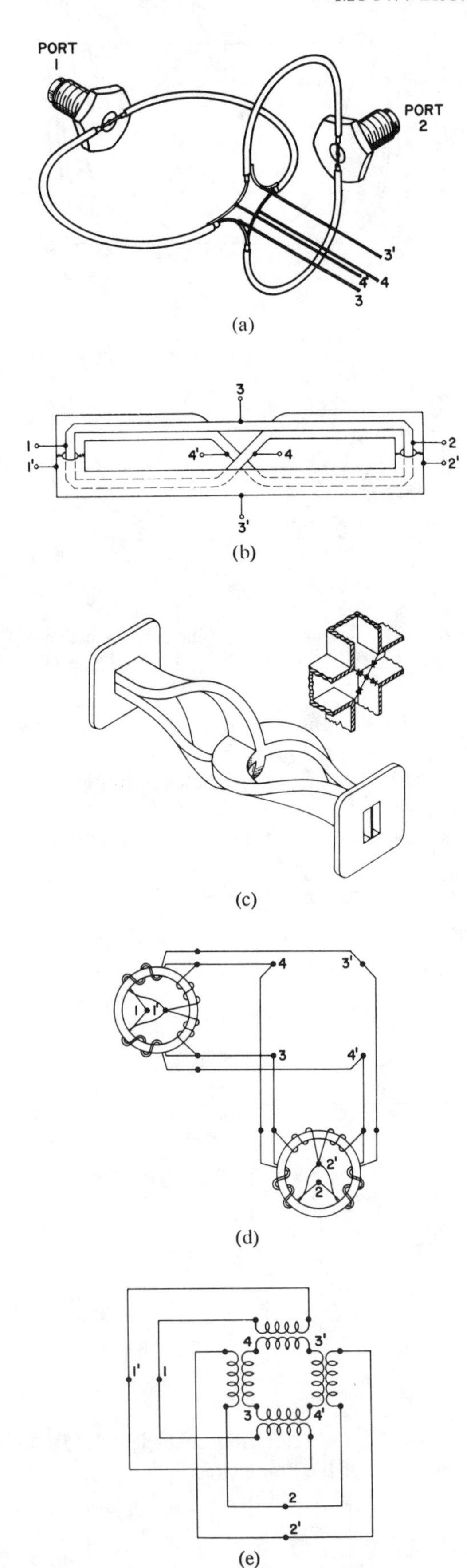

Fig. 7. Several possible realizations of the hybrid junction. (a) Coaxial line. (b) Microstrip. (c) Waveguide. (d) Bifilar windings on ferrite cores. (e) Magnetically coupled transformer windings.

Practical Realizations

Several possible physical realizations of the hybrid junction are shown in Fig. 7(a) through (e). The coaxial line version in Fig. 7(a) is one of the more practical constructions in that it is easily fabricated and radiation effects are not great. Terminals 33′ and 44′ may feed a four-wire line, a pair of two-wire lines, or may be directly connected to a semiconductor matrix. A possible construction in microstrip or stripline is shown in Fig. 7(b).

In Fig. 7(c), the waveguide version presents, without doubt, the most difficult construction problem. Terminals 33′ and 44′ may be extracted by means of electric probes in each of the diagonal corners of the central junction. More practical, at frequency ranges beyond the range of the other types, may be the placement of a four-element semiconductor matrix at the junction, with extraction of the center node through a conductor normal to the electric fields as shown in the insert.

In Fig. 7(d), balun transformers of the type described by Ruthroff [4] replace the two-wire transmission lines schematically shown in Fig. 1. At frequencies below approximately 1 GHz, where currently available ferrite cores do not have excessive loss, this construction is the most practical. The major advantages are small size, ease of fabrication, and attainable bandwidths in excess of 100:1.

The circuit using magnetically coupled transformer windings in Fig. 7(e) is an obvious extension of the hybrid junction to low-frequency applications. A dc return path for elements which may require it can be provided by center-tapping one or both pairs of secondary windings without effect on the ac performance.

Four-Diode Star Mixer/Modulator

A four-diode star is connected to conjugate terminal pairs 33′ and 44′ of the hybrid junction as shown in Fig. 8(a), which is a three-port network with port 1 (terminals 11′), port 2 (terminals 22′), and port 0 (terminals 00′). Terminals 1′, 2′, and 0′ are common. The diodes are arranged so that the direction of conduction is toward the center node from terminals 33′ and from the center node to terminals 44′. In all important respects, the operation of this network is identical to the four-diode ring mixer/modulator insofar as the external ports are concerned. Accordingly, the type of network shown in Fig. 8(a) is called the *four-diode star mixer/modulator*. An advantage of the star-diode network over the ring is the center node 0 which allows direct access to modulation/IF terminals and extends the frequency range of this port so that it may approach that of the other two. In the ring network, connection to these terminals must be made indirectly to points such as secondary winding center-taps or their microwave circuit equivalents. Either type of (double-balanced) four-diode mixer/modulator provides isolation of all ports with respect to each other as a result of inherent circuit balance independent of frequency, load impedance, or power, and without transformers or bypass elements associated with the modulation/IF port.

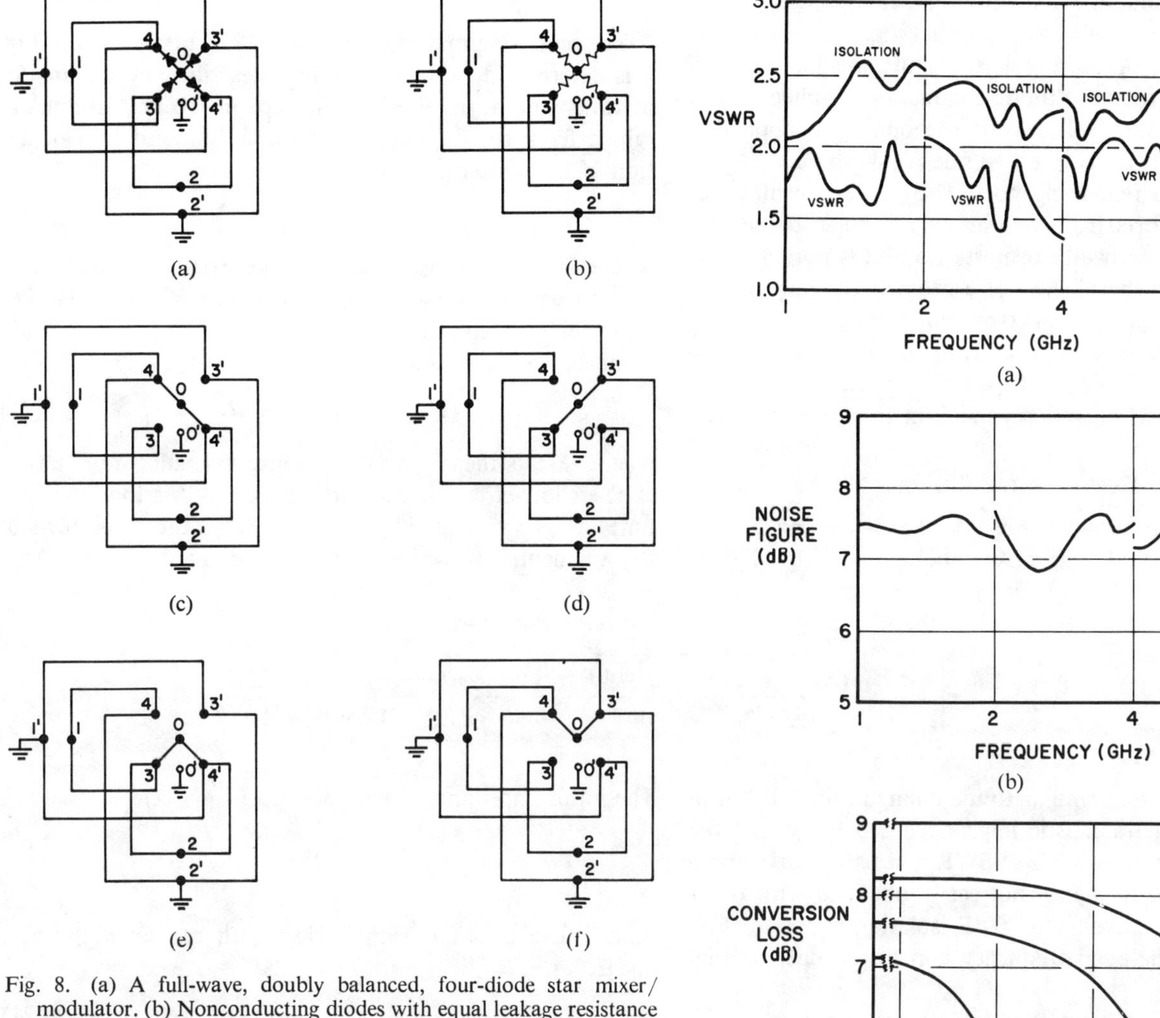

Fig. 8. (a) A full-wave, doubly balanced, four-diode star mixer/modulator. (b) Nonconducting diodes with equal leakage resistance assigned to each of the diodes. (c) Bias current entering port 0. (d) Bias current leaving port 0. (e) Excitation at port 1 or port 2: terminals 3 and 4 positive with respect to terminals 3′ and 4′. (f) Polarity reversed.

Circuit Operation

In Fig. 8(b) the equivalent circuit for all nonconducting diodes is shown with a nominal equal leakage resistance assigned to each of the diodes. It was shown previously that port 1 is isolated from port 2 under these conditions. As positive bias is applied to port 0, the network approaches the state shown in Fig. 8(c) where a direct transmission path is provided from port 1 to port 2. When negative bias is applied at port 0, the limiting condition is that shown in Fig. 8(d), and direct transmission is again provided; however, the phase of transmission is reversed. The circuit behavior described is that of a balanced modulator in which a carrier applied at port 1, for example, and a modulation signal applied at port 0 will produce two sidebands at port 2 with the carrier suppressed.

The operation is similar when the modulation (or local oscillator) is applied at port 1. For this excitation, the phase of transmission from port 2 to port 0 is reversed on successive half cycles of the signal at port 1. Fig. 8(e) is the limiting

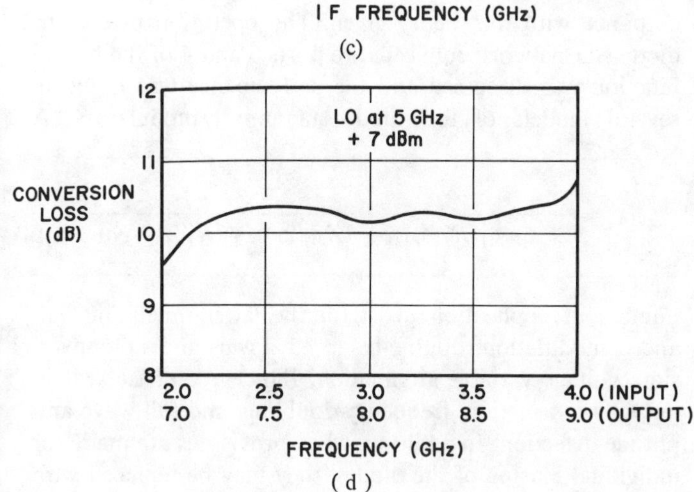

Fig. 9. Performance of L-, S-, C-, and X-band doubly balanced mixer/modulators. (a) Port-to-port isolation and VSWR. (b) Single-channel noise figure including the contribution of a 1.5-dB noise figure IF amplifier. (c) Conversion loss as a function of IF frequency. (d) Conversion loss of the X-band model used as an up-converter. (The S-band signal is applied at the junction of the four diodes.)

condition for terminals 3 and 4 instantaneously positive with respect to terminals 3′ and 4′, while Fig. 8(f) shows the condition for the opposite polarity. Thus, a signal of sufficient amplitude to cause diode conduction, applied at any one of the three ports will convert a second signal applied at either of the other ports into two sidebands which are available at the third remaining port. The property which distinguishes the ports from one another is their bandwidth capability. The bandwidth of ports 1 and 2 is nominally an octave or two in the vicinity of a center frequency f_o. The bandwidth capability of port 0 extends from dc to an upper 3-dB frequency which has been experimentally determined to be approximately $2/3\, f_o$ for the TEM transmission line types but may extend well beyond f_o in the waveguide versions.

Experimental Results

Typical data are shown in Fig. 9 for L-, S-, C-, and X-band silicon Schottky-diode star mixer/modulators using a hybrid junction construction similar to the configuration shown in Fig. 7(a). The characteristic impedance and lengths of the lines were chosen to optimize VSWR and bandwidth. Some aspects of the performance not easily obtained with other designs are the octave-band 20-dB isolation shown in Fig. 9(a) and the wide-band frequency conversion data in Fig. 9(c) and (d).

Conclusion

An elementary analysis was given for the hybrid junction of Fig. 1 and experimental results in an application of the hybrid junction verify that the behavior is in general accordance with the theory given. The operation of a four-diode star network connected to ports 3 and 4 of the hybrid junction was discussed and the performance was tested in several models of four-diode star mixer/modulators. A number of applications exist for the latter, including balanced modulation, double-balanced down- and up-conversion, voltage variable attenuation, 180-degree phase switching, phase detection, frequency doubling, and full-wave amplitude detection. In addition, when provisions are made for individual biasing of the diodes, they may be replaced with varactors, which suggests a variety of parametric applications. Another interesting prospect is the possibility of obtaining multioctave swept microwave power by extending to higher frequencies the broad-band down-conversion data of Fig. 9(c).

Appendix

The scattering matrix of the hybrid junction shown in Fig. 1 is determined using the method described by Carlin [5], in which the normalized admittance matrix Y_A is first obtained from the augmented network, followed by calculation of the scattering elements by

$$S = I - 2Y_A, \tag{20}$$

where I is the identity matrix. The network in Fig. 1 is augmented with resistors in series with each of the ports. The unnormalized admittance matrix Y_A' is obtained directly from this network, and

$$Y_A = \sqrt{R_0}\, Y_A' \sqrt{R_0} \tag{21}$$

where $\sqrt{R_o}$ is the diagonal resistance normalization matrix.

The short-circuit input admittance y_{111A}' is the input admittance at terminals 11_A in Fig. 3(a) with the other ports of the augmented network short-circuited:

$$y_{11A}' = \frac{1}{r\left[1 + \dfrac{Z_{01}}{2r}\, \dfrac{(Z_{01}/R) - j[\cot\theta + (Z_{01}/Z_{02})\cot\phi]}{1 - (Z_{01}/Z_{02})\cot\theta - \cot\phi - j(Z_{01}/R)\cot\theta}\right]}. \tag{22}$$

The input impedance (8) is obtained from (22):

$$Z_{i1} = Z_{i2} = \frac{1}{y_{11A}'} - r. \tag{23}$$

Since the network is symmetrical with respect to ports 1 and 2,

$$y_{22A}' = y_{11A}'. \tag{24}$$

From the conjugate property (1) of port 1 with respect to port 2,

$$y_{12A}' = y_{21A}' = 0. \tag{25}$$

The short-circuit transfer admittance from port 1 to port 3 is simply that of the equivalent half-circuit shown in Fig. 3(b).

$$y_{31A}' = \frac{j}{2r\sin\theta\left[(R/Z_{01}) + (Z_{01}/2r) - (R/Z_{02})\cot\theta\cot\phi - j\{[1 + (R/2r)]\cot\theta + (R/2r)(Z_{01}/Z_{02})\cot\phi\}\right]}. \tag{26}$$

Port 1 is symmetrical with respect to ports 3 and 4 and from (4)

$$y_{41A}' = y_{31A}'. \tag{27}$$

By reciprocity

$$y_{13A}' = y_{31A}' \tag{28}$$

and

$$y_{14A}' = y_{41A}'. \tag{29}$$

Port 3 is symmetrical with respect to ports 1 and 2, so that

$$y_{13A}' = y_{23A}' \tag{30}$$

and by reciprocity

$$y_{32A}' = y_{23A}'. \tag{31}$$

From (5)

$$y_{42A}' = -y_{32A}' \tag{32}$$

and by reciprocity,

$$y_{24A}' = y_{42A}'. \tag{33}$$

From (2)

$$y_{34A}' = y_{43A}' = 0. \tag{34}$$

The short-circuit input admittance y_{33A}' is obtained from the input admittance to the circuit of Fig. 4(f):

$$y_{33A}' = \frac{1}{R\left[1 + \frac{Z_{01}}{R}\frac{(Z_{01}/2r) - j\cot\theta}{1 - (Z_{01}/Z_{02})\cot\theta\cot\phi - j(Z_{01}/2r)[\cot\theta + (Z_{01}/Z_{02})\cot\phi]}\right]}. \tag{35}$$

The input impedance (12) is obtained from (35):

$$Z_{i3} = Z_{i4} = \frac{1}{y_{33A}'} - R. \tag{36}$$

From the symmetry of port 3 with respect to port 4,

$$y_{44A}' = y_{33A}'. \tag{37}$$

This completes the computation of the short-circuit admittance parameters. It is convenient to abbreviate these parameters as follows:

$$\begin{aligned} y_{11A}' &= y_{22A}' = a \\ \left.\begin{aligned} y_{31A}' &= y_{13A}' = \\ y_{41A}' &= y_{14A}' = \\ y_{32A}' &= y_{23A}' = \\ -y_{42A}' &= -y_{24A}' = \end{aligned}\right\} &c \\ y_{33A}' &= y_{44A}' = b. \end{aligned} \tag{38}$$

Then, from (24), (33), and (37), we have

$$Y_A' = \begin{bmatrix} a & 0 & c & c \\ 0 & a & c & -c \\ c & c & b & 0 \\ c & -c & 0 & b \end{bmatrix} \tag{39}$$

and from (20)

$$Y_A = \sqrt{R_0}\, Y_A' \sqrt{R_0}$$

$$= \begin{bmatrix} ar & 0 & c\sqrt{Rr} & c\sqrt{Rr} \\ 0 & ar & c\sqrt{Rr} & c\sqrt{Rr} \\ c\sqrt{Rr} & c\sqrt{Rr} & br & 0 \\ c\sqrt{Rr} & -c\sqrt{Rr} & 0 & ar \end{bmatrix}, \tag{40}$$

where

$$\sqrt{R_0} = \begin{bmatrix} \sqrt{r} & 0 & 0 & 0 \\ 0 & \sqrt{r} & 0 & 0 \\ 0 & 0 & \sqrt{R} & 0 \\ 0 & 0 & 0 & \sqrt{R} \end{bmatrix} \tag{41}$$

and $R_1 = R_2 = r$; $R_3 = R_4 = R$ are the respective port termination resistances.

The scattering matrix is now calculated from (20):

$$S = \begin{bmatrix} \alpha & 0 & \gamma & \gamma \\ 0 & \alpha & \gamma & -\gamma \\ \gamma & \gamma & \beta & 0 \\ \gamma & -\gamma & 0 & \beta \end{bmatrix} \tag{42}$$

where

$$\begin{aligned} \alpha &= 1 - 2ar \\ \gamma &= -2c\sqrt{Rr} \\ \beta &= 1 - 2br. \end{aligned} \tag{43}$$

ACKNOWLEDGMENT

The author wishes to thank F. M. Schumacher for suggestions and encouragement; P. Toft-Nielsen for contributions to the construction of the hybrid junction; S. R. Geraghty who supplied diode assemblies; S. M. Fukuchi who performed many of the tests.

REFERENCES

[1] L. J. Cutrona, "The theory of biconjugate networks," *Proc. IRE*, vol. 39, pp. 827–832, July 1951.
[2] G. A. Campbell and R. M. Foster, "Maximum output network for telephone substation and repeater circuits," *Trans. AIEE*, pt. 1, vol. 39, pp. 231–280, 1920.
[3] R. B. Mouw, "Broadband dc isolator—monitors," *Microwave J.*, vol. 7, pp. 75–77, November 1964.
[4] C. L. Ruthroff, "Some broad-band transformers," *Proc. IRE*, vol. 47, pp. 1337–1342, August 1959.
[5] H. J. Carlin, "The scattering matrix in network theory," *IRE Trans. Circuit Theory*, vol. CT—3. pp. 88–97, June 1956.

RECENT DEVELOPMENTS IN MICROWAVE SLOT-LINE MIXERS AND FREQUENCY MULTIPLIERS

J. K. Hunton & J. S. Takeuchi
Sylvania Electronic Systems--Western Division
Mountain View, California

A new type of planar balanced circuit, suitable for the realization of broadband microwave mixers and frequency multipliers, in integrated circuit form, was described in September 1969.[1] In this circuit, a hybrid junction was formed by joining a slot transmission line[2] and a coplanar transmission line.[3] Four diodes were connected at the junction in the form of a ring to realize a balanced mixer.

Balanced Mixer Circuits. A number of mixer circuits which are realizable in the planar form are shown in Table 1. In the analysis of these circuits, the diode model is an ideal switch which is an open circuit when the LO voltage is in the reverse direction and a fixed-loss resistance, R_d, when the LO voltage is in the forward direction. Through the switching action, which is a square-wave time function at the LO frequency, the signal terminals are connected in alternating fashion to the IF terminals, and the circuits can be classified as SPDT or DPDT.

The circuits which utilize four diodes have the essential characteristics of doubly balanced mixers provided the IF frequency band is much lower than the signal frequency band. The signal port is considered to be terminated at all frequencies in a generator resistance, R_s, while the IF port is resistively terminated (in R_o) only at the IF frequency, being short-circuited for all other frequency components. In the circuits which utilize two diodes, the IF and the LO connections are common, and the filter elements only serve as a frequency diplexer. In this case, the load on the IF port is considered to be resistive (R_o) at all frequencies. The minimum "ideal" conversion loss for all circuits is $(\pi^2/4)$ or 3.92 dB, assuming a perfect square-wave switching function. The effect of finite diode resistance is different for different circuits, as shown in the table. Here, the numbers are the values which would be calculated if $R_s = 50$ ohms and $R_d = 10$ ohms. Many diodes were measured, and typical values of R_d were found to be of the order of 6 to 8 ohms.

All of the circuits in Table 1 are realizable in planar form using the slot-line/coplanar-line junction concept. The photographs in Figures 1 and 2 show the MIC realization of the four-diode SPDT circuit (Table 1-C), which was designed for the 1 to 4 GHz frequency region. It consists of a chrome-gold metallization on one side of a 0.080-inch thick sheet of magnesium titanate ($\epsilon_r = 16$). The sheet is suspended in the center of a metal box. Connection to the slot-line is made by a small copper coaxi-tube cable at right angles to the axis while a coaxial connection to the coplanar-line is made directly along the axis. Excitation of the center conductor of the coplanar-line provides the unbalanced signal to the junction and excitation of the slot-line provides the balanced signal to the junction. The unbalanced signal from the coplanar-line cannot propagate past the junction. However, the balanced signal from the slot-line can propagate past the junction as far as the coaxial connection point, where it faces a short circuit. The balanced mode impedance of the coplanar-line is quite high, however, and a very broad frequency range can be achieved with the shorted section one-quarter wavelength long at mid-band. Four-beam lead Shottky barrier diodes are connected at the junction, in the firm of a ring. The filter capacitors are the silicon dioxide type and serve, also, as bonding pads for the tiny beam-lead diodes. The IF connecting wires are brought out through holes in the substrate to filter inductors located below. Connections from the other ends of the coils are brought up through the substrate to additional filter capacitors beyond the end of the slot-line, and thence to the ferrite toroidal IF balun.

The data plotted in Figure 3 shows the performance of the SPDT four-diode mixer (with 10 mw incident LO power) and indicates a frequency range capability of 6:1 with 6.5 to 7.0 dB conversion loss. The main bandwidth limiting factor is the shorted line stub which terminates the balanced (LO) port. This stub is a relatively high impedance by comparison with the LO drive impedance at the diode ring. The circuit losses amount to about 1.5 dB, most of which occurs in the IF balun. The four-diode circuit of Table 1-D has also been fabricated and tested. The frequency range capability in this case was less--about 3:1. This is because the signal is now coupled to the junction via the slot-line instead of the coplanar-line, and the loading effect of the shorted slot on the coplanar side is greater since the signal impedance at the diode ring is higher.

The realization of the two-diode circuit of Table 1-B is shown in the magnified photograph of Figure 4. Here, the silicon dioxide capacitor on the coplanar center conductor and the IF connecting wire serve to diplex the frequency spectrum, generated at the junction, between the IF load and the LO source resistance. Unlike the four-diode circuits, the two-diode circuits will generate the even harmonics of the LO at the signal port. Hence, the bandwidth, in a practical case, is limited to an octave. The mixer shown in Figure 4 was designed for the X-band range and exhibited good performance from 7 to 14 GHz.

Balanced Frequency Multiplier Circuits. The planar balanced circuit concept is equally applicable to the realization of balanced frequency doublers and quadruplers. The advantage of a balanced multiplier circuit is that the odd and even harmonics are separated--the odd harmonics at the input port and the even harmonics at the output port--thus allowing greater bandwidth. Balanced frequency doublers have been constructed with good efficiency over octave bandwidths, using coaxial line circuits.[4]

Reprinted from *1970 IEEE G-MTT Int. Microwave Symp.*, 1970, pp. 196–199.

Neglecting the tuning and filter elements, a planar balanced-circuit frequency doubler can take the form of either of the circuits in Table 1-A or 1-B with the mixer diodes replaced by punch-through varactors. The LO port becomes the fundamental input port; the signal port becomes the harmonic output port; and the IF port is eliminated, while appropriate bias connections are added. The unbalanced signal is supplied, as in the mixer case, by coplanar-line and the balanced signal via slot-line. Appropriate bandpass tuning and filtering elements are realizable on both sides of the junction.

REFERENCES

1. "A Microwave Integrated Circuit Balanced Mixer with Broad Bandwidth," J. K. Hunton, W. M. Kelly, J. S. Takeuchi, Proceedings of the 1969 Microelectronics Symposium, September 1969, pp. A3-1, 2.

2. "Slot-Line - An Alternative Transmission Medium for Integrated Circuits," S. B. Cohn, 1968 G-MTT Symposium Digest, May 1968, pp. 104-109. (See also 1969 G-MTT Symposium Digest, pp. 99-105.)

3. "Coplanar Waveguide, A Surface Strip Transmission Line," C. P. Wen, 1969 G-MTT Symposium Digest, May 1969, pp. 110-115.

4. "The Design of Broadband Frequency Doublers Using Charge-Storage Diodes," K. L. Kotzebue and G. L. Matthaei, 1969 G-MTT Symposiu Digest, May 1969, pp. 136-142.

	Circuit	OPTIMUM CONVERSION LOSS	OPTIMUM IF LOAD RES	OPTIMUM SIG REFL COEFF
A	2 DIODE SPDT, BALANCED LO	$\frac{\pi^2}{4}+\frac{\pi^2}{4}\frac{R_d}{R_s}$ (4.7 DB)	$4R_s+4R_d$ (240 Ω)	$\frac{\frac{R_d}{R_s}}{1+\frac{R_d}{R_s}}$ (0.17)
B	2 DIODE SPDT, BALANCED SIG	$\frac{\pi^2}{4}+\pi^2\frac{R_d}{R_s}$ (6.5 DB)	$\frac{R_s}{4}+R_d$ (22.5 Ω)	$\frac{4\frac{R_d}{R_s}}{1+4\frac{R_d}{R_s}}$ (0.40)
C	4 DIODE SPDT, BALANCED LO	$\frac{\pi^2}{4}+\frac{\pi^2}{8}\frac{R_d}{R_s}$ (4.4 DB)	$4R_s+2R_d$ (220 Ω)	$\frac{-(1-\frac{4}{\pi^2})+\frac{R_d}{2R_s}}{\frac{R_d}{2R_s}}$ (-0.45)
D	4 DIODE DPDT, BALANCED SIG	$\frac{\pi^2}{4}+\frac{\pi^2}{2}\frac{R_d}{R_s}$ (5.4 DB)	R_s+2R_d (70 Ω)	$\frac{-(1-\frac{4}{\pi^2})+\frac{2R_d}{R_s}}{1+\frac{2R_d}{R_s}}$ (-0.14)

Table 1 -- Mixer Circuits Realizable in Planar Form.

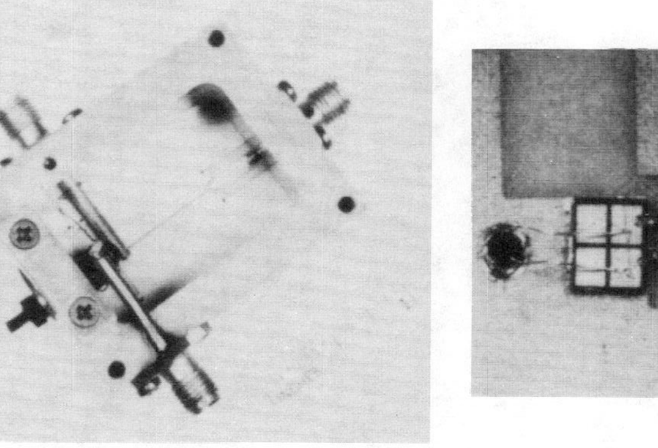

Figure 1 -- Slot-Line Mixer -- 1 to 4 GHz Band.

Figure 2 -- Slot-Line Mixer -- 1 to 4 GHz Band, Magnified Junction View.

Figure 3 -- Slot-Line Mixer -- 1 to 4 GHz Band, Measured Data.

Figure 4 -- Slot-Line Mixer Junction -- X-Band.

A NEW MIC DOUBLE-BALANCED MIXER WITH RF AND IF BAND OVERLAP

By

R. Pflieger
AIL, a division of Cutler-Hammer
Melville, New York

Abstract

A new type of MIC double-balanced mixer has been developed using a combination of slot-line, microstrip, and wire near a ground-plane transmission line. A number of these mixers have been built with octave RF bandwidths in the 2 to 18 GHz region with RF and IF band overlap.

Summary

The requirement for MIC double-balanced mixers in the 2 to 18 GHz range with all three ports operating in the microwave region prompted the development of the mixer described. Two new types of 180-degree hybrids are used to realize the three orthogonal means of driving the four mixer diodes necessary for double-balanced operation. Hunton and Takeuchi[1] used a combination of slot line and coplanar line to realize a MIC single-balanced mixer. Here, slot line, microstrip, and wire near a ground-plane transmission line[2] are combined to realize a MIC double-balanced mixer. This combination of transmission lines results in IF performance unique in the higher microwave region; IF band overlap with the RF band. A typical mixer for the RF band from 2.8 to 6.2 GHz provides an IF response from dc to 3.8 GHz at the 3-dB point, with port-to-port isolation greater than 20 dB.

This configuration also permits the three dimensional structure of the double-balanced mixer to be confined to the surfaces of a single alumina substrate; the connections between the surfaces are made only at the edges of the substrate. Figure 1 shows the microstrip and wire-line side of a 5.5 to 11.5 GHz unit. The slot line on the ground-plane side runs directly beneath the tuning-fork-shaped microstrip circuit to an SMA connector. (The connection cannot be seen in the figure.) Except for the wire line, the beam-lead diodes, and some gold wire and ribbon, all the circuits are etched on the two surfaces of a single alumina substrate.

Mixers of this design have been built operating over octave bands from approximately 2 to 18 GHz. The following general description applies to all the mixers.

Each of the three ports in the mixer is simply referred to by its transmission-line type. Although each of the three ports can be used for IF, LO, or RF, the microstrip port is usually called the IF port because its response extends down to dc. The 3-dB upper-frequency limit of the microstrip port can be made to extend into the operating band of the wire-line port. The wire-line port is an octave or more wide, and the slot-line port is 5 to 50 percent wide. The frequency range of the slot-line port falls within or above that of the wire-line port and can also overlap that of the microstrip port. Figure 2 shows an example of the frequency overlap of the three ports.

Both the slot-line port and the wire-line port can operate anywhere in the 2 to 18 GHz range. Port-to-port isolation runs 20 to 30 dB; the conversion loss ranges from 5 to 8 dB through X-band. Figure 3 shows a typical plot of conversion loss. At K_u-band, conversion losses have been running several dB higher because of circuit losses; however, these losses can be reduced by using better diodes and by refining the circuits.

The slot line to microstrip hybrid is based on the configuration of the slot-line electric field.[3] Figure 4 shows the slot line to microstrip hybrid. Two microstrips running above the slot to each side couple to the electric field 180 degrees out of phase with each other; they do not couple to the magnetic field. Joining the two microstrips forms a fourth port, which is orthogonal to the slot line. It is not obvious that this is a four-port hybrid, because the two ports that are 180 degrees out of phase are used to drive the diode ring without a ground connection. The normally weak coupling between the slot line and the two microstrips results in narrow operating bandwidths. To broadband the slot line to microstrip hybrid, the connection between the two parallel microstrips and the end of the slot line must be kept an equivalent quarter-wavelength from the diodes. This minimizes loading of the diodes by these circuits over the widest possible band. However, minimizing the loading also minimizes the coupling between the slot line and the microstrips. To obtain strong coupling and wide bandwidths, capacitive stubs must be added beyond the diodes.

The primary advantage of the slot line to microstrip hybrid is that it does not limit the IF response. The wire line to microstrip hybrid, in providing the IF grounding of the diodes, determines the upper IF operating frequency.

Theoretically, any two hybrids can serve in a double-balanced mixer. The use of a second slot line

Reprinted from *1973 IEEE G-MTT Int. Microwave Symp.*, 1973, pp. 301–303.

to microstrip hybrid was rejected because of the difficulty in achieving isolation between the two slot lines. A simple method of building the second hybrid was devised. Two parallel microstrips are driven 180 degrees out of phase with each other by a third transmission line. Joining the two microstrips forms a fourth port, which is orthogonal to the third transmission line. In the mixer, the third transmission line must not couple to the slot-line fields if proper orthogonality is to be achieved. A single wire, running above one of the microstrips and using the microstrip as its ground plane, satisfies these requirements because the microstrip effectively shields the wire line from the fields of the slot line. Figure 5 shows a transmission-line model of this circuit. The two microstrips are connected approximately one quarter-wavelength from both the diode connection and the wire-line connection. The wire uses one of the microstrips as a ground plane and is connected to the other microstrip.

The model in Figure 5 adequately describes the performance of the wire line to microstrip hybrid. As is to be expected from the quarter-wavelength stub circuitry, octave bandwidth operation is achieved.

Conclusion

A MIC mixer using a combination of slot line, microstrip, and wire near a ground-plane transmission line has been developed for the 2 to 18 GHz range. It is compact, reproducible, and rugged. The mixer is useful where octave-or-less RF bandwidths are required, especially where a very broadband IF bandwidth is required. The techniques used to build this mixer will prove valuable in other applications where elements on a substrate are required to be driven in a balanced circuit.

Acknowledgments

This work was supported by in-house funds. The author gratefully acknowledges the advice and encouragement of J. Taub and the technical assistance of his colleagues at AIL.

References

1. Hunton, J., and Takeuchi, J., "Recent Developments in Microwave Slot Line Mixers and Frequency Multipliers," G-MTT Symposium, 1970.

2. Jasik, H., "Antenna Engineering Handbook," McGraw-Hill Book Co., Inc., p 30-4, 1961.

3. Cohn, S. B., "Slot Line - An Alternate Transmission Medium for Integrated Circuits," G-MTT Symposium, 1968.

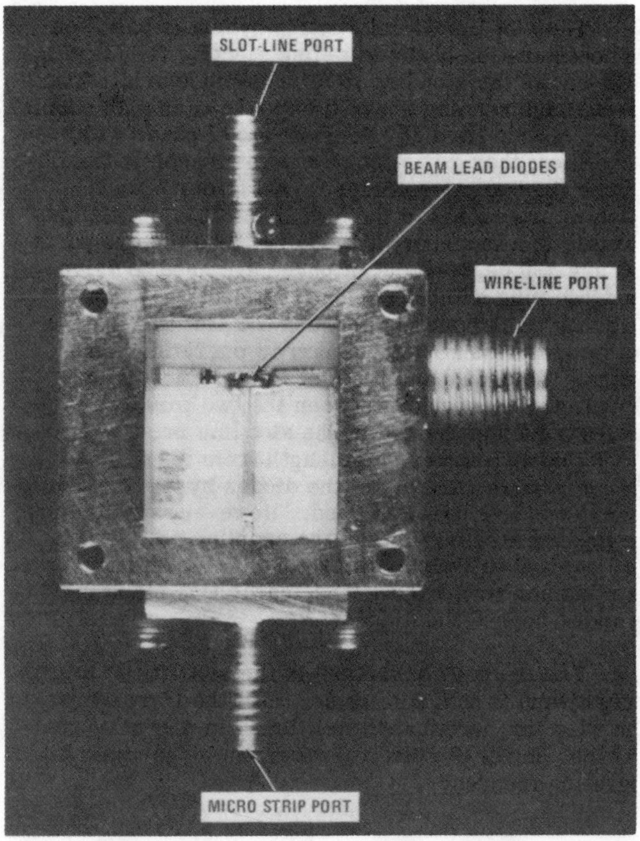

FIG. 1. X-BAND MIC DOUBLE BALANCED MIXER

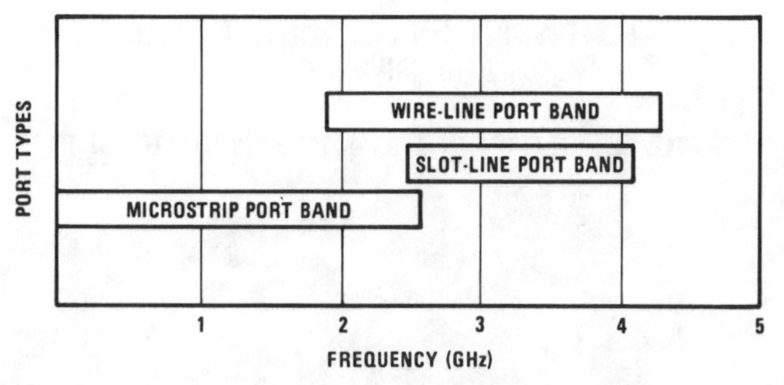

FIG. 2. OPERATING FREQUENCY BANDS OF S-BAND MIXER

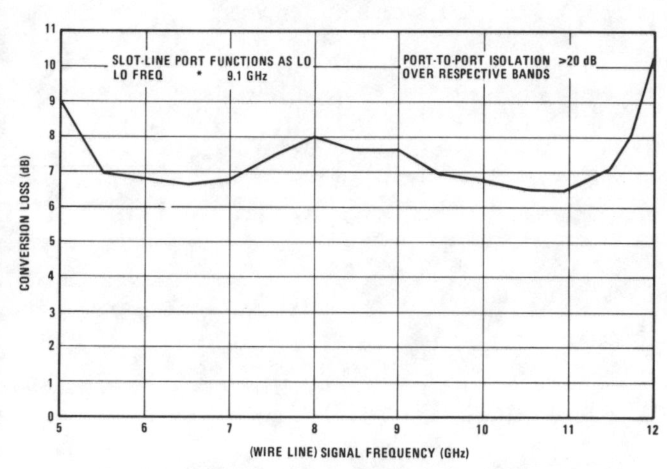

FIG. 3. X-BAND MIC DOUBLE BALANCED MIXER CONVERSION LOSS

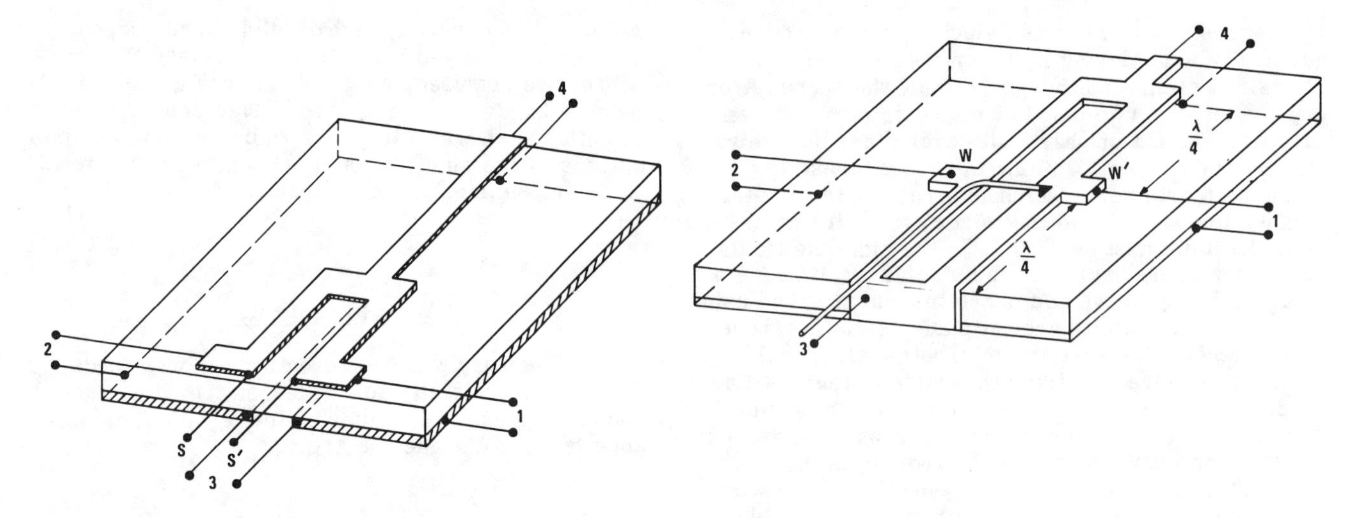

A. SLOT LINE TO MICROSTRIP HYBRID

A. WIRE LINE TO MICROSTRIP HYBRID

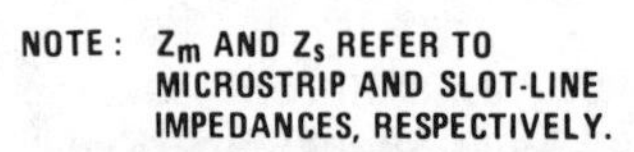

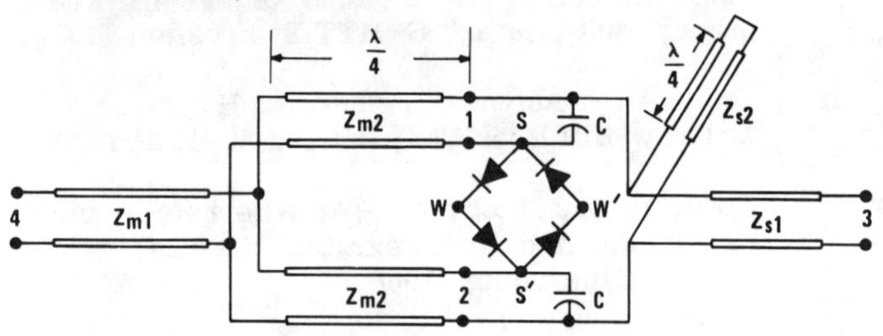

B. TRANSMISSION-LINE MODEL OF HYBRID AS USED TO DRIVE DIODES.

FIG. 4. SLOT LINE TO MICROSTRIP HYBRID

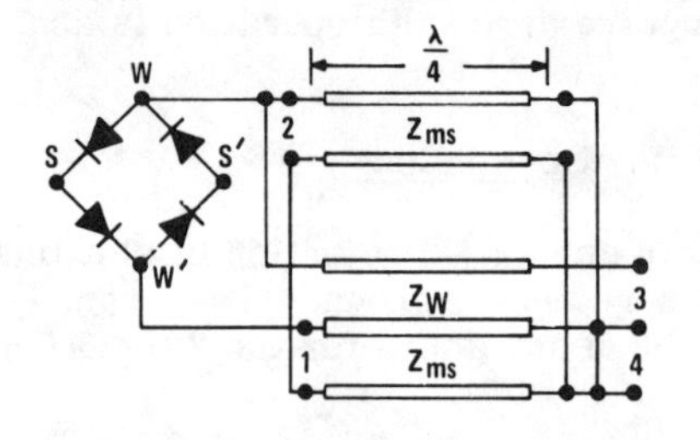

B. TRANSMISSION-LINE MODEL OF HYBRID AS USED TO DRIVE DIODES.

FIG. 5. WIRE LINE TO MICROSTRIP HYBRID

Simplified 12-GHz Low-Noise Converter with Mounted Planar Circuit in Waveguide

YOSHIHIRO KONISHI, SENIOR MEMBER, IEEE, KATSUAKI UENAKADA, NORIHIKO YAZAWA, NORIO HOSHINO, AND TADASHI TAKAHASHI

***Abstract*—A 12-GHz low-noise converter consisting of a planar circuit mounted in waveguide is described. This circuit consists of a metal sheet with proper patterns that is inserted in the middle of a waveguide parallel to the *E* plane. All circuit elements required for the converter are pressed or etched. This circuit is very useful for low-cost mass production and good performance. A measured noise figure of 4.5 dB was obtained with a 12-GHz signal frequency and a 420-MHz intermediate frequency.**

I. INTRODUCTION

A low-noise 100-GHz converter comprising a microwave integrated circuit mounted in a waveguide was reported by Konishi and Hoshino in 1971 [1]. A low-noise 12-GHz converter based on a similar principle is described in this short paper. It consists of a planar circuit mounted in a waveguide and is suitable for low-cost mass production.

Manuscript received July 31, 1973; revised December 4, 1973.
The authors are with the Technical Research Laboratories, Japan Broadcasting Corporation, Tokyo, Japan.

Millimeter components using integrated circuits mounted in a waveguide were also developed by Meier in 1972 [2], where unloaded Q of this transmission line takes a value less than 900 at X band [3].

A new type of filter, that is, a planar circuit mounted in waveguide, that we have developed has an unloaded Q factor of 2000–2500 at X band. This new type of filter is used in our 12-GHz converter to achieve low-noise performance.

The 12-GHz converter described in this short paper has a mixer conversion loss of 3.5 dB and a total noise figure of 4.5 dB, including the contribution of an intermediate frequency amplifier with a noise figure of 2.0 dB at 420 MHz.

II. 12-GHz LOW-NOISE CONVERTER WITH MOUNTED PLANAR CIRCUIT IN WAVEGUIDE

A high-sensitivity and low-cost converter was required that could be constructed simply and be mass produced. The construction of the circuit we developed is such that every necessary circuit element is arranged on a metal sheet merely by pressing or etching, and the metal sheet is inserted into a waveguide.

We will describe the results of the experiment that was carried out on our converter with a mounted planar circuit.

Fig. 1 shows the planar circuit pattern from left to right, a signal frequency bandpass filter, a Schottky barrier diode mount, a local oscillator frequency bandpass filter, and a Gunn diode mount for the local oscillator. A 0.3–0.5-mm-thick copper sheet is favorable for this pattern. When etching is used in forming the pattern, a 0.3-mm copper sheet is utilized for dimensional precision. When pressing is used, with a 0.5-mm-thick metal sheet, dimensions are maintained within 20 μm.

Reprinted from *IEEE Trans. Microwave Theory Tech.*, vol. MTT-22, pp. 451–454, Apr. 1974.

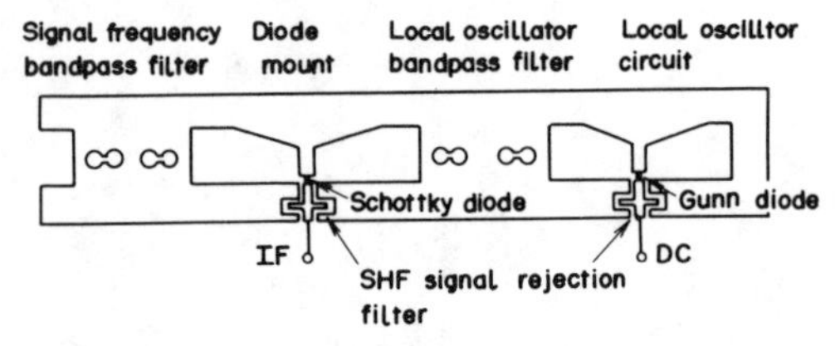

Fig. 1. Construction of 12-GHz converter with planar circuit in waveguide.

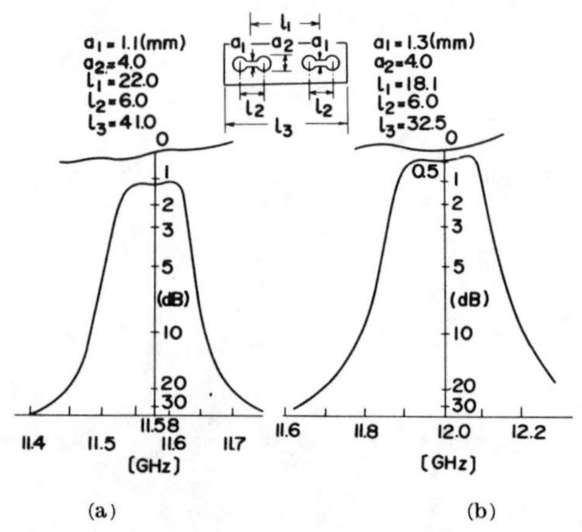

(a) (b)

Fig. 2. Bandpass filter (BPF). (a) Characteristic of oscillator BPF. (b) Characteristic of signal BPF.

A. Bandpass Filter

The construction of the bandpass filters for both the signal and the local oscillator frequencies is shown in Fig. 1. The diameter of the circle or the gap is adjusted to set the center frequency.

Fig. 2(a) and (b) shows the characteristics of the local oscillator frequency and the signal frequency bandpass filters. The signal frequency bandpass filter has a 12-GHz center frequency and a 3-dB bandwidth of 230 MHz. Its insertion loss is 0.3 dB. The attenuation is more than 30 dB for the local oscillator frequency and the image frequency. The local oscillator frequency bandpass filter has an 11.58-GHz center frequency and a 3-dB bandwidth of 140 MHz. Its insertion loss is 1.2 dB. Attenuation is more than 30 dB at the signal and image frequencies. It has a narrow band in comparison with the signal frequency bandpass filter. As a result, its insertion loss is higher. An unloaded Q of approximately 2000–2500 is provided for both the bandpass filters.

B. Schottky Diode Mount

Fig. 3(a) shows the Schottky diode mount circuit. A beam lead diode or a Schottky diode packaged for microwave integrated circuits is employed, allowing the diode to be directly mounted on the planar circuit.

Round and flat posts were considered as diode mounts. When round or flat posts are used alone, the height of the waveguide determines the length of the post, and unwanted reactance is included. Matching is more difficult, and the bandwidth will inevitably be narrower. When the diode is mounted at the foot of an open-tip probe that couples to the waveguide mode, the length of the antenna can be varied to match to impedances between 50 and 200 Ω. The width of the antenna determines the frequency bandwidth.

Fig. 3(a) shows the construction that allows the diode to be mounted directly with the metal sheet alone. As shown in Fig. 3(a), a distributed line transformer and a ridge taper are provided. The flat post that is located close to the $\lambda/4$ line is connected to the ridge taper, which has been set vertically in the middle of the H plane of the waveguide, and the diode is connected between the waveguide bottom wall and the flat post, as shown in Fig. 1. In this case, the flat post inserted along the E plane functions approximately as a distributed line transformer of $\lambda/4$. That is, where the diode impedance is Z_D and the $\lambda/4$ line impedance is Z, the impedance viewed toward the diode from the point A in Fig. 3(a) is Z^2/Z_D. As Z is varied by changing the post width, impedance matching becomes possible for a low-impedance diode, too.

The curves in Fig. 3(b)–(d) show the impedance viewed from the diode terminal when the post width W and the height of the ridge taper h are changed. Fig. 3(b) shows the impedance characteristics for various post widths W in the case of $\lambda/4$ post length. An approximate change is made as was described previously. Fig. 3(c) shows the impedance when W of the flat post remains constant at 4 mm and a change is made in the height of the taper. Fig. 3(d) shows the impedance when the width W is changed only by the flat post, the tip of which is shortened. In Fig. 3(d), in comparison with Fig. 3(b), there is an increase in reactance and the bandwidth proves to be narrower. According to those charts, we can obtain arbitrary impedance in wide bands by selecting appropriate W and h.

C. Consideration of Image Impedance

It is necessary to consider the image impedance in reducing the conversion loss [4]. At the image frequency, the signal frequency and the local oscillator frequency bandpass filters, which are arranged at both sides of the diode mount, produce a short circuit at an inside point a little off the end surface of the filter. Accordingly, the distance between the two filters varies the image frequency impedance that is presented to the diode. In general, where the distance between the filters is selected equal to $n\lambda/2$ at the image frequency, the image impedance is an open circuit. As the position of the signal frequency and the local oscillator frequency bandpass filters are adjusted to optimize the signal frequency and local oscillator frequency impedances, it is not possible to provide the exact image frequency impedance for minimum conversion loss.

For the converter described in this short paper, the image impedance remains a little off the open condition and capacitive.

D. Local Oscillating Circuits

The local oscillator source used a Gunn diode, and the construction described in Section II-B is used as a mount.

III. RESULTS OF OUR EXPERIMENT ON PROPOSED CONVERTER

We will describe the results of our measurements on the experimental converter with the metal sheet on which every foregoing pattern is formed, as shown in Fig. 1.

Fig. 4 shows the characteristics of the converter we have developed. The local oscillator drive is 8 mW. The signal frequency is 12 GHz and the intermediate frequency is 420 MHz. We were successful in obtaining a minimum real conversion loss of 3.2 dB. An intermediate amplifier with approximately a 2.0-dB noise figure was connected and a total noise figure 4.5 dB was obtained. This value coincides with the result of the noise figure equation according to the Appendix.

APPENDIX

Overall receiver noise figure F_t can be expressed as

$$F_t = L_r L[t + F_{if} - 1] \tag{1}$$

where

- L_r RF circuitry loss;
- L mixer conversion loss;
- F_{if} noise figure of IF amplifier;
- t mixer noise temperature.

In the case of the image reactive load, the noise figure F of the mixer diode is given by [6]

$$F = 1 - n + nL \tag{2}$$

where

- $n = \dfrac{e}{2kT_0\alpha}$ equals noise ratio;
- e charge on an electron;
- k Boltzmann's constant;
- T_0 temperature;
- α parameter of Schottky barrier diode (see [5]).

In the case of $\alpha = e/(kT_0)$, the noise ratio is equal to 0.5. There-

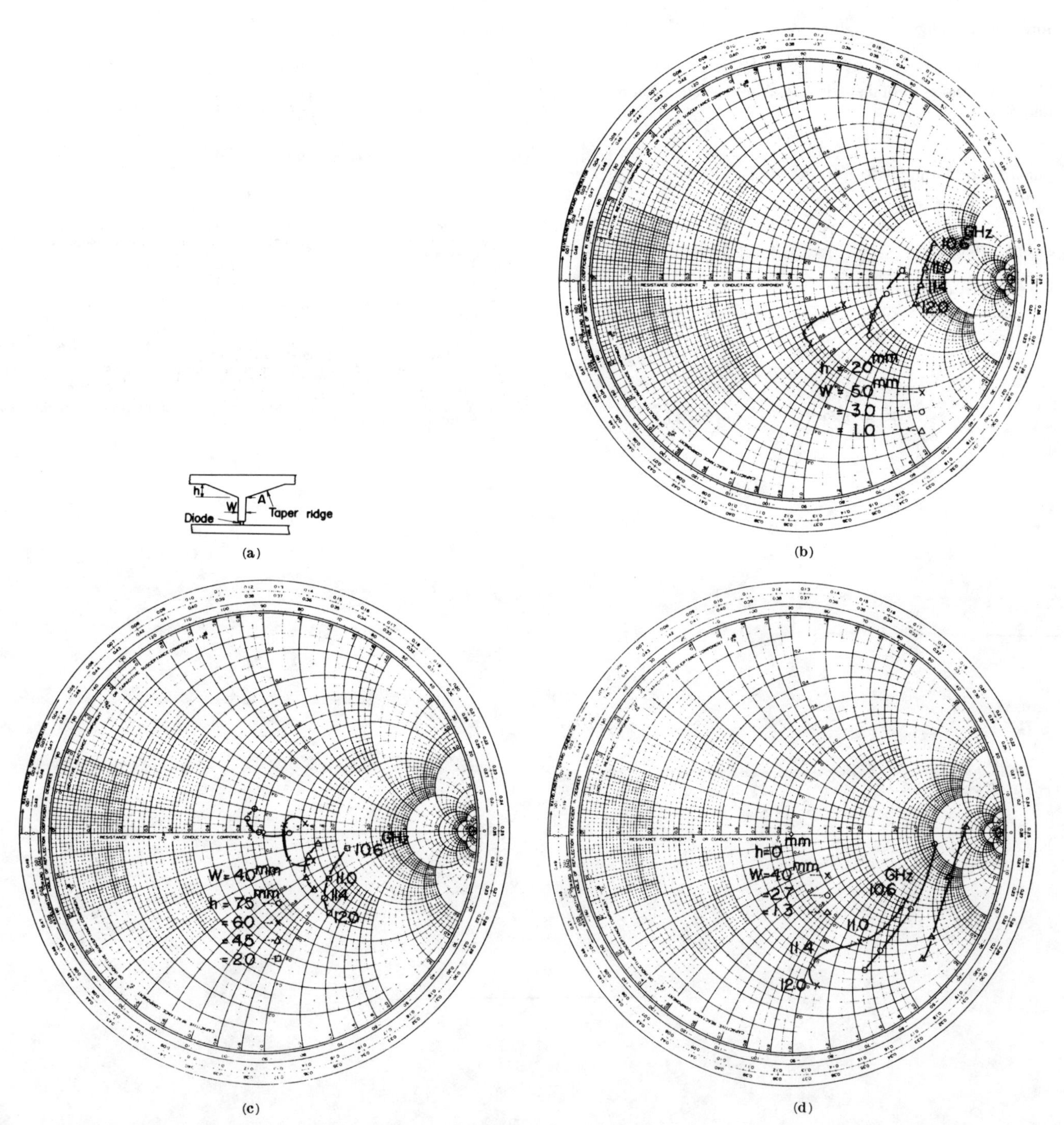

Fig. 3. Relationships between diode mount and impedance. (a) Diode mount. (b)–(d) Impedance or admittance coordinates.

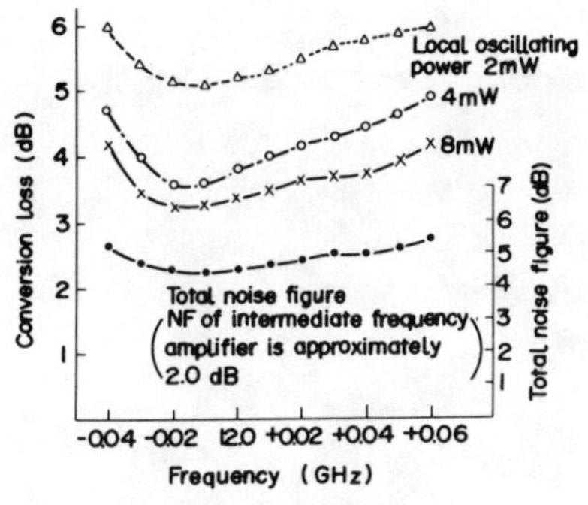

Fig. 4. Characteristics of 12-GHz converter with mounted planar circuit in waveguide.

fore, the noise figure F is given by

$$F = \tfrac{1}{2}(1 + L). \tag{3}$$

This is the same formula as in [5]. Also, the equation relating t, L, and F is

$$t = F/L. \tag{4}$$

Substituting formula (2) into (4) yields

$$t = \frac{1}{L}(1 - n) + n. \tag{5}$$

Measured values for the mixer were

$$L_t = L_r L = 2.09 \quad (3.2\ \text{dB})$$

$$F_{\text{if}} = 1.585 \quad (2.0\ \text{dB})$$

$$\alpha = 33.0$$

$$F_t = 2.82 \quad (4.5\ \text{dB}).$$

Calculated values were

$$n = 0.591$$

$$t \cong \frac{1}{L_t}(1 - n) + n = 0.7855$$

$$F_t = 2.86 \quad (4.57\ \text{dB}).$$

REFERENCES

[1] Y. Konishi and N. Hoshino, "100-GHz-band low-noise mixer," Inst. Electrical Communication Engineers Japan, Rep. MW 71-40, July 1971.

[2] P. J. Meier, "Two new integrated-circuit media with special advantages at millimeter wavelengths," in *IEEE G-MTT Symp. Dig.*, 1972, pp. 221–223.

[3] ——, "Equivalent relative permittivity and unloaded Q factor of integrated finline," *Electron. Lett.*, vol. 9, pp. 162–163, Apr. 1973.

[4] R. J. Mohr and S. Okwit, "A note on the optimum source conductance of crystal mixers," *IRE Trans. Microwave Theory Tech.*, vol. MTT-8, pp. 622–629, Nov. 1960.

[5] M. R. Barber, "Noise figure and conversion loss of the Schottky barrier mixer diode," *IEEE Trans. Microwave Theory Tech.*, vol. MTT-15, pp. 629–635, Nov. 1967.

[6] Y. Konishi, "Low noise amplifier" (in Japanese), Nikkankoogyo Shinbun, Ltd., 1969.

Low-Noise Microwave Down-Converter with Optimum Matching at Idle Frequencies

G. B. STRACCA, F. ASPESI, AND T. D'ARCANGELO

Abstract—**A low-noise balanced down-converter for microwave radio-link applications is described. Down-converters of this type have been realized with typical noise figures of 3.5 dB at 4 GHz, 4 dB at 7 GHz, and 5 dB at 13 GHz. These results are obtained mainly by taking into account high order sideband frequencies of the pump harmonics up to the third, by properly terminating the image frequency, by matching the input port of the mixer and by optimizing the mixer–preamplifier interface. The experimental results are compared with the theoretical ones obtainable with ideal purely resistive diodes.**

I. Introduction

Theoretical analysis has shown that mixer performance depends not only on the RF/IF mixer impedances and on the image-frequency termination [1]-[4], but also on the terminations at the various "idle frequencies,"[1] as well as on the LO waveform [5]-[8].

This short paper describes a mixer configuration designed to control the idle-frequency terminations up to the third harmonic and gives experimental results for mixers operating at different frequency ranges (i.e., 3.6–4.2 GHz; 7.1–7.7 GHz; 12.7–13.3 GHz) which are now used in the receiving section of commercial radio links. Down-converters of the same type, operating at frequencies between 11 and 12 GHz, have also been developed for radiometric applications, and are presently in operation in various stations of the European Space Research Organization.

Manuscript received April 4, 1972; revised March 26, 1973.

G. B. Stracca is with the Istituto di Elettrotecnica ed Elettronica, Trieste University, Trieste, Italy.

F. Aspesi and T. D'Arcangelo are with the GTE Telecommunications Research Laboratories, Cassina de Pecchi, Milan, Italy.

[1] Idle frequencies are defined as follows: $\omega_{\pm m} = m\omega_p \pm \omega_0$, where ω_p is the LO frequency, ω_0 is the IF frequency, and m is any positive integer. Here, the frequency ω_{+1} is the signal frequency and ω_{-1} is the image frequency. When m is an even (odd) integer, the corresponding frequencies are called even (odd) "idle frequencies."

Reprinted from *IEEE Trans. Microwave Theory Tech.*, vol. MTT-21, pp. 544–547, Aug. 1973.

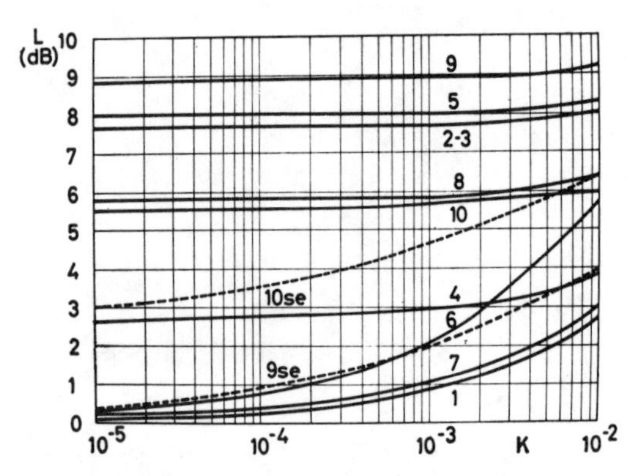

Fig. 1. Available conversion loss L_0 versus $k = g_{min}/g_{max}$ of resistive mixers. Solid lines—Balanced mixers with symmetrical rectangular LO drive waveform for the various cases of idle terminations given in Table I. Curve 9*se*—Single-ended mixer with optimum nonsymmetrical rectangular LO waveform (terminations as in case 9 of Table I). Curve 10*se*—Single-ended mixer with exponential resistive diode and sinusoidal voltage LO waveform (terminations as in case 10 of Table I).

TABLE I
IDLE-FREQUENCY TERMINATIONS FOR THE CASES OF FIG. 1

	Odd Idle Frequencies		Even Idle Frequencies		
Cases	ω_{-1}	all the others	ω_{-2}	ω_2	all the others
1	o.c.	o.c.	sh.c	sh.c	sh.c
2	sh.c	o.c.	sh.c	sh.c	sh.c
3	o.c.	o.c.	sh.c	o.c.	sh.c
4	o.c.	o.c.	o.c.	sh.c	sh.c
5	o.c.	o.c.	o.c.	o.c.	sh.c
6	sh.c	o.c.	sh.c	o.c.	sh.c
7	sh.c	o.c.	o.c.	sh.c	sh.c
8	sh.c	o.c.	o.c.	o.c.	sh.c
9	sh.c	sh.c	sh.c	sh.c	sh.c
10	o.c.	sh.c	sh.c	sh.c	sh.c

Note: Notice that the curves of Fig. 1 also hold for the dual termination cases, i.e., with short circuits (sh.c) replacing the corresponding open circuits (o.c.), and vice versa.

II. DESIGN CONSIDERATIONS

1) Single-ended and balanced[2] mixer configurations have been analyzed with various LO drive waveforms and idle-frequency terminations with purely resistive diodes [5]–[7]. For both configurations, the best conversion losses are the same and are obtained with a symmetrical rectangular LO drive and with dual terminations at even and odd idle frequencies, respectively (i.e., with all the even idle frequencies short circuited and all the odd idle frequencies open circuited, or vice versa).

In order to compare the best theoretical performance with the experimental results, the theoretical conversion loss L_0 calculated for input and output conjugate match conditions (i.e., "available" conversion loss) is plotted in curve 1 of Fig. 1 versus $k = g_{min}/g_{max}$, where g_{min} and g_{max} are the minimum forward and the maximum reverse differential diode conductance during the pump cycle, respectively.

This type of idle-frequency termination can easily be realized in balanced mixers, because all the odd idle frequencies appear at the input and all the even appear at the output,[2] which allows the corresponding constraints to be set, at least in theory, by means of two filters only at the input and output port, respectively.

Single-ended mixers do not present even and odd idle-frequency separations, and therefore an infinite number of filters is necessary to set the optimum terminations. Single-ended mixers, however, have been analyzed until now by imposing the same kind of constraints at all the idle frequencies (i.e., short circuit or open circuit), and only at a finite number of frequencies (ω_{-1} [1], [4], [6]; ω_2; and ω_{-2} [5], [7]) has the influence of different terminations been examined. In this case, in fact, the imbedding network may again be realized with a small number of filters. Suboptimal operating conditions have been found, which require nonsymmetrical rectangular LO waveforms and the same kind of terminations for all the idle frequencies [4], [5].[3] The curve 9*se* of Fig. 1 shows the behavior of L_0 in this case.

The balanced mixer configuration has finally been selected for the practical design described in Section III, because of the intrinsic simplicity in setting the optimum idle-frequency constraints as discussed above.

2) It is difficult to drive the diodes at microwave frequencies with a rectangular waveform. For the case of a sinusoidal waveform, theoretical analysis indicates that a sinusoidal current is better than a sinusoidal voltage [6], [7]. As an example, for $k = 10^{-3}$, the degradation of L_0 is about 1 dB for a sinusoidal LO current and 3 dB for a sinusoidal LO voltage with respect to curve 1 of Fig. 1. In the actual mixer, therefore, the LO signal is coupled to the diodes through a series resonator, which presents high impedance to LO harmonic frequencies and allows us to approximate better a sinusoidal current drive than a sinusoidal voltage drive.

3) In practical cases, mixer diodes are not purely resistive. However, measured results show good agreement with theory, provided the k value, chosen to characterize the diodes, is defined as the ratio of the minimum to the maximum real part of the diode small-signal admittance at the LO frequency, measured at 0 V and at full conduction bias conditions, respectively.

4) To show the influence of different idle-frequency terminations, we have also plotted in Fig. 1 L_0 versus k for the various cases defined in Table I, still for the same LO waveform. Notice that the dual cases are equivalent as far as conversion losses are concerned. However, they may exhibit different noise figures, as discussed in [7].

The behavior of L_0 at small k's for cases 2–5 and 8–10 can be justified heuristically by observing that some components of the currents circulating in the resistive diodes become very large as k approaches zero, in order to satisfy the constraints imposed at the input and output ports. It may be noticed that the idle-frequency terminations of cases 9 and 10 for balanced mixers are the same as those of the suboptimal cases for single-ended mixers mentioned before, and are presented in curves 9*se* and 10*se*.

Experiments have confirmed the influence of idle-frequency terminations in practical mixers with real diodes and nonrectangular LO waveforms, but the differences are not so pronounced, as shown in the theoretical cases of Fig. 1.

5) It is well known [3], [7] that the source impedance, which gives the minimum conversion loss, is not generally coincident with that giving the minimum down-converter noise figure. However, in practical situations, there is little difference in noise performance between the two cases [7], and therefore the mixer has been adjusted for a source impedance corresponding to that of minimum conversion loss.

6) To minimize the overall noise figure, particular care should be given to the mixer/IF preamplifier interface. For minimum noise, an impedance transformer should be used to transform the mixer output impedance into the optimum IF source impedance. To avoid the difficulties associated with the realization of a transformer network, an IF preamplifier has been realized which has an optimum noise source impedance close to the output impedance of the mixer.

III. DESCRIPTION OF THE EXPERIMENTAL MIXER

A schematic representation of the experimental balanced two-diode mixer is shown in Fig. 2. The mixer consists of a waveguide cavity containing an alumina substrate with two beam lead Schottky-barrier diodes mounted thereon.

The mixer configuration is designed to control the idle-frequency terminations up to the third harmonic. The image frequency ω_{-1} is terminated by properly positioning an image rejection filter at the input. This filter consists of two capacitively coupled bandstop coaxial resonators which are spaced a quarter wavelength at the band-

[2] The term "balanced" refers both to the four-diode configuration discussed in [9] and to the equivalent two-diode configuration of Fig. 2, i.e., for configurations in which there exist inherent even and odd idle-frequency isolations (and therefore inherent-signal IF isolation). Hybrid coupled two-diode mixers (although often called balanced mixers) do not present this property, and are considered as single-ended mixers.

[3] When a sinusoidal voltage LO waveform is applied to a diode with an ideal exponential V–I characteristic and all the idle frequencies higher than ω_{-1} are short-circuited, the optimum operation is achieved when ω_{-1} is open circuited (see curve 10*se* of Fig. 1).

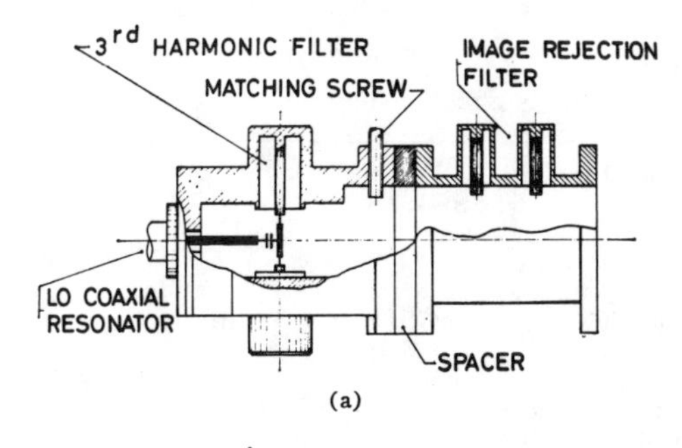

(a)

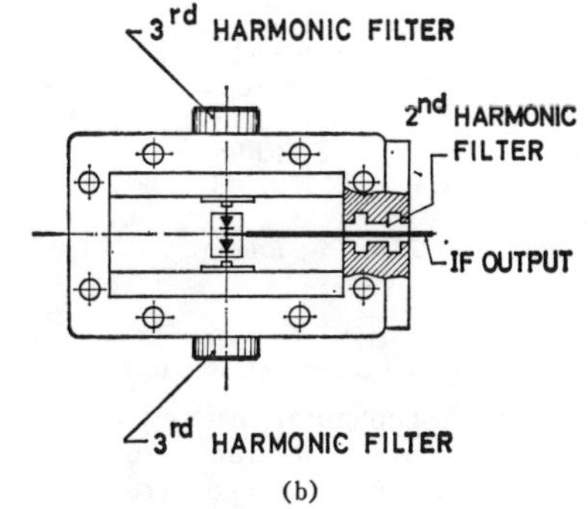

(b)

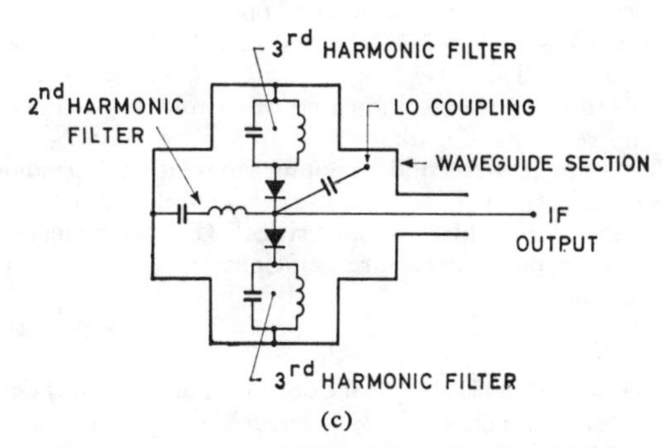

(c)

Fig. 2. (a) Mixer longitudinal cross-section view. (b) Mixer transverse cross-section view. (c) Mixer equivalent circuit.

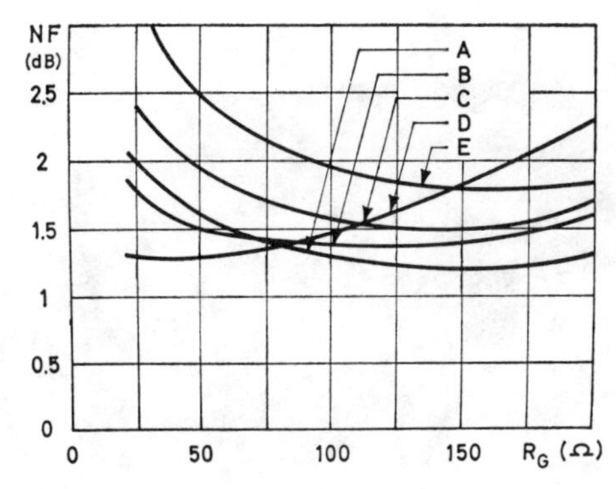

Fig. 3. Source impedance-dependence of IF preamplifier noise figure for various first-stage transistors. *A*—Texas Instruments MS 175 H; *B*—Avantek AT17A; *C*—Texas Instruments MS 173 H; *D*—KMC K6001; *E*—NEC V871.

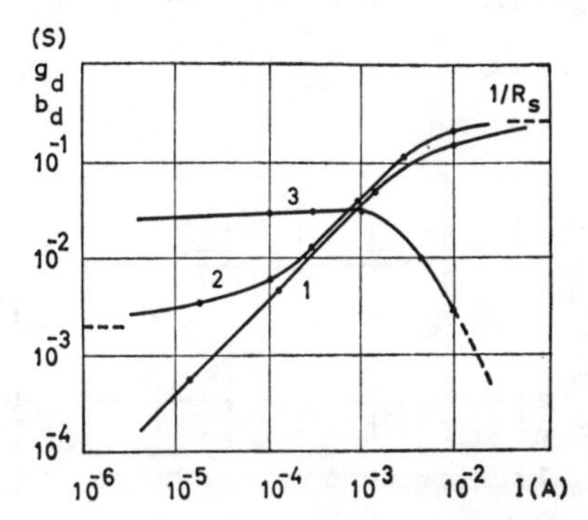

Fig. 4. Typical measured mixer diode dc and RF differential admittances as functions of diode dc current. 1—Differential dc conductance; 2—differential conductance at 7.350 GHz; 3—differential susceptance at 7.350 GHz.

stop center frequency. It can be tuned over a band of about 700 MHz and exhibits the following characteristics:

power reflected in the image frequency band (25 MHz) ≥ 80 percent;

insertion loss in the signal frequency band (25 MHz) ≤ 0.1 dB.

The electrical distance between the image rejection filter and the diodes' plane determines the actual admittance seen by the diodes at the image frequency, and is chosen to provide an open circuit at the diodes' plane. The distance can be varied by means of suitable spacers placed between the mixer cavity and the image rejection filter.

The currents at the third-harmonic idle frequencies $\omega_{\pm 3}$ are open circuited by means of fixed-tuned coaxial resonators connected to the diodes. The second-harmonic idle frequencies $\omega_{\pm 2}$ are properly short circuited at the output port by means of a suitable series resonator at the IF output port.[4]

The LO signal is capacitively coupled to the diodes from the rear of the waveguide. This type of coupling allows the use of a series resonator for achieving a sinusoidal current waveform.

To match the mixer-input impedance to the standard waveguide, the diodes are mounted in a reduced-height waveguide section.

The two diodes, which are in parallel at the mixer output, are connected directly to the IF-preamplifier input. The first stage of the IF preamplifier consists of a low-noise transistor in the common-emitter configuration. The transistor has been selected for an optimum noise source impedance approximately equal to the output impedance of the mixer when it is adjusted for minimum conversion loss. Fig. 3 shows the noise figure of different low-noise transistors as a function of the source impedance. The collector current level to which the curves of Fig. 3 refer has been chosen as a compromise to meet the intermodulation requirements of the overall receiving system due to the presence of adjacent channels and to minimize the IF noise figure.

Fig. 3 shows that the use of transistor *A* optimizes the IF noise figure with an output mixer impedance of about 125 Ω. This value of the output impedance is practically the same for all the Schottky-barrier diode types which have been tested.

The diode types are 1) the HP 5082-2709 for mixers operating in the 4- and 7-GHz ranges; and 2) the AEI DC 1306 for mixers operating in the 13-GHz range.

To show the difference between a practical nonlinear device and the ideal resistive diode, which the curves of Fig. 1 refer to, both the dc and RF admittances of the HP 5082-2709 diode are shown in Fig. 4. The results are plotted as a function of the diode current; curve 1 shows the differential dc conductance; curve 2 shows the differential conductance measured at 7350 MHz; and curve 3 shows the differential susceptance. Notice that the dc forward characteristic follows that of an exponential diode (i.e., $g = g_0 e^{-V/V_D}$, with $V_D = 31$ mV instead of the theoretical 25 mV) in series with a resistor R_S (with $R_S = 3.8$ Ω). The dc differential conductance approaches the limiting value $1/R_S$ for large currents ($I > 2$ mA). For reverse currents, the dc differential conductance does not decrease following the exponential law, but exhibits a minimum value of 10^{-7} Ω^{-1} at $I = -0.7$ μA. The diagram shows that at microwave frequencies, because of the presence of the diode junction capacity, the ratio g_{min}/g_{max} is markedly higher than the corresponding dc ratio ($3.8 \cdot 10^{-7}$). At 7350 MHz, for example, the ratio has been measured to be $5 \cdot 10^{-3}$.

It may be interesting to notice that, even though the curves in Fig. 1 refer to an idealized situation (purely resistive diodes and symmetrical rectangular LO waveform drive), the measured conversion losses are very close to the values given by the theoretical curve of case 1, when a k value is assumed equal to the ratio g_{min}/g_{max} measured at the appropriate microwave frequency.

As a typical example of practical realizations, Fig. 5 shows the 4-GHz version of the complete down-converter.

[4] A double-balanced four-diode mixer has also been tested, which uses an interdigitated capacitor to short circuit all the even idle frequencies at the output port of the mixer [10]. However, the two-diode balanced mixer configuration previously described has finally been selected because it allows a simpler implementation of the mixer–IF interface with no appreciable degradation of the conversion losses.

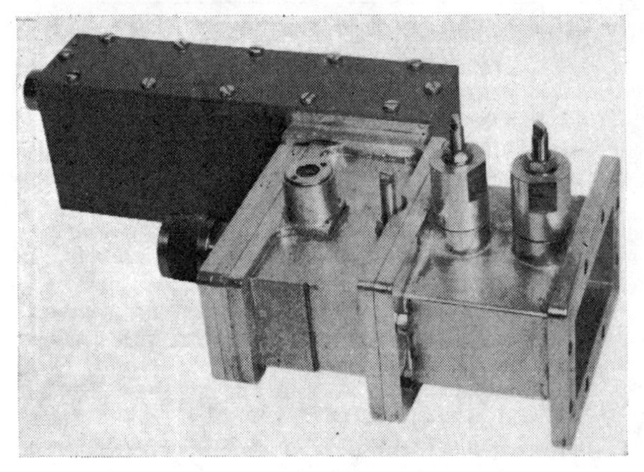

Fig. 5. Low-noise 4-GHz down-converter.

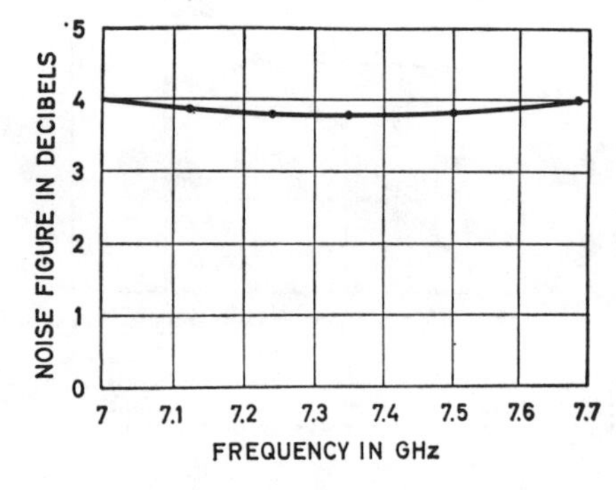

Fig. 6. Measured noise figure response of representative down-converter.

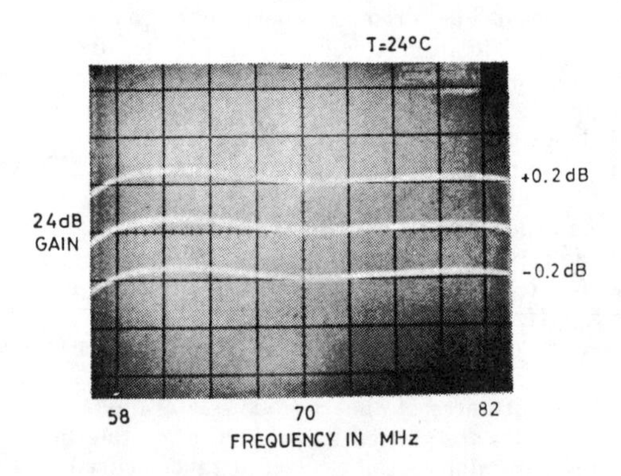

Fig. 7. Representative down-converter instantaneous overall gain-frequency response.

TABLE II

SUMMARY OF MEASURED PERFORMANCE OF RADIO-LINK MIXERS

RF input-signal tuning range	3.6–4.2 GHz	7.1–7.7 GHz	12.7–13.3 GHz
Overall noise figure at ambient temperature (including losses of image reject filter)	3.2 dB min 3.5 dB max	3.7 dB min 4 dB max	5 dB min 5.3 dB max
IF noise figure	1.5 dB	1.5 dB	1.5 dB
Conversion loss	2 dB max	2.6 dB max	4 dB max
Intermediate frequency		70 MHz	
Overall down-conversion gain of the receiver		24 dB	
Frequency response over any 25-MHz band	flat ±0.05 dB at ambient temperature ±0.2 dB from −5°C–55°C ±0.2 dB with ±2-dB pump power variation		
Diodes employed	HP 5082–2709	HP 5082–2709	AEI DC 1306
Rectified current	1 mA	1 mA	2 mA
AM/PM conversion	0.2°/dB at an input RF level of −20 dBm		

IV. EXPERIMENTAL RESULTS

Down-converters in the configuration described in Section III have been realized in the whole frequency range from 4 to 13 GHz. The measured performance of three down-converter versions, designed for microwave radio-link equipment, operating in the frequency ranges of 3.6–4.2 GHz, 7.1–7.7 GHz, and 12.7–13.3 GHz, respectively, are summarized in Table II.

The RF tunable range, indicated in Table II, is covered by mechanically retuning the image rejection filter only. No tuning is needed for both the second-harmonic and third-harmonic idle-frequency filters. It has been found experimentally that the position of the image rejection filter is not critical. Only two electrical distances (i.e., only one spacer) are sufficient to cover the entire RF input-signal tuning range.

Typical behavior of the noise figure of a down-converter unit within the RF frequency range is shown in Fig. 6. In addition to the results given in Table II, a typical example of an instantaneous gain-frequency response is shown in Fig. 7; such a response is practically the same in the whole RF frequency range.

REFERENCES

[1] H. C. Torrey and C. A. Whitmer, *Crystal Rectifiers* (M.I.T. Radiation Laboratory Series), vol. 15. New York: McGraw-Hill, 1948.
[2] P. D. Strum, "Some aspects of crystal mixer performance," *Proc. IRE*, vol. 41, pp. 875–889, July 1953.
[3] R. J. Mohr and S. Okwit, "A note on the optimum source conductance of crystal mixers," *IRE Trans. Microwave Theory Tech.*, vol. MTT-8, pp. 622–627, Nov. 1960.
[4] M. R. Barber, "Noise figure and conversion loss of the Schottky barrier mixer diode," *IEEE Trans. Microwave Theory Tech.*, vol. MTT-15, pp. 629–635, Nov. 1967.
[5] G. B. Stracca, "On frequency converters using non-linear resistors," *Alta Freq.*, vol. 38, pp. 318–331, May 1969.
[6] A. A. M. Saleh, *Theory of Resistive Mixer*. Boston, Mass.: M.I.T. Press, 1971.
[7] G. B. Stracca, "Noise in frequency mixers using non-linear resistors," *Alta Freq.*, vol. XL, no. 6, pp. 484–505, 1971.
[8] F. Aspesi and T. D'Arcangelo, "Low-noise down converter for radio link application," presented at the 1971 European Microwave Conf., Stockholm, Sweden.
[9] R. S. Caruthers, "Copper oxide modulators in carrier telephone systems," *Bell Syst. Tech. J.*, vol. XVIII, pp. 305–337, Apr. 1939.
[10] G. B. Stracca and C. Bassi, "Balanced and unbalanced frequency converters using non-linear resistor," presented at the Microwave Colloq., Budapest, Hungary, Apr. 1970.

Design and Performance Analysis of an Octave Bandwidth Waveguide Mixer

LLOYD T. YUAN

Abstract—**A new broad-band mixer capable of operating over two full adjacent waveguide bands (18 to 26.5 GHz and 26.5 to 40 GHz) is described. Within the octave bandwidth from 20 to 40 GHz, the maximum conversion loss is 6.5 dB with a corresponding average DSB noise figure of 5.7 dB. A theoretical analysis is given to treat quantitatively the performance of the octave bandwidth waveguide mixer.**

I. Introduction

RECENT TRENDS in millimeter-wave activities have shown increasing interest in broad-band receivers for use in communications, meteorology, and electronic warfare systems. One of the major requirements for a broad-band receiver design is that its front-end mixer be sufficiently broad-band. The need for an ultra-broad-band mixer design becomes more obvious when used in an ECM environment where maximum probability of intercept is required. Full waveguide bandwidth mixers with respectable performance have been reported recently [1], [2], but their operating frequency band is limited to that of the waveguide band. This paper describes a new approach for the design of a waveguide mixer which has an operating frequency band exceeding an octave bandwidth. The novel approach[1] utilizing a crossbar mixer configuration in a double-ridge waveguide mount, hereafter referred to as the double-ridge crossbar mixer, provides both low-noise and broad-band characteristics [3].

This paper describes the design and performance of an octave bandwidth mixer covering the frequency range from 18 to 40 GHz. Following a general discussion on the requirements for achieving more than full waveguide bandwidth performance, a theoretical analysis is given to treat quantitatively the performance of the double-ridge crossbar mixer over a broad range of operating frequencies. Finally, the detailed design and performance results of the octave bandwidth mixer are presented.

II. General Discussion

In a conventional waveguide mixer design, the usable bandwidth is generally restricted by the bandwidth-limiting elements, such as the mixer diodes, RF choke sections, impedance matching sections, etc., but ultimately it is limited by the operating frequency range of the waveguide used. Ordinarily, mixer performance degrades considerably when operating with a bandwidth exceeding 10 to 15 percent. It becomes totally unusable when operating near the waveguide cutoff frequency because of unacceptably high losses. In order to achieve a low-loss octave bandwidth mixer design, a new configuration utilizing a crossbar mixer together with a double-ridge waveguide mount was evolved. This mixer design eliminates practically the critical bandwidth-limiting elements. Fig. 1 shows the pictorial representation of the mixer design. As shown, the two mixer diodes are connected across the ridges of the waveguide mount and the IF signal is extracted via a center crossbar through the sidewall of the ridged waveguide mount where the RF fields are at a minimum. This virtually eliminates the requirements of RF chokes. In addition, the double-ridge waveguide mount provides a relatively low waveguide impedance to match the diode impedance. No bandwidth-limiting quarter-wave impedance transformers are required, thus further enhancing the broad-band performance of the mixer design.

For the design of a broad-band mixer with a bandwidth exceeding the operating frequency range of a standard waveguide, it is necessary that the waveguide cutoff limitation be removed. As an example, for an octave bandwidth (e.g., 18 to 40 GHz) mixer design, the operating frequency range covers two adjacent waveguide bands; i.e., K-band (18 to 26.5 GHz) and K_a-band (26.5 to 40 GHz). It is not possible to use the standard K_a-band waveguide (WR28) for the mixer design since the theoretical cutoff of the WR28 waveguide is at 21.08 GHz. The use of a double-ridge waveguide mount extends the cutoff frequency for the TE_{10} mode to below 18 GHz. Specifically, for the standard double-ridge waveguide (WRD 180 C24), the cutoff frequency is lowered to 15.25 GHz. The crossbar mixer design approach utilizing a standard WR28 waveguide has demonstrated excellent RF performance covering the full K_a-band waveguide bandwidth [4]. A conversion loss as low as 4 dB and an instantaneous RF bandwidth of over 13 GHz (from 26.5 to 40 GHz) was achieved. The present design, combining the advantages of the crossbar mixer and that of a double-ridge waveguide mount, exhibits both low loss and extremely broad-band characteristics.

III. Theoretical Analysis

In a mixer design, the performance parameters of primary concern are the operating bandwidth and the conversion loss. Basically, the requirement for a broad-band mixer design is that its frequency response be uniform and, more particularly, its conversion loss be uniformly low over a wide range of frequencies. For the purpose of analysis, the conversion loss of a mixer is considered to consist of three losses, namely:

Manuscript received May 11, 1977; revised July 10, 1977.

The author is with TRW Defense and Space Systems Group, Redondo Beach, CA 90278.

[1] Patent pending.

Reprinted from *IEEE Trans. Microwave Theory Tech.*, vol. MTT-25, pp. 1048–1054, Dec. 1977.

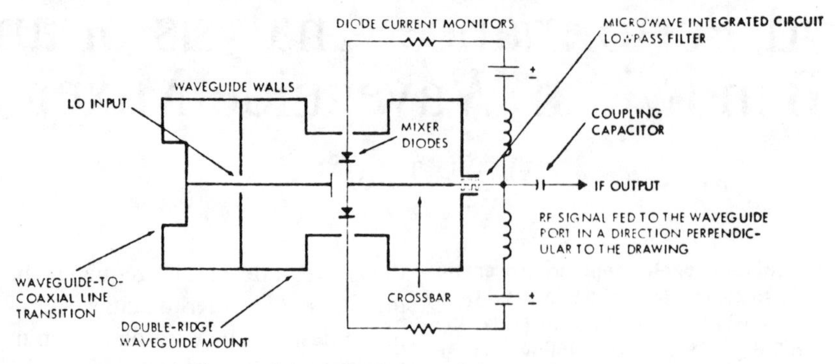

Fig. 1. Schematic diagram of the crossbar mixer.

L_1 mismatch loss due to impedance mismatches at the RF and IF ports

L_2 loss due to the diode junction capacitance and its series resistance

L_3 junction loss—the "intrinsic" loss at the diode junction which is dependent on the characteristic of the mixer diode and its terminating conditions.

Mathematically, these losses can be expressed as

$$L_1 = 10 \log (1 - |\Gamma_{RF}|^2)(1 - |\Gamma_{IF}|^2) \text{ dB} \tag{1}$$

where Γ_{RF} and Γ_{IF} are the reflection coefficients at the RF and IF ports.

$$L_2 = 10 \log \left| 1 + \frac{R_s}{R_j} + (\omega C_j)^2 R_s R_j \right| \text{ dB} \tag{2}$$

where

R_s diode series resistance

C_j diode junction capacitance

R_j diode junction resistance

ω operating frequency.

The "intrinsic" junction loss of a mixer depends on the terminating conditions of the image frequency and is not dependent on frequency. As an example, for the image matched condition, the junction loss is given as [5]

$$L_3 = \left[1 + \left\{ \frac{1 + \frac{g_2}{g_0} - 2\left(\frac{g_1}{g_0}\right)^2}{1 + \frac{g_2}{g_0}} \right\}^{1/2} \right]^2 \left(1 + \frac{g_2}{g_0}\right)\left(\frac{g_0}{g_1}\right)^2 \tag{3}$$

where g_0, g_1, and g_2 are the Fourier coefficients of the diode conductance. As seen from (1)–(3), the frequency-dependent losses are L_1 and L_2, while L_3 is frequency-independent. To achieve broad-band performance, i.e., low and uniform conversion loss over a wide range of frequencies, it is necessary that losses L_1 and L_2 be minimized and be uniformly low over the frequency range of interest.

For the purpose of analysis, a simplified equivalent circuit for the double-ridge crossbar mixer is developed, as shown in Fig. 2, assuming that the crossbar structure is transparent to the RF signals. The crossbar mixer is treated as two single-diode mixers connected in series with respect to the RF signals, except biased under different conditions.

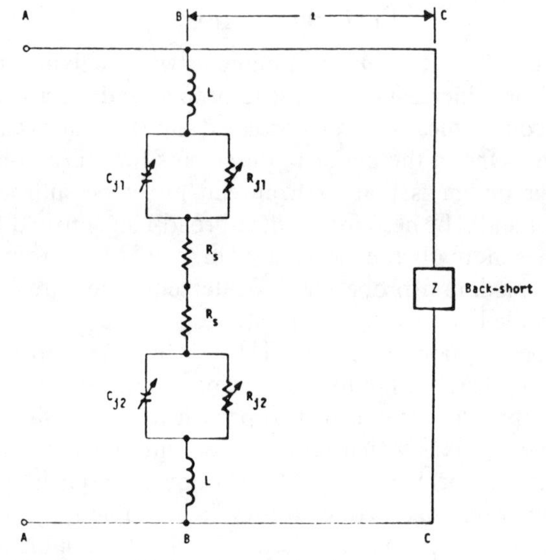

Fig. 2. Simplified equivalent circuit of the crossbar mixer.

Included in the equivalent circuit is a backshort section which is considered as an ideal contacting short located at a distance l from the diode pair.

The following abbreviations are used in Fig. 2.

Z backshort section impedance $= jZ_0 \tan (2\pi l/\lambda_g)$

Z_0 double-ridge waveguide characteristic impedance

λ_g wavelength in the double-ridge waveguide

l distance from diode pair to the backshort section

L whisker inductance

R_s diode series resistance

C_{j1}, C_{j2} junction capacitance for diode 1 and diode 2 under different bias conditions

R_{j1}, R_{j2} junction resistance for diode 1 and diode 2 under different bias conditions

For convenience, the whisker inductance and series resistance of the two diodes are considered to be equal.

The equivalent circuit of the crossbar mixer at plane *B–B* can be simplified by the impedance transformation as shown in Fig. 3.

By combining the equivalent circuit of Fig. 3(d) with the

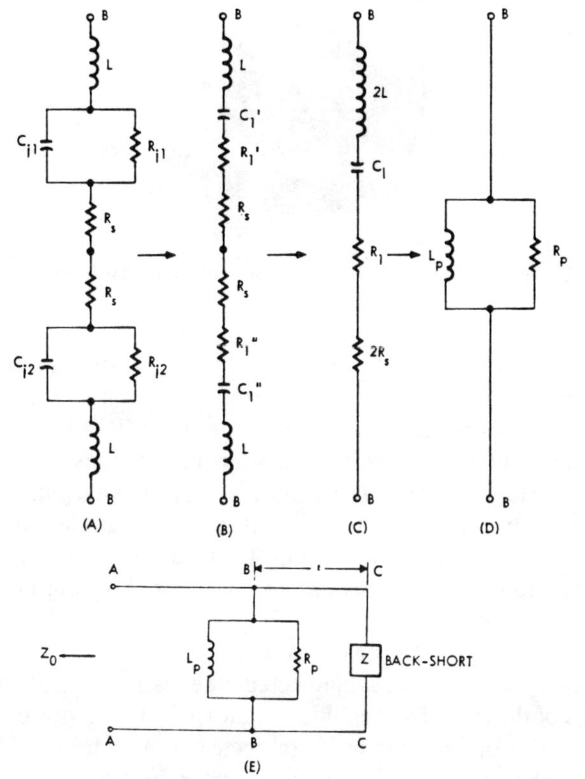

where

$$R_1' = \frac{R_{j1}}{1 + (\omega C_{j1} R_{j1})^2}; \quad \frac{1}{\omega C_1'} = \frac{1}{\omega C_{j1}\left[1 + \frac{1}{(\omega C_{j1} R_{j1})^2}\right]},$$

$$R_1'' = \frac{R_{j2}}{1 + (\omega C_{j2} R_{j2})^2}; \quad \frac{1}{\omega C_1''} = \frac{1}{\omega C_{j2}\left[1 + \frac{1}{(\omega C_{j2} R_{j2})^2}\right]},$$

$$R_p = (2R_s + R_1' + R_1'') + \frac{\left[2\omega L - \frac{1}{\omega}\left(\frac{1}{C_1'} + \frac{1}{C_1''}\right)\right]^2}{2R_s + R_1' + R_1''}$$

$$\omega L_p = 2\omega L - \frac{1}{\omega}\left(\frac{1}{C_1'} + \frac{1}{C_1''}\right) + \frac{(2R_s + R_1' + R_1'')^2}{\left[2\omega L - \frac{1}{\omega}\left(\frac{1}{C_1'} + \frac{1}{C_1''}\right)\right]};$$

$$\text{for } \omega L > \frac{1}{\omega}\left(\frac{1}{C_1'} + \frac{1}{C_1''}\right)$$

Fig. 3. Equivalent circuit transformation.

backshort section, a new equivalent circuit for the crossbar mixer is obtained as shown in Fig. 3(e). It is seen that for impedance matching, the following conditions should hold.

$$R_p = Z_0 \tag{4}$$

and

$$\omega L_p = -Z_0 \tan \frac{2\pi l}{\lambda_g}. \tag{5}$$

It is recalled that the frequency-dependent losses are the impedance mismatch loss L_1, and the diode parasitic loss L_2, while the junction loss L_3 is frequency-independent. For the calculation of bandwidth limitation of a mixer design, only the frequency-dependent losses L_1 and L_2 will be considered.

In order to simplify calculations, we assume that the impedance mismatch at the IF port is a negligibly small constant. Thus L_1 becomes

$$L_1 = -10 \log (1 - |\Gamma_{RF}|^2). \tag{6}$$

Using the equivalent circuit of Fig. 4, it can be shown that the reflection coefficient is given as

$$|\Gamma_{RF}| = \left[\frac{\left(1 - \frac{Z_0}{R_p}\right)^2 + \left(\frac{Z_0}{\omega L_p} + \frac{1}{\tan 2\pi l/\lambda_g}\right)^2}{\left(1 + \frac{Z_0}{R_p}\right)^2 + \left(\frac{Z_0}{\omega L_p} + \frac{1}{\tan 2\pi l/\lambda_g}\right)^2}\right]^{1/2} \tag{7}$$

The mismatch loss L_1 is given as

$$L_1 = -10 \log \frac{4Z_0}{R_p\left[\left(1 + \frac{Z_0}{R_p}\right)^2 + \left(\frac{Z_0}{\omega L_p} + \frac{1}{\tan 2\pi l/\lambda_g}\right)^2\right]}. \tag{8}$$

The diode parasitic loss L_2 not only is frequency-dependent but also depends on the values of R_s, R_j, and C_j. For minimum losses, the following condition exists

$$\frac{\partial L_2}{\partial R_j} = 0 \quad \text{or} \quad R_j = \frac{1}{\omega C_j}. \tag{9}$$

By substituting (9) into (2), we have

$$L_{2\,\min} = 10 \log \left[1 + 2\frac{f}{f_c}\right] \tag{10}$$

where

$$f_c = \frac{1}{2C_j R_s}$$

where f is the operating frequency.

Based on the equivalent circuit model developed for the crossbar mixer, the mismatch and parasitic losses are calculated over an octave bandwidth of operating frequency range, i.e., 18 to 40 GHz as shown in the Appendix. The calculated results are tabulated in Table I.

For the fully conducting diode having the parameters as described in the Appendix, the average "intrinsic" junction loss L_3 for a mixer under an image matched condition is approximately 3.8 dB over the frequency range of interest [6]. Assuming that the IF mismatch loss is 0.5 dB, the total conversion loss L of the mixer (i.e., $L = L_1 + L_2 + L_3 + 0.5$ dB) is plotted against frequency as shown in Fig. 7(a). It is seen that the frequency response of the mixer is remarkably uniform over an octave bandwidth frequency range, clearly indicating the broad-band characteristic of the mixer design.

IV. Octave Bandwidth Mixer Design

In the design of a low-noise broad-band mixer, careful consideration must be given to reducing parasitic losses in the circuit components and eliminating critical bandwidth-limiting elements. The key components of primary concern in a broad-band mixer design generally include the mixer

TABLE I
MIXER MISMATCH AND PARASITIC LOSS

Frequency (GHz)	Mismatch Loss, L_1 (dB)	Parasitic Loss, L_2 (dB)		Total Loss (dB) $L_1 + L_2$
		Diode 1	Diode 2	
18	1.2414	.6946	.0811	2.0171
19	.8478	.6950	.0860	1.6288
20	.5718	.6955	.0913	1.3586
21	.3767	.6960	.0968	1.1695
22	.2398	.6965	.1025	1.0388
23	.1455	.6970	.1086	.9511
24	.0828	.6975	.1148	.8951
25	.0431	.6981	.1214	.8626
26	.0200	.6987	.1281	.8468
27	.0083	.6993	.1352	.8428
28	.0039	.7000	.1425	.8464
29	.0037	.7006	.1500	.8543
30	.0050	.7013	.1578	.8641
31	.0062	.7020	.1659	.8741
32	.0061	.7028	.1742	.8831
33	.0043	.7035	.1827	.8905
34	.0016	.7043	.1915	.8974
35	.0000	.7051	.2006	.9057
36	.0033	.7060	.2099	.9192
37	.0183	.7068	.2194	.9445
38	.0556	.7077	.2291	.9924
39	.1317	.7086	.2391	1.0794
40	.2710	.7095	.2494	1.2299

diodes, the impedance matching elements, and the RF choke section for the extraction of the IF signals. The performance of these components is frequency-limited—particularly the stepped impedance transformer and the RF choke section ordinarily used in the conventional mixer design, which seriously degrades the bandwidth performance of the mixer. In the present mixer design approach, a crossbar mixer together with a double-ridge waveguide mount is utilized. This eliminates the bandwidth-limiting circuit elements, such as the stepped impedance transformer and the RF choke sections, and provides an inherent impedance-matching condition for both the RF and the IF signals, resulting in a mixer design with octave bandwidth performance.

The basic requirements for designing a mixer with an operating frequency range over an octave bandwidth are as follows.

1) The waveguide components used for the fabrication of the mixer must have an effective usable bandwidth well exceeding an octave bandwidth.

2) The circuit elements, such as the mixer diode and its embedding network, must be sufficiently broad-band and must be properly matched to the waveguide impedance over the octave bandwidth frequency range.

Based on these requirements, an octave bandwidth mixer, utilizing a crossbar configuration in a double-ridge waveguide mount, was designed, fabricated, and tested. This mixer is capable of operating over the frequency range from 18 to 40 GHz. Fig. 4 shows the mechanical configuration of the mixer design where the mixer mount is built from the standard double-ridge waveguide mount (WRD 180 C24).

Fig. 4. Mechanical configuration of the crossbar mixer.

The octave bandwidth crossbar mixer consists of a double-ridge waveguide mount, a ridged waveguide-to-coaxial line transition, two Schottky-barrier diodes, and a low-pass filter, as shown in Fig. 1. The double-ridge waveguide mount also serves as the backshort housing of the mixer mount.

The heart of the mixer is a pair of low-noise GaAs Schottky-barrier diodes connected in series across the broad walls of the ridged waveguide. As shown in Fig. 1, one of the electrodes of each diode is connected to a metal crossbar which serves as a mechanical support for the diodes and also as a transmission line for the incoming LO power and the IF output signals. In actual operation, the RF signal is fed directly to the ridged waveguide port. On one end of the crossbar, LO power is fed via a ridged waveguide-to-coaxial line transition and capacitively coupled to the diodes. The IF output is extracted from the opposite end of the crossbar via a microwave integrated-circuit low-pass filter (LPF). The LPF was fabricated on sapphire substrate with a cutoff frequency exceeding 12 GHz. The use of the LPF at the IF port is to prevent LO power leaking to the output. Electrically, the two mixer diodes are connected in series with respect to the RF signal and in parallel with the IF output. This provides a higher impedance level to the RF signal and a lower impedance level to the IF signal than a single diode mixer. Therefore, an inherent impedance match condition for both the RF and IF signals is achieved for broad-band performance. In addition, the relatively low impedance of the ridged waveguide further improves impedance matching of the mixer diodes. This results in an extremely broad-band mixer design with uniformly low losses over an operating frequency range exceeding an octave bandwidth.

The other microwave circuit components essential for the octave bandwidth mixer design are the double-ridge waveguide mount and the ridged waveguide-to-coaxial line transition. These components must provide sufficient bandwidth to cover the operating frequency range, e.g., 18 to 40 GHz.

The double-ridge waveguide used for the 18 to 40-GHz mixer design has the dimensions shown schematically in Fig. 5.

The key parameters essential for a ridged waveguide design are:

1) the effective usable bandwidth defined by the cutoff frequencies of the TE_{10} and TE_{20} modes

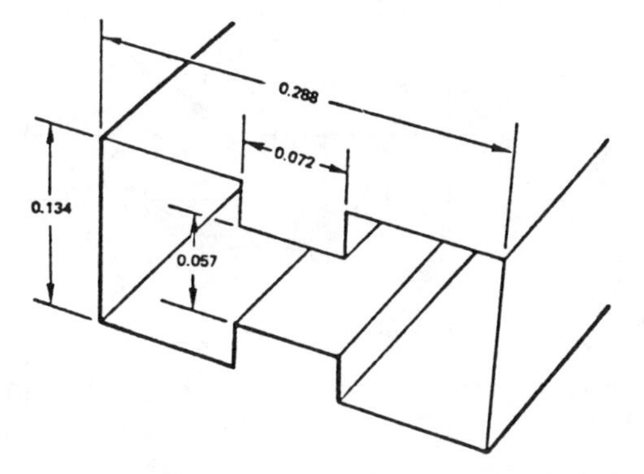

Fig. 5. Double-ridge waveguide (all dimensions in inches).

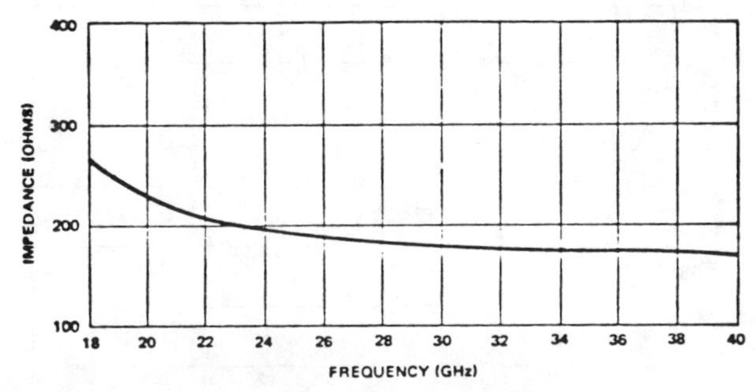

Fig. 6. Double-ridge waveguide impedance versus frequency.

2) the waveguide attenuation
3) the waveguide impedance.

According to Hopfer [7], with the dimensions shown in Fig. 5, the double-ridge waveguide has a usable bandwidth from 15.25 to 44.8 GHz, which is more than adequate for the 18 to 40-GHz operating bandwidth. The attenuation of the double-ridge aluminum waveguide is approximately 0.6 dB/ft at 18 GHz and 0.2 dB/ft at 40 GHz, clearly insignificant for this application.

The characteristic impedance for a double-ridge waveguide is given by Chen [8] as

$$Z_0 = \frac{Z_{0\infty}}{\sqrt{1 - \left(\frac{f'_c}{f}\right)^2}} \tag{11}$$

where

$Z_{0\infty}$ the characteristic impedance at infinite frequency for the TE_{10} mode
f'_c the TE_{10} mode cutoff frequency of the double-ridge waveguide
f the frequency of operation.

Fig. 6 shows a plot of the double-ridge waveguide impedance versus frequency. It is seen that the impedance level of a double-ridged waveguide is substantially lower than that of a standard K_a-band (26.5 to 40 GHz) waveguide (WR28), which varies from 443 to 623 Ω.

The ridged waveguide-to-coaxial line transition used in the 18 to 40-GHz mixer is a double-ridge aluminum waveguide construction (Model R45, MRC). This transition provides acceptable performance for the present design. Over the frequency range of 18 to 36 GHz, the typical VSWR is 1.3 : 1, with a maximum insertion loss of 1 dB within the 18 to 36 GHz range and 2 dB from 36 to 40 GHz. The higher insertion loss at the high end of the frequency band is not important for the present application because it is at the LO port. It will not affect the conversion loss of the mixer except that higher LO power is required to drive the diodes.

V. Mixer Diode Characterization

The Schottky-barrier diode plays the key role in the performance of the mixer. In fact, the ultimate bandwidth and noise performance of a mixer is determined by the quality of the diodes. It is, therefore, essential that the Schottky-barrier diodes be fully evaluated prior to their use in the mixer circuits. Evaluation is generally performed by measuring the "n-factor," R_s and f_c, of the mixer diode. The n-factor and R_s are defined by the I–V characteristics of a forward-biased diode as follows:

$$I = I_0 \exp \frac{q(V - IR_s)}{nkT} \tag{12}$$

where

I diode current
V applied voltage
I_0 diode saturation current
q electron charge
k Boltzmann's constant
T absolute temperature
n diode ideality factor
R_s diode series resistance.

It is seen from (12) that the n-factor and R_s can be determined by the I–V characteristic of the diode. The values of the n-factor and R_s are used to determine the dc quality of the mixer diode. Generally, for a good quality mixer diode, the n-factor and R_s should be less than 1.1 and 10 Ω, respectively.

The cutoff frequency f_c of a mixer diode is defined as

$$f_c = \frac{1}{2\pi C_j R_s} \tag{13}$$

where C_j is the diode junction capacitance. The cutoff frequency sets the ultimate limit of the mixer for use at high frequencies. It is desirable that the cutoff frequency be high for high-frequency applications. As can be seen from (13), for high f_c, both the C_j and R_s must be minimized. Minimization of C_j requires a small diode junction area as well as the use of low carrier concentration semiconductor materials. However, both of these (small junction area and low concentration material) increase the series resistance of the diode, which in turn lowers the cutoff frequency [see (13)]. To reach a compromise, an extremely thin epitaxial layer of moderately low carrier concentration is used. As an example, GaAs epitaxial layers with a carrier concentration of 6×10^{16} cm^{-3} and a thickness of less than 0.3 μm have been used successfully for the fabrication of Schottky-barrier diodes of 4-μm diameter. These mixer diodes have typical C_j's and R_s's of 0.01 pF and 5 Ω, respectively, under zero-bias conditions. This type of mixer diode has been used over the frequency range from 20 to 75 GHz with excellent performance.

VI. Performance Evaluation and Results

The bandwidth performance of a mixer is primarily restricted by the frequency-dependent elements associated with the mixer as well as its embedding network. In order to achieve ultra-broad-band and low-loss performance, considerations must first be given to minimizing the parasitic inductances, capacitances, and loss elements. These include the whisker inductance, the junction and package (if any) capacitances, and the series resistance. However, for a given mixer diode (i.e., C_j and R_s fixed), the ultimate broad-band performance of the mixer can be achieved by proper selection and implementation of the following critical bandwidth-limiting elements:

contact whiskers
adjustable backshort section
crossbar geometry.

An octave bandwidth mixer (Fig. 4) was designed and fabricated using GaAs Schottky-barrier diodes having $C_j = 0.01$ pF and $R_s = 5\ \Omega$. In the mixer design, performance optimization was achieved by precision design and fine tuning of every single element in the mixer circuit. The length of the contact whisker must be properly selected so that sufficient inductance will be provided for tuning out the parasitic capacitances of the mixer. A whisker length of approximately 0.02 in using a 0.001-in diameter wire was used in the present design. In addition, a contacting short of the plunger type was used to provide external tuning for the mixer. Ideally, the metal crossbar embedded in the waveguide should be transparent to the microwave signal and be lossless. In fact, for a properly designed crossbar, it can be considered as a microwave circuit with reactive elements only. It is, therefore, essential that the geometry and dimensions of the crossbar be properly designed such that its resonances do not occur within the operating frequency band of interest. In the present design, a 0.025-in diameter cylindrical crossbar of full waveguide width is used.

Evaluation of the octave bandwidth mixer was carried out using K-band (18 to 26.5 GHz) and K_a-band (26.5 to 40 GHz) test equipment with appropriate ridged waveguide-to-standard waveguide transitions. Fig. 7 shows the mixer conversion loss and noise figure performance over the 18 to 40-GHz range, measured with the mixer fixed tuned at 35 GHz. Also shown in Fig. 7 is the calculated conversion loss of the mixer obtained by the method and the equivalent circuit shown in Section III. It is seen that the measured conversion loss is in good agreement with the calculated value over the operating frequency range. The uniform distribution of the conversion loss over the frequency band fully demonstrates the broad-band characteristics of the mixer design. The double-sideband noise figure of the mixer was measured using an IF amplifier with a 1.5-dB noise figure. Over the octave bandwidth from 20 to 40 GHz, the measured minimum noise figure was 4.8 dB and the maximum 7 dB. It should be noted that from 22 to 40 GHz, the minimum and maximum noise figures are 4.8 and 5.7 dB. A differential of only 0.9 dB over a bandwidth of 18 GHz clearly indicated the excellent broad-band performance of the mixer design.

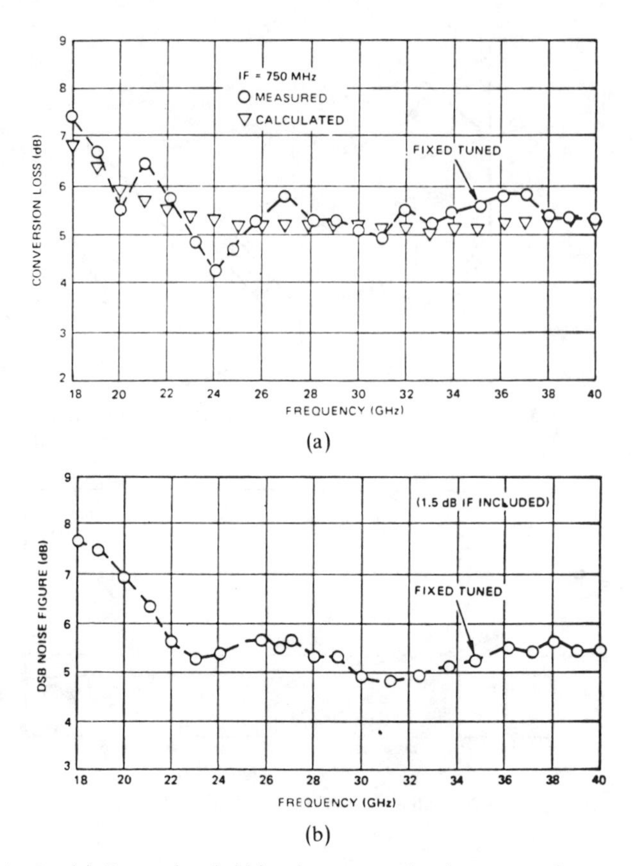

Fig. 7. (a) Octave bandwidth mixer conversion loss versus frequency. (b) Noise figure versus frequency.

VII. Conclusion

The performance of the octave bandwidth mixer is extremely encouraging, particularly, the bandwidth performance which exceeds an octave bandwidth. It was found that the most critical bandwidth-limiting parameter is the junction capacitance of the mixer diode which ultimately limits the operating bandwidth of the mixer design. An equivalent circuit model for the double-ridge crossbar mixer was developed, which treats quantitatively the bandwidth performance of the mixer over a wide range of frequencies. This circuit model provides a guideline for selecting the essential circuit parameters of the mixer design to achieve broad-band performance.

The design principle of the octave bandwidth mixer can definitely be extended to other frequency ranges by proper scaling. It is possible that the same design approach may be used for the design of mixers operating from 40 to 80-GHz and 80 to 160-GHz regions. This offers the technical feasibility of covering the 20 to 160-GHz frequency range with just three mixers.

Appendix
Calculation of Mismatch and Parasitic Losses of the Crossbar Mixer

To calculate the mismatch and parasitic losses of the mixer, we need to know the parameters, R_s's, R_j's, and C_j's of the mixer diodes. The exact values of these parameters cannot be determined without knowing the bias and local-oscillator drive conditions on the mixer diode. Since the local-oscillator drive level on the diode is not precisely known, in order to get an estimate of the value of these parameters, we make the following assumptions.

1) At each half cycle of the LO frequency, one of the diodes is fully conducting while the other one is partially conducting and being biased with a dc bias only.

2) Both of the diodes are operating under a small-signal condition such that

$$R_j = \left(\frac{\partial I}{\partial V}\right)^{-1} = \frac{nkT}{qI}, \quad \text{for } V > IR_s$$

$$= \frac{29}{I(ma)}, \quad \text{for } n = 1.1, \quad T = 298 \text{ K.} \quad \text{(A-1)}$$

Equation (A-1) is derived from the I–V characteristic of a mixer diode [see (12)].

The junction capacitance of the diode can be determined as follows [9]

$$C_j = C_{j0}\sqrt{\frac{\Phi}{(\Phi - V)}} \quad \text{(A-2)}$$

where

C_{j0} zero bias junction capacitance

$$= A\left[\frac{\varepsilon q N_d}{2\Phi}\right]^{1/2}$$

A junction area

ε semiconductor permittivity $= 1.11 \times 10^{-12}$ F/cm for GaAs with $\varepsilon = 12.5$

q electron charge

N_d donor concentration

assuming that GaAs Schottky-barrier diodes are used for the fabrication of the crossbar mixer operating over the frequency range from 18 to 40 GHz. The Schottky barrier diode has the following characteristics:

carrier concentration	$N_d = 6 \times 10^{16}$ cm^{-3}
barrier potential	$\Phi = 0.9$ V
diode area (4 μm diameter)	$A = 1.25 \times 10^{-7}$ cm^2
series resistance	$R_s = 5\ \Omega$.

Under the condition of minimum conversion loss, the measured bias currents were typically 1 mA for the fully conducting diode and 50 μA for the partially conducting diode. Based on the I–V characteristic of the GaAs Schottky-barrier diode used, this corresponds to a bias voltage of 0.75 V for the fully conducting diode and 0.6 V for the partially conducting diode. Using the preceding bias conditions, the parameters of the two diodes are calculated as follows:

$$C_{j0} = A\left[\frac{\varepsilon q N_d}{2\Phi}\right]^{1/2} = 9.6 \times 10^{-15} \text{ F.}$$

For the fully conducting diode,

$$C_{j1} = C_{j0}\sqrt{\frac{\Phi}{(\Phi - V)}} = 9.6 \times 10^{-15}\sqrt{\frac{0.9}{0.9 - 0.75}}$$

$$= 2.35 \times 10^{-14} \text{ F}$$

$$R_{j1} = \frac{29}{1} = 29\ \Omega.$$

For the partially conducting diode,

$$C_{j2} = 9.6 \times 10^{-15}\sqrt{\frac{0.9}{0.9 - 0.6}} = 1.66 \times 10^{-14} \text{ F}$$

$$R_{j2} = \frac{29}{50 \times 10^{-3}} = 580\ \Omega.$$

By substituting R_s, R_{j1}, R_{j2}, C_{j1}, and C_{j2} into the equation shown in Fig. 3 at a fixed tuned frequency of 35 GHz, we have

$$R_1' = 28\ \Omega,\ C_1' = 1.07 \times 10^{-12} \text{ F}$$

$$R_1'' = 106\ \Omega,\ C_1'' = 2.03 \times 10^{-14} \text{ F.}$$

Using the impedance match condition, $R_p = Z_0$, (4), where Z_0 is the double-ridge waveguide impedance at 35 GHz, and solving for L, we have $L = 6.66 \times 10^{-10}$ h. Then, we calculate ωL_p using the equation shown in Fig. 3, i.e., $\omega L_p = 386$ Ω. The backshort distance l can be obtained by solving (5) for l as follows:

$$\omega L_p = -Z_0 \tan \frac{2\pi l}{\lambda_g}$$

where

$$\lambda_g = \frac{\lambda_0}{\sqrt{1 - \left(\frac{f_c'}{f}\right)^2}} = 0.94 \text{ cm at 35 GHz.}$$

Thus $l = 0.298$ cm.

Now with the mixer fixed tuned at 35 GHz, i.e., the backshort position is fixed, the mismatch and parasitic losses from 18 to 40 GHz are calculated. The calculated results are tabulated as shown in Table I. The parasitic loss is the sum of the losses contributed from the two diodes under different biased conditions.

Acknowledgment

The author wishes to thank Dr. J. S. Honda and Dr. J. E. Raue for their support and many helpful discussions during the course of the work. Appreciation is also due to F. M. Garcia for carrying out the microwave measurements.

References

[1] A. Hislop and R. T. Kihm, "A broad-band 40–60 GHz balanced mixer," *IEEE Trans. Microwave Theory Tech.*, vol. MTT-24, pp. 63–64, Jan. 1976.

[2] N. Kanmuri and R. Kawasaki, "Design and performance of a 60–90 GHz broad-band mixer," *IEEE Trans. Microwave Theory Tech.*, vol. MTT-24, pp. 256–261, May 1976.

[3] L. T. Yuan, "Low noise octave bandwidth waveguide mixer," presented at the 1977 IEEE MTT-S Int. Microwave Symp., June 21–23, 1977.

[4] ——, "Millimeter wave technology and EHF receiver development," TRW Defense and Space Systems Group, Redondo Beach, CA, 1975 IR&D Final Report, Sec. 5, pp. 45–70, Feb. 1976.

[5] M. R. Barber, "Noise figure and conversion loss of the Schottky barrier mixer diode," *IEEE Trans. Microwave Theory Tech.*, vol. MTT-15, pp. 629–635, Nov. 1967.

[6] L. E. Dickens and D. W. Maki, "An integrated-circuit balanced mixer, image and sum enhanced," *IEEE Trans. Microwave Theory Tech.*, vol. MTT-23, pp. 276–281, Mar. 1975.

[7] S. Hopfer, "The design and ridged waveguide," *IRE Trans. Microwave Theory Tech.*, vol. MTT-3, pp. 20–29, Oct. 1955.

[8] T. S. Chen, "Calculation of the parameters of ridge waveguides," *IRE Trans. Microwave Theory Tech.*, vol. MTT-5, pp. 12–17, Jan. 1957.

[9] H. A. Watson, Ed., *Microwave Semiconductor Devices and Their Circuit Applications.* New York, NY: McGraw-Hill, 1969, ch. 11.

An X-Band Balanced Fin-Line Mixer

GÜNTHER BEGEMANN

***Abstract*—The fin-line technique has been used in a balanced 9–11-GHz mixer with a 70-MHz intermediate frequency. The mixer without an IF amplifier has an available conversion loss of less than 5 dB with a 3.8-dB minimum and a SSB noise figure of less than 6.9 dB with a 5.3-dB minimum. The mixer is tunable by variable shorts. It is possible to scale the device to millimeter-wave frequencies.**

I. Introduction

THIS PAPER describes the design and performance of a microwave integrated-circuit (MIC) balanced mixer that covers the bandwidth of 2 GHz within the X band with available conversion losses of less than 5 dB and a noise figure of less than 6.9 dB. Not included is the noise contribution from the IF amplifier. The mixer operates with an IF of 70 MHz, but the device is able to handle higher IF's up to some gigahertz. For this purpose, the low-pass filter coupling out the intermediate frequency must have a suitable cutoff frequency.

Manuscript received May 30, 1978; revised August 1, 1978. This work was supported in part by the Deutsche Forschungsgemeinschaft.

The author is with the Institut für Hochfrequenztechnik, Technische Universität, Braunschweig, Germany.

In the circuit considered here, a fin-line technique [1] has been used to realize a mixer which is capable to work well up to millimeter-wave frequencies. To this end the mixer is equipped with connections of rectangular waveguides both at the signal and the local oscillator input.

Because the fundamental mode of a fin-line (H_{10} mode) is the same as the one of a rectangular waveguide, transitions between these two guides are easy to handle and have a very small insertion loss and a VSWR over the entire waveguide bands. Parasitic radiation which often is a problem connected with planar waveguides especially at higher frequencies can be avoided. So the fin-line has very low losses. Moreover, it offers the same possibilities of integration as other planar circuits.

The most essential part of the mixer is a planar magic T completely integrated in a rectangular waveguide. The magic T proved itself as a rather broad-band and low-loss device. The purpose of the magic T is twofold. First, it distributes the signal and local oscillator voltages with their proper phase relationships to the two nonlinear elements, and, second, it blocks the local oscillator input from the signal frequency input and vice versa.

Reprinted from *IEEE Trans. Microwave Theory Tech.*, vol. MTT-26, pp. 1007–1011, Dec. 1978.

A main feature of the mixer is that it is tunable by variable shorts. Thus tuning of the signal input impedance is performed with low loss and as fast and accurate as in conventional waveguide circuitry.

Although the fin-line seems to be advantageous, especially for millimeter-wave applications, we have designed the mixer for a signal center frequency of 10 GHz because at this frequency measurements can be made exactly.

II. A Fin-Line Magic T

To realize balanced mixers, 90° or 180° hybrids are required. To do this with planar circuits, one can use branchline couplers, ring hybrids, parallel-coupled-line couplers, combinations of orthogonal transmission lines like slot lines and coplanar lines, or magic T's. There are several papers which deal with such mixers [2]–[4], and excellent results have been reported.

A hybrid junction similar to the one treated here is known from the literature [5]. The junction presented in [5] is but a pure planar device. We have designed a magic T of a fin-line–microstrip hybrid junction, as shown in Fig. 1. The fin-line magic T acts as follows. A wave incident from the left on the slot of a fin-line (its electric field is represented by solid arrows) excites two waves first guided by the slots between the microstrip antenna and the top and the bottom of the fin-line slot. These waves provide antisymmetric antiphase excitation of the cross arms which act as output ports. In case of perfect geometrical symmetry there occurs no excitation of the waves on the microstripline. A wave incident from the right on the microstripline (its electric field is represented by dashed arrows) excites symmetric inphase waves on the cross arms. The protruding part of the microstrip, i.e., the part of the microstrip without back-side metallization, acts as an antenna. Its optimum length, in order to match the microstrip input, was determined experimentally at 10 GHz. Typical dimensions are given in Fig. 2 but depend on the impedance level of the device. Data that is generally valid cannot be given because we did not execute theoretical investigations.

At nonlinear elements placed in the cross arms, as shown in Fig. 1, the incident waves beat once in phase and once in antiphase. If the microstrip side of the hybrid is used as a local oscillator (LO) input and the fin-line side as a signal (RF) input in a common IF output, the LO currents cancel and the IF currents are inphase. So this arrangement is capable of being used as a balanced mixer.

Because the cross arms are series-connected referred to the fin-line input but parallel-connected referred to the microstrip input, their impedance must be chosen to be half the impedance of the fin-line input and twice that of the microstrip input. Methods for calculating fin-line impedances are available from the literature [6], [7].

Fig. 2 essentially shows the planar structure of the complete magic T with its waveguide ports as used for optimization. One has to imagine this planar structure as being mounted parallel to the narrow sides of two cross-

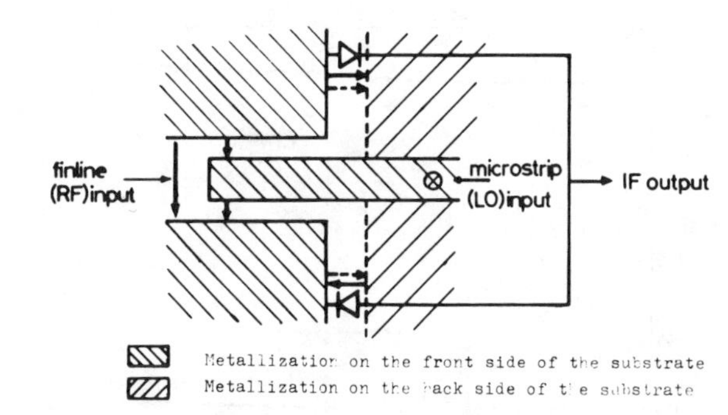

Fig. 1. The fin-line magic T.

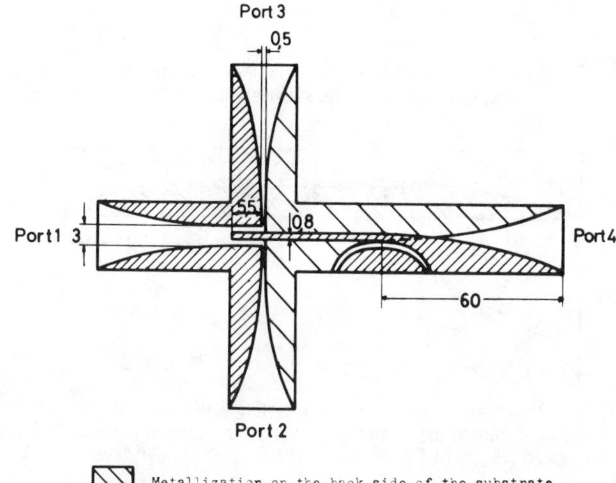

Fig. 2. Substrate of the complete fin-line magic T with its waveguide input and output ports; typical dimensions are in millimeters; thickness of the substrate is 0.254 mm; permittivity is 2.22 (RT/Duroid 5880).

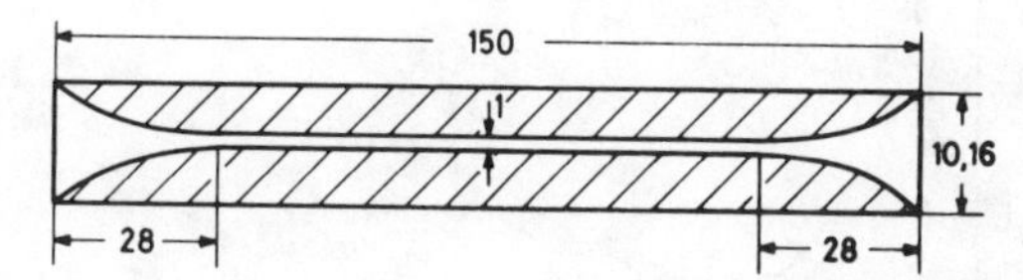

Fig. 3. Two fin-line tapers in a series connection; dimensions are in millimeters as used for the measured results of Fig. 4.

connected rectangular waveguides. Ports 1 and 4 serve as input ports, and ports 2 and 3 are used as output ports.

Port 1 is constructed with a fin-line taper. Such tapers are uncritical in their dimensions. Measurements at two tapers in a series connection, arranged as in Fig. 3 and printed on a 0.254-mm teflon substrate (RT/Duroid 5880, $\epsilon_r = 2.22$) which we used throughout the development of the mixer, showed a typical insertion loss of about 0.2 dB and a VSWR of about 1.2 over the entire X band, as shown in Fig. 4. Included are the losses of the cylindrical part of the device between the two tapers.

The transition at port 4 contains a fin-line taper, a short section of an antipodal fin-line [6], and a microstripline. An additional metallization prevents the metal-free space below the taper from resonating in the considered frequency band. Measurements at two of such transitions as shown in Fig. 5 yielded a typical insertion loss of less

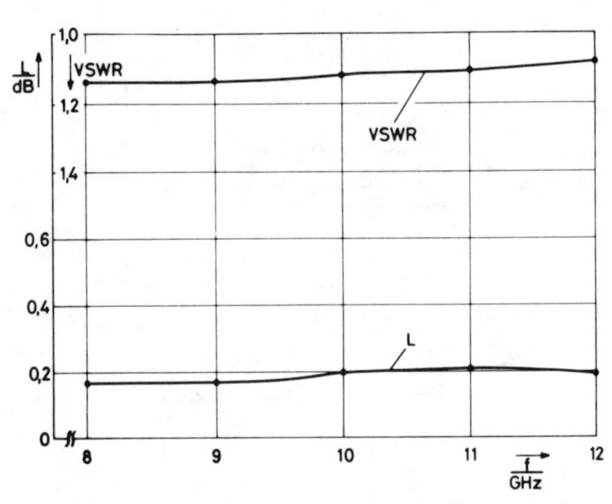

Fig. 4. VSWR and insertion loss L of two fin-line tapers in a series connection; dimensions of the substrate are as shown in Fig. 3.

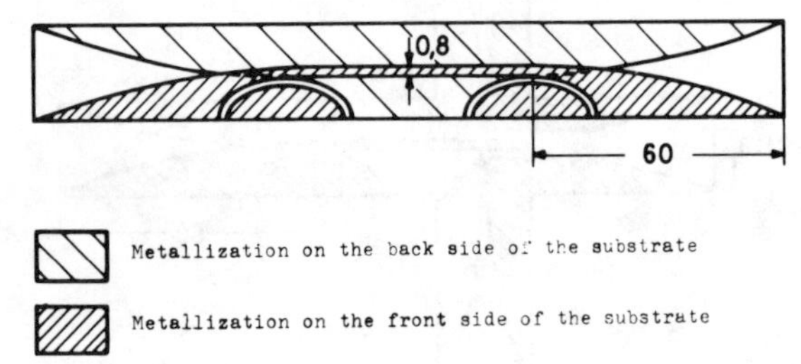

Fig. 5. Two transitions from a rectangular waveguide to a microstrip in a series connection; dimensions are in millimeters as used for the measured results of Fig. 6.

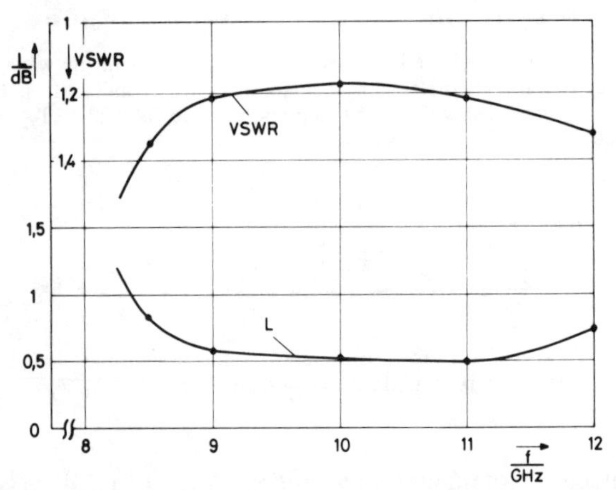

Fig. 6. VSWR and insertion loss L of two transitions from a rectangular waveguide to a microstrip; dimensions of the substrate are as shown in Fig. 5.

than 0.8 dB and a VSWR of less than 1.35 within the 8.5–12-GHz band, as shown in Fig. 6. All the measurements showed that a transition from a rectangular waveguide to a microstrip is more difficult to design than the simpler transition from a rectangular waveguide to a fin-line. Thus the useable bandwidth of a device using both types of transition mainly is determined by the transition from a rectangular waveguide to a microstrip. In general, the insertion loss and the VSWR of both the transitions increase with an increasing ratio of the waveguide input and the fin-line or microstrip output impedances and with a decreasing length of the transition.

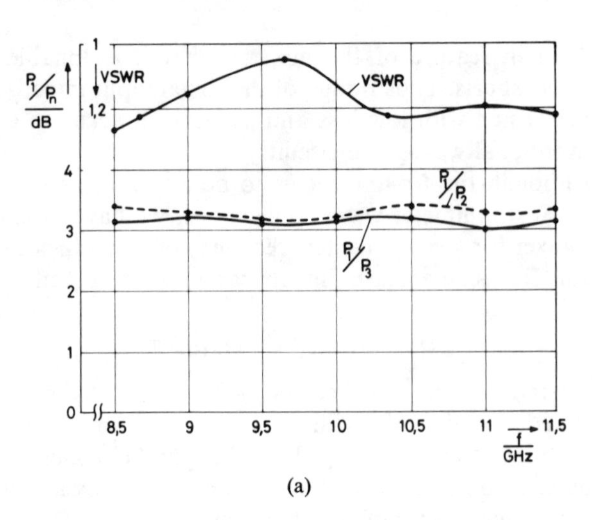

(a)

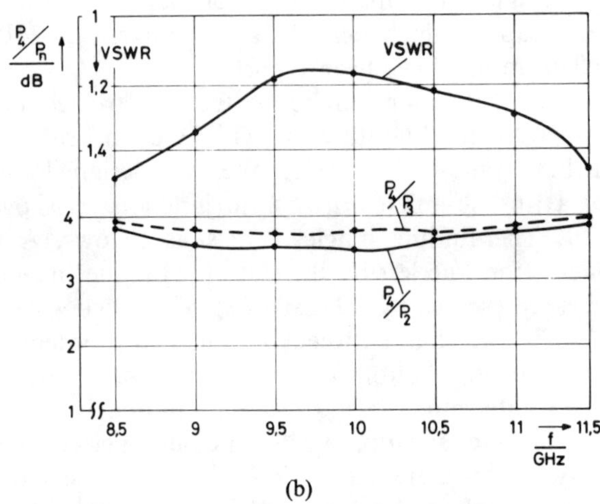

(b)

Fig. 7. VSWR at the input ports and ratios of the input and output ports powers measured at the fin-line magic T of Fig. 2. (a) Input at port 1. (b) Input at port 4. $P_n \cdots$ power at port n.

Measurements at the hybrid junction of Fig. 2 have yielded a maximum power imbalance of 0.3 dB and a maximum phase imbalance of 2° between the output ports 2 and 3. That was true for waves incident at port 1 as well as at port 4. The isolation between port 1 and port 4 and vice versa was 35-dB minimum with a 40-dB average. These results critically depend on the symmetry of the device. Fig. 7(a) shows the maximum insertion loss between port 1 and one of the output ports to be 0.4 dB with a 0.3-dB average within the 8.5–11.5-GHz band. The VSWR was better than 1.27. As Fig. 7(b) shows, the insertion loss between port 4 and one of the output ports was slightly higher due to a higher reflection coefficient; a 0.9-dB maximum with a 0.7-dB average has been observed within the same frequency band. The VSWR was less than 1.5.

III. The Complete Mixer Configuration

The complete mixer is shown in Fig. 8 [8]. The LO power is fed to the mixer from a rectangular waveguide via the transition-1 and a 50-Ω microstripline. A metallization-2 is added as in the case of the fin-line magic T. The RF power input is constructed of a fin-line taper-3. The

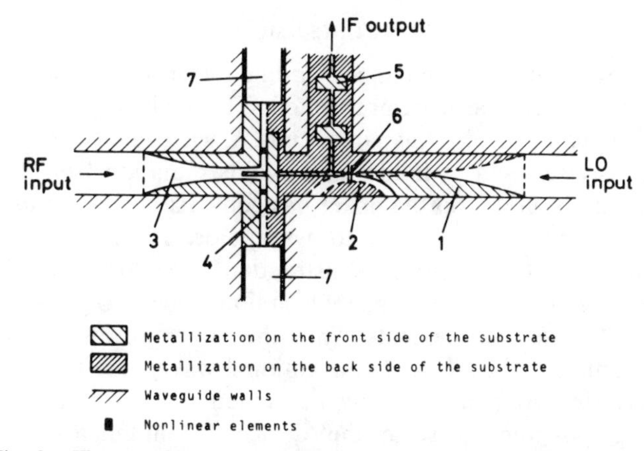

Fig. 8. The complete mixer configuration: 1 is the waveguide to microstrip transition, 2 is the additional metallization, 3 is the fin-line taper, 4 is the microstrip stub, 5 is the low-pass filter, 6 is the block capacitor, and 7 is the variable shorts.

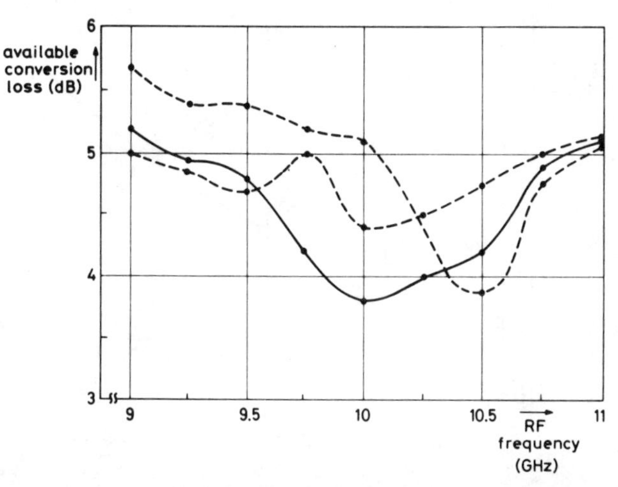

Fig. 9. Measured available conversion loss as a function of the RF frequency with a fixed LO power of 12 dBm and an intermediate frequency of 70 MHz. Solid line is the mixer optimized at 10 GHz. Dashed lines are the mixer optimized at 9.5 and 10.5 GHz, respectively.

impedance of the fin-line is 200 Ω such that it is matched to the 100-Ω cross arms.

Because the ratio of the microstrip to the cross arms and fin-line input impedances must be one to two to four, one is in a way forced to choose the impedance level as done here. If the microstrip impedance is chosen less than 50 Ω on the one hand, matching of this line to the waveguide input becomes more difficult; on the other hand, the cross-arms impedances have to be less than 100 Ω which is difficult to realize in a fin-line technique. On the other hand, a microstrip impedance of more than 50 Ω would require cross-arms impedances of more than 100 Ω which makes matching of the diodes more difficult. So the impedance level chosen here is a compromise.

The microstripline carrying the LO power is simultaneously used as the IF output. To do this, the nonlinear elements are connected with the microstrip by stubs-4 terminated in an open-circuit. Concerning the LO propagating on the microstripline, the stubs act as open-circuits and the LO energy gets to the nonlinear elements by way of the magic T and the cross arms. The nonlinear elements are soldered at the middle of the stubs such that one of the two connecting contacts is grounded at the RF and LO frequencies. The hot point is on the other side of the slot. The impedances of the stubs were chosen to be about 15 Ω in order to achieve a considerably broad-band resonance behavior.

A five-section conventional microstrip low-pass filter-5 acts as a diplexer between the LO and IF. The passband insertion loss of the filter which has a cutoff frequency of 3 GHz is less than 1 dB. The RF and LO isolation is better than 35 dB. An 8.2-pF capacitor-6 which blocks the IF from the LO input is connected in series with the microstrip. A remarkable advantage of the mixer is that it is tunable by variable shorts-7 contacting at the metallized substrate. So it is easy to match the RF input and to minimize the conversion loss for different LO power levels over a considerable wide frequency band without a laborious variation of the whole mixer configuration. After tuning, the shorts can be fixed by bridging the slots of the cross arms.

The mixer was printed on a 0.254-mm teflon substrate (RT/Duroid 5880). GaAs Schottky barrier diodes[1] with a typical series resistance of 1.7 Ω, a junction capacitance of 0.4 pF, an ideality factor of 1.06, and Si beam-lead Schottky barrier diodes[2] were used as nonlinear elements. A conventional X-band waveguide served to hold the planar structure.

IV. Mixer Performance

In order to measure the conversion loss, the mixer was fed with an RF input power directly measured at the mixer input and was pumped with different LO power levels. The IF output power was measured directly at the IF output with a selective microvoltmeter. The difference of the RF input and the IF output powers gives the conversion loss L. With the IF output reflection coefficient r_{ZF} directly measured with a network analyzer, the available conversion loss L_A results in

$$L_A = L(1-|r_{ZF}|^2).$$

Available conversion loss data for the mixer is presented in Fig. 9 for a fixed LO power of 12 dBm. For these measurements, we used the GaAs diodes. The Si diodes gave slightly worse results. When the shorts were fixed such that the mixer was optimized at an RF frequency of 10 GHz, the mixer had a minimum conversion loss of 3.8 dB (solid line). When the mixer was

[1]1SS11 of NEC.
[2]HP 5082–2264 of Hewlett–Packard.

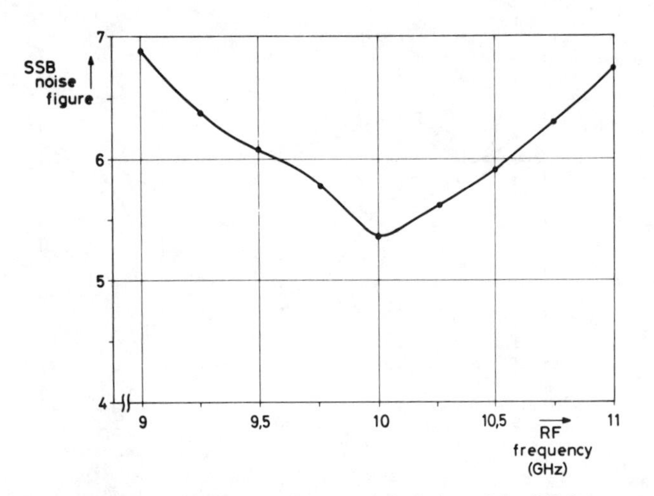

Fig. 10. Measured SSB noise figure as a function of the RF frequency with a fixed LO power of 12 dBm and an intermediate frequency of 70 MHz.

optimized at other frequencies (dashed lines), the overall behavior was slightly worse but better at specified frequencies. The RF VSWR was better than 1.3 in the 9–11-GHz band. The LO VSWR was less than 4.0 which can be improved by properly matching the LO input. The mixer output impedance was about 50 Ω. The LO–RF isolation was measured to be between 25 and 28 dB.

In order to determine the mixer's SSB noise figure, the DSB noise figure of the mixer connected with an IF amplifier (NF=2 dB) and the upper- and lower-sideband conversion of the mixer itself were measured separately. There was no external IF matching used, but the mismatch between the mixer and IF amplifier was considered. The maximum difference between the upper- and lower-sideband conversion loss was about 0.6 dB. The results calculated from the measured values are presented in Fig. 10, and they show the SSB noise figure of the mixer itself to be less than 6.9 dB over the 9–11-GHz band when the mixer was optimized at an RF center frequency of 10 GHz.

V. Conclusions

The features of a balanced fin-line mixer working at the *X* band have been demonstrated. The main part of the mixer consists of a planar circuit. A waveguide simultaneously serves as a holder and avoids radiation from the planar structure. So the mixer has conversion losses and a noise figure comparable to waveguide mixers but the advantage of a simpler construction. Low-loss variable shorts as used in a waveguide technique have proven to be excellent tuning elements. Thus the mixer profits by the advantages of MIC and conventional waveguide technology. Measurements have been executed at the *X* band, but it is possible to scale the device to millimeter-wave frequencies.

Acknowledgment

The author gratefully acknowledges J. Hartmann for many helpful ideas and the execution of the practical work.

References

[1] P. J. Meier, "Integrated finline millimeter components," *IEEE Trans. Microwave Theory Tech.*, vol. MTT-22, pp. 1209–1216, Dec. 1974.

[2] K. M. Johnson, "*X*-Band integrated circuit mixer with reactively terminated image," *IEEE Trans. Microwave Theory Tech.*, vol. MTT-16, pp. 388–397, July 1968.

[3] L. E. Dickens and D. W. Maki, "An integrated circuit balanced mixer, image and sum enhanced," *IEEE Trans. Microwave Theory Tech.*, vol. MTT-23, pp. 276–281, Mar. 1975.

[4] U. H. Gysel, "A 26.5-to-40-GHz planar balanced mixer," in *Proc. 5th European Microwave Conf.* (Hamburg, Germany), Sept. 1975, pp. 491–494.

[5] B. D. Geller and M. Cohn, "An MIC push-pull FET amplifier," in *IEEE Int. Microwave Symp. Dig.* (San Diego, CA), June 1977, pp. 187–190.

[6] H. Hofmann, "Finline dispersion," *Electron. Lett.*, vol. 12, pp. 428–429, Aug. 1976.

[7] A. M. K. Saad and G. Begemann, "Electrical performance of finline of various configurations," *Inst. Elec. Eng. Trans. Microwaves Opt. Acoust.*, vol. 1, no. 2, pp. 81–88, Jan. 1977.

[8] J. Hartmann, "Entwicklung eines Flossenleitungs-Gegentaktmischers," Masters thesis, Technical University, Braunschweig, Germany, 1978.

K-Band Integrated Double-Balanced Mixer

HIROYO OGAWA, MASAYOSHI AIKAWA, MEMBER, IEEE, AND KOZO MORITA

Abstract—**A novel microwave integrated circuit (MIC) double-balanced mixer with good isolation between the three ports is described. The mixer is fabricated using a combination of microstrip lines, slotlines, and coupled slotlines, together with four beam-lead Schottky-barrier diodes. The K-band magic-T has been developed for the double-balanced mixer. The minimum conversion loss measured at a signal frequency of 19.6 GHz is 4.7 dB. Isolation between RF and LO ports is greater than 20 dB from 18 to 21 GHz. The mixer can be expected to have wide applications in MIC receivers and transmitters up to the millimeter-wave band.**

I. Introduction

RECENTLY, microwave integrated circuit (MIC) balanced mixers have been produced for use in high-frequency bands [1]–[8]. Balanced-type mixers have several desirable features, such as good isolation and suppression of undesired signals. Though the hybrid circuits are easily fabricated on a dielectric substrate, the MIC high-Q filters which are needed for an unbalanced-type mixer cannot be easily fabricated for the millimeter-wave band. Consequently, the balanced-type mixer which does not require filters is suitable for MIC's in high-frequency bands.

MIC balanced mixers have been constructed using combinations of microstrip lines, slotlines, and coplanar lines. In particular, the cascade connection of slotlines and coplanar lines has recently been used for single-balanced mixers [3]–[7]. Four-port hybrid circuit have also been used for single-balanced mixers. Double-balanced mixers are not easily fabricated in high-frequency bands because of the complexity of their circuit configuration as compared with unbalanced or single balanced mixers.

In this paper, a new structure for a double-balanced mixer is proposed using a combination of microstrip lines, slotlines and coupled slotlines. Slotlines are employed as the main MIC transmission line. The mixer is composed of an MIC magic-T (180° hybrid), and a diode circuit with four beam-lead Schottky-barrier diodes. The magic-T is developed for K-band [9], [10] and is an essential part of the double-balanced mixer. Since this magic-T has a four-port configuration different from that of the conventional 180° hybrid, it is possible to construct a double-balanced mixer with no crossing of transmission lines. The high directivity of the magic-T will only result in good isolation in the mixer if the diodes and diode circuits are also well-matched to one-another. The diode circuit consists of slotlines, microstrip lines and four beam-lead diodes, and has low coupling of even LO harmonics to the RF input. The harmonic sidebands of $mf_{LO} \pm f_{RF}$ (m = an odd integer) are obtained at the IF signal port. This paper describes the circuit configuration, the basic behavior of the mixer using its equivalent circuit and the experimental results obtained in the 20-GHz band.

Manuscript received July 30, 1979; revised October 15, 1979.

The authors are with the Radio Transmission Section, Yokosuka Electrical Communication Laboratories, Nippon Telegraph and Telephone Public Corporation, Yokosuka, 238-03 Japan.

II. Circuit Configuration

The configuration of the double-balanced mixer is shown in Fig. 1. This circuit is composed of an MIC magic-T and a diode circuit. In this figure, solid lines indicate slotlines and coupled slotlines on the substrate, while dotted lines indicate microstrip lines on the reverse side of the substrate. Two circular marks "o" indicate cylindrical conductors used for connecting slotlines and microstrip lines through holes in the substrate. Ⓡ, Ⓛ, and Ⓘ denote RF input port, LO input port, and IF output port, respectively.

The RF and LO signals are fed into ports Ⓡ and Ⓛ, which correspond to the H-arm and the E-arm of a conventional waveguide magic-T. The magic-T couples the RF signal in-phase, and the LO signal out-of-phase to the two diode circuits. Matching between the magic-T and the diode circuits is accomplished by the slotlines. The IF signal is derived from port Ⓘ. The low-pass filter is connected to the IF port in order to suppress undesired signals.

III. Magic-T

To make double-balanced mixers, 90° or 180° four-port hybrids such as a branch line hybrid, parallel coupled line hybrid, a rat-rate, or a magic-T are required. These circuits are easily fabricated on a substrate and are applied to various types of balanced mixers. This section describes a K-band MIC magic-T with good isolation characteristics.

The configuration of the magic-T is shown in Fig. 2. This circuit is constructed with microstrip lines, slotlines and coupled slotlines. The magic-T is characterized by its special four-port configuration, that is, the two ports Ⓗ and Ⓔ, which correspond respectively to the H-arm and the E-arm of an ordinary wave-guide magic-T, are located on the same side opposite the other two-ports ① and ②. This port location has a significant advantage for practical applications, because this makes it possible to produce balanced-type mixers with no crossing of transmission lines. Crossing of lines is not suitable for a MIC structure, because it needs additional bonding and causes signal leakage which degrades the characteristics of balanced-type mixers.

Reprinted from *IEEE Trans. Microwave Theory Tech.*, vol. MTT-28, pp. 180–185, Mar. 1980.

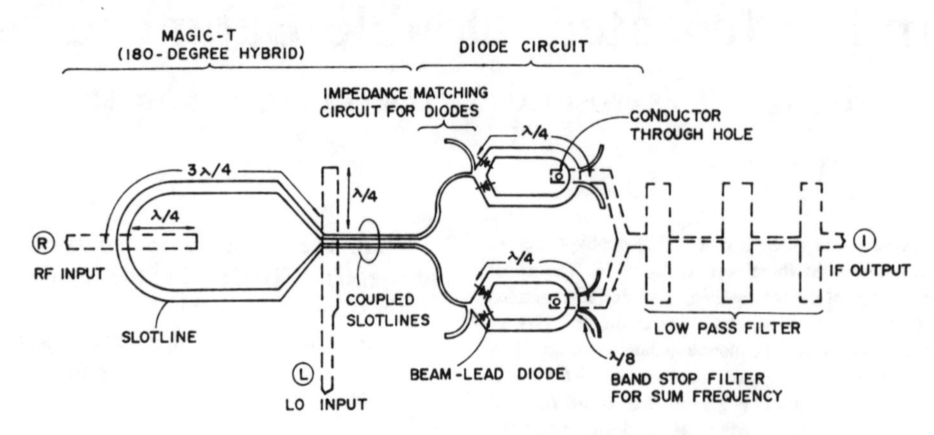

Fig. 1. Configuration of double-balanced mixer, where solid lines are slotlines and coupled slotlines on the substrate, dotted lines are microstrip lines on the reverse side of the substrate.

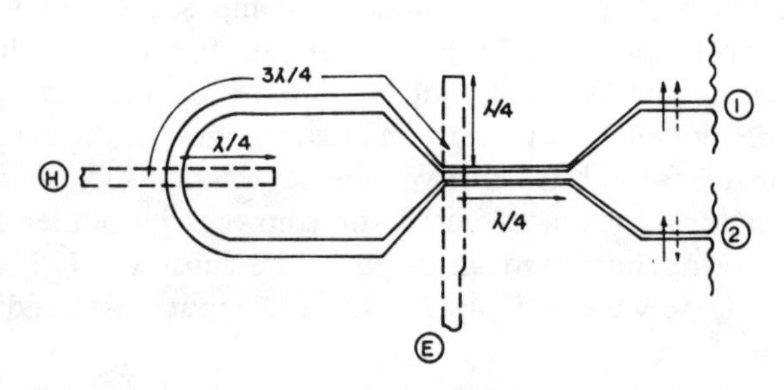

Fig. 2. Configuration of *K*-band MIC magic-T, where solid arrows and dotted arrows are schematic electric fields for the even-mode and the odd-mode of coupled slotlines, respectively.

The magic-T utilizes two orthogonal modes (even-mode and odd-mode [11]) of coupled slotlines. The signals fed into ports Ⓔ and Ⓗ are converted into the even-mode and the odd-mode, respectively. Thereafter, these signals are derived from ports ① and ②. The electric fields for the even-mode and the odd-mode are represented by solid arrows and dotted arrows in Fig. 2. Isolation between ports Ⓔ and Ⓗ is accomplished as follows: The signal from port Ⓗ is not coupled to the microstrip line of port Ⓔ because of the orthogonal characteristic of the coupled slotlines mode. On the other hand, for the signal from port Ⓔ, the three-quarter wavelength slotlines which facilitate matching of port Ⓔ behave as short-circuited stubs. Thus the signal from port Ⓔ is not propagated to port Ⓗ, because the signal is short-circuited at the intersection of the slotlines with the microstrip line port Ⓔ.

The magic-T is fabricated on a 0.3-mm thick alumina substrate with a center frequency of 20 GHz. To measure the performance of a magic-T, waveguide to microstrip transitions were necessary because the experimental equipment consisted of waveguide circuits. The waveguide to microstrip transition in this experiment is a ridged waveguide type [12], having six quarter-wave sections. Ports ① and ② are connected to microstrip lines using microstrip to slotline transitions, each having a quarter-wavelength. The insertion loss of the waveguide to microstrip transition is less than 0.2 dB and the return loss is greater than 20 dB in the 18- to 20.5-GHz band. Fig. 3 shows the experimental results of the MIC magic-T. In Fig. 3, dotted lines indicate the reference power level. The in-phase and out-of-phase coupling characteristics shown in Fig. 3(a) and (b), have some frequency sensitivity due to the effect of the waveguide to microstrip transitions. The couplings between Ⓗ↔①, ②, and Ⓔ↔①, ② are 4.0 ± 0.8 dB over a range of 19–21 GHz and 4.0 ± 0.4 dB in the 19.5- to 20.5-GHz band. The isolation between port Ⓗ and port Ⓔ is greater than 20 dB over a range of 18 to 21.0 GHz and greater than 30 dB in the 19- to 20.5-GHz band.

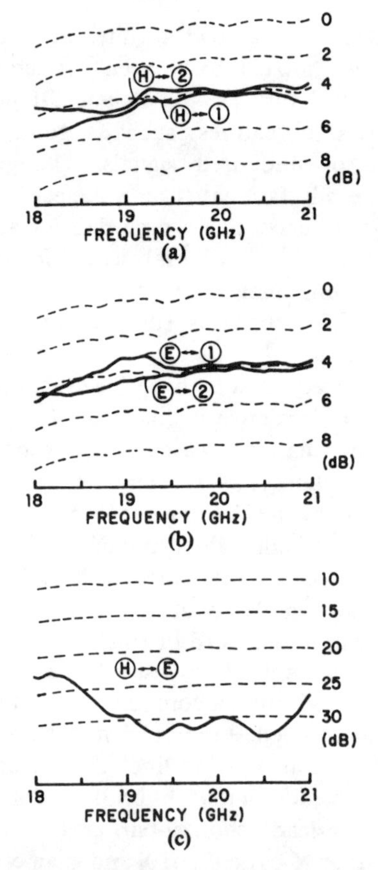

Fig. 3. Performance of MIC magic-T. (a) In-phase coupling characteristics. (b) Out-of-phase coupling characteristics. (c) Isolation characteristics.

IV. Diode Circuit

As shown in Fig. 1, the diode circuit of the MIC double-balanced mixer consists of two impedance-matching circuits, four quarter-wavelength slotlines, two pairs of beam-lead Schottky-barrier diodes, two cylindrical conductors with a diameter of 0.5 mm, and a microstrip line. The equivalent diode circuit without the impedance matching circuit is shown in Fig. 4. In this figure four quarter-wavelength slotlines (S_1), (S_2), (S_3), and (S_4) act as short-circuited stubs for the LO and RF signals because these signals are short-circuited at the point directly below the microstrip lines which are connected to port (I). In this type of double-balanced mixer these quarter-wavelength slotlines are used to utilize effectively the RF and LO powers fed to the diodes. The basic principle of the mixer is described as follows:

When the RF and LO powers are fed into the diodes, total current appearing at port (I), port (R) and port (L) is expressed as follows (see Appendix):

$$\begin{aligned} i_{(I)}(t) = &4\alpha i_s V_{RF} I_1(\alpha V_{LO}) \cos(\omega_{LO}-\omega_{RF})t \\ &+4\alpha i_s V_{RF} I_1(\alpha V_{LO}) \cos(\omega_{LO}+\omega_{RF})t \\ &+4\alpha i_s V_{RF} I_3(\alpha V_{LO}) \cos(3\omega_{LO}-\omega_{RF})t \\ &+\cdots \end{aligned} \tag{1}$$

$$\begin{aligned} i_{(R)}(t) = &4\alpha i_s V_{RF} I_0(\alpha V_{LO}) \cos \omega_{RF} t \\ &+4\alpha i_s V_{RF} I_2(\alpha V_{LO}) \cos(2\omega_{LO}-\omega_{RF})t \\ &+4\alpha i_s V_{RF} I_2(\alpha V_{LO}) \cos(2\omega_{LO}+\omega_{RF})t \\ &+4\alpha i_s V_{RF} I_4(\alpha V_{LO}) \cos(4\omega_{LO}-\omega_{RF})t \\ &+\cdots \end{aligned} \tag{2}$$

$$\begin{aligned} i_{(L)}(t) = &4\alpha i_s V_{LO}[I_0(\alpha V_{LO}) + I_2(\alpha V_{LO})] \cos {}_{LO} t \\ &+4\alpha i_s V_{LO}[I_2(\alpha V_{LO}) + I_4(\alpha V_{LO})] \cos 3 {}_{LO} t \\ &+4\alpha i_s V_{LO}[I_4(\alpha V_{LO}) + I_6(\alpha V_{LO})] \cos 5 {}_{LO} t \\ &+\cdots \end{aligned} \tag{3}$$

where

α diode slope parameter,

V_{RF} amplitude of the RF signal which is applied to each diode,

V_{LO} amplitude of the LO signal which is applied to each diode,

ω_{RF} RF angular frequency$(=2\pi f_{RF})$,

ω_{LO} LO angular frequency$(=2\pi f_{LO})$,

I_k modified Bessel function of the first kind of order k.

From (1), it can be seen that the total current at port (I) only contains frequency terms $mf_{LO} \pm f_{RF} (m \neq 0)$, when m is an odd integer. From (2) and (3), it can be seen that the frequency terms $mf_{LO} \pm f_{RF}$, where m is an even integer, and the fundamental RF signals appear at port (R), and the fundamental and odd harmonics of the LO signal appear at port (L).

Thus the IF, sum frequency and harmonic sidebands $(mf_{LO} \pm f_{RF}$; $m =$ odd integer) are obtained at port (I). The image frequency signal does not appear at port (I) and port (L) but at port (R). Therefore, the band stop filters constructed at the eighth-wavelength slotlines, as shown in Fig. 1, are connected to the slotlines in order to suppress the sum frequency. A five-section low-pass filter [13] formed by the microstrip line is also connected to the IF port in order to suppress those undesired harmonics. The cutoff frequency of the filter is 8 GHz. In this type of double-balanced mixer it is comparatively easy to reactively terminate the image signal if the LO frequency is far from the RF frequency.

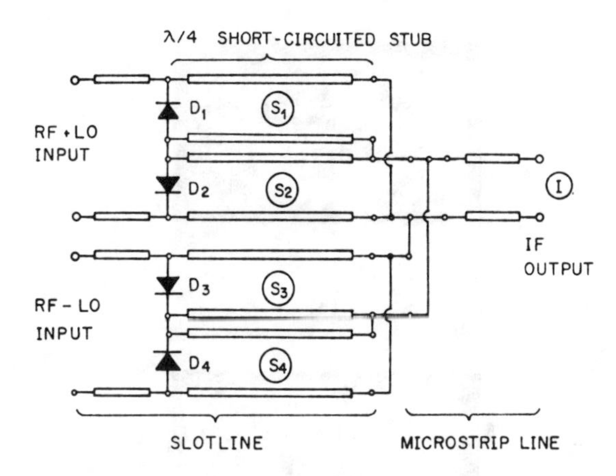

Fig. 4. Equivalent circuit of diode circuit.

The impedance of the beam-lead diode[1] is measured by a network analyzer, using the waveguide to microstrip transition mentioned in Section III. GaAs Schottky barrier diodes used here have a typical series resistance of 2.5 Ω, a junction capacitance of 0.05 pF at zero bias, a stray capacitance of 0.05 pF and an ideality factor of 1.17. The RF impedance of the diode at 20 GHz is $20 - j40$ (Ω), when the bias current is 0.5 mA at which the minimum conversion loss is obtained. The impedance-matching circuit for the diode is designed as shown in Fig. 1 using a quarter-wavelength impedance transformer and a short-circuited stub connected in series to the slotline in order to cancel the capacitance of diodes. It is noteworthy that short- or open-circuited stubs can be easily connected to the slotline. This configuration of the matching circuit is suitable for adjusting integrated circuits, because the stub length can be easily changed by bonding with gold wires or ribbons.

V. Experimental Results

The double-balanced mixer shown in Fig. 1 is fabricated by a photolithographic technique on a 0.3-mm thick alumina substrate with a relative permittivity of 9.6. A 500-Å thick nickel-chromium and 6000-Å thick gold are deposited on the substrate by a vacuum evaporation method. Fig. 5 shows photographs of the mixer pattern. In Fig. 5(a) is the pattern of the microstrip line on the reverse side of the substrate and Fig. 5(b) is the pattern of a slotline and coupled slotlines on the substrate. The gold thickness of the microstrip lines and slotlines in this

[1] V558 of NEC

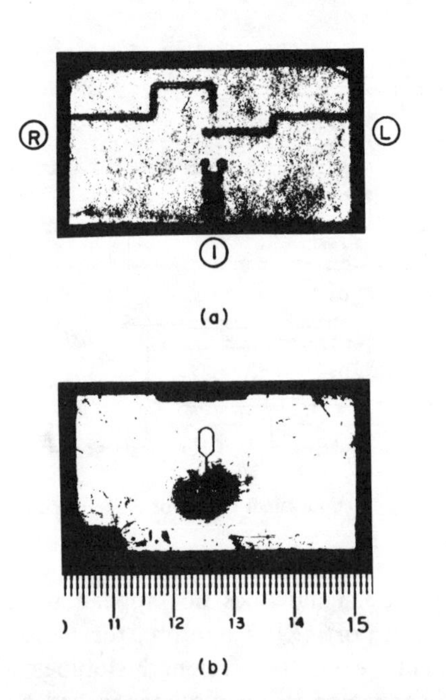

Fig. 5. Photograph of double-balanced mixer. (a) Microstrip pattern on the reverse substrate. (b) Slotline pattern.

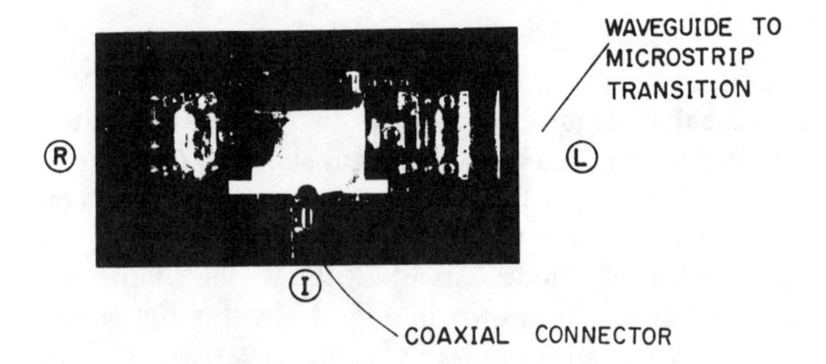

Fig. 6. Top view of an entire double-balanced mixer.

circuit is 3 μm, which is electrically plated after the photoetching process. A top view of the entire integrated mixer is shown in Fig. 6, including the waveguide to microstrip transitions at the RF and LO ports and the coaxial connector at the IF port. The microstrip pattern can be seen in this photograph.

The measured conversion losses of the mixer are presented in Fig. 7 for several LO frequencies. These losses include the insertion loss, that is, 0.2 dB of waveguide to microstrip transition and 0.5 dB of low-pass filter. As can be seen in Fig. 7, the minimum conversion loss attained at a signal frequency of 19.6 GHz is 4.7 dB. The cause for some frequency sensitivities in conversion loss is probably the positional inaccuracy of the holes on the substrate and the bonded diodes on the slotlines. Another reason is that four diodes used here are not completely matched. The isolation between ports Ⓛ and Ⓡ is greater than 20 dB, and between ports Ⓛ and Ⓘ greater than 30 dB, as shown in Fig. 8. As described above, the double-balanced mixer used for the 20-GHz band has good isolation between the three ports in addition to a low conversion loss, due to the effective combination of microstrip lines, slotlines, coupled slotlines and two pairs of Schottky-barrier diodes.

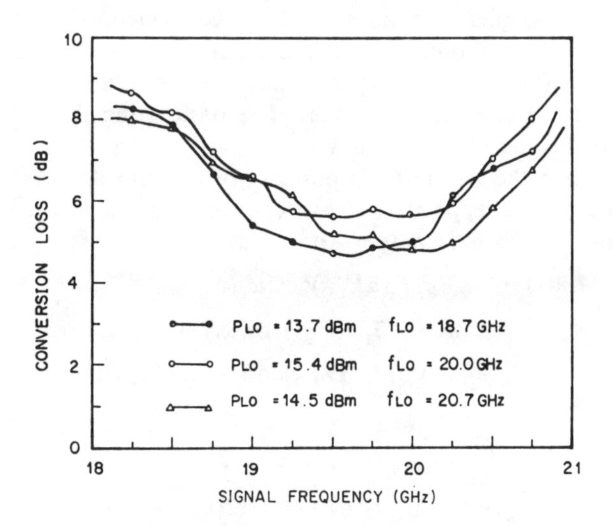

Fig. 7. Conversion loss of double-balanced mixer.

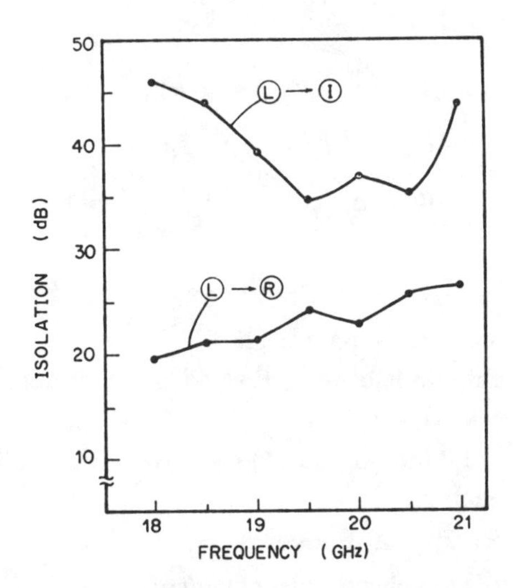

Fig. 8. Isolation characteristics of double-balanced mixer.

VI. Conclusion

A new MIC double-balanced mixer has been successfully developed, which is suitable for high frequencies up to the millimeter-wave bands. This mixer consists of an MIC magic-T and a diode circuit. These circuits are constructed by using combinations of microstrip lines, slotlines, coupled slotlines and beam-lead diodes. The coupling characteristics of the magic-T are typically 4 dB between 19 and 21 GHz and isolation is greater than 20 dB in the 18- to 21-GHz range. The minimum conversion loss of the mixer is 4.7 dB at a signal frequency of 19.6 GHz, and isolation between the three ports is greater than 20 dB in the 18- to 21-GHz range. This type of double-balanced mixer can be made without crossing transmission lines, and its configuration is very suitable for the beam-lead diode. The signals appearing at port Ⓘ are the IF signal and the harmonic sidebands. The image signal does not appear at port Ⓘ or port Ⓛ, but does appear at port Ⓡ. This mixer can be easily fabricated using ordinary MIC techniques and can be applied to other balanced type devices, such as balanced modulators and balanced upconverters.

Appendix

Equations (1)–(3) are derived as follows [14]–[16].

Fig. 9 shows the schematic circuit for illustrating the signal flow around the diodes. In this figure, the four quarter-wavelength slotlines in Fig. 4 is regarded as two coplanar lines for the current appearing at port Ⓘ. In Fig. 9, when the voltage V is fed into diodes D_1, D_2, D_3, and D_4, the instantaneous currents passing through diodes i_1, i_2, i_3, and i_4 are written in the usual expressions.

$$i_1 = -i_s(e^{-\alpha V}-1) \tag{4}$$

$$i_2 = i_s(e^{\alpha V}-1) \tag{5}$$

$$i_3 = i_s(e^{\alpha V}-1) \tag{6}$$

$$i_4 = -i_s(e^{-\alpha V}-1) \tag{7}$$

where α is the diode slope parameter and i_s is the saturation current. From (4)–(7), the differential conductance for each diode is expressed as follows:

$$g_1 = di_1/dV = \alpha i_s e^{-\alpha V} \tag{8}$$

$$g_2 = di_2/dV = \alpha i_s e^{\alpha V} \tag{9}$$

$$g_3 = di_3/dV = \alpha i_s e^{\alpha V} \tag{10}$$

$$g_4 = di_4/dV = \alpha i_s e^{-\alpha V} \tag{11}$$

We assume that the conductance of the diodes is modulated with periodic signals

$$V = V_{LO}\cos\omega_{LO}t, \qquad \text{for } D_1 \text{ and } D_2 \tag{12}$$

$$V = V_{LO}\cos(\omega_{LO}t+\pi), \qquad \text{for } D_3 \text{ and } D_4 \tag{13}$$

where $\omega_{LO} = 2\pi f_{LO}$ is LO angular frequency and V_{LO} is the amplitude of the LO signal applied to each diode. The conductances of the diodes are expressed as follows:

$$g_1(t) = \alpha i_s \exp[-\alpha V_{LO}\cos\omega_{LO}t] \tag{14}$$

$$g_2(t) = \alpha i_s \exp[\alpha V_{LO}\cos\omega_{LO}t] \tag{15}$$

$$g_3(t) = \alpha i_s \exp[\alpha V_{LO}\cos(\omega_{LO}t+\pi)] \tag{16}$$

$$g_4(t) = \alpha i_s \exp[-\alpha V_{LO}\cos(\omega_{LO}t+\pi)]. \tag{17}$$

Furthermore, the RF signal

$$v = V_{RF}\cos\omega_{RF}t \tag{18}$$

is fed into the above differential conductances, where $\omega_{RF} = 2\pi f_{RF}$ is RF angular frequency and V_{RF} is the amplitude of the RF signal applied to each diode. From the instantaneous currents passing through the diodes, $i_{12}(t)$ and $i_{34}(t)$ are expressed as follows:

$$\begin{aligned} i_{12}(t) &= i_2(t) - i_1(t) \\ &= 2\alpha i_s \sinh[\alpha V_{LO}\cos\omega_{LO}t] \\ &\quad \cdot [V_{LO}\cos\omega_{LO}t + V_{RF}\cos\omega_{RF}t] \\ &= 2\alpha i_s V_{LO} I_1(\alpha V_{LO}) \\ &\quad + 2\alpha i_s V_{LO}[I_1(\alpha V_{LO}) + I_3(\alpha V_{LO})]\cos 2\omega_{LO}t \\ &\quad + 2\alpha i_s V_{RF} I_1(\alpha V_{LO})\cos(\omega_{LO}-\omega_{RF})t \\ &\quad + 2\alpha i_s V_{RF} I_1(\alpha V_{LO})\cos(\omega_{LO}+\omega_{RF})t \\ &\quad + 2\alpha i_s V_{LO}[I_3(\alpha V_{LO}) + I_5(\alpha V_{LO})]\cos 4\omega_{LO}t \\ &\quad + 2\alpha i_s V_{RF} I_3(\alpha V_{LO})\cos(3\omega_{LO}-\omega_{RF})t + \cdots \end{aligned} \tag{19}$$

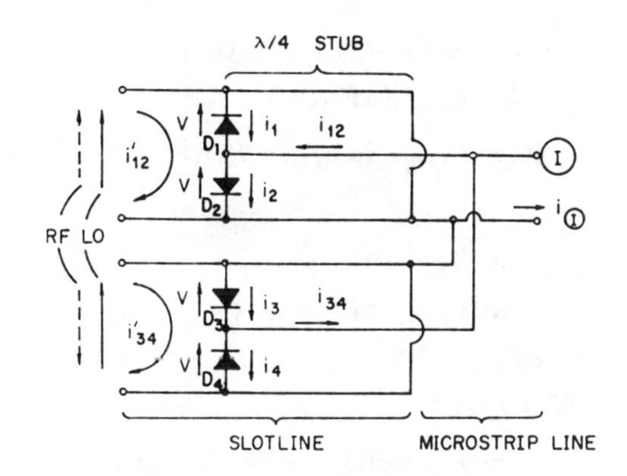

Fig. 9. Schematic circuit of Fig. 4.

$$\begin{aligned} i_{34}(t) &= i_3(t) - i_4(t) \\ &= 2\alpha i_s \sinh[\alpha V_{LO}\cos(\omega_{LO}t+\pi)] \cdot [V_{LO}\cos(\omega_{LO}t+\pi) \\ &\quad + V_{RF}\cos\omega_{RF}t] \\ &= 2\alpha i_s V_{LO} I_1(\alpha V_{LO}) \\ &\quad + 2\alpha i_s V_{LO}[I_1(\alpha V_{LO}) + I_3(\alpha V_{LO})]\cos(2\omega_{LO}t+2\pi) \\ &\quad + 2\alpha i_s V_{RF} I_1(\alpha V_{LO})\cos(\omega_{LO}t-\omega_{RF}t+\pi) \\ &\quad + 2\alpha i_s V_{RF} I_1(\alpha V_{LO})\cos(\omega_{LO}t+\omega_{RF}t+\pi) \\ &\quad + 2\alpha i_s V_{LO}[I_3(\alpha V_{LO}) + I_5(\alpha V_{LO})]\cos(4\omega_{LO}t+4\pi) \\ &\quad + 2\alpha i_s V_{RF} I_3(\alpha V_{LO})\cos(3\omega_{LO}t-\omega_{RF}t+\pi) \\ &\quad + \cdots. \end{aligned} \tag{20}$$

Total current expression at port Ⓘ is

$$\begin{aligned} i_{Ⓘ}(t) &= i_{12}(t) - i_{34}(t) \\ &= 4\alpha i_s V_{RF} I_1(\alpha V_{LO})\cos(\omega_{LO}-\omega_{RF})t \\ &\quad + 4\alpha i_s V_{RF} I_1(\alpha V_{LO})\cos(\omega_{LO}+\omega_{RF})t \\ &\quad + 4\alpha i_s V_{RF} I_3(\alpha V_{LO})\cos(3\omega_{LO}-\omega_{RF})t \\ &\quad + \cdots. \end{aligned} \tag{21}$$

In (19) and (20), the following formula is used:

$$\sinh(Z\cos\theta) = 2\sum_{k=0}^{\infty} I_{2k+1}(Z)\cos(2k+1)\theta \tag{22}$$

where I_k are modified Bessel functions of the first kind of order k.

It can be seen that the total current at port Ⓘ only contains frequency terms $mf_{LO} \pm f_{RF}$ ($m \neq 0$), where m is an odd integer.

In Fig. 9 sum current $i_1 + i_2$ and $i_3 + i_4$ flow toward the magic-T. These currents do not appear at port Ⓘ, but appear at port Ⓡ or port Ⓛ. These currents are expressed as follows:

$$\begin{aligned} i'_{12}(t) &= i_1(t) + i_2(t) \\ &= 2\alpha i_s \cosh[\alpha V_{LO}\cos\omega_{LO}t] \\ &\quad \cdot [V_{LO}\cos\omega_{LO}t + V_{RF}\cos\omega_{RF}t] \end{aligned}$$

$$=2\alpha i_s V_{LO}[I_0(\alpha V_{LO})+I_2(\alpha V_{LO})]\cos\omega_{LO}t$$
$$+2\alpha i_s V_{RF}I_0(\alpha V_{LO})\cos\omega_{RF}t$$
$$+2\alpha i_s V_{LO}[I_2(\alpha V_{LO})+I_4(\alpha V_{LO})]\cos 3\omega_{LO}t$$
$$+2\alpha i_s V_{RF}I_2(\alpha V_{LO})\cos(2\omega_{LO}-\omega_{RF})t$$
$$+2\alpha i_s V_{RF}I_2(\alpha V_{LO})\cos(2\omega_{LO}+\omega_{RF})t$$
$$+2\alpha i_s V_{RF}I_4(\alpha V_{LO})\cos(4\omega_{LO}-\omega_{RF})t$$
$$+\cdots. \quad (23)$$

$$i'_{34}(t)=i_3(t)+i_4(t)$$
$$=2\alpha i_s\cosh[V_{LO}\cos(\omega_{LO}t+\pi)]\cdot[V_{LO}\cos(\omega_{LO}t+\pi)+V_{RF}\cos\omega_{RF}t]$$
$$=2\alpha i_s V_{LO}[I_0(\alpha V_{LO})+I_2(\alpha V_{LO})]\cos(\omega_{LO}t+\pi)$$
$$+2\alpha i_s V_{RF}I_0(\alpha V_{LO})\cos\omega_{RF}t$$
$$+2\alpha i_s V_{LO}[I_2(\alpha V_{LO})+I_4(\alpha V_{LO})]\cos(3\omega_{LO}t+3\pi)$$
$$+2\alpha i_s V_{RF}I_2(\alpha V_{LO})\cos(2\omega_{LO}t-\omega_{RF}t+2\pi)$$
$$+2\alpha i_s V_{RF}I_2(\alpha V_{LO})\cos(2\omega_{LO}t+\omega_{RF}t+2\pi)$$
$$+2\alpha i_s V_{RF}I_4(\alpha V_{LO})\cos(4\omega_{LO}t-\omega_{RF}t+4\pi)$$
$$+\cdots. \quad (24)$$

In (23) and (24), the following formula is also used.

$$\cosh(Z\cos\theta)=I_0(Z)+2\sum_{k=1}^{\infty}I_k(Z)\cos 2k\theta. \quad (25)$$

From (23) and (24), the currents appearing at port Ⓡ and Ⓛ are expressed as follows:

$$i_{Ⓡ}(t)=i'_{12}(t)+i'_{34}(t)$$
$$=4\alpha i_s V_{RF}I_0(\alpha V_{LO})\cos\omega_{RF}t$$
$$+4\alpha i_s V_{RF}I_2(\alpha V_{LO})\cos(2\omega_{LO}-\omega_{RF})t$$
$$+4\alpha i_s V_{RF}I_2(\alpha V_{LO})\cos(2\omega_{LO}+\omega_{RF})t$$
$$+4\alpha i_s V_{RF}I_4(\alpha V_{LO})\cos(4\omega_{LO}-\omega_{RF})t$$
$$+\cdots. \quad (26)$$

$$i_{Ⓛ}(t)=i'_{12}(t)-i'_{34}(t)$$
$$=4\alpha i_s V_{LO}[I_0(\alpha V_{LO})+I_2(\alpha V_{LO})]\cos\omega_{LO}t$$
$$+4\alpha i_s V_{LO}[I_2(\alpha V_{LO})+I_4(\alpha V_{LO})]\cos 3\omega_{LO}t$$
$$+4\alpha i_s V_{LO}[I_4(\alpha V_{LO})+I_6(\alpha V_{LO})]\cos 5\omega_{LO}t$$
$$+\cdots. \quad (27)$$

It can be seen from (26) and (27) that the frequency terms $mf_{LO}\pm f_{RF}$, where m is an even integer, and the fundamental RF signal appear at port Ⓡ, and the fundamental and odd harmonics of the LO signal appear at port Ⓛ.

ACKNOWLEDGMENT

The authors wish to thank Dr. Yamamoto, Dr. Ohtomo, Dr. Akaike, and Dr. Kurita in Yokosuka Electrical Communication Laboratory for their encouragement and suggestions.

REFERENCES

[1] T. Araki and H. Hirayama, "A 20-GHz integrated balanced mixer" *IEEE Trans. Microwave Theory Tech.*, vol. MTT-19, pp. 638–643, July 1971.

[2] K. M. Johnson, "*X*-band integrated circuit mixer with reactively terminated image," *IEEE Trans. Microwave Theory Tech.*, vol. MTT-16, pp. 388–397, July 1968.

[3] L. E. Dickens and D. W. Maki, "An integrated-circuit balanced mixer, image and sum enhanced," *IEEE Trans. Microwave Theory Tech.*, vol. MTT-23, pp. 276–281, Mar. 1975.

[4] G. Begemann, "An X-band balanced fine-line mixer," *IEEE Trans. Microwave Theory Tech.*, vol. MTT-26, pp. 1007–1011, Dec. 1978.

[5] T. K. Hunton and T. S. Takeuchi, "Recent developments in microwave slotline mixers and frequency multipliers," in *1970 IEEE G-MTT Int. Symp. Dig. Tech. Papers*, pp. 196–199, May 1970.

[6] U. H. Gysel, "A 26.5-to-40-GHz planar balanced mixer," in *Proc. 5th European Microwave Conf.*, pp. 491–495, Sept. 1975.

[7] L. E. Dickens and D. W. Maki, "A new "phased-type" image enhanced mixer," in *1975 IEEE G-MTT Int. Symp. Digest Tech. Papers*, pp. 149–151, 1975.

[8] R. Pflieger, "A new MIC double-balanced mixer with RF and IF band overlap," in *1973 IEEE G-MTT Int. Sym. Digest ,Tech. Papers*, pp. 301–303, June 1973.

[9] M. Aikawa and H. Ogawa, "A new MIC magic-T using coupled slotlines," submitted to *IEEE Trans. Microwave Theory Tech.*, 1979.

[10] M. Aikawa and H. Ogawa, "2 Gb double-balanced PSK modulator using coplanar waveguides," in *1979 ISSCC Digest*, pp. 172–173, Feb. 1979.

[11] J. B. Knorr and K. D. Kuchler, "Analysis of coupled slots and coplanar strips on dielectric substrate," *IEEE Trans. Microwave Theory Tech.*, vol. MTT-23, pp. 541–548, July 1975.

[12] S. Hopfer, "The design of ridged waveguide," *IRE Trans. Microwave Theory Tech.*, vol. MTT-3, pp. 20–29, Oct. 1955.

[13] Matthaei, Young, and Jones, *Microwave Filters, Impedance-Matching Networks, and Coupling Structures*. New York: McGraw-Hill, 1964, pp. 83–104.

[14] M. R. Barber, "Noise figure and conversion loss of the Schottky-barrier diode," *IEEE Trans. Microwave Theory Tech.*, vol. MTT-15, pp. 629–635, Nov. 1967.

[15] M. V. Schneider and W. W. Snell, "Harmonically pumped stripline down-converter," *IEEE Trans. Microwave Theory Tech.*, vol. MTT-23, pp. 271–275, Mar. 1975.

[16] M. Cohn, J. E. Degnenford, and B. A. Newman, "Harmonic mixing with an antiparallel diode pair," *IEEE Trans. Microwave Theory Tech.*, vol. MTT-23, pp. 667–673, Aug. 1975.

E-PLANE COMPONENTS FOR A 94-GHz PRINTED-CIRCUIT BALANCED MIXER

Paul J. Meier

Eaton Corporation AIL Division
Melville, N.Y. 11747

Summary

E-plane components for a new form of a 94-GHz printed-circuit balanced mixer are described. The components include a low-loss printed-probe hybrid, advanced beam-lead diodes, and fin-line mounts. The E-plane approach features production economy, effective shielding, high (> 400) unloaded Q, light dielectric loading, and simple waveguide interfaces.

Introduction

Recent interest in the 3-mm atmospheric window has underscored the need for low-cost, high-performance receivers. Balanced mixers are of special interest in radiometers, where the IF is generally low, or in those applications where LO radiation must be minimized. A key component of a balanced mixer is the hybrid coupler, which can be constructed in various forms. At millimeter wavelengths, the most common forms of printed-circuit hybrids have been the ring hybrid[1], the branch-line coupler[2], and the slot/coplanar/microstrip junction[3, 4]. Although existing designs can be scaled into the 3-mm band, problems are to be expected in terms of radiation, stray coupling, Q limitations, manufacturing tolerances, and interaction with waveguide transitions.

As an alternative to the older forms of IC hybrids, the printed-probe E-plane coupler has been developed[5]. Through recent work, a high-performance E-plane hybrid has been developed in the 3-mm band[6]. By integrating such a hybrid with fin-line diode mounts, a balanced mixer can be constructed entirely from E-plane lines. Advantages of the E-plane approach at millimeter wavelengths include printed-circuit economy, negligible radiation and stray coupling, high unloaded Q (> 400 at 94 GHz), low-equivalent dielectric constant (for eased tolerances), and simple wideband transitions to standard waveguide instrumentation.

The following paragraphs describe the design and performance of E-plane components for a 94-GHz printed-circuit balanced mixer.

Components and Integration Goal

Figure 1 shows a preliminary layout of a balanced mixer assembly. The major components, a seven-probe hybrid and a pair of fin-line diode mounts, are printed on a single board which is suspended in the E-plane of a four-port housing. All four ports were required in early tests of the hybrid alone; in the illustrated assembly, only the two waveguide ports at the left are utilized. These ports serve as the RF and LO inputs. The IF outputs leave the housing through SMA connectors and are combined in a coax tee (not shown).

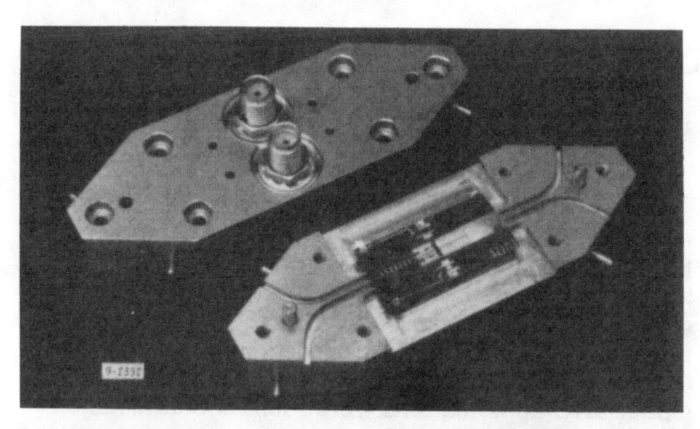

Figure 1. E-Plane Balanced Mixer

Figure 2 identifies the components of the printed-circuit assembly. Included are:

- Quarter-wave notches (a) which provide a match between the air-filled and slab-loaded waveguides.
- Mounting holes (b) which align the 5-mil, Duroid 5880 board within the housing.
- Foil tabs (c) where dc bias can be applied.
- Seven-element printed-probe coupler (d).
- Fin-line mounts for the beam-lead diodes (e); the mounts include RF transformers (f) and gold wire (g) which provides the ground return for the RF, LO, IF, and dc bias.
- IF output, containing high-impedance lines (h), dc-blocking capacitors (i), and connector contacts (j).
- Resistors (k) which block the IF from the dc bias circuit.
- Printed-circuit E-plane bifurcations (ℓ) which reactively terminate the diode mounts at RF and LO.

Details on the major components follow.

Printed-Probe Hybrid

At the center of the housing, illustrated in Figure 1, parallel waveguides share a common broadwall, which is slotted to accept a pair of dielectric boards. An array of coupling probes is printed on one board, and insulated from the common wall by a second board, fabricated from 2-mil Teflon.

Reprinted from *IEEE MTT-S Int. Microwave Symp. Dig.*, 1980, pp. 267–269.

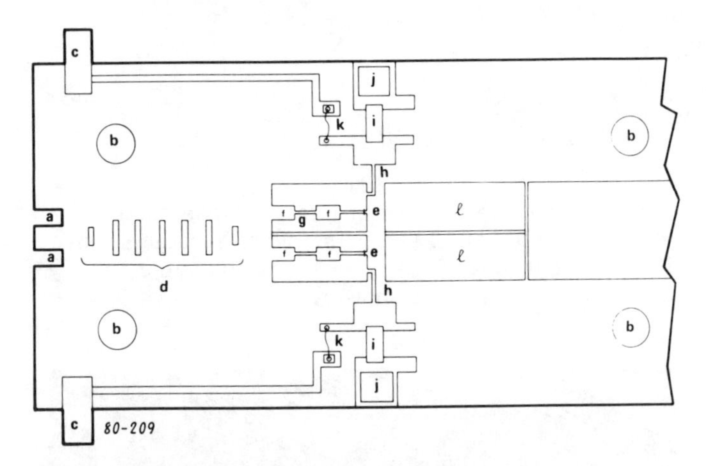

Figure 2. Balanced Mixer Components

The hybrid was designed with the aid of the equivalent circuit shown in Figure 3. The coupling probes were modeled as L-C branches, interconnected by lengths of slab-loaded waveguide. The characteristic impedance of the waveguide and terminations was calculated from the power-voltage definition. Based on a computer-aided technique, unique L-C combinations were found which modeled the measured coupling/frequency response for probes of various lengths and widths[6].

After studying a variety of array configurations, a seven-probe equal-element coupler was chosen for the final design. The configuration is simple and compact, and it avoids the large range in coupling levels found in more complex distributions (such as Tchebycheff). The L-C values for the final design are tabulated in Figure 3. The interprobe spacings were initially set a quarter-wavelength, and then optimized by a gradient-search technique.

Figure 4 shows the measured and calculated performance of the seven-probe coupler. Across a 3.7-GHz band, centered at 94 GHz, the measured isolation is 22 dB or better, and the coupling to either output port is 3.4 ±0.9 dB. In comparing the measurements with calculations, it should be noted that the measured isolation includes the effects of waveguide bends, flange discontinuities, and detector mismatch, whereas the model assumes perfect terminations.

At midband, the coupling is only 0.4 dB below the ideal 3-dB value. Since the housing alone has an insertion loss of 0.3 dB, the hybrid contributes only 0.1 dB to the total loss. Based on an axial length of 1.5 wavelengths, the unloaded Q in the printed probe region is greater than 400.

Diode Mounts

Other important parts of the assembly depicted in Figure 2 are the GaAs beam-lead mixer diodes and the fin-line circuit in which they are mounted. The diodes chosen for this program were developed by Calviello, et al[7], and have the following characteristics:

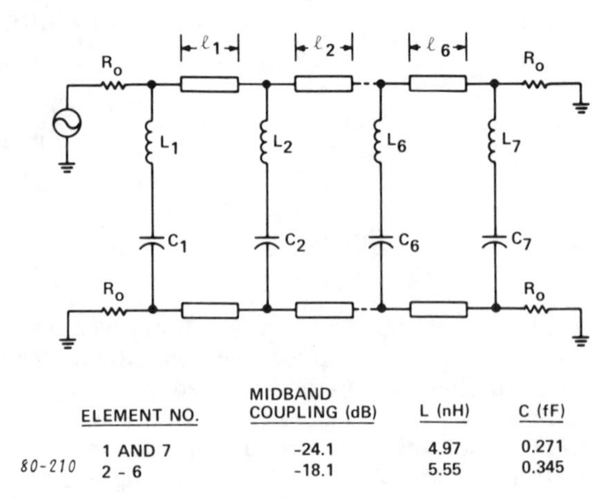

ELEMENT NO.	MIDBAND COUPLING (dB)	L (nH)	C (fF)
1 AND 7	-24.1	4.97	0.271
2 - 6	-18.1	5.55	0.345

80-210

Figure 3. Hybrid Equivalent Circuit

Zero-bias junction capacitance:	15 fF
Package capacitance:	20 fF
Series resistance:	2.5 ohms
Ideality factor:	1.07

The initial dimensions for the fin-line mount were scaled from a related design[4]. The circuit was then optimized as a single-ended mount (i.e., without the E-plane hybrid). By observing the reflected power with an external waveguide coupler, the circuit was adjusted to be matched, across the band of interest, under typical bias conditions.

Figure 5 shows the return loss versus frequency, for the single-ended fin-line mount. Across a 4-GHz band centered at 94 GHz, the return loss is 16.5 ±4.5 dB.

After obtaining a satisfactory RF/LO match, the conversion loss of the single-ended mount was measured. The RF and LO were coupled to the mount through the main and decoupled arms, respectively, of an external waveguide coupler. The RF and LO power levels were measured with a bolometer, calibrated against a wet calorimeter (TRG V981). The IF output was measured with a coax thermistor (HP478A).

Figure 6 shows the measured conversion loss as a function of the LO power at the input to the single-ended mount. The measurements were performed with an LO of 94 GHz and an RF of 93 GHz; the plotted performance is typical of that recorded in the lower part of the passband depicted in Figure 5. At each LO power level, the bias was adjusted to minimize the conversion loss. The optimum bias levels are also plotted in Figure 6. For an LO drive of 5.7 dBm (3.7 mW), the conversion loss is 7.2 dB. Although the final balanced-mixer design will require 3.4 dB more LO drive, this is well within the capability of existing, fundamental Gunn oscillators.

Conclusion

The major components for a new form of printed-circuit 94-GHz balanced mixer have been completed. Included are a low-loss printed-probe hybrid, low-parasitic GaAs beam-lead diodes, and fin-line matching circuits. This E-plane approach provides advantages in terms of production economy, effective shielding, high (> 400) unloaded Q, low equivalent dielectric constant, and simple waveguide interfaces. An integrated balanced mixer is now being optimized, and performance data will be available shortly.

Acknowledgements

The work reported was sponsored by AIL, under the direction of M. Lebenbaum, K. Packard, and J. Whelehan. The diodes were developed by J. Calviello and co-workers in our Central Research Laboratory, under the direction of J. Taub. Technical assistance was provided by A. Cooley, A. Kunze, J. Pieper, A. Rees, C. Thompson, and K. Walsh.

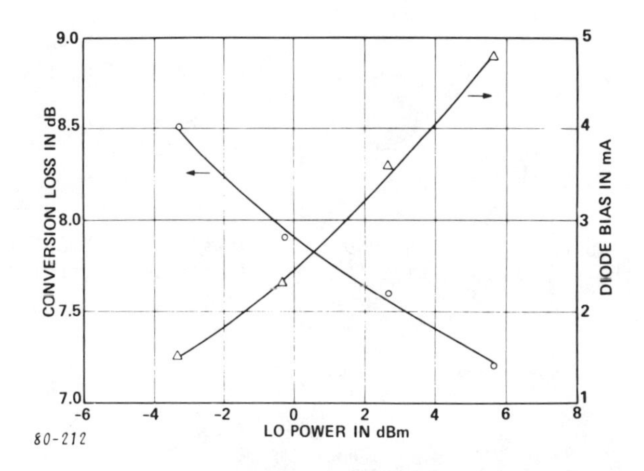

Figure 6. Conversion Loss of Single-Ended Mount

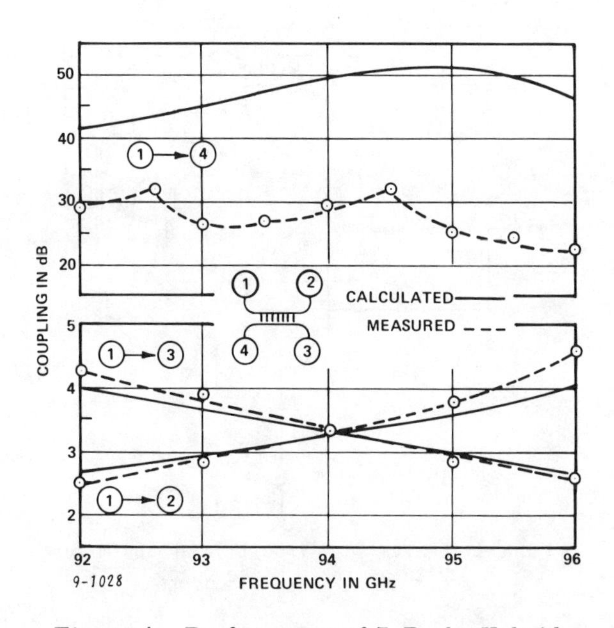

Figure 4. Performance of 7-Probe Hybrid

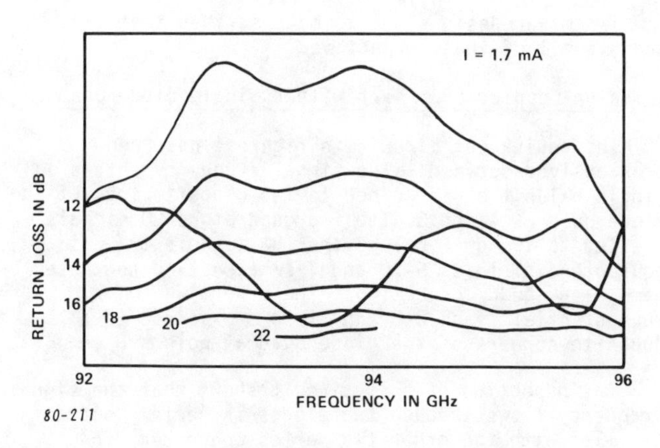

Figure 5. Return Loss of Diode Mount

References

1. T. H. Oxley, et al, "Hybrid Microwave Integrated Circuits for Millimeter Wavelengths," Digest of 1972 MTT Symposium, p 224-226, May 1962.

2. T. Araki and M. Hirayama, "A 20-GHz Integrated Balanced Mixer," IEEE Trans Vol MTT-19, p 638-643, July 1971.

3. A. K. Gorwara, et al, "K-Band Image-Reject and Ka-Band Balanced Mixers Constructed Using Planar Millimeter-Wave Techniques," Final Report on Contract N00123-74-C-1957, March 1975.

4. P. J. Meier, "Printed-Circuit Balanced Mixer for the 4- and 5-mm Bands," Digest of 1979 MTT Symposium, p 84-86, April 1979.

5. P. J. Meier, "Millimeter Integrated Circuits Suspended in the E-Plane of Rectangular Waveguide," IEEE Trans, Vol MTT-26, p 726-773, October 1978.

6. P. J. Meier, "Printed-Probe Hybrid Coupler for the 3-mm Band," Proceedings of Ninth European Microwave Conference, p 443-447, September 1979.

7. J. A. Calviello, J. L. Wallace and P. R. Bie, "High-Performance GaAs Beam-Lead Mixer Diodes for Millimeter and Submillimeter Applications," Electronics Letters, Vol 15, No. 17, p 509-510, August 1979.

SINGLE-SIDEBAND MIXERS FOR COMMUNICATIONS SYSTEMS

Ben R. Hallford
Rockwell International
Collins Transmission Systems Division
P.O. Box 10462
Dallas, Texas 75207

ABSTRACT

Two types of balun-coupled single-sideband (SSB) mixers suitable for communications systems will be presented in their equivalent circuit and RF planar configuration. A 6-GHz SSB mixer model will be fully described.

Summary

Introduction

Microwave components for radio relay systems are not required to perform over wide extremes of temperature and frequency, but they must meet peculiar requirements in their electrical specifications, reliability, and manufacturability. The mixer is still a key component in these systems. Mixer specifications have not been made any easier, but the design challenges have been shifted to other areas. The SSB mixer for today's communications systems should have the following characteristics:

- Return loss to signal frequency should exceed 20 dB to avoid a costly load isolator after a low-noise amplifier (LNA).
- Performance should not be changed for a local oscillator (LO) power variation of 4 dB.
- LO power for optimum performance should not exceed +7 dBm.
- Return loss at LO input connector should exceed 20 dB to avoid a load isolator.
- Image frequency response rejection should exceed 20 dB to reduce image frequency response filtering.
- Maximum of four diodes provides higher reliability. A single monolithic diode quad should be used, which would contain and shorten the image frequency return path and also lower cost.
- IF amplitude and delay response should be essentially flat over a 30-MHz bandwidth centered at 70 MHz.
- Dc block in the mixer eliminates output capacitor in LNA.
- RF circuits may be designed on a low dielectric constant plastic substrate to reduce cost.
- 50-ohm IF output impedance allows a direct connection to a 70-MHz 90° hybrid.
- RF to IF conversion loss plus 70-MHz hybrid loss should be 4 ±1 dB.
- Mixer should be compact in size, yet be field repairable.
- Total cost of mixer plus 70-MHz IF hybrid should be near $50.00.

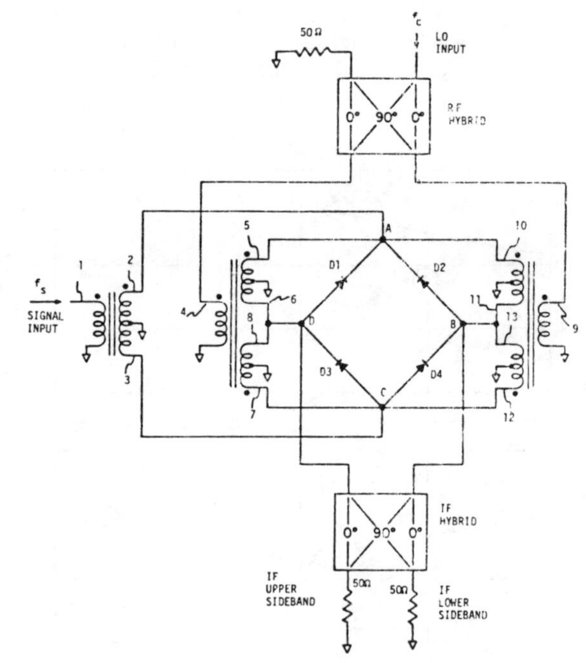

Figure 1. Equivalent Circuit of an SSB Mixer With a Single Diode Quad.

Two mixer designs are to be described that can achieve all of these objectives.

Balun-Coupled SSB Mixer With a Single Diode Quad

The equivalent circuit in figure 1 has been previously described.[1] The signal frequency enters a single balun that is joined to two opposite legs of the diode quad at A and C. The two quadrature LO signals are fed to two dual baluns that have their two equipotential leads 5-10 and 7-12 also tied to these same diode quad connections. The other two equipotential leads 6-8 and 11-13 are joined to opposite corners of the diode quad at points D and B.

An inspection of this circuit shows that the signal frequency flows through two diodes in series, but in parallel with the other two series-connected diodes. The LO carriers flow through only one diode that is connected to each of the four dual-balun outputs. Ideally perfect isolation exists between both LO inputs and the signal input.

Reprinted from *IEEE MTT-S Int. Microwave Symp. Dig.*, 1982, pp. 30–32.

The image frequency current flows in the diode quad only and cannot couple to any of the three RF baluns. The image frequency path length is defined by the line lengths that join the diodes together.

The two IF outputs are removed by filtering from the LO balun lines at points D and B. A 90° IF hybrid, then, separates the upper and lower sidebands.

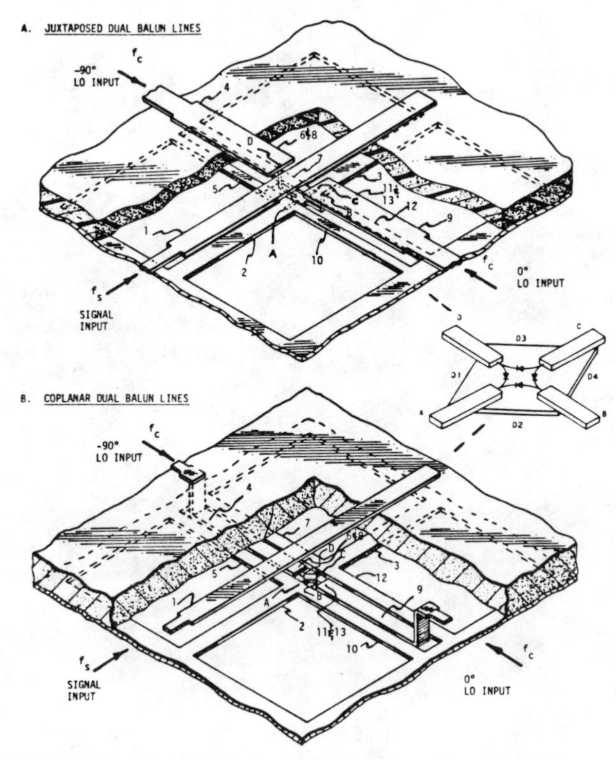

Figure 2. Planar Pictorial Diagram of an SSB Mixer With a Single Diode Quad.

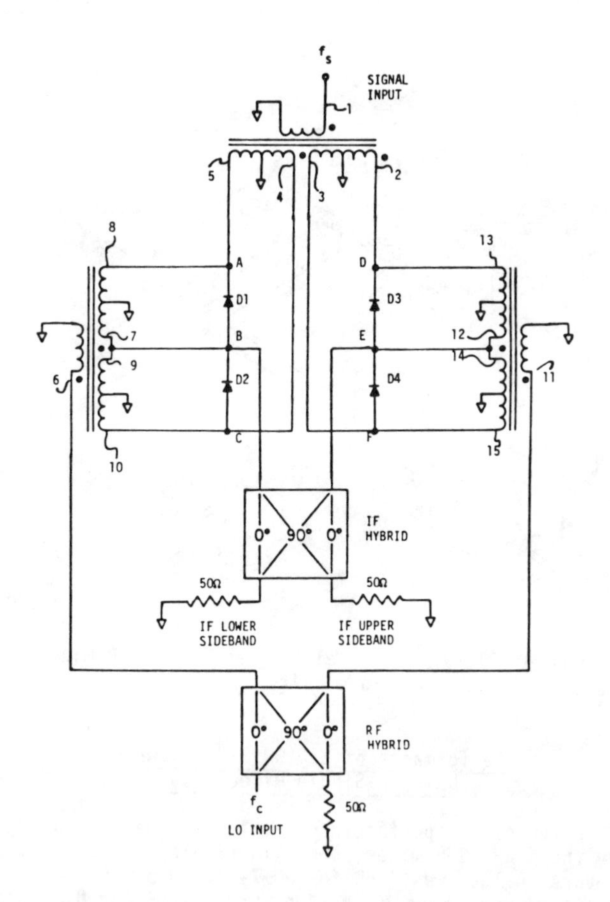

Figure 3. Equivalent Circuit of an SSB Mixer With Two Diode Parts.

The RF circuits shown in figures 2A and 2B are a direct representation of this equivalent circuit when the interconnecting transmission lines between the three balun circuits are allowed to shrink to a vanishingly small limit. A direct comparison between the equivalent circuit and the RF pictorial diagram is possible since equivalent lines bear the same numbers. The circuit in figure 2A requires two diode leads to be fed through the substrate, where they are connected at points D and B to the lines that are equivalent to the joined dual-balun output leads 6-8 and 11-13. The other alternative shown in figure 2B routes the leads 6-8 and 11-13 to the opposite side of the substrate, where it is nested between the other two balun output leads. The advantage of this latter circuit is that the two diode leads at D and B are now shortened and may be easily joined to the coplanar conductors.

The RF circuits in figure 2 use a bridge quad diode connection even though a ring quad theoretically may be used. This bridge quad diode reduces the loss caused by one LO carrier, which has been observed to feed through the diode quad to the other mixer section, where it combines with the signal frequency to generate an IF component that is 180° out of phase with the desired IF signal. This action is surprising because the isolation between points D and B in figure 2A has been measured and found to be in excess of 26 dB.

Balun-Coupled SSB Mixer With Two Diode Pairs

The loss of IF output signal that was caused by a feedthrough of the LO from the other mixer section can be greatly reduced by isolating the two mixer sections where the diodes are joined together. The circuit to accomplish this separation of the two mixer sections is shown in figure 3. The signal is fed into a dual-balun circuit that allows each balun half to feed the signal to each mixer section separately. The signal frequency flows through the two diodes in series in each of the dual-balun outputs. The quadrature local oscillator carriers enter dual baluns similar to figure 1. Again the dual balun secondary is connected to one diode only that is fed by a single local oscillator carrier.

The image frequency in this circuit is isolated from each LO balun input because of its 180° relationship and therefore does not couple into either LO line. Also, the image frequency currents flow 180° out of phase in the signal input dual balun, and therefore it does not couple into the signal input port. The two mixer sections are now completely separate, such that the image frequency current from one section does not flow to the diodes in the other section.

The pictorial diagram of an RF circuit using this equivalent circuit is shown in figure 4. This diagram is similar to the figure 2 pictorial diagrams with the exception of the dual balun for the signal input section and the separate diode pairs.

This equivalent circuit could also be constructed using the approach shown in figure 2B, where the primary balun lead was routed through the substrate and nested between the balun output secondary lines.

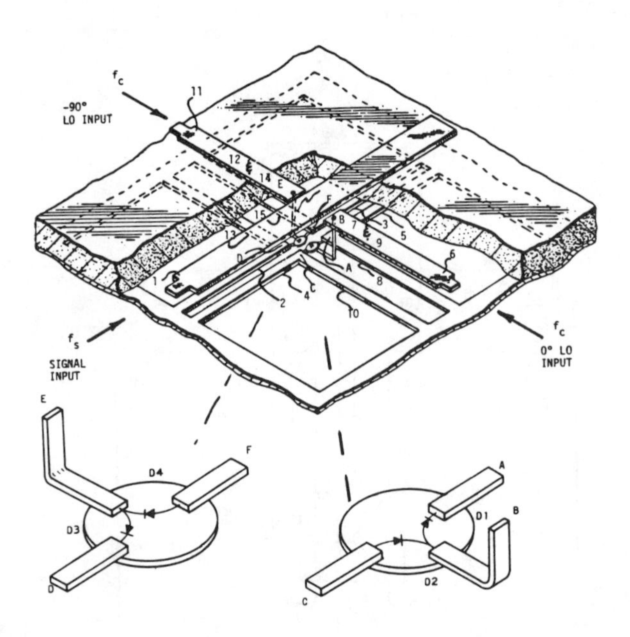

Figure 4. Planar Pictorial Diagram of an SSB Mixer With Two Diode Pairs.

Performance of a 6-GHz SSB Mixer Using a Single Diode Quad

A few of the performance levels will be mentioned for the 6-GHz SSB mixer, shown in figure 5. The conversion loss was 4 dB for a +7-dBm LO power. The amplitude and delay distortion were 0.1 dB and 0.25 ns, respectively, over a 30-MHz bandwidth centered at 70 MHz. The return loss at the signal and LO input ports exceeded 23 dB. Changes in performance were imperceptible for an LO power range of 5 to 9 dBm.

These mixers are exceedingly well balanced and suppress second harmonics to such an extent that the image frequency power level is low and image recovery is less than 1 dB. The image frequency response rejection typically ranges from 25 to 35 dB.

The four square copper areas that are adjacent to the balun primaries in the front view of figure 5 were used as ground planes in the early stage of development, when a loss of IF output was observed with a ring diode quad. The quarter-wave open studs that are at a 45° angle in the back view of figure 5 used these ground planes in an attempt to form a low impedance at terminals A and C in figure 2. After the problem was found to be an LO cross feedthrough in the ring quad, and a bridge quad was used to eliminate the loss, these microstrip stubs were removed to determine their effects. No noticeable change was observed, so the copper conductors were allowed to remain to stiffen the board.

[1]Ben R. Hallford, "Simple Balun Coupled Mixers," 1981 IEEE/MTT-S International Microwave Symposium Digest, June 15-19, pp 304-306.

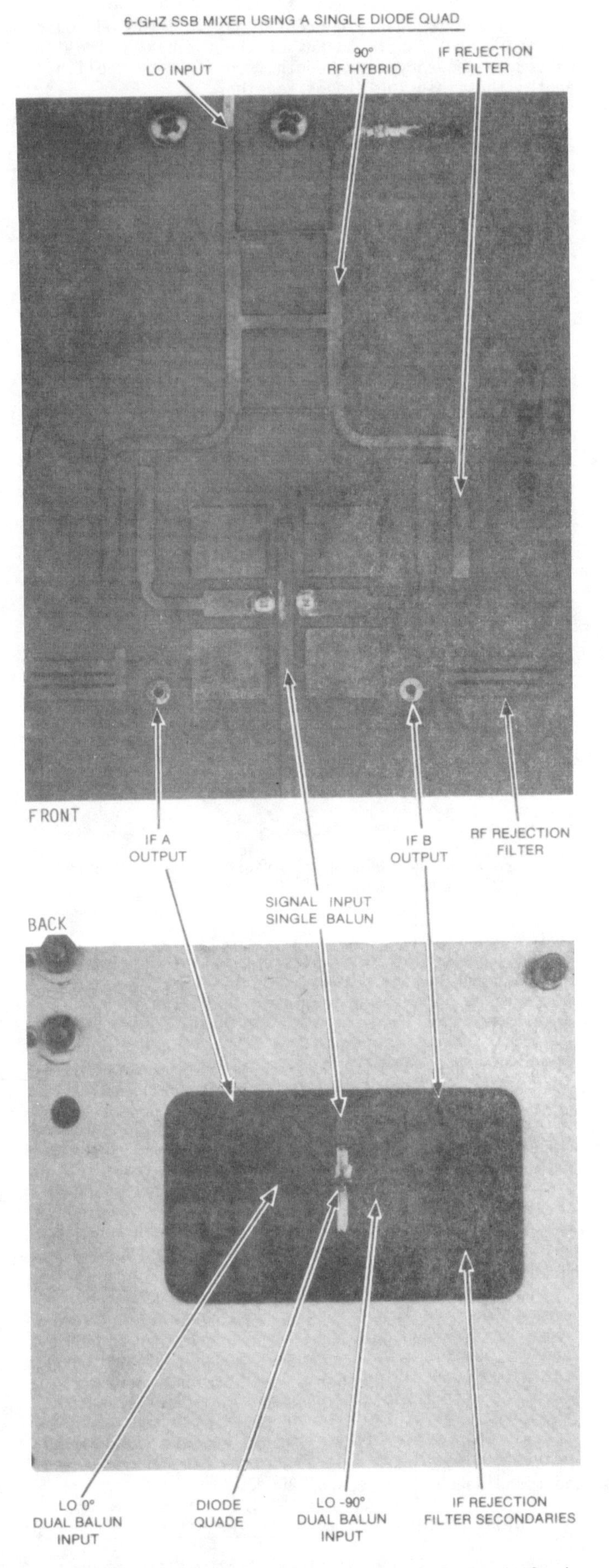

Figure 5. 6-GHz SSB Mixer Using a Single Diode Quad.

AN ANALYTIC DESIGN APPROACH FOR 2-18 GHz PLANAR MIXER CIRCUITS

Roy B. Culbertson and A.M. Pavio
Texas Instruments Incorporated
P. O. Box 226015
Dallas, Texas 75266

ABSTRACT

A design procedure for planar broadband mixers is proposed which employs filter synthesis for exact balun design and nodal analysis to check for deleterious port interactions. The method has been successfully used to obtain working mixers in one thin film fabrication cycle on alumina and fused quartz substrates. A representative design is presented with experimental data.

INTRODUCTION

Planar thin film mixers offer advantages in repeatability, ease of assembly, and amplitude and phase tracking performance. Good isolation characteristics and simultaneous impedance matching of the RF, LO and IF ports hinge on effective balun design. Hallford[1, 2] has reported various planar mixers employing resonant baluns which have relied on empirical design. This paper describes a design procedure which minimizes costly thin film iterations by applying two restrictions to resonant baluns, using filter synthesis techniques and checking port interactions with a nodal analysis of the mixer circuit.

MIXER CONFIGURATION

The basic form of the double balanced mixer to be described is illustrated in Figure 1. The diode ring is constructed with a monolithic beam lead pair on the top side of the suspended substrate connected to a diode pair on the bottom side via two plated through holes. Beam lead pairs offer good balance, repeatable mounting and low parasitics for broadband operation. This configuration promotes good isolation, since the LO energy fed by two edge coupled strips is orthogonal to the RF signal present on the two broadside coupled strips. The RF balun is a nonresonant tapered type, typically a quarter wavelength long at the lowest RF frequency. The bottom conductor follows a cosine taper while the top conductor varies in width to provide a DolfChebyshev impedance taper along the length of the balun.

RESONANT BALUN DESIGN

Matsumoto, has analyzed the coaxial Marchand balun as a three port, four conductor network as shown in Figure 2.[3] Restricting the immitance matrix representation to balanced currents and voltages at nodes 2 and 3 results in the following conditions: (1) the capacitance to ground of conductor 2 and conductor 4 must be zero and (2) the capacitance to ground of conductors 1 and 3 must be equal. Meeting these two requirements will provide a balanced output regardless of frequency. The VSWR performance of the network is restricted only by its filter characteristics. The balun design is reduced to a distributed high pass filter synthesis problem. Cloete, for example, has published element impedance values for a four element coaxial balun based on exact filter synthesis.[4] With the aid of a filter synthesis program a variety of topologies may be investigated to provide realizable element values, parasitic absorption, bandwidth, and impedance tranformation.

Figure 1 illustrates the realization of a typical balun. Z_1 and Z_2 are microstrip. Z_3 and Z_4 are inverted microstrip and must be equal in impedance. Their dimensions are calculated using the routines of Smith, making sure the strips are wide enough to shield the microstrip from the lower lid of the mixer housing.[5]

IF PORT DESIGN

The resonant balun provides a convenient low inductance IF ground. When coils are not desirable, the IF can be summed through a shunt shorted stub. To provide this stub, an RF bandpass filter is synthesized. Since the stub is a quarter wavelength at the RF band center frequency, it is quite short for even relatively high IF frequencies. The balanced filter elements are shown in Figure 3 along with even and odd mode impedances present when the housing lids are considered. These coupled mode values were determined from the equations of Bahl and Bahrtia[6].

The remaining design task is to determine the shunt effects of the RF filter and tapered balun on the IF port. To do this, the tapered balun is modeled as multiple sections of uniform asymmetric line. The two even mode and two odd mode impedances for each section are calculated ignoring the dielectric. The lowest of these impedances are used in the coupled line model shown in Figure 3 and analyzed on a microwave circuit analysis program. The aim of the calculation is not a highly accurate prediction of the shunting impedance, but a guide to avoid severe matching problems. In practice the IF impedance is measured and a simple match added, typically a series capacitor.

RESULTS

Several dual mixers have been built on both alumina and quartz substrates. The circuit of Figure 3 was used for 6 to 16 GHz operation with a 2 GHz IF. The alumina substrate dimensions were 1.1" x .85" x .010". The quartz mixer shown in Figure 4 uses coils to sum a 100 MHz IF signal. The conversion loss and amplitude matching appears in Figure 5. LO to RF port isolations are typically better than 25 dB and phase tracking within ±3 degrees over the 2 to 18 GHz band.

Reprinted from *IEEE MTT-S Int. Microwave Symp. Dig.*, 1982, pp. 425–427.

CONCLUSION

A simple mixer design technique has been outlined which is applicable to a variety of planar mixers which use a resonant balun. The method has proved useful in the design of thin film mixers where empirical approaches can be costly. Resonant planar baluns can be designed with virtually any high pass filter topology which includes two equal inverted microstrip shunt grounded stubs shielding the microstrip elements. This enables the realization of broadband impedance matching baluns.

References

1. B. R. Hallford, "Simple Balun-Coupled Mixers", 1982 IEEE MTT-S 1982 International Microwave Symposium Digest, pp 304-306.

2. B. R. Hallford, "A Designer's Guide to Planar Mixer Baluns", Microwaves, December 1979, Vol. 18, No. 12, pp. 53-57.

3. Akio Matsumoto, Advances in Microwaves Supplement 2, Adademic Press, New York, 1970, pp. 295-306.

4. J. H. Cloete, "Exact Design of the Marchand Balun", 1979 European Microwave Conference, pp. 480-484.

5. J. I. Smith, "The Even and Odd Mode Capacitance Parameters for Coupled Lines in Suspended Substrate", T-MTT, Vol. 9, No. 5 May 1971, pp. 424-431.

6. I. J. Bahl and P. Bhartia, "Characteristics of Inhomogeneous Broadside-Coupled Striplines", T-MTT, Vol. 28, No. 6, June 1980, pp. 529-535.

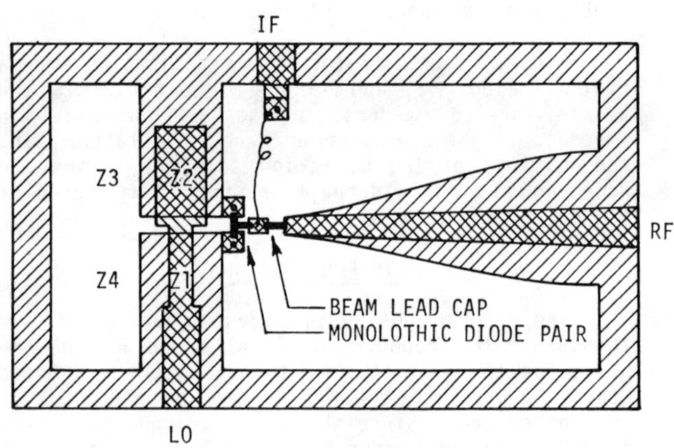
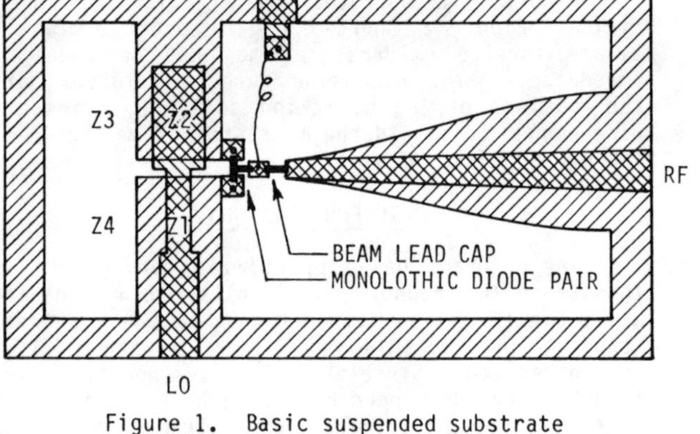

Figure 1. Basic suspended substrate double balanced mixer.

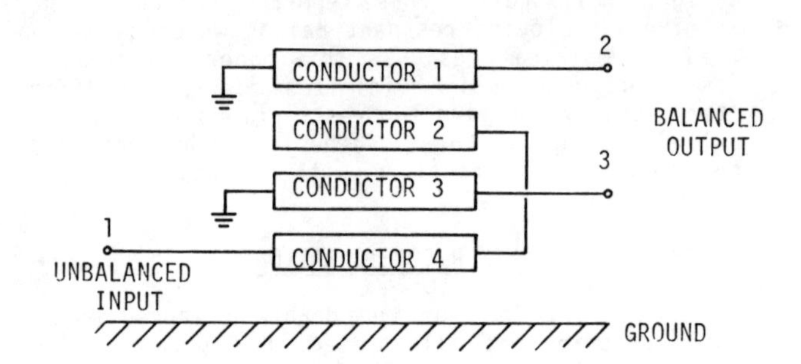

Figure 2. Generalized resonant balun (after Matsumoto).

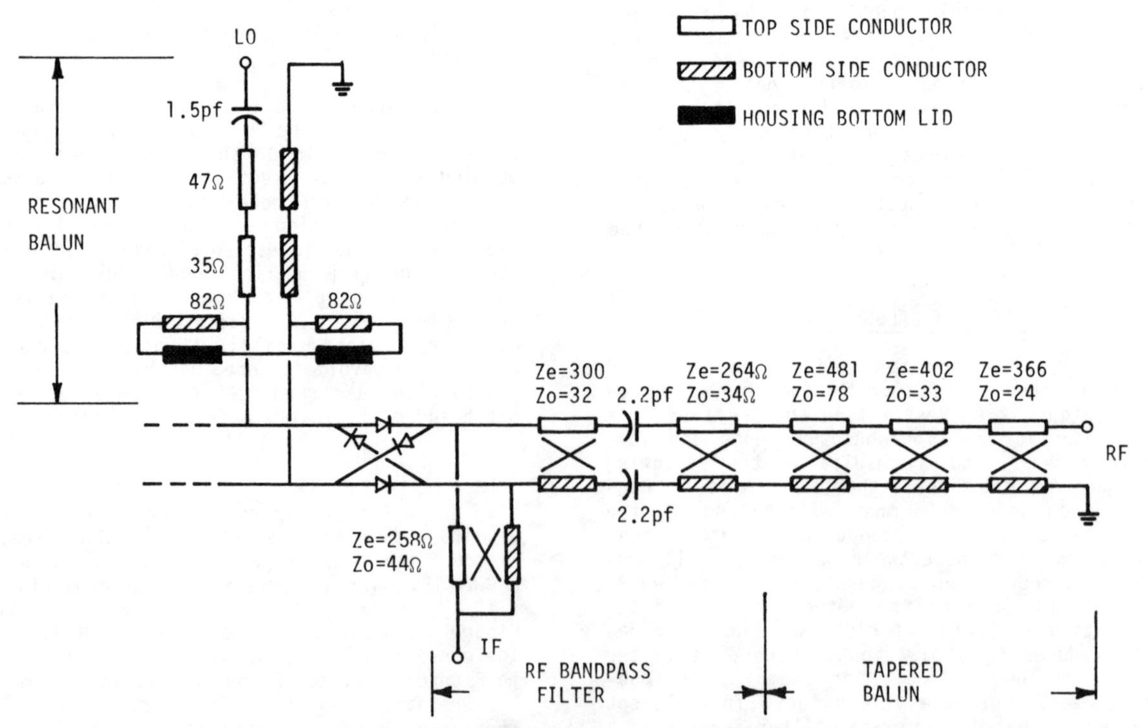

Figure 3. Nodal model for a dual alumina mixer.

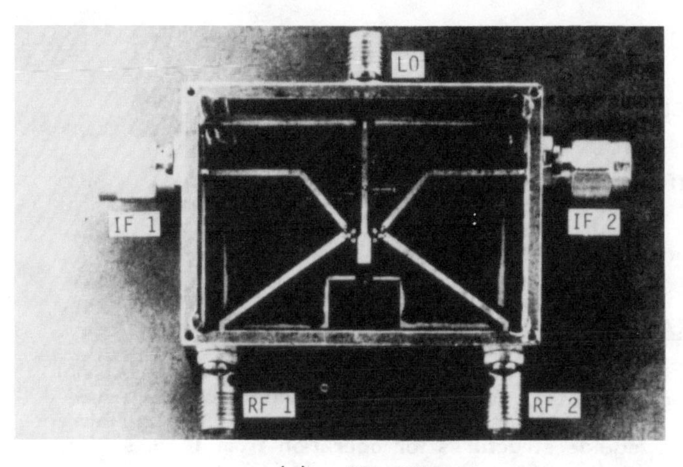

(a) TOP SIDE

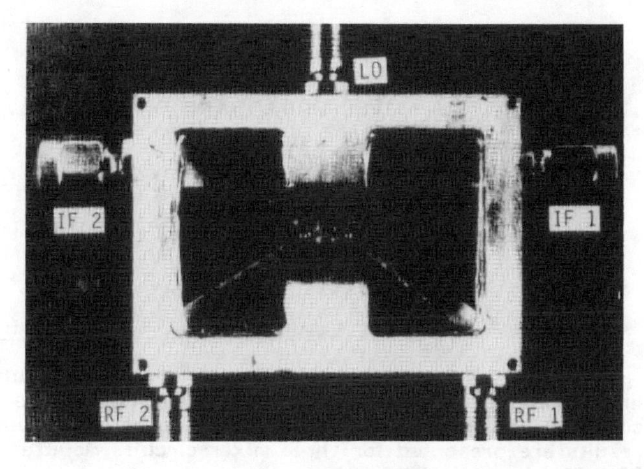

(b) BOTTOM SIDE

Figure 4. 2-18 GHz dual quartz mixer.

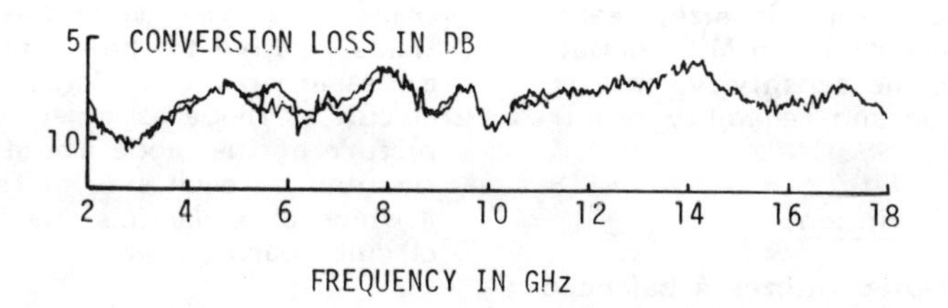

Figure 5. Amplitude tracking of dual quartz mixer.

A NOVEL BROADBAND DOUBLE BALANCED MIXER FOR THE 18-40 GHZ RANGE*

A. Blaisdell, R. Geoffroy, and H. Howe

M/A-COM Millimeter Products, Inc.
Burlington, Massachusetts 01803

ABSTRACT

Double balanced mixers have been constructed in the 18-40 GHz range, utilizing unique planar transmission line techniques** which permit easy integration of the mixers with other components to provide a compact, low-cost receiver assembly for band extension of EW systems. The design concepts are discussed and the performance results are presented for three mixer circuits mounted in waveguide structures for operation from 18-26.5 GHz, 26.5-40 GHz and 18-40 GHz.

Introduction

Much work has been published concerning broadband downconverters in the 18-40 GHz frequency range to extend EW capabilities to these high frequencies. The primary thrust behind the development of the mixers discussed in this presentation was to evolve a design which would be low-cost, small in size, readily integratable with other components in an MIC format and which would also provide the sensitivity, intermodulation product suppression and reliability required in a practical EW system.

Circuit Description

The design approach adopted utilizes a balanced slot line and a bifurcated, balanced microstrip line to couple the signal and L.O. to four beam lead Schottky diodes. The diodes are mounted in a star configuration as suggested by R. Mouw.[1] For ease in interfacing with test equipment, the development models were constructed on a substrate mounted in the E-plane of split block waveguide housings. Figure 1 illustrates the conceptual design. The RF signal is coupled to the diodes via a waveguide to bilateral fin-line which is then transformed to a balanced microstrip configuration which, in turn, is bifurcated to form a symmetrical four-wire slot line. The four beam lead mixer diodes are mounted centrally at the junction of the two four-wire lines. The IF is coupled orthogonally from the center of the junction area by means of a plated through hole which connects annular pads on either side of the substrate.

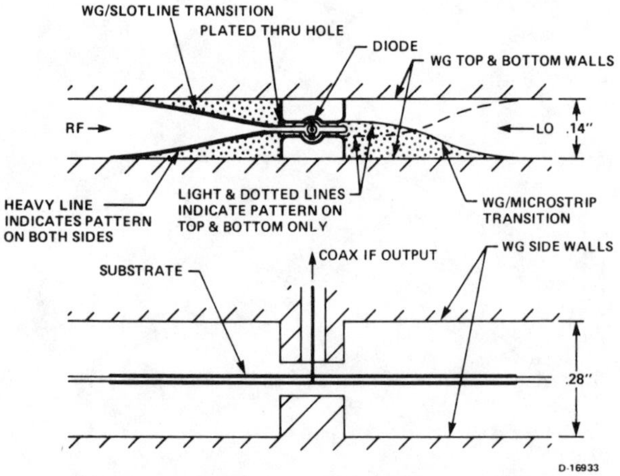

FIGURE 1. CONCEPTUAL DESIGN OF PLANAR BALANCED LINE DOUBLY BALANCED MIXER

Figure 2 illustrates the electric fields associated with the two four-wire lines. The top view depicts the electric field configuration of the four-wire slot line and the bottom view that of the bifurcated microstrip line. These modes are referred to as the parallel and transverse even modes based on the orientation of the fields relative to the original waveguide field. The odd modes are not intentionally excited but they can be by circuit or diode asymmetries. Figure 3 gives a clearer picture of the diode mounting detail. Two diodes are mounted on each side of the board as shown in Figure 4 which also clarifies the IF output connection to the circuit board.

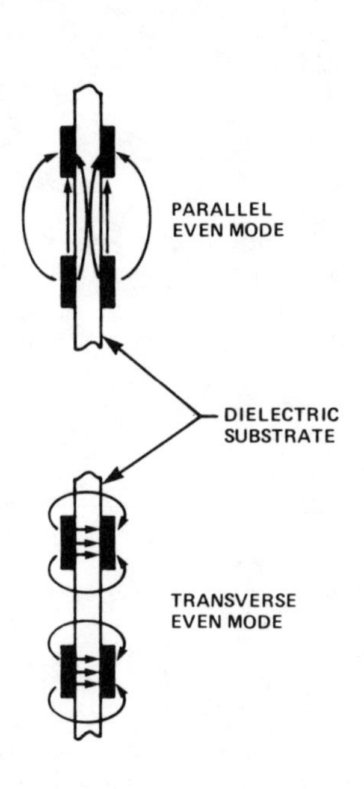

FIGURE 2. EVEN MODE PATTERNS ON BALANCED 4-WIRE LINE

Reprinted from *IEEE MTT-S Int. Microwave Symp. Dig.*, 1982, pp. 33–35.

Referring again to Figure 1, note that the parallel mode launches in the slot is terminated in a short circuit formed by the bifurcation of the balanced microstrip line. Locating the bifurcation approximately one quarter wavelength from the plane of the diodes results in an open circuit at the diodes for the parallel mode. The placement of a pair of plated-through holes through the slot line conductors, as shown in Figure 1, terminates the transverse mode in a short circuit. Locating these plated-through holes a quarter wavelength from the diodes transforms the short circuit to an open circuit at the plane of the diodes for the transverse mode. These quarter wavelength stubs are, in fact, the only structures which limit the L.O. and RF operating bandwidths other than the diodes themselves, so that this design approach should perform well over an instantaneous bandwidth of at least an octave for both the L.O. and the RF. Note that, although the slot line port and the microstrip port in Figure 1 have been arbitrarily labeled as the "RF" and "L.O." ports respectively, they are interchangeable.

The principal factor limiting the IF bandwidth is the length of the ground return from the diodes to the outer conductor of the coaxial IF output line. This path is along the four-wire lines to the waveguide walls and along the waveguide walls to the outer conductor of the IF line. As indicated in Figure 1 this path length can be made quite short by narrowing the waveguide "A" dimension in region occupied by the four-wire lines.

RF to L.O. isolation is inherent in this mixer design, since there is no cross-coupling between the transverse and parallel modes are also isolated from the IF output unless circuit or diode asymmetry is introduced.

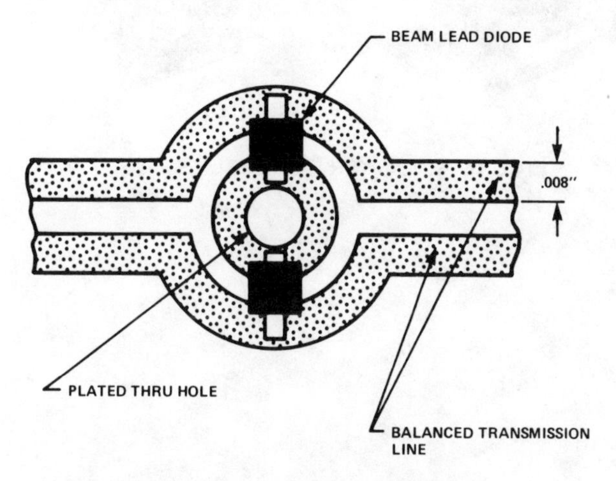

FIGURE 3. DIODE MOUNTING DETAIL

D-16938

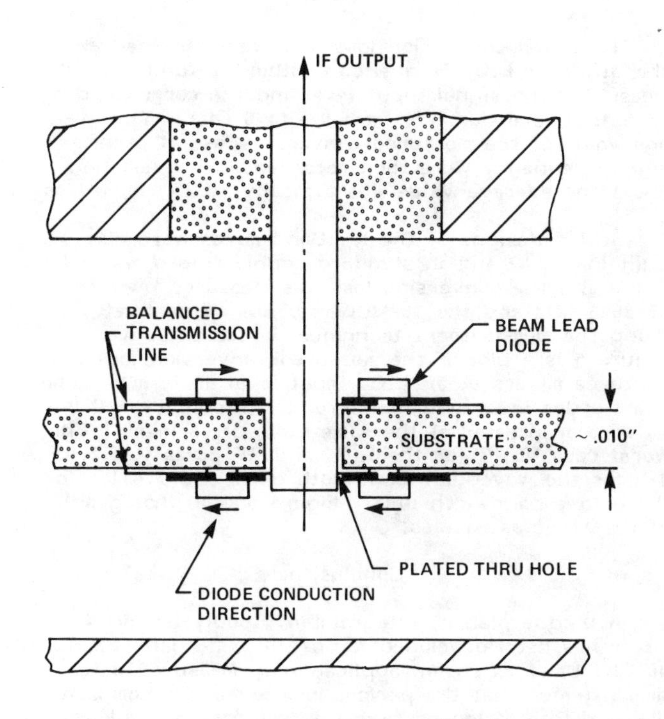

FIGURE 4. CROSS-SECTION THRU CENTRAL PLANE THRU DIODES

D-16939

Prototype Results

A mixer using these design concepts was constructed for operation from 26.5 to 40 GHz with transitions to WR-28 for the L.O. and RF signal ports and SMA connector at the IF output. RT Duroid type 5880 with copper clad on both sides and a dielectric thickness of 8 mils was used for the substrate material for low cost considerations. MA-40185 silicon Schottky beam lead diodes were selected for their intrinsic mechanical strength and reliability. A split-block aluminum housing was used to mount the substrate centrally in the waveguide. The transitions from waveguide to both the parallel and the transverse four-wire transmission line modes were designed for a typical return loss in excess of 20 dB.

Conversion loss was measured using fixed frequency L.O.'s at several frequencies throughout the 26.5 to 40 GHz band, without retuning. To preclude the possibility of measurement error due to L.O. or RF leakage a filter was used in the IF output port to provide a minimum of 40 dB attenuation in 26.5 to 40 GHz band. The match and loss of the filter was calibrated on a computer controlled network analyzer throughout the IF frequency range.

The actual conversion loss data were obtained at the different L.O. frequencies without retuning by measuring the signal input level and the corresponding IF output level for IF's from 0.5 to 8 GHz. The average value of the measured conversion loss at each signal frequency were then recorded. The variation about the average value was typically ±0.25 dB or less.

Scaled designs of the K_A Band mixer were also built in WR-42 and in standard double ridged waveguide and the conversion loss was measured over the 18-26.5 GHz and the 18-40 GHz bands respectively, using the measurement technique described above. Figure 5 is a plot of the measured conversion loss for all three mixers at an L.O. input level of 10 mW. The third order intermodulation product intercept point is 16 dBm or greater at the same L.O. drive level. The worst case VSWR for the L.O. and RF ports was 2.0:1.0 for the waveguide bandwidth units and 2.4:1.0 for the octave bandwidth unit. Figure 6 is a photograph of the mixer assemblies.

Conclusion

A unique planar, integratable, doubly-balanced mixer has been developed for use in wide band EW, 18-40 GHz band extension applications. Measured data demonstrates that the performance levels are comparable to doubly balanced mixers in common use at lower frequencies, so that the mixers should be useful for Radar and Communications Applications as well. The planar feature permits low cost integration of the mixer design into complex fin line and microstrip subassemblies without the need for waveguide transitions.

Acknowledgement

*This work was sponsored by AFAL, Wright-Patterson AFB, Ohio 45438, Contract No. F333615-80-C-1013. **Patented, US Patent No. 4,291,415. (Inventor, Dr. Charles D. Buntschuh).

Reference

1. R. Mouw, "A Broadband Hybrid Junction and Application to the Sideband Modulator," MTT-16, pp 911 ff., November 1968.

Typical Performance Curve

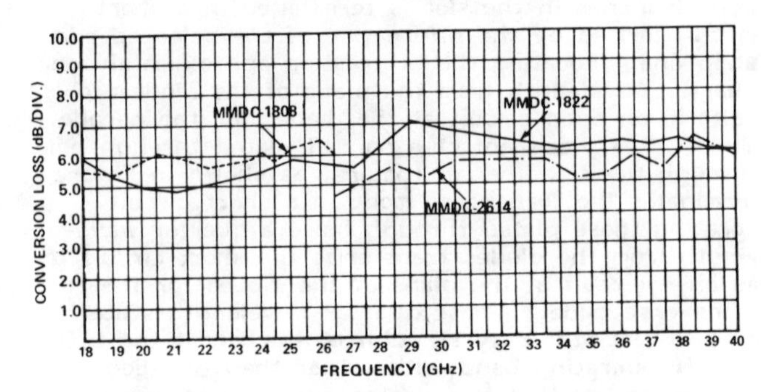

FIGURE 5. AVERAGE CONVERSION LOSS vs. FREQUENCY FOR VARIOUS IF'S AND LO FREQUENCIES

FIGURE 6. MIXER ASSEMBLIES

Design and Performance of *W*-Band Broad-Band Integrated Circuit Mixers

RAGHBIR S. TAHIM, GEORGE M. HAYASHIBARA, AND KAI CHANG, MEMBER, IEEE

***Abstract*—Broad-band integrated circuit mixers using a crossbar suspended stripline configuration and a finline configuration were developed with GaAs beamlead diodes. For the crossbar suspended stripline balanced mixer, less than 7.5-dB conversion loss for 15-GHz instantaneous IF bandwidth was achieved with the LO at 75 GHz and the RF swept from 76 to 91 GHz. With the LO at 90 GHz, a conversion loss of less than 7.8 dB was achieved over a 14-GHz instantaneous bandwidth as the RF is swept from 92 to 105 GHz. For the finline balanced mixer, a conversion loss of 8 to 12 dB over a 32-GHz instantaneous IF bandwidth was achieved as the RF is swept from 76 to 108 GHz. Integrated circuit building blocks, such as filters, broadside couplers, matching circuits, and various transitions, were also developed.**

I. Introduction

RAPIDLY EXPANDING activities in millimeter-wave hardware developments have created an urgent need for broad-band mixers for the receivers used in electronic warfare, surveillance, meterology, radiometer, and communication systems. During the past decade, significant improvements have been made in millimeter-wave mixers. Most of these, however, have been limited to narrow-band applications. Recently, several broad-band mixers have been built in waveguide circuits [1–3]. Among these is a crossbar configuration reported by Yuan [3]. This circuit was later realized at *W*-band using a stripline structure and pill type mixer diodes [4], and using a waveguide configuration and honeycomb mixer diodes [5].

Integrated circuit technologies provide the advantages of low cost, light weight, and small size. These also have the potential of direct translation into monolithic circuits and even large-scale integration.

The use of beamlead diodes in integrated circuit mixers avoids a mechanical diode contact in whisker-contacted mixers and therefore has better reliability and needs less assembly time. A 94-GHz mixer using beamlead diodes has recently been reported with a single sideband conversion loss of 6.2 dB [6] and 5 to 9 dB [7] over a 3-GHz RF bandwidth.

This paper describes the performance of a *W*-band wide-band integrated circuit mixer using beamlead diodes with a wide instantaneous IF bandwidth [8]. This mixer is especially important for millimeter-wave receivers which, in addition to accepting a wide range of RF frequencies, should provide wide instantaneous IF bandwidth with a fixed LO frequency.

Manuscript received June 25, 1982; revised October 14, 1982. This work was supported in part by the Naval Research Laboratory under Contract N000-14-C-1650.

The authors are with TRW Electronics and Defense, One Space Park, Redondo Beach, CA 90278.

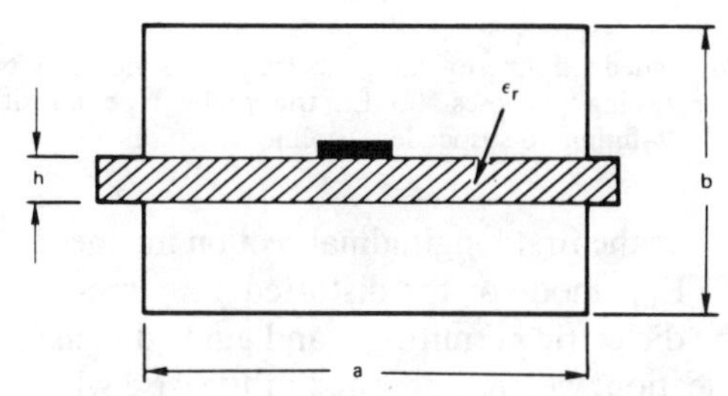

Fig. 1. Suspended stripline configuration.

A crossbar stripline mixer was developed to achieve less than 7.5-dB conversion loss for a 15-GHz instantaneous IF bandwidth with the LO at 75 GHz. A conversion loss of less than 7.8 dB with 14-GHz IF instantaneous bandwidth was achieved with the RF swept from 92 to 105 GHz. For narrow-band operation at 94 GHz, less than 5.5-dB conversion loss was achieved for a 500-MHz bandwidth and less than 6 dB for a 2-GHz bandwidth. A finline balanced mixer was also developed. It operates over a 32-GHz IF instantaneous bandwidth with a conversion loss of 8 to 12 dB, with the LO at 74 GHz as the RF is swept from 76 to 108 GHz. With the LO at 82 GHz, a conversion loss of less than 10 dB has also been achieved with the RF swept from 77 to 107 GHz. These results represent state-of-the-art performance in integrated circuit mixers.

To facilitate the mixer development, filters, broadside couplers, matching circuits, and various transitions were developed in integrated circuit form with low insertion loss. The performance of these components will also be discussed.

II. Suspended Stripline-to-Waveguide Transition

A suspended stripline-to-waveguide transition is essential for individual component testing before final integration. The transition should have low loss and a bandwidth sufficient for the application. Two types of suspended stripline-to-waveguide transitions used in mixers were investigated: 1) electric probe transition for crossbar suspended stripline mixer, and 2) waveguide-to-finline-to-stripline transition for finline mixer.

Theoretical studies have considered the case of an *H*-plane slab of dielectric, centrally located within the waveguide as shown in Fig. 1. The dominant waveguide mode

Reprinted from *IEEE Trans. Microwave Theory Tech.*, vol. MTT-31, pp. 277–283, Mar. 1983.

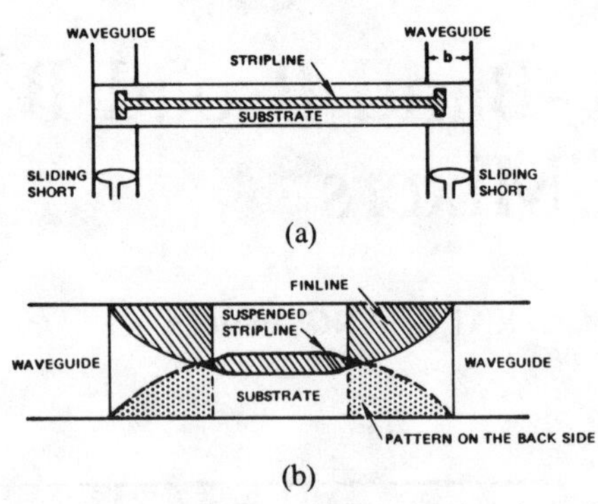

Fig. 2. Suspended stripline-to-waveguide transitions including two transitions for testing purposes. (a) Electric probe type transition. (b) Waveguide-to-finline-to-suspended stripline transition.

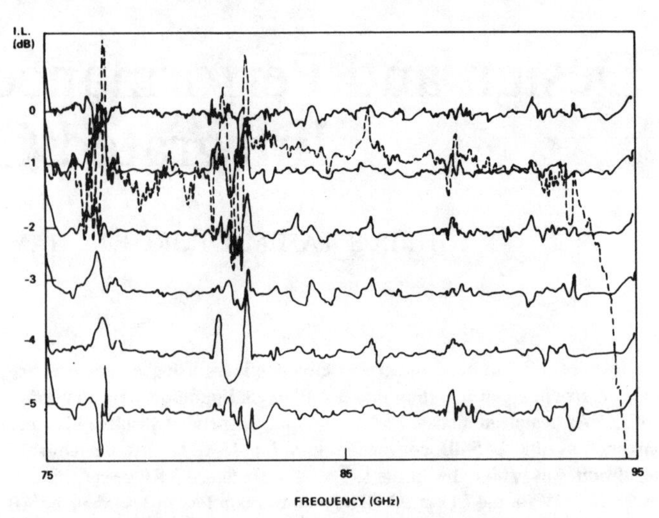

Fig. 3. Performance of W-band electric probe type transition (loss including two transitions and a 1-in line). The spikes shown are due to the sweeper.

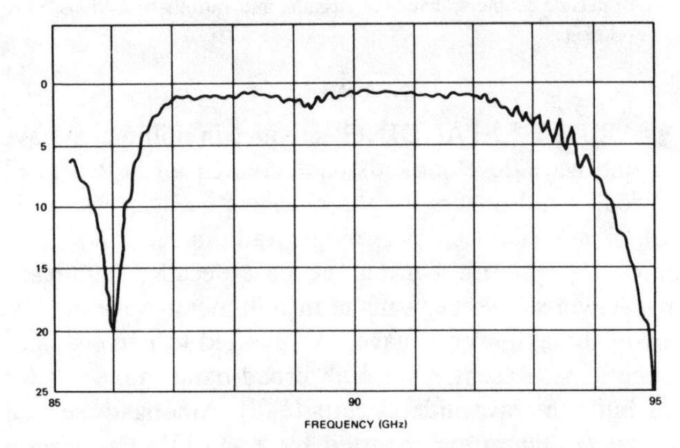

Fig. 4. Performance of waveguide-to-finline-to-stripline transition (including two transitions and a 0.5 in stripline).

can be either the first longitudinal section magnetic (LSM_{11} or quasi TE_{10}) mode or the distorted TE_{01} mode, depending on the dielectric permittivity and guide dimensions. For this application, we are interested in the case where $h/b \leqslant 0.25$ and $\epsilon_r \leqslant 10$. The cutoff frequency [9] of the dominant waveguide mode LSM_{11} can be approximated by

$$f_c = \frac{c}{2a}\sqrt{1 - \frac{h(\epsilon_r - 1)}{b\epsilon_r}} \tag{1}$$

where

- a width of channel;
- b height of channel;
- h thickness of the substrate;
- ϵ_r relative dielectric constant;
- c velocity of light in a vacuum.

The a dimension of suspended stripline is chosen so that the first higher order mode is much higher than the frequency band of interest. In the case of a channel cross section of 0.050×0.025 in^2 and a substrate thickness of 0.005 in, the cutoff frequency for duroid is 112 GHz, which guarantees a simple quasi-TEM mode of operation for W-band.

The electric probe type transition consists of an electric probe inserted into the waveguide formed by an extension of the suspended stripline or microstrip line beyond the ground plane (Fig. 2). The idea is very similar to the conventional waveguide-to-coaxial line transition. The transition was originally developed for narrow-band operation at Ka-band [10]. The advantage of this transition is that the probe can be fabricated as an integral part of the stripline and the difficulty of making reliable electrical contacts is avoided. By optimizing the probe shape, broad bandwidth has been accomplished with low insertion loss. As shown in Fig. 3, excellent results have been achieved at W-band using a rectangular probe shape. The total insertion loss for two transitions and a 1-in line is typically 1 dB. The insertion loss of each transition is about 0.25 dB over an 18-GHz range. The upper operating frequency can be easily adjusted by shaping the probe geometry. The assembly of this type of transition is relatively simple, and a sliding short can be used to optimize the performance at a specified frequency.

The second type transition [11] is designed to couple the power from waveguide-to-finline and then to stripline. As shown in Fig. 2, the finline is tapered down to a small gap. The performance of this transition is shown in Fig. 4. The total insertion loss for two transitions and a half-inch stripline is about 1 dB over a 5-GHz bandwidth. The insertion loss per transition is therefore about 0.25 dB. The narrow bandwidth is believed due to the discontinuities introduced at the fin and stripline interface.

Because of the low loss, wide bandwidth, and easy assembly, the electric probe type transitions were used for most of our component testing.

III. Filter Design

In the mixer, a low-pass filter is required to pass the IF frequency and reject the LO and RF signals. The RF and LO ports are isolated in the crossbar mixer because the input waveguides are orthogonal to each other. In the finline mixer, however, a bandpass filter is needed to increase the LO/RF isolation. This section discusses the development of these filters.

A Chebyshev IF filter was selected. The element values for a Chebyshev filter with a 0.2-dB ripple are computed

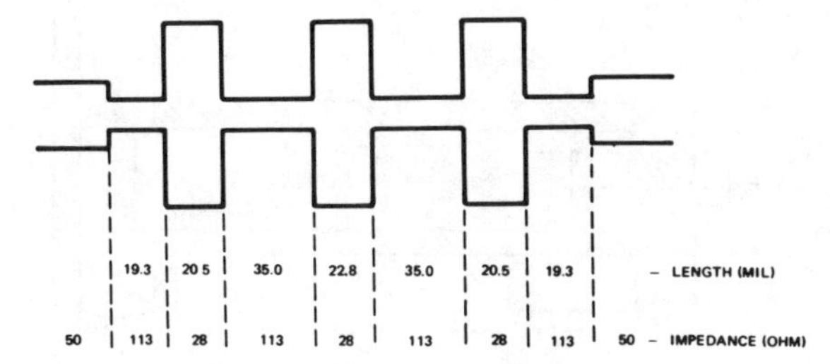

Fig. 5. Low-pass filter circuit layout.

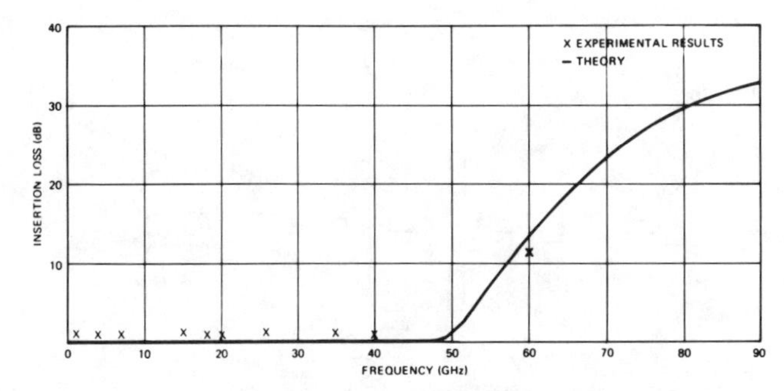

Fig. 6. Predicted and measured performance of a low-pass filter.

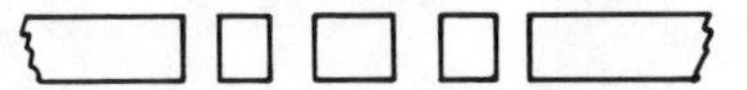

Fig. 7. Direct-coupled resonator filter.

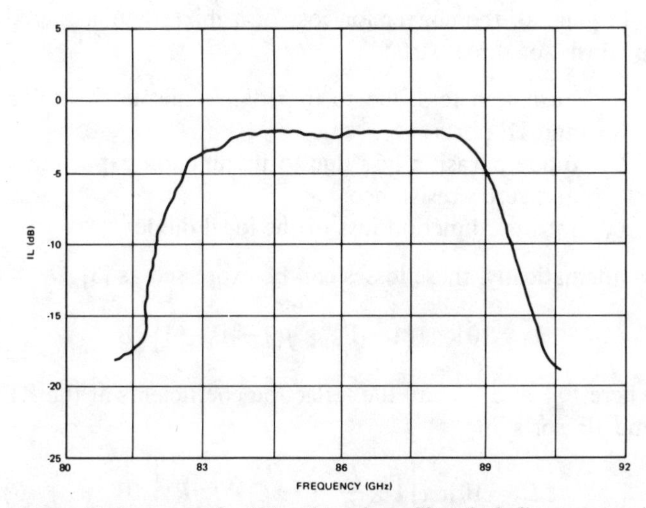

Fig. 8. Performance of a *W*-band bandpass filter (including two transitions and a 0.5 in extra length).

from tables given by Matthaei *et al.* [12]. The filter performance is optimized through the COMPACT computer program. The cutoff frequency is designed to be 50 GHz to provide an attenuation greater than 20 dB at the RF and LO frequencies.

A seven-element semi-lumped structure consisting of three capacitive and four inductive sections is shown in Fig. 5. The predicted and measured performances are given in Fig. 6.

The bandpass filter consists of multiple-coupled resonators and can be realized from end-coupled half-wavelength strips in a suspended stripline configuration, as shown in Fig. 7. The minimum gap is set at 3 mils for ease of fabrication.

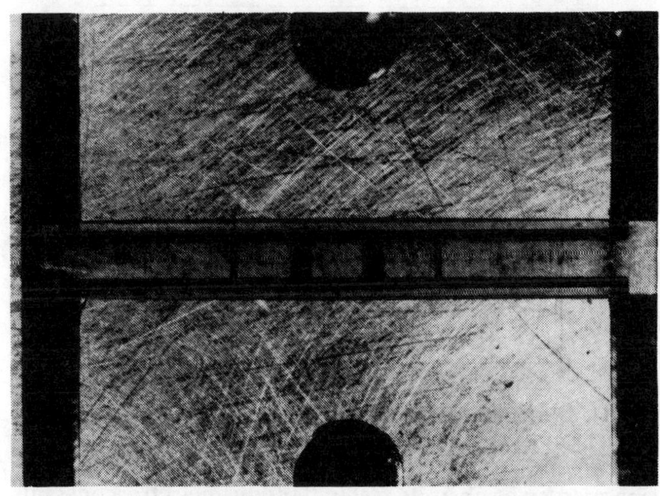

Fig. 9. *W*-band bandpass filter with two transitions.

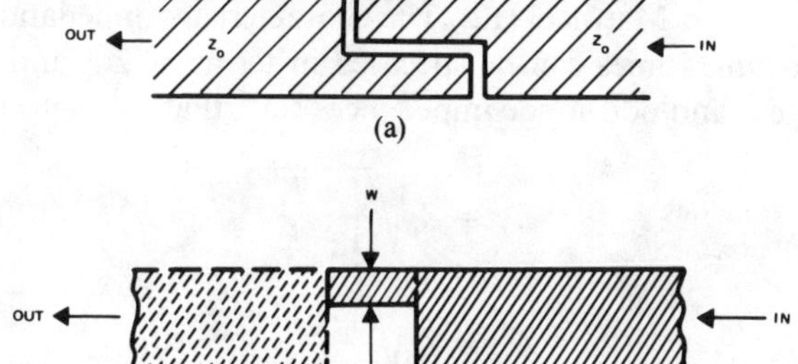

(a)

(b)

Fig. 10. Two different configurations of broadside coupler. (a) Edge coupler with lines on the same side of substrate. (b) Broadside coupler with lines separated by the substrate.

A bandpass filter has been designed using the COMPACT program to pass the LO and reject the RF frequency. This computer program calculates the required capacitances from which the gap dimensions can be determined [13], [14].

Fig. 8 shows typical performance of a *W*-band bandpass filter centered at 86 GHz. The insertion loss includes two transitions and a 0.5 in extra length of stripline. The insertion loss of the filter is thus about 1.5 dB over a 5-GHz bandwidth after the correction of transition loss. Fig. 9 is a photo of the filter with two transitions made for testing purposes.

IV. Broadside Coupler Development

The RF power propagating down a suspended stripline couples to the stripline on the other side of the substrate through a broadside coupler. The broadside coupler could serve as a dc block or present an open circuit to the IF frequencies.

At lower frequencies, the conventional edge coupler shown in Fig. 10(a) has been used quite successfully. At *W*-band, to achieve minimum insertion loss, it is necessary to have very tight coupling with a gap too small to realize practically. In this case, thin lines are superimposed and

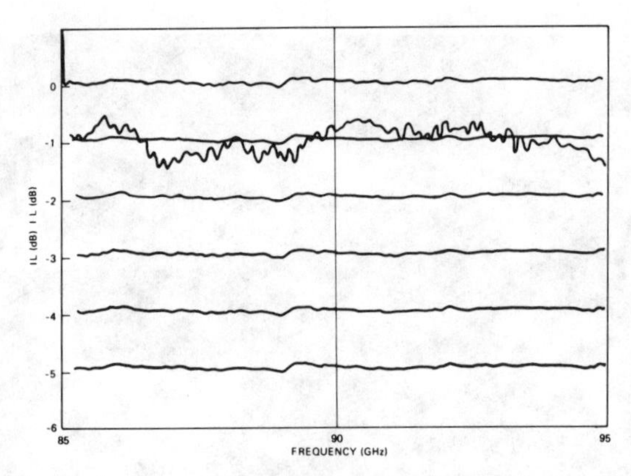

Fig. 11. Performance of a broadside coupler (including a two transitions and a 0.5-in section).

separated by a substrate material of low dielectric constant, with thickness on the order of 0.005 in, as shown in Fig. 10(b). The design of this coupler is straightforward and can be found in Matthaei [12]. The characteristic impedance of the coupled lines can be specified in terms of Z_{oe} and Z_{oo} as even- and odd-mode impedances such that

$$Z_{oe} = Z_o \sqrt{\frac{1+K}{1-K}} \tag{2}$$

$$Z_{oo} = Z_o \sqrt{\frac{1-K}{1+K}} \tag{3}$$

$$Z_o = \sqrt{Z_{oe} Z_{oo}} \tag{4}$$

where Z_o is the impedance of the terminating transmission lines and K is the coupling coefficient.

The width of the coupling lines is determined by Z_{oe} and Z_{oo}, and the substrate thickness as well as the distance between the ground planes above and below the thin substrate [12]. The length of the coupling section is about a quarter of a wavelength corrected for the compensation due to discontinuities. Typical performance is shown in Fig. 11 for the frequency range of interest. The insertion loss shown includes two transitions for testing purposes and a 0.5-in extra length of transmission line. The insertion loss of the coupler is thus less than 0.2 dB.

V. Crossbar Suspended Stripline Balanced Mixer

The circuit configuration of our crossbar stripline is shown in Fig. 12. The RF signal is applied to mixer diodes from a waveguide perpendicular to the circuit board. The crossbar configuration is formed by two mixer diodes with opposite polarity connected in series across the broadwall of the waveguide. The mixer diodes are thus in series with respect to the RF signal and in parallel with respect to the IF circuit. The IF signal is extracted via a low-pass filter and the LO signal is injected from the other side through a broadside coupler and an electric probe type transition.

A. Design Considerations

In a mixer design, the performance parameters of primary concern are the operating bandwidth and the conversion loss. To treat this analytically, we have developed a circuit

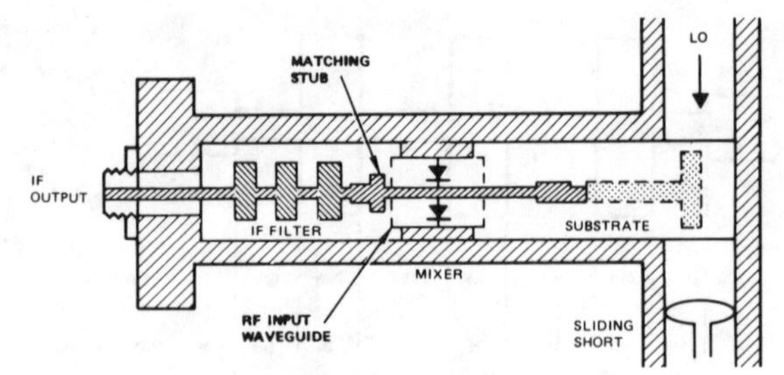

Fig. 12. Crossbar stripline mixer.

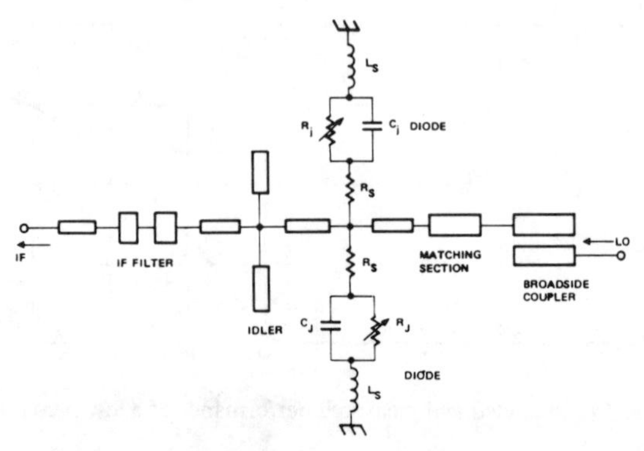

Fig. 13. Crossbar stripline mixer equivalent circuit.

model (Fig. 13) which deals quantitatively with the mixer performance.

In general, the conversion loss of a mixer is considered to consist of three parts:

L_1 mismatch loss due to impedance mismatch at RF and IF ports

L_2 diode parasitic loss due to its junction capacitance and series resistance

L_3 intrinsic junction loss of the ideal diode.

Mathematically, these losses can be expressed as [3]

$$L_1 = 10 \log \left[\left(1 - |\Gamma_{\rm RF}|^2\right)\left(1 - |\Gamma_{\rm IF}|^2\right) \right] \text{dB} \tag{5}$$

where $\Gamma_{\rm RF}$ and $\Gamma_{\rm IF}$ are the reflection coefficients at the RF and IF ports and

$$L_2 = 10 \log \left[1 + \frac{R_s}{R_j} + (\omega C_j)^2 R_s R_j \right] \text{dB} \tag{6}$$

where

R_s diode series resistance;
C_j diode junction capacitance;
R_j diode junction resistance;
ω operating frequency in radians.

The intrinsic junction loss of a mixer depends on the terminating conditions of the image frequency and nonlinearity of the diode conductance and is not dependent on frequency. As an example of the image matched condition, the junction loss is given as

$$L_3 = \left\{ 1 + \left[\frac{1 + \frac{g_2}{g_0} - 2\left(\frac{g_1}{g_0}\right)^2}{1 + \frac{g_2}{g_0}} \right]^{1/2} \right\}^2 \left[1 + \left(\frac{g_2}{g_0}\right) \right] \left(\frac{g_0}{g_1}\right)^2 \tag{7}$$

Fig. 14. Circuit layout of *W*-band crossbar stripline mixer.

where g_0, g_1, and g_2 are the Fourier coefficients of the diode conductance [15].

The diode parasitic loss L_2 depends on the values of the R_s, R_j, and C_j of the diode. For specified diodes, L_2 is fixed and cannot be improved. Consequently, circuit optimization was concentrated on the reduction of mismatch loss (L_1) and intrinsic junction loss (L_3).

L_1 is minimized by optimizing the RF and IF circuit matching. The junction resistance R_j of the mixer diode is varied with LO voltage and its value can be as low as 100 to 150 Ω under fully turned-on conditions. Waveguide impedance is in the range of 400 to 600 Ω and can be matched to the diode impedance by a reduced-height taper transformer. The sliding short on the opposite side of the RF port will tune out the reactance part of the diode.

IF and LO matching are aided by a computer analysis of the equivalent circuit shown in Fig. 13. An IF filter passes the IF frequency band and rejects the LO and RF signals. The connecting transmission line between the IF and LO ports can be optimized to provide matched conditions at the LO and IF ports.

A broadside coupler was designed to present an open circuit to the IF frequencies to prevent dissipation of IF power in the LO port. The coupler also serves as a dc block as the mixer is integrated with an MIC local oscillator. It can be designed to present an open circuit at the image frequency and thus reflect back the power for remixing with the LO to generate power at the IF frequency. A double open stub (Fig. 12) was used to facilitate the LO matching.

B. Mixer Performance

A photograph of a *W*-band crossbar stripline circuit layout is shown in Fig. 14. With the LO at 75 GHz, a conversion loss of less than 7.5 dB for 15-GHz instantaneous IF bandwidth was achieved with two beamlead diodes as the RF is swept from 76 to 91 GHz (Fig. 15(a)). The beamlead diodes are commercially available diodes with a C_j of approximately 0.04 to 0.05 pF and R_s of 5 to 7 Ω. With the LO at 90 GHz, a conversion loss of less than 7.8 dB was achieved over a 14-GHz instantaneous IF

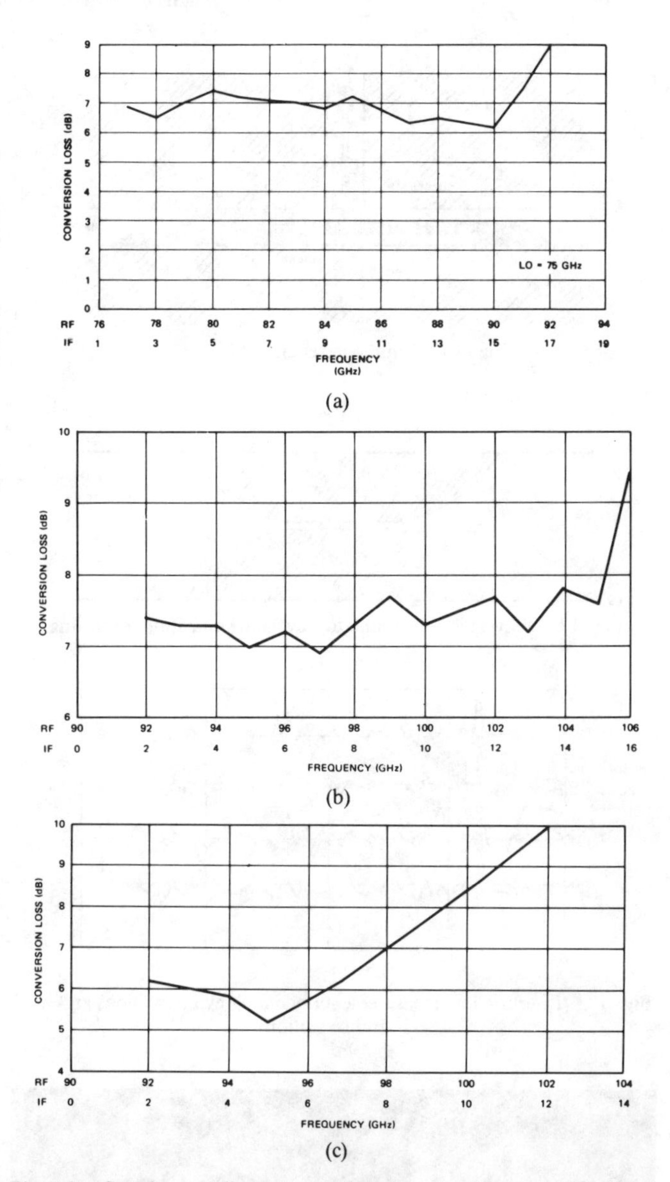

Fig. 15. Crossbar stripline mixer performance. (a) LO = 75 GHz, RF = 76–90 GHz. (b) LO = 90 GHz, RF = 92–105 GHz. (c) Narrow-band operation.

bandwidth as the RF is swept from 92 to 105 GHz (Fig. 15(b)). The mixer can be tuned for narrow-band operation at 94 GHz. Less than 5.5-dB conversion loss was achieved for 500-MHz bandwidth and less than 6 dB for 2-GHz bandwidth as shown in Fig. 15(c). RF/LO isolation of over 20 dB was achieved. These results represent state-of-the-art performance in this frequency range for an integrated circuit mixer.

VI. Finline Balanced Mixer

In conjunction with the crossbar stripline balanced mixer development, a finline balanced mixer was also developed. Finline has a simple geometry and is compatible with metal waveguides. A finline *W*-band balanced mixer has been reported by Meier for narrow-band operation [16] and recently realized for medium bandwidth operation by others [17], [18]. This paper presents a design which results in wide-band operation.

A circuit layout is given in Fig. 16 showing the basic configuration of this mixer. The RF is fed from the right

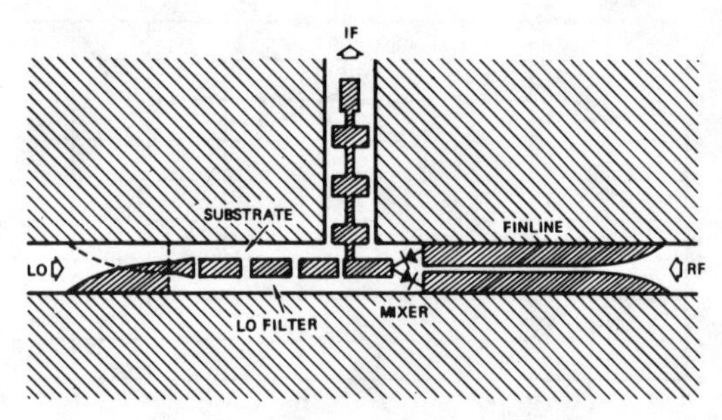

Fig. 16. Finline mixer circuit layout.

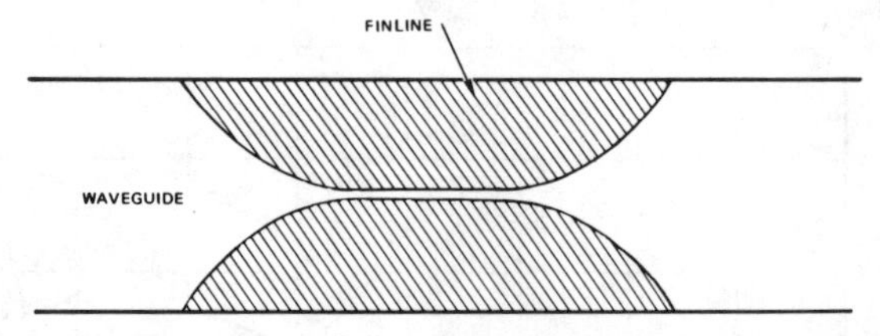

Fig. 17. Waveguide-to-finline-to-waveguide transition for testing.

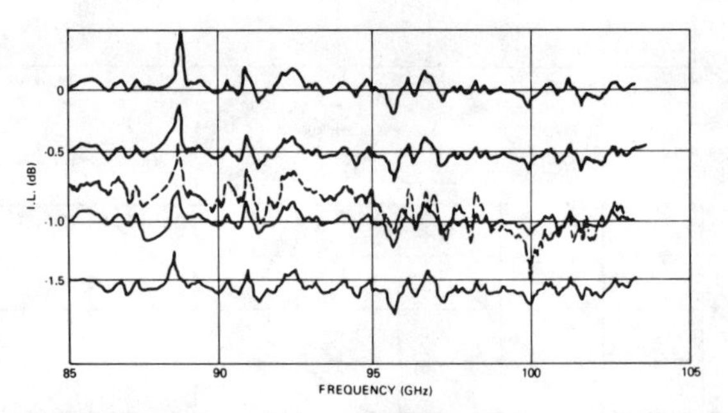

Fig. 18. Insertion loss measurement (including two transitions and a 0.5 in finline section).

Fig. 19. Photo of finline mixer.

through a tapered finline and LO power is coupled to the mixer diodes through a waveguide-to-finline-to-suspended stripline transition and a suspended stripline bandpass filter. The LO bandpass filter built on suspended stripline is implemented to achieve good RF to LO isolation. IF output is taken out via a low-pass filter.

To achieve good RF matching, a cosine taper finline transition was incorporated into the design. The transition

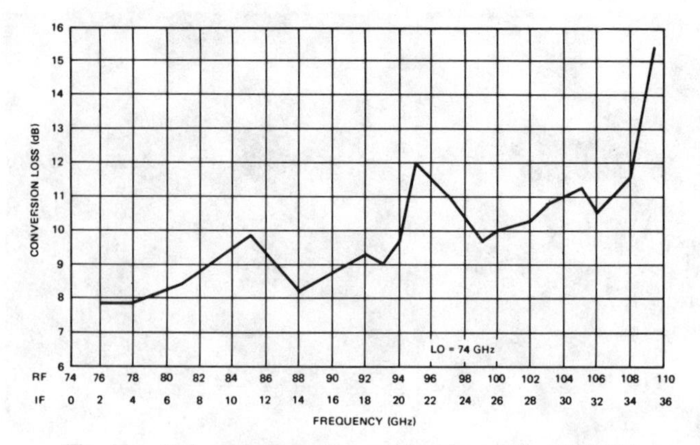

Fig. 20. *W*-band finline mixer performance (LO at 74 GHz).

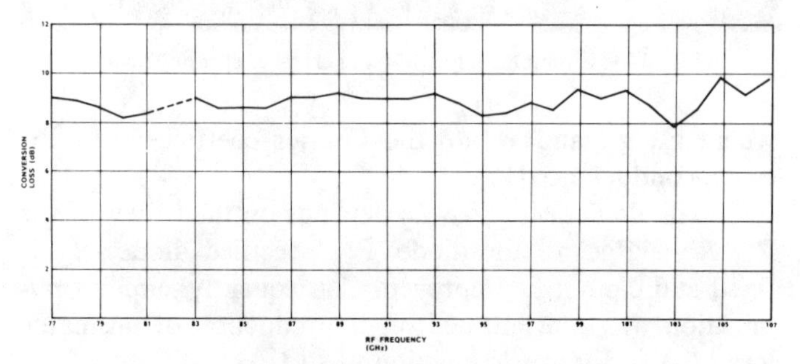

Fig. 21. *W*-band finline mixer performance (LO at 82 GHz).

was first built, tested, and optimized as an individual component (Fig. 17). Fig. 18 shows the characterization of this transition with typical insertion loss per transition of about 0.2 to 0.3 dB. It is believed that this transition has a much wider bandwidth than that shown in Fig. 18. The gap size between the tapered finlines is varied to achieve the optimum impedance matching to the mixer diodes. To match the low diode impedance (100 to 150 Ω), the gap size was designed to be less than 0.004 in, as optimized by a computer program.

A photograph of this mixer is shown in Fig. 19. A conversion loss of 8 to 12 dB was achieved over 32-GHz IF instantaneous bandwidth with the LO at 74 GHz as the RF is swept from 76 to 108 GHz (Fig. 20). Compared to the crossbar stripline mixer, the finline mixer operates at a wider bandwidth with slightly higher conversion loss. With the LO at 82 GHz, a conversion loss of less than 10 dB has been achieved with the RF swept from 77 to 107 GHz (Fig. 21).

VII. Conclusions

W-band broad-band integrated circuit mixers were developed using beamlead diodes. The mixer configurations are crossbar suspended stripline and finline. For crossbar suspended stripline mixers, a conversion loss of less than 7.5 dB over a 15-GHz instantaneous IF bandwidth was achieved with a fixed LO at either 75 or 90 GHz. For a finline mixer, a conversion loss of 8 to 12 dB was achieved over a 32-GHz instantaneous IF bandwidth with the LO at 74 GHz and the RF swept from 76 to 108 GHz. With the LO at 82 GHz, a conversion loss of less than 10 dB has also been achieved with the RF swept from 77 to 107 GHz.

All these results represent state-of-the-art performance in integrated circuit mixers.

Acknowledgment

The authors wish to thank Dr. C. Sun for helpful suggestions and discussions.

References

[1] A. Hislop and R. T. Kihm, "A broad-band 40–60 GHz balanced mixer," *IEEE Trans. Microwave Theory Tech.*, vol. MTT-24, pp. 63–64, Jan. 1976.

[2] N. Kanmuri and R. Kawasaki, "Design and performance of a 60–90 GHz broad-band mixer," *IEEE Trans. Microwave Theory Tech.*, vol. MTT-24, pp. 259–261, May 1976.

[3] L. L. Yuan, "Design and performance analysis of an octave bandwidth waveguide mixer," *IEEE Trans. Microwave Theory Tech.*, vol. MTT-25, pp. 1048–1054, Dec. 1977.

[4] C. P. Hu and A. Denning, "A broad-band, low-noise receiver at *W*-band," in *IEEE 1981 MTT-S Int. Microwave Symp. Dig.*, pp. 111–113.

[5] K. Louie, "A *W*-band wide-band crossbar mixer," in *IEEE 1982 MTT-S Int. Microwave Symp. Dig.*, pp. 369–371.

[6] P. T. Parrish, A. G. Cardiasmenos, and I. Galin, "94-GHz beam-lead balanced mixer," *IEEE Trans. Microwave Theory Tech.*, vol. MTT-29, pp. 1150–1157, Nov. 1981.

[7] J. Paul, L. Yuan, and P. Yen, "Beam lead dielectric crossbar mixers from 60 to 140 GHz," in *IEEE 1982 MTT-S Int. Microwave Symp. Dig.*, pp. 372–373.

[8] R. S. Tahim, G. M. Hayashibara, and K. Chang, "*W*-band broad-band integrated circuit mixers," *Electron. Lett.*, vol. 18, no. 11, pp. 471–473, May 27, 1982.

[9] M. Schneider, "Millimeter-wave integrated circuits," *IEEE 1973 MTT-S Int. Microwave Symp. Dig.*, pp. 16–18.

[10] B. Glance and R. Tramarulo, "A waveguide to suspended stripline transition," *IEEE Trans. Microwave Theory Tech.*, vol. MTT-21, pp. 117–118, Feb. 1973.

[11] L. J. Lavedan, "Design of waveguide-to-microstrip transitions specially suited to millimeter wave applications," *Electron. Lett.*, vol. 13, no. 20, pp. 604–605, Sept. 29, 1977.

[12] G. L. Matthaei, L. Young, and E. M. T. Jones, *Microwave Filters, Impedance Matching Networks, and Coupling Structures.* New York: McGraw-Hill, 1964.

[13] M. Maeda, "An analysis of gap in microstrip transmission lines," *IEEE Trans. Microwave Theory Tech.*, vol. MTT-20, pp. 390–395, June 1972.

[14] K. C. Gupta, R. Gary, and I. J. Bahl, *Microstrip Lines and Slotlines.* Dedham, MA: Artech, 1979.

[15] M. R. Barber, "Noise figure and conversion loss of the Schottky barrier mixer diode," *IEEE Trans. Microwave Theory Tech.*, vol. MTT-15, pp. 629–635, Nov. 1967.

[16] P. J. Meier, "*E*-plane components for a 94-GHz printed circuit balanced mixer," in *IEEE 1980 MTT-S Int. Microwave Symp. Dig.*, pp. 267–269.

[17] R. N. Bates *et al.*, "Millimeter-wave low noise *E*-plane balanced mixers incorporating planar MBE GaAs mixer diodes," in *IEEE 1982 MTT-S Int. Microwave Symp. Dig.*, pp. 13–15.

[18] L. Bui and D. Ball, "Broadband planar balanced mixers for millimeter-wave applications," in *IEEE 1982 MTT-S Int. Microwave Symp. Dig.*, pp. 204–205.

Part V
Harmonic Mixers

SUBHARMONICALLY pumped mixers using a *single* diode have been reported in the literature [1],[2]. In order to achieve a low conversion loss in a subharmonically pumped mixer, in particular the fundamental mixing between the signal and local oscillator should be suppressed. In a second harmonic mixer ($f_{IF} = f_s - 2f_{LO}$) this may be quite difficult, since for the fundamental mixing, $f_s - f_{LO}$ is approximately equal to the LO frequency. However, this problem is avoided in a mixer utilizing the symmetry properties of two diodes in antiparallel. Aside from the fact that the potential conversion loss is lower since no fundamental mixing can occur and direct detection is suppressed, lower noise figure is expected through suppression of local oscillator noise sidebands and inherent protection against large peak inverse voltage burn out.

The first paper in this part, by Schneider and Snell, and the second, by Cohn *et al.*, give the basic description of such mixers. Büchs and Begemann extend the general idea to harmonic mixers with more than two diodes. In particular, they discuss a fourth harmonic mixer using four diodes. Carlson *et al.* describe in their paper an elegant design for millimeter waves. The suppression of local oscillator noise is discussed in detail by Henry *et al.* in the fifth paper and they make a comparison to local oscillator noise expected in a single diode mixer.

In the final paper Malik *et al.* introduce a new mixing element, the planar doped barrier diode with symmetric *I–V* characteristic. Hence it is now possible to replace the antiparallel diode pair with a single element. Problems with a matched diode pair for exact symmetry should be diminished and the loop inductance problem recognized by Kerr [3] (see also paper 6 in Part II) is eliminated. Since the diode barrier height can be tailored to a low value, a low local oscillator power will result as well.

REFERENCES

[1] R. Meredith and F. L. Warner, "Superheterodyne radiometers for use at 70 GHz and 140 GHz," *IEEE Trans. Microwave Theory Tech.*, vol. MTT-11, pp. 397–411, Sept. 1963.

[2] F. A. Benson, *Millimetre and Submillimetre waves.* London, England: Iliffe, 1969, ch. 22.

[3] A. R. Kerr, "Noise and loss in balanced and subharmonically pumped mixers, Parts I and II, theory and applications," *IEEE Trans. Microwave Theory Tech.*, vol. MTT-27, pp. 938–950, Dec. 1979.

Harmonically Pumped Stripline Down-Converter

MARTIN V. SCHNEIDER, SENIOR MEMBER, IEEE, AND WILLIAM W. SNELL, JR.

***Abstract*—A novel thin-film down-converter which is pumped at a submultiple of the local-oscillator frequency has given a conversion loss which is comparable to the performance of conventional balanced mixers. The converter consists of two stripline filters and two Schottky-barrier diodes which are shunt mounted in a strip transmission line. The conversion loss measured at a signal frequency of 3.5 GHz is 3.2 dB for a pump frequency of 1.7 GHz and 4.9 dB for a pump frequency of 0.85 GHz. The circuit looks attractive for use at millimeter-wave frequencies where stable pump sources with low FM noise are not readily available.**

I. INTRODUCTION

THE PROCESS of frequency conversion and its applications are well known and have been extensively treated in the literature [1]–[7]. The conversion is usually performed by pumping a nonlinear resistive or reactive element embedded in a linear network and extracting the sum or difference frequencies which are generated by the signal and by the pump frequency. The purpose of this paper is to describe a novel thin-film converter which has the following properties.

1) The pump frequency required for efficient up-conversion or down-conversion is a submultiple of that needed in conventional frequency converters.

2) The circuit does not require a dc return path.

3) The separation of the signal and the local-oscillator frequency is readily obtained and the loss in the signal path is small.

The thin-film circuit consists of two strip transmission-line filters and two metal-semiconductor diodes which are shunt mounted in a stripline with opposite polarities as shown in the block diagram of Fig. 1. The strip transmission line is used because the conversion from the hybrid TEM mode to the first-order waveguide mode (longitudinal-section magnetic mode) is substantially reduced compared to the conversion obtained with other transmission-line circuits such as microstrip lines. This approach eliminates noise contributions from undesired bands near harmonics of the pump frequency.

The circuit design has the desired properties that the image frequency is reactively terminated and that the isolation of the pump oscillator is approximately 30 dB which is about 10 dB better than the isolation obtained with conventional circuits. The unit is therefore well suited for use in digital-communication receivers in the microwave and millimeter-wave frequency bands. The converter described herein is serving as a model for similar units being developed for use at higher frequencies. The individual components, the electrical characteristics of the diode pair, and the performance of the down-converter are described in the following paragraphs.

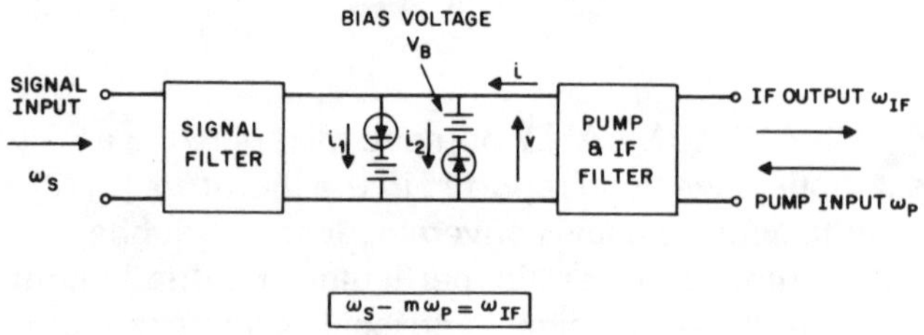

Manuscript received April 22, 1974; revised August 23, 1974.

The authors are with the Crawford Hill Laboratory, Bell Laboratories, Holmdel, N. J. 07733.

$\omega_S - m\omega_P = \omega_{IF}$

Fig. 1. Block diagram of harmonically pumped down-converter, including signal filter, pump and IF filter, and diode pair with opposite polarities and dc bias.

II. DESCRIPTION OF STRIPLINE CIRCUIT

The stripline conductor pattern of a harmonically pumped frequency converter is shown in Fig. 2. The pattern consists of a 50-Ω line section at the signal input, a half-wavelength resonator for the bandpass filter, a five-element low-pass filter, and a 50-Ω line section for the pump input and the IF output. Two Schottky-barrier diodes with opposite polarities are connected to the section between the filters at opposite sides of the stripline conductor. The conductor pattern is deposited on a quartz substrate which is suspended in a rectangular channel as shown in Fig. 3. Coupling to waveguide modes at harmonics of the pump frequency is substantially reduced because of the opposite polarity of the electric fields as indicated in Fig. 3. Undesired waveguide modes can also be suppressed if the slot which supports the substrate in each side wall has an electrical depth which is a quarter wavelength of a harmonic of the pump frequency [8]. The dimensions of the channel, the substrate, and the stripline conductor are given in Fig. 3 for a signal frequency of 3.5 GHz. They can be linearly scaled to higher frequencies using electromagnetic scaling laws [9]. The

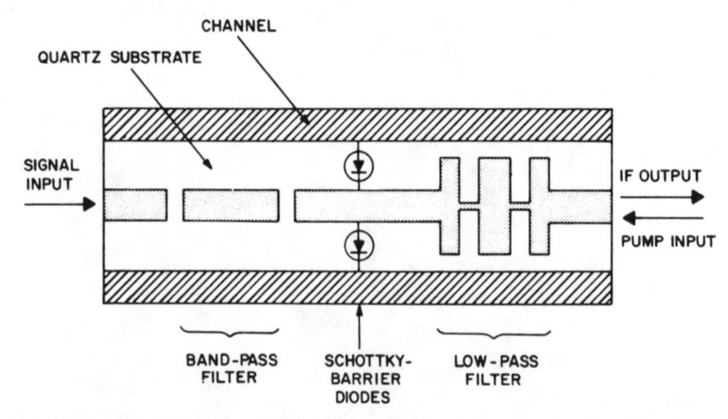

Fig. 2. Microstrip conductor pattern on quartz substrate in a metal channel of harmonically pumped down-converter. The diode pair is shunt mounted to the ground on opposite sides of the strip transmission line.

Reprinted from *IEEE Trans. Microwave Theory Tech.*, vol. MTT-23, pp. 271–275, Mar. 1975.

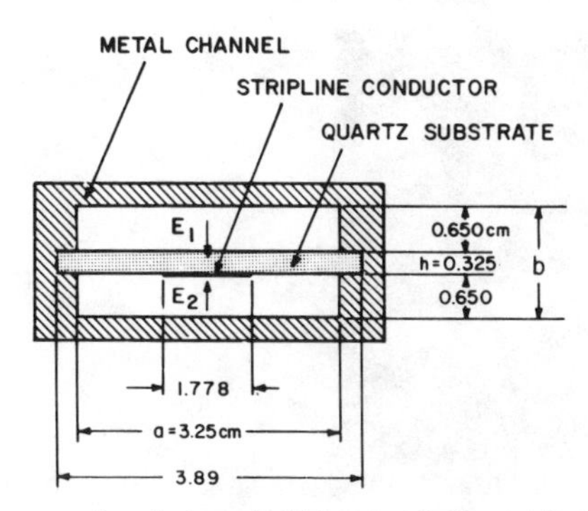

Fig. 3. Cross-sectional view of shielded stripline with symmetrically suspended quartz substrate and stripline conductor.

cutoff frequency for the first-order longitudinal-section magnetic mode for the channel dimensions a and b and a substrate thickness h is given by [10]

$$\omega_2 = \frac{\pi c}{a}\left(1 - \frac{h(\epsilon_r - 1)}{b\epsilon_r}\right)^{1/2} \tag{1}$$

where ϵ_r is the relative dielectric constant of the substrate. The signal frequency must satisfy the condition

$$\omega_1 < \omega_s < \omega_2 \tag{2}$$

where ω_1 is the cutoff frequency of the low-pass stripline filter. This condition insures that there is no signal loss through the low-pass filter and that waveguide modes are suppressed at the signal frequency. The cutoff frequency ω_2 calculated for the channel dimensions given in Fig. 3 with $\epsilon_r = 3.8$ is 4.25 GHz. The measured cutoff frequency of the channel is 4.20 GHz as shown in Fig. 4. The figure also gives the measured transmission loss of the low-pass filter as a function of frequency between 2 and 8 GHz. The usual low-pass characteristics are observed up to 4.2 GHz. Multimoding occurs at higher frequencies as shown by several peaks in the transmission curve which are caused by longitudinal LSM and LSE modes in the channel. The first-order LSM mode is partially suppressed if the mode conversion at a discontinuity is equal, but of opposite phase for the fields surrounding the stripline conductor shown in Fig. 3. This condition is fulfilled if the center conductor is symmetrically located between the top and the bottom plane of the surrounding shield. It is approximately fulfilled if the center conductor is deposited on a dielectric substrate with an appropriate dielectric constant for a fixed set of air gaps.

It should be noted that higher order mixer products are terminated by a complex reactance which is mainly determined by the transmission characteristics of the low-pass filter and the electrical length to the bandpass filter. An optimum conversion loss and noise figure for the converter is obtained by using an appropriate spacing between the diode pair and each filter, and by designing the filters in such a way that the excitation of undesired channel modes for the higher order mixer products is minimized.

A photograph of a harmonically pumped down-converter is shown in Fig. 5. Two Western Electric 497A GaAs Schottky-barrier diodes are used on opposite sides of the

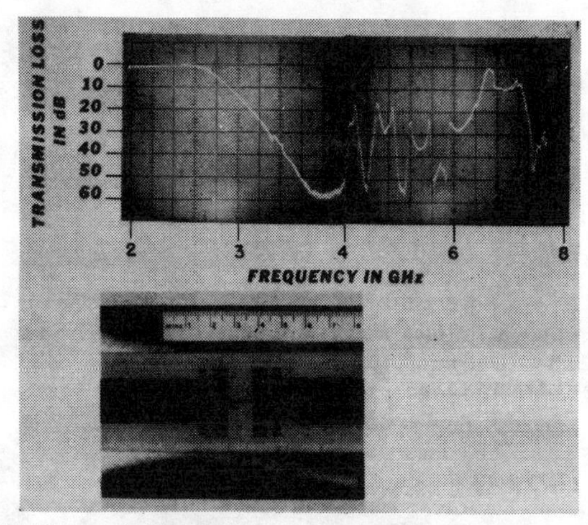

Fig. 4. Transmission loss of seven-element low-pass stripline filter from 2–8 GHz. The transmission spikes above 4 GHz are caused by resonances of the first-order LSM mode.

stripline conductor. The figure also shows the performance of the signal filter and the IF and pump filter measured with a network analyzer. The transmission loss of the signal filter at 3.5 GHz is normalized to zero. Its actual insertion loss at that frequency is 0.2 dB. The cutoff frequency of the low-pass filter is 2.6 GHz and the transmission loss at the signal frequency is 40 dB. The current-voltage characteristics of the diode pair are symmetrical with respect to the origin as shown in Fig. 6. This results in a current waveform which has only odd-order harmonics and a conductance waveform with even-order harmonics. The second feature combined with the low conversion to waveguide modes results in a converter which has a good conversion loss and a low noise figure for subharmonic pumping. The electrical characteristics of the diode pair and the converter characteristics are discussed in the two following paragraphs.

III. ELECTRICAL CHARACTERISTICS OF DIODE PAIR

The current-voltage characteristics of a Schottky-barrier diode can be computed from the thermionic-emission diffusion theory of Crowell and Sze [11], [12]. For a sufficiently small minority carrier injection ratio one obtains for the diode current

$$i = i_s\left\{\exp\left[\frac{q}{nkT}(v - iR_s)\right] - 1\right\} \tag{3}$$

where v is the applied voltage, R_s the diode series resistance, n the idealty factor, and $q/kT = 38.7\ \mathrm{V}^{-1}$ for 300 K. The saturation current i_s is a function of the diode area S, the barrier height ϕ_B, and the modified Richardson constant A^{**}. It is given by

$$i_s = A^{**}ST^2 \exp\left[-\frac{e\phi_B}{mkT}\right] \tag{4}$$

where $\phi_B = kT/e + V_D + \phi_F - \Delta\phi$. V_D is the diffusion voltage, ϕ_F the position of the Fermi level relative to the bottom of the conduction band, and $\Delta\phi$ the image-force lowering of the barrier.

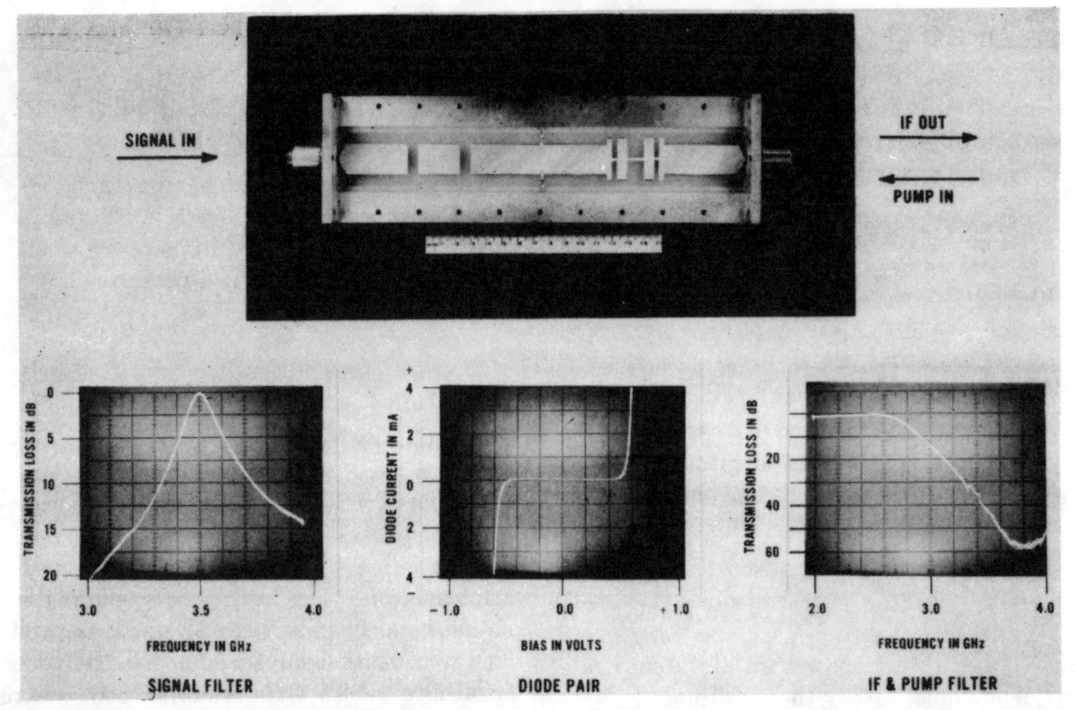

Fig. 5. Photograph of harmonically pumped down-converter showing top view of stripline conductor pattern in a rectangular channel. The characteristics of the bandpass filter, the low-pass filter, and the diode pair are displayed at the bottom of the photograph.

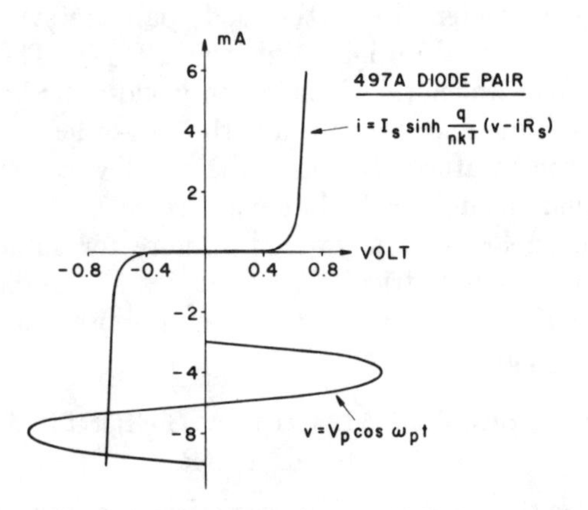

Fig. 6. Current-voltage characteristics of diode pair consisting of two Western Electric 497A Schottky-barrier diodes connected with opposite polarities. Two conducting switching states per RF cycle are obtained if the pair is driven by a sinusoidal pump frequency.

The current flowing through a diode pair shown in Fig. 1 is a function of the applied dc bias V_B, and is given by the sum $i = i_1 + i_2$ where

$$i_1 = i_s \left\{ \exp \frac{q}{nkT} (v - V_B - i_1 R_s) \right\} \tag{5}$$

$$i_2 = -i_s \left\{ \exp \frac{q}{nkT} (-v - V_B + i_2 R_s) \right\}. \tag{6}$$

We assume that the voltage drop i_1R_s and i_2R_s is only significant for $i_1 \gg i_2$ or $i_2 \gg i_1$. The resulting current becomes

$$i = 2i_s \exp \left(-\frac{qV_B}{nkT} \right) \sinh \frac{q}{nkT} (v - iR_s) \tag{7}$$

and for $iR_s \ll v$, which is fulfilled for most practical applications, one obtains

$$i = \frac{I_{sat} \sinh (qv/nkT)}{1 + (qI_{sat}R_s/nkT) \cosh (qv/nkT)} \tag{8}$$

where we define I_{sat} by

$$I_{sat} = 2i_s \exp (-qV_B/nkT). \tag{9}$$

In order to compute the conversion loss one has to know the conductance di/dv of the diode pair. From (7) one obtains

$$\frac{di}{dv} = \left(\frac{nkT}{q} \frac{1}{(i^2 + I_{sat}^2)^{1/2}} + R_s \right)^{-1}. \tag{10}$$

For $R_s = 0$ the resulting conductance is

$$\frac{di}{dv} = \frac{qI_{sat}}{nkT} \cosh \frac{qv}{nkT}. \tag{11}$$

IV. VOLTAGE PUMPING AND CONVERSION LOSS

Let us assume that the diode pair is pumped with a periodic signal

$$v = V_p \cos \omega_p t. \tag{12}$$

This means that the pumping voltage across the diode is constrained to be sinusoidal, while the current i can have all the possible harmonics. The current through the diode pair and the pair conductance becomes a periodic function which can be computed from the series expansions

$$\cosh (z \cos \theta) = I_0(z) + 2 \sum_{k=1}^{\infty} I_k(z) \cos 2k\theta \tag{13}$$

$$\sinh (z \cos \theta) = 2 \sum_{k=0}^{\infty} I_{2k+1}(z) \cos (2k+1)\theta \quad (14)$$

where I_k is the modified Bessel function of the first kind of order k. Using the abreviation $\alpha = q/nkT$ one obtains from (7), (11), and (12) for $R_s = 0$

$$i = 2I_{sat} \sum_{k=0}^{\infty} I_{2k+1}(\alpha V_p) \cos (2k+1)\omega_p t \quad (15)$$

$$di/dv = \alpha I_{sat} I_0(\alpha V_p) + 2\alpha I_{sat} \sum_{k=1}^{\infty} I_k(\alpha V_p) \cos 2k\omega_p t. \quad (16)$$

As expected, 1) there is no dc current flowing through the diode pair, i.e., the converter does not need a dc return path; 2) the diode-pair current contains only odd-order harmonics; and 3) the pair conductance has only even-order harmonics. Let us write the conductance waveform $g(t) = di/dv$ as follows

$$g(t) = y_0 + 2y_1 \cos 2\omega_p t + 2y_2 \cos 4\omega_p t + \cdots \quad (17)$$

$$y_n = \alpha I_{sat} I_n(\alpha V_p), \qquad n = 0,1,2,\cdots. \quad (18)$$

The conversion loss L of the down-converter operated at a pump frequency $2\omega_p$ is now given by

$$L = \Phi(Y_{image}, y_1/y_0, y_2/y_0) \quad (19)$$

where Φ is an irrational function of the image conductance Y_{image} and the first two normalized Fourier coefficients of the conductance waveform. The function Φ is exactly the same as the one which is used for computing the conversion loss of a conventional down-converter operated with a single diode and pumped with a local oscillator frequency which is close to the signal frequency ($\omega_{LO} = \omega_s \pm \omega_{IF}$). The properties of the function Φ are extensively treated in the work by Saleh [3]. For the special case $\alpha V_p \gg 1$ and for $Y_{image} = 0$, Y_{signal}, or ∞ the resulting conversion loss is

$$L(Y_{image} = 0) = 1 + (2/\alpha V_p)^{1/2} \quad (20)$$

$$L(Y_{image} = Y_{signal}) = 2(1 + \sqrt{2}/\alpha V_p) \quad (21)$$

$$L(Y_{image} = \infty) = 1 + 2/(\alpha V_p)^{1/2}. \quad (22)$$

This means that the minimum conversion loss which can be achieved with an open or short at the image frequency is 0 dB. The minimum conversion loss for a matched image is 3 dB. The minimum conversion loss for a series resistance $R_s \neq 0$ is a function of the ratio ω_s/ω_c where ω_s is the signal frequency and ω_c the cutoff frequency of the diode

$$\omega_c = (R_s C_0)^{-1} \quad (23)$$

where C_0 is the zero-bias capacitance of one single converter diode. Computations of the conversion loss L as a function of ω_s/ω_c have been performed by Dragone [13]. For a Western Electric diode WE 497A with $R_s = 2\ \Omega$ and $C_0 = 0.45$ pF pumped at a signal frequency of 3.5 GHz one obtains $f_c = 177$ GHz or $L = 2.1$ dB. Beam-leaded devices fabricated for use at millimeter-wave frequencies [14] have shown a cutoff frequency of approximately 1000 GHz. The corresponding diode conversion loss of an optimized converter at 50 GHz built with these diodes would be 3.1 dB.

V. PERFORMANCE OF THE STRIPLINE CONVERTER

The measured single-sideband noise figure for the stripline down-converter of Fig. 5 is plotted in Fig. 7 as a function of the signal frequency ω_s for $m = 2$ and $m = 4$. The harmonic integer m is defined by

$$m = \frac{\omega_s \pm \omega_{IF}}{\omega_p}. \quad (24)$$

The noise figure of the 100-MHz IF amplifier is 1.7 dB. The total single-sideband noise figure including the IF amplifier noise at a signal frequency of 3.455 GHz is 4.9 dB for $m = 2$ and 6.6 dB for $m = 4$. The corresponding conversion loss is 3.2 dB for $m = 2$. This result approaches the theoretically predicted loss of 2.1 dB obtained in the last paragraph. The loss for $m = 4$ is higher because the circuit does not contain a trap at the second subharmonic. The total noise figure for $m = 2$ is 6 dB or better over a bandwidth of 35 MHz which is 1 percent of the signal frequency. The bandwidth for $m = 4$ is much smaller because the circuit was designed for optimized performance at a pump frequency which is the second subharmonic of the conventional local-oscillator frequency. A well-designed harmonic down-converter which is pumped at the fourth subharmonic ($m = 4$) should include a trap at the second subharmonic consisting of an open-ended section of a strip transmission line which is connected to the part of each diode terminal at the stripline side, or an approximately chosen length for the stripline sections between the diode pair and the low-pass and the bandpass filter. The total noise figure as a function of a pump power for $m = 2$ is shown in Fig. 8. A first minimum is obtained at a power of approximately +3 dB. It can be shown that this minimum can be reduced further by using a small bias voltage V_B in series with each diode as shown in Fig. 1. It has also been found that for a down-converter

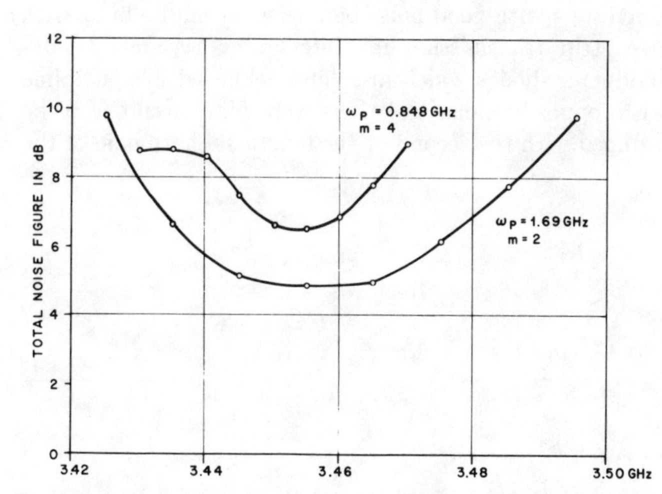

Fig. 7. Single-sideband noise figure including noise of IF amplifier in decibels as a function of signal frequency for harmonically pumped stripline down-converter pumped at the second subharmonic ($m = 2$) and at the fourth subharmonic ($m = 4$). The noise figure of the 100-MHz IF amplifier is 1.7 dB.

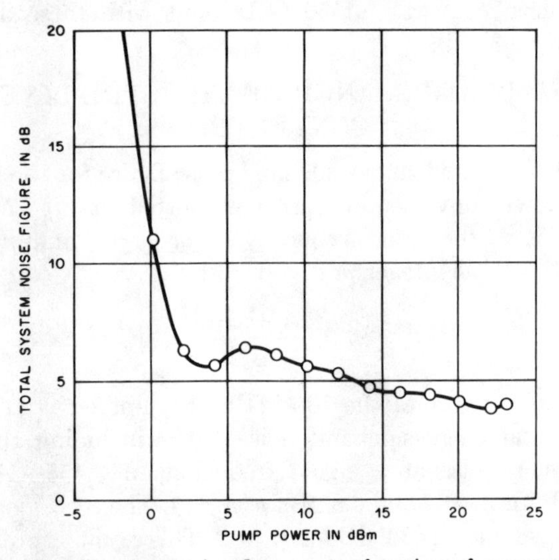

Fig. 8. Total system noise figure as a function of pump power for $m = 2$. The first minimum at +3 dBm can be further reduced using a small bias voltage V_B in series with each diode.

with a much higher signal frequency substantially less pump power is needed to obtain a minimum noise figure for the case $V_B = 0$.

A circuit which is similar to the one shown in Fig. 2 was also built using microstrip transmission-line filters and microstrip line sections. It was found that the noise figure of the microstrip down-converter is approximately 6 dB higher than the noise figure of the corresponding stripline circuit. This effect is not surprising since shielded microstrip circuits show relatively strong coupling to waveguide modes above the cutoff frequency of the shielding enclosure. Coupling to undesired modes can be suppressed by using waffle-iron filters or other traps; however, such a circuit is likely to be more complex and more costly than the simple stripline structure shown in Fig. 2.

VI. CONCLUSIONS

It has been shown that harmonically pumped down-converters with good noise performance can be built with two strip transmission-line filters and two metal-semiconductor diodes which are shunt mounted in a stripline with opposite polarities. The thin-film circuit can be pumped with the second or the fourth subharmonic of the local-oscillator frequency required in conventional down-converters. DC bias and a dc return path for the diodes is not required.

The new harmonically pumped stripline circuit can be readily scaled to higher microwave frequencies and particularly to millimeter-wave frequencies where solid-state oscillators are only available at subharmonics of the local-oscillator frequency. The basic design principles outlined in this paper can also be applied to other converters in the electromagnetic spectrum, such as up-converters, harmonic generators, and parametric amplifiers.

ACKNOWLEDGMENT

The authors wish to thank C. Dragone, A. A. Penzias, and V. K. Prabhu for a number of very helpful suggestions and discussions. They also wish to thank S. R. Shah for work performed on the stripline filters.

REFERENCES

[1] C. T. Torrey and C. A. Whitmer, *Crystal Rectifiers* (M.I.T. Radiation Lab. Series), vol. 15. New York: McGraw-Hill, 1948.
[2] C. Dragone, "Amplitude and phase modulations in resistive diode mixers," *Bell. Syst. Tech. J.*, vol. 48, pp. 1967–1998, July–Aug. 1969.
[3] A. A. M. Saleh, *Theory of Resistive Mixers* (Res. Monograph 64). Cambridge, Mass.: M.I.T. Press, 1971.
[4] M. R. Barber, "Noise figure and conversion loss of the Schottky barrier mixer diode," *IEEE Trans. Microwave Theory Tech.*, vol. MTT-15, pp. 629–635, Nov. 1967.
[5] J. W. Gewartowski, "Noise figure for a mixer diode," *IEEE Trans. Microwave Theory Tech.* (Corresp.), vol. MTT-19, p. 481, May 1971.
[6] C. A. Liechti, "Down-converters using Schottky-barrier diodes," *IEEE Trans. Electron Devices*, vol. ED-17, pp. 975–983, Nov. 1970.
[7] K. M. Johnson, "*X* band integrated circuit mixer with reactively terminated image," *IEEE Trans. Electron Devices* (*Special Issue on Microwave Integrated Circuits*), vol. ED-15, pp. 450–459, July 1968.
[8] M. V. Schneider and B. S. Glance, "Suppression of waveguide modes in strip transmission lines," *Proc. IEEE* (Lett.), vol. 62, p. 1184, Aug. 1974.
[9] J. A. Stratton, *Electromagnetic Theory.* New York: McGraw-Hill, 1941, pp. 488–490.
[10] M. V. Schneider, "Millimeter-wave integrated circuits," in *1973 IEEE G-MTT Int. Symp. Digest Tech. Papers* (Univ. Colorado, Boulder), June 4–6, 1973, pp. 16–18.
[11] C. R. Crowell and S. M. Sze, "Current transport in metal-semiconductor barriers," *Solid-State Electron.*, vol. 9, pp. 1035–1048, Nov./Dec. 1966.
[12] S. M. Sze, *Physics of Semiconductor Devices.* New York: Wiley, 1969, pp. 363–409.
[13] C. Dragone, to be published.
[14] A. Y. Cho and W. C. Ballamy, "GaAs planar technology by molecular beam epitaxy," *J. Appl. Phys.*, vol. 46, Feb. 1975.

Harmonic Mixing with an Antiparallel Diode Pair

MARVIN COHN, FELLOW, IEEE, JAMES E. DEGENFORD, MEMBER, IEEE, AND BURTON A. NEWMAN, MEMBER, IEEE

***Abstract*—An analytical and experimental investigation of the properties of an antiparallel diode pair is presented. Such a configuration has the following unique and advantageous characteristics as a harmonic mixer: 1) reduced conversion loss by suppressing fundamental mixing products; 2) lower noise figure through suppression of local oscillator noise sidebands; 3) suppression of direct video detection; 4) inherent self protection against large peak inverse voltage burnout. These results are obtained without the use of either filters or balanced circuits employing hybrid junctions.**

Manuscript received October 17, 1974; revised March 17, 1975.
The authors are with the Electromagnetic Technology Laboratory, Westinghouse Defense and Space Center, Advanced Technology Laboratories, Baltimore, Md. 21203.

I. INTRODUCTION

HISTORICALLY harmonic mixing has been used primarily at the higher millimeter wave frequencies where reliable stable LO sources are either unavailable or prohibitively expensive. However, the conversion loss obtained by harmonic mixing has been typically 3 to 5 dB greater than that which could be obtained by fundamental mixing at the same signal frequency [1], [2]. An analysis [3], [4] has shown that such a large degradation should not exist, but it assumes that fundamental mixing between the signal and LO is suppressed. Fundamental mixing

Reprinted from *IEEE Trans. Microwave Theory Tech.*, vol. MTT-23, pp. 667–673, Aug. 1975.

will, however, take place unless the harmonic mixer provides a reactive termination for these mixer products. In general, that is difficult to accomplish, e.g., in the case of second-harmonic mixing the fundamental mixing difference frequency ($f_s - f_{LO}$) is close to the LO frequency.

In this paper, an analytical and experimental investigation of the properties of an antiparallel diode pair is presented [5]. Such a configuration has unique and advantageous characteristics as a harmonic generator or harmonic mixer. In the latter application, which is treated in this paper, it will be shown that this circuit provides:

1) reduced conversion loss by suppressing the fundamental mixing products;
2) lower noise figure through suppression of local oscillator noise sidebands;
3) suppression of direct video detection;
4) inherent self-protection against large peak inverse voltage burnout.

These results are obtained without the use of either filters or balanced circuits employing hybrid junctions.

II. ANALYSIS

In a conventional single diode mixer as shown in Fig. 1(a), application of a voltage waveform

$$V = V_{LO} \sin \omega_{LO} t + V_s \sin \omega_s t$$

to the usual asymmetric diode characteristic results in the diode current having all frequencies $mf_{LO} \pm nf_s$. It will be shown in this section that the total current of the antiparallel diode pair shown in Fig. 1(b) contains only frequencies for which $m + n$ is an odd integer. The terms in which $m + n$ is even, i.e., even harmonics, fundamental mixing products ($\omega_s - \omega_{LO}$ and $\omega_s + \omega_{LO}$), and the dc term flow only within the diode loop.

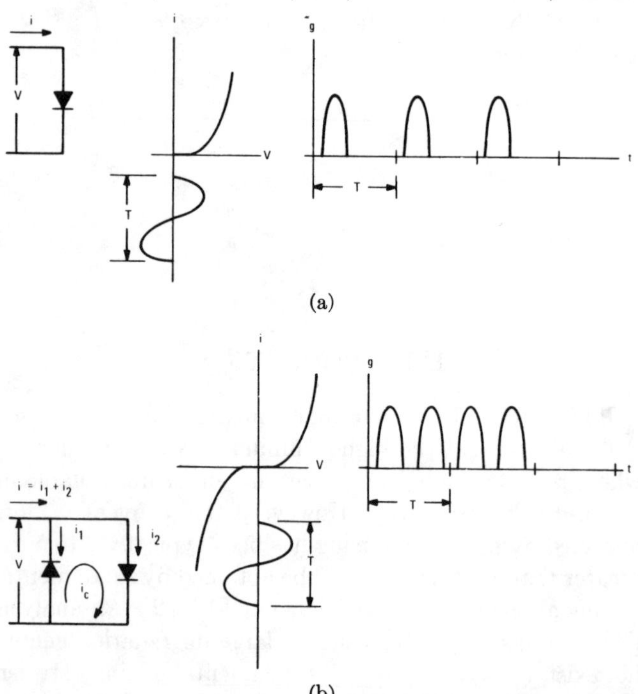

Fig. 1. Mixer circuit. (a) Single diode mixer. (b) Antiparallel diode pair mixer.

The basic antiparallel diode pair circuit is shown in Fig. 1(b). The instantaneous currents through the diodes i_1 and i_2 may be written in the usual fashion

$$i_1 = -i_s(e^{-\alpha V} - 1) \tag{1}$$

$$i_2 = i_s(e^{\alpha V} - 1) \tag{2}$$

where α is the diode slope parameter ($\alpha \approx 38$ V^{-1} for typical high-quality gallium arsenide Schottky barrier diodes). Similarly, the differential conductance for each diode may be written as

$$g_1 = \frac{di_1}{dV} = \alpha i_s e^{-\alpha V} \tag{3}$$

and

$$g_2 = \frac{di_2}{dV} = \alpha i_s e^{\alpha V}. \tag{4}$$

The composite time varying differential conductance g is simply the sum of the individual differential conductances.

$$g = g_1 + g_2 = \alpha i_s(e^{\alpha V} + e^{-\alpha V}) = 2\alpha i_s \cosh \alpha V. \tag{5}$$

Examination of this expression reveals that g has even symmetry with V and, as illustrated in Fig. 1(a) and (b), double the number of conductance pulses per LO cycle as compared to a single diode mixer.

For the usual case in which only the LO modulates the conductance of the diodes we may substitute

$$V = V_{LO} \cos \omega_{LO} t$$

into (5) with the following result

$$g = 2\alpha i_s \cosh (\alpha V_{LO} \cos \omega_{LO} t) \tag{6}$$

which may be expanded in the following series:

$$g = 2\alpha i_s [I_0(\alpha V_{LO}) + 2I_2(\alpha V_{LO}) \cos 2\omega_{LO} t + 2I_4(\alpha V_{LO}) \cos 4\omega_{LO} t + \cdots] \tag{7}$$

where $I_n(\alpha V_{LO})$ are modified Bessel functions of the second kind. Notice that the conductance components consist of a dc term plus even harmonics of the LO frequency, ω_{LO}. For the applied voltage, $V = V_{LO} \cos \omega_{LO} + V_s \cos \omega_s t$, the current expression is

$$i = g(V_{LO} \cos \omega_{LO} t + V_s \cos \omega_s t) \tag{8}$$

$$i = A \cos \omega_{LO} t + B \cos \omega_s t + C \cos 3\omega_{LO} t + D \cos 5\omega_{LO} t + E \cos (2\omega_{LO} + \omega_s) t + F \cos (2\omega_{LO} - \omega_s) t + G \cos (4\omega_{LO} + \omega_s) t + H \cos (4\omega_{LO} - \omega_s) t + \cdots. \tag{9}$$

It can be seen that the total current only contains frequency terms $mf_{LO} \pm nf_s$ where $m + n$ is an odd integer; i.e., $m + n = 1,3,5,\cdots$.

In Fig. 1(b) a circulating current i_c is also indicated.

This current arises from the fact that Fourier expansions of the individual currents i_1 and i_2 reveal that certain components of each current are oppositely phased. Because of their opposite polarity, these components cancel as far as the external current i is concerned and simply circulate within the loop formed by the two diodes. From Fig. 1, one can mathematically describe this circulating current as

$$i_c = (i_2 - i_1)/2 = i_s[\cosh \alpha V - 1]. \quad (10)$$

Substituting

$$V = V_{LO} \cos \omega_{LO} t + V_s \cos \omega_s t \quad (11)$$

into the expansion for the hyperbolic cosine yields

$$i_c = i_s\left[1 + \frac{(V_{LO} \cos \omega_{LO} t + V_s \cos \omega_s t)^2}{2!} + \cdots - 1\right]$$

$$= \frac{i_s}{2}[V_{LO}^2 \cos^2 \omega_{LO} t + V_s^2 \cos^2 \omega_s t + 2V_{LO}V_s \cdot \cos \omega_{LO} t \cos \omega_s t + \cdots] \quad (12)$$

$$= \frac{i_s}{2}\left\{\frac{V_{LO}^2 + V_s^2}{2} + \frac{V_{LO}^2}{2} \cos 2\omega_{LO} t + \frac{V_s^2}{2} \cdot \cos 2\omega_s t + V_{LO}V_s[\cos(\omega_{LO} - \omega_s)t + \cos(\omega_{LO} + \omega_s)t] + \cdots\right\} \quad (13)$$

from which it can be seen that the circulating current only contains frequencies $mf_0 \pm nf_s$, where

$$m + n = \text{even integer}. \quad (14)$$

Thus the antiparallel pair has the advantage of suppressing fundamental and other odd harmonic mixing products as well as even harmonics of the LO.

This natural suppression is lessened, of course, by diode unbalance. If we first consider the case where the saturation currents i_s are different for the two diodes, then we may let

$$i_{s1} = i_s + \Delta i_s \quad \text{and} \quad i_{s2} = i_s - \Delta i_s. \quad (15)$$

Substitution of the above expressions into (3), (4), and (5) yields the following equation for the total conductance g:

$$g = 2\alpha i_s\left[\cosh \alpha V + \frac{\Delta i_s}{i_s} \sinh \alpha V\right]. \quad (16)$$

Similarly, if the diode slope parameters are different, we may let

$$\alpha_1 = \alpha + \Delta\alpha \quad \text{and} \quad \alpha_2 = \alpha - \Delta\alpha \quad (17)$$

which yields the following expression for the total conductance:

$$g = 2\alpha i_s e^{(\Delta\alpha)V}\left[\cosh \alpha V + \frac{\Delta\alpha}{\alpha} \sinh \alpha V\right]. \quad (18)$$

Notice that in both cases, the conductance function contains the desired hyperbolic cosine term plus a hyperbolic sine term whose coefficient is proportional to either $\Delta i_s/i_s$ or $\Delta\alpha/\alpha$. This hyperbolic sine term introduces conductance variations at the fundamental and other odd harmonics of the LO. We may find the ratio of the conductance component at the fundamental $g^{(1)}$ to the conductance component at the second harmonic $g^{(2)}$ by simply substituting $V = V_s \cos \omega_s t$ into (16) and (18) and expanding (16) and (18) with the following result:

$$\frac{g^{(1)}}{g^{(2)}} = \frac{\Delta i_s}{i_s} \cdot \frac{I_1(\alpha V_0)}{I_2(\alpha V_0)}, \quad \text{for } i_s \text{ unbalance} \quad (19)$$

and

$$\frac{g^{(1)}}{g^{(2)}} = \frac{\Delta\alpha}{\alpha} \frac{I_1(\alpha V_0)}{I_2(\alpha V_0)}, \quad \text{for } \alpha \text{ unbalance.} \quad (20)$$

In Fig. 2 the ratio $g^{(1)}/g^{(2)}$ is plotted versus $\Delta\alpha/\alpha$ or $\Delta i_s/i_s$. It can be seen that if one were to operate under LO "starved" conditions ($V_{LO} < 0.1$ V) the effect of the unbalance can be severe. However, for typical LO voltages of 0.7 V or greater, (19) and (20) reduce approximately to

$$\frac{g^{(1)}}{g^{(2)}} = \frac{\Delta i_s}{i_s}, \quad \text{for } i_s \text{ unbalance} \quad (21)$$

and

$$\frac{g^{(1)}}{g^{(2)}} = \frac{\Delta\alpha}{\alpha}, \quad \text{for } \alpha \text{ unbalance.} \quad (22)$$

Thus the percentage unbalance in either i_s or α translates directly into the percentage of $g^{(1)}$ as compared to $g^{(2)}$. This does not mean, however, that the ratio of undesired fundamental conversion loss to the desired second-harmonic conversion loss will be directly proportional to $g^{(1)}/g^{(2)}$, since the conversion loss will depend on exactly what load impedance is presented to the diode at the fundamental mixing frequency. To accurately predict the fundamental conversion loss, detailed characterization of the embedding circuit at the fundamental mixing frequency would be necessary.

The second-harmonic mixing conversion loss may be estimated by noting (Fig. 1) that the pulse duty ratio (PDR) [6] for the antiparallel diode pair will be essentially double that attainable at the fundamental LO frequency since the period at the second harmonic is halved. Referring to Barber's paper [6, fig. 4], a doubling of

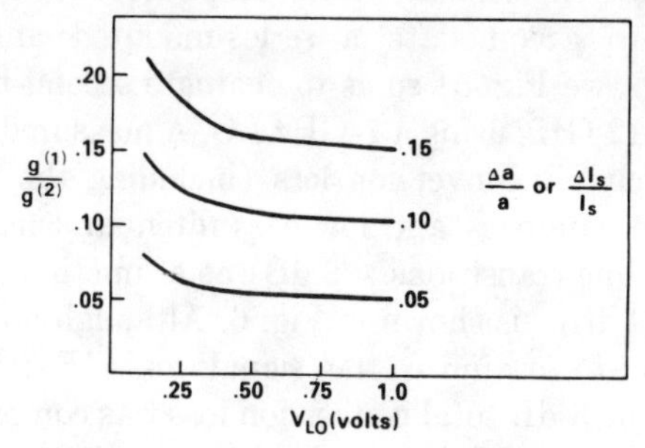

Fig. 2. Unbalance versus LO power.

the PDR (for the matched image case) from a typical value of 15 percent to 30 percent represents a degradation of approximately 1.5 dB for the second-harmonic conversion loss as compared to fundamental mixing. Such an estimate, of course, is only approximate since: 1) the conversion loss depends on the terminations presented to the diode at the various frequencies, and 2) the cited curve from Barber's paper is for an ideal diode with no series resistance or shunt junction capacitance. Nonetheless, such an estimate is useful in assessing the merits of the harmonic mixing approach as compared to fundamental mixing.

It should also be pointed out that image enhancement techniques can be used with an antiparallel diode pair mixer to improve conversion loss. Such an experiment is described in Section III of this paper, and also in a recent paper by Schneider [7]. Schneider has investigated an image enhanced stripline mixer with a very similar diode configuration and achieved a conversion loss = 3.2 dB at 3.5 GHz.

The degradation of receiver noise figure due to LO noise sidebands is also reduced in even harmonic mixing (m even, $n = 1$) in an antiparallel diode pair as shown in Fig. 3. LO noise sidebands (f_{NL} and f_{NH}) whose separation from the LO (f_{LO}) equals the IF (f_{IF}) generate IF noise which only circulates within the diode loop when they mix fundamentally with the LO. Second-harmonic mixing of these noise sidebands with the virtual LO ($2f_{LO}$) produces noise which is not within the IF amplifier passband. Like a conventional balanced mixer, of course, the degree of suppression is affected by the balance between the diodes.

Finally, the circuit has inherent self-protection against large peak inverse voltage burnout since a reverse biased junction is always in parallel with a forward biased junction. This limits the maximum reverse voltage excursion to a value much less than the reverse breakdown voltage of the diodes.

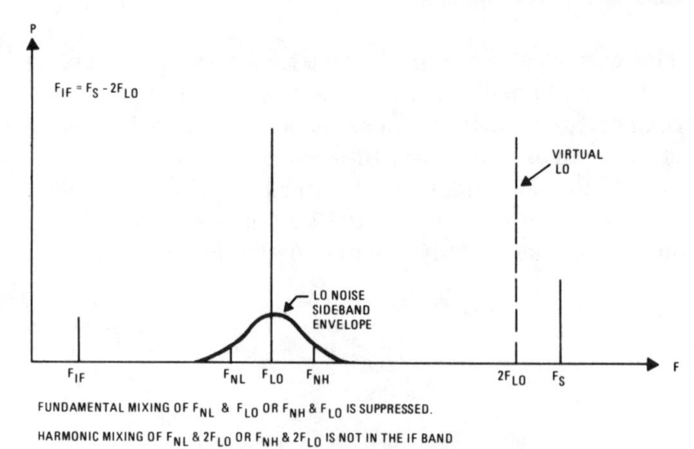

Fig. 3. Noise sideband mixing products.

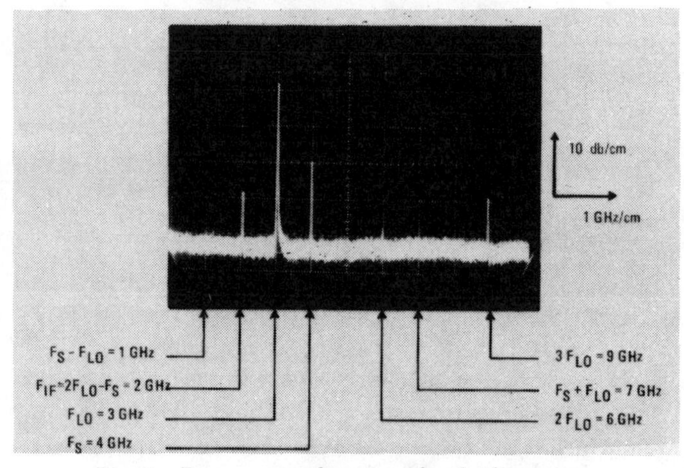

Fig. 4. Power versus frequency for slot line mixer.

III. EXPERIMENTAL INVESTIGATION

In order to verify many of the predicted characteristics, an antiparallel pair of GaAs Schottky barrier diodes were shunt mounted across a slot line. A 3-GHz LO input and a 4-GHz low-level signal were impressed at the slot line input. A photograph of the output spectrum is shown in Fig. 4. Note that the output at $3f_{LO}$ is much greater than that at $2f_{LO}$, and the absence of fundamental mixing products, $f_s - f_{LO}$ and $f_s + f_{LO}$, and the relatively large 2-GHz IF output due to second-harmonic mixing ($2f_{LO} - f_s$).

In another experiment, an existing microstrip mixer was modified to accomodate a series-mounted antiparallel diode pair (see Fig. 5) so as to evaluate second-harmonic mixing at 12 GHz using a 7-GHz LO. A measured curve of the total circuit conversion loss (including the insertion loss of the bandpass and low-pass filters and microstrip-to-coaxial line transitions $\approx$2 dB) as a function of fundamental LO drive is shown in Fig. 6. Although no attempt was made to optimize the signal and IF impedance matches, the 8-dB total conversion loss was comparable to that obtained by fundamental mixing at 12 GHz.

An experiment was also conducted into image enhanced harmonic mixing using the circuit shown in Fig. 7. This circuit was originally designed for fundamental image enhanced mixing at the following frequencies [8]:

$$\text{signal}: f_s = 9.5 \text{ GHz}$$
$$\text{LO}: f_0 = 8.5 \text{ GHz}$$
$$\text{IF}: f_{IF} = 1.0 \text{ GHz}.$$

In order to modify the circuit for the harmonic mixing experiment, a second diode was mounted antiparallel fashion across the first diode and the following frequencies were applied:

$$\text{LO}: f_0 = 4.25 \text{ GHz}$$
$$\text{signal}: f_s = 9.5 \text{ GHz}$$
$$\text{IF}: f_{IF} = 1.0 \text{ GHz}.$$

The LO was injected via a directional coupler since the resonant ring would only pass a narrow range of frequencies about 8.5 GHz. The measured performance of this mixer is shown in the table below.

Fundamental mixing conversion loss	$f_s - f_{LO} = 5.25$ GHz	>45 dB
Harmonic image enhanced mixing conversion loss	$f_s - 2f_{LO} = 1.0$ GHz	< 5 dB
LO harmonic suppression	$2f_{LO} = 8.5$ GHz	>50 dB

For comparison, the measured fundamental image en-

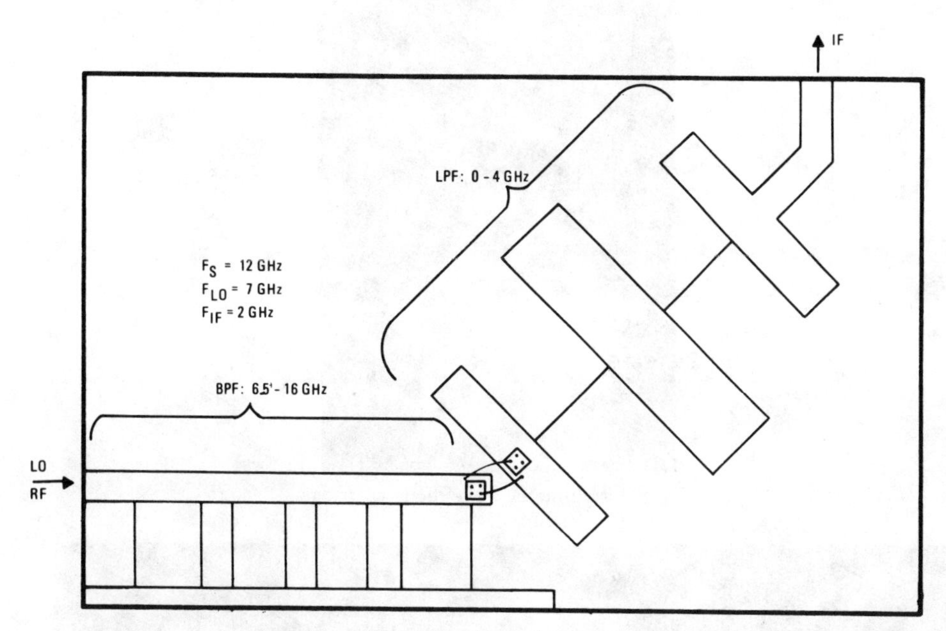

Fig. 5. *X*-band MIC harmonic mixer layout.

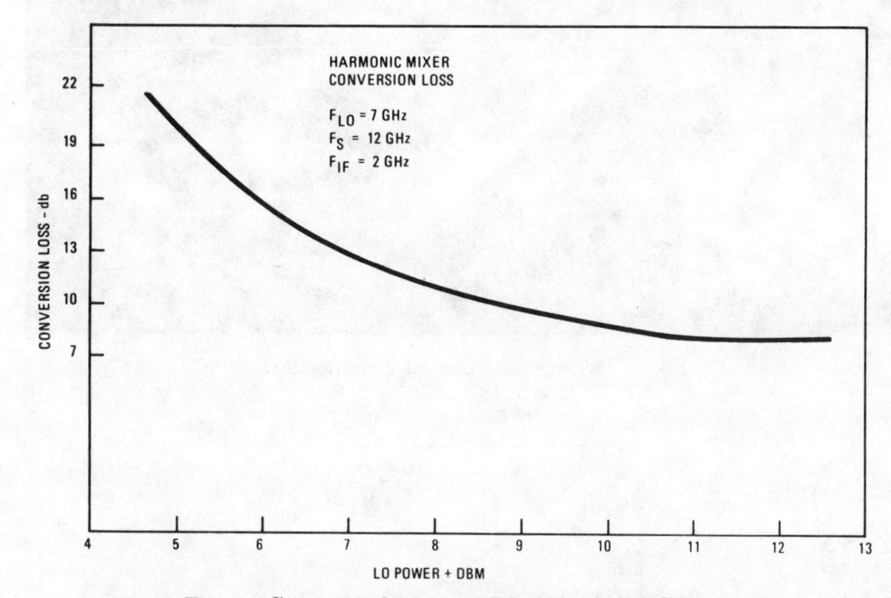

Fig. 6. Conversion loss versus LO power for MIC mixer.

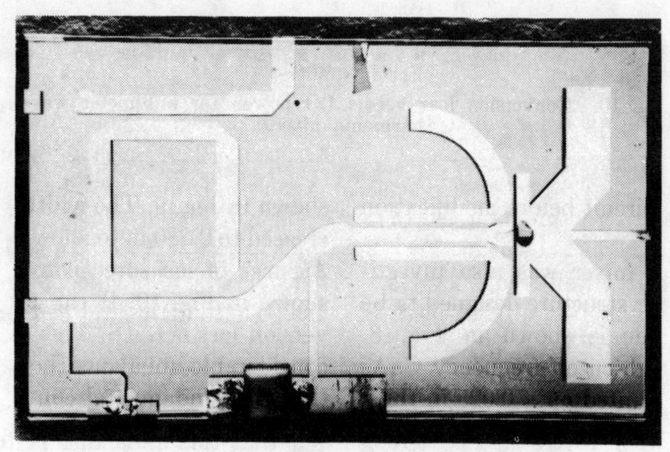

Fig. 7. Image enhanced harmonic mixer.

Fig. 8. Millimeter wave diode cartridge.

Fig. 9. Waveguide harmonic mixer mount.

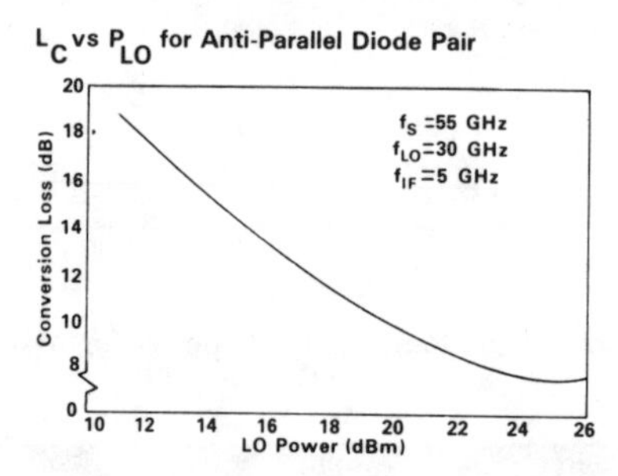

Fig. 10. Conversion loss versus LO power for millimeter wave harmonic mixer.

hanced conversion loss for this circuit before modification was typically 3.5 dB.

A millimeter-wave harmonic mixer was also investigated. The antiparallel diode pair structure designed to be inserted in a WR-15 waveguide is shown in Fig. 8. The Westinghouse developed high-cutoff-frequency GaAs Schottky barrier diodes are mounted on a 0.025-in-thick sapphire substrate which is metallized only on the surface shown. The smaller metallic end cap connects to the center conductor of the coaxial IF output port.

The disassembled mount with cartridge inserted is shown in Fig. 9. The width of the waveguide has been increased to 0.280 in to allow propagation of the 30-GHz LO. A curve of measured conversion loss versus LO power is shown in Fig. 10. It can be seen that the minimum conversion loss is 8 dB. This diode structure proved to have considerable unbalance, however, as evidenced by the fact that the fundamental conversion loss was also 8 dB. The dc current flowing in the IF circuit was 4 mA versus 0 for the perfect balance case. By way of comparison the dc current in the X-band mixers never exceeded 0.2 mA. It is felt that this unbalance is due to slight differences in the length of the bonding wires which at this frequency have an inductive reactance of approximately 6 Ω/0.001 in of length. Future efforts on this mixer will be concentrated on keeping these wires as short and as equal as possible.

These experiments have confirmed the theoretical predictions of Section II and demonstrated the usefulness of the antiparallel diode pair as a harmonic mixer. Potentially the most useful application of this circuit will be at millimeter wavelengths although careful balancing of the diodes will be required to realize its full potential.

REFERENCES

[1] M. Cohn, F. L. Wentworth, and J. C. Wiltse, "High-sensitivity 100- to 300-Gc radiometers," *Proc. IEEE*, vol. 51, pp. 1227–1232, Sept. 1963.

[2] R. J. Bauer, M. Cohn, J. M. Cotton, Jr., and R. F. Packard, "Millimeter wave semiconductor diode detectors, mixers, and frequency multipliers," *Proc. IEEE* (*Special Issue on Millimeter Waves and Beyond*), vol. 54, pp. 595–605, Apr. 1966.

[3] R. Meredith and F. L. Warner, "Superheterodyne radiometers for use at 70 Gc and 140 Gc," *IEEE Trans. Microwave Theory Tech.*, vol. MTT-11, pp. 397–411, Sept. 1963.

[4] F. A. Benson, *Millimetre and Submillimetre Waves.* London, England: Iliffe Books, 1969, ch. 22.

[5] M. Cohn, J. E. Degenford, and B. A. Newman, "Harmonic mixing with an anti-parallel diode pair," in *1974 MTT Int. Symp. Dig.*, pp. 171–172, June 12–14, 1974.

[6] M. R. Barber, "Noise figure and conversion loss of the Schottky barrier mixer diode," *IEEE Trans. Microwave Theory Tech.*, vol. MTT-15, pp. 629–635, Nov. 1969.

[7] M. V. Schneider, "Harmonically pumped stripline down converter," presented at the European Microwave Conf., Montreux, Switzerland, Sept. 10–13, 1974.

[8] J. B. Cahalan, J. E. Degenford, and M. Cohn, "An integrated X-band, image and sum frequency enhanced mixer with 1 GHz IF," in *1971 IEEE Int. Microwave Symp. Dig.* (Washington, D. C.), May 17–19, 1971.

Frequency conversion using harmonic mixers with resistive diodes

Just-Dietrich Büchs and Günther Begemann

Indexing terms: *Frequency convertors, Mixers (circuits)*

Abstract: In the paper, mixers are described that consist of M resistive diodes, e.g. Schottky-barrier diodes in a parallel connection. When such mixers are used in frequency convertors, the diodes are pumped by signals having a frequency M times lower than the frequency of conventional local oscillators. A phase shift equal to multiples of $2\pi/M$ is necessary between the pump signals. The frequency multiplication, which is usually carried out in the local-oscillator circuit, is done in the mixer. For this reason it is possible to use pump sources that are much simpler than conventional local oscillators. Often the harmonic mixers proposed in the paper are more complicated than the mixers used so far. Sometimes, however, circuits of moderate complexity can be found, so that a simplification of the frequency convertors is possible. This will be shown by mixers that are pumped at the first or second subharmonic, respectively. Theoretical investigations show that the conversion losses of harmonic mixers are slightly higher than those of conventional mixers.

1 Introduction

In heterodyne radio relay systems, the carriers are converted from the intermediate frequency (i.f.) range to the radio frequency (r.f.) range before being transmitted, and reconverted from the r.f. to the i.f. range after reception. The up- and down-conversions are usually carried out with the aid of mixers that are pumped by local oscillators (l.o.s). The pumped mixer diodes present periodically time-dependent admittances to the signals to be converted. They produce new signals shifted in frequency by an amount equal to multiples of the local-oscillator frequency.

In most cases, the local oscillator of such a system consists of a highly stable fundamental oscillator and a frequency multiplier. This type of local oscillator is widely used and works well. Unfortunately, with increasing radio frequency the multiplier becomes more and more complicated and expensive, and the f.m. noise behaviour of the local oscillator deteriorates.

These problems are mitigated when the multiplication factor is reduced. This, in fact, can be achieved by using harmonic mixers instead of conventional ones. Since 1971 a new type of harmonic mixer has been known, in which the pump frequency is only half the l.o. frequency f_s of a conventional frequency convertor.[1] One can look upon this circuit as being a mixer and a frequency doubler in one.

In this paper, circuits are proposed in which not only part but all of the multiplication is carried out in the mixer. Thereby the use of multipliers in the l.o. circuit can be avoided. The circuits proposed here consist of several two-poles in parallel. Their periodically time-dependent admittances have a fundamental frequency that is M times below f_s and a time shift equal to multiples of f_s^{-1} between each other. Frequency convertors containing this sort of harmonic mixer are often called subharmonic pumped convertors.

The conversion losses of a mixer pumped at the first subharmonic can be calculated using Barber's formulas.[2] The current/voltage characteristics of the nonlinear elements are assumed to be purely exponential. Parasitic effects are neglected. The investigations show the conversion loss to be about 0·5 dB greater than a conventional mixer.

Paper T177M, first received 12th December 1977 and in revised form 10th March 1978

Dr.-Ing. Büchs is with AEG-Telefunken, Geschäftsbereich Weiterverkehr und Kabeltechnik, Fachbereich Richtfunk, Gerberstraße 34, D-7150 Backnang, West Germany. Dipl.-Ing. Begemann is with the Institut für Hochfrequenztechnik, Technische Universität Braunschweig, Postfach 3329, D-3300 Braunschweig, West Germany

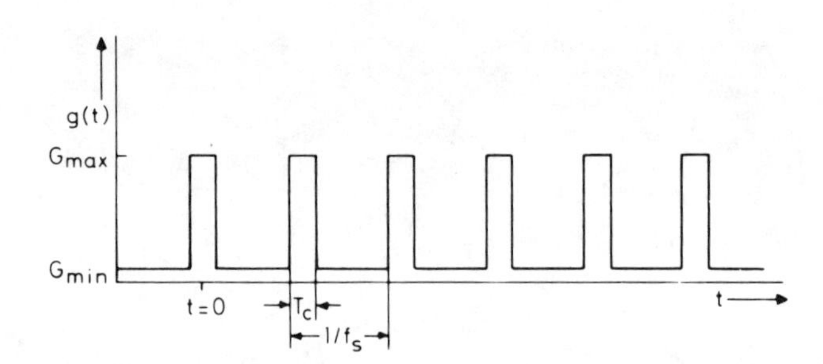

Fig. 1 *Small-signal conductance of a pumped mixer diode as a function of time: approximation by rectangular pulses*

2 Frequency conversion by means of a resistive time-dependent two-pole

The small-signal admittance of a pumped mixer diode can in most cases be approximated by rectangular conductance pulses of length T_c, as is shown in Fig. 1.[2] It can be written as

$$g(t) = \sum_{n=-\infty}^{\infty} \underline{G}_n \exp(jn2\pi f_s t) \tag{1}$$

where

$$G_0 = G_{min} + \Delta G\theta \tag{2}$$

and

$$\underline{G}_n = \frac{\Delta G}{n\pi} \sin(n\pi\theta) \tag{3}$$

with

$$\Delta G = G_{max} - G_{min} \tag{4}$$

and

$$\theta = T_c f_s \tag{5}$$

$\underline{G}_n$ are the Fourier coefficients of $g(t)$, ΔG is the difference between the maximum and minimum of the small-signal conductance and θ is the pulse duty ratio.

Reprinted with permission from *IEE J. Microwave, Opt. Acoust.*, vol. 2, pp. 71–76, May 1978.

The small-signal voltage u existing at the diode and the small-signal current i flowing through it are related as follows:

$$i = g(t)u \tag{6}$$

If the voltage u contains a component with frequency f_0 it may, according to eqn. 6, also contain components at

$$f'_k = k'f_s + f_0, \qquad k' = 0, 1, 2 \ldots \tag{7}$$

$$f''_k = k''f_s - f_0, \qquad k'' = 1, 2, \ldots \tag{8}$$

All these voltage components are characterised by complex quantities, which are defined as

$$\underline{U}_k = \hat{u}'_k \exp(j\varphi'_k) \tag{9}$$

and

$$\underline{U}_{-k} = \hat{u}''_k \exp(-j\varphi''_k) \tag{10}$$

where $\hat{u}'_k$, φ'_k are the amplitude and phase, respectively, of the voltage at frequency f'_k, and $\hat{u}''_k$, φ''_k are the amplitude and phase of the voltage at f''_k.

Equivalent definitions are introduced for the current.

Now the relation between the small-signal current and voltage components can be expressed as follows:

$$\boldsymbol{I} = \boldsymbol{Y} \cdot \boldsymbol{U} \tag{11}$$

Here, $\boldsymbol{I}$ and $\boldsymbol{U}$ are matrices with one column. They are defined as

$$\boldsymbol{I} = (\underline{I}_i) \tag{12}$$

and

$$\boldsymbol{U} = (\underline{U}_k) \tag{13}$$

$\boldsymbol{Y}$ is a matrix of the form

$$\boldsymbol{Y} = (\underline{Y}_{ik}) = (\underline{G}_{i-k}) \tag{14}$$

Obviously, if an i.f. signal with frequency f_0 is injected into the convertor a voltage $\underline{U}_0$ is generated. According to eqn. 11, it causes currents at all frequencies defined by eqns. 7 and 8 to flow through the diode. Unless these currents are short-circuited, voltages with the same frequencies also exist. Among these is a signal at f_{+1}, which can be interpreted as an r.f. signal. Thus we have demonstrated how up-conversion works. In a similar way it may be shown that down-conversion is possible as well.

It is important to note that, in all cases, undesired frequencies are generated. The most important one is the image frequency f_{-1}.

3 Frequency conversion using harmonic mixers with several resistive two-poles

In conventional frequency convertors, as described in the previous Section, the local-oscillator frequency has to be as high as the fundamental frequency of the time-dependent conductance. As was mentioned earlier, this means that multipliers are necessary in the local-oscillator circuits, which become complicated and expensive when the radio frequency increases. Therefore it is desirable to derive the time-dependent conductance of eqn. 1 from a pump source oscillating at a frequency M times below the fundamental frequency of $g(t)$.

It is proposed here to accomplish this by connecting M equal mixers in parallel and pumping each one at f_s/M. Then each mixer represents a time-dependent conductance with a fundamental frequency f_s/M (Fig. 2). If, in addition, a time shift of f_s^{-1} is introduced between adjacent conductances, the resultant conductance function resembles that of Fig. 1. This is shown in Fig. 3.

We assume the pulses of the individual conductance functions to be as narrow as for the conventional frequency convertor. Then $g(t)$ may be written as

$$g(t) = \sum_{m=1}^{M} \left\{ G_0 + \sum_{\substack{n'=-\infty \\ n' \neq 0}}^{\infty} \exp\left(j\frac{m2\pi}{M}n'\right) \frac{\Delta G}{n'\pi} \sin\left(n'\pi\frac{\theta}{M}\right) \exp\left(jn'2\pi\frac{f_s}{M}t\right) \right\} \tag{15}$$

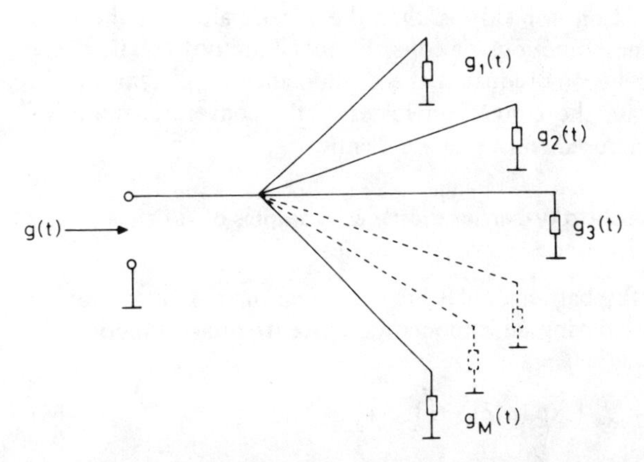

Fig. 2 *Parallel connection of M equal mixers with small-signal conductances $g_{1 \ldots M}$, each pumped with a fundamental frequency of f_s/M (schematic)*

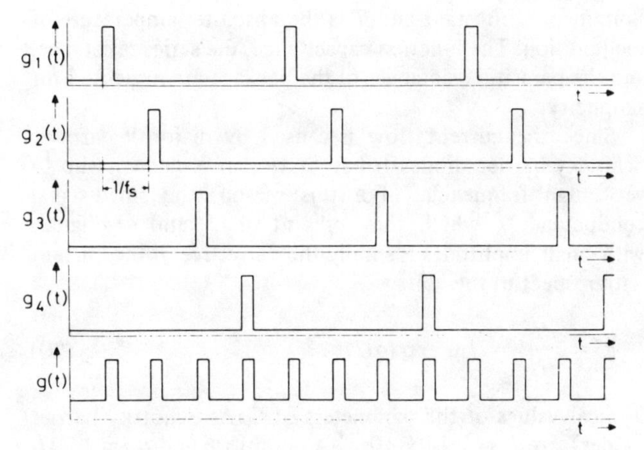

Fig. 3 *Resulting small-signal conductance $g(t)$ when four equal mixers with small-signal conductance $g_{1 \ldots 4}$ are connected in parallel*

Each mixer is pumped with a frequency of f_s/M
The time shift between adjacent mixers is f_s^{-1}

Summing over m yields

$$g(t) = MG_0 + \sum_{\substack{n'=-\infty \\ n' \neq 0 \\ n'=kM}}^{\infty} M \frac{\Delta G}{n'\pi} \sin\left(n'\pi \frac{\theta}{M}\right) \exp\left(jn'2\pi \frac{f_s}{M} t\right) \quad (16)$$

where k is an integer, $-\infty \leqslant k \leqslant \infty$.

Finally, substituting for n'/M with n leads to

$$g(t) = (M-1)\,G_0 + \sum_{n=-\infty}^{\infty} \frac{\Delta G}{n\pi} \sin(n\pi\theta) \exp(jn2\pi f_s t) \quad (17)$$

Because this expression has basically the same form as eqn. 1, the circuit proposed may indeed be used as a frequency convertor.

The conversion losses of the harmonic mixer are, however, somewhat higher than those of conventional mixers. The reason for this is that the mean values of the conductance functions of eqns. 1 and 17 are not equal. These values become equal if the conductance G_{min} vanishes. In this case the conversion losses of the conventional and of the proposed convertor are identical.

4 Schottky-barrier diodes as examples of resistive frequency-conversion devices

Schottky-barrier diodes are near-ideal metal-semiconductor devices having an exponential current/voltage function, as follows:

$$I = I_s \{\exp(\alpha U) - 1\} \quad (18)$$

$$\alpha = \frac{e}{mkT} \quad (19)$$

where I is the current flowing through the diode, U is the voltage at the diode, I_s is the saturation current, e is the charge of an electron, m is the ideality factor, k is Boltzmann's constant and T is the absolute temperature of the junction. The junction capacitance, the series resistance, and the parasitic reactances of the package are neglected for simplicity.

Since the current flow is caused by majority carriers, Schottky-barrier diodes behave as resistive elements up to very high frequencies. For this reason, the small-signal conductance g which they present to i.f. and r.f. signals with small amplitudes is simply the derivative of the current with respect to the voltage:

$$g = \frac{dI}{dU} = I_s \alpha \exp(\alpha U) \quad (20)$$

Typical values of the parameters of GaAs Schottky-barrier diodes* are $I_s = 2{\cdot}14 \times 10^{-14}$ A, $m = 1{\cdot}06$ and $R_s = 1{\cdot}7\,\Omega$. These values were obtained by measuring the current/voltage function and evaluating it with a computer program.

*1SS11 type, NEC

5 Conventional frequency convertors with Schottky-barrier diodes

We assume that a sinusoidal pump signal is applied to the diode. Then a time-dependent small-signal conductance of the following form results:

$$g(t) = I_s \alpha \exp(\alpha U_{dc}) \exp(\alpha \hat{u}_s \cos 2\pi f_s t) \quad (21)$$

where U_{dc} is the d.c. voltage at the diode and $\hat{u}_s$ is the pump voltage amplitude. This expression can be rewritten to give

$$g(t) = I_s \alpha \exp(\alpha U_{dc}) \sum_{n=-\infty}^{\infty} \underline{G}_n \exp(jn2\pi f_s t) \quad (22)$$

with

$$\underline{G}_n = I_n(\alpha \hat{u}_s) \quad (23)$$

where $I_n(\alpha\hat{u}_s)$ is a modified Bessel function of order n.[2]

Fig. 4 shows the conductance waveform of the pumped Schottky-barrier diode. The pulse duty ratio θ is defined here as the percentage of time, where

$$g(t) > \tfrac{1}{2}(G_{max} + G_{min}) \quad (24)$$

The conductance pulses do not differ much from the idealised pulses of Fig. 1.

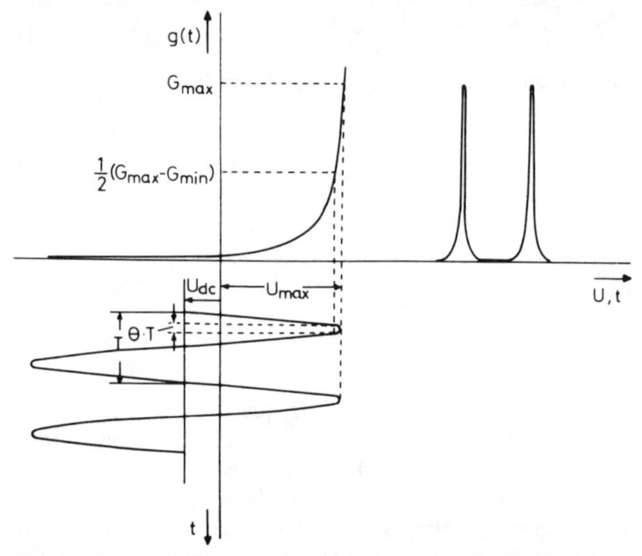

Fig. 4 *Conductance waveform g(t) of a pumped Schottky-barrier diode*

$T = 1/f_s$ is the period of the pump voltage

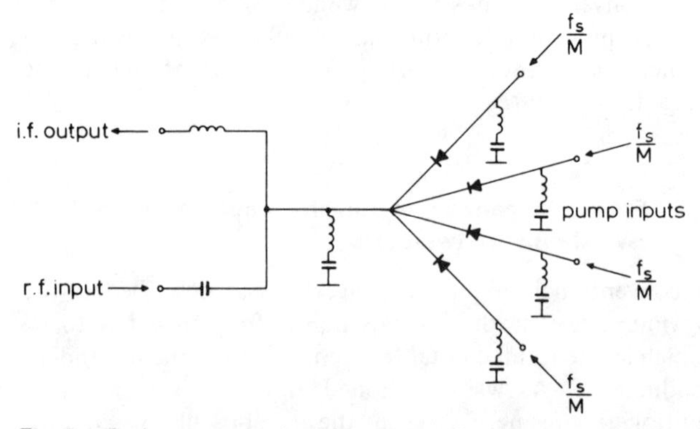

Fig. 5 *Basic structure of a subharmonically pumped mixer (M = 4)*

M Schottky-barrier diodes are connected in parallel and pumped with a frequency of f_s/M and phase difference of $2\pi/M$ from each other

6 Harmonic mixers with Schottky-barrier diodes

Next, we assume M Schottky-barrier diodes to be connected in parallel and to be pumped by sinusoidal signals having a frequency of f_s/M and phase differences of $2\pi/M$ from each other.

Fig. 5 shows the basic structure of the whole configuration. Not included are the networks that distribute the pump power from the source to the diodes.

The series resonant circuits between the diodes and the pump power inputs short the r.f. signals to ground. The series resonator on the other side of the diode is a short for the pump signals. The i.f. and the r.f. signal are separated with a branching network. The network shown in Fig. 5 is a 1st-order branching filter consisting of an inductance and a capacitance.

The conductance functions of the diodes and the resultant one are pulse trains with pulse rates of f_s/M and f_s, respectively. The shape of the pulses is derived in Fig. 4. When the pump power and the d.c. bias voltage of the diodes are adjusted, two conditions have to be fulfilled. The first is that the conductance $g(t)$ has to reach a certain value G_{max}. The second condition arises from the need not to exceed that pulse duty ratio at which the conversion loss has its minimum. This minimum arises whenever the mixer is loaded by fixed impedances at the input and output. This will be shown in Section 6.1 for a mixer pumped at the first subharmonic.

Assuming that the first condition is fulfilled, the second can be put into the following form:

$$\theta = \frac{M}{\pi} \cos^{-1}\left[1 + \frac{1}{\alpha\hat{u}_s}\left\{\ln \frac{1 + \exp(-2\alpha\hat{u}_s)}{2}\right\}\right] \quad (25)$$

Here the pulse duty ratio θ is given as a function of M and the pump voltage amplitude $\hat{u}_s$. It would be more comfortable if this expression were solved for $\hat{u}_s$, but this is only possible if

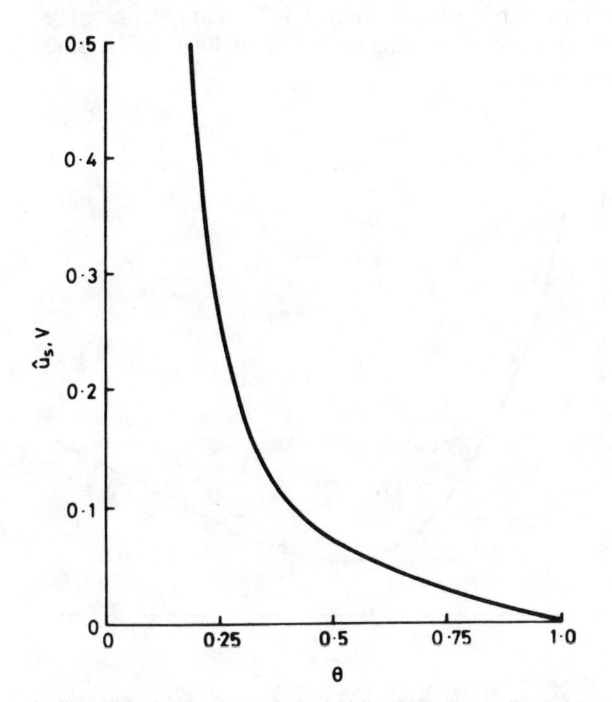

Fig. 6 *Pump voltage amplitude $\hat{u}_s$ as a function of the pulse duty ratio θ; evaluation of eqn. 25*

$$\hat{u}_s \gg \frac{1}{\alpha} \quad (26)$$

Provided eqn. 26 holds, eqn. 25 can be written as

$$\hat{u}_s = \frac{(1/\alpha)\ln 2}{1 - \cos(\theta\pi/M)} \quad (27)$$

From Fig. 6 it can be seen that, with $\hat{u}_s$, the d.c. bias voltage is also known:

$$U_{dc} = U_{max} - \hat{u}_s \quad (28)$$

where U_{max} is given as

$$U_{max} = \frac{1}{\alpha} \ln \frac{G_{max}}{\alpha} \quad (29)$$

Eqn. 25 is evaluated in Fig. 6 for $M = 2$ and the diode parameters given earlier. The diagrams show that large pump voltage amplitudes are necessary to obtain small pulse duty ratios.

When the number M of the diodes connected in parallel grows, the pulse duty ratio increases, as can be seen from eqn. 25. For compensation, the pump voltage amplitude has to be enlarged. This sets a natural limit to the frequency-conversion principle proposed in this paper, because the pump power is restricted and the breakdown voltage of the diodes is not infinite.

In practice, the main problem is to find simple circuits that fulfil the pump power-distribution and phase-shifting functions. In the following two Sections we give two examples of such circuits, one for $M = 2$ and the other for $M = 4$.

6.1 *Second-harmonic frequency convertors with Schottky-barrier diodes*

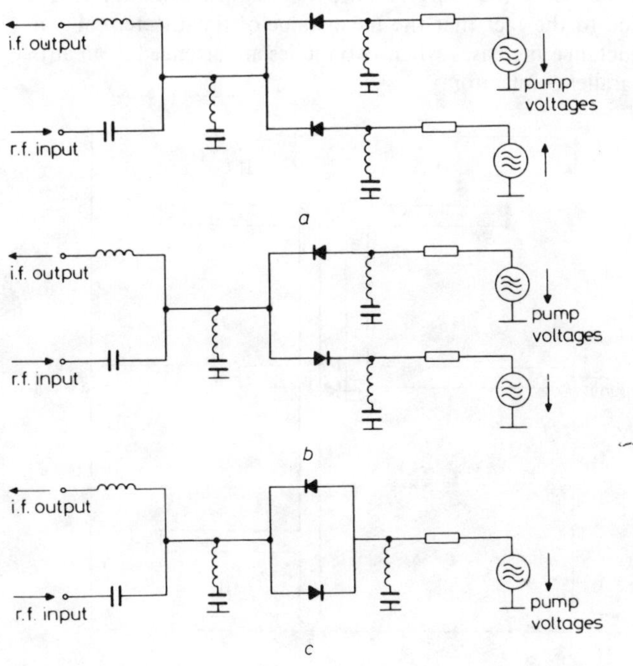

Fig. 7 *Basic structure of a mixer pumped at the first subharmonic ($M = 2$)*

a Phase difference of π between the pump voltages
b Pump voltage in phase, one of the diodes has to be reversed
c Reduced network

If M equals 2 the pump frequency must be $f_s/2$, and the phase difference between the pump signals of the two diodes must equal π. In this case, the circuit of Fig. 5 reduces to the one shown in Fig. 7*a*.

This circuit can be simplified significantly. Because the small-signal conductance does not depend on the polarity of a diode, no essential change occurs if one of the two diodes and the associated pump source are reversed (Fig. 7*b*). After this change the two pump sources are in phase. Therefore one of them can be eliminated, so that finally one single pump source remains (Fig. 7*c*).

The simplification leads to a network that consists essentially of two diodes in an antiparallel connection. This configuration was first published in 1971 and has spread since then.[1, 3-6]

A microstrip layout of this circuit is shown in Fig. 8. The shorted stub I does not impede the pump signal but shorts the r.f. signal to ground. The open stub II, on the other hand, shorts the pump signal without affecting the r.f. signal. The branching filter III presents an infinite impedance to the r.f. signal, looking from the diodes to the i.f. load. The capacitor IV keeps the i.f. power from the r.f. input. Therefore the i.f. power can be extracted without significant loss of r.f. power.

For the mixer shown in Figs. 7 and 8, the conversion losses have been calculated. For details, see Appendix 9. As a result of the investigations, Fig. 9 shows the conversion losses L of a normally (L_1) and a subharmonically (L_2) pumped mixer. All the curves hold for the same loading at signal and image frequency (matched image). The solid lines show the conversion losses L_{1m} and L_{2m} when the mixers are matched at the r.f. input and the i.f. output; the dashed lines (L_{1r}, L_{2r}) hold for fixed loads (50 Ω) at the input and the output. Fig. 9 shows that the conversion loss of a subharmonic mixer with $M = 2$ is about 0·5 dB higher than the conversion loss of a normal mixer. This is due to the fact that the mean value of the differential conductance increases when two diodes are arranged in an antiparallel connection.

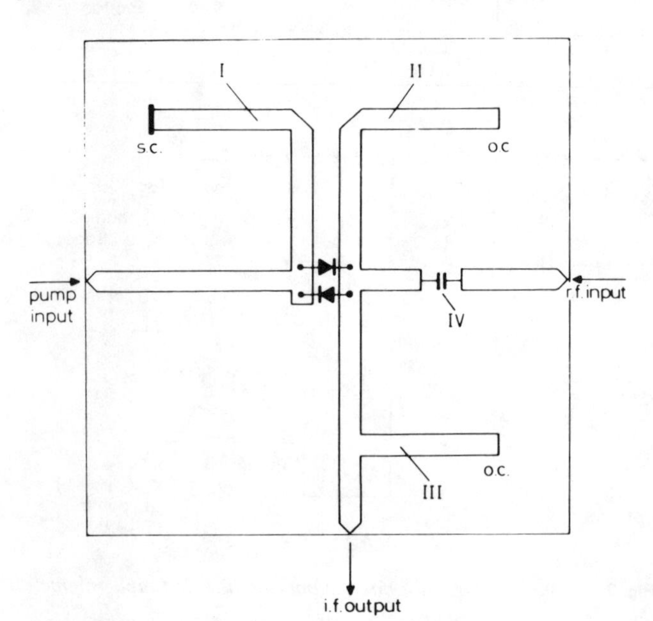

Fig. 8 *Microstrip layout of a 2nd-harmonic mixer*

s.c. = short circuit, o.c. = open circuit

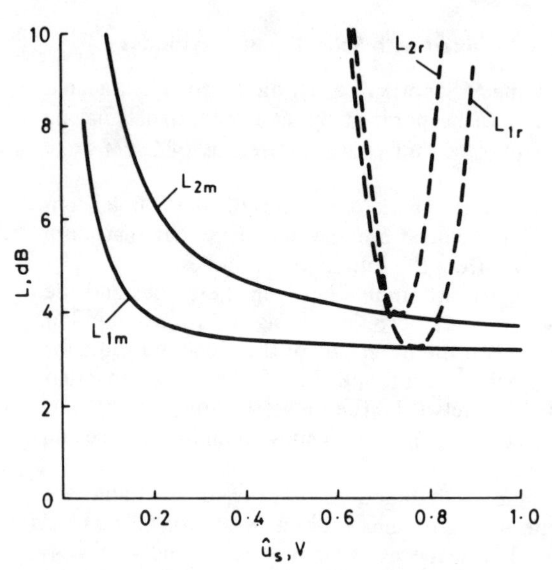

Fig. 9 *Calculated conversion losses L of a conventionally (L_1) and subharmonically (L_2, $M = 2$) pumped mixer (image matched) as a function of the pump voltage amplitude $\hat{u}_s$*

Solid lines: conversion losses when the r.f. inputs and i.f. outputs of the mixer are matched

Dashed lines: conversion losses when the mixers are loaded with fixed loads (50 Ω) at the input and output

Experiments with the circuit of Fig. 8 have yielded a minimum conversion loss of 6·1 dB. Fig. 10 shows the conversion loss L as a function of the pump power P_s. The measurements have been performed at a pump frequency of about 2·5 GHz and a signal frequency (r.f.) of about 5 GHz. The intermediate frequency was about 70 MHz. The conversion loss is about 2–3 dB higher than that of a well designed conventional mixer with reactive terminations at the image and sum frequency, but the mixer considered here is a so-called broadband mixer with the same (real) loading at image and r.f. Such mixers have a theoretical minimum loss of 3 dB (generally 4·5–5 dB in practice). The loss of our mixer is even higher, by about 1 dB. The reasons for this are:

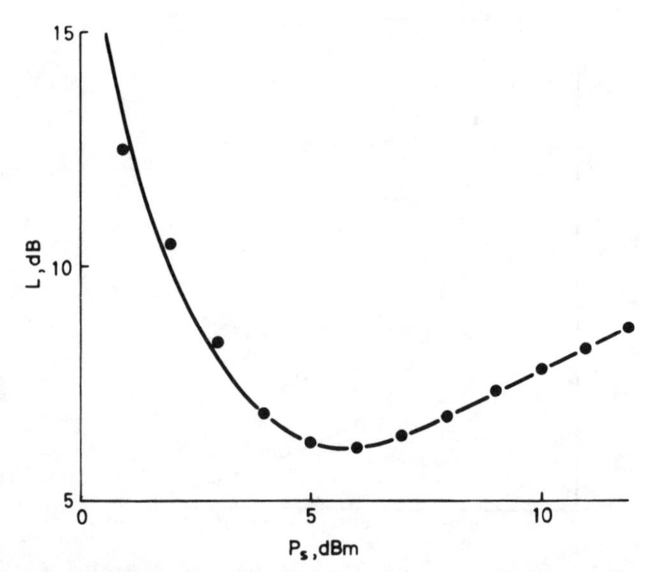

Fig. 10 *Measured conversion loss L of a 2nd-harmonic mixer as a function of the pump power P_s*

(*a*) The theoretical increase of loss, which is about 0·5 dB for a subharmonic mixer with $M = 2$.

(*b*) No attempt was made to reach a lower conversion loss by applying a d.c. bias voltage.

(*c*) If the diode parameters are not the same, there occurs a mixing process between the signal and the fundamental pump frequency. These mixing products are lost for conversion because they do not fall within the i.f. band. So one has to take care for equal diode parameters.

6.2 *Fourth-harmonic frequency convertors with Schottky-barrier diodes*

For $M = 4$, the pump frequency needed is only $f_s/4$, and phase difference of $\pi/2$, π and $3\pi/2$ occur. The circuit of Fig. 7 consists of four parallel branches in this case.

Again, in two of the branches the diode and the associated pump source can be reversed and, consequently, two pump sources can be eliminated. The resultant circuit contains two mixers with antiparallel diodes, as described in Section 6.1. Fig. 11 shows a microstrip layout of this circuit.

The two pump sources that are still necessary must have a phase difference of $\pi/2$ from each other. They can therefore be derived from one pump source, the power of which is divided with a 3 dB directional coupler or with a 3 dB hybrid. As is well known, 3 dB couplers and hybrid rings consisting of four quarterwave line sections have output signals of equal amplitudes and a phase difference of $\pi/2$ from each other.

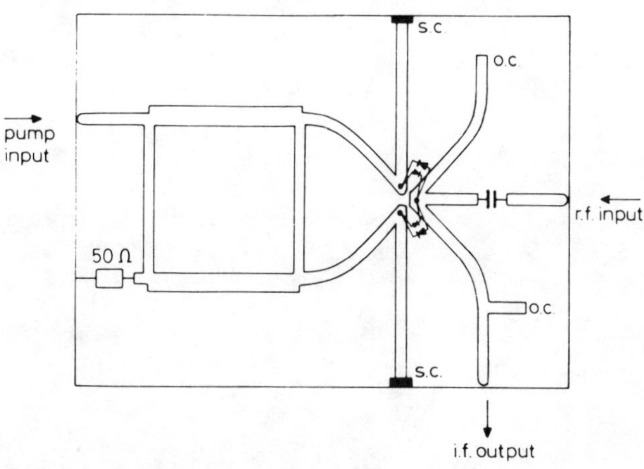

Fig. 11 *Microstrip layout of a 4th-harmonic mixer*
s.c. = short circuit, o.c. = open circuit

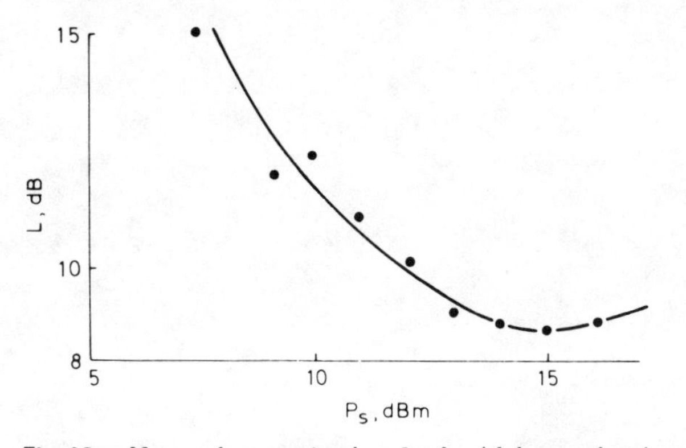

Fig. 12 *Measured conversion loss L of a 4th-harmonic mixer as a function of the pump power P_s*

Essential parts of the 4th-harmonic mixer are two 2nd-harmonic mixers. On the pump side they are connected with a 3 dB hybrid. On the signal side they are connected directly or with lines which are multiples of a half wavelength long.

Fig. 12 shows the conversion loss L as a function of the pump power P_s measured with this circuit. Since the pump frequency was about 2·5 GHz, we used a signal frequency of about 10 GHz. The minimum conversion loss was about 8·6 dB.

7 Conclusions

It has been shown that harmonic mixers with several diodes in a parallel connection can replace the mixer and the l.o. multiplier of a conventional frequency convertor. This is advantageous, expecially in the millimetre-wave region. Whereas theoretical investigations showed the conversion losses of a mixer pumped at the first subharmonic to be about 0·5 dB greater than a conventional mixer, the losses measured at a subharmonically pumped microstrip mixer with an r.f. of about 5 GHz were about 1 dB higher than that of a conventional broadband mixer. In addition, there are other advantages that have not been discussed so far. One of them is the noise behaviour of the harmonic mixers, which is very favourable because the noise of the pump source does not contribute to the noise of the signals involved in the conversion process. For these reasons, harmonic mixers with a subharmonic pump are promising alternatives for conventional frequency convertors.

8 References

1 BÜCHS, J.-D.: 'Ein subharmonisch gesteuerter Schottkydioden-mischer in Streifenleitungstechnik', *Arch. Elekt. & Übertragungstech.*, 1975, **25**, pp. 52–53
2 BARBER, M.R.: 'Noise figure and conversion loss of the Schottky barrier mixer diode', *IEEE Trans.*, 1967, **MTT-15**, pp. 629–635
3 SCHNEIDER, M.V.: 'Harmonically pumped stripline down-convertor'. Proceedings of the European microwave conference, Montreux, 1974, pp. 599–603
4 SCHNEIDER, M.V., and SNELL, W.W.: 'Stripline down-convertor with subharmonic pump', *Bell Syst. Tech. J.*, 1974, **53**, pp. 1179–1183
5 SCHNEIDER, M.V., and SNELL, W.W.: 'Harmonically pumped stripline down-convertor', *IEEE Trans.*, 1975, **MTT-23**, pp. 271–275
6 COHN, M., DEGENFORD, J.E., and NEWMAN, B.A.: 'Harmonic mixing with an antiparallel diode pair', *ibid.*, 1975, **MTT-23**, pp. 667–673
7 SALEH, A.A.M.: 'Theory of resistive mixers' (Research monograph 64, MIT Press, 1971)

9 Appendix

Neglecting parasitic effects, the current/voltage characteristic of two Schottky diodes in an antiparallel connection reads[1]

$$I = I_s \sinh(\alpha U) \tag{30}$$

With a pump-voltage of

$$U = U_{dc} + \hat{u}_s \cos(2\pi f_s t) \tag{31}$$

the Fourier expansion of the differential conductance

$$g(t) = 2\alpha I_s \cosh(\alpha U) \quad (32)$$

results in

$$g(t) = \alpha \cdot I_s e^{\alpha U_{dc}} (1 + e^{-2\alpha U_{dc}}) I_0(\alpha \hat{u}_s) + 2\alpha I_s e^{\alpha U_{dc}} \sum_{n=1}^{\infty} I_n(\alpha \hat{u}_s) \{1 + (-1)^n e^{-2\alpha U_{dc}} \cos(n 2\pi f_s t)\} \quad (33)$$

Assuming U_{dc} to be zero, eqn. 33 can be simplified to give

$$g(t) = 2\alpha I_s \{I_0(\alpha \hat{u}_s) + 2I_2(\alpha \hat{u}_s) \cos(4\pi f_s t) + 2I_4(\alpha \hat{u}_s) \cos(8\pi f_s t) + \ldots\} \quad (34)$$

Using the same assumptions, eqn. 22 can be rewritten to give

$$g(t) = \alpha I_s \{I_0(\alpha \hat{u}_s) + 2I_1(\alpha \hat{u}_s) \cos(2\pi f_s t) + 2I_2(\alpha \hat{u}_s) \cos(4\pi f_s t) + \ldots\} \quad (35)$$

Since the pump frequency of a subharmonic mixer with $M = 2$ is half that of a conventional mixer, one can rewrite eqn. 34, which finally results in

$$g(t) = 2\alpha I_s \{I_0(\alpha \hat{u}_s) + 2I_2(\alpha \hat{u}_s) \cos(2\pi f_s t) + 2I_4(\alpha \hat{u}_s) \cos(4\pi f_s t) + \ldots\} \quad (36)$$

Except for the mean values of the differential conductances, there exists a formal identity between eqns. 35 and 36 if one substitutes I_n by I_{2n}, and so one can apply the methods of References 2 and 7 to calculate the conversion loss and the input and output impedances.

Subharmonically Pumped Millimeter-Wave Mixers

ERIC R. CARLSON, MEMBER, IEEE, MARTIN V. SCHNEIDER, FELLOW, IEEE, AND THOMAS F. McMASTER, MEMBER, IEEE

Invited Paper

Abstract—**The two-diode subharmonically pumped stripline mixer has a pair of diodes shunt mounted with opposite polarities in a stripline circuit between the signal and local oscillator inputs. The circuit has low noise and conversion loss and substantial AM local oscillator noise cancellation. The local oscillator frequency is about half the signal frequency. A novel diode chip, the notch-front diode, which has ohmic contacts on the chip faces adjacent the face containing the diode junctions, was developed for these circuits. The notch-front diode permits the low parasitic reactance of the waveguide diode mount to be achieved in stripline circuits. The best performance for a two-diode subharmonically pumped mixer with notch-front diodes was a 400 K mixer noise temperature, obtained at 98 GHz, which is comparable to the best fundamental mixers in this frequency range. The performance over a 47–110-GHz frequency range for this circuit with commercial beam-lead diodes is also presented.**

I. INTRODUCTION

THE TWO-DIODE subharmonically pumped hybrid integrated downconverter [1], [2], referred to here as the two-diode mixer, has many desirable properties which make it an interesting alternative to conventional mixers, especially at millimeter wavelengths. The local oscillator requirements are easier to meet in a two-diode mixer because the LO frequency is about half that in the corresponding conventional mixer, and because the two-diode mixer has substantial AM local oscillator noise suppression [3]. The large difference between the signal and LO frequencies simplifies the design of a filter to separate these frequencies and permits elimination of the potentially lossy diplexing arrangement used at the signal input of conventional mixers. The position of this stripline filter can be readily changed to optimize the impedance seen by the diode looking toward the local oscillator port. The two-diode mixer described here can be tuned for either single-sideband or double-sideband response, and covers an entire waveguide band. Due to the symmetry of the circuit, the diode-pair current contains no even harmonics of the LO [4], so that no dc return path is needed, and emission from the mixer in the signal frequency band is suppressed. Passive elements are incorporated in the strip transmission line circuit with high precision by conventional photolithographic techniques rather than by intricate machining. And, as discussed below, the noise and loss performance of the two-diode mixer is comparable to the best fundamental mixers.

When compared with second-harmonic mixers employing a single diode [5], [6], the two-diode mixer has another important advantage. The conductance waveform of the diode pair contains only even harmonics of the LO [4], so the two-diode mixer is inherently insensitive to frequencies in the band corresponding to fundamental mixing with the local oscillator. Response at these frequencies must be suppressed by the embedding network of the single-diode second-harmonic mixer, and a nonreactive termination will result in a contribution to the mixer noise. Furthermore, in practice, in single-diode second-harmonic mixers, a filter structure to separate the signal and LO frequencies has not been used, so that the impedance at the signal frequency seen by the diode looking toward the LO port has not been optimized.

A novel diode chip geometry, which we call the notch-front diode [7], was developed for use in strip transmission line circuits. This chip has ohmic contact metallization on the chip faces adjacent to the face containing the diode junctions. Notch-front diodes can be readily soldered to millimeter-wave thin-film circuits. They are particularly suited for use in conventional and subharmonically pumped millimeter-wave mixers because the reduced parasitic capacitance in comparison with beam-lead diodes results in a better switching waveform and because the chip geometry has a lower RF series resistance than conventional millimeter-wave diode structures.

Results have been obtained over a 66–110-GHz frequency range for notch-front and beam-lead diodes in the two-diode downconverter. The best performance with notch-front diodes, a single-sideband (SSB) mixer noise temperature of 400 K at 98 GHz, is comparable to the best results of fundamental mixers in this frequency range. In comparison, beam-lead diodes in this mixer at 98 GHz had a SSB mixer noise temperature of 1600 K and required 10 dB more local oscillator power. At lower frequencies, the performance of the beam-lead diode mixer improves as parasitic effects become less important.

These two-diode downconverters look promising for scaling to other frequencies. Potential applications include

Manuscript received November 3, 1977.

E. R. Carlson and M. V. Schneider are with Bell Laboratories, Crawford Hill Laboratory, Holmdel, NJ 07733.

T. F. McMaster is with Bell Laboratories, Holmdel, NJ 07733.

Reprinted from *IEEE Trans. Microwave Theory Tech.*, vol. MTT-26, pp. 706–715, Oct. 1978.

microwave and millimeter-wave communications, plasma diagnostics, collision avoidance radar, and radio astronomy.

This paper mainly discusses results in the 66–110-GHz frequency range for two-diode mixers with a WR-10 input waveguide. Preliminary reports of these results have appeared [4], [8]. The design and performance of two-diode mixers at 5 GHz and 50 GHz have also been published [4] and are briefly reviewed here.

II. Circuit Description

Stripline, i.e., suspended-substrate strip transmission line, was used in the design of this mixer because it has lower loss and less excitation of higher modes than conventional microstrip. The mixer circuit was optimized in a large scale model at a signal frequency of about 5 GHz, and then all circuit dimensions were reduced by the ratio of the model and millimeter-wave signal frequencies to make the millimeter-wave mixers [4].

A schematic view of the stripline conductor pattern is shown in Fig. 1, and a photograph of the millimeter-wave mixer with the cover removed is presented in Fig. 2. The circuit consists of a signal waveguide input section, a waveguide to stripline transition, a stripline conductor pattern including mounting pads for a pair of Schottky barrier diodes and two low-pass filters, and a transition from the LO waveguide to the stripline. The signal waveguide to stripline transition, illustrated in Fig. 3, can be tuned by adjusting the waveguide backshort and the *H*-plane waveguide short so that the downconverter can be operated either as a single-sideband or as a double-sideband mixer.

The input signal is coupled to a pair of Schottky barrier diodes which are shunt mounted on the stripline with opposite polarities. The notch-front diodes were soldered to the stripline conductor. The commercial beam-lead diodes [9] were thermocompression bonded to the circuit.

Two low-pass filters are needed to separate the signal frequency $\omega_s = 2\omega_p \pm \omega_{IF}$, the LO frequency ω_p and the intermediate frequency ω_{IF}. The filter adjacent to the pair of diodes has a cutoff frequency of 59 GHz in order to reject the signal (66–110 GHz) while transmitting the LO (32.3–54.3 GHz) and the IF (1.4 GHz). In the lumped-element approximation, it is a seven-element *L–C* ladder type low-pass filter with a Chebyscheff response. The second low-pass filter was designed empirically to reject the LO frequency and transmit the IF.

The stripline conductor pattern is fabricated using standard thin-film and photolithographic techniques. The substrates are polished optical-grade fused quartz 0.12 mm thick. The conductor metallization, deposited by evaporation, is 2.0–2.5 μm of gold on top of 75–150 Å of chromium. The surface finish of the metallization was found to be important: mixers made on substrates with matte finish metallization had conversion losses almost one dB higher and mixer noise temperatures several hundred degrees higher than mixers made on substrates with mirror finish metallization.

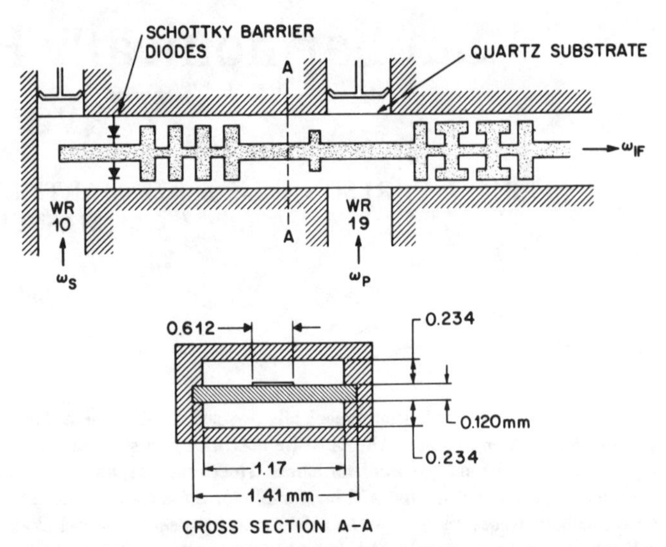

Fig. 1. Top view and cross-sectional view of the stripline circuit with signal and LO waveguide input ports. The LO waveguide was WR-15 for the first version of the circuit, including mixer 113.

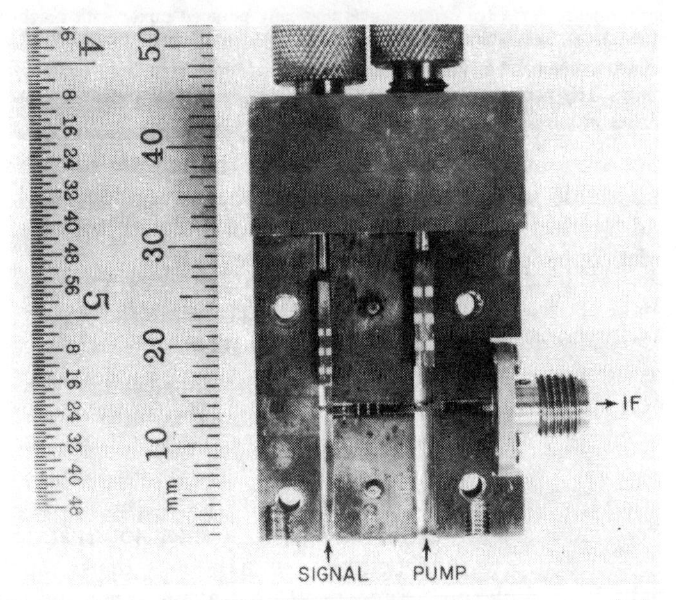

Fig. 2. Photograph of the low noise mixer. The top cover of the housing is removed to show the conductor pattern on the quartz substrate and the mounted diodes.

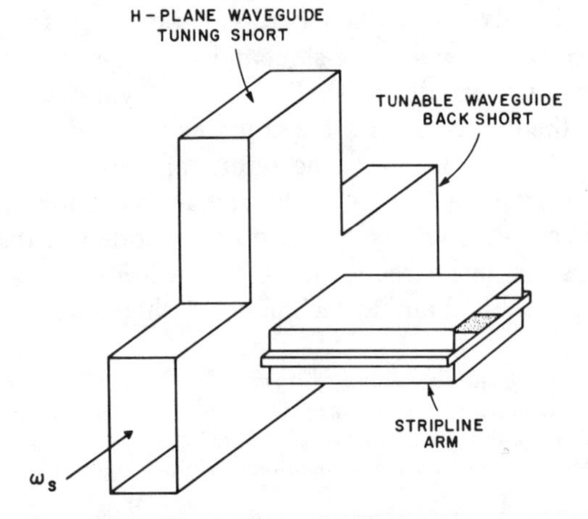

Fig. 3. Transition from signal waveguide to stripline circuit with a tunable waveguide backshort and an *H*-plane tunable short.

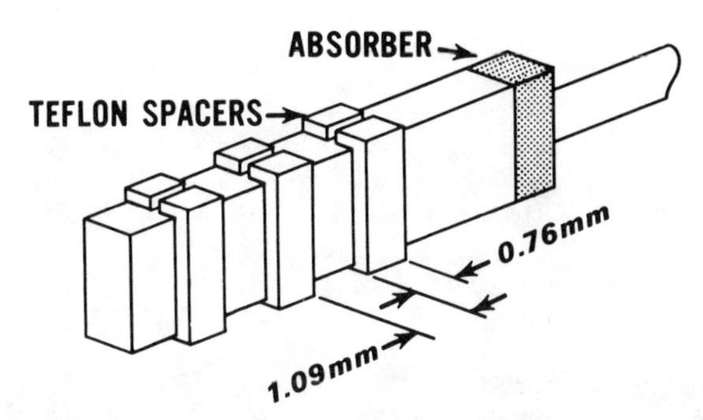

Fig. 4. Sliding noncontacting waveguide short used in these mixers. The dimensions shown are for WR-10 waveguide in which the narrow sections are 0.20 mm wide and the wide sections are 1.19 mm.

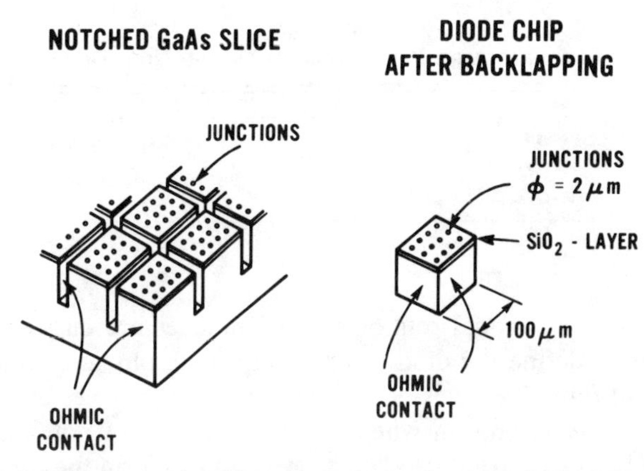

Fig. 5. Schematic view of notched GaAs slice and diode chip after backlapping.

The circuit housing was machined from OFHC copper. The first housing had a WR-15 local oscillator waveguide scaled from the low-frequency model, and the cutoff of this waveguide placed a lower limit of about 85 GHz on the circuit operation. Subsequently it was found that the LO waveguide could be changed to WR-19 and circuit operation extended below 70 GHz without any penalty in performance at higher frequencies.

The sliding noncontacting shorts used in both the signal and LO waveguides had alternating high and low impedance sections as shown in Fig. 4. The shorts were smooth and stable in operation and across the waveguide band had a reflection coefficient greater than or equal that of a solid copper plate terminating the waveguide.

III. Notch-Front Diodes

A novel diode chip geometry was developed which has low parasitics and can readily be mounted in strip transmission line circuits. This structure, called the notch-front diode [7], has ohmic contacts and an array of junctions fabricated on adjacent sides of a chip as shown in Fig. 5. A silicon dioxide layer with a thickness of 5000 Å is first deposited on the epitaxial layer side of a gallium arsenide slice. An array of notches is cut into the GaAs slice with a diamond saw blade. The slice has a thickness of approximately 300 μm and the notches are cut to a depth of 100 μm. An ohmic contact is formed in the notches by electroplating Sn–Ni, Ni, and Au into the notches and subsequently alloying the slice at a temparature of 400°C for 60 s. The ohmic contact metallization is about 1 μm thick. The notches are filled with photoresist and an array of Schottky barrier diodes is fabricated on the front of the slice by a sequence of processing steps involving masking with photoresist, plasma etching of holes into the SiO_2 layer which covers the top surface of the slice, and electroplating of platinum and gold on the exposed GaAs surface in the hole areas. A pulse plating technique developed by Burrus [10] is used to obtain uniform deposition of the metal films in all the hole areas. A description of the junction formation steps is given in a separate paper [11].

N-type vapor phase epitaxy material was used in this work. The epitaxial layer was about 1200 Å thick and was doped with sulphur to a carrier concentration of about 2×10^{17} cm^{-3}. The dopants for the buffer layer and substrate were sulphur and silicon, respectively, and the carrier concentration was about 3×10^{18} cm^{-3}.

Fig. 6. Scanning electron micrograph of notch-front diodes with a dice size of 75 μm×80 μm and a notch depth of 100 μm. The junction diameter is 2 μm with a center-to-center spacing of 5 μm.

Fig. 6 is a scanning electron micrograph of the slice after fabricating the notches with ohmic contacts and the junctions on top of the slice. The width of the notches is about 70 μm. While some overplating of gold is visible around some parts of the top periphery of each chip, this overplating does not affect the performance of the final device. The individual junctions on the top surface of the slice have a diameter of 2 μm and a center-to-center spacing of 5 μm. With this array of junctions, a diode may be contacted anywhere on the surface of the chip [12]. Individual diode chips are fabricated by mounting the notched slice on a glass slide with wax with the top surface down and backlapping the slice to a thickness which is smaller than the depth of the notches. The chips are separated by dissolving the wax in a solvent. An individual diode chip after the backlapping of the slice is illustrated in Fig. 5. A typical length of the side of the chip is 75 μm. The minimum length is determined by the depth of the damage created by the diamond saw blade which is about 10 μm.

TABLE I
LOW-FREQUENCY PARAMETERS FOR NOTCH-FRONT DIODES

Batch	R_s (Ω)	C_o (fF)	n	V_B(V)
B20-99	8	7	1.13	9
B20-119	7	8	1.12	8
B17-122A	5	8	1.15	8
Conventional Chips	4	8	1.15	8

Two orthogonal sets of parallel notches are cut in the slice to define the diode chips. Chips with ohmic contacts on all four side faces or on two opposite side faces can be made depending on whether the second set of notches is cut before or after the slice is processed to form the ohmic contacts and the diodes. Both types have been made; better diodes were obtained on chips with ohmic contacts on two side faces due to details of the processing techniques.

Typical parameters for three batches of notch-front diodes are shown in Table I. The series resistance R_s measured at 10-mA forward current, and the n-factor are taken from dc $I-V$ characteristics for chips mounted in circuits, The reverse breakdown voltage V_B and the zero-bias junction capacitance C_0 were measured in unmounted chips. A 1-MHz capacitance bridge was used to measure C_0. The series resistance was reduced as processing technology improved through experience. Also listed in Table I are the parameters for one of our best conventional diodes with the ohmic contact on the back of the chip. The low-frequency parameters for the notch-front diodes are not significantly different from those for the coventional chip. The commercial beam-lead diodes used in these circuits typically had $R_s \approx 3.5$ Ω and a total capacitance of about 60–75 fF.

Although these low-frequency measurements may not be sufficient to predict completely the performance of a mounted device at millimeter wavelengths [13], [14], they are useful as the most widely compared characteristics of Schottky barrier diodes. Sources of the series resistance for a millimeter-wave diode are considered in Appendix A, where it is shown that the series resistance contributed by the skin effect in the notch-front diode is about 2/3 that in a conventional chip. However, the total series resistance in both cases is dominated by effects near the junction.

Notch-front diodes are mounted on the stripline conductor of the millimeter-wave mixers as shown in Fig. 7. The chip is soldered on one side of a gap in the conductor with an indium-based solder [15], and contact is made to a diode on the chip by a pointed spring wire soldered on the other side of the gap with another, lower temperature, indium-based solder [16]. The wire used for the diode contact is 12-μm diameter Phosphor Bronze A. A point is etched on the end of the wire at 1.2-V dc in an electrolyte consisting of 23-ml H_2SO_4, 80-ml H_3PO_4, 1.5-g CrO_3, and 50-ml H_2O [17], with a final dilution of about one part H_2O to ten parts of the above solution to allow for constituent concentration variations. A typical point, shown in Fig. 8, has a tip radius of about 1 μm to contact the 2-μm diameter junctions without penetrating to the epilayer, and a large included angle to reduce parasitic inductance and minimize susceptibility to damage during the contacting process. A spring having the shape shown in Fig. 7 is bent with a micromanipulator, and the wire is then stress relieved for one hour at 218°C in a forming gas atmosphere. The two chips are simultaneously soldered to the substrate on a strip heater. The substrate with the

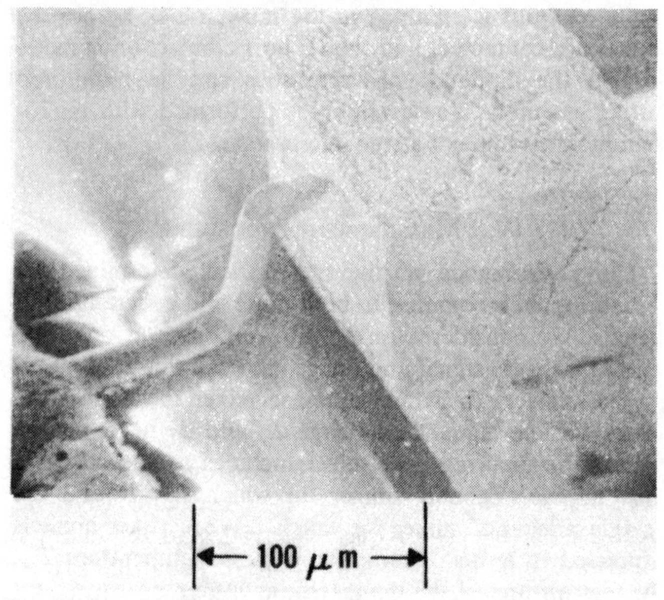

Fig. 8. Scanning electron micrograph of a pointed 12-μm diameter phosphor bronze wire used for making contact to Schottky barrier diodes.

Fig. 8. Scanning electron micrograph of a pointed 12-μm diameter phosphor bronze wire used for making contact to Schottky barrier diodes.

mounted chips is clamped in the mixer block before the diodes are contacted. and the IF port connection is made so that the diode dc characteristics can be monitored during assembly. The assembly is performed with micromanipulators under a stereo microscope.

IV. Measurement Procedure

The performance of the mixers was determined by measuring their response to both noise and coherent input signals. We consider a mixer with responses at only two frequencies, the signal and image, and call this a "dual-response" mixer. In a dual-response mixer the conversion losses for the signal and image L_s and L_i need not be equal. The dual-response mixer includes as special cases the "double-sideband" mixer for which $L_s = L_i$ and the "single-sideband" mixer for which $L_i = \infty$. Mixer noise is expressed in terms of mixer input noise temperature T_M, the temperature of the input termination on an equivalent noise-free mixer which would produce the same output noise power as the actual mixer with a noise-free input termination [18]. Similarly, a receiver input noise temperature T_R can be defined which includes the noise contributed by both the mixer and the IF amplifier.

For a dual-response receiver, the result of a Y-factor measurement [19] using a broad-band noise source is a "dual-sideband" receiver noise temperature T_R (DSB). If the signal of interest appears in the signal sideband only, then the appropriate measure of receiver noise is the single-sideband receiver noise temperature, given by

$$T_R(\mathrm{SSB}) = \left(1 + \frac{L_s}{L_i}\right) T_R(\mathrm{DSB}). \tag{1}$$

A similar equation applies to mixer noise temperature. The mixer and receiver noise temperatures are related by

$$T_R(\mathrm{SSB}) = T_M(\mathrm{SSB}) + L_s T_{\mathrm{IF}} \tag{2a}$$

or by

$$T_R(\mathrm{DSB}) = T_M(\mathrm{DSB}) + \frac{L_s}{1 + \frac{L_s}{L_i}} T_{\mathrm{IF}} \tag{2b}$$

where T_{IF} is the noise temperature of the IF amplifier. Some results are expressed in terms of the receiver noise figure given by [20]

$$F_R(\mathrm{SSB}) = \left(1 + \frac{L_s}{L_i}\right)\left[1 + \frac{T_R(\mathrm{DSB})}{290}\right]. \tag{3}$$

The relationships between single-sideband and dual-sideband noise temperature and noise figure are summarized in Fig. 9.

The noise power at the IF output P_n was measured for three different conditions [18]:

1) mixer terminated with a hot load:

$$P_n(H) = \left[(T_H + T_M)\left(\frac{1}{L_s} + \frac{1}{L_i}\right) + T_{\mathrm{IF}}\right] kBG_{\mathrm{IF}} \tag{4}$$

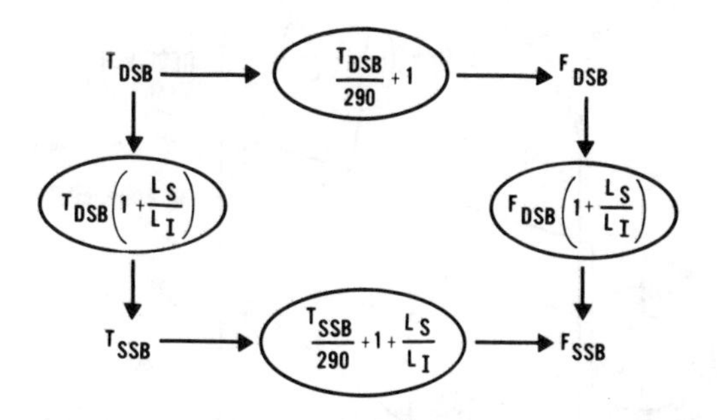

Fig. 9. Noise relationships in dual-response receivers for single-sideband, SSB, and dual-sideband, DSB, noise temperature, T, and noise figure, F, where L_s and L_i are the signal and image conversion losses. The relationships apply to both receiver and mixer noise temperature or noise figure.

2) mixer terminated with a cold load:

$$P_n(C) = \left[(T_C + T_M)\left(\frac{1}{L_s} + \frac{1}{L_i}\right) + T_{\mathrm{IF}}\right] kBG_{\mathrm{IF}} \tag{5}$$

3) IF amplifier terminated with a hot load:

$$P_n(\mathrm{IF}) = [T_H + T_{\mathrm{IF}}] kBG_{\mathrm{IF}} \tag{6}$$

where T_M is the DSB mixer noise temperature, T_H is the temperature of the hot load, T_C is the temperature of the cold load, k is Boltzmann's constant, B is the bandwidth, and G_{IF} is the IF gain. The mixer termination was a piece of millimeter-wave absorber at room temperature, $T_H = 299$ K, or soaked with liquid nitrogen, $T_C = 77$ K. The IF amplifier termination was a coaxial load at room temperature. The IF noise temperature was measured separately with hot and cold loads [21]. Transistor amplifiers with $T_{\mathrm{IF}} = 330$ K and $T_{\mathrm{IF}} = 438$ K and an uncooled paramp for which $T_{\mathrm{IF}} = 108$ K were used in the 1.4-GHz IF system. The IF output power was measured with a thermocouple power meter [22] coupled to a digital voltmeter for increased resolution. The Y factors $Y_H = P_n(H)/P_n(\mathrm{IF})$ and $Y_C = P_n(C)/P_n(\mathrm{IF})$ can be solved for the signal conversion loss:

$$L_s = \frac{T_H - T_C}{(Y_H - Y_C)(T_H + T_{\mathrm{IF}})}\left(1 + \frac{L_s}{L_i}\right) \tag{7}$$

and the SSB mixer noise temperature:

$$T_M(\mathrm{SSB}) = L_s[T_H Y_H + T_{\mathrm{IF}}(Y_H - 1)] - T_H\left(1 + \frac{L_s}{L_i}\right). \tag{8}$$

These solutions are indicated graphically in Fig. 10.

Coherent power measurements of L_s and L_i were made to obtain the image rejection ratio L_s/L_i needed to evaluate (7) and (8). The millimeter-wave signal generator shown in Fig. 11 was used to inject into the mixer a known amount of power at the signal or image frequency. The conversion loss was derived from the ratio of the RF input power and the IF output power measured on the thermocouple power meter. Although the accuracy of the image rejection ratio measurement depends primarily on the frequency response of the thermistor head, the entire system was carefully calibrated so that a comparison of the direct measurement of L_s and the noise power

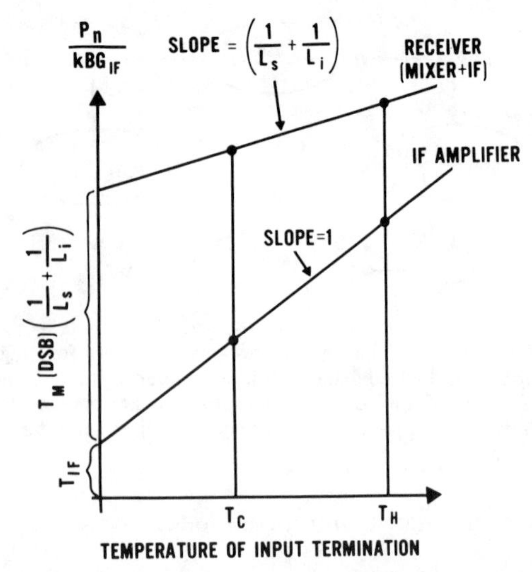

Fig. 10. Noise temperature, referred to the IF amplifier input, as a function of input termination temperature for a receiver and the incorporated IF amplifier. Noise temperature measurements at termination temperatures T_H and T_C are used to obtain the conversion loss and receiver and mixer noise temperatures.

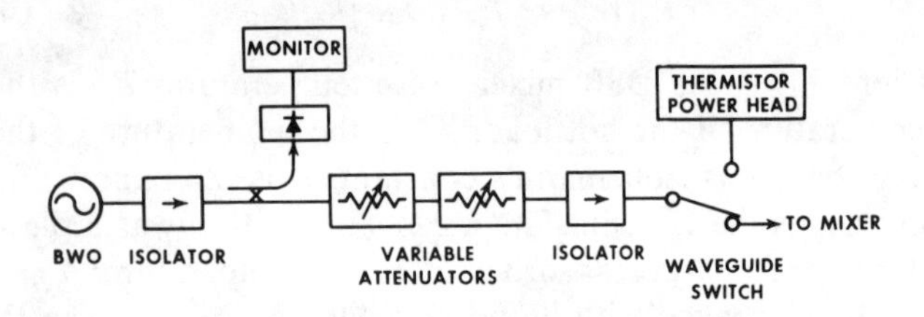

Fig. 11. Block diagram of the millimeter-wave signal generator system used in coherent signal conversion loss measurements.

measurement might serve as an accuracy check. The thermistor head was calibrated across the frequency band against a dry calorimeter [23]. The attenuator linearity was checked with the calorimeter and against the mixer linearity. The IF gain was measured using a set of precision attenuators. The measurement accuracy of the direct conversion loss is estimated to be ±0.5 dB and of the image rejection ratio to be ±0.3 dB.

A "simple procedure" (as opposed to the "full procedure" described above) was used to measure some of the data. The receiver noise temperature was determined from the Y factor $Y=P_n(H)/P_n(C)$. The conversion loss and image rejection ratio were obtained from coherent signal measurements, and T_M was calculated from (2). In both cases, the signal generator was used in the initial tuning of the mixer to obtain minimum L_s and the desired L_s/L_i, and a noise tube was then used in the final tuning for minimum F_R before measurement.

For some mixer tunings with large image rejection ratios, it was found that the conversion loss of (7) was about one dB lower then L_s measured directly. Measurement of F_R (SSB) with the signal generator by means of the small-signal method [24] and noise response measurements ((4) and (5)) made with a 2.5-cm length of WR-4 waveguide on the mixer input indicated that the discrepancy was caused by mixer response at higher harmonics. Such data were rejected.

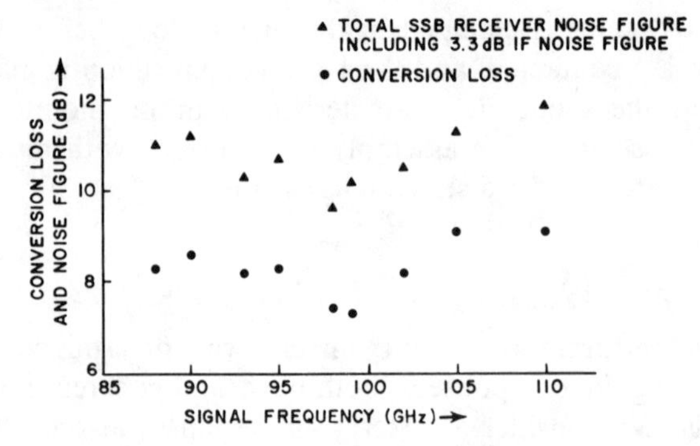

Fig. 12. Total SSB receiver noise figure, including the contribution from a 3.3-dB IF noise figure, and conversion loss of mixer 113 with notch-front diodes as a function of frequency from 85 to 110 GHz. The data points are obtained with the circuit adjusted for optimum receiver noise figure at each frequency. Typical error bars are estimated to be ±0.2 dB for receiver noise figure and ±0.5 dB for conversion loss.

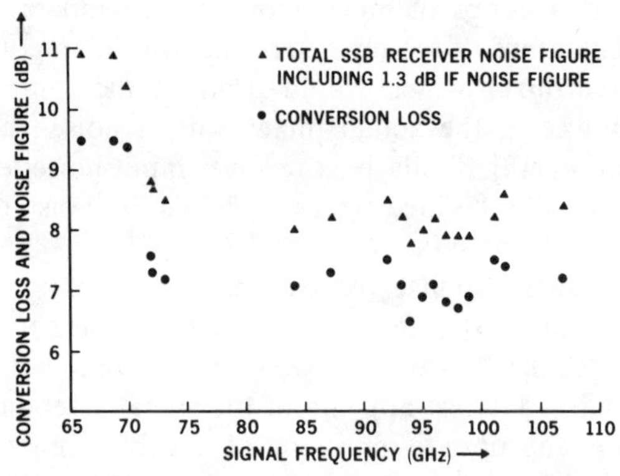

Fig. 13. Total SSB receiver noise figure, including the contribution from a 1.3-dB IF noise figure, and conversion loss of mixer 225 with notch-front diodes as a function of frequency from 66 to 110 GHz. Typical error bars are estimated to be ±0.1 dB for receiver noise figure and ±0.2 dB for conversion loss. The circuit was adjusted for minimum receiver noise figure at each frequency.

V. Results

The receiver noise figure and conversion loss for two mixers with notch-front diodes, labeled 113 and 225, are presented in Figs. 12 and 13 and for a mixer with commercial beam-lead diodes labelled TM58 in Fig. 14. Mixer noise temperatures for mixers 113 and 225 and for another mixer with notch-front diodes, labelled 223, are shown in Fig. 15.

The lowest mixer noise temperature measured in a room temperature two-diode mixer was obtained in mixer 113. The comparison of mixers 223 and 225 in Fig. 15 shows the reproducibility obtained with diodes from a slice which has demonstrated uniform properties and substrates from the same batch. The data were taken with the mixer tuned for minimum receiver noise figure at each frequency.

Mixers 223 and 225 were measured by the "full procedure" described in Section IV, and mixer 113 and the beam-lead diode mixer by the "simple procedure." The typical error bars listed are evaluated from the expressions in Appendix B. Mixers 223 and 225 have lower error bars for T_M principally because the IF amplifier used with these mixers had a much lower noise temperature.

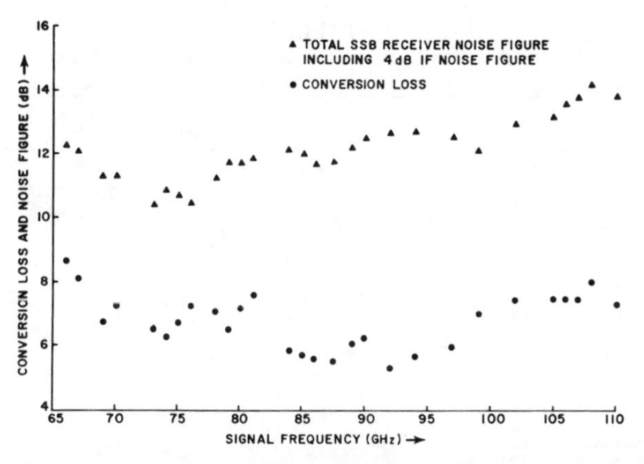

Fig. 14. Total SSB receiver noise figure, including the contribution from a 4-dB IF noise figure, and conversion loss of a two-diode mixer TM58 using commercial beam-lead diodes as a function of frequency. The data points are obtained with the circuit adjusted for optimum receiver noise figure at each frequency. Typical error bars are estimated to be ±0.2 dB for receiver noise figure and ±0.5 dB for conversion loss.

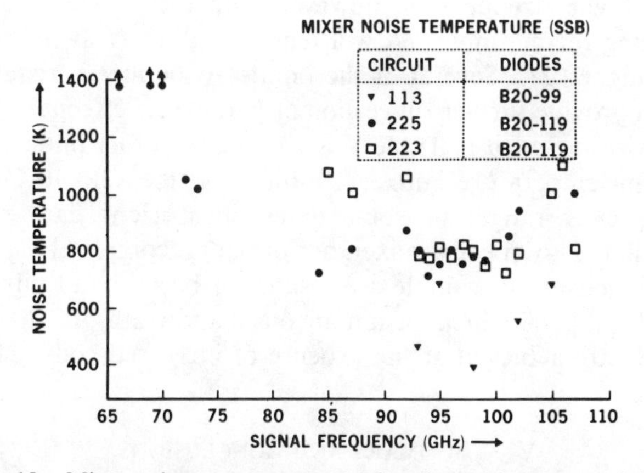

Fig. 15. Mixer noise temperature as a function of frequency of three mixers with notch-front diodes. Typical error bars are ±250 K for mixer 113 and ±60 K for mixers 223 and 225.

Corrections for input loss, LO noise, and IF mismatch are not subtracted from the mixer noise temperature because these effects are unavoidable in making a mixer. The input loss is small because no diplexing arrangement is needed for LO injection as in a conventional mixer. Loss in the input waveguide is estimated to be ~0.1 dB and the corresponding contribution to the mixer noise temperature $\lesssim$25 K. Significant LO noise cancellation occurs in the two-diode mixer [3]. The IF impedance of these two-diode mixers is close to 50 Ω, with return losses greater than 15 dB being measured at the IF port. A tuner between the mixer and IF amplifier eliminated this small mismatch. These corrections are sometimes found to be necessary in characterizing other mixers.

The data are summarized in Table II which includes the lowest mixer noise temperature obtained in each of the three notch-front diode mixers and the lowest noise temperature in the beam-lead diode mixers at both the low end of the signal frequency band (Fig. 14) and the high end [4, Fig. 11]. For comparison, the published performance of several conventional mixers is presented in

TABLE II
SUMMARY OF RESULTS IN TWO-DIODE MIXERS, 67–110 GHz

MIXER	DIODES	f (GHz)	F_R(SSB)	L_s (dB)	L_i/L_s (dB)	T_M(K)	P_{LO}(dBm)
113	B20-99	98	9.6	7.4	2.9	390±250	+8
223	B20-119	101	7.7	6.5	1.4	720±60	-
225	B20-119	94	7.8	6.5	0.7	710±60	-
TM28	BEAM-LEAD	98	11.3	7.7	4.0	1600±600	+18
TM58	BEAM-LEAD	76	10.4	7.2	2.3	590±310	-

TABLE III
CONVERSION LOSS AND MIXER NOISE TEMPERATURE IN CONVENTIONAL ROOM TEMPERATURE MIXERS AT 70–115 GHz

REFERENCE	f(GHz)	L_s(dB)	T_M(K)
Kerr [25]	85	4.6	420
	115	5.5	500
Wilson [26]	90	5.4	600
	115	6.2	700
Zimmermann & Haas [27]	107	6.2	600
	112	6.1	760
Linke [28]	72	6.2	700
	117	6.5	840

Table III. For two of these conventional mixers, the data reported [25], [27] includes corrections for IF mismatch. The most important conclusion from these results is that the noise performance of the two-diode mixer with notch-front diodes is comparable to that of the best conventional mixers. Furthermore, the beam-lead diodes can produce competitive results at lower frequencies, although their performance does deteriorate substantially above 85–90 GHz due to device parasitics.

The noise temperature and conversion loss of a resistive mixer are theoretically related by the expression [29], [30]

$$T_M = \frac{nT_p}{2}\left[L_s - 1 - L_s/L_i\right] \tag{9}$$

where T_p is the physical temperature of the diode(s) and where it is assumed that the diode has no series resistance or time-varying junction capacitance. Observed noise temperatures greater than the values given by (9) have been attributed to noise contributions from diode series resistance and the "parametric" effect of the time-varying diode capacitance acting on the correlated components of the diode shot noise [13], [31]. The noise temperature in Table II for mixer 113 is less than, although still consistent with, the theoretical value $T_M = 650$ K predicted by (9). Noise temperatures less than the theoretical value have also been observed in lower frequency two-diode mixers. Two sources could produce such a result, an uncorrected higher harmonic response of the mixer or parametric effects. Although no check for higher

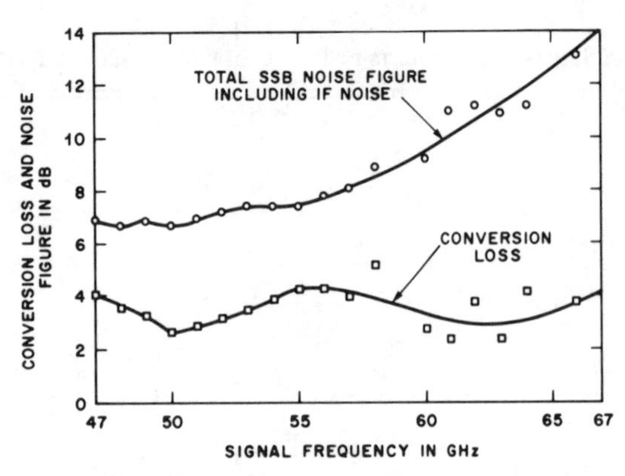

Fig. 16. Total SSB receiver noise figure, including the contribution from a 3.7-dB IF noise figure, and conversion loss of a two-diode mixer using commercial beam-lead diodes as a function of frequency from 47 to 66 GHz. Typical error bars are estimated to be ±0.3 dB for both receiver noise figure and conversion loss.

TABLE IV
PERFORMANCE OF COOLED TWO-DIODE MIXER AT 98 GHz

PHYSICAL TEMPERATURE	L_s(dB)	T_M(K) (SSB)
T=299K	6.8	890
T∿77K	7.0	540

harmonic responses was made with mixer 113, these responses were not observed in other mixers with the type of tuning for which this data was taken. Parametric effects have been seen in two-diode mixers with beam-lead diodes at lower frequencies in the form of conversion gain and also in millimeter-wave fundamental mixers [13]. The operation of a mixer with both variable resistance and variable capacitance has not been fully analyzed, but the general formulation of the problem [31] appears to permit values of T_M either smaller than or larger than that given by (9) for the purely resistive mixer.

The bandwidth of a two-diode mixer with notch-front diodes was measured with the LO frequency and the tuning adjustments fixed. A variation of 0.3 dB in the coherent signal L_S was observed over a 650-MHz frequency band centered at 98 GHz. The instantaneous bandwidth was ultimately limited by the IF amplifier rather than the mixer.

A notch-front diode mixer was cooled to a temperature approaching 77 K by bolting it to a copper bar immersed in liquid nitrogen. The mixer was successfully cooled and returned to room temperature ten times to test the stability of the diodes against thermal stress. The performance of the cooled mixer is given in Table IV and is similar to the results seen in conventional mixers [11], [25].

In Fig. 16 the receiver noise figure and conversion loss for a 47–66-GHz two-diode mixer with beam-lead diodes are reproduced [4, Fig. 9]. In these measurements of F_R, a waveguide noise tube [32] calibrated by the National Bureau of Standards was used. Two points, chosen to represent the low and high end of the frequency band, are

TABLE V
MIXER PERFORMANCE IN 30–70-GHz FREQUENCY RANGE

Reference	f (GHz)	F_R(dB) SSB	L_s(dB)	T_M(K) or F_m(dB) SSB
Two-diode mixer	49	6.9	3.3	290±140 K 3.1±1.0 dB
	60	9.2	2.8	1360±190 K 7.6±0.5 dB
Weinreb and Kerr [33]	33	-	5.9	623 K
Calviello, et al. [34,35]	35	5.9	3.3	3.9 dB
Akaike, et al. [36]	50	10.1	5.3	-
Cohn, et al [2]	55	-	8	-
Meier [35,37]	60	9.0	-	-

listed in Table V. Mixer noise temperature deteriorates at the higher frequencies; however, with the high T_{IF} used, tuning for minimum F_R will tend to reduce L_s at the cost of higher T_M. Several of the points at the lower frequencies provide further suggestion of parametric effects.

Also included in Table V is a sample of other published room-temperature mixer results in the 30–70-GHz frequency range. In some cases, insufficient data was available to specify mixer performance completely. Not included in this sample are results for broad-band mixers [38], [40], in which instantaneous bandwidths of 20–30 GHz are achieved at the expense of increased conversion loss.

VI. SUMMARY AND CONCLUSIONS

This paper describes the design of the two-diode subharmonically pumped stripline mixer, the advantages of this circuit in comparison with conventional mixers, and its performance over the 47–110-GHz frequency range. The notch-front diode, which has ohmic contacts on the chip faces adjacent to the face containing the diode junctions, is described along with an outline of the processing steps for making this chip and mounting it on a strip transmission line circuit. The geometry and size of the notch-front diode results in a lower series resistance than for a conventional millimeter-wave diode, and can bring to stripline circuits the benefits of low parasitic reactance found in the waveguide wafer diode mount.

The performance of the two-diode subharmonically pumped mixer with notch-front diodes was found to be comparable to that of the best fundamental mixers. Even with commercial beam-lead diodes, the performance of the two-diode subharmonically pumped mixer compares favorably with other mixers over much of the millimeter-wave region.

Finally, evidence was seen in these results of parametric effects producing a reduction in mixer noise temperature to below the generally recognized limit for pumped resistive mixers.

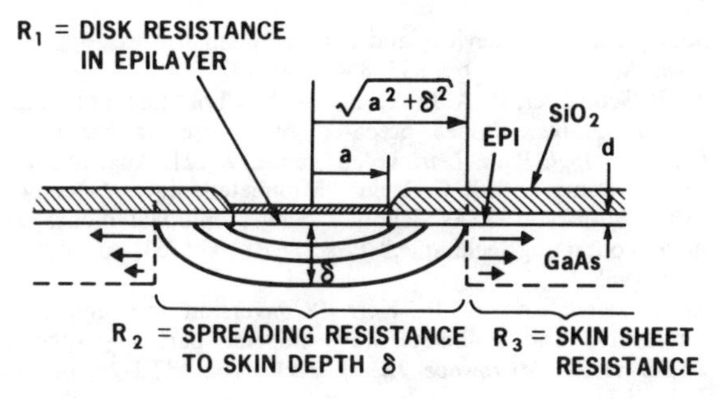

Fig. 17. Sources of RF series resistance in a millimeter-wave diode.

APPENDIX A

Sources of Series Resistance

The contributions to the diode millimeter-wave series resistance are shown in Fig. 17. The disk resistance in the epilayer is given by

$$R_1 = \frac{\rho_{\text{epi}} d}{\pi a^2} \tag{A-1}$$

where ρ_{epi} is the epilayer resistivity, d is the epilayer thickness, and a is the junction radius. The spreading resistance from the disk to one skin depth δ in the substrate is [40]

$$R_2 = \frac{\rho_s}{2\pi a} \arctan \frac{\delta}{a} \tag{A-2}$$

where ρ_s is the substrate resistivity.

Calculation of the skin sheet resistance is simplified if a circular geometry is assumed with the junction at the center of a chip of circular cross section with radius b and height h. The skin sheet resistance for a conventional chip is then [15]

$$R_3 = \frac{\rho_s}{2\pi\delta}\left[\ln \frac{b}{\sqrt{a^2+\delta^2}} + \frac{h}{b}\right]. \tag{A-3}$$

The sides of mechanically separated chips are not smooth, so the second term in (A-3) underestimates the actual sheet resistance of the sides. The sides of a notch-front diode are plated with gold and the second term in (A-3) is negligible.

Some typical values for the millimeter-wave diodes parameters are

$$\rho_{\text{epi}} = 8 \times 10^{-3}\ \Omega\cdot\text{cm}$$

$$\rho_s = 1 \times 10^{-3}\ \Omega\cdot\text{cm}$$

$$a = 1\ \mu\text{m}$$

$$d = 0.1\ \mu\text{m}$$

$$h = 100\ \mu\text{m}$$

$$\delta(100\ \text{GHz}) = 5\ \mu\text{m}$$

$$b = \begin{cases} 50\ \mu\text{m}, & \text{notch-front diode} \\ 125\ \mu\text{m}, & \text{conventional chip.} \end{cases}$$

The contributions to the series resistance are then

$$R_1 = 2.5\ \Omega$$

$$R_2 = 2.2\ \Omega$$

$$R_3 = \begin{cases} 0.7\ \Omega, & \text{notch front diode} \\ 1.2\ \Omega, & \text{conventional chip.} \end{cases}$$

The size and geometry of the notch-front diode lead to a 40-percent reduction in the sheet resistance term. This reduction is about 10 percent of the total series resistance, which is dominated by the resistance at the junction. The junction diameter and epilayer thickness vary with processing parameters, however, and the resulting variations in R_1 are larger than the calculated reduction in R_3 for the notch-front diode, so that experimental verification of this reduction would be difficult.

APPENDIX B

Error Estimates

Consider a quantity y which is a function of variables x_i so that $y = y(x_i)$. Then the variance of y is given by

$$\sigma_y^2 = \sum_i \left(\frac{\partial y}{\partial x_i}\right)^2 \sigma_{x_i}^2 \tag{B-1}$$

where $\sigma_{x_i}^2$ is the variance of x_i and it is assumed that the x_i's are uncorrelated. Equation (B-1) will be applied to the various quantities which describe mixer performance. All quantities used below are defined in the main text except $R = L_s/L_i$. In the examples, noise temperatures are rounded to 10 K and conversion losses to 0.1 dB.

Single-Sideband Receiver Noise Temperature:

$$T_R = \frac{T_H - YT_C}{Y-1}(1+R) \tag{B-2}$$

$$\sigma_{T_R}^2 = \left(\frac{1+R}{Y-1}\right)^2\left[\sigma_{T_H}^2 + Y^2\sigma_{T_C}^2 + \left(T_C + \frac{T_R}{1+R}\right)^2 \sigma_Y^2\right] + \left(\frac{T_R}{1+R}\right)^2 \sigma_R^2. \tag{B-3}$$

Mixer Noise Temperature:

$$T_M = T_R - L_s T_{\text{IF}} \tag{B-4}$$

$$\sigma_{T_M}^2 = \sigma_{T_R}^2 + T_{\text{IF}}^2\sigma_{L_s}^2 + L_s^2\delta_{T_{\text{IF}}}^2. \tag{B-5}$$

Conversion Loss Measured from Noise Power:

$$L_s = \frac{(T_H - T_C)(1+R)}{(Y_H - Y_C)(T_H + T_{\text{IF}})} \tag{B-6}$$

$$\sigma_{L_s}^2 = \left[\frac{L_s}{T_H - T_C}\right]^2 \sigma_{T_C}^2 + \left[\frac{L_s}{1+R}\right]^2 \sigma_R^2 + \left[\frac{L_s}{Y_H - Y_C}\right]^2 \left(\sigma_{Y_H}^2 + \sigma_{Y_C}^2\right) + \left[\frac{L_s}{T_H + T_{\text{IF}}}\right]^2 \sigma_{T_{\text{IF}}}^2 + \left[\frac{L_s(T_C + T_{\text{IF}})}{(T_H + T_{\text{IF}})(T_H - T_C)}\right]^2 \sigma_{T_H}^2. \tag{B-7}$$

Examples:

Example I. Mixer 113, $f=98$ GHz

$Y=1.145\pm.004$ $\quad T_H=299\pm2$ K

$L_s=7.4\pm0.5$ dB $\quad T_C=77\pm3$ K

$R=2.9\pm0.3$ dB $\quad T_{IF}=330\pm5$ K

$1+R=1.51\pm0.04$ $\quad T_R(\text{SSB})=2204$ K

$$\sigma_{T_R}=100\text{ K}((\text{B-3}))$$

$$\sigma_{T_M}=250\text{ K}((\text{B-5})).$$

Example II. Mixer 225 $f=94$ GHz

$Y=1.306\pm.004$ $\quad T_H=229\pm2$ K

$Y_H=0.961\pm.004$ $\quad T_C=77\pm3$ K

$Y_C=0.735\pm.004$ $\quad T_{IF}=108\pm5$ K

$R=0.7\pm0.3$ dB $\quad T_R(\text{SSB})=1199$ K

$1+R=1.85\pm0.06$ $\quad L_s=6.5$ dB

$$\sigma_{T_R}=50\text{ K}((\text{B-3}))$$

$$\sigma_{L_s}=0.2\text{ dB}((\text{B-7}))$$

$$\sigma_{T_M}=60\text{ K}((\text{B-5})). \tag{B-8}$$

Acknowledgment

The authors acknowledge the assistance of A. C. Chipaloski in measuring receiver performance and of A. A. Olenginski in processing notch-front diodes and useful discussions with E. T. Harkless, W. W. Snell, Jr., and R. F. Trambarulo. R. A. Linke suggested the form of the noise relationships in Figs. 9 and 10 and contributed to our understanding of (4)–(6). The authors also wish to thank A. R. Kerr for providing a preprint of a paper discussing the mixer noise theory expressed in (9).

References

[1] M. V. Schneider and W. W. Snell, Jr., "Harmonically pumped stripline down-converter," *IEEE Trans. Microwave Theory Tech.*, vol. MTT-23, pp. 271–275, Mar. 1975.

[2] M. Cohn, J. E. Degenford, and B. A. Newman, "Harmonic mixing with an antiparallel diode pair," *IEEE Trans. Microwave Theory Tech.*, vol. MTT-23, pp. 667–673, Aug. 1975.

[3] P. S. Henry, B. S. Glance, and M. V. Schneider, "Local-oscillator noise cancellation in the subharmonically pumped down-converter," *IEEE Trans. Microwave Theory Tech.*, vol. MTT-24, pp. 254–257, May 1976.

[4] T. F. McMaster, M. V. Schneider, and W. W. Snell, Jr., "Millimeter-wave receivers with subharmonic pump," *IEEE Trans. Microwave Theory Tech.*, vol. MTT-24, pp. 948–952, Dec. 1976.

[5] R. J. Bauer, M. Cohn, J. M. Cotton, Jr., and R. F. Packard, "Millimeter wave semiconductor diode detectors, mixers, and frequency multipliers," *Proc. IEEE*, vol. 54, pp. 595–605, Apr. 1966.

[6] P. F. Goldsmith and R. L. Plambeck, "A 230-GHz radiometer system employing a second-harmonic mixer," *IEEE Trans. Microwave Theory Tech.*, vol. MTT-24, pp. 859–861, Nov. 1976.

[7] M. V. Schneider and E. R. Carlson, "Notch-front diodes for millimeter-wave integrated circuits," *Electron. Lett.*, vol. 13, pp. 745–747, Nov. 1977.

[8] T. F. McMaster, E. R. Carlson, and M. V. Schneider, "Subharmonically pumped millimeter-wave mixers built with notch-front and beam-lead diodes," presented at 1977 *IEEE-MTT-S Int. Microwave Symp. Digest*, San Diego, CA, June 21–23, 1977, pp. 389–392.

[9] AEI Semiconductors, Ltd., type DC 1308.

[10] C. A. Burrus, "Pulse electroplating of high-resistance materials, poorly contacted devices and extremely small areas," *J. Electrochem. Soc.*, vol. 118, pp. 833–834, May 1971.

[11] M. V. Schneider, R. A. Linke, and A. Y. Cho, "Low-noise millimeter-wave mixer diodes prepared by molecular beam epitaxy (MBE)," *Appl. Phys. Lett.*, vol. 31, pp. 219–221, Aug. 1, 1977.

[12] D. T. Young and J. C. Irvin, "Millimeter-frequency conversion using Au-n-type GaAs Schottky barrier epitaxial diodes with a novel contacting technique," *Proc. IEEE*, vol. 53, pp. 2130–2131, Dec. 1965.

[13] D. N. Held and A. R. Kerr, "Conversion loss and noise of microwave and millimeter-wave mixers: part 2—experiment," *IEEE Trans. Microwave Theory Tech.*, vol. MTT-26, pp. 55–61, Feb. 1978.

[14] J. A. Calviello, J. L. Wallace, and P. R. Bie, "High performance GaAs quasi-planar varactors for millimeter waves," *IEEE Trans. Electron. Devices*, vol. ED-21, pp. 624–630, Oct. 1974.

[15] Indium Corporation of America, Indalloy No. 2.

[16] Indium Corporation of America, Indalloy No. 8.

[17] F. Rosebury, *Handbook of Electron Tube and Vacuum Techniques*. Reading, MA: Addison-Wesley, 1965, p. 20.

[18] W. W. Mumford and E. H. Scheibe, *Noise Performance Factors in Communications Systems*. Dedham, MA: Horizon House—Microwave, 1968, pp. 20–21.

[19] [18, pp. 25–32].

[20] [18, p. 42].

[21] AILTECH Type 70 Hot-Cold Standard Noise Generator.

[22] Hewlett-Packard Models 435A and 8481A.

[23] Hitachi Model E-3904.

[24] [18, pp. 61–64].

[25] A. R. Kerr, "Low-noise room-temperature and cryogenic mixers for 80–120 GHz," *IEEE Trans. Microwave Theory Tech.*, vol. MTT-23, pp. 781–787, Oct. 1975.

[26] W. J. Wilson, "The Aerospace low-noise millimeter-wave spectral line receiver," *IEEE Trans. Microwave Theory Tech.*, vol. MTT-25, pp. 332–335, Apr. 1977.

[27] P. Zimmermann and R. W. Haas, "A broadband low noise mixer for 106–116 GHz," *Nachrichtentech. Z.*, vol. 30, pp. 721–722, Sept. 1977.

[28] R. A. Linke and M. V. Schneider in *Workshop on Mixers at Millimeter Wavelengths*, Max-Planck-Institut für Radioastronomie, Bonn, Apr. 26–28, 1977.

[29] C. Dragone, "Analysis of thermal and shot noise in pumped resistive diodes," *B.S.T.J.*, vol. 47, pp. 1883–1902, Nov. 1968.

[30] A. A. M. Saleh, *Theory of Resistive Mixers*, Cambridge, MA: M.I.T. Press, 1971, p. 170.

[31] D. N. Held and A. R. Kerr, "Conversion loss and noise of microwave and millimeter-wave mixers: Part 1—theory," *IEEE Trans. Microwave Theory Tech.*, vol. MTT-26, pp. 49–55, Feb. 1978.

[32] Signalite Model TN-164.

[33] S. Weinreb and A. R. Kerr, "Cryogenic cooling of mixers for millimeter and centimeter wavelengths," *IEEE J. Solid-State Circuits*, vol. SC-8, pp. 58–63, Feb. 1973.

[34] J. A. Calviello and J. L. Wallace, "Performance and reliability of an improved high temperature GaAs Schottky junction and native-oxide passivation," *IEEE Trans. Electron Devices*, vol. ED-24, pp. 698–704, June 1977.

[35] J. J. Whelehan, "Low-noise millimeter-wave receivers," *IEEE Trans. Microwave Theory Tech.*, vol. MTT-25, pp. 268–280, Apr. 1977.

[36] M. Akaike, N. Kanmuri, H. Kato, and K. Hiyama, "Millimeter-wave solid-state circuits," *Rev. Elect. Commun. Laboratories*, vol. 23, pp. 904–918, July–August 1975.

[37] P. J. Meier, "Low-noise mixer in oversized microstrip for 5-mm band," *IEEE Trans. Microwave Theory Tech.*, vol. MTT-22, pp. 450–451, Apr. 1974.

[38] A. Hislop and R. T. Kihm, "A broad-band 40-60-GHz balanced mixer," *IEEE Trans. Microwave Theory Tech.*, vol. MTT-24, pp. 63–64, Jan. 1976.

[39] L. T. Yuan, G. M. Yamaguchi, and J. E. Raue, "Design, implementation, and performance analysis of a broad-band V-band network analyzer," *IEEE Trans. Microwave Theory Tech.*, vol. MTT-24, pp. 981–987, Dec. 1976.

[40] L. T. Yuan, "Low noise octave bandwidth waveguide mixer," 1977 *IEEE MTT-S Int. Microwave Symp. Dig.* (San Diego, CA), June 21–23, 1977, pp. 480–482.

[41] R. Holm, *Electric Contacts*. New York: Springer-Verlag, 1967, pp. 15–16.

Local-Oscillator Noise Cancellation in the Subharmonically Pumped Down-Converter

P. S. HENRY, MEMBER, IEEE, B. S. GLANCE, MEMBER, IEEE, AND M. V. SCHNEIDER, SENIOR MEMBER, IEEE

Abstract—**The noise power at the IF output of a superheterodyne mixer which is caused by local-oscillator noise can be significantly reduced by using the recently developed subharmonically pumped down-converter. In many cases this reduction is so large that even noisy sources, such as IMPATT oscillators, can be used to pump low-noise mixers without causing significant degradation of noise figure.**

I. Introduction

A new type of mixer called a subharmonically pumped down-converter has recently been developed [1]–[4]. The mixer is particularly attractive for use at millimeter-wave frequencies because it can be pumped at one-half the frequency required for a conventional mixer, yet shows negligible increase in conversion loss.

An important property of this mixer, first discussed by Cohn *et al.* [1], [4], is the strong attenuation of down-converted local-oscillator noise available at the IF output. The amount of attenuation is determined by the noise spectrum of the local oscillator. In many cases it is sufficiently large so that the noise power at IF due to a noisy local oscillator, such as an IMPATT diode, is negligible. In this short paper we present a detailed discussion of this phenomenon, and report measurements on a 60-GHz mixer (local oscillator at 30 GHz), which show a noise reduction of at least 19 dB relative to the noise expected from a conventional, single-ended mixer using the same local oscillator.

Usually, the contribution of local-oscillator noise to IF noise power output can be reduced by using a balanced mixer, but at millimeter-wave frequencies the balanced mixer presents cumbersome mechanical problems. The new mixer, however, is relatively simple to build. We conclude that this down-converter provides an attractive solution to the problem of using noisy sources as local oscillators at millimeter-wave frequencies.

II. Response of Mixers to Local-Oscillator Noise

It is well known that conventional single-ended mixers are susceptible to degradation by local-oscillator noise [5], [6]. The cause of the problem is illustrated in Fig. 1. We assume in Fig. 1, and throughout this paper, that $\omega_{IF} \ll \omega_{LO}$. Noise components in the local-oscillator sidebands at $\omega_{LO} \pm \omega_{IF}$ beat with the local oscillator to produce noise power at IF. These noise sidebands are of nearly equal amplitude, resulting in a signal-to-noise ratio of

$$\text{SNR} = \frac{\frac{1}{2}V_s^2}{2n^2(\omega_{LO} + \omega_{IF}) \cdot B} \tag{1}$$

where V_s is the peak input signal voltage, $n^2(\omega)$ is the mean-squared noise voltage per unit bandwidth at ω, and B is the IF bandwidth.

The subharmonically pumped down-converter provides a way to reduce the signal-to-noise degradation which is caused by local-oscillator noise. This property is directly related to the fact that the *I–V* curve of the nonlinear element in the mixer is an antisymmetric function of voltage. The relationship between symmetry and noise reduction is illuminated by the following discussion.

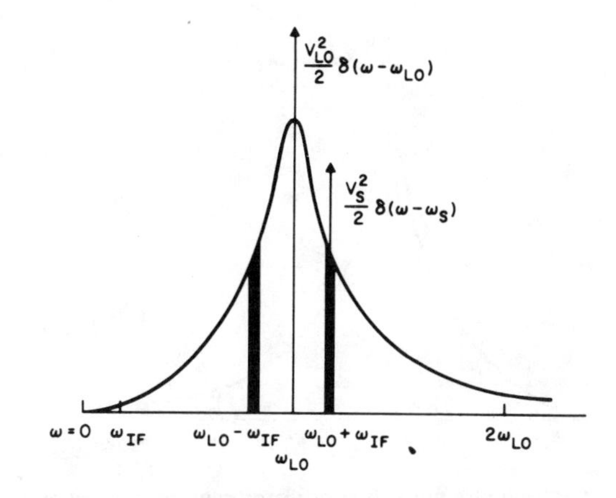

Fig. 1. Spectrum of a noisy local oscillator, plus input signal at $\omega_{LO} + \omega_{IF}$. The size of ω_{IF} relative to ω_{LO} has been exaggerated for clarity. The indicated noise sidebands are appropriate for a conventional mixer.

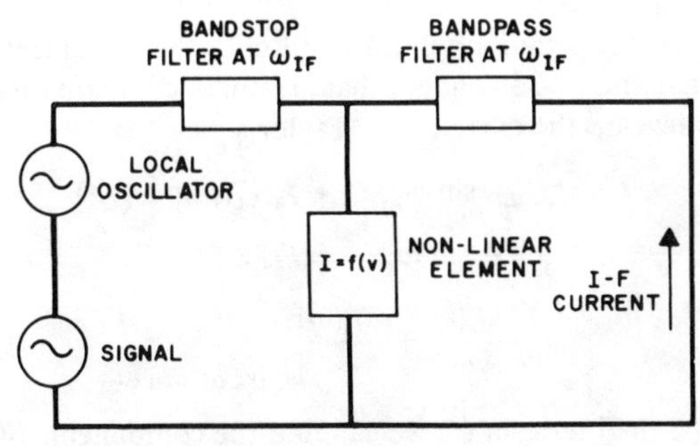

Fig. 2. A simple mixer.

Consider the mixer shown in Fig. 2. The voltages from the local oscillator and signal are impressed across the nonlinear element. The filters separate the current flowing at ω_{IF} from the other components of diode current. Assume the current through the nonlinear element, I, to be a function only of the voltage across it, v:

$$I = f(v). \tag{2}$$

For a conventional mixer the function f is the sum of symmetric and antisymmetric parts, $f_s(v)$ and $f_a(v)$

$$I(v) = f_s(v) + f_a(v) \tag{3}$$

where

$$f_s(v) = [f(v) + f(-v)]/2$$

and

$$f_a(v) = [f(v) - f(-v)]/2.$$

For this calculation we will be interested in local-oscillator noise components at frequencies far from the carrier, i.e., in the wings of the local-oscillator spectrum. In this case we can represent the local oscillator as the sum of a pure sinusoid of amplitude V_{LO}, plus a small noise voltage $n(t)$ [7]. The total

Manuscript received June 27, 1975; revised December 3, 1975.
The authors are with the Crawford Hill Laboratory, Bell Laboratories, Holmdel, NJ 07733.

Reprinted from *IEEE Trans. Microwave Theory Tech.*, vol. MTT-24, pp. 254–257, May 1976.

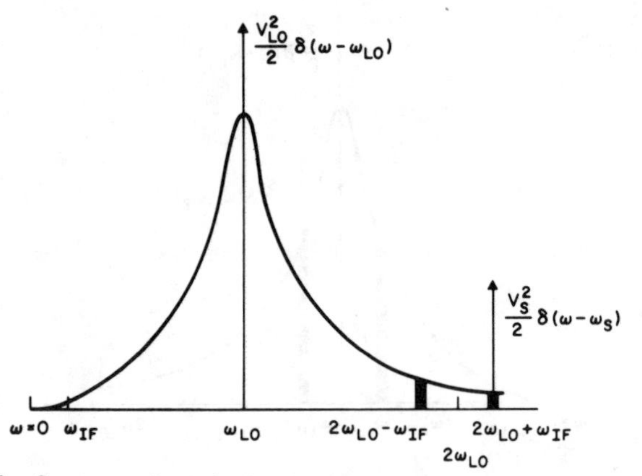

Fig. 3. Spectrum of a noisy local oscillator, plus input signal at $2\omega_{LO} + \omega_{IF}$. The indicated noise sidebands are appropriate for an antisymmetric mixer.

voltage across the nonlinear element is then

$$v(t) = V_{LO} \sin \omega_{LO}t + n(t) + V_s \sin \omega_s t \quad (4)$$

where V_s is the signal amplitude, and ω_s is the signal frequency. For signal and noise voltages that are small compared with V_{LO} we can develop the current in a Taylor series

$$I = f_s(V_{LO} \sin \omega_{LO}t) + f_a(V_{LO} \sin \omega_{LO}t) + \{V_s \sin \omega_s t + n(t)\} \cdot \left\{\frac{df_s(v)}{dv} + \frac{df_a(v)}{dv}\right\}_{v=V_{LO}\sin\omega_{LO}t} \quad (5)$$

From the final term in (5) we can find the components of signal and noise currents flowing at ω_{IF}. Since $f_s(v)$ is symmetric in v, $df_s(v)/dv$ is antisymmetric. It is well known that if a function of a variable with period T is antisymmetric in that variable, the Fourier series representation of that function contains only terms at frequencies $\omega, 3\omega, 5\omega, \cdots$, where $\omega = 2\pi/T$. Therefore the Fourier expansion of

$$\left(\frac{df_s(v)}{dv}\right)_{v=V_{LO}\sin\omega_{LO}t}$$

contains only odd harmonics of ω_{LO}: ω_{LO}, $3\omega_{LO}$, $5\omega_{LO}$, etc. The frequencies of the time-varying terms in the expansion of

$$\left(\frac{df_a(v)}{dv}\right)_{v=V_{LO}\sin\omega_{LO}t}$$

on the other hand, will be only the even harmonics of ω_{LO}: $2\omega_{LO}$, $4\omega_{LO}$, etc. Thus in a conventional mixer the signal current at IF is caused by the "beat" between the signal at $\omega_s = \omega_{LO} \pm \omega_{IF}$ and the component of $df_s(v)/dv$ at ω_{LO}. There is no IF signal contribution from $df_a(v)/dv$ because it has no component at ω_{LO}. The IF signal-to-noise ratio is determined primarily by the relative strengths of signal and local-oscillator noise at $\omega_{LO} \pm \omega_{IF}$, as shown in Fig. 1.

Now consider a mixer where $f(v)$ is purely antisymmetric, so that $f_s = 0$. In this case the IF signal current is the beat between the component of $df_a(v)/dv$ at $2\omega_{LO}$ and the signal at $2\omega_{LO} \pm \omega_{IF}$, as shown in Fig. 3. The noise components at $\omega_{LO} \pm \omega_{IF}$ cannot contribute to the IF current. Rather, it is the components at $2\omega_{LO} \pm \omega_{IF}$ that are important. Since local-oscillator noise tends to drop off rapidly away from the carrier [7], the relevant oscillator noise components are much smaller at $2\omega_{LO} \pm \omega_{IF}$ than at $\omega_{LO} \pm \omega_{IF}$. Hence, for a given input signal, the signal-to-noise ratio will be much higher for the purely antisymmetric mixer than for the conventional mixer.

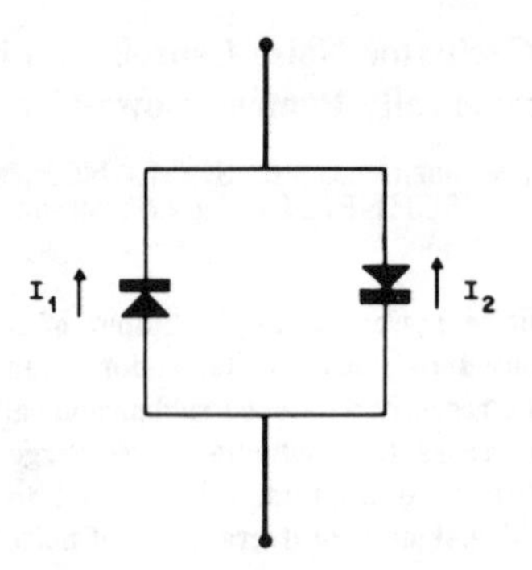

Fig. 4. The nonlinear element of a subharmonically pumped down-converter.

In the following section we will study an example of a mixer which displays these desirable symmetry properties.

III. The Subharmonically Pumped Down-Converter

The subharmonically pumped down-converter is a mixer in which the nonlinear element is a pair of antiparallel diodes, as shown in Fig. 4. The total current through the diode pair is given by the algebraic sum of the separate currents. For ideal diodes with negligible parasitic resistance and capacitance

$$I = I_1 + I_2 = I_s(e^{\alpha v} - 1) - I_s(e^{-\alpha v} - 1) = 2I_s \sinh \alpha v \quad (6)$$

where I_s is the diode reverse saturation current and $\alpha = q/nkT \approx (25 \text{ mV})^{-1}$ [3]. From (6) the antisymmetry of the I–V characteristic is clear. For the rest of this short paper we will refer to the subharmonically pumped down-converter as the hyperbolic-sine-law mixer.

To study the response of the hyberbolic-sine-law mixer to local-oscillator noise, we assume, as before, that the total applied voltage is the sum of the local oscillator, signal, and noise voltages. The total diode current is

$$I = 2I_s \sinh \alpha\{V_{LO} \sin \omega_{LO}t + V_s \sin \omega_s t + n(t)\}. \quad (7)$$

For small signal and noise, i.e.,

$$\alpha\{V_s \sin \omega_s t + n(t)\} \ll 1 \quad (8)$$

we find the current to be

$$I \simeq 2\alpha\{V_s \sin \omega_s t + n(t)\}I_s \cosh(\alpha V_{LO} \sin \omega_{LO}t) + 2I_s \sinh(\alpha V_{LO} \sin \omega_{LO}t). \quad (9)$$

All the components of current flowing at ω_{IF} are contained in the first term on the right-hand side of (9). Therefore we neglect the second term and expand the hyperbolic cosine in a Fourier series

$$I \simeq 2I_s\alpha\{V_s \sin \omega_s t + n(t)\} \cdot \{I_0(\alpha V_{LO}) + 2\sum_{k=1}^{\infty} (-1)^k I_{2k}(\alpha V_{LO}) \cdot \cos 2k\omega_{LO}t\} \quad (10)$$

where the I_{2k} are the hyperbolic Bessel functions [8]. The argument of the functions, αV_{LO}, is approximately 24, because V_{LO} must be $\sim$0.6 V in order to get appreciable current to flow through the mixer diodes [3]. For this argument the first few terms of the I_{2k} form a slowly decreasing sequence. Thus the hyperbolic-sine-law mixer is capable of high-order harmonic mixing [3], [4]. In normal operation, however, the mixer is used to receive signals at $\omega_s = 2\omega_{LO} \pm \omega_{IF}$. If we assume, as we did in Section II, that local-oscillator noise falls off rapidly away from the carrier, then the dominant noise contribution is made by noise components near $2\omega_{LO} \pm \omega_{IF}$, as shown in Fig. 3. Since these are nearly equal for $\omega_{IF} \ll \omega_{LO}$, we have

$$\text{SNR} = \frac{\frac{1}{2}V_s^2}{2n^2(2\omega_{LO} + \omega_{IF}) \cdot B}. \tag{11}$$

Compared with the performance of the conventional mixer given by (1), the hyperbolic-sine-law mixer, with "effective" local oscillator at $2\omega_{LO}$, offers a noise improvement of

$$\eta = 10 \log \frac{n^2(\omega_{LO} + \omega_{IF})}{n^2(2\omega_{LO} + \omega_{IF})} \text{ dB}. \tag{12}$$

This is the principal result of our analysis. It is valid provided that oscillator noise at higher order even harmonics is negliglble compared with that at $2\omega_{LO}$.

It should be remarked that the antisymmetry of the I–V curve for a mixer with two identical antiparallel diodes is unaffected by the presence of diode series resistance. Since the argument following (5) in Section II shows that noise suppression is related to antisymmetry and not to the details of the shape of the I–V curve, the noise improvement given by (12) also applies to diodes with series resistance.

An estimate of the noise improvement available with the hyperbolic-sine-law mixer requires a model for the noise spectrum of the local oscillator. However, it is often difficult to know accurately the behavior of the oscillator circuit elements at frequencies near $2\omega_{LO}$. To make an order-of-magnitude estimate of oscillator noise, let us assume that the oscillator is represented by a frequency-independent negative conductance in parallel with a singly tuned resonant circuit. For such an oscillator the noise spectral density far from the carrier has the same frequency dependence as the response of an RLC circuit to white noise [7][1]

$$n^2(\omega) \propto \left[S^2 + Q^2\left(\frac{\omega}{\omega_{LO}} - \frac{\omega_{LO}}{\omega}\right)^2\right]^{-1} \tag{13}$$

where S is the saturation parameter of the oscillator [7], and Q is its external Q. In those cases where $\omega_{IF} \gtrsim S\omega_{LO}/Q$, so that we need not be concerned with noise components near the resonant peak, (12) and (13) can be combined to yield a simple result

$$\eta' = 20 \log (3\omega_{LO}/4\omega_{IF}). \tag{14}$$

Equation (14) is useful as a guide to the noise reduction expected with the hyperbolic-sine-law mixer.

IV. Experimental Results

Measurements of the noise performance of a 60-GHz hyperbolic-sine-law mixer with IF at 1.4 GHz show clearly the reduction of the IF noise contribution from the local oscillator. For these experiments the required local-oscillator power of ~5 mW was supplied by a 30-GHz microstrip IMPATT oscillator [10]. The IMPATT local oscillator caused the mixer noise figure to increase not more than 1 dB over the 10-dB value measured using a klystron local oscillator with negligible noise [11]. Thus we have

$$10 \log \left(\frac{N_{sinh}}{kT} + 10\right) \leq 11 \text{ dB} \tag{15}$$

[1] Reference [7, eq. (41)] is an approximate expression valid only near $\omega = \omega_{LO}$. See [9].

where N_{sinh} is the contribution to IF noise from the local oscillator referred to the mixer input, and $kT \simeq 4 \times 10^{-18}$ mW/Hz. Therefore,

$$N_{sinh} \leq 10^{-17} \text{ mW/Hz}. \tag{16}$$

Let us compare this with the effective noise that would be expected using this same local oscillator with a conventional mixer (IF = 1.4 GHz) to receive a signal near 30 GHz. At 5-mW power level, the single-sideband noise power spectral density of the local oscillator is measured to be 4×10^{-16} mW/Hz at 1.4 GHz away from the carrier. From (1) we see that the noise bands at $\omega_{LO} \pm \omega_{IF}$ both contribute to IF noise power, giving an effective noise density, referred to the mixer input, of

$$N_{conv} = 8 \times 10^{-16} \text{ mW/Hz}. \tag{17}$$

Comparing (16) and (17), we find the ratio of IF noise powers for the conventional and hyperbolic-sine-law mixers to be

$$\frac{N_{conv}}{N_{sinh}} \geq 19 \text{ dB}. \tag{18}$$

That is, we have observed a noise reduction of at least 19 dB. The value of η' calculated from (14) is 24 dB.

The noise-reduction capability of the hyperbolic-sine-law mixer can be degraded if the diodes are not identical. This destroys the antisymmetry of the mixer and allows local-oscillator noise at $\omega_{LO} \pm \omega_{IF}$ to contribute to the IF noise current. If we call this contribution P_1, and let P_2 be the IF contribution due to local-oscillator noise at $2\omega_{LO} \pm \omega_{IF}$, then (see Appendix)

$$\frac{P_1}{P_2} \approx \left[2\frac{I_{dc}}{I_{LO}} \cdot \frac{\omega_{LO}}{\omega_{IF}}\right]^2 \tag{19}$$

where I_{dc} is the net dc current flowing through the diodes ($I_{dc} = 0$ for identical diodes), and I_{LO} is the amplitude of the diode current flowing at ω_{LO}. In the mixer described in this section I_{dc} is ~30 μA for 5-mW local-oscillator power. Since the voltage across the diodes is ~0.6 V [3], and the power absorbed is ~5 mW, the current flowing through them, I_{LO}, is of order 10 mA. Substituting these values of I_{dc} and I_{LO} into (19), we find

$$\frac{P_1}{P_2} \approx 0.02. \tag{20}$$

Thus, for the mixer described here, diode imbalance causes a negligible increase in total mixer noise.

V. Reception of Angle-Modulated Signals

Phase noise (FM noise) in the local oscillator causes random phase fluctuations in the IF output signal, which can produce radio system degradation if the signal is angle modulated. However, even though the hyperbolic-sine-law mixer doubles the rms phase fluctuations of the local oscillator [see (10)], its output phase noise is not significantly greater than that of a conventional mixer receiving the same signal. This is because the local oscillator for the hyperbolic-sine-law mixer, running at about half the frequency of the local oscillator for the conventional mixer, will show about half the rms phase fluctuations, provided the two oscillators are of similar design [12]. The doubling of these fluctuations by the hyperbolic-sine-law mixer results in IF fluctuations roughly equal to those of a conventional mixer.

Experimentally, we observed that an IMPATT local oscillator with phase noise of ~400 Hz/$\sqrt{\text{kHz}}$ produces phase noise of

~ 700 Hz/$\sqrt{\text{kHz}}$ at the IF output of a hyperbolic-sine-law mixer. Considering our probable error of 10–20 percent on each of those measurements, the data are consistent with the hypothesized doubling of local oscillator FM noise by the hyperbolic-sine-law mixer.

APPENDIX
EFFECT OF DIODE IMBALANCE ON MIXER NOISE

An expression relating diode imbalance to hyperbolic-sine-law mixer noise can be derived in a straightforward fashion. Assume that the diodes have identical saturation current but slightly different α parameters. The diode current is given by

$$I(v) = I_s(e^{\alpha v} - e^{-(\alpha+\delta\alpha)v}) \tag{A-1}$$

where $\delta\alpha/\alpha \ll 1$. If V_{LO} is much greater than the signal and noise voltages, we can obtain an expansion of (A-1) similar to (10). From this we find the components of the total current at dc (which is due to diode imbalance) and ω_{LO}, as well as the contributions of local-oscillator noise at $\omega_{\text{LO}} \pm \omega_{\text{IF}}$ and $2\omega_{\text{LO}} \pm \omega_{\text{IF}}$. Using the notation of Sections III and IV we have

$$\frac{I_{\text{dc}}}{I_{\text{LO}}} = \frac{\delta\alpha}{\alpha} \cdot \frac{\alpha V_{\text{LO}}}{8} \tag{A-2}$$

and

$$\frac{P_1}{P_2} = \left[\frac{(\delta\alpha/\alpha) \cdot \alpha V_{\text{LO}} \cdot I_0(\alpha V_{\text{LO}})}{4 I_2(\alpha V_{\text{LO}})}\right]^2 \cdot \frac{n^2(\omega_{\text{LO}} \pm \omega_{\text{IF}})}{n^2(2\omega_{\text{LO}} \pm \omega_{\text{IF}})}. \tag{A-3}$$

Equations (A-2) and (A-3) can be combined and simplified by noting that $I_0(\alpha V_{\text{LO}}) \approx I_2(\alpha V_{\text{LO}})$, and assuming that $n^2(\omega)$ falls off roughly like $(\omega_{\text{LO}} - \omega)^{-2}$ away from the carrier. Then

$$\frac{P_1}{P_2} \simeq \left[2 \frac{I_{\text{dc}}}{I_{\text{LO}}} \cdot \frac{\omega_{\text{LO}}}{\omega_{\text{IF}}}\right]^2. \tag{A-4}$$

Equation (A-4) shows that when $\omega_{\text{IF}} \ll \omega_{\text{LO}}$, extremely good diode balance is required to ensure $P_1 \ll P_2$.

A calculation assuming identical α parameters but slightly different values of the saturation current also yields the result given in (A-4).

ACKNOWLEDGMENT

The authors wish to thank W. W. Snell, Jr., who provided them with the low-noise 60-GHz hyperbolic-sine-law mixer, and also L. J. Greenstein, V. K. Prabhu, and H. E. Rowe for offering helpful insights and advice.

REFERENCES

[1] M. Cohn, J. E. Degenford, and B. A. Newman, "Harmonic mixing with an anti-parallel diode pair," in *1974 IEEE S-MTT Int. Microwave Symp. Dig. Tech. Papers*, 1974, pp. 171–172.
[2] M. V. Schneider and W. W. Snell, Jr., "Stripline down-converter with subharmonic pump," *Bell Syst. Tech. J.*, vol. 53, pp. 1179–1183, July–Aug. 1974.
[3] ——, "Harmonically pumped stripline down-converter," *IEEE Trans. Microwave Theory Tech.*, vol. MTT-23, pp. 271–275, Mar. 1975.
[4] M. Cohn, J. E. Degenford, and B. A. Newman, "Harmonic mixing with an antiparallel diode pair," *IEEE Trans. Microwave Theory Tech.*, vol. MTT-23, pp. 667–673, Aug. 1975.
[5] R. V. Pound, *Microwave Mixers*. New York: McGraw-Hill, 1948, ch. 5 and 6.
[6] M. I. Skolnik, *Introduction to Radar Systems*. New York: McGraw-Hill, 1962, ch. 8.
[7] W. A. Edson, "Noise in oscillators," *Proc. IRE*, vol. 48, pp. 1454–1466, Aug. 1960.
[8] S. M. Selby, Ed., *Handbook of Tables for Mathematics*. Cleveland, OH: Chemical Rubber Co., 1970.
[9] G. E. Valley, Jr., and H. Wallman, *Vacuum Tube Amplifiers*. New York: McGraw-Hill, 1948, ch. 4.
[10] B. S. Glance and M. V. Schneider, "Millimeter-wave microstrip oscillators," *IEEE Trans. Microwave Theory Tech.* (Short Papers), vol. MTT-22, pp. 1281–1283, Dec. 1974.
[11] W. W. Snell, Jr., private communication.
[12] K. Kurokawa, "Injection locking of microwave solid-state oscillators," *Proc. IEEE*, vol. 61, pp. 1386–1410, Oct. 1973.

A Subharmonic Mixer Using a Planar Doped Barrier Diode with Symmetric Conductance

R. J. MALIK, MEMBER IEEE, AND S. DIXON, MEMBER, IEEE

Abstract—**Planar doped barrier (PDB) diodes with symmetric *I-V* characteristics have been successfully used in a subharmonically-pumped coplanar stripline mixer circuit. A conversion loss around 6 dB has been measured using a 1.2 GHz local oscillator at a pump power of 7 dBm and a 2 GHz signal frequency for large area, 100 μm diameter PDB mesa diodes. The particular advantages of this device structure are that a single PDB replaces two critically matched Schottky barrier diodes in conventionally balanced mixers, and the PDB's designable barrier height reduces the local oscillator power requirement.**

INTRODUCTION

THE FIRST demonstration of a subharmonically-pumped mixer using an antiparallel conducting pair of Schottky barrier diodes was reported by Schneider and Snell [1]. The subharmonic mixer has a number of advantages in comparison to conventional mixers, especially at millimeter-wave frequencies. The local oscillator (LO) power requirement is easier to meet since the LO operates at an even submultiple (about one half) of the signal frequency. The separation of the LO and signal frequencies through bandpass filtering is also simplified due to the large difference in these frequencies. In addition, the symmetric *I-V* characteristic of the subharmonic mixer affords substantial AM noise suppression of the LO [2]. However, the performance of subharmonic mixers which utilize a pair of Schottky barrier diodes is critically dependent upon matching the electrical characteristics of the two diodes as well as their placement in the microwave circuit. Conductance or phase mismatches between the diodes at any of the mixing product frequencies leads to degraded conversion loss and noise figure in the mixer. A much more attractive alternative to the two-diode mixer is the use of a single diode element with a symmetrical *I-V* characteristic. Schottky barriers employing a reverse tunneling current and graded bandgap heterojunctions have been proposed to approximate this characteristic [3]. However, this article reports the use of a planar doped barrier (PDB) to obtain a truly symmetric *I-V* characteristic in a single diode. Additionally, the ability to design the barrier height and capacitance in the PDB diode makes it ideally suited for use in subharmonic mixers.

PDB DIODE THEORY

The planar doped barrier concept has been previously described [4, 5]. The PDB is a majority carrier device structure with an n^+-i-p^+-i-n^+ doping configuration in which an extremely thin, fully depleted acceptor layer (20-100 Å) is used to form a triangular potential profile of predetermined shape and height. For the particular case of a symmetric *I-V* curve for a subharmonic mixer, the acceptor layer is positioned in the middle of the undoped region and separated by the distance L from each of the respective donor regions (Fig. 1).

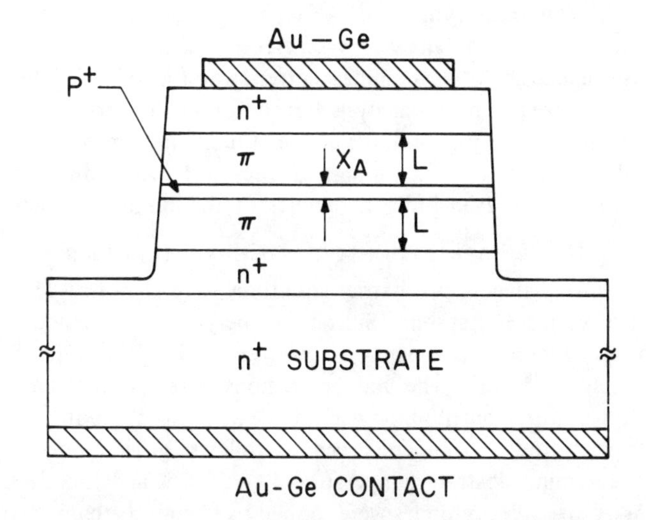

Fig. 1. Schematic cross-section of a GaAs PDB diode with a symmetric current-voltage characteristic (not drawn to scale).

Assuming that the ionized impurity widths of the acceptor layer and at the edges of the donor layers are much less than L (typically < 100 Å), and the potential due to the ionized impurities in the undoped regions is negligible, then the zero-bias barrier height ϕ_{BO} is given approximately by

$$\phi_{BO} = \frac{qN_A X_A L}{2\epsilon} \quad (1)$$

where q is the unit electron charge, N_A the volume density, X_A the acceptor width, and ϵ is the dielectric permitivity of the semiconductor. The capacitance of the diode which is constant with applied voltage and is approximately equal to

$$C = \frac{\epsilon a}{2L} \quad (2)$$

where a is the diode area.

The symmetric *I-V* characteristic of the diode can be expressed by a hyperbolic sine function [3] of the applied voltage V, whereby

$$I = 2I_s \operatorname{Sin}h\left(\frac{qV}{nkT}\right) \quad (3)$$

The saturation current I_s is related to the zero-bias barrier height by

$$I_s = aA^*T^2 \exp\left(\frac{-q\phi_{BO}}{kT}\right) \quad (4)$$

In (3) and (4), k is Boltzmann's constant, T the absolute temperature, and A^* the effective Richardson constant. The n-

Manuscript received April 23, 1982; revised May 19, 1982.

The authors are with the US Army Electronics Technology and Devices Laboratory (EARDCOM) Fort Monmouth, NJ 07703.

Reprinted from *IEEE Electron Devices Lett.*, vol. EDL-3, pp. 205–207, July 1982.

factor in (3) is determined by the geometry of the diode which effectively divides the applied voltage across the barrier region. Ideally, $n = 2$ for a truly symmetric PDB diode. The n-factor should not be confused with the so-called ideality factor which is an empirical parameter used to describe the effects of image force lowering and interface traps in Schottky barriers.

Schneider [3] has reported on the electrical properties of diodes exhibiting symmetric I-V characteristics for frequency conversion. An ac analysis was performed through a Taylor series expansion of the nonlinear current waveform. The key features derived from this analysis are: there is no dc current flowing through the junction, the device current contains only odd order harmonics of the pump frequency, and the conductance contains only even order harmonics of the pump frequency.

Diode Fabrication and dc Characteristics

GaAs planar doped barrier structures were grown by MBE in a Varian-360 system. Silicon and beryllium were used as n-type and p-type dopants, respectively, at levels of approximately 10^{18} cm^{-3}. The undoped regions were low 10^{14} cm^{-3} p-type, thus contributing negligible charge to the potential. Typical growth parameters were as follows: growth rate r = 200 Å/min, substrate temperature T_s = 580°C, and flux ratio As_4/Ga = 2. Mesa diodes were formed by chemical etching and alloying of evaporated Au/Ge contacts to result in the structure shown in Fig. 1.

The parameters for PDB Diode 265 are the following: planar acceptor density $N_A X_A$ = 3.6 x 10^{11} cm^{-2}, undoped region widths L = 2000 Å, and mesa diameter of 100 μm (a = 7.85 x 10^{-5} cm^2). A photograph of the dc I-V curve for this diode is shown in Fig. 2. The excellent symmetry for the I-V curve was checked with a dc electrometer. Within experimental error, there was no measurable difference between the forward and reverse I-V characteristics.

A logarithmic plot for the current dependence upon applied voltage for PDB Diode 265 is shown in Fig. 3. The equations for determining the n-factor and zero-bias barrier height ϕ_{BO} are found by taking the natural log of (3) and (4).

$$\frac{d(\ln I)}{dV} = \frac{q}{nkT} \tag{5}$$

$$\phi_{BO} = \frac{-kT}{q} \ln\left(\frac{I_s}{aA^*T^2}\right) \tag{6}$$

Thus, n and ϕ_{BO} can be determined from the slope and intercept of the Log I-V curve. From the inset box in Fig. 3, there is seen to be excellent agreement between the theoretical and experimental calculations of these parameters. These results demonstrate that MBE provides the requisite control necessary to form PDB device structures.

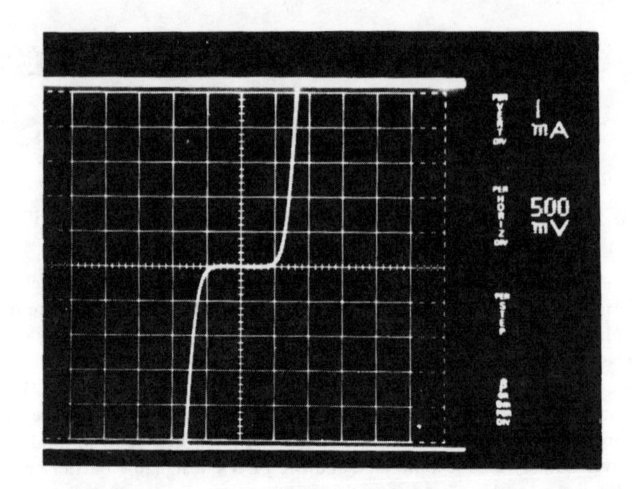

Fig. 2. Current-voltage characteristic of PDB Diode 265 with the following diode parameters: $N_A X_A$ = 3.6 x 10^{11} cm^{-2}, L = 2000 Å, area = 7.85 x 10^{-5} cm^2.

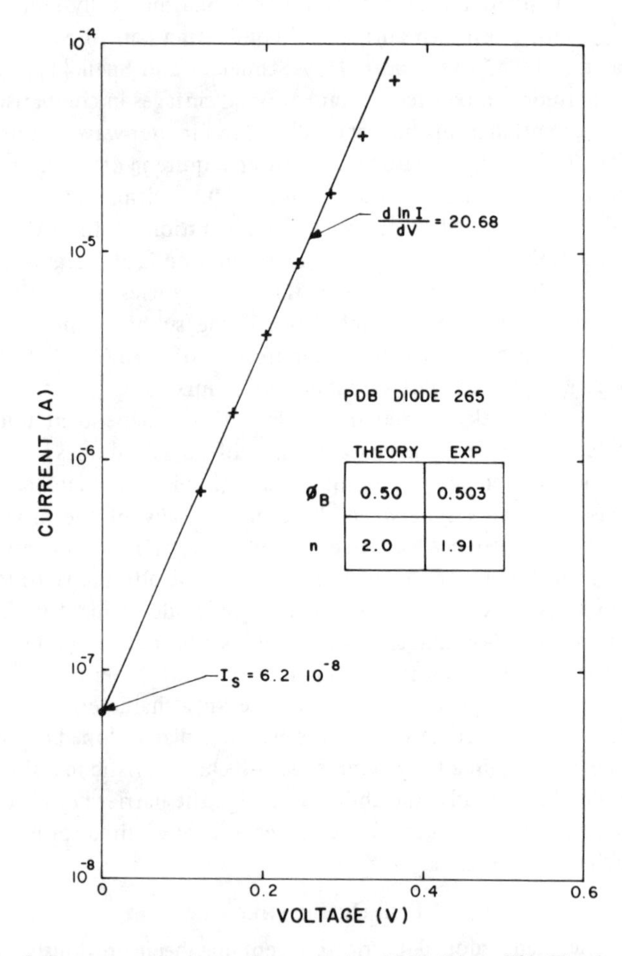

Fig. 3. Log current-voltage characteristic of PDB Diode 265. The slope and intercept are used in determining the diode n-factor and barrier height, respectively.

Subharmonic Mixer Results

The summetric PDB diodes were evaluated in a subharmonic mixer using a coplanar waveguide fixture fabricated on a 1" x 1" x 0.015" Al_2O_3 substrate. The ground planes and center conductor were formed by photolighography and Au evaporation. The diode wafer was sliced into 0.080" square chips and one diode on a chip was wire-bonded to the coplanar waveguide. A photograph of the resultant test fixture for the subharmonic mixer is seen in Fig. 4. A series resistance of R_s = 7 ohms and capacitance of C = 1.8 pF was measured for PDB Diode 265. This corresponds to a cut-off frequency of f_c = 12.6 GHz which, therefore, restricted RF measurements to a few GHz. The high value of the capacitance was due to the large area of these diodes. The results reported here serve to demonstrate the potential use of symmetric PDB diodes in subharmonic

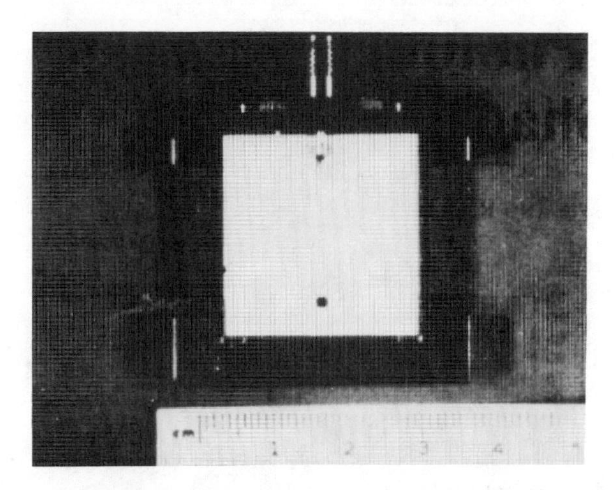

Fig. 4. Photograph of subharmonic mixer test fixture.

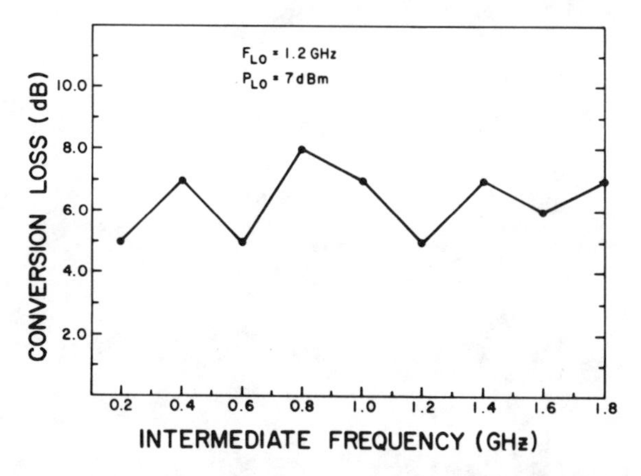

Fig. 5. Conversion loss as a function of the intermediate frequency with fixed local oscillator of 1.2 GHz and pump power of 7 dBm.

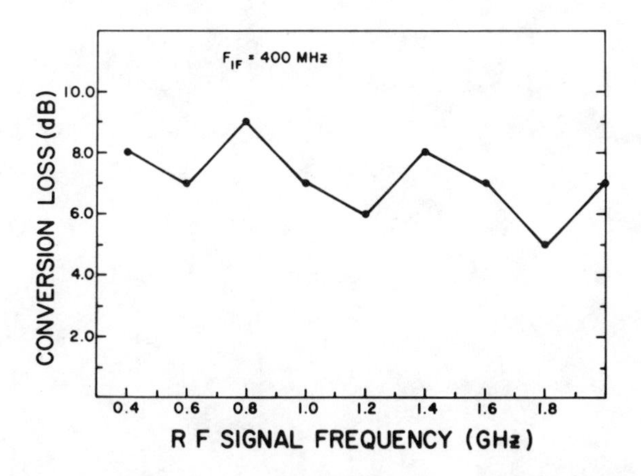

Fig. 6. Conversion loss as a function of the RF signal frequency with fixed intermediate frequency of 400 MHz.

mixers. High cut-off frequency ($f_c > 1000$ GHz) PDB diodes can easily be obtained by reducing the diode area and optimizing the ohmic contact metallization. The performance of a millimeter wave PDB subharmonic mixer will be reported in a subsequent publication.

Using separate signal generators as the local oscillator and signal source, the conversion loss was measured as a function of several different parameters. Figure 5 shows the dependence of the conversion loss upon the intermediate frequency (*IF*). The conversion loss varies from 5.0 to 8.0 dB over an *IF* range of 0.2 to 1.8 GHz. It should be noted that this conversion loss is achieved at a pump power of only 7 dBm for the LO, which is about a factor of one half the required pump power for GaAs Schottky barrier diode mixers. This demonstrates that a substantial reduction in the LO power can be achieved with the designable barrier height in the PDB. Diode 265 had a barrier height of 0.5 V, so that a further decrease in the barrier height should lead to an even lower required pump power for the mixer. Figure 6 shows the variation of the conversion loss as a function of the RF signal frequency with a fixed *IF* = 400 MHz. The conversion loss varies between 5.0 and 9.0 dB over a RF signal frequency range of 0.4 to 2.0 GHz. The conversion loss is relatively flat with a maximum deviation of 4.0 dB over this entire frequency range. Pumping of the mixer diode has also been achieved at submultiples of less than one half the signal frequency without a significant degradation of the conversion loss. This possibility is attractive for use in very-high-frequency mixers.

These preliminary results clearly demonstrate the potential use of PDB diodes in subharmonic and conventional mixer circuits. Work is presently under way in the development of PDB diodes for microwave and millimeter-wave mixers.

Conclusion

A novel subharmonic mixer has been demonstrated which uses a PDB diode with a symmetric *I–V* characteristic. A minimum conversion loss of 5 dB has been obtained for a RF signal of around 2 GHz and a local oscillator at 1.2 GHz with a pump power of 7 dBm. In this configuration, a single PDB diode replaces two well-matched Schottky barrier diodes in conventional balanced mixers and the designable barrier height of the PDB reduces the local oscillator power requirement. These results are presently being applied in the design of very high frequency millimeter-wave subharmonic mixers.

Acknowledgment

The authors express their appreciation to V. E. Rible, R. L. Ross, M. J. Wade, W. Goodreau, J. H. Kwiatkowski, and E. Malecki for technical assistance, and to M. V. Schneider for many useful discussions on subharmonic mixers.

References

[1] M. V. Schneider and W. W. Snell, Jr., "Harmonically pumped stripline down-converter," *IEEE Trans. Microwave Theory Tech.*, MTT-23, p. 271, 1975.

[2] P. S. Henry, B. S. Glance, and M. V. Schneider, "Local-oscillator noise cancellation in the subharmonically pumped down-converter," *IEEE Trans. Microwave Theory Tech.*, MTT-24, p. 254, 1976.

[3] M. V. Schneider, "Electrical characteristics of metal-semiconductor junctions," *IEEE Trans. Microwave Theory Tech.*, MTT-28, p. 1169, 1980.

[4] R. J. Malik, T. R. AuCoin, R. L. Ross, K. Board, C. E. C. Wood, and L. F. Eastman, "Planar-doped barriers in GaAs by molecular beam epitaxy," *Electron. Lett.*, vol. 16, p. 836, 1980.

[5] R. J. Malik, K. Board, L. F. Eastman, C. E. C. Wood, T. R. AuCoin, and R. L. Ross, "Rectifying variable planar-doped-barrier structures in GaAs," *Inst. Phys. Conf. Ser.*, vol. 45, p. 697, 1981.

Part VI
Intermodulation in Microwave Mixers

WHEN the power at the signal frequency (or frequencies) becomes large, i.e. (nearly) of the same order as the power of the local oscillator, saturation and intermodulation will become a problem. In this part some aspects of these problems are discussed in some detail.

The paper by Orloff is one of the early papers often referred to in the literature. An expression for the amplitude of the intermodulation products ($f_{IF} = mf_{LO} - nf_s$, $m,n \geq 1$) and harmonics is presented assuming a purely resistive and exponential diode. In the next paper Ernst *et al.* consider the problem of when two strong signals (f_{s1} and f_{s2}) are present simultaneously and spurious signals at $2f_{s1} - f_{s2} - f_{LO}$ and $2f_{s2} - f_{s1} - f_{LO}$ occur at the IF port. Suggestions for reducing saturation and intermodulation are discussed. In the third paper Beane discusses why intermodulation increases with LO power under some circumstances, which is contrary to what is expected from the simple models presented in the earlier papers.

The two final papers discuss intermodulation in multidiode mixers. Gardiner analyzes intermodulation distortion effects in diode modulators with a technique which is a significant improvement over the "modulation function" techniques used earlier. Finally, in the last paper Maiuzzo and Cameron present results of a computer assisted analysis of commonly used double-balanced mixers.

Intermodulation Analysis of Crystal Mixer*

L. M. ORLOFF†, MEMBER, IEEE

Summary—An expression for the amplitude of the intermodulation products and harmonics produced in a crystal mixer is derived using the coefficients of the power series expansion of the device. Using this expression and an exponential approximation of the current through a diode $[i=i_0(\epsilon^{\alpha v}-1)]$, the amplitude of intermodulation produced in a crystal mixer is found to be

$$2i_0\epsilon^{\alpha V_0}R_0I_s(\alpha V_1)I_b(\alpha V_2).$$

$I_n(x)$ is an nth order modified Bessel function of the first kind. The quantities s and b are the signal harmonic and the oscillator harmonic, R_0 is the output resistance, and V_0, V_1, and V_2 are the bias, signal, and oscillator voltages, respectively.

The quantities i_0, α, R_0, and V_2 are found from the dc E-I diode characteristics, the mixer bias current, and the loss in the desired signal. Experimental tests on a mixer operating from 450 Mc to 850 Mc show that the signal input power necessary to produce a given intermodulation output power can be predicted within 6 db.

I. Introduction

In the design of a receiving system, the engineer usually must reduce the intermodulation products appearing at the output below a certain design specification. If the output powers of the intermodulation products are not known, the designer must assume a conservative (high) value and then design filters to meet the specification. Such a design technique invariably results in an over-designed system.

The receiver-system designer should know in advance what spurious responses can occur in the mixer. Signals at undesired frequencies can produce intermodulation responses if the following relationship is true:

$$|s\omega_1 \pm b\omega_2| = \omega_{if} \tag{1}$$

where

ω_1 = signal frequency,
ω_2 = local-oscillator frequency,
ω_{if} = intermediate frequency,
s and b = integers ≥ 1.

This paper shows how the output power of intermodulation products is determined. An exponential expression, similar to one suggested by Tucker[1,2] describes the diode dc E-I characteristic. The results of experimental tests are presented to verify the accuracy of the calculated intermodulation output powers.

* Received November 5, 1962; revised manuscript received May 29, 1963. The work presented in this paper was sponsored by the Electromagnetic Vulnerability Laboratory of Rome Air Development Center, Griffiss AFB, N. Y., under Contr. AF 30(602)-2690.

† Airborne Instruments Laboratory, A Division of Cutler-Hammer, Inc., Deer Park, Long Island, N. Y.

[1] D. G. Tucker, "Rectifier resistance laws," *Wireless Engineer*, pp. 117–128; April, 1948.

[2] D. G. Tucker, "Intermodulation distortion in rectifier modulators," *Wireless Engineer*, pp. 145–152; June, 1954.

II. Mixer Intermodulation Products

When a voltage is impressed across a nonlinear resistance, the current through the nonlinear resistance, assuming no reactance elements, can be expressed as a power series of the voltage

$$i = \sum_{k=0}^{\infty} g_k v^k \tag{2}$$

where

v = voltage,
i = current,
g_k = factor related to diode conductance.

When a crystal is used in a conventional mixer circuit, the output voltage is equal to the crystal current times the load resistance. Thus,

$$E_0 = f(v) = iR_0 = R_0 \sum_{k=0}^{\infty} g_k v^k \tag{3}$$

where

E_0 = output voltage,
i = crystal current,
R_0 = load resistance.

The voltage is the sum of the local oscillator and signal voltages, v_1 and v_2. Letting $v_1 = A_1 \cos \omega_1 t$ and $v_2 = A_2 \cos \omega_2 t$, then

$$v = v_1 + v_2 = A_1 \cos \omega_1 t + A_2 \cos \omega_2 t \tag{4}$$

where

A_1 = amplitude of signal voltage,
A_2 = amplitude of local oscillator voltage,
ω_1 = signal frequency,
ω_2 = local-oscillator frequency.

Assuming $A_2 > A_1$, (3) can be expanded into a Taylor series about v_2

$$E_0 = \sum_{n=0}^{\infty} \frac{A_1{}^n}{n!} (\cos^n \omega_1 t) f^{(n)}(v_2) \tag{5}$$

where $f^{(n)}(v_2)$ is the nth derivative of $f(v_2)$.

The derivatives of the right side of (3) yield

$$f^{(n)}(v_2) = \frac{d^n}{dt^n} R_0 \sum_{k=0}^{\infty} g_k v_2{}^k$$

$$= R_0 \sum_{k=n}^{\infty} \frac{k!}{(k-n)!} g_k v_2{}^{k-n}. \tag{6}$$

Reprinted from *Proc. IEEE*, vol. 52, pp. 173–179, Feb. 1964.

Substituting $p=k-n$,

$$f^{(n)}(v_2) = R_0 \sum_{p=0}^{\infty} \frac{(p+n)!}{p!} g_{p+n} v_2^{p}$$

and using $v_2 = A_2 \cos \omega_2 t$

$$f^{(n)}(v_2) = R_0 \sum_{p=0}^{\infty} \frac{(p+n)!}{p!} g_{p+n} A_2^{p} \cos^{p} \omega_2 t. \quad (7)$$

Substituting this expression into (5):

$$E_0 = R_0 \sum_{n=0}^{\infty} \frac{A_1^n}{n!} \cos^n \omega_1 t \sum_{p=0}^{\infty} \frac{(p+n)!}{p!} g_{p+n} A_2^{p} \cos^{p} \omega_2 t$$

$$E_0 = R_0 \sum_{n=0}^{\infty} \sum_{P=0}^{\infty} \frac{(p+n)!}{p!n!} A_1^n A_2^p g_{n+p} \cdot \cos^n \omega_1 t \cos^p \omega_2 t. \quad (8)$$

The terms $\cos^n \omega_1 t$ and $\cos^p \omega_2 t$ can be evaluated in terms of their harmonic components.

$$\cos^k x = \left(\frac{e^{jx} + e^{-jx}}{2} \right)^k = \left(\frac{1}{2} \right)^k \sum_{y=0}^{k} \frac{k!}{(k-y)!y!} e^{j(k-2y)x}. \quad (9)$$

Arranging terms of the sum to form cosine functions

$$\cos^k x = \begin{cases} \left(\frac{1}{2}\right)^{k-1} \sum_{y=0}^{Y} \frac{k!}{(k-y)!y!} \cos (k-2y)x + \left(\frac{1}{2}\right)^k \frac{k!}{\left[\left(\frac{k}{2}\right)!\right]^2} & \text{for } k \text{ even} \\ \left(\frac{1}{2}\right)^{k-1} \sum_{y=0}^{Y} \frac{k!}{(k-y)!y!} \cos (k-2y)x & \text{for } k \text{ odd} \end{cases} \quad (10)$$

where $Y=(k-2)/2$ for k even and $Y=(k-1)/2$ for k odd. Substituting the results of (10) into (8) and using the identity

$$\cos a \cos b = \tfrac{1}{2}[\cos (a+b) + \cos (a-b)] \equiv \tfrac{1}{2}[\cos (a \pm b)]$$

yields

$$E_0 = R_0 \sum_{n=0}^{\infty} \sum_{p=0}^{\infty} \sum_{c=0}^{C} \sum_{d=0}^{D} \frac{(p+n)! A_1^n A_2^p g_{n+p}}{(n-c)!c!(p-d)!d!2^{p+n-1}} \times \cos [(n-2c)\omega_1 \pm (p-2d)\omega_2]t + R_0 \sum_{n=0}^{\infty} \sum_{P=0}^{\infty} \frac{(p+n)! A_1^n A_2^p g_{n+p}}{2^{p+n} \left(\frac{n}{2}!\right)^2 \left(\frac{p}{2}!\right)^2} \quad (11)$$

where

$$C = \frac{n-2}{2} \text{ for } n \text{ even and } C = \frac{n-1}{2} \text{ for } n \text{ odd,}$$

$$D = \frac{p-2}{2} \text{ for } p \text{ even and } D = \frac{p-1}{2} \text{ for } p \text{ odd.}$$

Eq. (11) contains all the harmonics of both the signal and oscillator frequencies and the amplitude of all the intermodulation products. To find the output for any particular intermodulation product E_{sb} where s is the signal harmonic and b is the local oscillator harmonic, let $(n-2c)=s$ and $(p-2d)=b$. Then, $n=s+2c$ and $p=b+2d$. Substituting for n and p in (11) yields:

$$E_{sb} = R_0 \sum_{c=0}^{\infty} \sum_{d=0}^{\infty} \frac{(2c+2d+s+b)! A_1^{2c+s} A_2^{2d+b} g_{2c+2d+s+b}}{(c+s)!c!(d+b)!d!2^{2c+2d+s+b-1}} \cdot \cos (s\omega_1 \pm b\omega_2)t \quad (12)$$

and

$$E_{00} = R_0 \sum_{c=0}^{\infty} \sum_{d=0}^{\infty} \frac{(2c+2d)! A_1^{2c} A_2^{2d} g_{2c+2d}}{2^{2c+2d}(c!)^2(d!)^2}.$$

If the coefficients of the power series of (2) are known, this double summation can be used to evaluate the intermodulation products in (12); however, this method is cumbersome. To simplify the analysis, one can consider the dc E-I characteristic to be of exponential form. Let

$$i = i_0[\epsilon^{\alpha(v+V_0)} - 1] = i_0 \epsilon^{\alpha V_0} \sum_{k=1}^{\infty} \frac{\alpha^k}{k!} v^k \quad (13)$$

where

α = function of the diode nonlinearity,
v = time-varying voltage across the diode,
V_0 = dc bias across the diode,
i_0 = leakage current.

Comparing (2) with (13)

$$g_k = \begin{cases} i_0 \epsilon^{\alpha V_0} \frac{\alpha^k}{k!} & \text{for } k \neq 0 \\ 0 & \text{for } k = 0. \end{cases} \quad (14)$$

Substituting g_k for $g_{2c+2d+s+b}$ and k for $2c+2d+s+b$ in (12) and rearranging terms

$$E_{sb} = 2 i_0 \epsilon^{\alpha V_0} R_0 \sum_{c=0}^{\infty} \frac{\left(\frac{\alpha A_0}{2}\right)^{2c+s}}{(c+s)!c!} \sum_{d=0}^{\infty} \frac{\left(\frac{\alpha A_2}{2}\right)^{2d+b}}{(d+b)!d!} \times \cos (s\omega_1 \pm b\omega_2)t. \quad (15)$$

The two summations in this expression are modified Bessel functions of the first kind (Bessel function for a pure imaginary argument[3]). That is

$$\sum_{c=0}^{\infty} \frac{\left(\frac{\alpha A_1}{2}\right)^{2c+s}}{(c+s)!c!} = I_s(\alpha A_1)$$

$$\sum_{d=0}^{\infty} \frac{\left(\frac{\alpha A_2}{2}\right)^{2d+b}}{(d+b)!d!} = I_b(\alpha A_2) \quad (16)$$

where I_s and I_b are the modified Bessel functions of the first kind with arguments αA_1 and αA_2. Thus,

$$E_{sb} = 2i_0\epsilon^{\alpha V_0} R_0 I_s(\alpha A_1) I_b(\alpha A_2) \times \cos(s\omega_1 \pm b\omega_2)t. \quad (17)$$

If the value of α is known, the amplitude factor $2i_0\epsilon^{\alpha V_0}R_0 I_s(\alpha A_1)I_b(\alpha A_2)$ can be evaluated using tables of modified Bessel functions of the first kind.

III. Intermodulation Output Power

P_{sb} is the intermodulation output power at a frequency whose s harmonic mixes with the b harmonic of the local oscillator. It is expressed as

$$P_{sb} = \frac{\overline{(E_{sb})^2}}{R_0} \quad (18)$$

where $\overline{(E_{sb})^2}$ is the mean square value of E_{sb}. Using (17) and averaging over periods of the local oscillator and signal frequencies gives

$$\overline{(E_{sb})^2} = \tfrac{1}{2}[2i_0\epsilon^{\alpha V_0}R_0 I_s(\alpha A_1)I_b(\alpha A_2)]^2,$$

then

$$P_{sb} = \frac{2}{R_0}(i_0\epsilon^{\alpha V_0}R_0)^2 I_s^2(\alpha A_1) I_b^2(\alpha A_2). \quad (19)$$

Since the signal levels are usually quite small, the value of αA_1 is much less than 1. This permits $I_s(\alpha A_1)$ to be approximated by the first term of its series

$$I_s(\alpha A_1) = \frac{\alpha^s A^s}{2^s s!}.$$

Thus,

$$P_{sb} = \frac{2}{R_0}(i_0\epsilon^{\alpha V_0}R_0)^2 \frac{\alpha^{2s}A_1^{2s}}{2^{2s}(s!)^2} I_b^2(\alpha A_2). \quad (20)$$

The input power is expressed as

$$P_1 = \frac{A_1^2}{2R_1}$$

where R_1 is the input resistance. If an impedance match at the input is assumed, then $R_1 = R_d$ where R_d is the diode resistance. Also, if a match is assumed at the output then $R_d = R_0 = R_1$. Expressing A_1 in terms of P_1 and R_0 gives

$$A_1^2 = 2P_1R_0.$$

Substituting this relationship into (19) and rearranging terms

$$P_{sb} = \left[\frac{(\alpha i_0 R_0 \epsilon^{\alpha V_0}) I_b(\alpha A_2)}{s!}\right]^2 \alpha^{2s-2}\left(\frac{R_0}{2}\right)^{s-1} P_1^s. \quad (21)$$

Using (21), the intermodulation output power as a function of the signal input power can be determined from the local oscillator voltage, the dc bias voltage, and the diode characteristics α, i_0, and R_0.

IV. Determination of Parameters

It is inconvenient and sometimes impossible to measure the value of ac oscillator voltage across the diode. Also, the values of α, i_0, and R_0 that best fit the diode E-I curve must be determined. Measurements that can be taken easily are the insertion loss of the desired mixing product, the diode bias current, and the dc E-I characteristic of the diode. The derivation of the desired parameters (using parameters that can be conveniently measured) follows.

A. Local-Oscillator Voltage

The manufacturer's published specifications of diode resistance can be used to approximate the local-oscillator voltage across the diode. The square of the voltage is

$$A_2^2 = 2PR \quad (22)$$

where P is the power delivered to the diode and R is the resistance. The factor of 2 appears because A_2 is the peak value of the voltage.

A resistance value cannot truly describe diode operation, and there are large variations between diodes of the same type. For these reasons, a method was devised to determine the local-oscillator voltage from the diode dc E-I characteristic and the dc bias current through the diode in the mixer.

This method consists of plotting the dc E-I characteristic of the diode and then graphically determining the current waveform that will flow for a few values of sinusoidal input voltage.

Fig. 1 shows the diode bias current for three trial values of local-oscillator voltage versus the angle of local-oscillator voltage. The correct local-oscillator voltage corresponds to the current waveform whose average value is equal to the dc bias current. Fig. 1 shows, for example, that a 0.3-volt local-oscillator voltage A_2 produces an average current of 1.12 ma over a quarter cycle. Since current only flows for one-half cycle, this corresponds to a bias current of 0.56 ma. This

[3] E. Jahnke and F. Emde, "Tables of Functions with Formulae and Curves," Dover Publications, New York, N. Y., pp. 232–233; 1945.

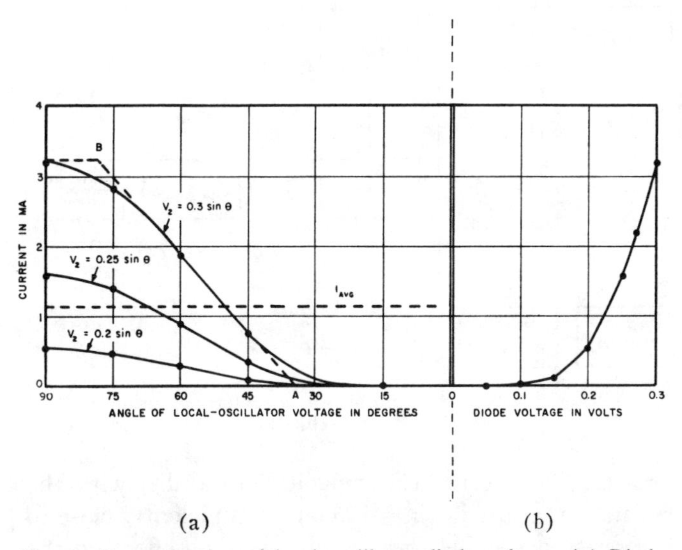

Fig. 1—Determination of local oscillator diode voltage. (a) Diode bias current for sinusoidal voltage inputs. (b) Measured diode dc E-I characteristic.

value compares with the average value of 0.5 ma for the diode used in the mixer for experimental verification. An advantage of this method is that it does not depend upon a good impedance match between the oscillator and the diode for accurate results. The bias current is a measure of the actual voltage at the diode, independent of the circuits used.

If an external dc bias is used, the current should be calculated for the ac voltage impressed over the dc bias. The method of determining the local-oscillator voltage remains the same.

B. Output Resistance (R_0)

Since an impedance match was assumed, calculation of the diode resistance R_d is equivalent to finding R_0. The diode resistance is an average value determined from the average power flow into the diode. The average signal power flow is equal to the time average of the instantaneous power over one cycle of the bias oscillator

$$P_i = E_i^2\left(\frac{dI}{dE}\right) \tag{23}$$

where

P_i = instantaneous power,
E_i = instantaneous signal voltage,
(dI/dE) = slope of diode E-I characteristic at that time.

Because the signal voltage is very small compared with the local-oscillator voltage, (dI/dE) can be assumed to come entirely from the local-oscillator voltage. Thus,

$$P_i = E_i^2\left(\frac{dI}{dE}\right)_{\mathrm{LO}}. \tag{24}$$

To find the average signal power at the diode, the instantaneous power P_i must be averaged for the local oscillator voltage and the signal voltage. Since the local oscillator and signal frequencies are noncoherent, the average value of the product is equal to the product of the average values. The average value of E_i^2 is the rms signal voltage squared E_s^2. The average power can be expressed by

$$P_{\mathrm{avg}} = E_s^2\frac{1}{2\pi}\int_0^{2\pi}\left(\frac{dI}{dE}\right)_{\mathrm{LO}} d\theta, \tag{25}$$

letting

$$E_{\mathrm{LO}} = A_2 \sin\theta,$$

$$dE_{\mathrm{LO}} = A_2 \cos\theta d\theta$$

also,

$$I_{\mathrm{LO}} = f(\theta),$$

$$dI_{\mathrm{LO}} = f'(\theta)d\theta.$$

Thus, (24) becomes

$$P_{\mathrm{avg}} = \frac{E_s^2}{2\pi A_2}\int_0^{2\pi}\frac{f'(\theta)}{\cos\theta}d\theta. \tag{26}$$

The expression $f'(\theta)$ can be determined from the measured data by curve fitting, and the use of a computer is almost mandatory. However, within the accuracy of the measured data and considering the variations that are likely to occur in the diode characteristics, a straight-line approximation to the curve in Fig. 1 will suffice. This approximation has a constant slope, K from A to B and a zero slope in the rest of the quarter cycle. Since the next quarter cycle has the same power flow, the value of the integral over the period is twice that from A to B. Thus,

$$P_{\mathrm{avg}} = \frac{E_s^2K}{A_2\pi}\int_A^B \sec\theta d\theta, \tag{27}$$

since

$$\int_A^B \sec\theta d\theta = \log\left[\frac{\tan\left(\frac{B}{2}+\frac{\pi}{4}\right)}{\tan\left(\frac{A}{2}+\frac{\pi}{4}\right)}\right]$$

and

$$R_d = \frac{E_s^2}{P_{\mathrm{avg}}}.$$

Then,

$$R_d = R_0 = \frac{\pi A_2}{K\log\left[\frac{\tan\left(\frac{B}{2}+\frac{\pi}{4}\right)}{\tan\left(\frac{A}{2}+\frac{\pi}{4}\right)}\right]} \tag{28}$$

C. Diode Characteristic (α)

The current through the diode is assumed to be an exponential function of voltage (13).

Taking any two points on the E-I characteristic, the ratio of the currents can be used to evaluate α

$$\frac{i_2}{i_1} = \frac{e^{\alpha(v_2+V_0)} - 1}{e^{\alpha(v_1+V_0)} - 1} \cdot \qquad (29)$$

It has been found, however, that the value of α varies considerably depending upon the two voltages chosen. The current only approximates an exponential over a reasonably small region. Therefore, it is necessary to determine α in the region where most of the current is flowing. This region is near the peak value of the local-oscillator voltage. The most accurate values of α are obtained when the diode current is measured at voltages equal to the peak local-oscillator voltage ($v_2 = A_2$) and at a second point 10 per cent lower than this voltage. Thus, (29) becomes

$$\frac{i_2}{i_1} = \frac{\exp\left[\alpha(A_2 + V_0)\right] - 1}{\exp\left[\alpha(0.9A_2 + V_0)\right] - 1} \cdot \qquad (30)$$

D. Leakage Current (i_0)

The leakage current can be evaluated from a measurement of the insertion loss of the desired response P_{11}/P_1. Using (21) with $s = b = 1$ and solving for i_0 yields

$$i_0 = \frac{\sqrt{\dfrac{P_{11}}{P_1}}}{\alpha R_0 \epsilon^{\alpha V_0} I_1(\alpha A_2)} \cdot \qquad (31)$$

Taking the previously calculated values of α, R_0, and A_2 with the known values of V_0 and P_{11}/P_1 uniquely determines i_0.

V. Experimental Verification

The theoretical values for the ratio of the intermodulation output power as a function of signal power (21) were checked using a mixer that covered frequencies from 450 to 850 Mc. A 1N21WE diode was used with no external dc bias. The curve of Fig. 1(b) is the measured characteristic of the diode. The significant parameters were

1) Local oscillator voltage, $A_2 = 0.3$ v,
2) Average diode resistance, $R_0 = 126$ ohms,
3) Diode characteristic $\alpha = 14.6$ v^{-1},
4) Leakage current, $i_0 = 1.58 \times 10^{-5}$ a.

The stray capacitance of this diode in its mount is about 1.0 pf. At a frequency of 850 Mc, the highest frequency used, this capacitance represents a reactance of 190 ohms. Although the reactance is significant when compared with the diode resistance of 126 ohms, the magnitude of the diode impedance is not changed greatly. Considering the capacity in parallel with the resistance results in $|z| = 105$ ohms, sufficiently close to the resistance value to have only a small effect on the intermodulation amplitudes.

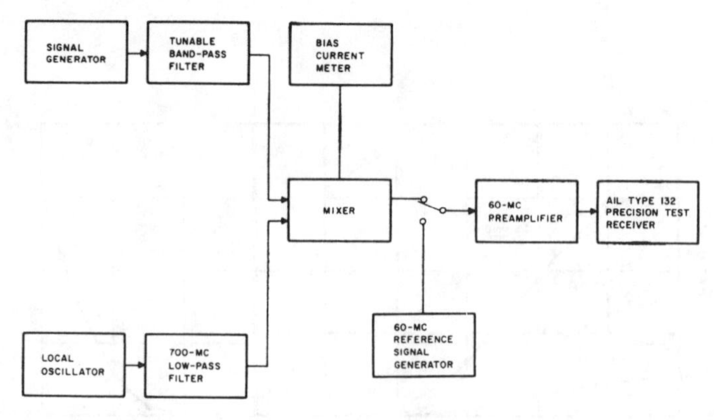

Fig. 2—Block diagram of test setup for measuring mixer intermodulation.

Data were taken using the test setup shown in Fig. 2. The output of the signal generator was sent through a frequency meter used as a tunable band-pass filter to eliminate any harmonic distortion present in the inputs, and the output of the local oscillator was sent through a low-pass filter. Both the signal and the oscillator voltages then went to the mixer. (The mixer and the 60-Mc preamplifier following it were designed at AIL as matching units.) The output of the preamplifier was then amplified further by the AIL Type 132 Precision Test Receiver and the detected output was indicated on its output meter.

The intermodulation power and input signal power were measured as follows:

1) The 60-Mc reference signal source was connected to the 60-Mc preamplifier. For a given input power a reference was established on the meter of the AIL Precision Test Receiver.
2) The mixer was connected to the 60-Mc preamplifier and the signal and local-oscillator frequencies adjusted to obtain a 60-Mc output for a given intermodulation product. The frequency of the signal and local oscillator depended upon the intermodulation product being measured. For example, to measure P_{22}, the signal was at 530 Mc and the local oscillator at 500 Mc (2×530 Mc $- 2 \times 500$ Mc $= 60$ Mc). To measure P_{12}, the signal was at 840 Mc and the local oscillator at 450 Mc (2×450 Mc $- 840$ Mc $= 60$ Mc). Frequencies for other intermodulation products were selected similarly.
3) The local-oscillator level was set for 0.5 ma of bias current.
4) The input-signal power was taken as the power setting on the signal generator. The mixer output power was obtained from the precision attenuator in the AIL receiver, which was varied to maintain a constant output-power reference level as in Step 1).

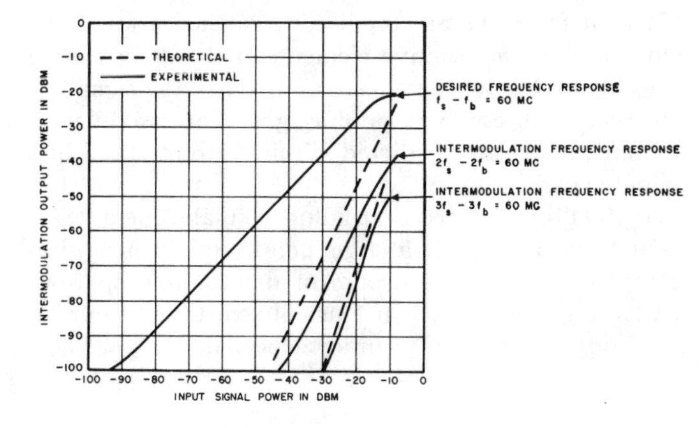

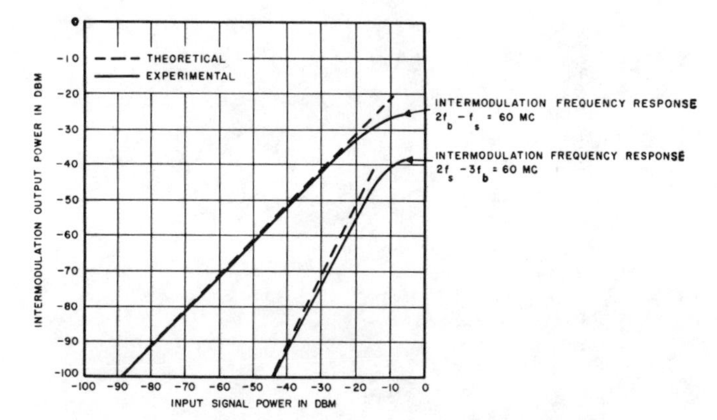

Fig. 3—Intermodulation output power vs signal power.

5) Steps 2) to 4) were repeated for various intermodulation products up to sixth order.

The insertion loss of the frequency meter was measured later and the power level of the signal generator applied to the diode was corrected accordingly.

Fig. 3 shows the measured intermodulation powers and the powers predicted by (21). No predicted power is shown for the primary response P_{11}, since the loss of the primary response is required to predict the intermodulation powers (calculation of i_0). There is excellent agreement for input-signal power levels less than -10 dbm. Above this level, saturation effects in the mixer cause increasing errors. This analysis is not expected to be valid for higher levels, since the small-signal assumption is no longer true.

VI. Illustrative Example

The calculation of R_0, α, i_0, and the response P_{12} follows to illustrate the method more clearly. The local oscillator voltage A_2 of 0.3 volt was obtained in Section IV-A.

The current waveform for a quarter-cycle voltage input is shown in Fig. 1(a). Using (28) and values from Fig. 1(a) for points A and B, $A=36$ degrees $=0.63$ radian; $B=82$ degrees $=1.43$ radians

$$K = \frac{3 \times 10^{-3}}{1.43 - 0.63} = 3.75 \times 10^{-3} \text{ amperes/radian}$$

and

$$R_0 = \frac{\pi(0.3)}{(3.75 \times 10^{-3}) \log \left[\dfrac{\tan\left(\dfrac{1.43}{2} + \dfrac{\pi}{4}\right)}{\tan\left(\dfrac{0.63}{2} + \dfrac{\pi}{4}\right)} \right]}$$

$$R_0 = 126 \text{ ohms}.$$

The value of α can be obtained from (30), and the values $i_2 = 3\times 10^{-3}$ a for $A_2 = 0.3$ v and $i_1 = 1.90\times 10^{-3}$ a for $A_1 = 0.27$ v, then

$$\frac{i_2}{i_1} = \frac{3 \times 10^{-3}}{1.9 \times 10^{-3}} = \frac{\epsilon^{0.3\alpha} - 1}{\epsilon^{0.27\alpha} - 1}.$$

There is no way of solving this equation explicitly for α, but it only takes a few trial values using a slide rule to determine the value of α at 14.6 $(\text{v})^{-1}$.

The value of i_0 is found by using (31). The insertion loss (P_{11}/P_1) was found to be 8 db $=0.16$. Therefore,

$$i_0 = \frac{(0.16)^{1/2}}{(14.6)(126)(13.8)}$$

$$i_0 = 1.58 \times 10^{-5} \text{a}.$$

Substituting the values for R_0 and i_0 into (21) and letting $V_0 = 0$, yields

$$P_{sb} = \left[\frac{(14.6)(1.58 \times 10^{-5})(126) I_b(4.48)}{s!} \right]^2 (14.6)^{2s-2} \cdot \left(\frac{126}{2}\right)^{s-1} P_1{}^s.$$

For the product involving the second harmonic of the LO mixing with the signal frequency ($s=1$, $b=2$) and $I_2(4.48) = 9.45$,

$$P_{12} = \left[\frac{(14.6)(1.58 \times 10^{-5})(126)(9.45)}{1} \right]^2 P_1$$

$$P_{12} = 0.075 P_1$$

or, expressed in dbm

$$10 \log \left(\frac{P_{12}}{0.001}\right) = 10 \log \left(\frac{P_1}{0.001}\right) - 11.2 \text{ db}$$

$$P_{12}(\text{dbm}) = P_1(\text{dbm}) - 11.2 \text{ db}.$$

VII. Conclusions

For the most part, (21) can be used to predict spurious responses in mixers at frequencies where reactive components of the diode are negligible. One need only know the diode dc E-I characteristic, the bias current, the desired response insertion loss, and the dc bias voltage. Such information can be used to determine what filtering is necessary in receivers using mixers, or what intermodulation is to be expected by changing diodes or adjusting the local-oscillator level. The unnecessary waste of overdesign can thereby be eliminated.

The analysis has been made as general as possible and can be used in single-ended mixers with and without external dc bias. In fact, the first part of the analysis applies to any nonlinear device. Eq. (12) can be used to predict intermodulation products in any nonlinear device if the proper values of g_k are used.

Inaccuracies in this analysis are caused mainly by the following assumptions: 1) that the diode E-I curve is an exponential function, 2) that the voltages across the diode, at the signal and oscillator frequencies, are sinusoidal, and 3) the reactive elements in the diode can be neglected. Despite these sources of error, the signal input powers necessary to produce given intermodulation output powers are predicted within 6 db of the values found experimentally.

It should be noticed that failure to make one of these assumptions results in a large increase in mathematical complexity. For the purpose of determining spurious levels this complexity is not usually worthwhile in relation to the increase in accuracy expected.

Designing Microwave Mixers for Increased Dynamic Range

ROBERT L. ERNST, MEMBER, IEEE, PETER TORRIONE, WEN Y. PAN, FELLOW, IEEE, AND MELVIN M. MORRIS, SENIOR MEMBER, IEEE

***Abstract*—The principal factors limiting dynamic range in microwave frequency mixers are reviewed, and recently developed methods of extending the dynamic range are summarized. The principles are shown to be applicable to a wide variety of nonlinear elements at all frequency ranges up to and including the microwave range. Specially developed equipment for measuring mixer performance is described, and typical experimental results given. The significance of the suggested techniques to communications and video systems is demonstrated.**

Introduction

MIXERS commonly used in the front ends of microwave receivers are often subjected to large-signal conditions. These signals, comprising either the desired information carrier or unwanted interfering signals, can cause distorted output due to limited mixer dynamic range. In particular, when more than one signal exists, intermodulation and cross-modulation distortion can be prevalent. Thus, it is important to know the sources of limited dynamic range and spurious responses. Once the sources are identified, they can be minimized or eliminated by proper device selection and circuit design.

Manuscript received June 3, 1969. This work was supported by USAECOM under Contracts DA36-039 AMC-02345(E) and DAAB07-67-C-0317.

L. R. Ernst and W. Y. Pan are with the RCA Defense Advanced Communications Laboratory, Somerville, N. J.

P. Torrione was with the RCA Defense Advanced Communications Laboratory, Somerville, N. J. He is now with ITT Laboratories, Nutley, N. J.

M. M. Morris is with the U. S. Army Electronics Laboratories, Fort Monmouth, N. J.

The factors which limit dynamic range and methods to minimize their detrimental effects are discussed in the following section. Examples of performance improvements and a measurement system capable of evaluating these improvements are shown. Improvements possible in multichannel communication and video systems are also discussed.

Sources of In-Band Distortion

Intermodulation (IM) distortion will be considered in detail because this type of distortion cannot be eliminated by filtering or by suppressing out-of-band responses. In particular, the input will be considered to consist of two signals of radian frequencies ω_{s1} and ω_{s2}, with $\omega_{s1} \approx \omega_{s2}$. The mixer is pumped at radian frequency ω_{LO}, so that the expected outputs at $\omega_{o1} = \omega_{s1} - \omega_{LO}$ and $\omega_{o2} = \omega_{s2} - \omega_{LO}$ result. However, IM distortion occurs in the form of spurious signals occurring at frequencies $\omega_{im_1} = 2\omega_{s1} - \omega_{s2} - \omega_{LO}$ and $\omega_{im_2} = 2\omega_{s2} - \omega_{s1} - \omega_{LO}$. This frequency spectrum is illustrated in Fig. 1, along with hypothetical values of frequency. It will be shown that interfering signals can be generated at these frequencies by several processes.

Assume that the mixer network is such that power flows only in the frequency bands associated with the signal, pump, and output. In other words, no power exists at frequencies such as $2\omega_{s1}$ and $2\omega_{s2}$. IM products will still be generated by any one of several sources, either singly or in combination. These sources will now be considered in detail.

Reprinted from *IEEE Trans. Electromagn. Compat.*, vol. EMC-11, pp. 130–138, Nov. 1969.

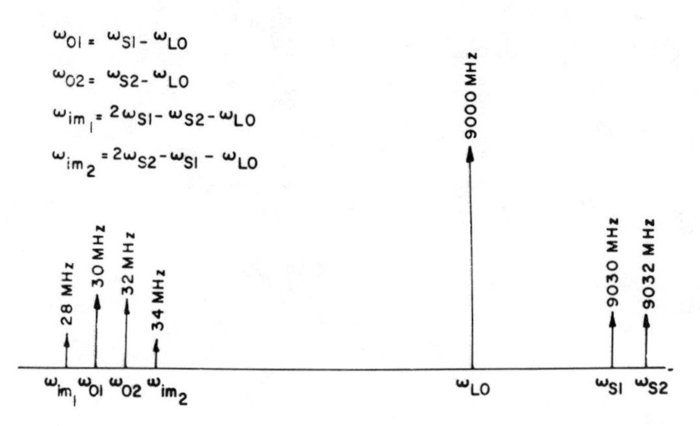

Fig. 1. Typical spectrum of microwave frequency converter.

High-Order Nonlinearities

Most devices used as the mixing element in a frequency-conversion circuit do not have a "square-law characteristic." In other words, the transfer function of the device contains terms higher than the second order. For example, the resistive diode, i.e., a diode in which energy storage effects are minimal, is typically characterized by the expression [1]

$$i = I_{\text{sat}}\,(e^{\alpha v} - 1), \tag{1}$$

when spreading resistance losses and parasitic elements are neglected.[1] Expanding to a Taylor series leads to the conventional expression

$$i = I_{\text{sat}}\left[\alpha v + \frac{(\alpha v)^2}{2!} + \frac{(\alpha v)^3}{3!} + \frac{(\alpha v)^4}{4!} + \cdots\right]. \tag{2}$$

Mixer analyses use the principle that the second-order term is responsible for generation of the desired converted output, while the higher even-order terms contribute IM and cross modulation terms.[2] The contribution from the odd-order terms can be effectively eliminated by filtering.

An understanding of the nonlinear nature of the mixing element and how this nonlinearity is affected by biasing and circuit configuration is essential to the design of high-dynamic-range mixers.

Gain Saturation

If a "perfect square-law" device were used for mixing, interfering responses at ω_{im_1} and ω_{im_2} would still be generated by virtue of a gain-saturation mechanism [2]. What happens is that, as the signal is increased, more power is needed from the local oscillator (LO) to provide conversion to the output. However, due to limited available pump power and changes in impedance presented to the pump generator, the gain from signal to output begins to decrease. As a result, the mixer transfer characteristic becomes nonlinear, and spurious responses are produced.

[1] Mathematical symbols are defined in the Glossary.
[2] This principle is used in [6] and [7].

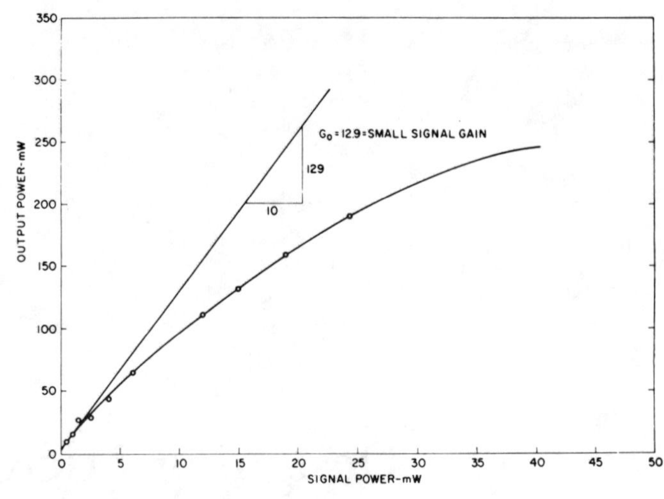

Fig. 2. Gain saturation characteristics of varactor upconverter.

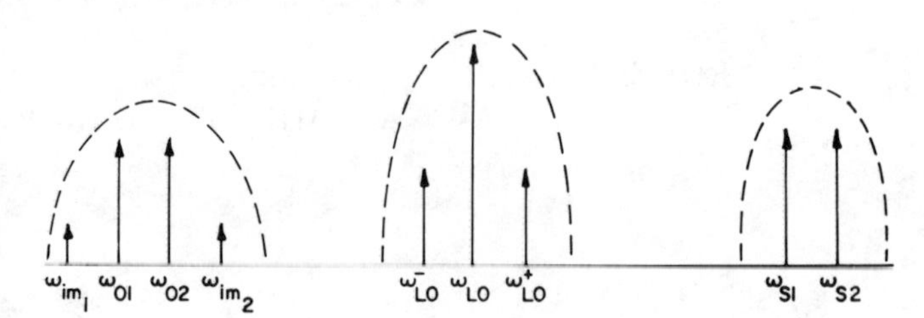

Fig. 3. Interference created by finite pump bandwidth.

This saturation is shown in Fig. 2 for the case of a parametric upconverter, in which the varactor is characterized by the equation [3]

$$v + \phi = \frac{V_B + \phi}{4} + \frac{S_m}{2}\,q + \frac{S_m}{Q_m}\,q^2. \tag{3}$$

Thus, to reduce IM distortion, it is important to know how to extend the dynamic range of the mixer by minimizing the effect of gain saturation.

Finite Pump Bandwidth

Even when the highest Q components are used to filter the signal, pump, and output frequencies, perfect filtering cannot be realized. Spurious frequencies can still be generated very close to the desired frequencies, and result in the generation of in-band interference [4]. Consider, for example, the spectrum shown in Fig. 3. The various frequency components are generated in the following manner.

1) The two desired outputs are produced by the conventional mechanism; i.e.,

$$\omega_{o1} = \omega_{s1} - \omega_{\text{LO}}, \tag{4}$$

$$\omega_{o2} = \omega_{s2} - \omega_{\text{LO}}. \tag{5}$$

2) Sidebands are generated about the pump frequency by an internal mixing within the diode of the outputs with the inputs:

$$\omega_{\text{LO}}^- = \omega_{s1} - \omega_{o2} = \omega_{\text{LO}} + \omega_{s1} - \omega_{s2}, \tag{6}$$

$$\omega_{\text{LO}}^+ = \omega_{s2} - \omega_{o1} = \omega_{\text{LO}} + \omega_{s2} - \omega_{s1}. \tag{7}$$

3) The pump sidebands, in turn, beat with the signal inputs to produce the in-band interference terms:

$$\omega_{\text{im}_1} = \omega_{s1} - \omega_{\text{LO}}^+ = 2\omega_{s1} - \omega_{s2} - \omega_{\text{LO}}, \tag{8}$$

$$\omega_{\text{im}_2} = \omega_{s2} - \omega_{\text{LO}}^- = 2\omega_{s2} - \omega_{s1} - \omega_{\text{LO}}. \tag{9}$$

The importance of this mechanism depends on the nonlinear device used as the mixing element. In some devices, some of these mixing processes might result in a gain mechanism when producing the spurious response. As a result, the sidebands generated by this multiple-conversion technique may be significant.

Methods of Extending Dynamic Range and Reducing Distortion

Having established the mechanism by which dynamic range is limited by distortion, techniques which effect a greater dynamic range can be considered. The various techniques can be roughly categorized by the process which is predominantly responsible for the distortion.

High-Order Nonlinearities

Distortion caused by high-order nonlinearities may be reduced by either of two methods: 1) reduction of high-order nonlinearity by applying proper bias and LO levels, and 2) cancellation of a high-order nonlinearity by a complementary nonlinearity.

With any nonlinear device it is important to guarantee that the applied voltage never enters a region of high nonlinearity, such as conditions of breakdown, saturation, cutoff, and (in devices such as the varactor diode) forward conduction. An example of this effect at UHF can be shown in the FET mixer. If the device were a perfect square-law device, and the gate voltage limited to excursions between zero and the pinchoff voltage V_{po} no distortion would be caused by high-order nonlinearities. The optimum bias voltage on the gate would be $V_{\text{go}} = {}^1/_2 V_{\text{po}}$ and the maximum LO would be $V_{\text{po}}/2$. An experimental curve showing this result when performing a cross-modulation test is shown in Fig. 4. As can be seen, the best performance occurs at the predicted bias point [5]. An almost identical result would be observed with the current-pumped abrupt-junction varactor.

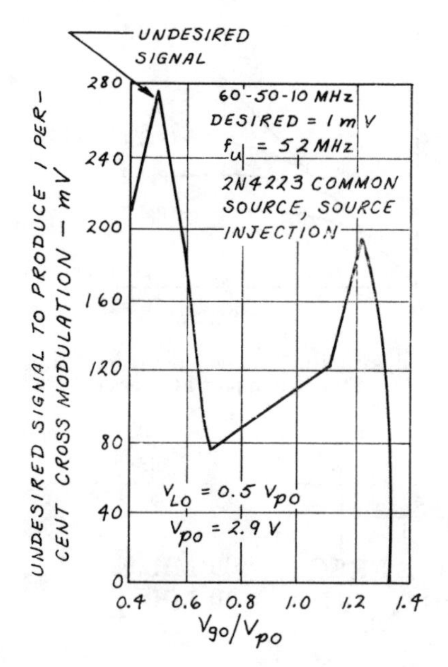

Fig. 4. Cross-modulation performance of FET mixer.

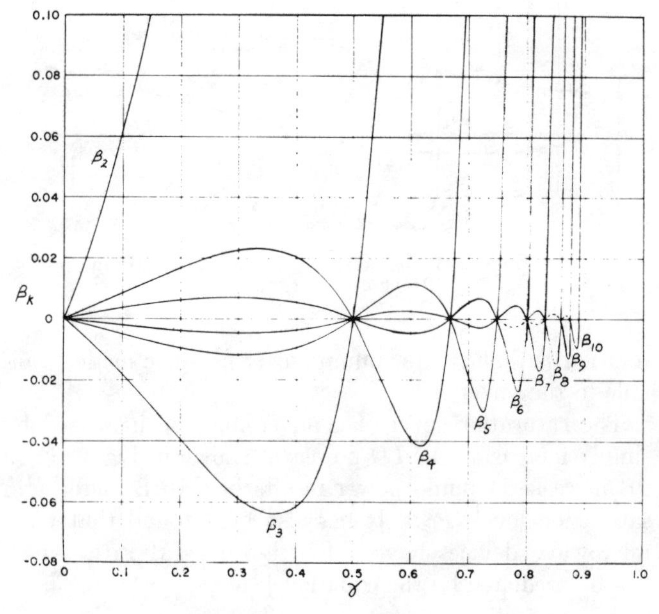

Fig. 5. β_k versus γ ($\beta_k \leq 0.1$).

The cancellation of IM distortion by using complementary nonlinearities can be demonstrated by a varactor upconverter. The varactor characteristic is given by

$$C(v) = \frac{dq}{dv} = C_0 \left(\frac{V_B + \phi}{V + \phi} \right)^\gamma. \tag{10}$$

The voltage-charge expansion is given by

$$v(q) = V_B(1 + \beta_1 q + \beta_2 q^2 + \cdots + \beta_n q^n + \cdots) \tag{11}$$

where the values of β are shown in Fig. 5. It can be seen that the coefficients of the terms of higher than second-order reverse sign as γ increases through the value $\gamma = 0.5$. Therefore, by using a balanced type of frequency converter, in which one diode has a γ less than 0.5 and the other diode has a γ in the range 0.5–0.667, partial cancellation of IM and cross modulation can be obtained. Experimental upconverters of this type have demonstrated an improvement of IM ratio of from 10 to 30 dB, as compared with a similar circuit using identical diodes [6].

Another result has been reported in the literature in which a combination of bias adjustment and nonlinearity cancellation is used to lower IM distortion by 50 dB with Schottky barrier diodes [7]. The diodes are biased differently, so that the nonlinearity of one diode complements the high-order nonlinearity of the other.

Gain Saturation

To lower spurious responses caused by gain saturation, it is necessary to be able to increase the upper limit of dynamic range without a sacrifice in the lower limit. The

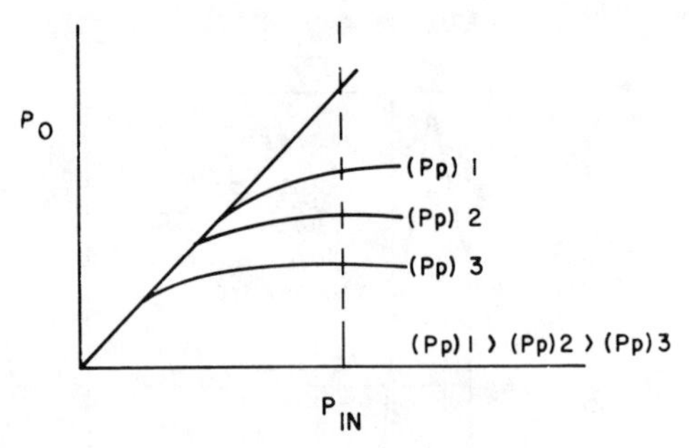

Fig. 6. Typical saturation characteristics of frequency converter (P_P = pump or LO power).

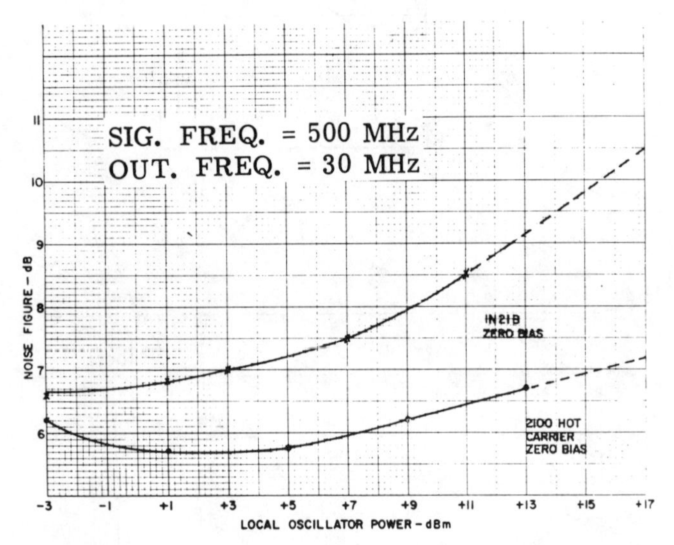

Fig. 7. Noise figure versus LO power for 1N21B and 2100 hot-carrier mixers.

solution is that greater pump power must be made available to the mixer.

The saturation level of a mixer may be increased by simply increasing the LO power, as shown in Fig. 6. Each dB increase in pump power results in a 1-dB increase in saturation level (P_{om}). It has also been found that most microwave devices have a fourth-order IM ratio which can be predicted by the equation [8]

$$\text{IMR} = 20 \log_{10} \left(\frac{P_o}{P_{om}}\right) - 19.5 \text{ (dB)}. \tag{12}$$

From this relation, it is seen that a 1-dB increase in pump power produces a 2-dB improvement in IM ratio. However, the improvement that can be obtained by simply increasing the LO power is limited by other factors, such as an increase in noise figure and thermal dissipation capabilities.

When a conventional point-contact diode is used, the amount of pump power which may be applied without severe degradation in noise figure is limited to about 2 mW [8] as shown in Fig. 7. A better choice for mixing appears to be the Schottky barrier diode. This diode can have approximately the same nonlinear characteristic as the point-contact diode, while the series spreading or loss resistance can be lower by roughly a factor of $^1/_4$. When properly chosen, and compared in the same frequency range, the hot-carrier diode has lower conversion loss, lower noise figure, relatively constant noise ratio, and the capability of handling 50 mW LO power while maintaining a noise figure approximately equal to that of the point-contact diode at 2 mW.

When upconversion or near-unity frequency ratio downconversion is desired, the varactor diode is presently the best choice at microwave frequencies. Diodes having a value of γ close to $^1/_2$ can be used to minimize high-order nonlinearities. Because low-loss diodes are available with high breakdown voltages, pump powers exceeding 1 watt can be applied while maintaining good noise performance. Outstandingly high dynamic range has been obtained in this manner [9].

Once the optimum mixing element is selected, the dynamic range of the mixing circuit can, in principle, be extended at all frequencies by incorporating several diodes in a single circuit [10]. As more diodes are added, greater LO power can be applied safely and, since the signal power is unchanged, a reduction in IM occurs. The reduction in the lowest order IM, compared to a single-diode mixer, is given by the expression

$$\Delta\text{IMR(dB)} = 20 \log_{10} N \tag{13}$$

where N is the number of diodes used. It makes no difference which circuit is used, because any technique used to cancel an IM response by circuit symmetry will also cancel the desired output. Hence, any balanced, push-pull, or multiple-diode mixer in which the number of diodes is unchanged will have the same spurious response.

Unsolved Problems

Finite Pump Bandwidth

As indicated earlier, IM products generated by multiple conversions within the limited passbands of pump, signal, and output circuits can never be completely eliminated. However, the significance of this mechanism depends upon the nonlinear element used for mixing. If a resistive diode is used, this process can be neglected, because each of the three mixing processes generating the IM, as outlined earlier, is lossy. On the other hand, if a reactive diode is used, some of these mixing steps can have significant gain associated with them, and the resulting spurious responses can be significant [11].

Out-of-Band Responses

With the increasing use of integrated microcircuits at microwave frequencies, adequate filtering of out-of-band signals becomes very difficult. It will be necessary in the near future to determine how the results discussed in this paper can be extended to include this particular case.

Measurement Techniques

Equipment Requirements

Measurement of gain phenomena in a microwave frequency converter may be done in a straightforward manner using conventional high-quality instrumentation.

Fig. 8. Microwave mixer analyzer, front view.

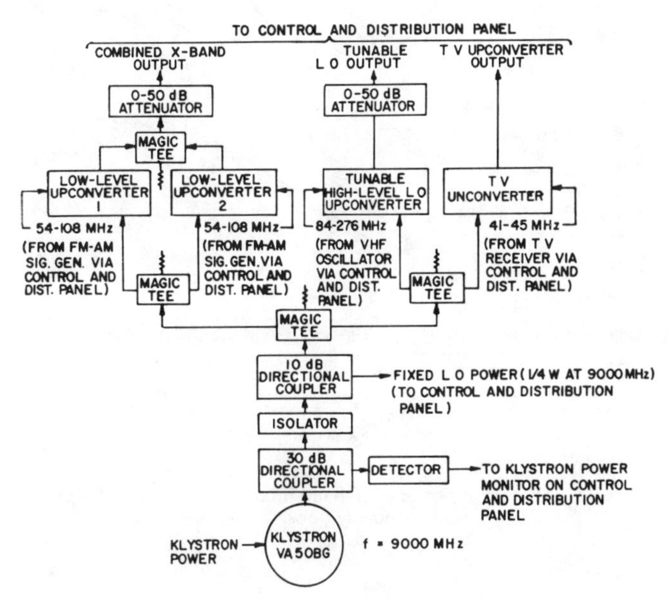

Fig. 9. Upconverter assembly block diagram.

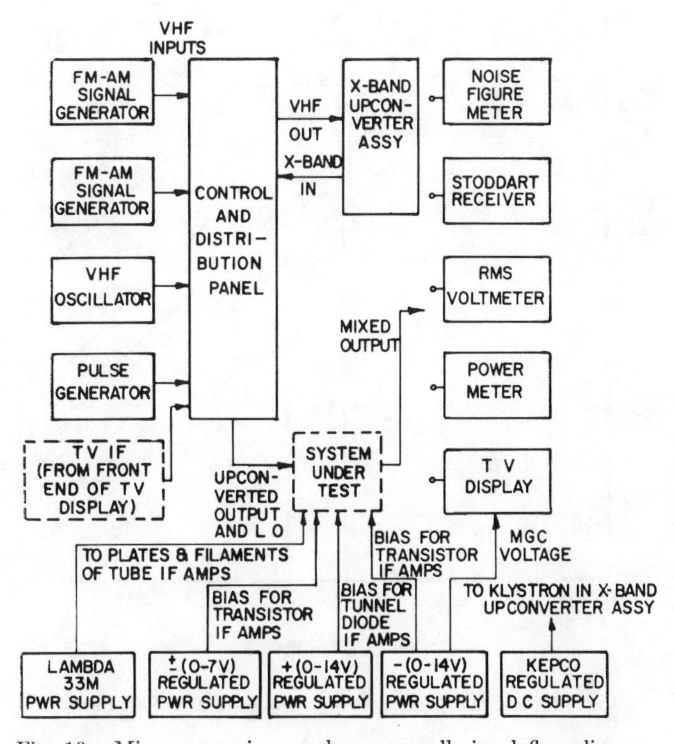

Fig. 10. Microwave mixer analyzer, overall signal flow diagram.

However, when performing a two-tone test of IM and cross-modulation characteristics, it is important that the test equipment satisfy several stringent requirements. These requirements are as follows.

1) A very-narrow-band tunable receiver must be used to measure the IM products and still adequately filter out the adjacent stronger signal.

2) The two tones applied to the mixer must have a high order of stability so that the IM product measured at the output does not drift out of the receiver passband.

3) The two tones must be generated in a way which will not create IM products before their application to a mixer.

Description of Equipment

RCA has developed a console of equipment which has been called a microwave mixer analyzer. This equipment is specially designed to measure dynamic range, spurious responses including IM and cross modulation, noise figure, and desensitization of X-band mixers. This equipment, shown in Fig. 8, meets all the requirements stated above, and has the capability to display the effect of mixer design on both video and audio systems.

The microwave portion of this system is shown in functional form in Fig. 9. Signal drift of the mixer output is virtually eliminated by an upconverter system in which the pump is made available for use as an LO with the mixer under test. In this fashion, any drift in pump frequency is eliminated, and only the drift of the VHF oscillators remains. Because high-level upconverters are used, distortion is minimized. By using a separate upconverter for each translation from VHF to X band and isolating these upconverters with hybrid junctions, IM within the system is virtually nonexistent.

A block diagram of the entire microwave mixer analyzer is shown in Fig. 10. The system is designed to permit tests with CW, AM, FM, and pulse signals. By selecting the right connection of equipment, it is possible to test virtually any type of mixer for gain, dynamic range, noise figure, spurious responses including IM and cross-modulation distortion, and desensitization.

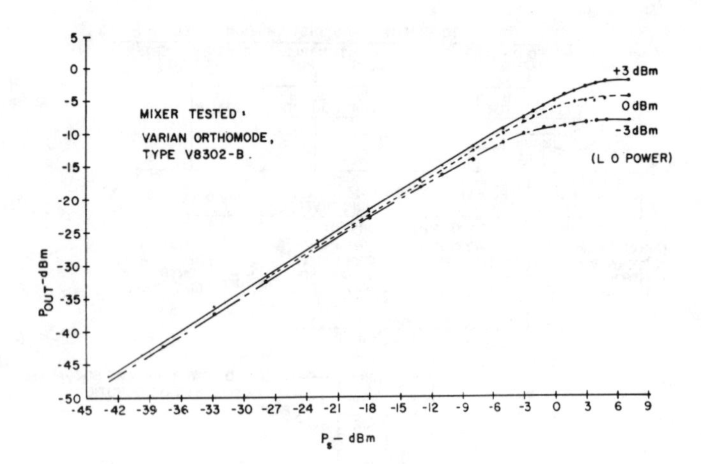

Fig. 11. IF output power of X-band mixer as a function of signal and LO power.

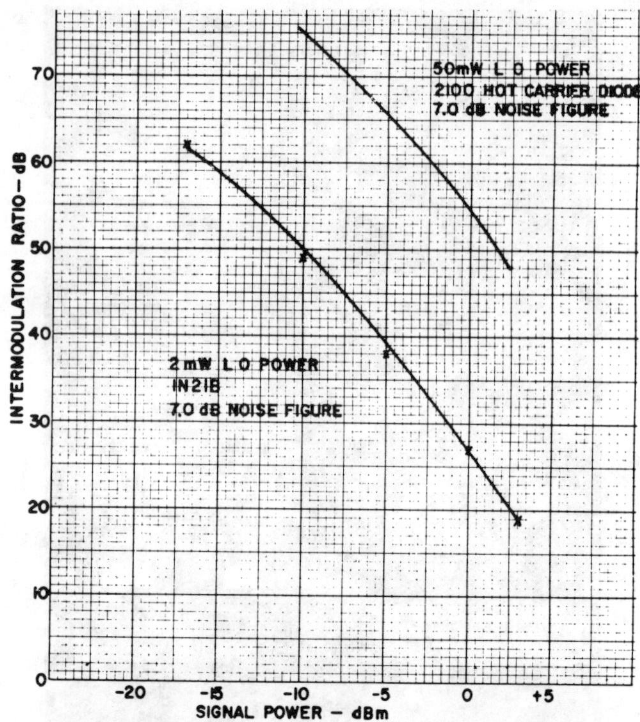

Fig. 13. IM distortion in 1N21B and 2100 hot-carrier mixers.

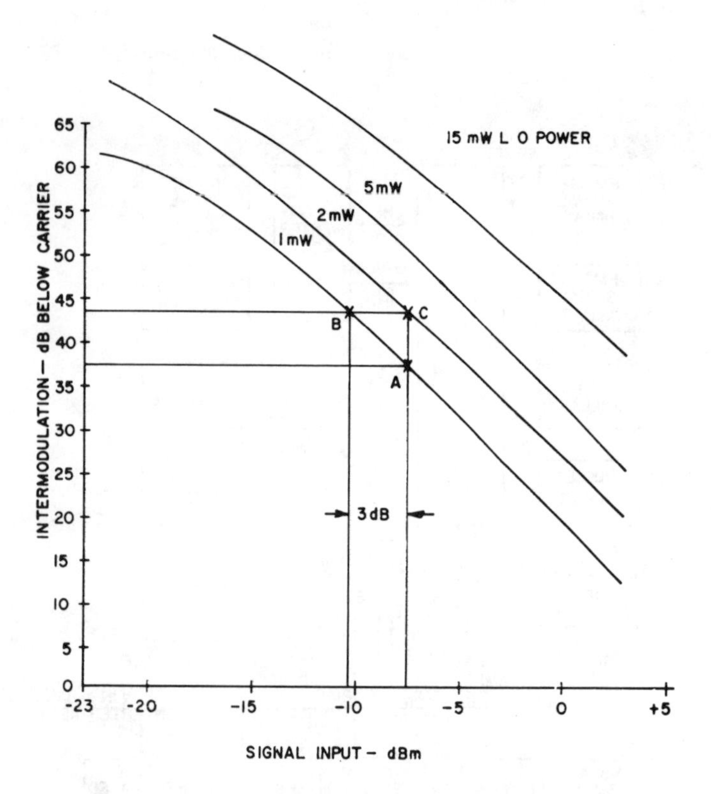

Fig. 12. IM characteristics of a point-contact diode mixer.

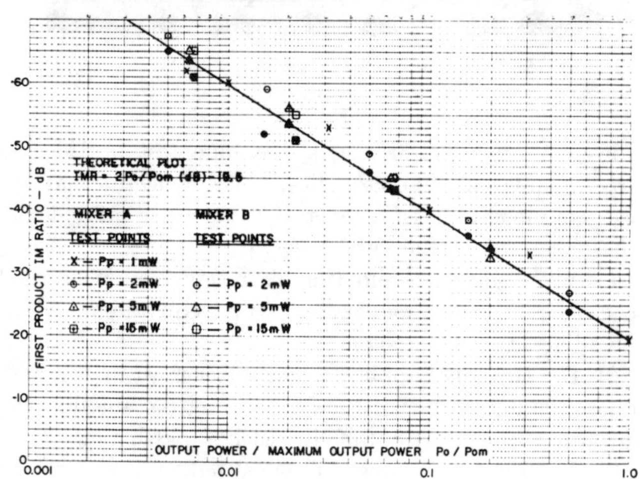

Fig. 14. IM distortion in point-contact diode mixers.

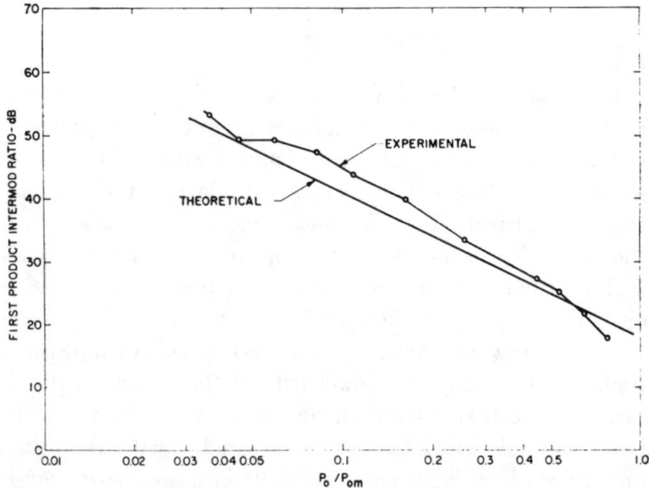

Fig. 15. IM distortion in a varacter diode upconverter.

Measurement Results

To verify the basic principles discussed in this paper, some typical experimental results are presented here. The gain saturation of a point-contact diode mixer and its variation with LO power is shown in Fig. 11. Here it can be seen that each dB increase in pump power extends the saturation level by 1 dB. Hence, the effective range of linearity is extended, and the associated in-band distortion is expected to be less.

The measurement of the resulting IM distortion is shown in Fig. 12. A standard two-tone technique was performed using signals of equal amplitude. The results verify that a 1-dB increase in signal power causes a 2-dB degra-

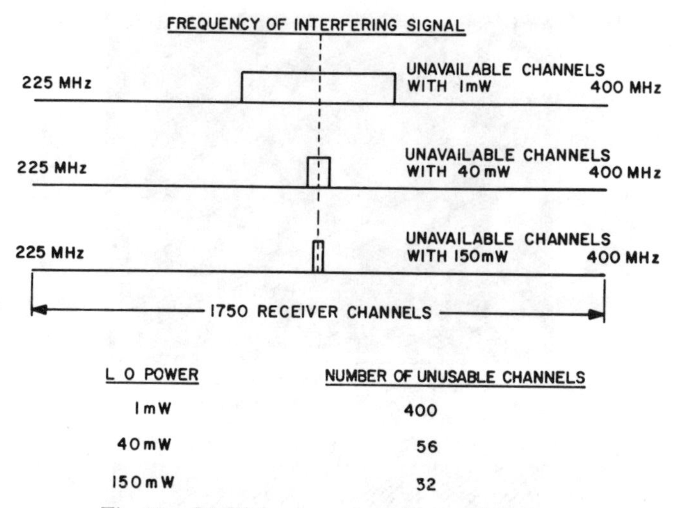

Fig. 16. Multiple-channel receiver performance.

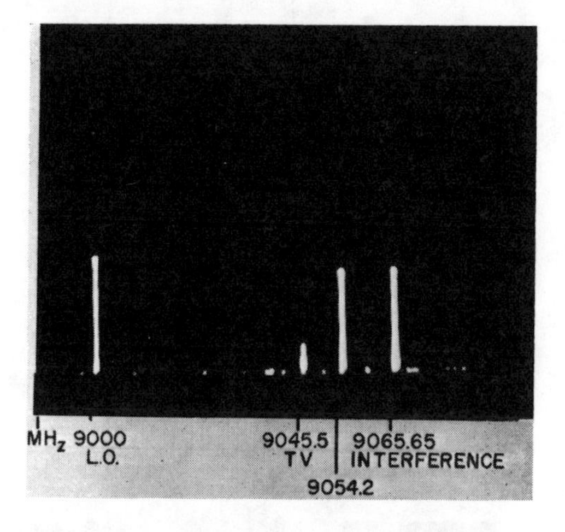

Fig. 17. Spectrum display of X-band input to mixer, $P_{\mathrm{LO}} = -3$ dBm.

dation of IM ratio, while a 1-dB increase in LO power causes a 2-dB improvement.

The improvement possible with a hot-carrier diode is shown in Fig. 13. Because this diode can operate with an LO level 14 dB greater than that of the point-contact diode, while maintaining the same noise figure, and improvement of IM distortion ratio approaching 28 dB results.

Experimental verification of (12) is shown in Fig. 14 for point-contact diode mixers, and in Fig. 15 for a varactor diode upconverter. Similar experiments with other types of mixers have provided even further verification of these principles.

The effects on TV signals of IM distortion in a microwave mixer are discussed in detail in the Appendix.

The effects of reduced distortion on system performance have been demonstrated with a multi-channel UHF receiver. As shown in Fig. 16, an increase in LO power permitted a greater number of channels to be used without excessive distortion from cross-modulation and IM responses.

Conclusions

A designer of mixer circuits frequently hears the following objection to increasing the LO power applied to a mixer, "It is desirable to reduce the oscillator amplitude as far as possible without affecting sensitivity or low-voltage operation (to reduce spurious responses especially images)" [12]. This paper has shown that when in-band spurious responses are considered, it is desirable to *increase* the LO as much as practical considerations will permit. This has the effect of extending the range of linear operation and reducing interfering signals.

The selection of a nonlinear element for the frequency converter depends on the ratio of output frequency to input frequency. For ratios greater than unity, varactors are most desirable because of their capacity for gain at large pump powers while maintaining low noise figure. When mixing to a lower frequency, Schottky barrier diodes are most attractive in the present state of the art. As the down-conversion ratio approaches unity, the advantage in terms of dynamic range goes to the varactor diode.

The observations made in this paper are valid for most devices used as mixers for the higher frequencies. It has been found that these devices may be placed in a hierarchy of preference where high dynamic range is desirable. This order is as follows:

1) varactor diode (for $\omega_o/\omega_s \gtrapprox 1$);
2) Schottky barrier diode;
3) point-contact diode;
4) tunnel and back diodes.

Once the optimum mixing element has been selected, further dynamic-range extension may be realized with a multiple-diode circuit. As more diodes are properly added to the circuit, the power-handling capacity is increased, permitting an extension of gain saturation. Since the noise figure is unchanged, the region of linear operation is greater, and interfering spurious responses are reduced.

Appendix

Visual Display of Intermodulation Effects in an X-Band Mixer

To visually demonstrate the effects of IM distortion in an X-band mixer, a standard TV signal was upconverted to 9045.5 MHz and used as a test signal. A commercially available balanced mixer incorporating 1N23WE diodes was used. The spectrum display of all tones applied to the mixer is shown in Fig. 17. The interference signals were generated through the upconverter system of the microwave mixer analyzer. The power levels of the tones from left to right are as follows: LO, -3 dBm; TV signal, -23 dBm, and interfering signals, -6 dBm each.

The reference condition obtained with no interference present is shown in Fig. 18. Shown is the spectrum display of the mixer output in which the TV signal is about -30 dBm. To obtain this "perfect" picture required a LO of -3 dBm.

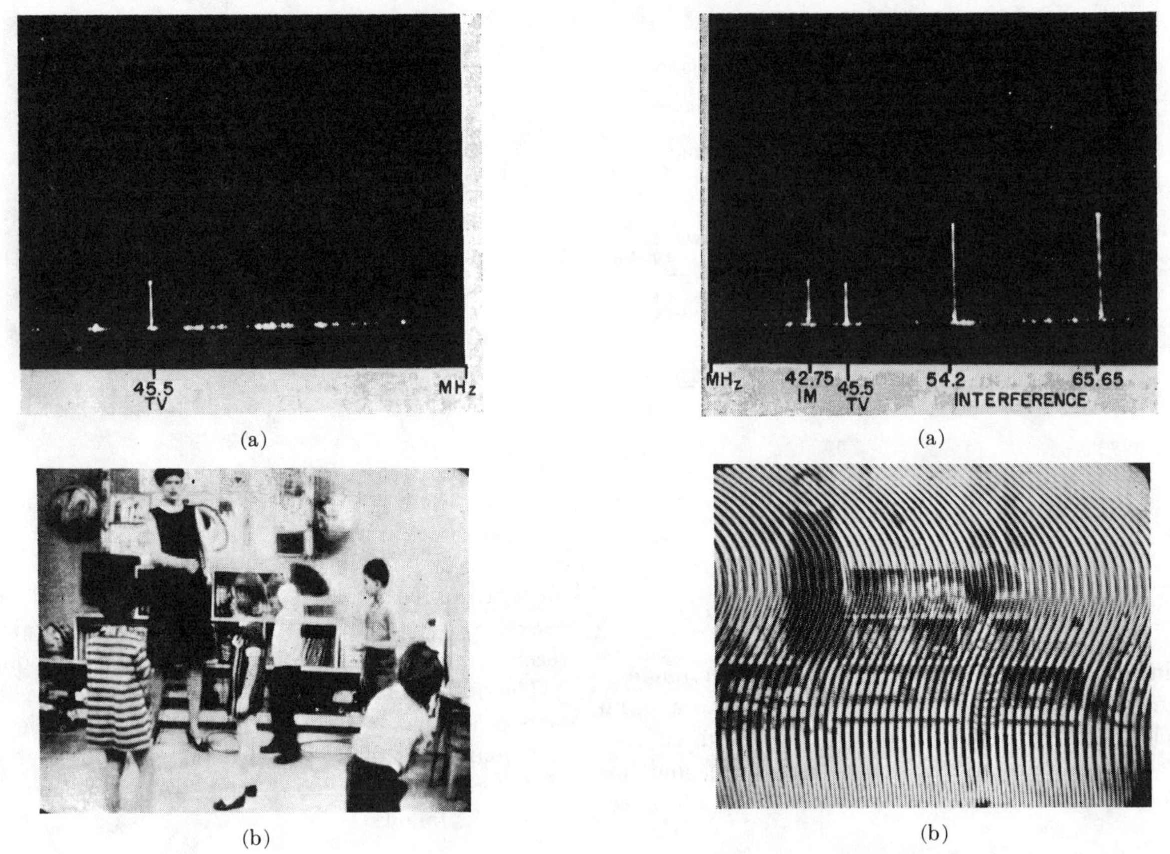

Fig. 18. Reference condition, no interference. (a) Spectrum display. (b) TV display.

Fig. 20. Effects of 3-dB increase in LO power. (a) Spectrum display. (b) TV display.

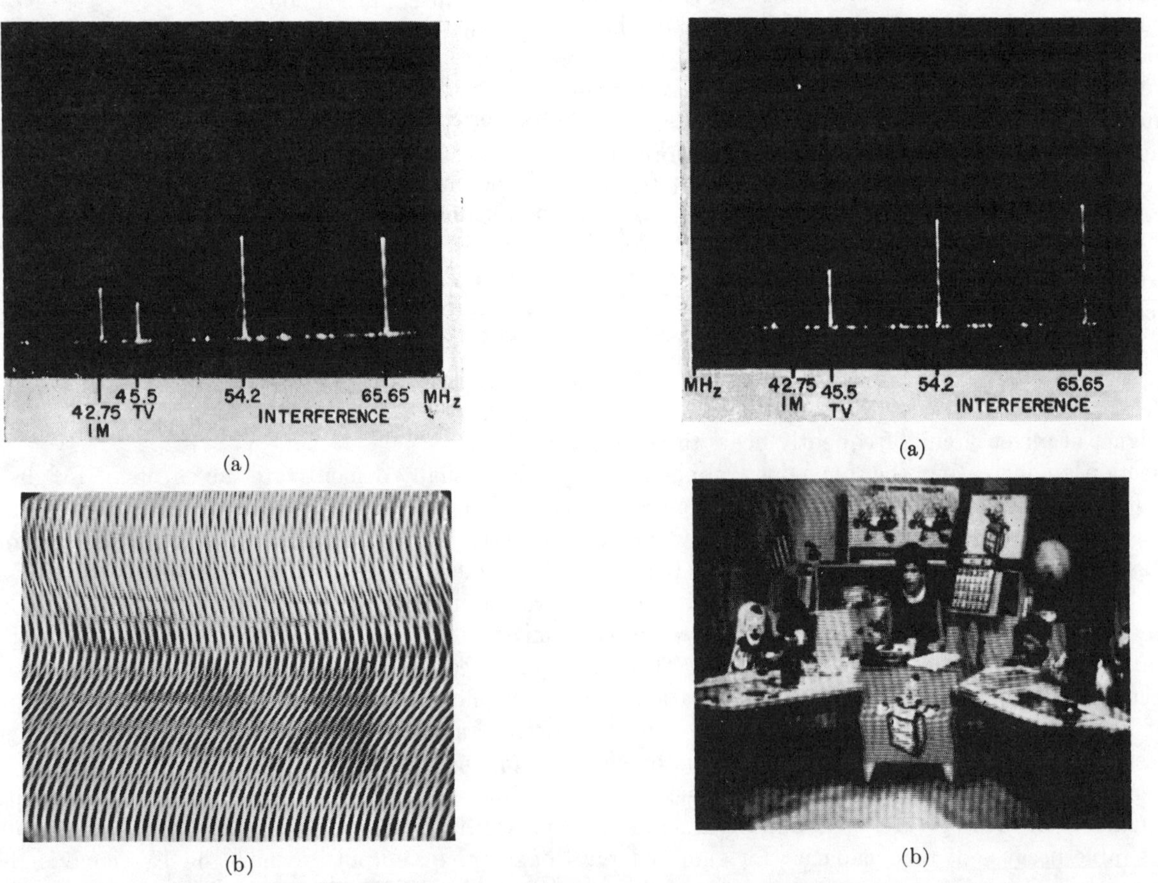

Fig. 19. Results of fourth-order IM product interference. (a) Spectrum display. (b) TV display.

Fig. 21. Effects of 6.5-dB increase in LO power. (a) Spectrum display. (b) TV display.

The result of a fourth-order IM product completely obscuring the signal is shown in Fig. 19. The TV signal is reduced by gain compression, and the difference in the two interfering signal strengths is believed to be caused by different impedances at different frequencies. The IM product exceeds the TV signal by about 2 dB.

Fig. 20 shows the effect of increasing the LO by 3 dB. The IM is at a lower level, while the TV signal and two interfering signals are larger. The signal-to-IM ratio is about 5 dB.

Increasing the LO power by another 6.5 dB causes the IM to be undetectable on the spectrum display as shown in Fig. 21. Some interference is still present in the TV display, but does not prevent most of the desired information from being seen. At the time of this writing, the cause of this residual interference has not definitely been determined.

Glossary of Mathematical Symbols

General Symbols

ω_{s1}, ω_{s2}	radian frequencies of two signals
ω_{o1}, ω_{o2}	radian frequencies of outputs associated with the respective signals
ω_{LO}	local oscillator or pump radian frequency
$\omega_{LO}^+, \omega_{LO}^-$	radian frequencies of upper and lower sidebands, respectively, within the pump circuit
$\omega_{im_1}, \omega_{im_2}$	radian frequencies of lower and upper "fourth-order" IM products, respectively
N	number of diodes in a mixer circuit
IMR	IM ratio
P_o	output power of a frequency converter
P_{om}	maximum (saturated) output power of a frequency converter

Resistive Diode Symbols

i	diode instantaneous current
i_{sat}	diode saturation current
α	exponential nonlinearity coefficient
v	instantaneous voltage

Varactor Diode Symbols

v	instantaneous voltage
ϕ	contact potential
V_B	breakdown voltage
S_m	maximum elastance (occurs at $v = V_B$)
Q_m	maximum diode charge (occurs at $v = V_B$)
q	instantaneous diode charge
$C(v)$	diode capacitance as a function of voltage
γ	varactor nonlinearity coefficient
β_k	coefficient of kth term in Taylor series expansion for $v(q)$

Field-Effect Transistor Symbols

V_{po}	pinchoff voltage
V_{go}	quiescent gate voltage.

References

[1] C. Kittel, *Introduction to Solid State Physics*, 2nd ed. New York: Wiley, 1962, p. 393.

[2] S. M. Perlow and B. S. Perlman, "A large signal analysis leading to intermodulation distortion prediction in abrupt junction varactor upconverters," *IEEE Trans. Microwave Theory and Techniques*, vol. MTT-13, pp. 820–827, November 1965.

[3] B. B. Bossard, *et al.*, "Interference reduction techniques for receivers," Quart. Rept. 2, Contract DA 36-039 AMC-02345(E), January 1964.

[4] A. I. Grayzel, S. M. Perlow, and B. S. Perlman, "Comments on 'A large signal analysis leading to intermodulation distortion prediction in abrupt junction varactor upconverters,' " *IEEE Trans. Microwave Theory and Techniques* (Correspondence), vol. MTT-15, pp. 183–184, March 1967.

[5] S. P. Kwok, "Field effect transistor RF mixer techniques," *1967 WESCON Conv. Dig.*, session 8, pp. 1–9.

[6] E. Markard, P. Levine, and B. B. Bossard, "Intermodulation distortion improvement in parametric upconverter," *Proc. IEEE* (Letters), vol. 55, pp. 2060–2061, November 1967.

[7] J. H. Lepoff and A. M. Cowley, "Improved intermodulation rejection in mixers," *IEEE Trans. Microwave Theory and Techniques*, vol. MTT-14, pp. 618–623, December 1966.

[8] B. B. Bossard *et al.*, "Interference reduction techniques for receivers," Quart. Rept. 5, Contract DA 36-039 AMC-02345(E), January 1965.

[9] B. S. Perlman, "Current-pumped abrupt-junction varactor power-frequency converters," *IEEE Trans. Microwave Theory and Techniques*, vol. MTT-13, pp. 150–161, March 1965.

[10] R. L. Ernst, "Multiple diode theorems," *Proc. IEEE* (Correspondence), vol. 53, p. 417, April 1965.

[11] D. R. Chambers and D. K. Adams, "A technique for the rapid calculation of distortion effects in varactor parametric amplifiers," *1968 G-MTT Internatl. Microwave Symp. Digest*, pp. 173–178.

[12] F. Langford-Smith, *Radiotron Designer's Handbook*, 4th ed 1952, p. 987.

Prediction of Mixer Intermodulation Levels as Function of Local Oscillator Power

ELI F. BEANE, MEMBER, IEEE

***Abstract*—During intermodulation testing with diode mixers an increase of intermodulation interference was observed due to an increase of LO power incident to the mixer. This phenomenon conflicted with the theory that increase of LO power reduces intermodulation output of the diode mixer. In these tests the intermodulation decreased as expected when the LO power was further increased. Results of a theoretical and experimental study of how the level of incident LO power affects the intermodulation output levels emanating from the mixer are presented. The predicted results lead to the following experimentally verified conclusions.**

1) A drop in power at some intermodulation frequencies occurs for an increase of LO power, depending on LO operating point and order of intermodulation.

2) Power at each intermodulation frequency will repeatedly increase, reach a maximum, and then decrease as power in LO signal increases, where the number of repetitions follows the orders of intermodulation.

3) The maximum intermodulation power at low-order intermodulation frequencies occurs for higher LO power than higher order intermodulation frequencies.

LO power operating point is shown to be a significant factor in mixer intermodulation consideration. Application of these results to receiver intermodulation improvement is discussed.

Manuscript received May 15, 1970.

The author was with the Electronic Warfare Laboratory, U. S. Army Electronics Command. He is now with the Ground Radar Team, Combat Surveillance Acquisition and System Integration Laboratory, U. S. Army Electronics Command, Fort Monmouth, N. J. 07703.

Introduction

THE EFFECT of LO level on the spurious intermodulation product output levels of a nonlinear resistance mixer was studied in [1]–[4]. These studies concluded that a 3-dB increase of LO power to a positive resistance point contact diode mixer will improve the intermodulation ratio (IMR) by 6 dB. IMR is defined as

$$\text{IMR} = \frac{P_{\text{IM}}}{P_{\text{IF}}} \tag{1}$$

where P_{IF} is the power output of the mixer at the desired IF and P_{IM} is the power output of the mixer at undesired intermodulation frequency. Figs. 1(a) and (b) show the input and output spectrums of the mixer, respectively. We are limiting our consideration at this point to the IF outputs: S_{IF_1} at f_{IF_1} and S_{IF_2} at f_{IF_2} (S_{IF_1}, S_{IF_2}, etc., are signal identifiers and do not specify amplitudes). These are due to the desired second-order mixing of input signals S_1 at frequency f_1, S_2 at frequency f_2, S_{LO}, the local oscillator, at f_{LO}, and undesired fourth-order intermodulation, S_{IM} at f_{IM} and S_{IM}' at f_{IM}' in the following manner:

$$f_{\text{IF}_1} = f_1 - f_{\text{LO}}$$

$$f_{\text{IF}_2} = f_2 - f_{\text{LO}} \quad \text{(second order)} \tag{2a}$$

Reprinted from *IEEE Trans. Electromagn. Compat.*, vol. EMC-13, pp. 56–63, May 1971.

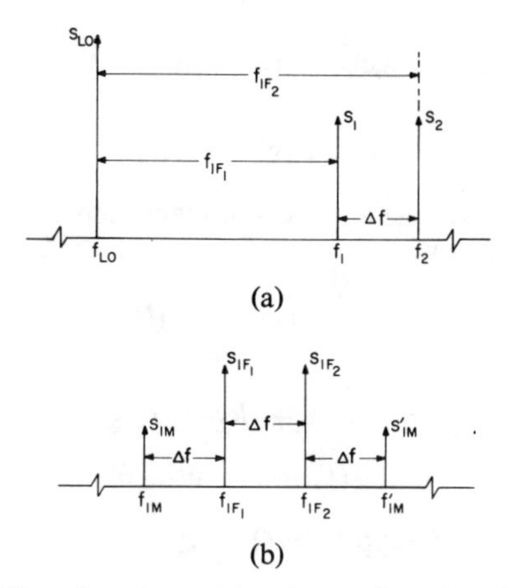

Fig. 1. (a) Mixer input spectrum. $f_{IF1} = f_1 - f_{LO}$; $f_{IF2} = f_2 - f_{LO}$; $\Delta f = f_2 - f_1$. (b) Mixer output spectrum. $f_{IM} = 2f_1 - f_2 - f_{LO}$; $f_{IM}' = 2f_2 - f_1 - f_{LO}$.

and

$$f_{IM} = 2_{f_1} - f_2 - f_{LO}$$

$$f_{IM}' = 2_{f_2} - f_1 - f_{LO} \quad \text{(fourth order).} \qquad (2b)$$

The orders of mixing are related to the order of the term in the power series expansion (given in (4)) of the nonlinear device from which a particular mix results [4] or from the sum of the absolute values of the integer coefficients of the frequencies involved in a particular "mix," as given in (2a) and (2b).

Laboratory observations in TV display intermodulation tests [4] of an X-band mixer that have been made with a microwave mixer analyzer [4] verify the conclusion [1]–[4] that increased LO power does eliminate to a great extent the effect of the fourth-order intermodulation product. However, this conclusion did not account for an *increase* of intermodulation and IMR that occurred for an initial increase of LO power incident to the mixer, although the intermodulation and IMR decreased when the LO power was further increased. This phenomenon seemed to suggest that intermodulation level is a function of LO power operating point. For certain LO operating points the following may occur: 1) an increase in LO power will decrease intermodulation and IMR; 2) a decrease in LO power decreases intermodulation and IMR; 3) an increase or decrease of LO power will decrease intermodulation and IMR (see Figs. 8–10).

A theoretical explanation of this effect seemed to be shown in an article by Rutz-Phillip [5] which deals with interference phenomena related to corroded joints aboard ships and other vehicles. She found by analysis and experiment 1) that the power at intermodulation frequencies generated in a nonlinear resistive element in an actual circuit can become smaller although the power in some of the incident waves increases; 2) that the power at higher order intermodulation frequencies generated in a nonlinear resistive element in an actual circuit can become smaller although the power in some of the incident waves increases; and 3) that the power at higher order intermodulation frequencies (e.g., fourth-order f_{IM} in (2b) and Fig. 1(b)) will reach its maximum at a lower level in the incident waves than lower order intermodulation frequencies (e.g., second-order f_{IF} in (2a) and Fig. 1(a)).

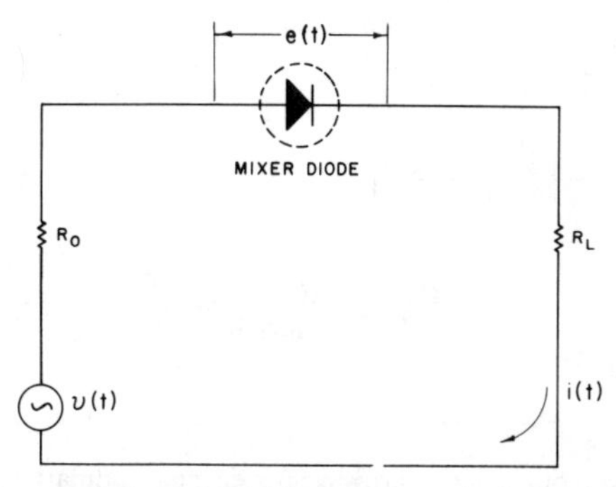

Fig. 2. Lumped-element low-frequency equivalent mixer circuit.

Rutz-Phillip derives power conversion relations for a nonlinear resistive element in series with linear resistors. She applies the analysis based on a lumped element low-frequency circuit (Fig. 2) to a microwave circuit where the dc current-voltage characteristic [6] functionally represents the diode or nonlinear resistive element given by the equation

$$i = i_0(e^{\alpha v} - 1) \qquad (3)$$

where

- v Voltage across nonlinear element.
- α Nonlinearity coefficient.
- i_0 Saturation current.

This current-voltage characteristic is expressed as a power series of a time dependent voltage $v = v(t)$:

$$i(t) = a_0 + a_1 v(t) + a_2 v(t)^2 + a_3 v(t)^3 \cdots \qquad (4)$$

where the coefficients $a_0, a_1, a_2 \cdots$ are given in the form of recursive polynomials as derived by Mills [7]. These coefficients can thus be obtained numerically for any order power series no matter how high. In previous analyses [2], [3] of nonlinear resistive elements coefficients higher than the tenth order were not obtained. Rutz-Phillip indicated that the prediction of intermodulation rise and fall which depends on signal level incident to the nonlinear element could be arrived at only by evaluating a considerable number of terms of the power series representation of the current-voltage characteristic. As indicated, this approach has a strong connection to the study of mixer intermodulation, and hence its application to an X-band single-ended resistive diode mixer was undertaken.

Analysis

Consider the block diagram in Fig. 3 of a microwave mixer circuit for a two equal tone intermodulation test of a single-ended 2-port microwave (X-band) mixer. Now, represent this microwave circuit by a lumped element low-frequency equivalent circuit (idealized approximation) as

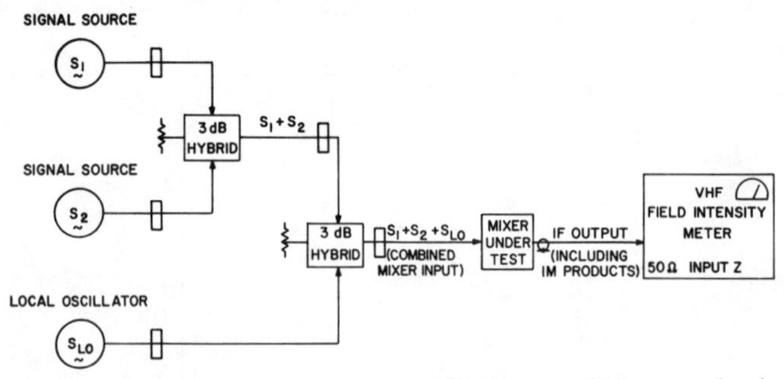

Fig. 3. Two equal tone microwave mixer intermodulation test circuit.

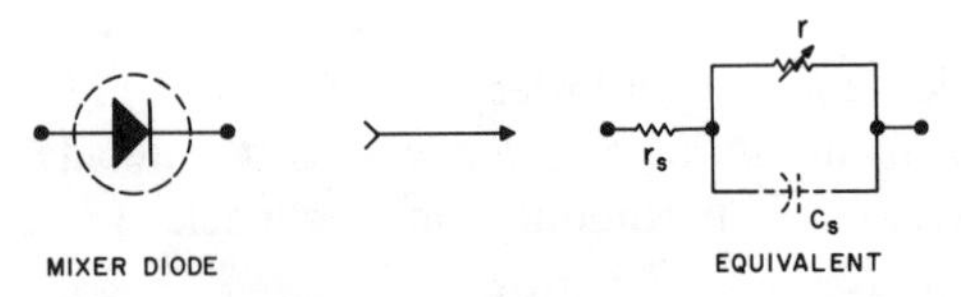

Fig. 4. Equivalent circuit of nonlinear resistance mixer diode.

shown in Fig. 2, where

- $v(t)$ Instantaneous input voltage of signals S_1, S_2, and S_{LO}.
- $e(t)$ Instantaneous voltage across diode.
- $i(t)$ Instantaneous current induced in equivalent circuit.
- R_0 Linear resistance equivalent of input impedance.
- R_L Linear resistance equivalent of output load impedance.

To complete the circuit in Fig. 2, the equivalent circuit for the semiconductor nonlinear resistance mixer diode is shown in Fig. 4, where

- r Nonlinear resistance.
- r_s Linear spreading or bulk resistance of semiconductor.
- c_s Shunt capacitance.

In our analysis reactive effects attributed to c_s will be ignored; we are taking into consideration only the linear and nonlinear resistances in the circuit. This is similar to the approach taken by Rutz-Phillip [5]. She justifies the use of the low-frequency circuit model (Fig. 2) for a microwave circuit, disregarding frequency dependent impedance mismatches by having both input and output (intermodulation) frequencies all in the same spectral vicinity. We follow a similar analysis and disregard frequency dependent impedance mismatches in order to facilitate measurements and obtain results for comparison with theoretical calculations which does not significantly alter the result.

This input voltage $v(t)$ consists of three CW sinusoidal signals, S_1, S_2, and S_{LO}, which are two equal amplitude input signals and a local oscillator signal, respectively. Corresponding to these signals are their respective voltage amplitudes V_1, V_2, and V_{LO} and angular frequencies ω_1, ω_2, and ω_{LO} such that

$$v(t) = V_1 \cos \omega_1 t + V_2 \cos \omega_2 t + V_{LO} \cos \omega_{LO} t. \quad (5)$$

The nonlinear resistive diode in Figs. 2 and 4 is represented by the dc current-voltage characteristic in (3), where the dc current and voltage are replaced by time-varying $i(t)$ and $e(t)$:

$$i(t) = i_0(e^{\alpha e(t)} - 1) \quad (6)$$

where $e(t)$, the voltage across nonlinear element resistance r in Fig. 4, is given by

$$e(t) = v(t) - i(t)R \quad (7)$$

where

$$R = R_0 + R_L + r_s.$$

Replacing the value of $e(t)$ in (7) into (6), we find

$$i(t) = i_0\{\exp [\alpha(v(t) - i(t)R)] - 1\}. \quad (8)$$

We express $i(t)$ in (8) as a power series in $v(t)$ (see (4)) about the point $v = 0$. The derivatives

$$\left.\frac{d^n i}{dv^n}\right|_{v=0}$$

needed for a Taylor series, as obtained from the analysis of Mills [7], are expressed as polynomials in Z, $H(Z)$, where

$$Z = \left.\frac{di}{dv}\right|_{v=0} = \frac{i_0\alpha}{1 + i_0\alpha R} \quad (9)$$

and the power series expression for $i(t)$ is given by

$$i(t) = \sum_{k=1}^{m} \frac{1}{k!} H_k(Z)v^k(t). \quad (10)$$

The polynomials $H(Z)$ in (10) are expressed as

$$H_k(Z) = \frac{\alpha^{k-1}}{R} \sum_{j=0}^{2k-1} a_{k,j}(ZR)^j \quad (11)$$

for $k = 1,2,3 \cdots$. The $a_{k,j}$ in (11) are computed by the following recursion relation:

$$a_{k+1,j} = ja_{k,j} - 2(j-1)a_{k,j-1} + (j-2)a_{k,j-2} \quad (12)$$

where

$$a_{k,j} = 0, \qquad \text{for } j < 0 \text{ or } j > 2k - 1$$

$$a_{1,1} = 1 \text{ and } a_{1,j} = 0, \qquad \text{for } j \neq 1.$$

Note from (10) that the kth-order component of current $i_k(t)$ is given by

$$i_k(t) = \frac{1}{k!} H_k(Z)v^k(t). \quad (13)$$

The general Fourier series expansion of the current $i(t)$ as a function of the generator voltage $v(t)$, where $v(t)$ is composed of three sinusoids as in (5), will have the form

$$i(t) = \sum_{x=-\infty}^{\infty} \sum_{y=-\infty}^{\infty} \sum_{z=-\infty}^{\infty} I_{xyz} \exp [j(x\omega_1 + y\omega_2 + z\omega_{LO})t] \quad (14)$$

where x, y, and z are integers and I_{xyz} is the current component at intermodulation frequency,

$$\omega_{xyz} = x\omega_1 + y\omega_2 + z\omega_{LO}. \quad (15)$$

The power output P_{xyz} at frequency f_{xyz} $(= \omega_{xyz}/2\pi)$ is shown, following Rutz-Phillip [5], for simplified equivalent

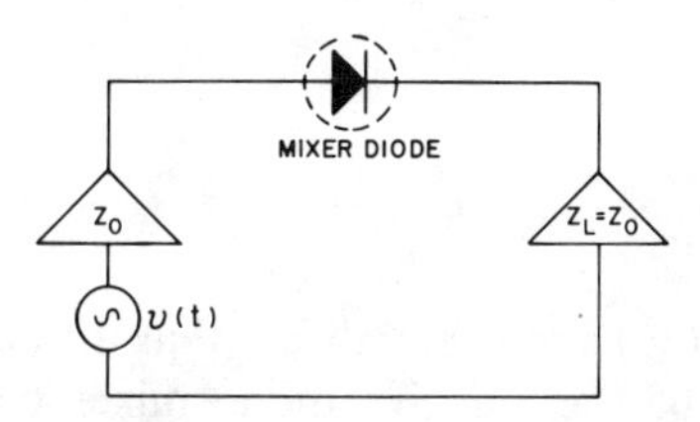

Fig. 5. Simplified equivalent microwave mixer circuit.

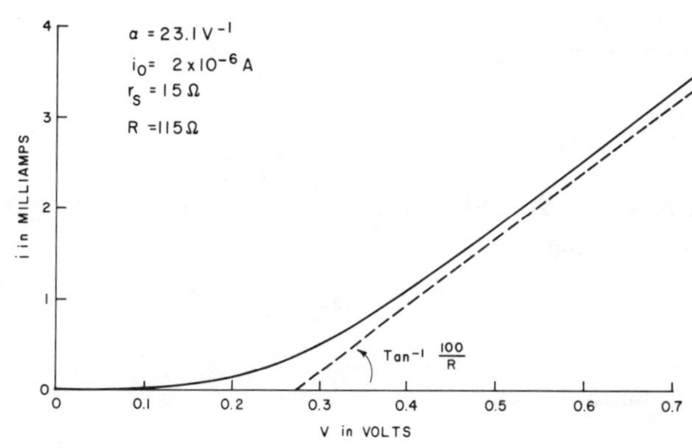

Fig. 6. Plot of diode mixer current-voltage characteristic.

microwave circuit in Fig. 5, where Z_O is the characteristic impedance of the circuit and $Z_L = Z_O$ is the matched terminating impedance to be

$$P_{xyz} = 2I_{xyz}^2 Z_L. \tag{16}$$

Identify Z_O and $Z_L = Z_O$ with R_O and R_L of Fig. 4, $Z_L = Z_O = R_L = R_O$; hence

$$P_{xyz} = 2I_{xyz}^2 R_O. \tag{17}$$

Each $v^k(t)$ in the expression for $i_k(t)$ in (13) is given by

$$v^k(t) = (V_1 \cos \omega_1 t + V_2 \cos \omega_2 t + V_{\text{LO}} \cos \omega_{\text{LO}} t)^k \tag{18}$$

which upon expansion yields an explicit expression for the amplitude of each frequency component generated by $v^k(t)$. The amplitude or the Fourier coefficient of the frequency

$$f_{xyz} = \frac{x\omega_1 + y\omega_2 + z\omega_{\text{LO}}}{2\pi}$$

due to $v^k(t)$, $(V_{xyz})_k$ (following the work of Wass [11]) is given by

$$(V_{xyz})_k = \sum_{b=0}^{k-1} \sum_{c=0}^{k-b} \frac{V_1{}^a V_2{}^b V_{\text{LO}}{}^c k!}{2^{k-1}[(a+x)/2]!\,[(a-x)/2]!\,[(b+y)/2]!\,[(b-y)/2]!\,[(c+z)/2]!\,[(c-z)/2]!} \tag{19}$$

where a,b,c are positive integers such that

$$a + b + c = k$$

or

$$a = k - b - c. \tag{20}$$

Using $(V_{xyz})_k$ from (19), we may write $v^k(t)$ as a Fourier series

$$v^k(t) = \sum_{x=-\infty}^{\infty} \sum_{y=-\infty}^{\infty} \sum_{z=-\infty}^{\infty} (V_{xyz})_k \times \exp\left[j(x\omega_1 + y\omega_2 + z\omega_{\text{LO}})t\right]. \tag{21}$$

Substituting this expression for $v^k(t)$ into (10), we obtain

$$i(t) = \sum_{k=1}^{m} \frac{1}{k!} H_k(Z) \sum_{x=-\infty}^{\infty} \sum_{y=-\infty}^{\infty} \sum_{z=-\infty}^{\infty} (v_{xyz})_k. \tag{22}$$

Comparing (22) with (14), we find

$$I_{xyz} = \sum_{k=1}^{m} (V_{xyz})_k \frac{H_k(Z)}{k!}. \tag{23}$$

Replacing $(V_{xyz})_k$ by its equivalent from (19),

$$I_{xyz} = \sum_{k=1}^{m} \sum_{b=0}^{k-1} \sum_{c=0}^{k-b} \frac{V_1{}^a V_2{}^b V_{\text{LO}}^c H_k(Z)}{2^{k-1}[(a+x)/2]!\,[(a-x)/2]!\,[(b+y)/2]!\,[(b-y)/2]!\,[(c+z)/2]!\,[(c-z)/2]!} \tag{24}$$

where $H_k(Z)$ is given by (11). Thus the Fourier coefficient I_{xyz} of the current associated with the output intermodulation frequency,

$$f_{xyz} = \frac{x\omega_1 + y\omega_2 + z\omega_{\text{LO}}}{2\pi}$$

should be computable by the formula in (24) for any order $k = m$ of the power series in (4).

The power output at any intermodulation frequency should now be obtained by substituting I_{xyz}, computed from (24), into (17), the only limiting factor being the capacity of the computer used to evaluate the triple summation indicated in (24). In fact serious computer limitations were encountered in the numerical evaluation of the $H_k(Z)$ polynomials, given in (11). The computer evaluation of $H_k(Z)$ involves terms of such immense magnitude, for all but the smallest of the k ($k \simeq 8$) that it is numerically unfeasible to use the series in (10) to represent the current-voltage characteristic of the mixer circuit under investigation. Thus the theoretical analysis which leads to the Rutz-Phillip conclusions regarding intermodulation in a resistive diode circuit based upon evaluation of higher order terms ($k > 8$) of the current-voltage relationship cannot be regarded as valid.

A simpler analytic examination of behavior of the current-voltage relationship (8) representing the diode mixer circuit leads to a basic understanding of the correlation of local oscillator power with the intermodulation output of a resistive diode mixer without a complicated and numerically cumbersome power series. Fig. 6 shows a plot of the time independent (i.e., dc) form of the current-voltage characteristic in (8):

$$i = i_O\{\exp\left[\alpha(v - iR)\right] - 1\}. \tag{25}$$

First, examine the behavior of the current i as the voltage v becomes very large. In (25), separating terms involving i

and v to opposite sides of the equation gives

$$\frac{i + i_0}{i_0} \cdot e^{\alpha iR} = e^{\alpha v}. \tag{26}$$

Let $v \to \infty$. We find that $e^{\alpha v} \to \infty$. Then by virtue of the equality in (26)

$$\frac{i + i_0}{i_0} \cdot e^{iR} \to \infty. \tag{27}$$

This cannot be, unless $i \to \infty$. Hence

$$i \to \infty, \qquad \text{as } v \to \infty. \tag{28}$$

Solving (25) which is implicit in i explicitly for v we obtain

$$v = \frac{1}{\alpha} \ln \frac{i + i_0}{i_0} + iR. \tag{29}$$

Differentiate (29) with respect to i:

$$\frac{dv}{di} = \frac{1}{\alpha(i + i_0)} + R. \tag{30}$$

Take the reciprocal of both sides of (30):

$$\frac{di}{dv} = \frac{\alpha(i + i_0)}{1 + \alpha(i + i_0)R}. \tag{31}$$

Let $i \to \infty$, which is equivalent to $v \to \infty$ (by (28)). We find

$$\frac{di}{dv} \to \frac{1}{R} \text{ (constant)}. \tag{32}$$

Next examine (25) as $v \to 0$. From (26) we see that $e^{\alpha v} \to 1$ as $v \to 0$ from which it follows that

$$\frac{i + i_0}{i_0} \cdot e^{\alpha iR} \to 1$$

as $i \to 0$. Hence

$$i \to 0 \text{ as } v \to 0. \tag{33}$$

Differentiate both sides of (31) with respect to v; then

$$\frac{d^2 i}{dv^2} = \alpha \frac{di}{dv} \left[\frac{1}{1 + (i + i_0)R} - \frac{\alpha(i + i_0)R}{[1 + R(i + i_0)]^2} \right]. \tag{34}$$

In (34) let $i \to 0$, which is equivalent to $v \to 0$ by (33):

$$\frac{d^2 i}{dv^2} = \frac{\alpha}{(1 + \alpha i_0 R)^2} \frac{di}{dv}. \tag{35}$$

Substituting

$$k = \frac{\alpha}{(1 + i_0 R)^2} \text{ (constant)}$$

in (35) we find:

$$\frac{d^2 i}{dv^2} - k \frac{di}{dv} = 0. \tag{36}$$

Therefore

$$i \sim e^{kv}, \qquad \text{as } v \to 0 \tag{37}$$

and from (32):

$$\frac{di}{dv} \sim \frac{1}{R} \text{ (constant)}, \qquad \text{as } v \to \infty. \tag{38}$$

The results in (37) and (38) indicate that the current-voltage characteristic of the resistive diode mixer circuit behaves like an exponential for small $v (v \simeq 0)$ and like a straight line of slope $1/R$ for large $v (v \gg 0)$. (See Fig. 6.)

In terms of a power series representation of the current-voltage characteristic, higher order coefficients would be relatively large for low voltage. As v is increased, the higher order coefficients of a power series representation of the function become smaller relative to the lower order coefficients, and the current-voltage characteristic tends to become a linear function. This functional behavior relates to intermodulation of a resistive diode mixer and leads to the previous conclusions, namely, power at intermodulation frequencies, especially of higher orders, can become smaller although power in some of the incident waves (i.e., LO) increases, and power at higher order intermodulation frequencies (e.g., fourth, sixth, etc.) will reach its maximum at lower levels in the incident waves than lower order intermodulation (e.g., second order). This is so since increasing incident LO power to the resistive diode mixer corresponds to the effect of increasing the voltage v, i.e., the bias in the dc current-voltage relationship. Hence higher LO power causes the incident waves to intercept a portion of the diode characteristic less nonlinear than that intercepted by lower LO powers.

Further understanding of the intermodulation behavior of the point contact diode mixer may be obtained by successive differentiation of (29) yielding the derivatives of v with respect to i. From these derivatives algebraic expressions for the coefficients, a_k of the kth-order terms of the Taylor series expansion of the diode mixer current-voltage relationship expanded about a voltage $v = v_0$ (i.e., bias) can be obtained by reversion of a power series as shown in [10, appendices I and II]. The algebraic expression for the first four Taylor series expansion coefficients is shown to be

$$a_1 = \left[\frac{1}{x + 1}\right] (i + i_0)\alpha \tag{39a}$$

$$a_2 = \left[\frac{1}{(x + 1)^3}\right] \frac{(i + i_0)}{2!} \alpha^2 \tag{39b}$$

$$a_3 = \left[\frac{-2x + 1}{(x + 1)^5}\right] \frac{(i + i_0)}{3!} \alpha^3 \tag{39c}$$

$$a_4 = \left[\frac{6x^2 - 8x + 1}{(x + 1)^7}\right] \frac{(i + i_0)}{4!} \alpha^4 \tag{39d}$$

where $x = \alpha(i + i_0)R$. These coefficients may now be computed as a function of i which in turn may be correlated to the bias voltage v_0 by means of (29).

In Fig. 7 the magnitudes of the coefficients a_k, for $k = 1,2,3,4$, are plotted (semilogarithmically) as a function of the bias voltage v_0 computed from (39) and (29). Observe how the magnitudes of the coefficients oscillate (circled nodes indicate zero crossings) about the v_0 axis with the

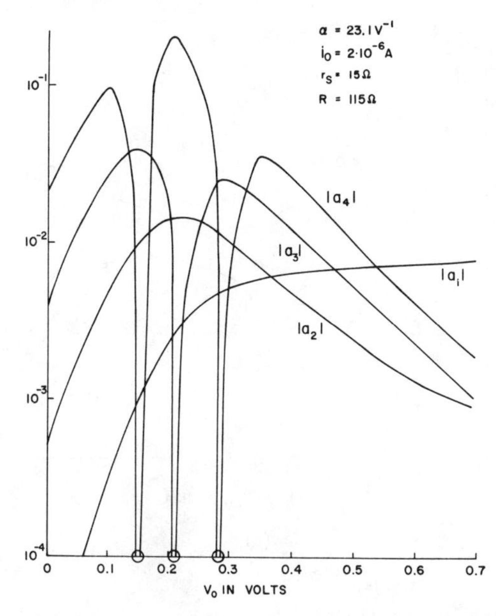

Fig. 7. Diode mixer coefficient magnitudes as functions of bias voltage (semilogarithmic plot).

number of oscillations increasing with increasing k. The behavior of these coefficients relates directly to the mixer intermodulation output product levels. It can be shown [10] that the most significant contribution to a kth-order intermodulation product of small input signals is due to the coefficient of the kth-order term of the Taylor series expansion. Hence variation (i.e., peaks and valleys) in power levels of kth-order mixer intermodulation products following the oscillatory behavior of the kth-order coefficient k are to be expected as the value v_0 is varied across the diode.

Although the variation of the kth-order coefficient a_k is being considered as a function of v_0 (the dc bias), similar results are to be expected if the LO voltage incident to the diode were varied. This is reasonable since V_{LO} creates an effective dc bias voltage across (or dc bias current through) the diode due to rectification. The results of measurements shown in Figs. 8–10 indicate that this conclusion corresponds very well with experiment. Note that the oscillatory behavior of the series coefficients comes about only when the diode mixer's linear circuit resistance ($R = R_0 + R_L + r_s$) is taken into account. Were these resistances neglected in the analysis, the resulting coefficients would be independent of the bias voltage v_0 or the effective LO bias level.

Results

Intermodulation measurements of a microwave X-band single-ended mixer, utilizing a 1N23WE crystal diode were made with circuit represented (power monitoring and calibration instrumentation not shown for simplicity) in Fig. 3, where

$$S_1 \text{ is at } f_1 = 9055 \text{ MHz}$$
$$S_2 \text{ is at } f_2 = 9065 \text{ MHz}$$
$$S_{LO} \text{ is at } f_{LO} = 9000 \text{ MHz.} \quad (40)$$

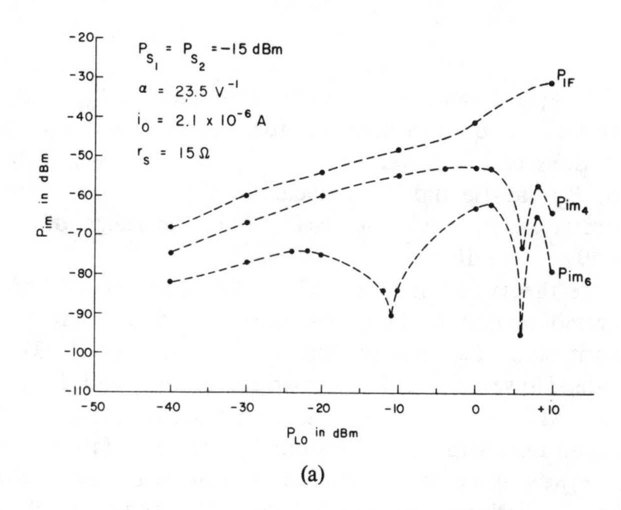

(a)

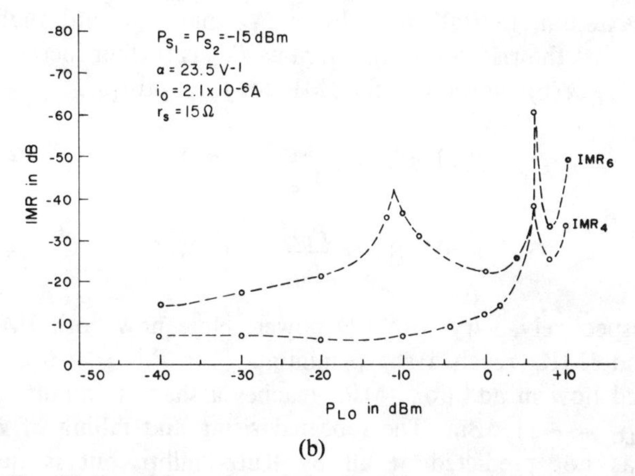

(b)

Fig. 8. (a) P_{IF} at $f_{IF} = f_1 - f_{LO}$ as function of P_{LO} at f_{LO} (second-order intermodulation). P_{IM4} at $f_{IM4} = 2f_1 - f_2 - f_{LO}$ as function of P_{LO} at f_{LO} (fourth-order intermodulation). P_{IM6} at $f_{IM6} = 3f_1 - 2f_2 - f_{LO}$ as function of P_{LO} at f_{LO} (sixth-order intermodulation). (b) IMR_4 at f_{IM4} as function of P_{LO} at f_{LO} (fourth-order intermodulation). IMR_6 at f_{IM6} as function of P_{LO} at f_{LO} (sixth-order intermodulation).

In all measurements the power P_{LO} of S_{LO} was varied from −40 to +10 dBm (+10 dBm was not exceeded due to saturation effects). In each test the power levels of S_1 and S_2 remained fixed, i.e., they served as parameters. Power outputs of the mixer were measured at frequencies which are in the vicinity of the desired IF which result from mixing signals (S_1 and S_2) within the mixer input bandwidth. These intermodulation products turn out to be all of even order, as follows:

$$\begin{aligned} f_{IF} &= f_1 - f_{LO} \\ &= 9055 \text{ MHz} - 9000 \text{ MHz} \\ &= 55 \text{ MHz} \quad \text{second-order IM, desired IF} \end{aligned} \quad (41a)$$

$$\begin{aligned} f_{IM_4} &= 2f_1 - f_2 - f_{LO} \\ &= 2(9055 \text{ MHz}) - 9065 \text{ MHz} - 9000 \text{ MHz} \\ &= 45 \text{ MHz} \quad \text{fourth-order IM, undesired} \end{aligned} \quad (41b)$$

$$\begin{aligned} f_{IM_6} &= 3f_1 - 2f_2 - f_{LO} \\ &= 3(9055 \text{ MHz}) - 2(9065 \text{ MHz}) - 9000 \text{ MHz} \\ &= 35 \text{ MHz} \quad \text{sixth-order IM, undesired.} \end{aligned} \quad (41c)$$

Fig. 8(a) shows the results of equal-tone (i.e., $P_{S_1} = P_{S_2}$) intermodulation measurements for P_{IF}, P_{IM_4}, and P_{IM_6} as functions of LO power P_{LO} with $P_{S_1} = P_{S_2} = -15$ dBm. For P_{IF} the rise and fall predicted by the analysis with the increase of P_{LO} did not occur within the range of P_{LO} (-40 to $+10$ dBm).

The theory is not necessarily wrong since second-order intermodulation might be reduced if the diode were able to withstand P_{LO} greater than $+10$ dBm on the IN23WE diode. However, the data presented for P_{IM_4} and P_{IM_6} in Fig. 8(a) does show the strong dependence of intermodulation on LO operating point. Both P_{IM_4} and P_{IM_6} fall sharply for $P_{LO} = +6$ dBm and rise for an increase or decrease of P_{LO} about this point. Furthermore, P_{IM_6} rises, reaches a maximum, and falls for a lower P_{LO} than P_{IM_4}, although it repeats the rise and fall pattern as P_{LO} is further increased.

Fig. 8(b) shows how the IMR for f_{IM_4} and f_{IM_6},

$$\mathrm{IMR}_4 = \frac{P_{IM_4}}{P_{IF}} \quad \text{(dB)} \tag{42a}$$

$$\mathrm{IMR}_6 = \frac{P_{IM_6}}{P_{IF}} \quad \text{(dB)} \tag{42b}$$

respectively, vary with LO power. Note how both IMR_4 and IMR_6 reach a sharp minimum for $P_{LO} = +6$ dBm, and how in addition IMR_6 reaches a sharp minimum for $P_{LO} = -11$ dBm. The repeated rising and falling of P_{IM} was not predicted at all by Rutz-Phillip, but is quite apparent from the simple analytical approach which we employ in the preceding section.

Note that the number of maxima and minima are related to the order of intermodulation as our analysis predicts. Thus for P_{IM_4} (fourth order) we have two peaks at $P_{LO} = 0$ dBm and $P_{LO} = +8$ dBm and two sharp falls at $P_{LO} = +6$ dBm and $P_{LO} \geq +10$ dBm. For P_{IM_6} (sixth order) we have three peaks: $P_{LO} = -23$ dBm, $P_{LO} = +2$ dBm, and $P_{LO} = +8$ dBm; we also have three minima: $P_{LO} = -11$ dBm, $P_{LO} = +6$ dBm, and $P_{LO} \geq +10$ dBm. We can thus relate the total number of maxima and minima to the order of intermodulation product, e.g., fourth-order intermodulation would have four maxima and minima, while sixth order has six maxima and minima. A similar phenomenon has been observed in measurements of spurious response level as a function of dc bias voltage by Donaldson and Moss [12] and as referred to by Herishen in his study [10] of mixer response versus bias current of a nonlinear diode mixer. They [12] indicate from their data that spurious responses of the same order display similar properties as bias is changed and that the number of minima relative to the absolute maximum is equal to the order of the mix product. They also noted the influence of local oscillator power in increasing and decreasing spurious response level.

Both the effects of LO power and bias on spurious response levels or intermodulation are related in that different bias values and LO input levels subtend different portions of the mixer nonlinear characteristic (3) and thereby subject input signal voltages to different mixer nonlinearities. In our study we considered only the effect of LO power levels without any external bias.

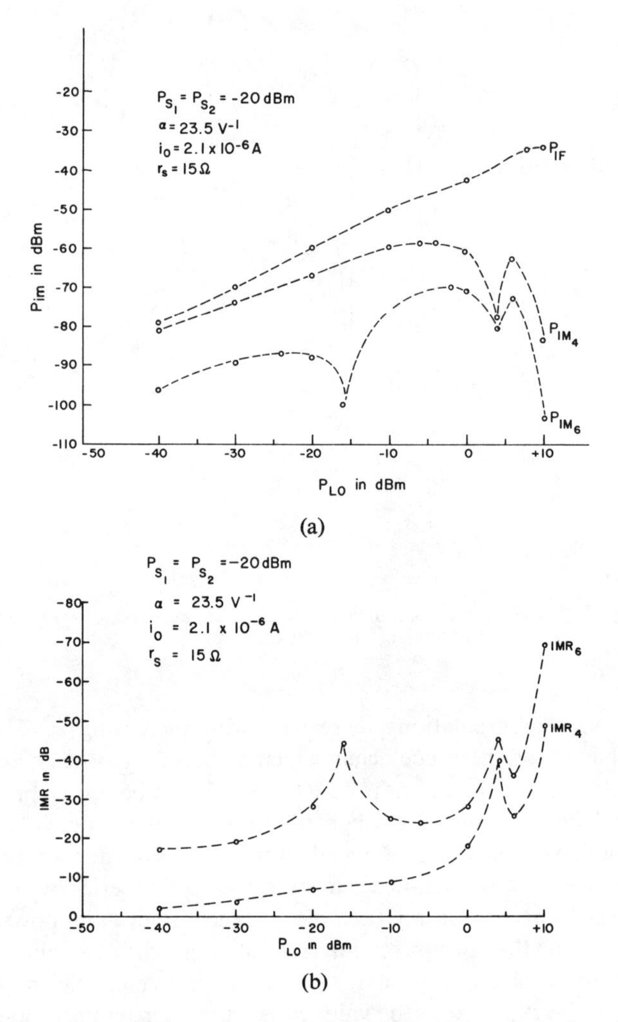

Fig. 9. (a) P_{IF} at $f_{IF} = f_1 - f_{LO}$ as function of P_{LO} at f_{LO} (second-order intermodulation). P_{IM4} at $f_{IM4} = 2f_1 - f_{LO}$ as function of P_{LO} at f_{LO} (fourth-order intermodulation). P_{IM6} at $f_{IM6} = 3f_1 - 2f_2 - f_{LO}$ as function of P_{LO} at f_{LO} (sixth-order intermodulation). (b) IMR_4 at f_{IM4} as function of P_{LO} at f_{LO} (fourth-order intermodulation). IMR_6 at f_{IM6} as function of P_{LO} at f_{LO} (sixth-order intermodulation).

Figs. 9 and 10 show results of equal tone intermodulation and IMR measurements for $P_{S_1} = P_{S_2} = -20$ dBm and $P_{S_1} = P_{S_2} = -30$ dBm, respectively. Data for P_{IM_6} and IMR_6 do not appear in Fig. 10 since the level of P_{IM_6} was too low for accurate measurements. Observe how for lower levels of input signals S_1 and S_2 the intermodulation maxima and IMR minima are shifted to lower levels (i.e., the intermodulation maxima and IMR minima occur for lower LO levels). Although the IMR for lower level input signals, -20 and -30 dBm at $P_{LO} = +10$ dBm, are less than those at the preceding minima (i.e., at $+4$ dBm for $P_{S_1} = P_{S_2} = -20$ dBm and -6 dBm for $P_{S_1} = P_{S_2} = -30$ dBm), the likelihood of "burnout" at $P_{LO} = +10$ dBm precludes operation of the single-diode mixer at such a high LO level. (This problem pertains only to the single point contact diode mixer configuration used in the preceding tests; multiple diode mixers or mixers using Schottky barrier diodes may overcome this power limitation.)

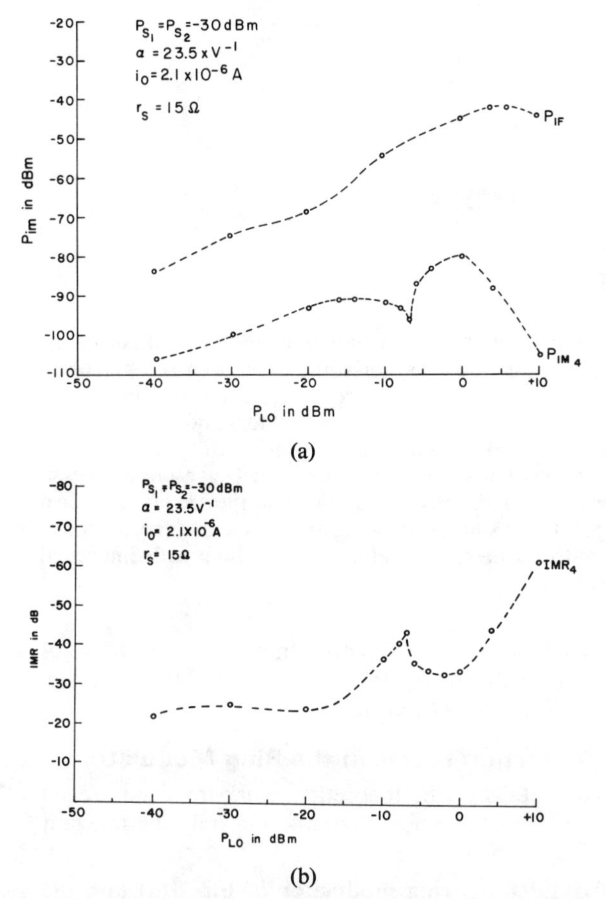

Fig. 10. (a) P_{IF} at $f_{IF} = f_1 - f_{LO}$ as function of P_{LO} at f_{LO} (second-order intermodulation). P_{IF} at $f_{IM4} = 2f_1 - f_2 - f_{LO}$ as function of P_{LO} at f_{LO} (fourth-order intermodulation). (b) IMR_4 at f_{IM4} as function of P_{LO} at f_{LO} (fourth-order intermodulation).

A study of the P_{IF} in Figs. 8(a) and 9(a) in the vicinity of $P_{LO} = +10$ dBm shows a drop in P_{IF} with increased P_{LO}. (This may be related to the down shifting in the P_{LO} of the maxima and minima.) This lends credence to the assertion that a drop of P_{IF} would occur for higher (+10 dBm) P_{LO}, as shown in Fig. 8(a), where $P_{S_1} = P_{S_2} = -15$ dBm. Also, as to be expected, from Taylor series expansion of the diode characteristic, our data shows that higher orders of intermodulation generally have lower power levels.

Conclusions

The critical nature of the LO power level operating point of a point contact diode mixer in reducing intermodulation has been shown. Depending on signal strength of intermodulating signals, there exists a critical LO operating point at which even ordered undesired intermodulation products and IMRs are at a minimum and any excursion of LO power about this point results in a sharp rise in intermodulation. This optimum operating point ranged from -12 to $+6$ dBm for mixer input interfering signals with levels from -30 to -15 dBm.

Although even smaller IMRs are indicated for higher LO power levels, they are not presently feasible for a single point contact diode due to "burnout" considerations. The similarity between dc bias and LO drive with regard to intermodulation rejection is shown both analytically and experimentally.

Further studies under consideration are the following: 1) develop an intermodulation prediction computer program based on the present indicated simple analytic approach; 2) study combined effects of bias and LO power operating point on intermodulation and IMR; 3) study intermodulation and IMR of a Schottky barrier diode mixer with regard to LO and dc bias operating points; 4) investigate multiple diode mixer configurations; and 5) study effect of mixer output impedance on intermodulation and IMR.

Acknowledgment

The author wishes to acknowledge the work of M. Morris, L. Doyle, R. Poulos, and S. Ippolito, whose assistance was employed in the preparation of this paper.

References

[1] B. B. Bossard *et al.*, "Interference reduction techniques for receivers," U. S. Army Electronics Lab., Fort Monmouth, N. J., Contract DA36-039 AMC-02345(E), Tech. Rep., June 1964–July 1965.
[2] B. B. Bossard, P. Torrione, and S. Yuan, "Theory and improvement of intermodulation distortion in mixers," in *Proc. 10th Tri-Service Conf. Electromagnetic Compatibility*, Nov. 1964.
[3] L. Becker and R. L. Ernst, "Nonlinear admittance mixers," *RCA Rev.*, Dec. 1964.
[4] R. L. Ernst, P. Torrione, W. Y. Pan, and M. M. Morris, "Designing microwave mixers for increased dynamic range," *IEEE Trans. Electromagn. Compat.*, vol. EMC-11, Nov. 1969, pp. 130–138.
[5] E. M. Rutz-Phillip, "Power conversion in nonlinear resistive elements related to interference phenomena," *IBM J.*, Sept. 1967.
[6] H. C. Torrey and C. A. Whitmer, *Crystal Rectifiers* (M. I. T. Radiation Lab. Series), vol. 16. New York: McGraw-Hill, 1948.
[7] H. D. Mills, "On the equation, $i = i_0[\exp \alpha(v - Ri) - 1]$," *IBM J.*, Sept. 1967.
[8] L. M. Orloff, "Intermodulation analysis of crystal mixer," *Proc. IEEE*, vol. 52, Feb. 1964, pp. 173–179.
[9] L. D. Neidleman, "An application of FORMAC," *Commun. Ass. Comput. Mach.*, vol. 10, 1967, pp. 167–168.
[10] J. T. Herishen, "Diode mixer coefficients for spurious response prediction," *IEEE Trans. Electromagn. Compat.*, vol. EMC-10, Dec. 1968, pp. 355–363.
[11] C. A. A. Wass, "A table of intermodulation products," *J. Inst. Elec. Eng.* (London), pt. III, Jan. 1948, pp. 31–39.
[12] E. E. Donaldson, Jr., and R. W. Moss, "Study of receiver mixer characteristics," Eng. Experiment Station, Georgia Inst. Technol., Atlanta, Final Rep., ECOM-01426-F, Sept. 1966.

An Intermodulation Phenomenon in the Ring Modulator

By

J. G. GARDINER,
Ph.D., B.Sc. (Graduate) †

Analysis of intermodulation distortion effects in diode modulators by 'modulating function' techniques has, for some time, been recognized to be applicable where the diodes can be represented by a bi-linear d.c. characteristic in which the change of state from blocking to conduction takes place at zero bias. Experiments on ring modulators have shown, however, that predictions based on this diode model give unduly pessimistic results under some conditions. It is shown that the use of a modified diode model incorporating a suitable offset voltage in the d.c. characteristic permits prediction of the hitherto anomalous results with greatly improved accuracy.

1. Introduction

The development of Schottky barrier diodes has made possible the use of switching diode modulators in h.f. communications applications where low intermodulation distortion is a major criterion of design. As a result there has been a considerable revival of interest in distortion analysis of these circuits, the 'ring' mixer in particular, using the switching or 'modulating' function analysis developed many years ago by Belevitch[1] and Tucker.[2]

This analysis assumes the diode to be a bi-linear device changing from high impedance to low at zero bias; distortion is generated as a result of interaction between the input signals and the applied local-oscillator at the point of transition from high impedance to low, the input signals influencing the time of switching.

A difficulty arises when it is required to take account of the finite curvature of the diode forward characteristic and analytical techniques have recently been proposed by Savin[3] which can treat a discontinuous characteristic for the diode, namely a high linear impedance under reverse bias and an exponential characteristic under forward bias. However, the mathematical procedures tend to be somewhat involved and it has been demonstrated by the present author[4] that the simple bi-linear approach can produce acceptable predictions of some aspects of distortion performance in Schottky barrier ring modulators, notably the relationship between intermodulation and cross-modulation distortion. However, experimental investigations into intermodulation distortion in the ring circuit indicate that the bi-linear diode model as proposed by Tucker does not contain sufficient information about the nature of the diode forward characteristic. A simple modification to include the 'offset' voltage in the diode d.c. characteristic is sufficient to permit greatly improved accuracy of predicting distortion levels under conditions of low distortion product output.

† Postgraduate School of Electrical and Electronic Engineering, University of Bradford.

2. Distortion Levels in the Ring Modulator

The details of the theoretical work involved are set out fully in References 2 and 4, and only the relevant results will be quoted here.

Consider the ring modulator of Fig. 1(a) and the equivalent local-oscillator circuit of Fig. 1(b). Suppose the local-oscillator generator to have a source resistance R_0 and an e.m.f. V_0. Then, using the bi-linear model of Tucker shown in Fig. 2, it is seen that in the absence of any input signal to the modulator, two diodes are always conducting in parallel with two diodes turned off. Thus if the forward resistance of one diode is r_f the voltage appearing across the local-oscillator input port to the mixer (V_0') is

$$V_0' = \frac{r_f/2}{R_0 + r_f/2} \cdot V_0 \qquad \text{......(1)}$$

It is shown by Belevitch[1] that this situation is substantially maintained when an input signal is present and interferes with the diode switching. No condition can arise where less than two diodes are conducting at any one time. Therefore, we may say that the local-oscillator voltage appearing across any one diode is given by $V_0' \cos \omega_0 t$ while the signal voltage across any 'off' diode is $V_s \cos \omega_s t$ (as defined in Fig. 1). Distortion occurs when the input signal voltage across an 'off' diode exceeds the instantaneous value of V_0' and is of opposite polarity. (Subsequent overload conditions are relatively unimportant.[1]) Consider now a practical two-tone intermodulation test with frequencies as indicated in Fig. 3. The voltage across the diode may be assumed to take the form

$$V_d = V_0' \cos \omega_0 t + V_{s1} \cos \omega_{s1} t + V_{s2} \cos \omega_{s2} t \qquad \text{......(2)}$$

A wide spectrum of significant output products results

Reprinted with permission from *Radio Electron. Eng.*, vol. 39, pp. 193–197, Apr. 1970.

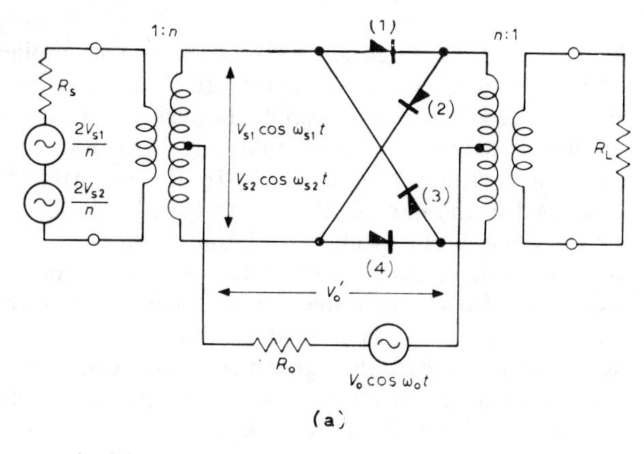

(a) Ring modulator terminated for minimum loss.

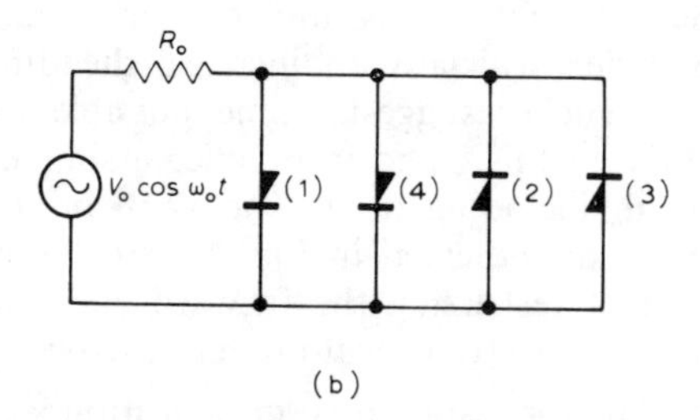

(b) Diode circuit as seen by local oscillator.

Fig. 1.

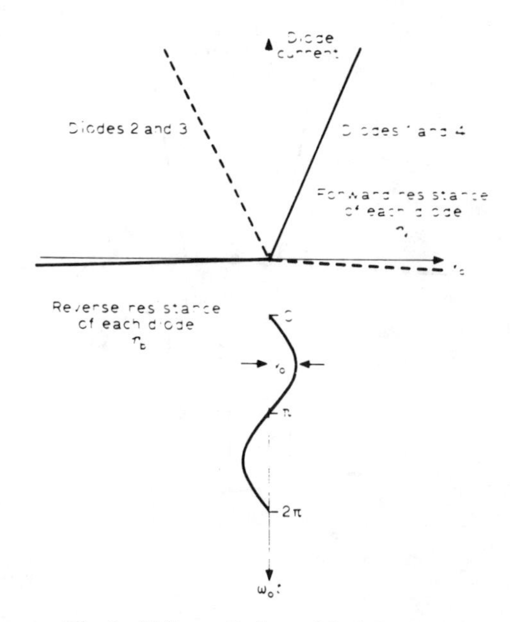

Fig. 2. Bi-linear diode model of Tucker.[2]

but only two are of major importance in assessing the modulator performance by this type of test. These are the products $\omega_0+2\omega_{s1}-\omega_{s2}$ and $\omega_0+2\omega_{s2}-\omega_{s1}$ for an upper-sideband modulator and $\omega_0-2\omega_{s1}+\omega_{s2}$ and $\omega_0-2\omega_{s2}+\omega_{s1}$ for lower-sideband conversion. As is demonstrated in Reference 4 the relative magnitude of these distortion products and the large-signal sideband outputs at $\omega_0\pm\omega_{s1}$; ω_{s2} is given by

$$\left|\frac{v_i}{V_{0\pm 1}}\right| = \frac{k^2}{8[1-\frac{3}{8}k^2]} \quad \text{......(3)}$$

where v_i is the intermodulation product voltage at the modulator output,

$V_{0\pm 1}$ is the large-signal sideband product voltage and $k = k_1 = k_2$

$$\text{where} \quad \left. \begin{aligned} k_1 &= \frac{V_{s1}}{V_0'} \\ k_2 &= \frac{V_{s2}}{V_0'} \end{aligned} \right\} \quad \text{......(4)}$$

3. Incorporating the Diode Offset Voltage

Consider now the diode characteristics indicated in Fig. 4. In this case it has been assumed that conduction does not commence until a certain value of forward bias has been attained. In silicon p–n junction devices this voltage may be of the order of 0·5–0·7 V, in Schottky barrier diodes 0·25–0·35 V. The effect of this offset voltage on the voltage at the local-oscillator input port is shown; a very brief interval exists twice per local-oscillator cycle when all the diodes are turned off and the voltage at the port changes at the same rate as the local-oscillator e.m.f. (assuming that, as would almost invariably be the case, the diode 'off' resistance is very large in comparison with the local-oscillator source resistance R_0).

Returning to the mechanism by which distortion is generated, it is seen that a comparison can be made between the effective shift in switching time which is produced by a given V_{s1}, V_{s2} in the modulator using the diodes of Fig. 4 on the one hand and those of Fig. 2 on the other. It is apparent that in the modulator using the diodes of Fig. 4 two distinct situations can be described.

(1) V_{s1} and V_{s2} are sufficiently small for their sum to be always smaller than the diode offset voltage. This means that even when V_{s1} and V_{s2} are interfering with the diode switching to the maximum extent, the shift produced in the time of switching is small; this overload condition always arises during the interval when all the diodes are turned off and the local-oscillator voltage at the diode is undergoing rapid transition in polarity, i.e. the level of distortion

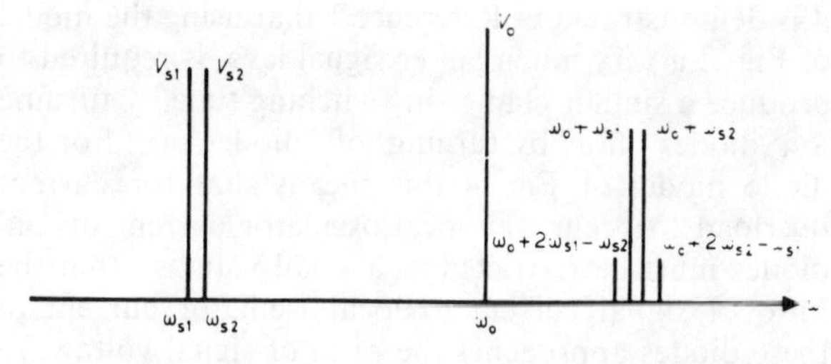

Fig. 3. Two-tone intermodulation test, relevant frequencies.

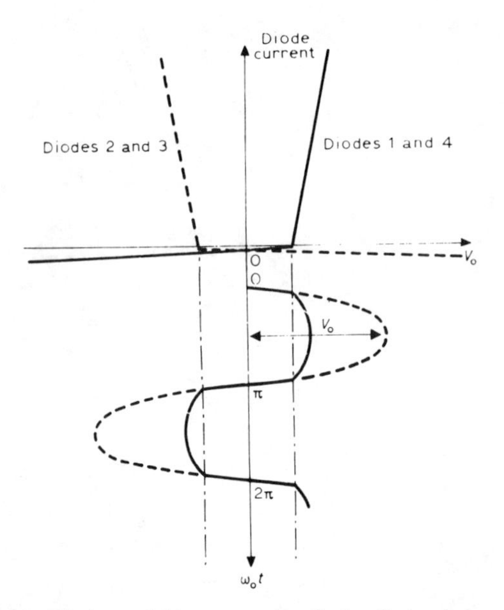

Fig. 4. Diode model incorporating diode offset voltage.

produced is determined not by V_0' but by V_0. Thus

$$\left.\begin{aligned} k_1 &= \frac{V_{s1}}{V_0} \\ k_2 &= \frac{V_{s2}}{V_0} \end{aligned}\right\} \quad \ldots\ldots(5)$$

(2) Once the combined peak amplitude of V_{s1} and V_{s2} has exceeded the diode offset voltage, the effect of the input signals on the diode switching time becomes increasingly marked, since, as indicated by the diagram, a small further increase in signal level results in a large change in the switching time. Ultimately, for very large input-signal levels the existence of the offset voltage becomes masked by these effects and the performance of the modulator tends to that predicted by the simple model of Fig. 2.

It is apparent that the diode model of Fig. 4 permits, in principle, other overload conditions to exist than are possible in the simple model; for instance, since all the diodes are turned off for a short interval it is possible that as local-oscillator current falls to zero in an 'on' diode the signal current flowing in this element may turn it off prematurely and so produce a larger change in the switching time of the diode than the mechanism described so far. However, it is demonstrated in Reference 1 that using the model of Fig. 2 a very much larger signal level is required to produce a similar change in switching time by turning on diodes off as by turning 'off' diodes on. For the diode model of Fig. 4 this means that for current overload to occur, the local-oscillator current in 'on' diodes must be restricted to a small value so that the ratio of signal current to local-oscillator current in these diodes approaches the ratio of signal voltage to local-oscillator voltage across 'off' diodes. This implies either low-level drive or drive from a very high resistance local-oscillator supply, both situations being readily avoided if intermodulation distortion is a significant criterion of design. In the experiments described in the next Section a local-oscillator supply of 4 V open circuit from a 50 Ω source was used and results predicted on the assumption that only voltage overload effects contributed. Whilst small departures from predicted results were observed these were not sufficient to suggest that any mechanism other than voltage overload need be taken into account with currently available diodes except possibly under some conditions of artificially enhanced offset voltage.

It will be seen from the experimental results of the next Section that it may, under some circumstances, be desirable artificially to increase the diode offset voltage. Tucker suggests some possible techniques for achieving this[5], and in practice a simple solution consists in including a bias network in series with each diode as indicated in Fig. 5. However this has the effect of restricting the forward current through the diode for a given local-oscillator drive voltage.

A modulator using this modification was tested with the local-oscillator supply described above but again predictions of distortion based on voltage overload proved adequate over the range of levels investigated.

4. Discussion of Experimental Results

To illustrate the arguments of the previous Section, suppose that a two-tone intermodulation test is carried out over a wide range of input-signal levels. At low signal levels distortion product outputs will be determined by equations (3) and (5), and at high signal levels by equations (3) and (4), with a region of signal levels over which a transition can be expected. The beginning of this transition will occur when the combined peak input signal levels equal twice the diode

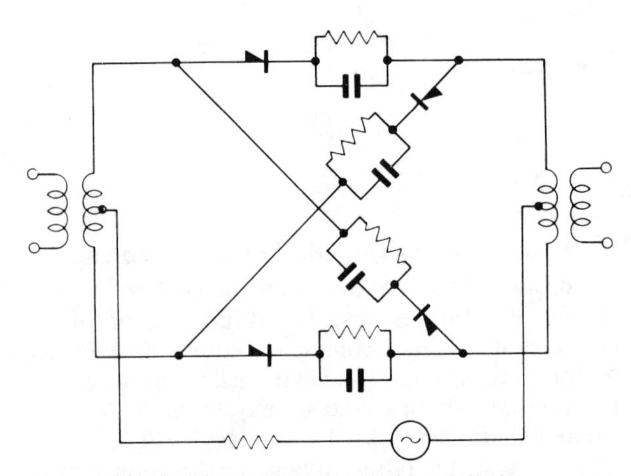

Fig. 5. Ring modulator with biased diodes.

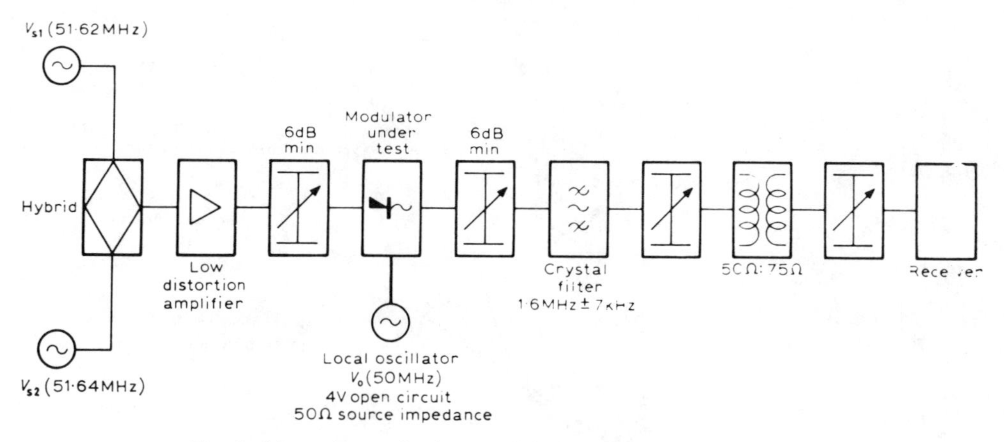

Fig. 6. Measuring set for intermodulation measurement.

offset voltage, since this is the voltage which must be developed across an 'off' diode to turn it on.

Three modulators were examined: (1) a conventional ring as in Fig. 1 with gallium arsenide diodes type CAY 11; (2) as (1) but with typical silicon Schottky-barrier diodes; and (3) a modulator as in Fig. 5 using the Schottky barrier diodes of (2). The measuring set is shown in Fig. 6.

Figure 7 shows a comparison between the modulators (1) and (2). The predicted results for the first phase of distortion are obtained from equations (3) and (5) and for distortion under high-level input conditions from (3) and (4). The diode offset voltages were 550 mV for the gallium arsenide devices and 250 mV for the Schottky-barrier types. These values result in maximum input levels for phase 1 distortion as shown in Fig. 7, i.e. 195 mV for the gallium arsenide, 88 mV for the Schottky barrier.

Figure 8 shows a comparison between the modulators (2) and (3) to illustrate the effect of bias on the maximum input-signal level for phase 1 distortion. The measuring set output level to the modulator was limited to 0 dBm (220 mV in 50 Ω) so that the nature of overload ultimately occurring in the modulator using biased diodes could not be determined. However, it is apparent that a useful extension of phase 1 distortion is possible by this technique. The bias elements used were 100 Ω and 10 000 pF.

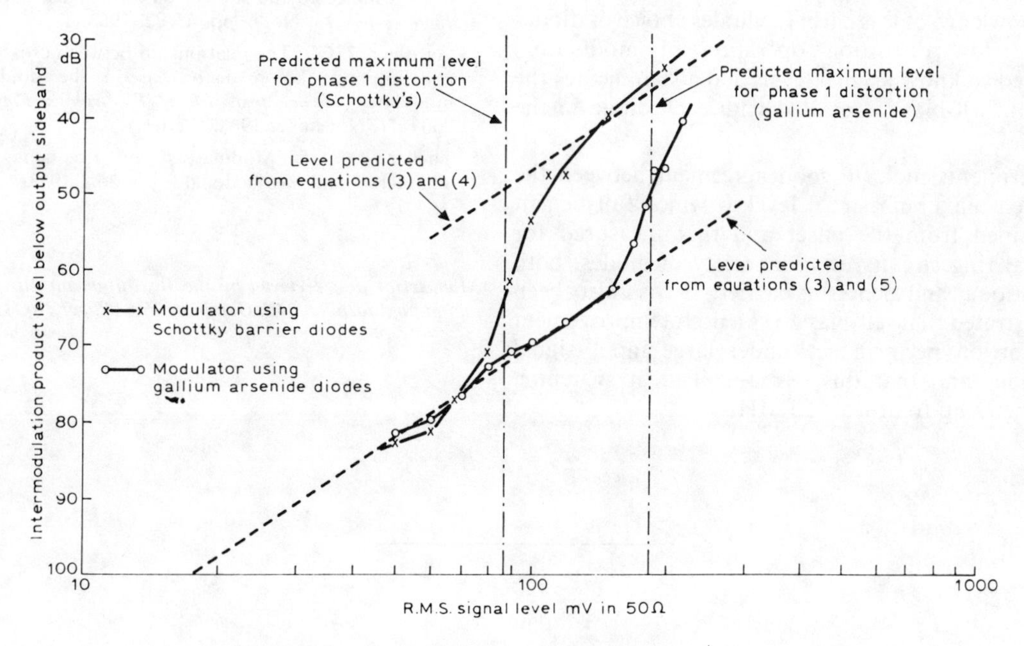

Fig. 7. Comparison of diodes with differing offset voltages.

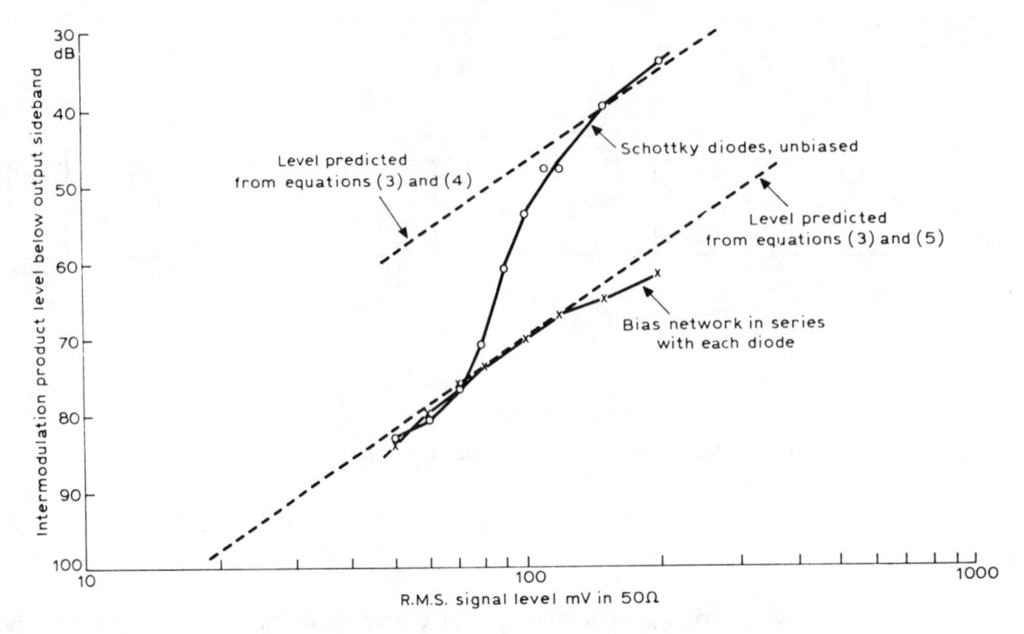

Fig. 8. Artificial enhancement of diode offset voltage.

5. Conclusions

It has been demonstrated that the offset voltage in the diode d.c. characteristic is an important parameter in the prediction of distortion levels in the ring modulator. The existence of the offset voltage results in the modulator achieving a condition, twice per local-oscillator cycle, in which all the diodes are turned off. If interfering signals generate voltages across the diodes during this period less than twice the value of the offset, then switching interference is reduced since the rate of change of the local-oscillator voltage across any diode in the ring is increased to substantially that of the open circuit local-oscillator source voltage.

A knowledge of this effect facilitates choice of diodes for the low distortion operation of modulators subjected to known signal levels and also indicates the value of 'self-bias' associated with each diode in the ring.

Experiments indicate good agreement between the predicted maximum signal level at which full benefit is obtained from the offset and that measured for practical ring circuits using a variety of diodes, both conventional and Schottky barrier. It has also been demonstrated that self-bias gives a useful improvement in distortion performance under large input signal conditions and that this is maintained at switching frequencies up to at least 50 MHz.

6. Acknowledgments

The experimental results reported here were obtained while the author was working with Racal Communications Ltd., Tewkesbury; permission from the Board of Directors to publish these is gratefully acknowledged.

7. References

1. Belevitch, V., 'Non-linear effects in rectifier modulators', *Wireless Engineer*, **27**, pp. 130–1, April 1950.
2. Tucker, D. G., 'Intermodulation distortion in rectifier modulators', *Wireless Engineer*, **31**, pp. 145–52, June 1954.
3. Savin; S. K., 'Response of a non-linear unilaterally conducting resistance to the sum of sinusoidal oscillations', *Radio Engineering*, **23**, No. 6, pp. 65–72, 1968.
4. Gardiner, J. G., 'The relationship between cross-modulation and intermodulation distortions in the double-balanced modulator', *Proc. Inst. Elect. Electronics Engrs*, **56**, pp. 2069–71, November 1968. (Letter.)
5. Tucker, D. G., 'Modulators and Frequency Changers', Chapters 5 and 8 (Macdonald, London, 1953).

Manuscript first received by the Institution on 4th August 1969 and in final form on 24th October 1969. (Paper No. 1315/CC72.)

Response Coefficients of a Double-Balanced Diode Mixer

M. A. MAIUZZO AND S. H. CAMERON

Abstract—**This paper presents the results of a computer-assisted analysis of a commonly used double-balanced diode mixer circuit. It is shown that the magnitudes of some of the responses to one or more input radio frequency signals are sharply dependent on the degree to which the diodes comprising the mixer are balanced. The average rejections imparted to a particular undesired intermodulation (or spurious) response by a random sample of such mixers are plotted as a function of the tightness of the control of diode characteristics of the population from which the four diodes comprising a single mixer are selected.**

A computer program based on a Fourier-series expansion of the time-dependent coefficients of a Taylor-series representation of the four diode currents was employed to compute the response coefficients of the mixer [1]–[3]. The response coefficients are defined as the magnitudes of the intermodulations of sinusoidal-input radio-frequency signals and the local-oscillator-frequency signal. The method treats the switching action characteristic of the large local-oscillator power levels, rather than relying on the more usual and invalid [1] assumption of "mild" nonlinearity. Orders of nonlinearity up to the tenth are treated.

Key Words—**Double-balanced diode mixer, response coefficients, large LO signals.**

I. INTRODUCTION

WHEN RECEIVERS are to be deployed in a closely spaced environment with one or more transmitters, the degree to which the receivers respond to spurious signals becomes important. A major determiner of the receiver's spurious/intermodulation response levels is the first mixer. In recent years, the use of balanced mixers has become widespread. The degree to which the components of these circuits are balanced can have a great effect on the levels of many of these undesired responses.

To arrive at this result, the following procedure was employed and is described in additional detail in the following paragraphs.

1) The Ebers-Moll model was selected to represent the individual diodes. Ebers-Moll model parameter values for a "typical" diode were determined. Possible variations in parameter values were investigated empirically.

2) Current-loop antennas for a common double-balanced diode mixer circuit were derived, leading to equations for the Taylor-Fourier-series coefficients.

3) A computer program was prepared which computes the Taylor-Fourier-series coefficient values and then uses these to compute the mixer rejection to spurious responses. A Monte Carlo technique was employed, whereby diode parameters were selected from a random distribution. The process is automatically repeated the desired number of times in order to provide a distribution of response rejection values.

Manuscript received July 6, 1978; revised July 2, 1979. This work was supported by Contract F-19628-78-C-0006 for the Department of Defense at the Electromagnetic Compatibility Analysis Center, Annapolis, MD 21402. This paper was to have been presented at the 1979 IEEE International Symposium on Electromagnetic Compatibility, San Diego, CA.

The authors are with the IIT Research Institute ECAC, North Severn, Annapolis, MD 21402.

A. The Taylor-Fourier Method

The output current is related to the input RF voltage, v_{RF} using

$$i = \sum_{Q=0} \sum_{P=0} (v_{RF})^Q a_{P,Q} \cos (P 2\pi f_{LO} t) \tag{1}$$

where

- P, Q integers representing the harmonic number of the LO and RF inputs, respectively, contributing to the response being analyzed,
- $a_{P,Q}$ the Taylor-Fourier-series response coefficients,
- f_{LO} the local-oscillator frequency,
- t time.

B. Using the Response Coefficients, $a_{P,Q}$

Once the response coefficients are known, the mixer output levels may be calculated from (1) for any desired or undesired input voltage. However, (1) can be put in a more convenient form for calculation purposes if the form of v_{RF} is given. For example, suppose the input waveform consists of a single tone at a frequency that will cause an IF-amplifier response. Let

$$V_{RF} = V_{RF} \cos 2\pi f_{RF} t \tag{2}$$

where

- V_{RF} peak voltage,
- f_{RF} radio frequency.

A response is said to occur when

$$f_{RF} = \frac{P f_{LO} \pm f_{IF}}{Q} \tag{3a}$$

where

- f_{IF} receiver intermediate frequency,
- f_{LO} receiver local-oscillator frequency.

Note, all but one of these responses (i.e., the desired one) can be termed spurious.

For example, setting $P = 2$ and $Q = 3$ and sign = +, an input of frequency

$$f_{RF} = \frac{2 f_{LO} + f_{IF}}{3} \tag{3b}$$

will cause a response at the mixer output at the IF-stage tuned

Reprinted from *IEEE Trans. Electromagn. Compat.*, vol. EMC-21, pp. 316–319, Nov. 1979.

frequency f_{IF}. The amplitude of this response may be calculated by considering the appropriate, $P = 2, Q = 3$ term of (1)

$$i_{SPUR.} \cong (V_{RF} \cos 2\pi f_{RF} t)^3 a_{2,3} \cos 4\pi f_{LO} t \tag{4a}$$

which may be expanded yielding the following term of interest:

$$\cong (\tfrac{1}{2})^3 V_{RF}{}^3 a_{2,3} \cos [2\pi(2f_{LO} - 3f_{RF} t]. \tag{4b}$$

When (3) is used and terms at frequencies other than IF are ignored,

$$i_{SPUR} \cong (\tfrac{1}{2})^3 V_{RF}{}^3 a_{2,3} \cos 2\pi f_{IF} t. \tag{4c}$$

Thus $a_{2,3}\ V_{RF}{}^2/8$ is the conversion transconductance of the mixer for a $P = 2, Q = 3$ spurious response. A more general expression for (4c) is

$$i_{SPUR} \cong (\tfrac{1}{2})^Q V_{RF}{}^Q a_{BQ} \cos 2\pi f_{IF} t, \qquad p > 0 \tag{5a}$$

and

$$i_{SPUR} \cong (\tfrac{1}{2})^{Q-1} V_{RF}{}^Q a_{P,Q} \cos 2\pi f_{IF} t, \qquad P = 0. \tag{5b}$$

Similar expressions may be employed to cover two-signal mixer intermodulation and other nonlinear effects.

In evaluating the coefficients, the diodes must be first modeled.

II. REPRESENTING THE DIODES

Fig. 1 is an equivalent-circuit diagram of the Ebers-Moll diode model [1]. The nonlinearity is represented by the relationship between $v_D{}'$ and i_D using

$$i_D = i_0 [e^{\alpha v_D{}'} - 1] \tag{6}$$

where

i_D	current through the ideal junction, in amperes
i_0	reverse leakage current, in amperes
$v_D{}'$	voltage across the ideal-diode junction, in volts
R_b	diode bulk resistance
α	diode parameter $\cong q/kT$, V^{-1}
q	electric charge of an electron
k	Boltzman's constant
T	temperature in K.

Direct-current (dc) measurements were made on six IN82A diodes. From these measurements, "average" diode parameter values of $\alpha = 25\ V^{-1}$, $R_b = 13\ \Omega$, and $i_0 = 2\ \mu A$ were estimated. Also, a definite correlation between α and the diode bulk resistance R_b was noted for the diodes. In fact it was noted that

$$\alpha \cong 36.77 - 0.867 R_b. \tag{7}$$

Equation (7) was used in the mixer model as discussed later in this paper. It was also noted that the shunt resistance R_{SH} was extremely large. This agrees with the results of Ebers-Moll [4]. For this reason and for the sake of simplicity, the current through this resistance was assumed negligible. In addition, the reverse leakage current on all diodes measured varied little from the 2-μA level.

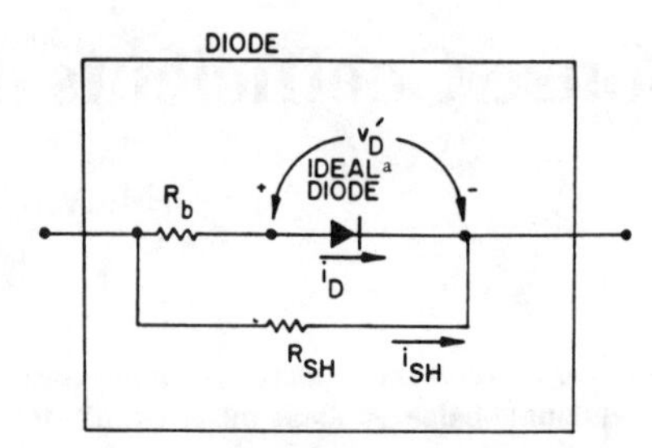

Fig. 1. Ebers-Moll diode model.
[a] An ideal diode is governed by (6).

Therefore, it was decided to simulate the situation where four diodes for the mixer are selected from a sample described by R_b and α values having Gaussian distributions and subject to (7) with i_0 constant.

III. REPRESENTING THE MIXER CIRCUIT

Fig. 2 is the circuit diagram for the DBDM mixer. For this circuit, four loop equations are necessary; ideal transformers are assumed. These equations are given in matrix notation by

$$\begin{bmatrix} \frac{1}{\alpha_1} \ln\left(\frac{i_1}{i_{01}} + 1\right) + k_3 E_0 - k_1 e_s \\ \frac{1}{\alpha_2} \ln\left(\frac{i_2}{i_{02}} + 1\right) - k_3 E_0 - k_2 e_s \\ \frac{1}{\alpha_3} \ln\left(\frac{i_3}{i_{03}} + 1\right) + k_4 E_0 + k_1 e_s \\ \frac{1}{\alpha_4} \ln\left(\frac{i_4}{i_{04}} + 1\right) - k_4 E_0 + k_2 e_s \end{bmatrix} = [A]\ [i] \tag{8}$$

where

k_1	n_{ae}/n_s,
k_2	n_{ed}/n_s,
k_3	n_{bc}/n_0,
k_4	n_{cf}/n_0,
K_n	indicated turns ratio, secondary to primary,
E_0	instantaneous LO voltage,
e_s	instantaneous RF voltage,
R_0	LO source resistance,
R_s	RF source resistance,
i_{0j}	reverse leakage current for jth diode,
α_j	diode parameter for jth diode,
R_{bj}	bulk resistance for jth diode.

The members of [i] are i_j where i_j is the forward current through diode j. The members of $[A]$ are loop resistances. For example,

$$A_{1,1} = -k_1{}^2 R_s - k_3{}^2 R_0 - R_L - R_{b1} \tag{9}$$

$$A_{1,2} = -k_1 k_2 R_s + k_3{}^2 R_0 + R_L. \tag{10}$$

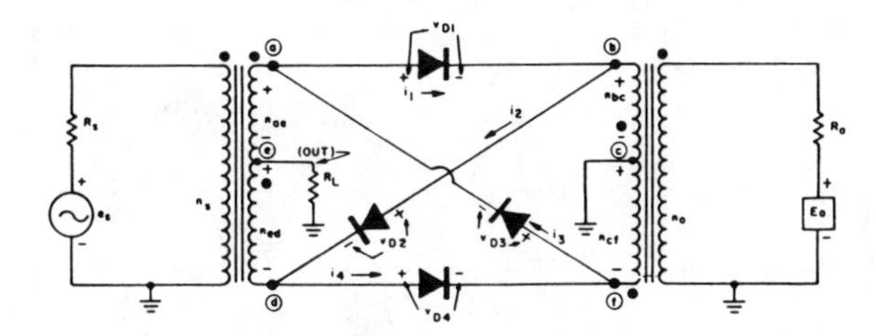

Fig. 2. Modeled DBDM circuit.

NOTE: D1 = diode #1 D2 = diode #2
D3 = diode #3 D4 = diode #4.

Expressions for the derivatives of the load-resistor current with respect to the RF source voltage were derived. From (8), it follows that

$$\begin{bmatrix} -k_1\alpha_1 J_1 \\ -k_2\alpha_2 J_2 \\ +k_1\alpha_3 J_3 \\ +k_2\alpha_4 J_4 \end{bmatrix} = [B]\left[\frac{\partial i}{\partial e_s}\right] \tag{11}$$

where

$$J_j = (i_j + i_{0,j}) \tag{12}$$

$$B_{j,j} = A_{j,j}\alpha_j J_j - 1 \tag{13}$$

$$B_{j,k} = A_{j,k}\alpha_j J_j, \qquad j \neq k. \tag{14}$$

The expressions for the Qth derivative have the general form

$$[C^Q] = [B]\left[\frac{\partial^Q i}{\partial e_s{}^Q}\right] \tag{15}$$

(See [1] for additional detail.) The load resistor current is given using

$$i_L = -i_1 + i_2 + i_3 - i_4. \tag{16}$$

Consequently

$$\frac{\partial^Q i_L}{\partial e_s{}^Q} = -\frac{\partial^Q i_1}{\partial e_s{}^Q} + \frac{\partial^Q i_2}{\partial e_s{}^Q} + \frac{\partial^Q i_3}{\partial e_s{}^Q} - \frac{\partial^Q i_4}{\partial e_s{}^Q}. \tag{17}$$

The Taylor series coefficients are

$$a_Q = \frac{1}{Q!}\frac{\partial^Q i_L}{\partial e_s{}^Q}. \tag{18}$$

Each a_0 value is dependent on the instantaneous value of the LO voltage; the a_Q values are periodic waveforms whose fundamental frequency is the LO frequency. The Taylor-Fourier coefficients are given by

$$a_{P,Q} = \frac{1}{2\pi}\int_0^{2\pi} a_Q(t)d(\omega_{LO}t), \qquad P = 0 \tag{19}$$

$$a_{P,Q} = \frac{1}{\pi}\int_0^{2\pi} a_Q(t)\cos(P\omega_{LO}t)d(\omega_{LO}t), \qquad P > 0 \tag{20}$$

where

$$\omega_{LO} = 2\pi f_{LO}.$$

These coefficients are obtained by sampling the LO-voltage time waveform. For each sample, the current and a_0 values are calculated using (8) and (18); numerical integration is employed to compute (19) and (20). For (20) to suffice, the LO waveform must be represented as an even function.

IV. RESULTS

The Taylor-Fourier representation of the DBDM was first tested by comparison of the predicted values of spurious-response rejection with spurious-response measurements conducted on a DBDM consisting of 4 diodes, whose Ebers-Moll parameters had been previously measured and used in the prediction model. The general conclusion was that, for frequencies less than about 300 MHz, the model is excellent, but that for higher frequencies energy-storage elements neglected in the formulation of the circuit model become significant.

The dependence of the spurious-response coefficients on the degree of mixer balance was investigated by the following procedure. The Ebers-Moll parameters of the individual diodes comprising the DBDM model were selected by random sampling techniques from a population having known characteristics. The average behavior of a number (10) of such mixers was established by executing the computer model described above. The parameters of the diode distribution were then altered and the computation repeated in order to display the relationships between the degree of diode matching and the spurious-response performance.

Measurements of the Ebers-Moll parameters of a series of diodes demonstrated the fact that the variations in leakage current from diode to diode were sufficiently small as to have negligible effect on the DBDM performance. Leakage current was, therefore, held constant in the diode distributions. The measurements also indicated the previously mentioned strong condition between R_b and α of the diodes. The relationship of (7) was, therefore, assumed and the diode distribution was characterized by a single degree of freedom, namely R_b. In all other respects, the DBDM circuit parameters were balanced.

Mathematical arguments based on the symmetry of the DBDM loop equations can be constructed to show that the reponds coefficients of a perfectly balanced DBDM will be zero if either P or Q is an even integer.

Computations results are shown in Fig. 3 for the case of Q = 1 and P = 1-10. Rather than plot the spurious-response coefficient for each (P, Q), the value plotted is the input RF-signal power (in dBm) required to produce a specified P, Q mixer output at the IF frequency, a quantity which can be conveniently compared with measurements. Fig. 3 shows the behavior of an actual DBDM based on laboratory measurements of the spurious-response characteristics, together with

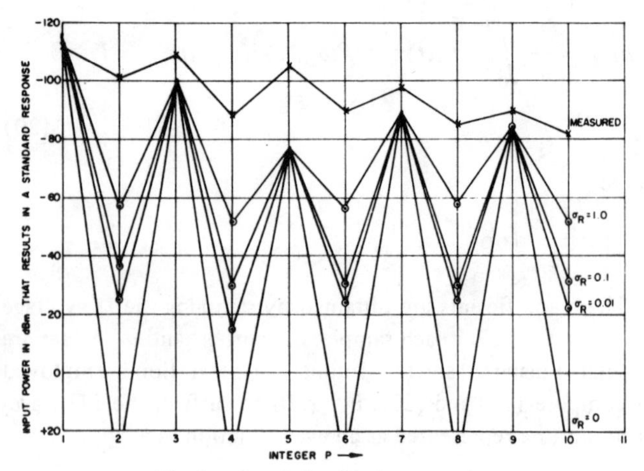

Fig. 3. Levels for $Q = 1$ responses.

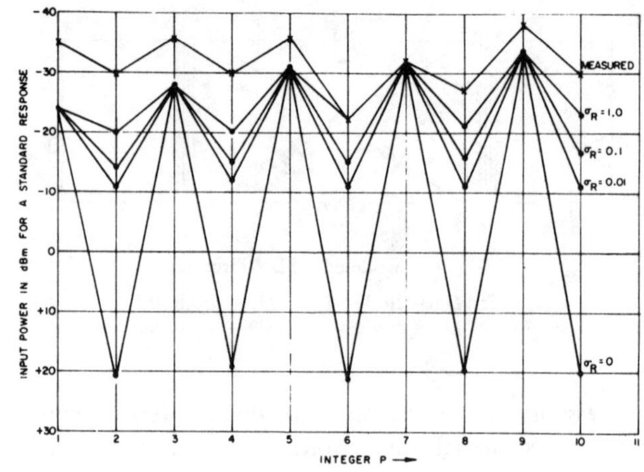

Fig. 5. Levels for $Q = 3$ responses.

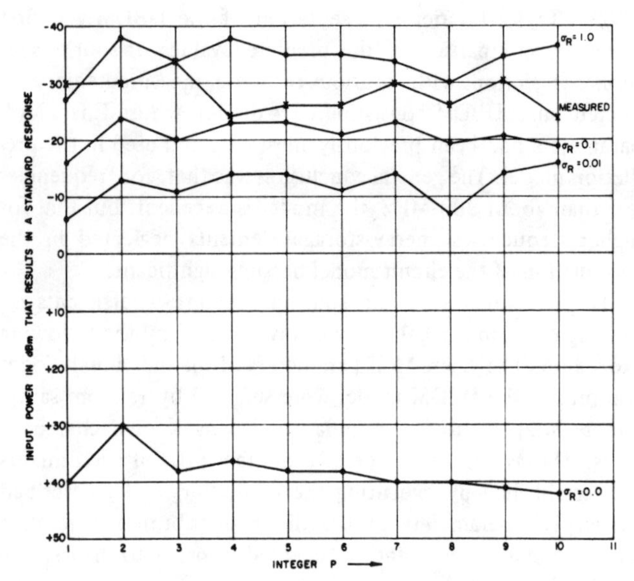

Fig. 4. Levels for $Q = 2$ responses.

the predicted behavior of a number of DBDM's of various degrees of balance. The predicted performance curves are labeled with the standard deviation (σ_R) of the normal distribution from which the diodes were randomly selected. The mean value of each of the R_b distributions was 15 Ω. A "perfectly balanced mixer" corresponds to the case of $\sigma_R = 0$. As can be seen, the predicted values for odd P are in good agreement with the measurements, good at least in terms of previous success in predicting mixer behavior involving high orders of nonlinearity. While the measured values for even P are, in general, smaller than the neighboring values for odd P, the deep nulls for even P shown by the perfectly balanced mixer are not in evidence in the measured values, presumably because of imperfect balance among the diodes and possibly other circuit components.

Note that even the $\sigma_R = 0$ case shows a finite value for P even. This may be attributed to the quantization noise and other errors associated with the digital-computer realization of the Taylor-Fourier-series model. Fig. 3 also illustrates the manner in which the calculated performance approaches the measured performance as the degree of unbalance increased.

Plots comparable to Fig. 3 are given in Figs. 4 and 5, illustrating the DBDM spurious-response performance for $Q = 2$ and $Q = 3$, respectively. As expected from the symmetry arguments, all of the response coefficients for $Q = 2$ are generally small, while the case of $Q = 3$ resembles $Q = 1$ in that values for P even are suppressed.

V. CONCLUSIONS

On the basis of the above results, the following conclusions would seem appropriate

1) The Taylor-Fourier-series representation provides a tool for the prediction of high-order spurious-response performance of a DBDM at frequencies up to 300 MHz.

2) The spurious-response performance for P and Q odd is relatively independent of the degree of diode balance.

3) Substantial diode unbalance is possible before response for P even approaches the performance for P odd.

4) Since many inband intermodulation problems occur primarily as a result of the ($P = 1$, Q odd) mixer responses, little benefit in these cases can be gained from carefully balanced mixers.

REFERENCES

[1] M. Maiuzzo, "Analysis of receiver mixers," ESD-TR-78-100, ECAC, Annapolis, MD, Mar. 1978.

[2] F. E. Terman, *Electronic and Radio Engineering*, New York: McGraw-Hill, Fourth Ed., 1955, pp. 574-575.

[3] *Nonlinear system modeling and analysis with applications to communications receivers*, RADC-TR-73-178, June 1973.

[4] J. J. Ebers and J. L. Moll, "Large signal behavior of junction transistors," *Proc. IRE*, Dec. 1954.

Part VII
Cryogenic Schottky Barrier Diode Mixers

FOR millimeter-waves the workhorse low noise receiver is the cooled Schottky diode mixer. The first extensive investigation of millimeter-wave cryogenic mixers was published by Weinreb and Kerr 1973 [1]. The first paper in this part, by Kerr, compares properties of cooled and uncooled mixers and describes in detail the design of such mixers. A somewhat different design is described in the next paper by Linke *et al.* The low noise operation of their mixer is seen to be a result of the short-circuiting of the noise entering the image port, and a diode specially designed for cryogenic operation.

In the third paper Kollberg and Zirath discuss the importance of the embedding impedance and point out that the harmonic response is of particular importance for low noise receivers. The problem of correct modeling diodes at cryogenic temperatures is pointed out. In the fourth paper Schneider *et al.* conclude that in fact cryogenically cooled diodes may show properties that are not necessarily the same from diode to diode, suggesting that the technology of making diodes is not yet perfect and that it may pay off to try contacting different diodes on the same chip for optimizing the mixer performance. In the last paper of this part Archer discusses low noise solid-state receiver systems for room temperature and cryogenic operation at near 230 GHz. Not only is the mixer itself described in detail, but also the local oscillator design and the quasioptical circuit for coupling the signal and local oscillator power to the diode are described.

In a paper to be published mid-1984 [2] Predmore *et al.* will not only describe an extremely low noise cooled mixer, but also will give a survey of results obtained by different groups working in this particular field.

References

[1] S. Weinreb and A. R. Kerr, "Cryogenic cooling of mixers for millimeter and centimeter wavelengths," Special Issue on Microwave Integrated Circuits, *IEEE J. Solid-State Circuits,* vol. SC-8, pp. 58–63, Feb. 1973.

[2] C. Read Predmore, A. V. Räisänen, J. L. R. Marrero, N. L. Erickson, and P. F. Goldsmith, "A broad-band, ultra low-noise Schottky mixer receiver from 80–115 GHz," *IEEE Trans. Microwave Theory Tech.,* May 1984.

Low-Noise Room-Temperature and Cryogenic Mixers for 80-120 GHz

ANTHONY R. KERR, ASSOCIATE MEMBER, IEEE

***Abstract*—A description is given of two new mixers designed to operate in the 80–120-GHz range on the 36-ft radio telescope at Kitt Peak, Ariz. It is shown that for a hard-driven diode the parasitic resistance and capacitance are the primary factors influencing the design of the diode mount. A room-temperature mixer is described which achieves a single-sideband (SSB) conversion loss (L) of 5.5 dB, and a SSB noise temperature (T_m) of 500 K (excluding the IF contribution) with a 1.4-GHz IF. A cryogenically cooled version, using a quartz structure to support the diode chip and contact whisker, achieves values of L = 5.8 dB and T_m = 300 K with a 4.75-GHz IF. The mixers use high-quality Schottky-barrier diodes in a one-quarter-height waveguide mount.**

I. INTRODUCTION

THIS PAPER describes the results of a program of mixer development aimed at producing more sensitive millimeter-wave receivers for the National Radio Astronomy Observatory's 36-ft radio telescope at Kitt Peak, Ariz.

The most significant development in millimeter-wave mixers since Sharpless [1] introduced the wafer diode mount in 1956, has been the introduction of the Schottky-barrier diode. The nearly ideal exponential characteristic of the Schottky diode led Barber [2] to approximate the device by a switch in series with a small resistance; the conversion loss of a mixer is then a function of the pulse duty ratio (PDR) of the switch. Dickens [3] has achieved good agreement between Barber's theory and experimental results at 60 and 95 GHz. Leedy *et al.* [4] demonstrated good agreement between theory and experiment when they assumed, following Torrey and Whitmer [5], a sinusoidal LO voltage at the diode. Although this assumption is unlikely to be strictly valid [6], [7], it is consistent at high LO levels with Barber's switching model.

More recently, nonlinear analysis techniques have been applied to the mixer problem in an effort to achieve a more accurate understanding of the mixing process [6]–[9]. However, these attempts have been limited to cases in which the diode has a fairly simple embedding network. The difficulty of characterizing the embedding network at the harmonics of the LO frequency has so far prevented these methods from being used to give an accurate solution for the case of a waveguide-mounted diode.[1]

In this paper the approach taken to mixer design is to consider the mixer as three interconnected networks as shown in Fig. 1:

- N_1 the embedding network, or diode mount;
- N_2 the network containing the diode's parasitic capacitance and resistance, which connects the ideal diode to the embedding network;
- N_3 the ideal exponential diode.

These three networks are optimized to obtain maximum power transfer between the embedding network and the periodically varying junction resistance at the input (RF) and output (IF) frequencies. The embedding network is assumed reactive at the harmonics of the LO frequency.

The single-sideband (SSB) noise temperature of a mixer receiver can be written as

$$T_R = T_M + LT_{\text{IF}} \qquad (1)$$

where T_M is the noise contribution of the mixer itself, L is the SSB conversion loss of the mixer, and T_{IF} is the noise temperature of the IF amplifier. Following the argument given by Weinreb and Kerr [12], T_M can be expressed in terms of an average temperature associated with the diode $T_{D_{\text{AV}}}$ and the conversion loss, thus

$$T_M = (L - 2)T_{D_{\text{AV}}}. \qquad (2)$$

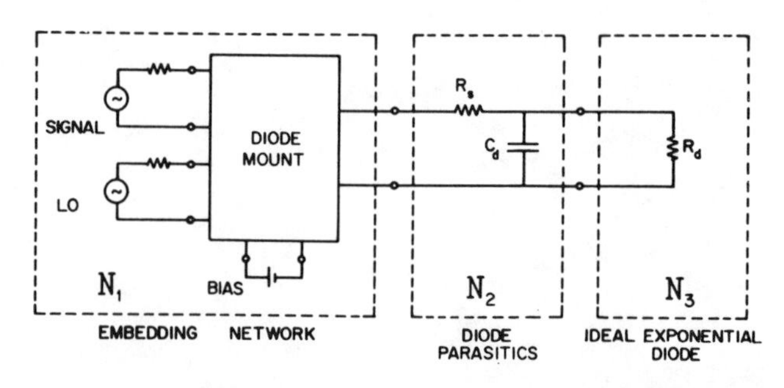

Fig. 1. A mixer represented as three interconnected circuits: N_1—the embedding network; N_2—the diode parasitic resistance and capacitance; and N_3—the ideal exponential diode.

Manuscript received January 13, 1975; revised April 28, 1975. This work was supported by Associated Universities, Inc., under contract with the National Science Foundation.

The author was with the National Radio Astronomy Observatory, Charlottesville, Va. 22901. He is now with the NASA Goddard Institute for Space Studies, New York, N. Y. 10025.

[1] Eisenhart and Khan [10] and Eisenhart [11] have made an analysis of the simple waveguide mount, which is accurate up to many times the normal operating frequency of the waveguide. In practice, however, the waveguide mount deviates from the simple model in such details as the nonideal RF choke, an input waveguide transformer, and a nonplanar short circuit behind the diode.

Reprinted from *IEEE Trans. Microwave Theory Tech.*, vol. MTT-23, pp. 781–787, Oct. 1975.

The mixer is assumed to be of the broad-band type for which the conversion loss is the same at the signal and image frequencies. It has been found in practice that if the tuning, LO drive, or bias of a mixer is varied, $T_{D_{AV}}$ generally changes to some extent, but that the change in $(L - 2)$ dominates the right-hand side of (2). Reducing L therefore reduces both the mixer and IF contributions to the receiver noise temperature as given by (1).

The object of this paper is to show that with careful attention to the mixer design, it is possible to achieve low noise and conversion loss at frequencies up to 120 GHz. Two mixers are described which are tunable from 80 to 120 GHz; one is intended for room-temperature operation, and the other, a variation of the first, uses a quartz diode mount which is suitable for cryogenic operation.

II. MIXER THEORY

A. Ideal Exponential Diode

The junction resistance of a practical Schottky-barrier diode behaves as an ideal exponential element over many decades of current. The current i and voltage v are related by

$$i = i_0(e^{\alpha v} - 1) \tag{3}$$

where $\alpha = q/\eta kT \simeq 35\ \mathrm{V}^{-1}$ at room temperature. For practical purposes $i_0 \ll i$, and the incremental conductance of the diode may be written as

$$g = \frac{\partial i}{\partial v} \simeq \alpha i. \tag{4}$$

The behavior of the diode as a mixer depends both on the waveform of $g(t)$ produced by the LO, and on the embedding network seen by the diode.

If a transformer of ratio n is inserted between an ideal exponential diode, operating as a mixer, and its embedding network, the properties of the mixer will remain unchanged provided the dc bias and LO power are changed according to

$$V_{\text{bias}} \rightarrow V_{\text{bias}} - \ln(n^2)/\alpha \tag{5a}$$

and

$$P_{\text{LO}} \rightarrow P_{\text{LO}}/n^2. \tag{5b}$$

It follows that the ideal exponential diode has no preferred impedance level and can perform equally well as a mixer at any impedance. Thus, in optimizing the three networks of Fig. 1 for maximum signal-frequency power transfer to the ideal diode, N_3 imposes no constraint on the impedance levels of N_1 and N_2. If the parasitic elements of N_2 are fixed for the available diodes, the impedance levels of N_1 and N_3 can be chosen to minimize the conversion loss.

B. Diode Capacitance and Series Resistance

The signal frequency equivalent circuit of a Schottky-barrier diode, operating as a mixer, is shown in Fig. 1. R_d is the input impedance of the time-varying junction resistance, and C_d and R_s are the mean values of the junction capacitance and series resistance, assumed equal to their values at the dc bias voltage when no LO power is applied. For a given semiconductor sample C_d and R_s depend primarily on the area A of the diode and on the doping and thickness of the epitaxial layer on which the diode is formed [13]. R_s includes contributions from skin effect in the semiconductor and contact wire. The cutoff frequency is defined as $\omega_c = 1/(R_sC_d)$.

The effects of R_s and C_d on the mixer performance are threefold.

1) They contribute to the conversion loss because of power dissipated in R_s at the RF and IF frequencies.

2) They affect the waveform $g(t)$ of the mixing element by changing the termination of the LO harmonics.

3) They affect the terminations seen by the frequencies $nf_{\text{LO}} \pm f_{\text{IF}}$, $n > 1$.

These effects may be further elaborated as follows.

1) The degradation of the conversion loss caused by power dissipated in R_s at the signal frequency ω, is

$$\delta_{\text{RF}} = 1 + \frac{R_s}{R_d} + \frac{R_d}{R_s}\frac{\omega^2}{\omega_c^2} \geq 1. \tag{6}$$

At the IF frequency R_s appears in series with the output impedance R_o of the exponential element. The loss due to R_s is

$$\delta_{\text{IF}} = 1 + \frac{R_s}{R_o} \geq 1. \tag{7}$$

The combined RF and IF loss due to R_s and C_d is $\delta = \delta_{\text{RF}} \times \delta_{\text{IF}}$; this is shown in Fig. 2 as a function of normalized frequency. The parameter $K = R_o/R_d$ is the quotient of the output (IF) impedance and the input (RF) im-

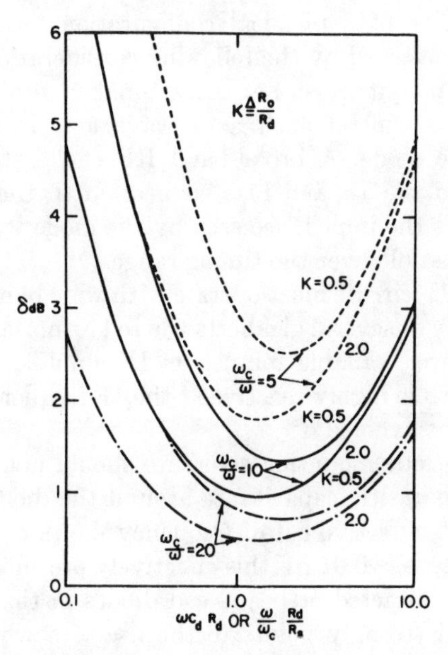

Fig. 2. Loss δ due to RF and IF dissipation in the diode series resistance R_s. The RF (signal) frequency is ω, and $K \triangleq R_o/R_d$ is the ratio of IF impedance to RF impedance.

pedance. It will be shown in the following that for a broad-band mixer K is expected to lie between 0.5 and 2.

2) Barber [2] has used the concept of an equivalent PDR to characterize the mixer properties of a diode with a conductance waveform $g(t)$. The PDR is a function of LO power and bias voltage, and can be maintained constant along with the conversion loss despite changes in $g(t)$ caused by variation of the embedding impedance at frequencies nf_{LO}, $n > 1$.

3) Saleh [14] has shown that for Barber's equivalent PDR to uniquely define the conversion loss it must be dependent not only on the $g(t)$ waveform but also on the embedding impedance seen by the diode at frequencies $nf_{LO} \pm f_{IF}$, $n \geq 1$. A change in the reactive termination at some sideband frequency $nf_{LO} \pm f_{IF}$, $n > 1$, affects both the PDR, which can be restored by appropriate LO and bias adjustments, and the optimum RF and IF impedances of the mixer. It is assumed here that loss in R_s at these sideband frequencies is small, an assumption which is likely to hold for a practical mixer.

C. IF Impedance

Saleh [14] has made an extensive investigation of the effects on mixer performance of the diode's conductance waveform $g(t)$ and of the embedding impedances at the harmonics of the LO. It is observed from his results that for a broad-band mixer the optimum source (RF) and load (IF) impedances never differ by a factor of more than 2, regardless of LO drive level or bias. Although this is not generally proven for all combinations of terminations of the higher frequency sidebands, $nf_{LO} \pm f_{IF}$, $n > 1$, it is consistent with observed mixer performance, and is a useful aid to design.

III. THE DIODE MOUNT

A. Mount Configuration

The choice of a physical configuration for the diode mount is governed by the following considerations.

1) The mount must be easily tunable, preferably by means of a control such as a waveguide short circuit behind the diode. A broad-band RF choke structure is required in the IF and bias connection to the diode to ensure that the impedance seen by the diode will vary as little as possible over the tuning range.

2) The IF circuit must operate with wide bandwidth at a frequency of several gigahertz where low-noise cryogenic paramps are available for use as IF amplifiers. An RF choke which is highly reactive at the IF frequency should therefore be avoided.

3) The diode mounting structure should not introduce excessive parasitic capacitance around the diode thereby reducing its effective cutoff frequency. For a diode whose capacitance is ~0.01 pF this effectively precludes the use of ribbon-contacted or beam-lead diodes in their present forms, and strongly points to the use of a whisker-contacted diode.

These requirements can be fulfilled by a waveguide mount similar in some respects to the wafer mount introduced by Sharpless [1], but using very much reduced-height waveguide, and a different RF choke structure.

B. Mount Equivalent Circuit

Eisenhart and Khan [10] and Eisenhart [11] have made a detailed investigation of the driving-point impedance seen by a small device connected across the gap G in a waveguide mount as shown in Fig. 3(a). The approximate equivalent circuit of the mount is shown in Fig. 3(b), where

$$Z_g = 2\left(\frac{\mu}{\epsilon}\right)^{1/2} \frac{b}{a} \frac{\lambda_g}{\lambda} \tag{8}$$

is the TE_{10}-mode guide impedance, L_s is the post inductance due to the evanescent TE_{m0} modes ($m > 1$), C is the gap capacitance due to the evanescent TE_{mn} and TM_{mn} modes, $n > 1$, and C_1 and L_1 are the capacitance and inductance due to the TE_{m1} and TM_{m1} modes. This equivalent circuit characterizes the mount in the normal operating range of the waveguide for which $f_c < f < 2f_c$, where f_c is the cutoff frequency for the TE_{10} mode.

The gap impedance, Z_{gap} in Fig. 3(b) is strongly affected by elements C_1 and L_1 which are series resonant at the frequency f_1 for which the waveguide height $b = \lambda/2$. For full-height waveguide $b \simeq a/2$, and the resonance f_1 occurs close to $2f_c$. Over most of the useful waveguide band L_1 and C_1 cause a rapid variation of Z_{gap}, both real and imaginary parts, which is clearly undesirable for a mixer in which broad tunability must be simply achieved. By reducing the waveguide height, however, it is possible to raise the resonant frequency f_1 until L_1 and C_1 are equivalent to a small capacitance C', which is independent of frequency for $b^2 \ll \lambda^2/4$.

The element C of Fig. 3(b) is independent of frequency for $b^2 \ll \lambda^2$, and can be considered together with C' as a single capacitance C'', provided $b^2 \ll \lambda^2/4$. In the case of a mixer, the gap of Fig. 3(a) is the depletion region of the diode. C'' is then the junction capacitance C_d of the diode, and can conveniently be measured by a capacitance bridge connected to the IF port of the mixer while the diode is being contacted.

C. Mount Analysis

We now investigate the reduced-height waveguide mount of Fig. 4(a) whose equivalent circuit is shown in

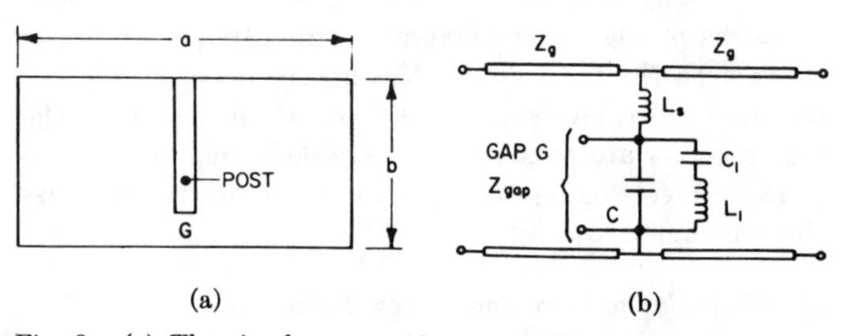

Fig. 3. (a) The simple waveguide mount. The diode is mounted across the gap G. (b) The equivalent circuit of the mount for frequencies $f_c < f < 2f_c$.

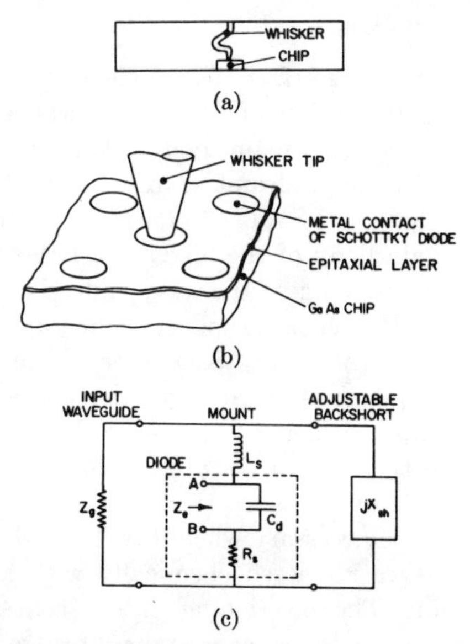

Fig. 4. (a) Reduced-height waveguide mount with a whisker-contacted diode. (b) Diode contact details. (c) Equivalent circuit of the mount and diode parasitics as seen by the junction resistance of the diode.

Fig. 4(c). The impedance Z_e is the embedding impedance seen by the junction resistance R_j of the diode. For efficient mixing, Z_e must be real and equal to some optimum source impedance. It is of interest to examine the real values of Z_e that are possible for this circuit. In particular we shall determine the values of X_{sh} (i.e., the backshort settings) for which Z_e is real, and the effect of L_s, C_d, R_s, Z_g, and frequency on these real values.

The equivalent circuit of Fig. 4(c) was analyzed by computer to determine the values of X_{sh} for which Z_e is real. Fig. 5 shows the real values of Z_e and corresponding values of X_{sh} as functions of $\omega L_s/Z_g$. The main curves are for $R_s = 0$, and typical points are indicated for $R_s = 0.05Z_g$. It is seen that there are, in general, two values of X_{sh} for which Z_e is real, and those real values may differ by a factor of 10 or more.

IV. MOUNT DESIGN FOR 80–120 GHz

A. Diodes

The Schottky-barrier diodes used in this work [15], [16] were formed by electroplating a platinum anode, followed by gold, on epitaxial gallium arsenide. Typical characteristics are shown in Table I. The parameters η and R_s are defined by the diode equations

$$i = i_0 \left\{ \exp\left(\frac{qv'}{\eta kT}\right) - 1 \right\} \tag{9a}$$

$$v = v' + iR_s. \tag{9b}$$

The diodes were supplied by Dr. R. J. Mattauch of the University of Virginia.

B. Electrical Design

The first step in the mount design is to use the loss curves of Fig. 2 to determine the optimum value of R_d,

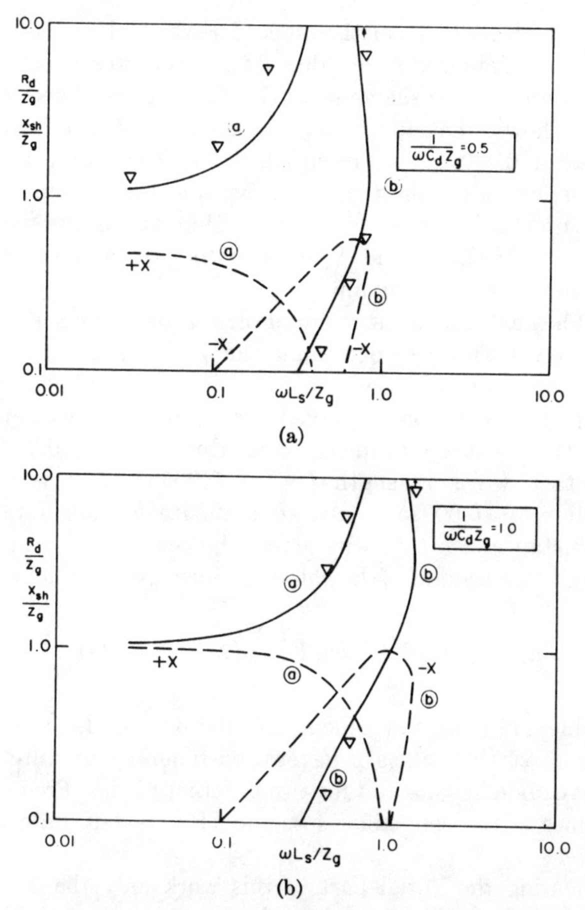

Fig. 5. Mount-matching curves showing values of diode impedance R_d which can be matched (solid curves), and the corresponding backshort reactance X_{sh} (broken curves), both as functions of whisker reactance ωL_s. Normalized reactance of diode capacitance, $1/(\omega C_d Z_g)$ = (a) 0.5, (b) 1.0. Diode series resistance is assumed zero; points (∇) are for $R_s = 0.05\, Z_g$.

TABLE I
CHARACTERISTICS OF THE GALLIUM ARSENIDE SCHOTTKY-BARRIER DIODES AT ROOM TEMPERATURE

EPITAXIAL LAYER	Doping	3×10^{17} cm^{-3}	
	Thickness	0.5 ± 0.25 µ	
SUBSTRATE	Orientation	(1 0 0)	
	Type	n	
	Doping	$2\text{-}3 \times 10^{8}$ cm^{-3}	
DIODE DIAMETER		2.5 µ	3.5 µ
MEASURED PARAMETERS			
		1.11	1.10
R_s (measured at DC)		8.0 Ω	3.6 Ω
C_d (at 0.0V, 1 MHz)		0.007 pF	0.012 pF
V_b (at −0.1 µA)		−8 V	−8 V
CALCULATED PARAMETERS			
C_d at V_{bias}		~0.011 pF	~0.020 pF
$\frac{1}{\omega C_d}$ at 100 GHz		145 Ω	80 Ω
R_s at 100 GHz[a]		10 Ω	6 Ω
f_c at V_{bias} and 100 GHz		1450 GHz	1330 GHz
$\frac{f_c}{f_{sig}} = \frac{\omega_c}{\omega}$		14.5	13.3
FROM FIG. 2			
Optimum $\omega C_d R_d$		1 - 2	1 - 2
R_d for minimum δ		145 - 290 Ω	80 - 160 Ω
δ		0.7 - 1.1 dB	0.8 - 1.2 dB

[a] The values of R_s at 100 GHz include contributions from skin effect in the whisker and diode substrate material.

the RF impedance of the diode, for which the power loss in R_s is minimized. The value of C_d used in this calculation is assumed to be the value at the bias voltage. Experience has shown that for gallium arsenide diodes a forward bias of 0.4–0.7 V is required. Table I gives the values of R_d and δ for the two diode types available. Since the IF impedance is known only within the limits set in Section II-C, R_d and δ can only be determined to lie within corresponding limits.

The next step in the mount design is to use the matching curves of Fig. 5 to determine the value(s) of diode impedance R_d which can be matched in the mount shown in Fig. 4. Dimensions assumed are as follows: waveguide width[2] $a = 0.100$ in, diode chip thickness $t = 0.006$ in, contact whisker length $l = b - 0.006$ in, and whisker radius $r = 0.00025$ in. An approximate formula for the inductance of a thin wire across the center of a reduced-height waveguide of height b is given by Sharpless [1]

$$L_s = 2 \times 10^{-7}\, l \log_e \left(\frac{2a}{\pi r}\right) \text{ (MKSA units)}. \quad (10)$$

Table II gives the salient calculations in determining the matchable values of R_d for three mounts with different waveguide heights and for two different diodes. Predicted values of the conversion loss and IF impedance are also given.

During the initial part of this work only the 3.5-μm diodes were available, and for these the one-quarter-height mount provides the best match. This mount was used for all the mixers described in this paper. For the 2.5-μm diodes the impedance level of the one-quarter-height mount is somewhat lower than the optimum value; however, Fig. 2 indicates a degradation in conversion loss of less than 0.1 dB.

TABLE II
CALCULATION OF MATCHABLE R_d VALUES AND CORRESPONDING CONVERSION LOSS AND IF IMPEDANCE FOR VARIOUS WAVEGUIDE HEIGHTS AND DIODES

DIODE DIAMETER	2.5 μ			3.5 μ		
Waveguide Height as a Fraction of Full Height	1/2	1/3	1/4	1/2	1/3	1/4
Z_g at 100 GHz, eq. 8	233 Ω	156 Ω	117 Ω	233 Ω	156 Ω	117 Ω
$\frac{1}{\omega C_d Z_g}$ using Table I	0.6	0.9	1.2	0.3	0.5	0.7
$\frac{\omega L_s}{Z_g}$ using eq. 10	1.4	1.2	0.9	1.4	1.2	0.9
$\frac{R_d}{Z_g}$ from Fig. 5	no match	1.5	0.9	no match	no match	1.0 (or~10.0)
R_d	---	230 Ω	110 Ω	---	---	117 Ω
Loss δ dB when diode is matched -- from Fig. 2	---	0.7 - 1.0 dB	0.7 - 1.3 dB	---	---	0.7 - 1.1 dB
$L_{SSB} = 3 + \delta$ dB	---	3.7 - 4.0 dB	3.7 - 4.3 dB	---	---	3.7 - 4.1 dB
Expected IF impedance -- from Section II-C	---	115 - 560 Ω	55 - 220 Ω	---	---	58 - 234 Ω

[2] The choice of $a = 0.100$ in allows the possibility of TE_{20}-mode propagation above 118 GHz. For a centrally mounted diode, however, there is no asymmetry to excite this mode. Our measurements have indicated no higher mode problems.

C. Mechanical Design

Room-Temperature Mixer: The room-temperature mixer, shown in Fig. 6, consists of two main parts, a waveguide transformer and the main body. The transformer is electroformed copper, shrunk into a brass block, and is designed to have a VSWR < 1.06 from 80 to 120 GHz [17]. The main body of the mixer is a brass block, split across the narrow walls of the waveguide. The upper part contains the RF choke supported in Stycast 36-DD dielectric,[3] and the lower part accepts an accurately machined copper post supporting the contact whisker. The diode chip is soldered in place on the end of the RF choke before the two halves of the block are finally assembled. The aluminum insert shown around the choke in Fig. 6 became necessary when it was found that during curing the Stycast reacted chemically with any copper-bearing metal. The positioning of the contact whisker was monitored with a capacitance bridge connected between the diode and the body of the mixer. This ensured that excessive capacitance was not introduced in parallel with the diode due to deformation of the whisker tip after contacting the diode. The whisker position was controlled to a fraction of a micron by a differential micrometer. The backshort is of the contacting finger type, milled from a single piece of beryllium–copper shim stock. Contact between the SMA connector and the RF choke is made by a small bellows spring.

The RF choke was designed to give low loss over 80–120 GHz while having low capacitance as seen at the IF. It consists of four coaxial sections of, alternately, 12- and 70-Ω characteristic impedance, inside an outer conductor of 0.027-in diameter. The cutoff frequency of the TE_{11} mode on the high impedance sections of the choke is ~170 GHz. Calculation of the choke impedance Z_c as seen from inside the waveguide gives $\text{Re}[Z_c] < 0.2\ \Omega$ and $\text{Im}[Z_c] < 5\ \Omega$ in the frequency range 80–120 GHz.

Cryogenic Mixer: The room-temperature mixer described in the preceding was found to be unstable when cooled because of movement between the diode and contact whisker. This was caused by differential contraction of the Stycast dielectric with respect to the metal body of the mount. To eliminate differential contraction poses a

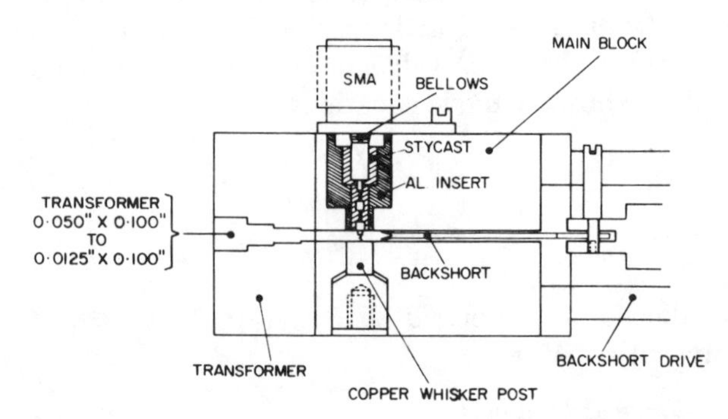

Fig. 6. Cross section of the room-temperature mixer.

[3] Emerson Cuming Company. $\epsilon_r = 1.7$.

difficult materials problem, but its effect can be controlled by using the quartz diode package shown in Fig. 7(b). Differential contraction between the contact whisker and the quartz is small enough to be taken up by the spring of the wisker. Fused quartz was chosen as the structural material because it has high mechanical strength and rigidity, relatively low dielectric constant and loss tangent, is easily cut by scribing and breaking, and is easily metallized with gold over a thin chromium adhesion layer.

It was desired to keep the electrical properties of the mount as close as possible to those of the room-temperature design, and for this reason the mount configuration shown in Fig. 7(a) was used. The main electrical difference between this and the room-temperature design is the quartz member across the waveguide adjacent to the diode. The additional shunt susceptance of this member can be tuned out by adjustment of the backshort.

The quartz diode mount is constructed from three strips of 0.006 × 0.015-in quartz as shown in Fig. 7(b). Two strips are metallized with the RF choke pattern, and the longer unmetallized third strip forms the mechanical support between the choke strips. On one choke strip two 0.001-in gold brackets are ultrasonically bonded, one to contact the IF connector, the other to support the diode which is soldered to it. The contact whisker is soldered to one end of the second choke strip. The three strips are assembled using Eastman 910 adhesive: first-the strip carrying the diode is glued to the long support strip, and then the strip carrying the whisker is slid into contact with the diode and glued. The positioning of the whisker point on the chip is observed through a high-power microscope and monitored with an I–V curve tracer. A differential screw is used to control the position of the whisker strip within a fraction of a micron.

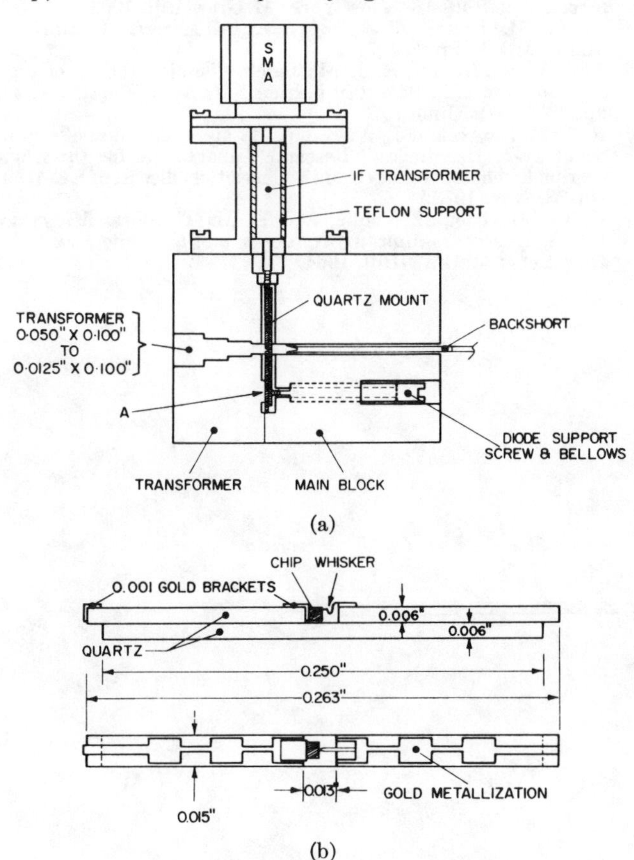

Fig. 7. The cryogenic mixer. (a) Cross section of the mixer. (b) Details of the quartz diode mount. Not to scale.

The quartz diode assembly is supported across the waveguide, as shown in Fig. 7(a), by the pressure of two springs. One spring holds the assembly against a raised part (A) of the block, ensuring a dc return path, and RF and IF grounds. The second spring, on the end of the IF transformer, contacts the gold bracket at the end of the quartz structure. The diode structure is then free to expand relative to the brass housing.

V. PERFORMANCE

The noise and conversion loss measurements given in the following were made using the IF noise radiometer/reflectometer described by Weinreb and Kerr [12]. This instrument enables the mixer performance to be determined without matching the IF port, which is expedient when a large number of measurements are to be made under conditions of varying IF port impedance. Results obtained in this way have been in good agreement with measurements made by the Y-factor method with the IF port matched using an appropriate transformer.

Typical performance figures for the room-temperature mixers are shown in Table III. The considerable superiority of the smaller diode is believed to be due to its smaller capacitance, enabling it to behave more nearly as an ideal switching mixer.

Table IV gives typical figures for the cooled mixers. These mixers were all constructed with 2.5-μm diodes. The cooled measurements made at 77 K were found to be close to those at 18 K; laboratory measurements were therefore generally made at 77 K for convenience. The mixers had 0.2–0.5-dB greater conversion loss when operating at 4.75-GHz IF than at 1.4-GHz IF. This was probably due to the following: 1) higher IF transformer losses at 4.75 GHz, and 2) the wider spacing (9.5 GHz)

TABLE III
MEASURED CHARACTERISTICS OF THE ROOM-TEMPERATURE MIXERS (f_{IF} = 1.4 GHz)

LO Frequency	85 GHz		115 GHz	
Diode	2.5 μ	3.5 μ	2.5 μ	3.5 μ
L_{SSB}	4.6 dB	6.2 dB	5.5 dB	6.7 dB
$T_{M_{SSB}}$	420°K	700°K	500°K	1400°K
Bias	0.4 v	0.6 v	0.4 v	0.4 v
	2.0 mA	4.0 mA	2.0 mA	4.0 mA

TABLE IV
MEASURED CHARACTERISTICS OF THE CRYOGENIC MIXERS (f_{LO} = 115 GHz)

TEMP.	IF FREQ.	L_{SSB}	$T_{M_{SSB}}$
298°K	1.4 GHz	5.4 dB	740°K
77°K[a]	4.75 GHz	5.8 dB	300°K

[a] Similar results were obtained at 18 K.

between the signal and image bands resulting in a poorer RF match.

The measured IF impedance levels all lie within the limits predicted in Section II-C.

VI. CONCLUSION

An approach to mixer design has been presented for cases where the diode is driven hard by the LO and can be approximated by a switch whose duty cycle depends on the basis voltage and LO level. The ideal diode is connected through a parasitic network, containing the diode's series resistance and capacitance, to the embedding network (mount). The optimum impedance of the embedding network is shown to depend primarily on the parasitic resistance and capacitance. For the particular diodes used in this work it was necessary to reduce the height of the waveguide in the mount to $\sim\frac{1}{4}$ of the standard height.

Two mixers have been described. One is for room-temperature operation, and the other, a modification of the first with a quartz diode mounting structure, is suitable for cryogenic cooling. Typical values of the SSB conversion loss and SSB mixer noise temperature [defined in (1)], measured at 115 GHz, are 5.5 dB and 500 K operating at room temperature with a 1.4-GHz IF, and 5.8 dB and 300 K when cryogenically cooled to 77 or 18 K with a 4.75-GHz IF. The difference between the measured conversion loss and the predicted value is due to nonideal switching behavior of the diode, and to dissipation of signal power converted to higher order sidebands, $nf_{\rm LO} \pm f_{\rm IF}$, $n \geq 2$, which were assumed to be reactively terminated.

The mixers described in this paper are currently in use on the National Radio Astronomy Observatory's 36-ft radio telescope at Kitt Peak, Ariz.

ACKNOWLEDGMENT

The author wishes to thank Dr. S. Weinreb of NRAO, whose support and inspiration sustained this work, and Dr. R. J. Mattauch of the University of Virginia for his patience and persistence in developing the diodes. He also wishes to thank J. E. Davis, T. J. Viola, W. Luckado, G. Green, J. Cochran, N. Horner, Jr., and J. Lichtenberger for their significant contributions to the work.

REFERENCES

[1] W. M. Sharpless, "Wafer-type millimeter wave rectifiers," *Bell Syst. Tech. J.*, vol. 35, pp. 1385–1402, Nov. 1956.
[2] M. R. Barber, "Noise figure and conversion loss of the Schottky barrier mixer diode," *IEEE Trans. Microwave Theory Tech.*, vol. MTT-15, pp. 629–635, Nov. 1967.
[3] L. E. Dickens, "Low conversion loss millimeter wave mixers," in *IEEE G-MTT Int. Microwave Symp. Proc.*, June 1973, pp. 66–68.
[4] H. M. Leedy *et al.*, "Advanced millimeter-wave mixer diodes, GaAs and silicon, and a broadband low-noise mixer," presented at the Conf. High Frequency Generation and Amplification, Cornell Univ., Ithaca, N. Y., Aug. 17–19, 1971.
[5] H. C. Torrey and C. A. Whitmer, *Crystal Rectifiers* (M.I.T. Radiation Lab. Ser., vol. 15). New York: McGraw-Hill, 1948.
[6] D. A. Fleri and L. D. Cohen, "Nonlinear analysis of the Schottky-barrier mixer diode," *IEEE Trans. Microwave Theory Tech.*, vol. MTT-21, pp. 39–43, Jan. 1973.
[7] A. R. Kerr, "A technique for determining the local oscillator waveforms in a microwave mixer," this issue, pp. 828–831.
[8] S. Egami, "Nonlinear, linear analysis and computer-aided design of resistive mixers," *IEEE Trans. Microwave Theory Tech.*, vol. MTT-22, pp. 270–275, Mar. 1974.
[9] W. K. Gwarek, "Nonlinear analysis of microwave mixers," M.S. thesis, Mass. Inst. Technol., Cambridge, Sept. 1974.
[10] R. L. Eisenhart and P. J. Khan, "Theoretical and experimental analysis of a waveguide mounting structure," *IEEE Trans. Microwave Theory Tech.*, vol. MTT-8, pp. 706–719, Aug. 1971.
[11] R. L. Eisenhart, "Understanding the waveguide diode mount," in *Dig. Tech. Papers, 1972 IEEE G-MTT Int. Microwave Symp.* (May 1972), pp. 154–156.
[12] S. Weinreb and A. R. Kerr, "Cryogenic cooling of mixers for millimeter and centimeter wavelengths," *IEEE J. Solid-State Circuits* (*Special Issue on Microwave Integrated Circuits*), vol. SC-8, pp. 58–63, Feb. 1973.
[13] H. A. Watson, *Microwave Semiconductor Devices and Their Circuit Applications.* New York: McGraw-Hill, 1968.
[14] A. A. M. Saleh, *Theory of Resistive Mixers.* Cambridge, Mass.: M.I.T. Press, 1971.
[15] T. J. Viola, Jr., and R. J. Mattauch, "Unified theory of high frequency noise in Schottky barriers," *J. Appl. Phys.*, vol. 44, pp. 2805–2808, June 1973.
[16] R. J. Mattauch and J. W. Kamps, "Lateral coupling effects in Schottky-barrier diodes," Research Laboratories for the Engineering Sciences, Univ. Virginia, Charlottesville, Rep. EE-4769-101-73, Nov. 1973.
[17] G. L. Matthaei, L. Young, and E. M. T. Jones, *Microwave Filters, Impedance-Matching Networks, and Coupling Structures.* New York: McGraw-Hill, 1964.

Cryogenic Millimeter-Wave Receiver Using Molecular Beam Epitaxy Diodes

RICHARD A. LINKE, MARTIN V. SCHNEIDER, FELLOW, IEEE, AND ALFRED Y. CHO

***Abstract*—A millimeter-wave cryogenic receiver has been built for the 60–90-GHz frequency band using GaAs mixer diodes prepared by molecular beam epitaxy (MBE). The diodes are mounted in a reduced-height image rejecting waveguide mixer which is followed by a cooled parametric amplifier at 4.5–5.0 GHz. At a temperature of 18 K the receiver has a total single-sideband (SSB) system temperature of 312 K at a frequency of 81 GHz. This is the lowest system temperature ever reported for a resistive mixer receiver. The low-noise operation of the mixer is seen to be a result of 1) the short-circuiting of the noise entering the image port and 2) an MBE mixer diode with a noise temperature which is consistent with the theoretical shot noise from the junction and the thermal noise from the series resistance.**

I. Introduction

A CRYOGENICALLY cooled receiver has been designed and built in the 60–90-GHz (WR-12) waveguide band for the 7-m offset Cassegrainian antenna at Bell Laboratories, Crawford Hill, NJ. The receiver is used for studies of spectral-line radiation from molecules in the interstellar medium. The mixer diode is fabricated from a slice of epitaxial gallium arsenide which is prepared by molecular beam epitaxy (MBE). The doping concentration in the epitaxial layer is specifically designed for low-temperature operation [1]. The diode is incorporated into a reduced-height waveguide block mount using a coaxial RF choke and a noncontacting backshort. The intermediate frequency (IF) is amplified by a cryogenic parametric amplifier at 4.5–5.0 GHz which is followed by room-temperature transistor amplifiers. The mixer and the IF amplifier are both cooled to 18 K by a helium closed-cycle refrigeration system. Signal and local oscillator injection is provided by a quasi-optical injection system described in [2]. The injection system provides 18 dB of image rejection. The total single-sideband (SSB) receiver noise temperature including the feed system is below 480 K from 62 to 92 GHz and has a minimum of 312 K at a signal frequency of 81 GHz.

II. The Mixer

Fig. 1 is a schematic cross-sectional view of the mixer block, and Fig. 2 is a photograph of the open mixer block with its components. The transition from a full-height waveguide (0.122 in × 0.061 in) to a one-fifth height waveguide (0.122 in × 0.012 in) is made by a linear taper with a length of 0.980 in which was measured to match 97 percent of the incident power to the reduced-height waveguide. This low waveguide height was chosen to reduce the total length and, therefore, the inductance of the contact whisker/post combination. The contact wire is a phosphor-bronze wire with a diameter of 12 μm. The wire is mounted on a nickel post which is pressed into the waveguide block.

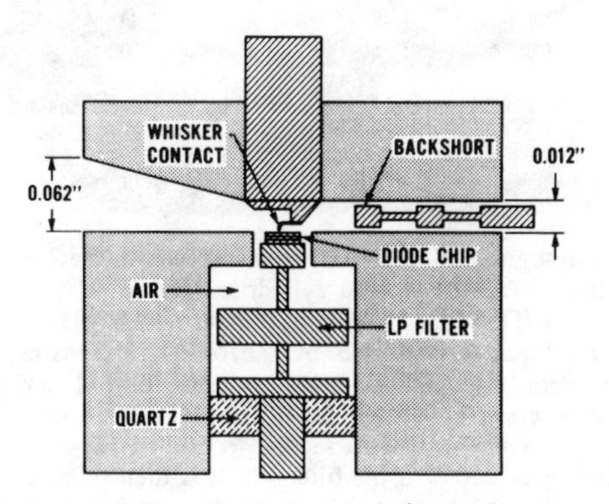

Fig. 1. Cross-sectional view of mixer block with a Schottky-barrier diode mounted in a reduced-height waveguide section.

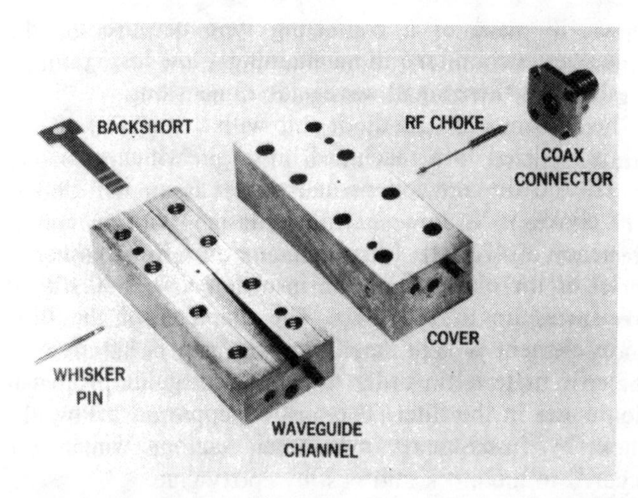

Fig. 2. Photograph of a mixer block with associated components. The waveguide channel with a linear taper is milled into the block shown in the lower left part of the figure.

Tuning is accomplished by means of a phosphor-bronze noncontacting waveguide backshort which consists of alternating low- and high-impedance sections (each 1/4 wavelength long), and it is insulated from the waveguide by a 19-μm thick mylar tape. A noncontacting short was

Manuscript received June 7, 1978; revised July 10, 1978.
R. A. Linke and M. V. Schneider are with the Bell Laboratories, Crawford Hill Laboratory, Holmdel, NJ 07733.
A. Y. Cho is with the Bell Laboratories, Murray Hill, NJ 07974.

Reprinted from *IEEE Trans. Microwave Theory Tech.*, vol. MTT-26, pp. 935–938, Dec. 1978.

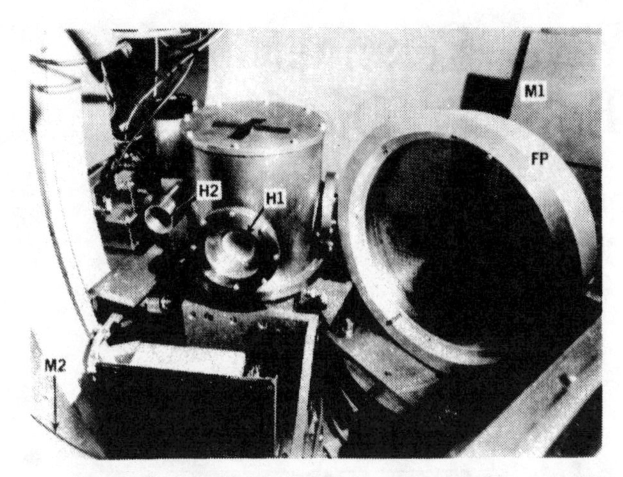

Fig. 3. Photograph of millimeter-wave receiver in the 7-m antenna cab. The mixer is located in the cylindrical Dewar in the center of the photograph. The signal path is as follows—the signal enters from the lower right and is reflected by mirror M1. It passes through the Fabrey–Perot filter FP and is then reflected by M2 toward the dewar window where it is focused on the receiver feed horn, H1. The LO power is transmitted from a second feed horn H2 toward M2 and is reflected by M2 toward the filter FP. It is then reflected by the filter and joins the signal on its path to the receiver feed horn H1.

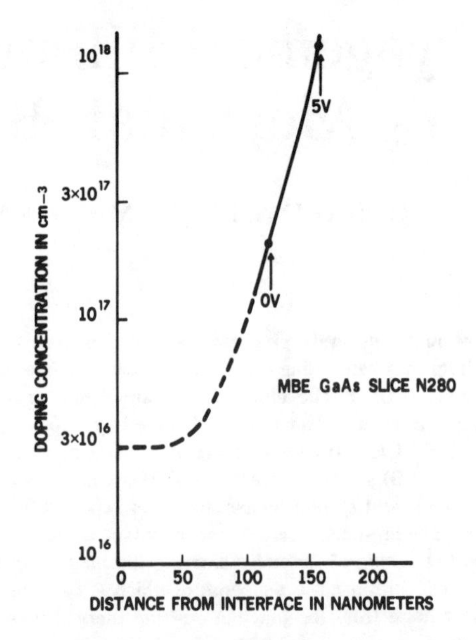

Fig. 4. Doping profile of an epitaxial GaAs slice grown by MBE as a function of depth from the surface. The dashed portion of the curve is inferred from measurements on a thicker sample.

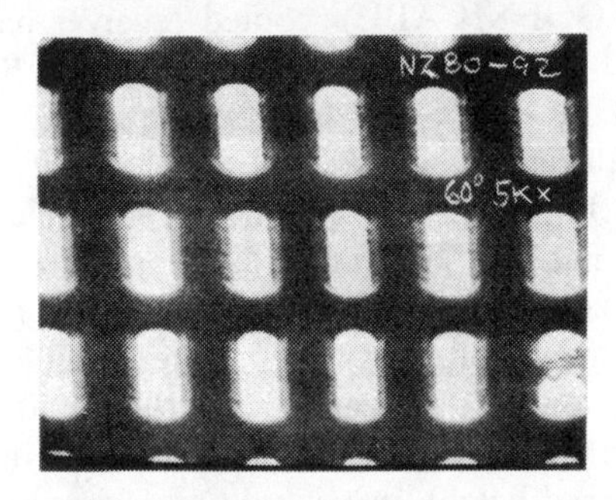

Fig. 5. Electron micrograph of bathtub diodes fabricated on an epitaxial gallium-arsenide slice. (MBE-GaAs batch N280-92. Junction size 1.8×6.2 μm.)

chosen in place of a contacting type because of the difficulties encountered in maintaining a low-loss contacting short for these small waveguide dimensions.

The 250 μm×250 μm diode chip with a thickness of 100 μm is soldered to a machined brass pin which provides the IF and dc connections and serves as an RF choke. The choke is a low-pass filter design with a cutoff frequency of 47 GHz. Measurements on a low-frequency model of the choke imply an insurtion loss >40 dB at frequencies up to 180 GHz. The diameter of the first choke element is kept small, and the gap is left free of dielectric material in order to avoid waveguide propagation modes in the filter. The post is supported below the choke by fused-quartz cylindrical sections which are secured to the block and post by epoxy glue.

Fig. 3 is a photograph of the receiver dewar and the quasi-optical components mounted in the 7-m antenna cab.

III. Mixer Diode

The nonlinear device which is used for the mixing process is a moat-etched elliptical junction (bathtub diode) prepared on a heavily doped GaAs substrate by MBE [3]. The doping profile achieved for the deposited epitaxial gallium arsenide on a heavily doped n-type substrate with a carrier concentration of 3×10^{18} cm^{-3} is shown in Fig. 4. The surface doping density is quite low (3×10^{16} cm^{-3}) in order to minimize conduction by electron tunneling and thereby to minimize shot noise [4], [5]. In spite of this low doping, the series resistance is kept low by virtue of the extremely small thickness of the low-doped region (increasing to 10^{17} cm^{-3} at ~1000 Å).

The growth process and the fabrication technique for preparing an array of these junctions are described in [1]. The junctions are fabricated by means of contact photolithography using a pattern on a chromium mask which is generated with an electron beam exposure system [6], [7]. An electron micrograph of the junctions is shown in Fig. 5. Each individual junction has a series resistance of 4 Ω and a zero bias capacitance of 14.7 fF at room temperature. The breakdown voltage of each junction is 7.6 V at a current of 10 μA, and the idealty factor is 1.08. The characteristics of the diode used in this mixer are given in Table I for ambient temperatures of 295, 88, and 18 K. In this table, R_s is the series resistance, n is the idealty factor, and V_0 is defined by

TABLE I
I Versus *V*: Characteristics of Mixer Diode

T_{amb}	R_s (Ω)	V_0 (mV)	n
295	4.5	28.2	1.11
77	6.8	11.3	1.70
18	7.4	11.9	7.67

$$I(V)=I_s(e^{V/V_0}-1)$$

where I_s is the saturation current.

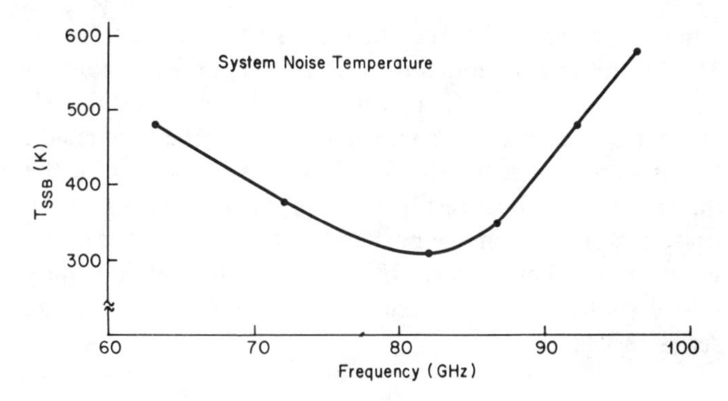

Fig. 6. Total SSB system noise temperature of cooled receiver as a function of frequency. The minimum system temperature of 312 K is obtained at a frequency of 81 GHz.

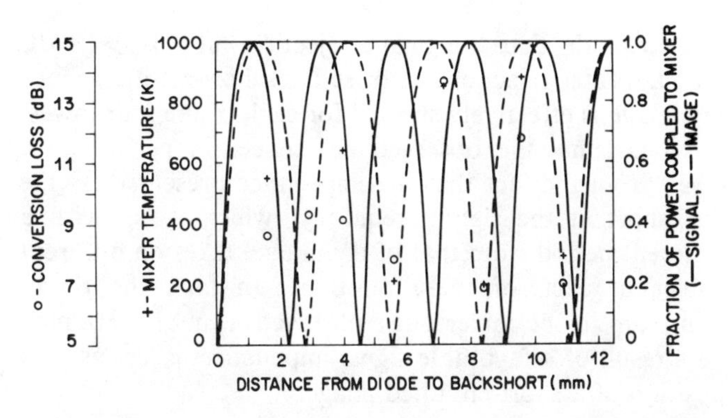

Fig. 7. The mixer temperature minima (filled circles) and the associated conversion loss (open circles) are plotted versus the backshort position. The curves give the efficiency of the coupling power to the diode at the signal frequency (solid) and image frequency (dashed) for the simple case of a diode shunted by a shorted waveguide section. Note that the lowest noise operation occurs for the positions of the backshort which give a short-circuit at the image frequency.

IV. Receiver Performance

A cold (18 K) reference termination on the refrigerator cold station and an external noise diode coupled to the paramp through a 30-dB directional coupler are used to determine the IF-system noise temperature which is found to be 22±3 K. Hot (ambient temperature) and cold (liquid-nitrogen temperature) millimeter-wavelength reference signals are coupled into the mixer by the quasi-optical injection system through a 0.100-thick (one wavelength at 75 GHz) rexolite window in the Dewar. The window has a <0.1-dB loss at 75 GHz and a 0.8-dB reflection loss at 60 and 90 GHz. The signal is converted to waveguide propagation by means of a corrugated scalar feed horn which is cooled to 18 K along with the mixer and IF amplifier. The mixer's IF port was measured to have a VSWR of 2.5 at room temperature. This results in a reflection loss of ~0.85 dB. The data reported here have not been corrected for this loss.

Using the calibrated IF system to measure noise from the mixer's IF port for the two different RF termination temperatures, the conversion loss, mixer temperature, and total receiver noise temperature are determined as a function of the frequency and backshort position. The receiver noise temperature is shown in Fig. 6 for signal frequencies from 63 to 98 GHz. A broad minimum in receiver temperature is seen at ~81 GHz. At this frequency, T_{rec} is 312 K, the conversion loss is 6.7 dB, and the mixer noise temperature is 209 K. Here T_{rec} is the total SSB system temperature including feed system losses and IF noise. The mixer temperature and conversion loss are related to T_{rec} by

$$T_{rec} = T_m + LT_{IF}$$

where all quantities are SSB.

Measured mixer parameters for $f_{sig} = 81.9$ GHz are given in Table II for the first eight positions of the backshort which give a minimum in system noise. It is apparent that the best operating point is the sixth noise minimum rather than the usual first position of the backshort. It is interesting to note that the mixer alternates between high- and low-noise operation with successive positions of the backshort. In Fig. 7 we have plotted the

TABLE II
Mixer Parameters Versus Backshort Position for Minima in T_{REC} ($f_{sig} = 81.9$ GHz)

Diode to Backshort Distance (mm)	L_s (dB)	L_i/L_s (dB)	T_M^{SSB} (K)	T^*	$\frac{T_m}{L_s}$ (K)
1.63	8.6	~0.5	550	41	75
2.90	9.3	>18	290	38	35
3.99	9.1	~0.5	640	43	78
5.61	7.8	>18	210	41	35
7.19	13.7	>18	860	36	37
8.38	6.9	>18	190	48	38
9.60	11.8	~3	880	52	58
10.90	7.0	>18	290	71	59

$$T^* = \frac{T_m}{L_s - 1}\left(1 - \frac{T_0}{T_m}\frac{L_s}{L_i}\right)$$

mixer noise temperature minima and the associated conversion loss as a function of the backshort position. In addition, we have plotted theoretical curves for the efficiency of coupling power into the diode at the signal and image frequencies as a function of the backshort position. These curves correspond to a simple mixer circuit model in which a real diode impedance, equal to the guide impedance, is shunted by a shorted waveguide section of varying length. It is clear from Fig. 7 that the best operation of the mixer occurs at the backshort positions which correspond to a short-circuited diode at the image frequency. Thus the improvement in mixer noise performance is a result of short-circuiting the noise entering the mixer through the image port.

In order to check this interpretation, we measured the image rejection of the mixer itself by tuning the Fabrey–Perot filter in the quasi-optical injection system to pass the image frequency and block the signal frequency. At the backshort positions corresponding to a shorted image, the DSB system temperature was found to be consistent with the known leakage through the filter at the signal frequency and no mixer sensitivity at the image frequency.

We conclude that the image rejection ratio is >18 dB (i.e., the filter rejection) at these backshort settings. Since the image is effectively shorted for each of the four lowest noise minima, the observed differences in performance result from the fact that the impedance presented by the backshort at the signal frequency (which varies with a different period from that of the image) takes on different values at each of these positions. In fact, the broad minimum in the mixer noise observed near 81 GHz may be a result of a favorable signal impedance occurring at a position of a short-circuited image.

We have attempted to determine the effective temperature of the mixer diode when it is viewed as a lossy element at some assignable temperature [8], [9]. In order to do this, we must subtract the noise contribution from the image termination which is taken to be at ambient temperature T_0 (295 K) since the feed system terminates the image in a room-temperature attenuator. The resulting effective diode temperatures, given in the column headed T^* in Table II, are seen to be nearly independent of image rejection while the value labeled T_m/L, which is the equivalent noise temperature of the mixer's IF port, is seen to double when the image rejection is poor indicating that noise from the image termination is in fact converted to the IF when the image is not shorted. We feel that the average T^* of 47 K is a reasonable measure of the mixer diode's effective temperature. It is interesting to compare this temperature with the value [8], [10], [11]

$$\frac{1}{2}nT = 69\text{ K}$$

obtained from the n and T given in Table II. When we allow for the fact that a significant portion of the conversion loss results from the diode's series resistance which is at a temperature of 18 K, we see that our measured mixer noise is consistent with a purely theoretical shot noise. That is, we see no evidence for excess noise as discussed in [9]. We attribute this to an absence of parametric effects since the Mott barrier used in this mixer shows very little change in capacitance with the bias voltage.

V. Conclusions

We have designed and built a cryogenic millimeter-wave resistive mixer receiver which operates with a system noise temperature (SSB) below 480 K from 62 to 92 GHz. The system is observed to have a broad minimum in noise temperature near 81 GHz where the system temperature is 312 K, the mixer temperature is 209 K, and the conversion loss is 6.7 dB. If a correction is made for the IF mismatch measured at room temperature, the conversion loss is found to be 5.9 dB. The low-noise operation of the mixer is seen to be a result of 1) the short-circuiting of the noise entering the image port and 2) the MBE mixer diode with a noise temperature which is consistent with a theoretical shot noise in the junction and thermal noise in the series resistance.

Acknowledgment

We are indebted to A. A. M. Saleh for numerous valuable discussions on mixer theory. The authors are grateful to W. W. Snell, Jr., for the design of the coaxial RF filter and to K. Ganson for fabricating precision components. We also would like to thank C. Radice, Jr., and A. Olenginski for assistance in producing the MBE layers and mixer diodes. Precision masks to fabricate the devices were provided by J. P. Ballantyne.

References

[1] M. V. Schneider, R. A. Linke, and A. Y. Cho, "Low-noise millimeter-wave mixer diodes prepared by molecular beam epitaxy (MBE)," *Appl. Phys. Lett.*, vol. 31, pp. 219–221, Aug. 1977.

[2] P. F. Goldsmith, "A quasi-optical feed system for radioastronomical observations at millimeter wavelengths," *Bell Syst. Tech. J.*, vol. 56, pp. 1483–1501, Oct. 1977.

[3] A. Y. Cho and J. R. Arthur, *Progress in Solid-State Chemistry, Vol. 10*, G. Somerjai and J. McCaldin, Ed. New York: Pergamon, 1975, p. 157.

[4] T. J. Viloa, Jr. and R. J. Mattauch, in *J. Appl. Phys.*, vol. 44, p. 2805, 1973.

[5] F. A. Padovani and R. Stratton, in *Solid-State Electron.*, vol. 9, pp. 695–707, 1966.

[6] D. R. Herriott, R. J. Collier, D. S. Alles, and J. W. Stafford, "EBES: A practical electron lithographic system," *IEEE Trans. Electron Devices*, vol. ED-22, pp. 385–392, July 1975.

[7] R. F. W. Pease, J. P. Ballantyne, R. C. Henderson, A. M. Voshenkov, and L.D. Yau, "Applications of the electron beam exposure system," *IEEE Trans. Electron Devices*, vol. ED-22, pp. 393–399, July 1975.

[8] C. Dragone, "Analysis of thermal and shot noise in pumped resistive diodes," *Bell Syst. Tech. J.*, vol. 47, pp. 1883–1902.

[9] S. Weinreb and A. R. Kerr, "Cryogenic cooling of mixers for millimeter and centimeter wavelengths," *IEEE J. Solid-State Circuits*, vol. SC-8, pp. 58–63, Feb. 1973.

[10] A. A. M. Saleh, *Theory of Resistive Mixers*. Cambridge, MA: M.I.T. Press, 1971, p. 170.

[11] A. R. Kerr, "Shot-noise in resistive-diode mixers, and the attenuator noise model," *IEEE Trans. Microwave Theory Tech.*, to be published.

A Cryogenic Millimeter-Wave Schottky-Diode Mixer

ERIK L. KOLLBERG, MEMBER, IEEE,
AND HERBERT H. G. ZIRATH

***Abstract*—We report theoretical calculations and measurements on cryogenic millimeter-wave Schottky-diode mixers. Measurements of the embedding impedances at the signal and image frequencies have been used for the theoretical predictions of the mixer performance, and an excellent agreement with measured performance was obtained. Measurements of embedding impedances for various waveguide structures are reported, and the choice of configuration for optimum single-sideband performance is discussed.**

I. Introduction

Cooled millimeter-wave Schottky-diode mixers are frequently used in low-noise radiometer systems [1]–[4]. Although the superconducting semiconductor–insulator–semiconductor (SIS) mixer has still lower noise [5], it needs cooling to 4.2 K or below, while there is no use cooling the diode mixer below about 20 K [6]. Therefore, for most practical purposes, cooled diode mixers are at present the best choice for low-noise receivers. Below, we will report on some results obtained when developing the mixer receivers now in use at Onsala Space Observatory.

II. Mixer Design

The mixers used are designed in a straightforward way (Fig. 1). An impedance transformer at the IF-output port matches the 50-Ω IF amplifier to the diode when dc-biased to have a differential dc-resistance $r = (di/dv)^{-1}$ equal to 175 Ω. This impedance level has been found to be a good compromise yielding a low-power reflection (less than about 1–2 dB extra conversion loss) for most bias conditions and the backshort settings where low-noise operations are obtained. The diode is mounted in a reduced-height waveguide in order to facilitate impedance matching. In Fig. 2, a typical locus of the diode impedance (the complex conjugate) versus dc- and local-oscillator bias is shown. The waveguide cross-section dimensions, the length and the shape of the whisker, the low-pass filter impedance, and the position of the diode chip in the waveguide are the most important parameters that influence the locus of the impedance circle, representing the embedding impedance seen by the diode, obtained when the backshort (l_b) is moved over $\lambda_g/2$ (see Fig. 2).

In Fig. 3, an equivalent embedding circuit is shown. The inductive (X_L) and capacitive (X_{C1}, X_{C2}) circuit elements, however, cannot be represented by simple inductances and capacitances [7]. For the fundamental frequency band of the waveguide (no higher modes can propagate), the embedding impedance can be measured as described in [8] and [9]. In Figs. 4 and 5, measurements on some experimental mixers made using the method described in [8] are depicted. From these measurements, X_L, X_{C1}, and X_{C2} of Fig. 3 can be determined.

It is interesting to see how shortening the whisker not only moves the impedance circle downwards in the impedance diagram (corresponding to smaller inductance X_0 and X_L), but also makes the diameter of the impedance circle smaller (corresponding to a decrease in X_{C1}). In Fig. 5, the effect of going from an ordinary reduced-height waveguide to a ridged waveguide by making grooves is illustrated (Fig. 5(a) and (b)). The decrease in the waveguide impedance is observed as a decrease in the impedance circle diameter. By further decreasing the whisker length, the position of the circle moves, corresponding to a lower inductance X_L.

Short enough whisker wires (Figs. 4 and 5(c)) will leave impedance circles with diameters close to the impedance of the waveguide, indicating that X_{C1} is large compared to X_L. Similar observations have been made by Pospieszalski and Weinreb [9], [10].

Manuscript received May 5, 1982; revised July 28, 1982. This work was supported (in part) by the Swedish Board for Technical Development.

The authors are with the Department of Electron Physics I and Onsala Space Observatory, Chalmers University of Technology, Göteborg, Sweden.

Reprinted from *IEEE Trans. Microwave Theory Tech.*, vol. MTT-31, pp. 230–235, Feb. 1983.

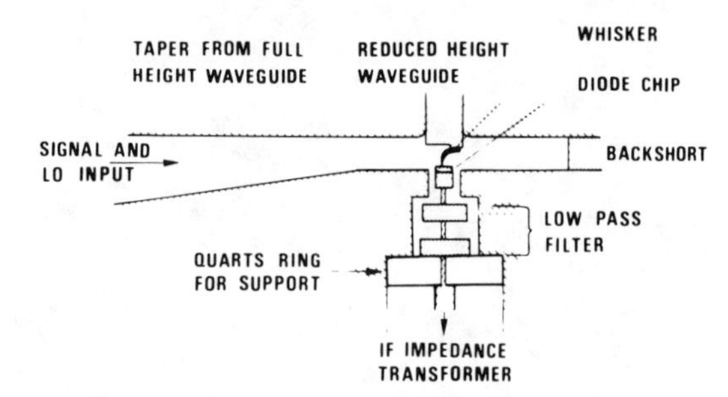

Fig. 1. Schematic diagram of the waveguide mixer mount.

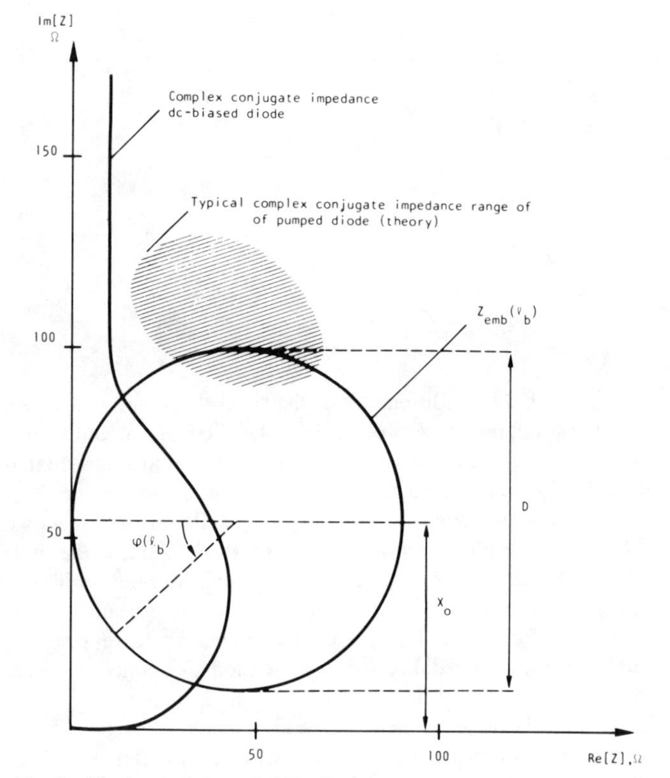

Fig. 2. The locus of the embedding impedance as seen from the diode versus backshort setting, and the locus of the complex conjugate impedance of the dc-biased diode at 100 GHz, with $\phi = 1080$ mV, $C_0 = 14.5$ fF, $kT/q = 28.5$ mV, $i_0 = 4.9 \cdot 10^{-14}$ mA, and $R_s = 10\ \Omega$. $\varphi(I_b)$ is increasing when the backshort is moved towards the diode mount. The shaded area indicates the approximate impedance of the pumped diode.

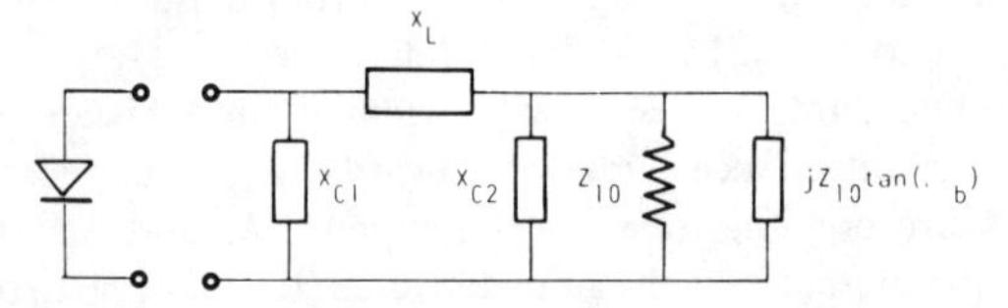

Fig. 3. The equivalent circuit of the embedding impedance for fundamental waveguide mode frequency range. The inductive reactance X_L is related to the whisker inductance. X_{C1}/X_L is important for the diameter and location of the impedance circle (Fig. 2) (see [8]). For very reduced height waveguide mount X_{C1} and X_{C2} can, to a first approximation, be assumed large.

III. Theoretical Evaluation

In order to evaluate a mixer properly, one has to know the diode impedance and noise properties, and the embedding circuit impedances at all harmonics of the local oscillator frequency f_{LO}, as well as at all harmonic sidebands ($nf_{LO} \pm f_{IF}$). In the theory by Held and Kerr, the diode is modeled in the usual way, i.e., an ideal exponential diode is in parallel with a variable capacitance and in series with a current independent series resistance. The

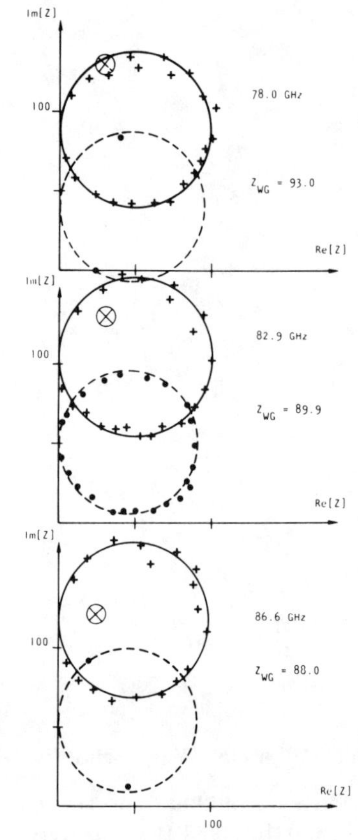

Fig. 4. Impedance loci for the 70–95-GHz mixer. The full line circles are for the longest whisker, while the broken line circles are for the shortest whisker.

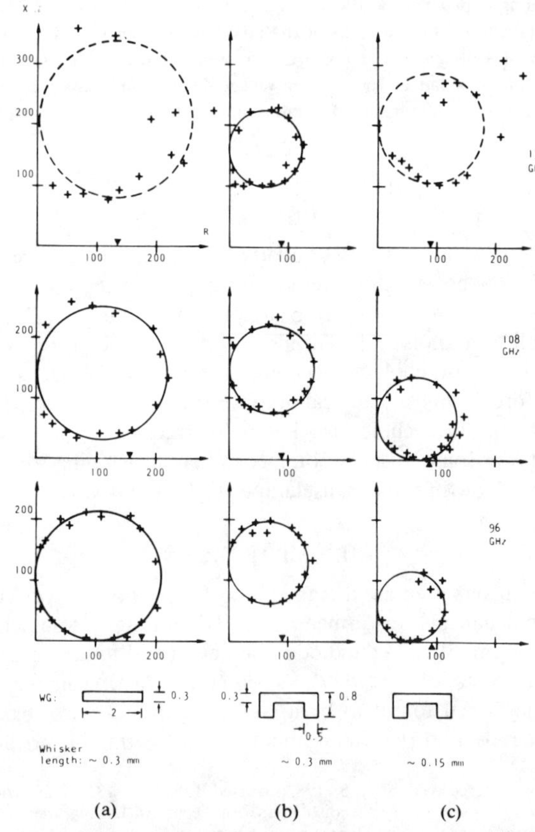

Fig. 5. Impedance loci for a successively modified 90–140-GHz mixer mount. Each column is for one mount and three different frequencies (96, 108, and 137 GHz). The cross section and approximate whisker length are given below each column.

diode is assumed to give shot noise, and the series resistance pure thermal noise. In the theory, the properties of the embedding network are also accounted for. However, modeling the diode at high forward currents is not straightforward since the exponential voltage dependance of the current is not constant due to microcluster effects at the metal–semiconductor interface [13], and a current dependent excess noise is generated [11]–[13]. Examples are given in Figs. 6 and 7.

The embedding impedances at the signal and the image frequency are the most important ones, and have been measured for the mixers described in this paper. However, the harmonic sideband impedances cannot be determined for the actual mounts unless a scaling experiment, as described by Held and Kerr [11], is used. In the computer runs below, we have estimated the harmonic impedances, assuming X_L to scale approximately linearly with frequency, and X_{C1}, X_{C2}, and $\tan(\beta l) Z_{wg} \approx \infty$. This is to some extent justified by the results by Held and Kerr [11], and by the fact that a reasonable agreement is obtained between theory and experiments. It was also found that in order to describe theoretically the main features of the measured properties, taking into account two harmonics is usually sufficient.

The Pt–GaAs diodes used in the experiments described here were made from molecular beam epitaxy GaAs and supplied by Dr. A. Y. Cho and Dr. M. Schneider at the Bell Telephone Laboratories, Crawford Hill, NJ. The diode properties necessary for the theoretical evaluation were investigated [13], and a shot-noise temperature of $nT_0/2 = 40$ K at low currents (Fig. 6) and a zero bias capacitance of 14.5 fF was found. The Mott character of the diode capacitance versus bias-voltage was accounted for, although using an ordinary Schottky-diode capacitance $C_0 \cdot (1 - v/\phi)^{-0.45}$ or the Mott character of the diode did not give a considerably different result, as long as the capacitance versus voltage was correct for large forward bias (where it has the Schottky character anyway). The potential ϕ was assumed equal to 1080 mV at 20 K [13].

Since the $I-V$ characteristic has a somewhat peculiar shape (Fig. 6), one has to be careful when defining the series resistance R_s from the equation

$$i = i_0 \left[\exp\left(\frac{v - iR_s}{v_0} \right) - 1 \right] \tag{1}$$

where $v_0 = \eta T \cdot k/e$. We have obtained R_s as shown in Fig. 6, and used the experimental $I-V$ characteristic in the computations. The dc series resistance is typically 8–10 Ω, which increases to 12–15 Ω at 100 GHz due to the skin effect. The noise temperature increase at high forward currents (Fig. 7) has been suggested to be caused by hot electron noise in the epilayer of the diode [11], [12]. Due to the difficulties in properly taking this extra noise contribution into account theoretically, we have omitted this in the calculations reported below. It should be pointed out that the computer calculations show that 20-K thermal noise from the series resistance contributes to the mixer noise with between 10 and 20 K typically (see Fig. 12), and that the computed mixer noise is normally a few tens of degrees lower than the measured mixer noise.

IV. Measurements of Conversion Loss and Noise

The IF amplifier of the cooled (18-K) mixer receiver is a 3.7–4.2-GHz (18-K) parametric amplifier followed by a cooled (60-K) FET amplifier. The overall IF-noise temperature at the diode is 24 K. An impedance transformer between the low-pass filter and the 50-Ω input port of the parametric amplifier makes the diode see 175 Ω at 4 GHz. The noise temperature of the

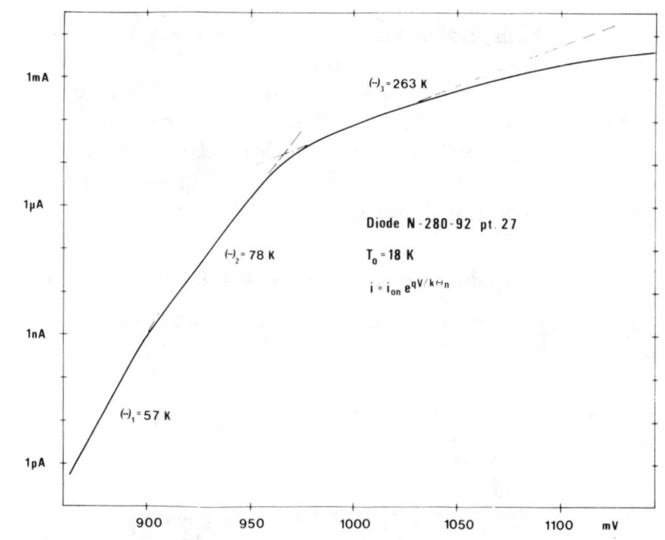

Fig. 6. Typical $\log(I)-V$ characteristic at 20 K for the N280–92 diode.

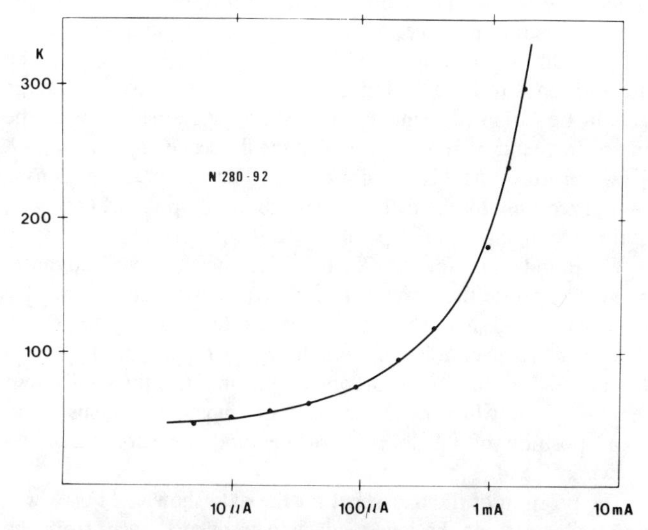

Fig. 7. Typical effective diode noise temperature versus current measured at 4 GHz for the dc-biased N280–92 diode at 20 K.

system was measured using a room-temperature and a liquid nitrogen cooled load in front of the window of the receiver dewar. The noise temperature was measured over the 3.7–4.2-GHz IF band using a narrow-band (1-MHz) precision receiver.

The noise temperature versus IF frequency was found to vary typically with about 20 percent. In the experiments discussed below, we quote a mean over the IF band.

The single-sideband response was measured using another mixer mount used to multiply the output of a 8–12-GHz sweeper to the actual frequency range. We found that the (noncontacting) backshort could suppress one sideband better than −20 dB. The single-sideband noise temperature versus backshort position was measured as follows.

a) Set the backshort to cancel one sideband (e.g., the lower sideband) and measure the noise temperature of the system, which is then considered to be the single-sideband noise temperature $T_{SSB,0}$ (upper sideband).

b) Inject the coherent signal and measure the ratio of the coherent signal power to the noise power. Hence, the coherent signal power is *calibrated in degrees Kelvin*.

In this way, a "narrow-band noise source" is obtained which

may be used for evaluating the single-sideband system noise temperature. In order to evaluate the conversion loss, we must know the IF-amplifier noise (T_{IF}), which consists not only of the noise from the IF-amplifier itself, but also from noise from the IF-amplifier input circulator (T_0), which is reflected back from the mixer ($|\Gamma_{mix}|^2 \cdot T_0$). More details of the full measuring procedure will be published elsewhere.

V. Measurements on the 60–90-GHz Mixer

The system noise temperature of a single-sideband mixer receiver can be written as

$$T_{syst} = T_A\left(1 + \frac{L_s}{L_i} + L_s \cdot \sum_{n=2}^{\infty}\left(\frac{1}{L_{n+}} + \frac{1}{L_{n-}}\right)\right) + T_{MXR} + L_s \cdot T_{IF} \tag{2}$$

where T_A is the antenna temperature, here assumed to be the same at the image frequency with conversion loss L_i and at the upper and lower harmonic sidebands, with conversion loss L_{n+} and L_{n-}, respectively. L_s is the signal conversion loss, T_{MXR} the inherent single-sideband noise of the mixer itself, and T_{IF} the noise temperature of the intermediate frequency amplifier. From (2), it can be seen that in practical applications it may be advantageous to have L_i and $L_{n\pm}$ large compared to L_s. A large L_i can be obtained using the backshort positioned so that the image frequency is short-circuited at the diode, i.e., $l_{bs} = n \cdot \lambda_{gi}/2$. However, this puts a constraint on T_{MXR} and L_s that can then be minimized only by the dc bias, the amount of applied LO power, and by changing l_{bs} in steps of $\lambda_{gi}/2$.

We found that, for the 70–100-GHz mixer, it was advantageous to operate the mixer over the frequency range 75–95 GHz having the backshort short-circuiting the lower sideband. Fig. 8 depicts the receiver noise temperature $T_{rec} = T_{MXR} + L_s \cdot T_{IF}$ versus frequency for this mode of operation, and for three different lengths of the whisker. The embedding impedance versus backshort position of the longest whisker and the shortest one is shown in Fig. 4.

The position of the backshort for the data shown in Fig. 8 was one wavelength at the lower sideband frequency away from the diode, i.e., the lower sideband was short-circuited.

The theoretical curves of Fig. 8 were obtained assuming the harmonic impedances equal for the three cases and using the nonlinear analysis made for 85 GHz for the whole frequency range. The result clearly demonstrates two effects. First, increasing the length of the whisker, i.e., in the first approximation increasing the whisker inductance, tunes the mixer to lower frequencies. It is also seen that, at least for this particular mode (short-circuited lower sideband), the shortest whisker yields the better performance over the examined frequency range (the corrugated feed horn used was cut off at 70 GHz).

Fig. 9 shows the system noise temperature for the system as used in the 20-m dish at the Onsala Space Observatory. The short whisker is used, but the backshort position is chosen for lowest system noise, i.e., either $\lambda_{gi}/2$, λ_{gi}, or $3\lambda_{gi}/2$. The full lines demonstrate the theoretical results, which are in very good agreement with the measured performance.

VI. Measurements on the 90–140-GHz Mixer

The advantages were less pronounced operating this mixer with a short-circuited image frequency. The diodes used had a capacitance a factor of 1.3 lower than that for the 70–95-GHz mixer,

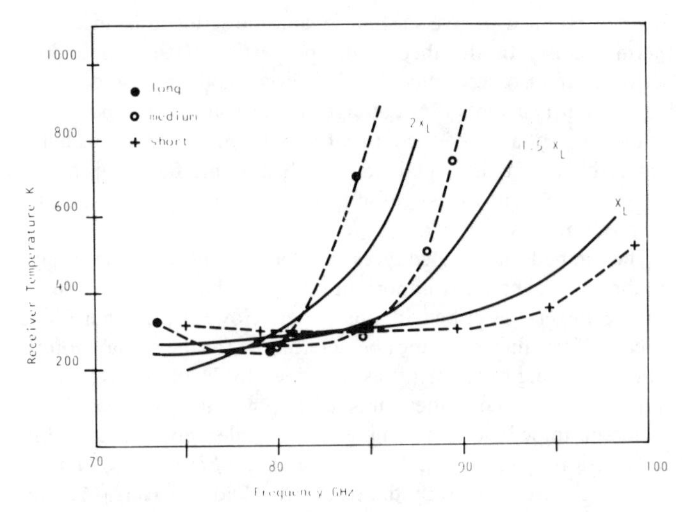

Fig. 8. Receiver noise temperature versus frequency for the 70–95-GHz mixer for three different lengths of the whisker. X_L and $2X_L$ correspond to the short and long whisker, respectively, discussed in Fig. 4. The broken line indicates the experimental result, and the full line the theoretical result. Only the two first harmonic impedances were considered.

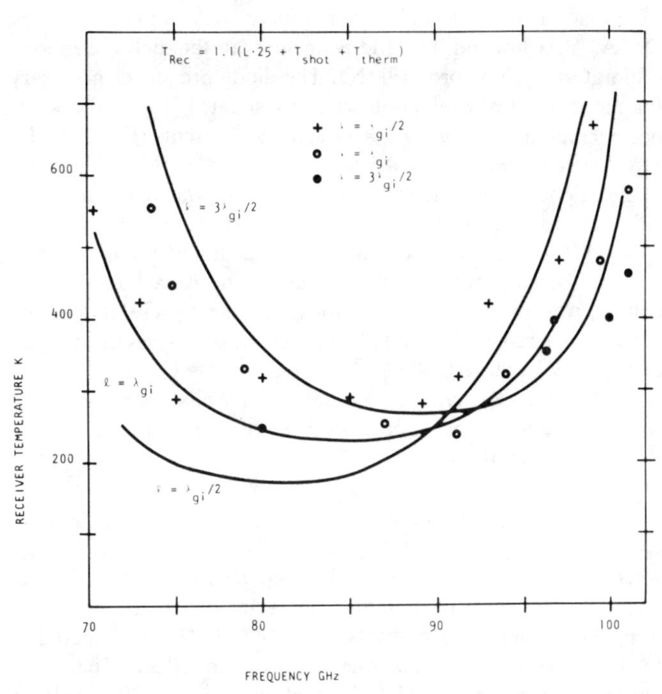

Fig. 9. Receiver noise temperature of the 70–95-GHz Onsala system. The whisker is "short" and the full lines indicate the theoretically derived noise temperature $T_{Rec} = 1.1(L_s \cdot 24 + T_{shot} + T_{therm})$. Only the two first harmonic and harmonic sidebands are accounted for theoretically.

which should roughly compensate for the higher frequency. The waveguide height (0.2 mm) and the waveguide width (2.0 mm) were decreased with a factor 1.5, and the whisker was made very short. The resulting embedding impedance circles then very much corresponded to those for the short whisker in Fig. 4.

The input reactance of the low-pass filter at the signal frequency will affect the embedding impedance, and should, in the first place, affect the "whisker inductance" reactance X_L. We changed the filter in order to increase the theoretical capacitive reactance of the filter from $-j10\ \Omega$ to $-j30\ \Omega$ at 106 GHz. A decrease in X_L of $15 \pm 5\ \Omega$ was then measured, and a somewhat better performance at higher frequencies was observed.

Although several design parameters of the 60–90-GHz mixer

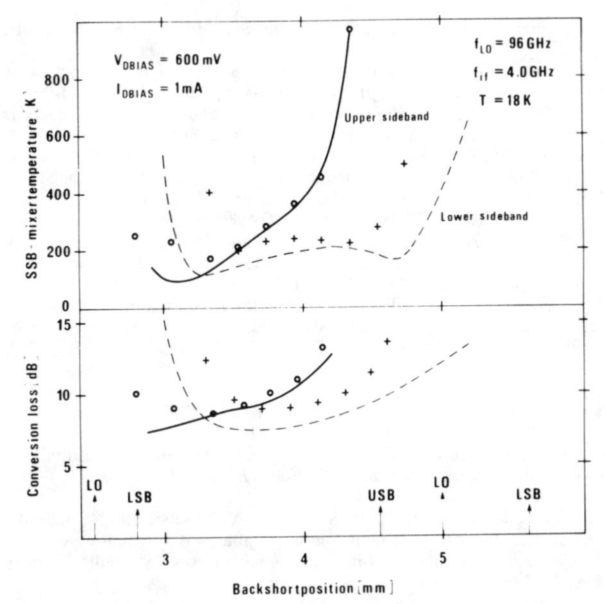

Fig. 10. An example on noise temperature and conversion loss for the 90–140-GHz mixer cooled to 20 K. The theoretical conversion loss has been adjusted 2 dB to account for circuit losses.

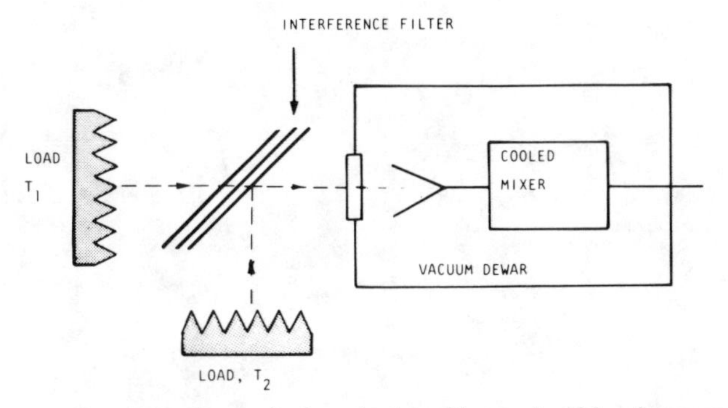

Fig. 11. Experimental setup for investigation of harmonic sideband response (at 200 GHz).

could be scaled to the 90–140-GHz frequency band (waveguide dimensions, whisker length) it was necessary, in this case, to stay with the 4-GHz IF frequency, and to introduce the smooth step in the waveguide (Fig. 1) in order to have a larger waveguide height of the backshort section, which facilitated the manufacturing of the noncontacting short. This step causes the backshort reactance in the diode plane not to vary like $jZ_0 \tan(\beta l)$, as in the 75–95-GHz mixer where no step was necessary.

Fig. 10 shows conversion-loss and mixer noise temperature versus back-short position at 96 GHz obtained experimentally as well as theoretically. There is a discrepancy for the conversion loss between theory and experiments which can be explained by losses in the experimental setup. For similar experiments at 110 GHz and 140 GHz, the agreement between theory and experiment was also good.

One experiment was also made to evaluate the harmonic response, i.e., the ratio

$$L_s \cdot \sum_{n=2} \left(\frac{1}{L_{n+}} + \frac{1}{L_{n-}} \right)$$

appearing in (1). The experimental technique is shown in Fig. 11. A quasioptical interference filter is frequency selective, i.e., the signal at 200 GHz is propagating straight through the filter while 100 GHz is reflected almost entirely. Measuring the system noise temperatures from the two different positions A and B using a

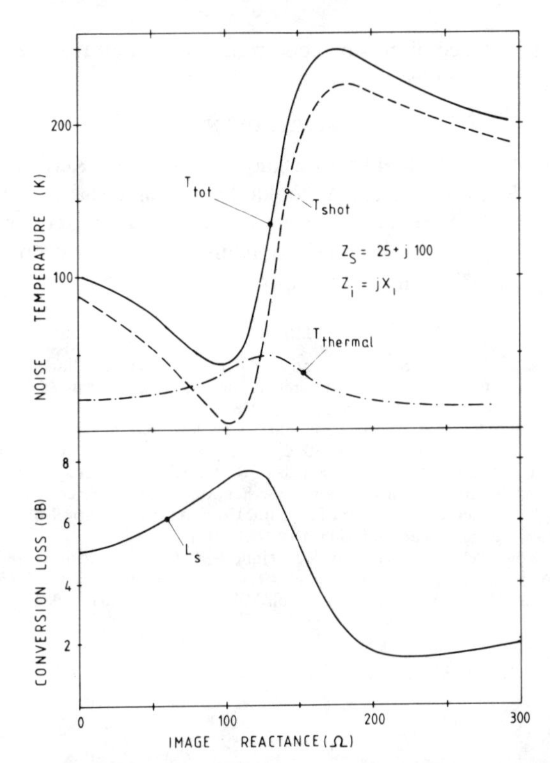

Fig. 12. Theoretical mixer noise and conversion loss for an 80-GHz diode mixer cooled to 20 K versus image reactance. The diode zero bias capacitance is 10.6 fF and series resistance 15 Ω. The signal input impedance $25 + j100$ Ω. The harmonic frequency impedances are assumed equal to $50 + j250$ Ω.

room-temperature absorber and a liquid nitrogen cooled absorber suggested that

$$L_s \cdot \sum_{n=2}^{\infty} \left(\frac{1}{L_{n+}} + \frac{1}{L_{n-}} \right) \simeq 0.08 \text{ at 96 GHz.}$$

VII. Optimizing the Mixer Performance

From the results reported above, we can draw the conclusion that knowing the signal and image impedances is far more important than knowing the harmonic and harmonic sideband impedances. The agreement demonstrated between theory and the experimental results also gives us some confidence in using the theory for optimizing the circuit. Again, an interesting case exists when the image frequency is reactively loaded, i.e., when the backshort short-circuits the diode mount (compare Figs. 3 and 4). Fig. 12 shows how the mixer noise temperature and the conversion loss ($T_{IF} = 0$) varies versus the image reactance. The impedances at the signal frequency are chosen to $25 + j100$, which is close to optimum, while at the harmonics and harmonic sidebands the impedances are assumed to be highly inductive, viz., $Z_2 = 50 + j200$, $Z_n = 50 + j250$, $n = 3$ to 6. It is interesting to notice the minimum in the mixer noise near $jX_i = j100$ Ω where the shot noise is as low as 4 K. Actually, for X_i larger than 100 Ω, parametric effects due to the pumped capacitance become important, which is why this region should be avoided (notice the low-conversion loss for $X_i = 200$ Ω, which is due to parametric effects). The result suggests that an image reactance equal to or smaller than the signal reactance is essential. This case can be achieved if the whisker inductance is low enough and for the lower sideband short-circuited (compare Fig. 2). It is interesting to point out that this is exactly the case for the 70–95-GHz mixer.

A more detailed discussion concerning the optimization procedure will be presented in a forthcoming report.

Acknowledgment

The authors would like to acknowledge Dr. A. Kerr at the Goddard Space Institute, NY, for kindly lending us his computer program, and T. Ståhlberg for some calculations on mixer performance. They would also like to thank C. O. Lindström for carrying through many of the experiments.

References

[1] R. A. Linke, M. V. Schneider, and A. Y. Cho, "Cryogenic millimeter-wave receiver using molecular beam epitaxy diodes," *IEEE Trans. Microwave Theory Tech.*, vol. MTT-26, pp. 935–938, 1978.

[2] N. Keen, R. Haas, and E. Perchtold, "Very low noise mixer at 115 GHz using a Mott diode cooled to 20 K," *Electron. Lett.*, vol. 14, pp. 825–826.

[3] A. V. Räisänen, N. R. Erickson, J. L. R. Marrero, P. F. Goldsmith, and C. R. Predmore, "An ultra low-noise Schottky mixer receiver at 80–120 GHz," presented at the 1981 IEEE Int. Conf. on Infrared and Millimeter Waves, Miami Beach, Florida, Dec. 7–12, 1981.

[4] B. Vowinkel, J. K. Peltonen, W. Reinhert, K. Grüner, and B. Aumiller, "Airborne imaging system using a cryogenic 90-GHz receiver," *IEEE Trans. Microwave Theory Tech.*, vol. MTT-29, pp. 535–541, 1981.

[5] M. Feldman and S. Rudner, *Reviews Infrared Millimeter Waves*, vol. II, 1982.

[6] T. J. Viola and R. J. Mattauch, "Unified theory of high-frequency noise in Schottky barriers," *J. Appl. Phys.* vol. 44, pp. 2805–2808, 1973.

[7] R. L. Eisenhart and P. J. Kahn, "Theoretical and experimental analysis of a waveguide mounting structure," *IEEE Trans. Microwave Theory Tech.*, vol. MTT-19, pp. 706–719, 1971.

[8] C. E. Hagström and E. L. Kollberg, "Measurements of embedding impedance of millimeter-wave diode mounts," *IEEE Trans. Microwave Theory Tech.*, vol. MTT-28, pp. 899–904, 1980.

[9] M. Pospieszalski and S. Weinreb, "A method for measuring an equivalent circuit for waveguide-mounted diodes," in *Proc. 10th Eur. Microwave Conf.* (Warszawa, Poland), Sept. 1980.

[10] M. Pospieszalski and S. Weinreb, "A method for measuring an equivalent circuit for waveguide-mounted diodes," National Radio Astronomy Observatory, Charlottesville, VA, Electronics Division Internal Rep. 201, Oct. 1979.

[11] N. H. Held and A. R. Kerr, "Conversion loss and noise of microwave and millimeter-wave mixers: Parts 1 and 2," *IEEE Trans. Microwave Theory Tech.*, vol. MTT-26, pp. 55–70, Feb. 1978.

[12] N. J. Keen, "Low-noise millimeter-wave mixer diodes, results and evaluation of a test programme," *Proc. IEEE*, vol. 127, pp. 188–198, Aug. 1980.

[13] H. Zirath, E. Kollberg, M. V. Schneider, A. Y. Cho, and A. Jelenski, "Characteristics of metal-semi-conductor junctions for mm-wave detectors," in *Proc. 7th Int. Conf. Infrared millimeter waves*, (Marseille, France), Feb. 1983.

Characteristics of Schottky diodes with microcluster interface

M. V. Schneider
Bell Laboratories, Crawford Hill Laboratory, Holmdel, New Jersey 07733

A. Y. Cho
Bell Laboratories, Murray Hill, New Jersey 07974

E. Kollberg and H. Zirath
Chalmers University of Technology, Goteborg, Sweden

(Received 11 April 1983; accepted for publication 27 June 1983)

We present experimental evidence that a single Schottky diode on GaAs is an agglomorate of paralleled microjunctions with different barrier heights and saturation currents. The current-voltage characteristic of the cluster breaks up into sections of exponentials with different slopes as one cools the diode from 300 to 10 K. Noise measurements performed on cooled diodes at 4 GHz also confirm that a single device is a cluster of paralleled diodes.

PACS numbers: 85.30.Mn, 73.30. + y, 73.40.Ns

The classical model of a metal-semiconductor contact assumes that the junction can be uniquely described by a fixed barrier height, a saturation current, a depletion layer thickness, and other secondary parameters which do not vary as one moves along the interface.[1,2] This theory predicts the observed data with sufficient accuracy for conventional diodes. However, the model does not explain the observed facts for cooled devices which are designed for use as fast switches and low-noise detectors at microwave and millimeter-wave frequencies.

The purpose of this letter is to present experimental results on Pt-GaAs diodes which show that a single diode is a microcluster of paralleled junctions with different barrier heights and saturation currents. Our measurements are in agreement with recently proposed models for GaAs contacts by Freeouf and Woodall[3,4] and for silicon contacts by Ohdomari and Tu.[5] The results also confirm observations reported by Aydinli and Mattauch[6] that the noise temperature of a junction is critically dependent on the chemical nature of the interface between the metal and the semiconductor. In our simplified model we assume that the parallel subdiodes are characterized by areas S_n, barrier heights Φ_n, and series resistances R_n. The total current i through a single diode is

$$i = \sum_{n=1}^{m} i_n, \tag{1}$$

where the current i_n through the nth subdiode is given by

$$i_n = S_n A^* \theta^2 \exp\left(-\frac{q\Phi_n}{k\theta}\right)\exp\left\{\frac{q(v - R_n i_n)}{k\theta}\right\} \quad v \leqslant \phi_n, \tag{2}$$

$$I_n = S_n A^* \theta^2 \quad v > \phi_n. \tag{3}$$

In these equations the effective temperature of the junction, θ, is given by $\theta = \eta(T)\, T$, where η is the ideality factor and T the physical temperature of the diode. A^* is the Richardson constant, v the voltage applied to the diode, q the electron charge, and k the Boltzmann constant. For GaAs the Richardson constant for low fields[2] is $A^* = 8.16\ \text{A/cm}^2/\text{K}^2$. A schematic plot of the logarithm of the total current, log i vs applied voltage v, for the simple case of two subdiodes is shown in Fig. 1. We assume in this specific case that two stable and different chemical phases exist at the interface between the metal and the semiconductor. The two subdiodes are characterized by different areas and barrier heights. It is to be noted that the room-temperature plot for the total current versus voltage remains a straight line, while the corresponding plot at low temperatures shows a distinctive knee. This knee becomes evident at low temperatures because the log i vs v characteristic has a much steeper slope as shown in Fig. 1(b).

Measurements performed on a diode for both room temperature and 20 K are displayed in Fig. 2. The diode was

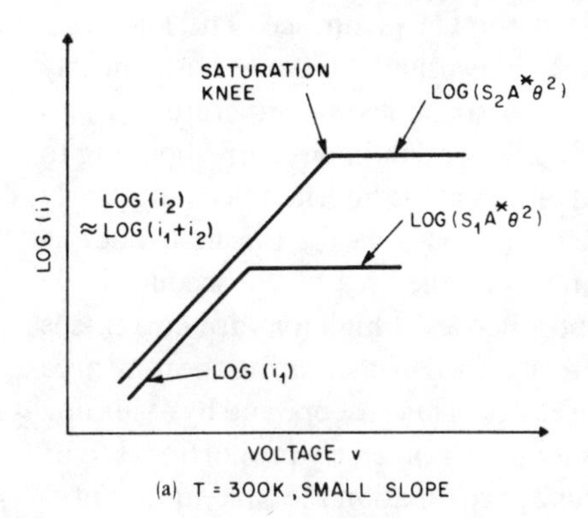

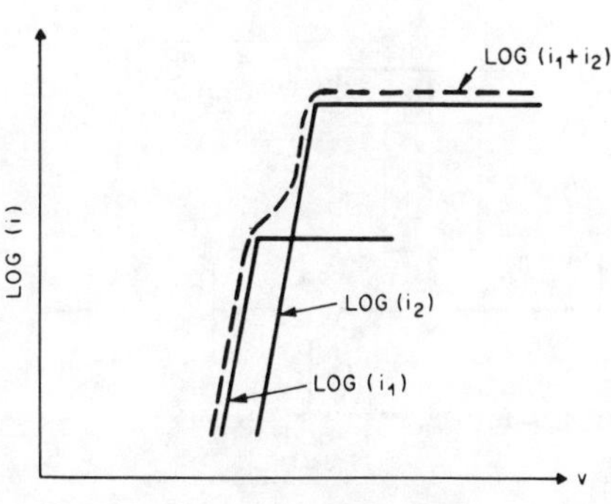

FIG. 1. Schematic drawing of the current-voltage characteristic of a metal-semiconductor junction with two distinct phases at the interface. (a) Room temperature; (b) cryogenically cooled diode. It is assumed in this model that the voltage drop caused by the series resistance of each subdiode is small compared to the forward bias, and that the field in the bulk semiconductor is zero.

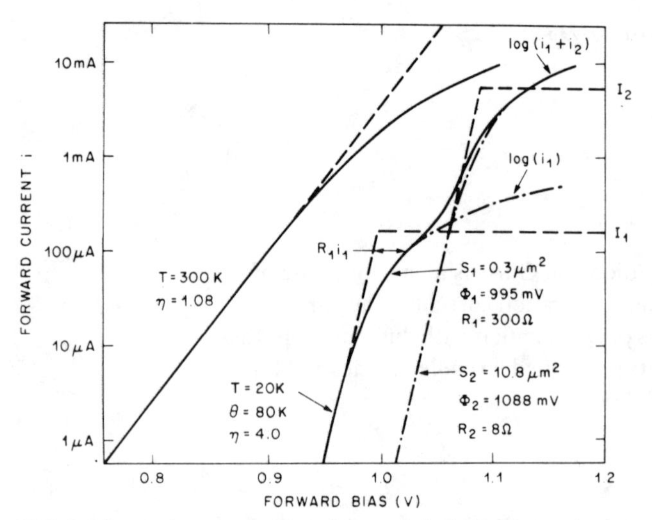

FIG. 2. Measured current-voltage characteristic for Pt-GaAs microjunction at 300 and 20 K. The solid lines are the measured data. The existence of two distinct phases at the interface can be clearly seen from the low-temperature data.

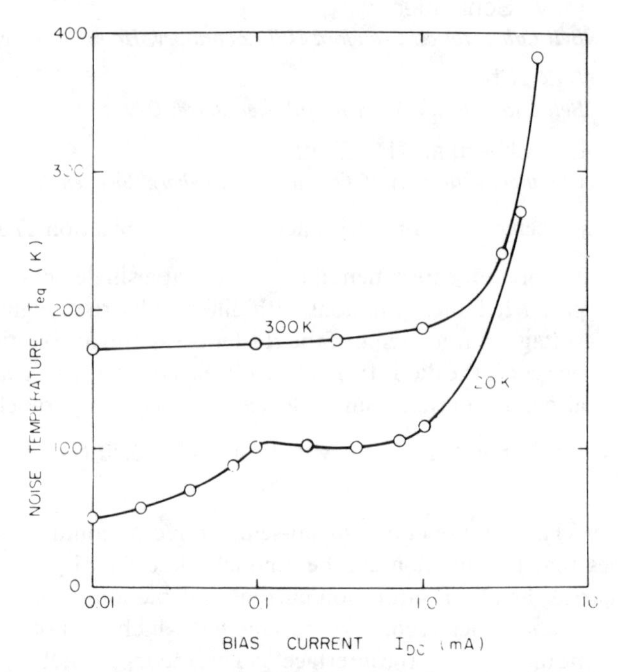

FIG. 4. Equivalent diode noise temperature T_{eq} at 4 GHz measured for electroplated MBE junction N280-92. The mesa-shaped characteristic obtained for a junction temperature of 20 K is caused by electron transport through the subdiode with the smaller junction area.

fabricated on n-type GaAs prepared by molecular beam epitaxy (MBE).[7] The doping concentration of the n^+ substrate was 3×10^{18} cm^{-3} and the Sn-doped epitaxial layer had a carrier concentration of 3×10^{16} cm^{-3} and a thickness of 100 nm. The junctions with an area of 11 μm^2 were formed by pulse plating Pt on the GaAs surface. The junction area was defined by fabricating holes in a 400-nm-thick SiO_2 layer which was deposited on the GaAs surface. The log i vs v characteristic of Fig. 2 shows that the diode has a nearly ideal exponential characteristic at room temperature. For a device temperature of 20 K two subdiodes corresponding to two different chemical phases at the interface become clearly visible. The calculated junction area for the first diode is approximately 0.3 μm^2 while the area of the second diode controlling the device properties at high forward currents is 10.8 μm^2. The areas are obtained by measuring the total area of the junction with an electron microscope and by assuming that the area of each subdiode is inversely proportional to its series resistance. In this approximation the currents are normal to the surface and fringe effects are neglected. This is valid because the epitaxial layer is thin with respect to the diameter of the junction. The calculated knee currents from Eq. (3) for a device temperature of 20 K are $I_1 = 156\,\mu$A and $I_2 = 5.6$ mA. The measured current shows the expected leveling off in these ranges as shown in Fig. 2. The effect of different phases at the interface on the log i–v characteristic is enhanced by fluctuations of the doping concentration in the epitaxial layer which increase the carrier transport through the barrier in the higher doped areas at low forward voltage. This type of cluster will be discussed in more detail in a subsequent publication.

A detailed analysis of the temperature-variable current-voltage characteristics of a number of junctions of the type shown in Fig. 2, shows that the barrier heights increase by 80 mV as the junctions are cooled from 300 to 15 K. The measurements give room-temperature barrier heights of $\Phi_1 = 915$ mV and $\Phi_2 = 1008$ mV for the junction of Fig. 2. The apparent barrier height calculated from the formalism

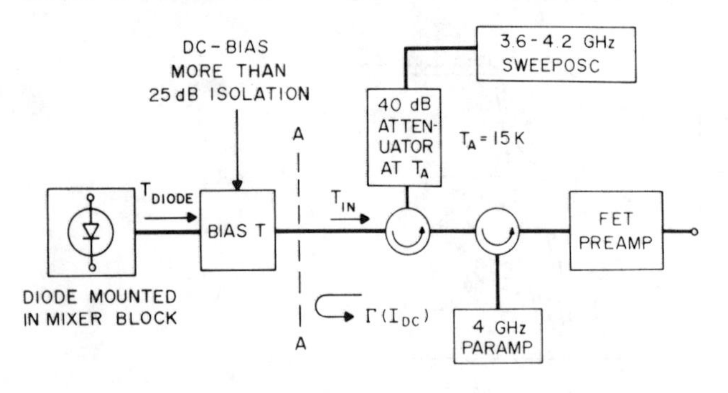

FIG. 3. Schematic drawing of 4-GHz noise measurement apparatus for measuring the noise temperature of a diode as a function of applied dc bias. The noise power available from the diode is $P_{diode} = kT_{eq}\Delta f$. The corrections due to the reflection coefficient $\Gamma_D(I_{DC})$ at the interface A-A are indicated in the figure.

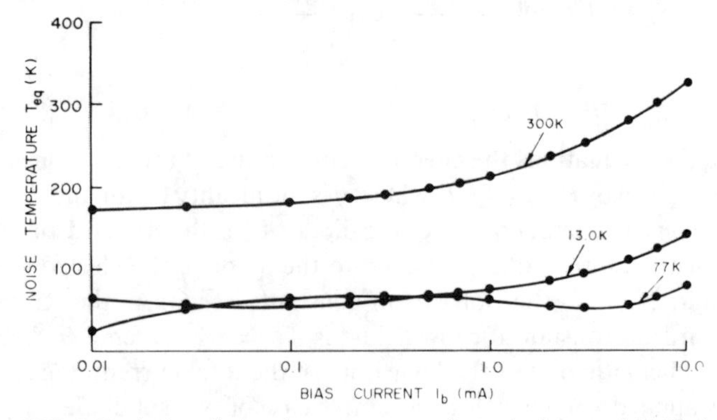

FIG. 5. Equivalent diode noise temperature of new single-crystal Al/GaAs MBE junction at 4 GHz, batch A244A-55. The device geometry and the processing steps are described in Ref. 8.

of Ohdomari and Tu[5] for a ratio of the junction areas $S_1/S_2 = 0.03$ is 992 mV and the apparent barrier height deduced from the room-temperature data shown in Fig. 2 is 1006 mV. Thus, the difference between predicted and measured data is 18 mV or 1.8% of the total barrier height, which is relatively small percentage.

A schematic drawing of the experimental apparatus for measuring the microwave noise generated by the diode as a function of temperature is shown in Fig. 3. The various components of the circuit are described in the drawing. The equivalent noise temperature measured for a Pt-GaAs diode as a function of forward current with the apparatus shown in Fig. 3 is displayed in Fig. 4. The microwave noise measured at room temperature is independent of bias up to a dc bias current of approximately one milliampere where a sharp increase occurs due to electron noise of heated carriers which are not in equilibrium with the lattice temperature. A different noise characteristic is measured for the same diode for a junction temperature of 20 K. The relatively high noise temperature at a forward current above 0.1 mA is caused by a high current density in the subdiode which has reached the range of the maximum knee current as indicated in Fig. 2.

Experiments performed on a novel single-crystal metal-semiconductor junction prepared by MBE[8] show that the electrical effects of multiple phases at the interface are less pronounced than those observed for an electroplated junction. The single-crystal diode has a noise temperature which remains constant over a large bias range and increases slightly at forward currents approaching 10 mA. Measurements of the noise characteristics at room temperature, 77 K, and 13 K are shown in Fig. 5 for a single-crystal Al on GaAs microjunction for which the metal is interface lattice matched to the semiconducting substrate. It is believed that this device is an improved structure for building low-noise microwave detectors operating from 10 to 300 K.

We would like to thank A. A. Olenginski and P. A. Verlangieri for preparing the GaAs diodes, R. F. Trambarulo for helpful discussions, and C. O. Lindstrom for assistance with the experiments. The authors would also like to acknowledge financial support from the Swedish Board for Technical Development.

[1]W. Schottky and E. Spenke, Wissenschaftliche Veroeffentlichungen aus den Siemens Werken **18**, 225 (1939).
[2]S. M. Sze, *Physics of Semiconductor Devices*, second ed. (Wiley, New York, 1981), pp. 245–311.
[3]J. L. Freeouf and M. J. Woodall, Appl. Phys. Lett. **39**, 727 (1981).
[4]J. M. Woodall and J. L. Freeouf, J. Vac. Sci. Technol. **21**, 574 (1982).
[5]I. Ohdomari and K. N. Tu, J. Appl. Phys. **51**, 3735 (1980).
[6]A. Aydinli and R. J. Mattauch, Solid-State Electron. **25**, 551 (1982).
[7]A. Y. Cho and J. R. Arthur, Prog. Solid State Chem. **10**, 157 (1975).
[8]A. Y. Cho, E. Kollberg, H. Zirath, W. W. Snell, and M. V. Schneider, Electron. Lett. **18**, 424 (1982).

All Solid-State Low-Noise Receivers for 210–240 GHz

JOHN W. ARCHER

Abstract — Low-noise all solid-state receiver systems for room temperature and cryogenic operation between 210 and 240 GHz are described. The receivers incorporate a single-ended fixed tuned Schottky barrier diode mixer, a frequency-tripled Gunn source as local oscillator and a GaAsFET IF amplifier. Single sideband receiver noise temperatures are typically 1300 K (7.39-dB noise figure) for a room temperature system and 470 K (4.18-dB noise figure) for a cryogenically cooled receiver operating at 20 K.

I. Introduction

A NUMBER OF researchers have reported the development of heterodyne receiver systems operating at frequencies near 230 GHz [1]–[4]. However, receiver noise figures achieved have been relatively high (typically about 10 dB). Furthermore, the lack of a convenient and reliable local oscillator source with adequate output power has limited the receiver performance, and in many cases necessitated the use of relatively noisy harmonic mixers or complex dual-diode subharmonically pumped devices.

High-performance 210- to 240-GHz receiver systems have recently become practical as a result of significant improvements in single-ended mixer design [5] and the development of efficient frequency multipliers as LO sources [6]. Although receiver noise temperatures can be reduced with mixers and IF amplifiers cooled to 20 K, in many applications it is desirable that the receiver be readily portable and operate at 300 K ambient without the necessary complicated closed cycle helium refrigerators and vacuum systems required for cooled operation. The primary emphasis of this paper concerns the realization of a portable low-noise receiver for room temperature operation between 210–240 GHz. One of the prerequisites for portability was the development of practical solid-state local oscillator sources for this frequency range. Results are also presented which indicate that about a factor of three improvement in receiver noise temperature can be achieved by cooling mixer and IF amplifier to 20 K, but with a necessary increase in complexity and reduced portability.

II. Description of the Receiver and Components

Fig. 1 shows a photograph and block diagram of the ambient temperature receiver. The cooled system is similar except for the inclusion of a small vacuum dewar and closed cycle helium refrigerator[1] in which mixer and IF amplifier are mounted.

A lightweight compact polarizing interferometer diplexer [7] is used for LO/RF combining and filtering. The modular construction of the diplexer (each module forms an 88.9-mm sided aluminum cube) readily enables the implementation of single or dual linearly polarized receivers. The

[1] CTI Inc., Model 21.

Manuscript received February 2, 1982; revised March 18, 1982. The National Radio Astronomy Observatory is operated by Associated Universities, Incorporated, under contract with the National Science Foundation.

The author is with the National Radio Astronomy Laboratory, Charlottesville, VA 22903.

Reprinted from *IEEE Trans. Microwave Theory Tech.*, vol. MTT-30, pp. 1247–1252, Aug. 1982.

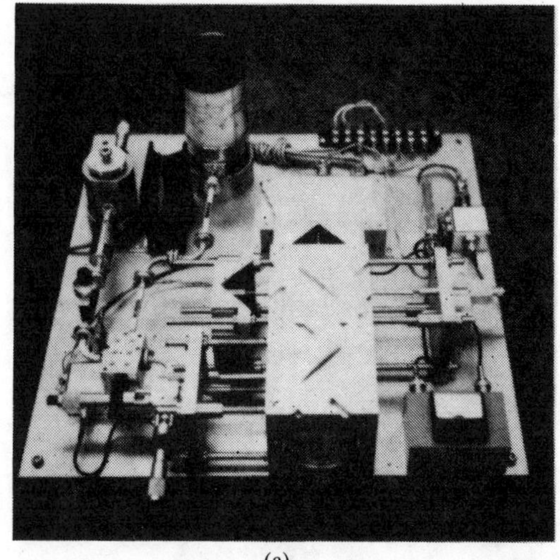

(a)

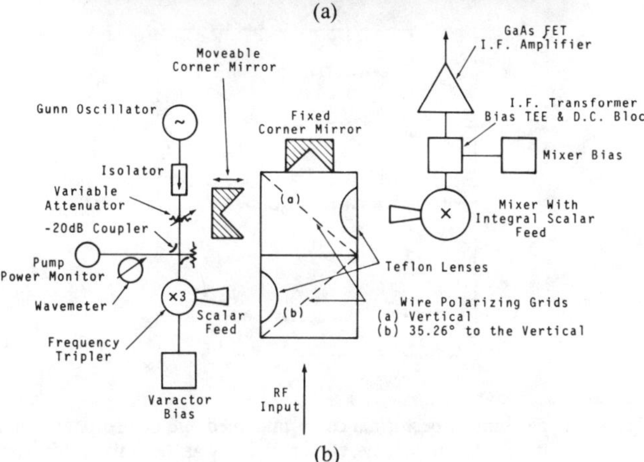

(b)

Fig. 1. (a) Photograph and (b) block diagram of the portable 210- to 240-GHz receiver designed for room temperature operation.

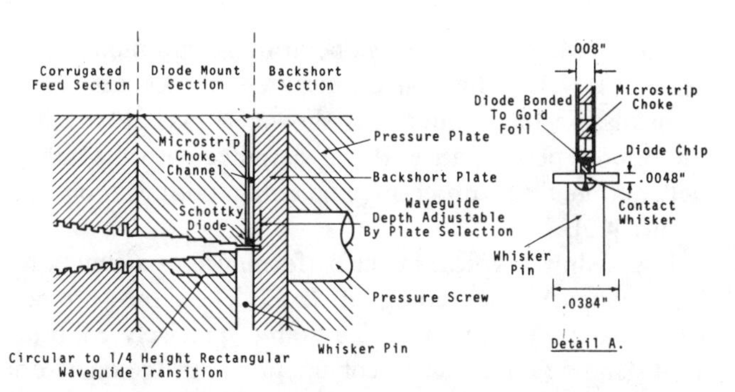

Fig. 2. A sketch of the mixer mount in partial cross-section (not to scale) shows the principal features of the design. Detail A shows the diode mounting and contacting geometry.

polarizing grids used in the diplexer are free-standing 0.05-mm diameter BeCu wire grids with 75 wires per centimeter mounted on removable circular cylindrical forms.

When adjusted for operation at any frequency between 200 and 240 GHz with a 1.5-GHz IF center frequency, the diplexer provides a theoretical -1-dB passband width for each RF sideband of the LO, of 980 MHz and measured rejection for the LO noise sidebands of greater than 20 dB. Teflon dielectric lenses are employed to match the LO and mixer scalar feed patterns to the quasi-collimated beam within the diplexer. The lenses have a focal length of 45.7 mm, a diameter of 55.9 mm, and are grooved on the surfaces to reduce reflection losses. In the cooled receiver, the lens also serves as a window at the dewar vacuum/air interface. The far field -11-dB full beamwidth of the lens corrected mixer feed pattern is 4.2°, independent of frequency between 200 and 240 GHz. The total diplexer loss, including lens reflection and feed coupling losses, is 0.40 dB over this RF frequency range when operating with a 1.5-GHz IF.

The mixer, shown schematically in Fig. 2, is a single-ended fixed tuned mount with integral scalar feed, developed from an earlier design [5]. It comprises a whisker contacted Schottky barrier diode mounted in reduced height waveguide of dimension 0.122 mm$\times$0.978 mm. A five-section circular to rectangular waveguide step transition couples power from the feed to the reduced height waveguide. The GaAs Schottky barrier diode, fabricated by R. Mattauch at the University of Virginia (designated type 2P8-400), has a zero bias capacitance of 6.5 fF and a dc series resistance of 10 Ω at 300 K. The diode is bonded to a 0.076 mm thick crystalline quartz dielectric microstrip RF choke which is epoxied[2] in a 0.203-mm$\times$0.203-mm channel in the diode block so that the diode is recessed into the channel, itself forming the first capacitive section of the choke. The diode mounting structure is similar to that described by Kerr *et al.*, [8] but with the orientation with respect to the waveguide described by Cong *et al.* [9]. The choke comprises a 10-section high/low impedance design with transmission line characteristic impedances of 99 Ω and 28 Ω, respectively. The lengths of the sections were optimized with the aid of a network analysis program [10] to ensure that the choke presents a reactive termination to the diode at frequencies up to at least the third harmonic of the local oscillator. The diode is contacted with a 0.0127-mm diameter phosphor bronze whisker of 0.178-mm unbent length, attached to a 0.51-mm diameter gold-plated BeCu alloy pin which is an interference fit in the mixer body. The whisker is bent so that it just spans the guide and the pin sits flush with the guide wall after contacting.

The mixer used in these receivers employs a novel fixed backshort structure for mount tuning. The backshort is implemented with the aid of a section of short circuited waveguide electroformed into a backing plate 1.760 mm thick. The backshort plate is held in place adjacent to the block containing the diode by a clamping block attached to screws which pass through the mixer body. Backshort plates with a range of diode to short spacings are available to facilitate initial mixer tuning. This method of mount tuning has distinct advantages over that used in an earlier design [5] in that it is much more mechanically stable, exhibits repeatable performance, and offers improved mixer bandwidth.

The mixer IF output is connected to an integral IF matching transformer, bias tee, and dc block which results in an IF output VSWR of less than 1.2:1 between 1.2 and

[2]Sears, Roebuck and Co., 2-part epoxy, stock number 9 8059.

1.8 GHz, when the mixer is operating at optimum LO and dc bias levels. The transformer/tee/block uses coaxial transmission line sections, as shown in Fig. 3, to achieve the desired performance. It was designed with the aid of a network analysis program employing optimization techniques [10].

Fig. 4 shows typical mixer performance as a function of LO frequency with tuning left fixed. These measurements were made with the aid of a stable, precisely calibrated 1.5-GHz IF radiometer in conjunction with the quasi-optical LO combining system described above. The results have been corrected for the signal loss associated with the diplexer, but not for residual IF mismatch and were determined by conventional techniques with 300 K and 77 K RF loads provided using Eccosorb AN-72 formed into a pyramidal shape for minimal error due to reflections from the terminations. All single sideband values quoted assume equal sideband gains and are based on double sideband measurements. The sideband gains for the mixers used here have been measured and found to be equal to within 5 percent [11].

It can be seen, for mixer 1A, with diode and tuning optimized for 300 K operation, that the SSB mixer temperature is less than 950 K between 210 and 240 GHz, reaching a minimum of 800 K at 228 GHz. Conversion loss, which includes corrugated feed horn losses, estimated to be about 0.25 dB, is less than 7.25 dB between 210 and 240 GHz with a minimum of 6.60 dB at 228 GHz. For room temperature operation, the dc bias voltage was held fixed at a typical value of 0.620 V during the measurements, and the LO power was adjusted to give a mixer current of 1.10 mA. For mixer 3, optimized for cooled operation at 20 K, the SSB cooled mixer temperature is less than 350 K with a minimum of 250 K extending from 220–230 GHz. The conversion loss response is very similar to that obtained with mixer 1A. DC bias conditions were typically 0.89 V at 0.40 mA for cooled operation.

In the receiver, the IF signal is normally fed to a low-noise GaAsFET amplifier [12]. At 300 K ambient typical amplifier noise temperature is less than 70 K between 1.2 and 1.8 GHz with a gain of 28 ± 1 dB and an input VSWR of less than 1.4:1 over the same range. For an amplifier designed to be cooled to 20 K, the noise temperature falls to less than 15 K over this frequency band, with a small increase in gain and little change in input match compared with the room temperature version.

The local oscillator source is a frequency tripled, mechanically and electrically tuneable Gunn oscillator. The harmonic generator employs a split block crossed waveguide design [6] shown in Fig. 5. Power incident in the full height input waveguide is fed via a tuneable transition to a seven-section suspended substrate low-pass filter which passes the pump frequency with low loss, but is cutoff for higher harmonics. The low-pass filter transforms the impedance of the pumped varactor at the input frequency to a convenient value at the plane of the waveguide to stripline transition. Pump circuit impedance matching is achieved using two adjustable waveguide stubs with sliding

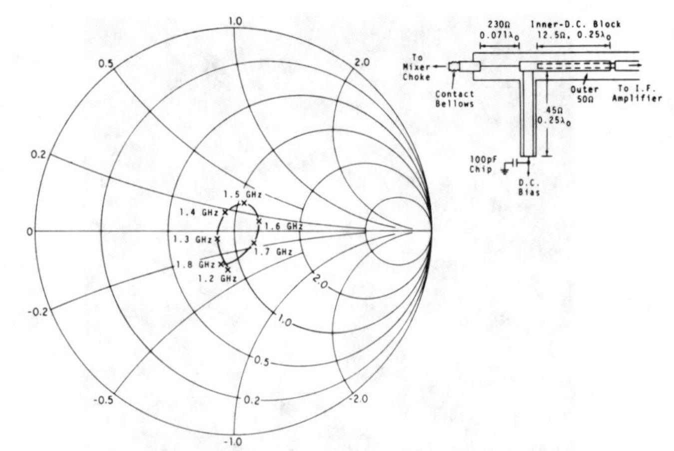

Fig. 3. Schematic and performance of the 1.2- to 1.8-GHz IF transformer, bias tee and dc block.

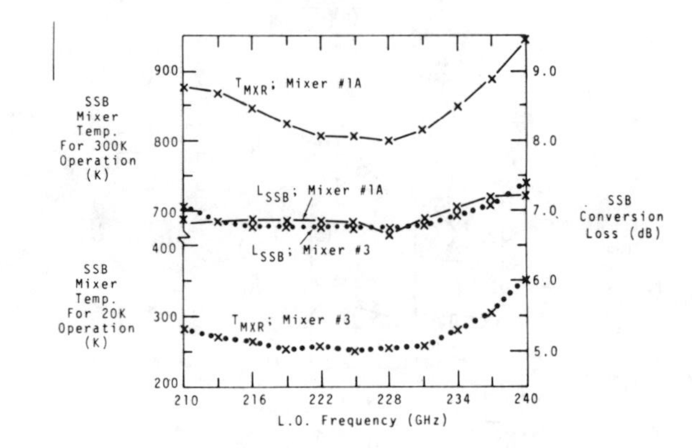

Fig. 4. Typical mixer performance for uncooled and cooled operation as a function of LO frequency. Mixer #1A was optimized for room temperature operation; mixer #3 was optimized for cooled operation.

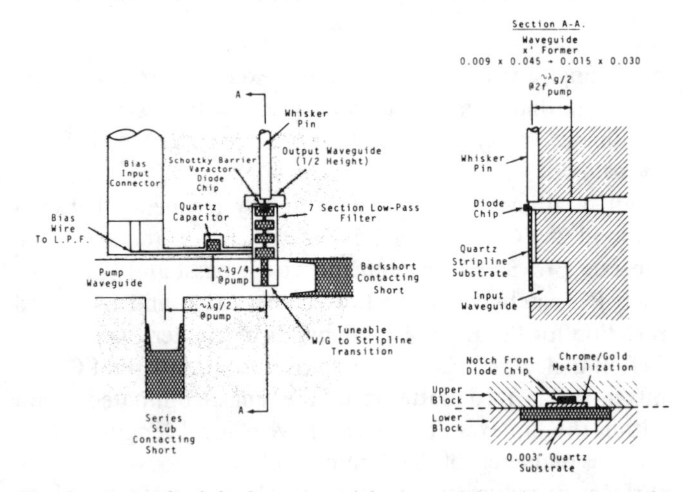

Fig. 5. A sketch of the frequency tripler in partial section (not to scale) shows the principal features of the design.

contacting shorts. One stub acts as a backshort for the probe type waveguide to stripline transition and a second as an E-plane series stub located $\lambda_g/2$ (at the pump wavelength) towards the source from the plane of the transition. Mechanical adjustment of these tuners typically enables the input to be well matched to the diode impedance at any frequency within the operating bandwidth of the pump waveguide.

The whisker-contacted varactor chip is mounted on the

filter substrate in the reduced height output waveguide. Output tuning is accomplished with the aid of an adjustable backshort in this guide. DC bias is brought to the device via a transmission line bias filter. The bias circuit comprises a 140-Ω transmission line, consisting of a 0.025-mm diameter gold wire center conductor bonded at one end to a low impedance section of the low-pass filter and at the other end to a 100-fF quartz dielectric bypass capacitor and enclosed in a rectangular shield machined into the mount. The bias line approximates, at the chosen mount center frequency, a quarter-wave short-circuited stub.

A quarter-wave three-section impedance transformer couples the reduced height guide to the full height output guide. Power can flow in the reduced height guide at the second harmonic whereas the output guide is cut off at this frequency. The transformer is thus used to implement a second harmonic idler termination by spacing it approximately $\lambda_g/2$ (at the second harmonic wavelength) from the plane of the diode.

The varactor diode is a Schottky barrier device fabricated by R. Mattauch at the University of Virginia (designated 5M2) with a zero bias capacitance of 21 fF and a dc series resistance of 8.5 Ω. The breakdown voltage is 14 V at 1 μA. These devices have a highly nonlinear capacitance versus voltage law which approximates the inverse half-power behavior of the ideal abrupt junction varactor to within about 2 V of the breakdown limit.

The frequency tripler exhibits typical conversion efficiency greater than 3 percent for output frequencies between 210 and 240 GHz, with 50-mW pump power, when dc bias and tuning are optimized at each operating frequency. Peak conversion efficiency of 6 percent is attained at an output frequency of 222 GHz.

The 70- to 80-GHz Gunn oscillator employs a commercial GaAs Gunn diode[3] in a quarter-wave radial line resonator structure, as shown in Fig. 6. The 2.667-mm diameter resonator is mounted in a waveguide with nonstandard dimensions, which were adjusted to optimize the coupling between resonator and guide in the 70–80-GHz range (3.81 mm × 1.55 mm). A taper transformer couples this waveguide to the standard WR-12 output guide (3.10 mm × 1.55 mm). DC bias and ground return are brought to the diode via multisection coaxial choke structures made from tellurium copper. The choke center conductors are coated[4] with a spray-on fluorocarbon material to insulate them from the outer wall. The bias and heatsink assembly is designed so that the vertical position of the diode in the guide may be continuously varied by approximately ±0.125 mm with dc bias applied to the device.

The operating frequency of the oscillator is primarily determined by the resonant frequency of the radial line/Gunn diode combination. A contacting backshort is used to optimize the coupling between resonator and guide at each operating frequency, but the adjustment of this

[3]Hughes Aircraft Company, Model 47205H-0305.
[4]Whitford Corp., Xylan 1000 Fluorocarbon Coating.

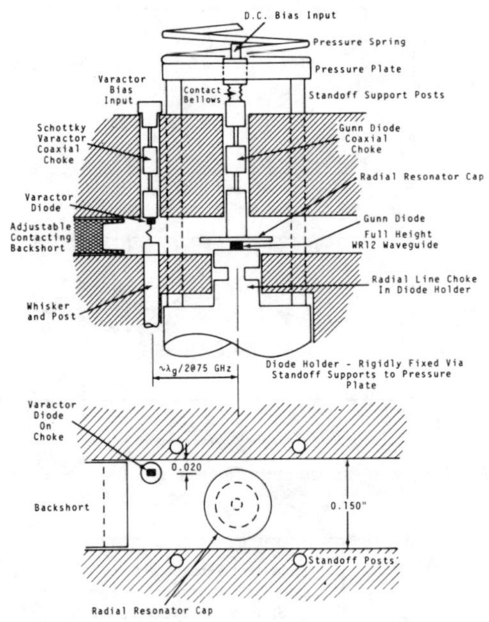

Fig. 6. A sketch of the Gunn oscillator mount in partial section (not to scale) shows the principal features of the design.

short has only a second order effect on oscillator frequency. Varying the spacing between the guide wall and the radial hat changes the fringing field strength at the edges of the resonator and alters the fringing capacitance. Hence, it is possible to tune the oscillator by varying the vertical position of the hat in the guide [13]. For the present design, a reduction in height from about 1.25 mm–1.00 mm corresponds to a frequency change from 80–70 GHz.

Since the backshort adjustment has a small but measurable effect on operating frequency (for approximately 1-dB reduction in output power, the frequency can be pulled on the order of 100 MHz), it has been found possible to provide limited electrical tuning of the oscillator with a varactor diode mounted about $\lambda g/2$ from the Gunn diode plane, in the backshort guide. The diode, a Schottky barrier, whisker-contacted device similar to those used in the frequency tripler, is mounted on a coaxial choke in a position offset from the guide axis. The degree of offset, which determines the coupling between guide and varactor circuit, has been experimentally optimized to give an electrical tuning range of about ±50 MHz at any operating frequency with less than 1.5-dB change in output power. The electrical tuneability of the oscillator should allow it to be phase locked, although this has not yet been attempted.

Oscillator performance is illustrated in Fig. 7. Output power varies between a minimum of 45 mW and a maximum of 55 mW as the oscillator is tuned over the range 70–80 GHz. DC bias requirements are approximately 5 V at 1.2 A. As is also shown in Fig. 7, the Gunn oscillator/tripler combination can provide a minimum of 2.0 mW at any frequency between 210 and 240 GHz.

III. Receiver Parameters and Performance

The complete room temperature receiver package occupies a volume of less than 0.015 m^3 and weighs less than 9.0 kg (excluding power supplies). Typical mixers and

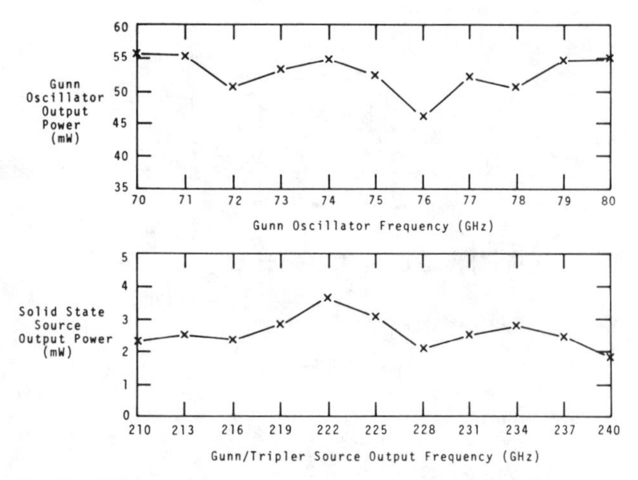

Fig. 7. Solid state source output power as a function of frequency.

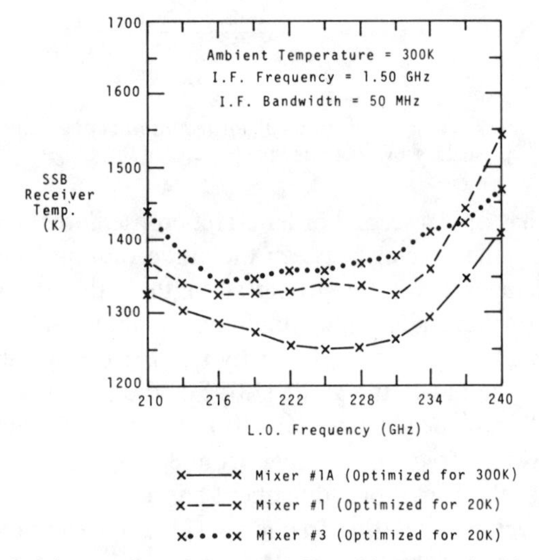

Fig. 8. Receiver performance at 300 K ambient as a function of LO frequency, for three different mixers. Mixer #1A was optimized for 300 K operation; mixers #1 and #3 for operation at 20 K.

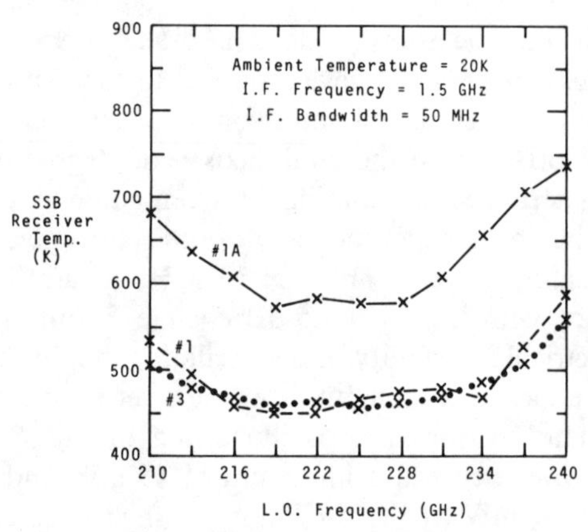

Fig. 9. Receiver performance at 20 K ambient as a function of LO frequency, for three different mixers.

frequency multipliers used in the receiver have undergone extensive mechanical shock and vibration testing to evaluate their performance under extreme environmental conditions. They can withstand for an indefinite period, continuous vibration with a sinusoidal acceleration of up to 4-G peak at 60 cycles per minute rate. However, permanent

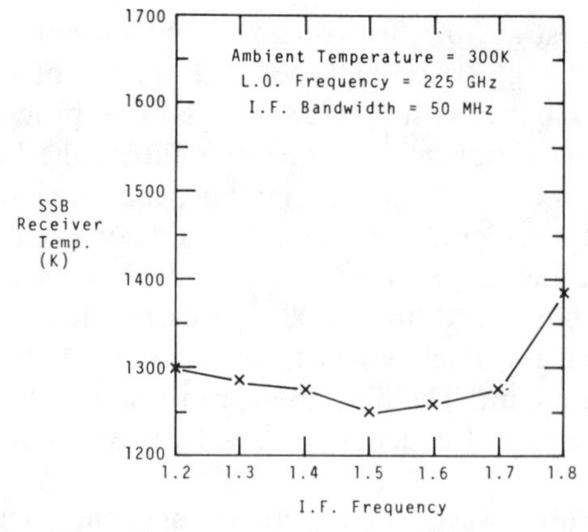

Fig. 10. Receiver performance as a function of IF center frequency for mixer #1A at 300 K.

changes in diode $I-V$ characteristics are observed after five minutes at 10-G peak at the same repetition rate. The mixers have also been tested under thermal shock conditions. A typical mixer can withstand a minimum of five sequential immersions in liquid nitrogen, each followed by warming to room temperature in an evacuated desiccator, without detectable change in performance.

Receiver RF performance was determined using similar techniques to those used in evaluating mixer behavior. As the LO frequency was varied between 210 and 240 GHz and the diplexer tuning adjusted accordingly, the performance curves shown in Figs. 8 and 9 were obtained at 300 K and 20 K ambient, respectively, for three different mixers. For the room temperature receiver, the single sideband receiver noise temperature with mixer 1A varies between 1400 K at the band edges to a minimum of 1250 K at band center. The IF center frequency for these measurements was 1.5 GHz with a bandwidth of 50 MHz. Typical LO power required was less than 1.5 mW at the tripler output.

When the mixer and IF amplifier are cooled to 20 K, the SSB receiver temperature varies between 575 K at the band edges and a minimum of 470 K at band center (mixers 1 and 3). The IF center frequency and bandwidth remain the same for these measurements. The required LO power drops to about 300 μW at the tripler output.

For a fixed LO frequency of 225 GHz, the instantaneous response shown in Fig. 10, as a function of IF frequency, was measured for the 300 K receiver. The usable instantaneous bandwidth is at least 500 MHz with less than 100 K degradation in receiver noise temperature relative to the value at 1.5 GHz.

IV. Conclusion

The results described in this paper demonstrate the feasibility of building all solid-state high performance receiver systems to utilize the 200- to 300-GHz atmospheric transmission window. It has been shown that portable, mechanically robust receivers, which exhibit SSB noise temperatures of the order of 1300 K, can be built for room temperature operation. Cooling the mixer and IF amplifier to 20 K results in about a factor of three reduction in receiver noise temperature, but with necessary penalties of increased weight and complexity and limited portability. Current research is aimed at developing components that would enable all solid-state low-noise receivers to be constructed for operation at frequencies up to 350 GHz.

Acknowledgment

The author wishes to thank N. Horner, who assembled the mixers and multipliers, G. Taylor, who fabricated the diplexer, Dr. S. Weinreb and R. Harris, for supplying the GaAsFET amplifiers, and Dr. R. Mattauch for providing the Schottky barrier diodes.

References

[1] N. R. Erikson, "A 200–350 GHz heterodyne receiver," *IEEE Trans. Microwave Theory Tech.*, vol. MTT-29, pp. 557–562, June 1981.

[2] S. Lidholm and T. DeGraauw, "A heterodyne receiver for submillimeter wave astronomy," presented at 4th Int. Conf. Infrared and Millimeter Waves, (Miami, FL), Dec. 1979.

[3] P. F. Goldsmith and R. Plambeck, "A 230-GHz radiometer system employing a second harmonic mixer," *IEEE Trans. Microwave Theory Tech.*, vol. MTT-24, pp. 859–867, Nov. 1976.

[4] D. R. Carlson and M. V. Schneider, "Subharmonically pumped millimeter wave receivers," in *Conf. Dig., 4th Int. Conf. Infrared Millimeter Waves*, (Miami, FL,) Dec. 1979, pp. 82–83.

[5] J. W. Archer and R. J. Mattauch, "Low noise, single-ended mixer for 230 GHz," *Electron. Lett.*, vol. 17, pp. 180–181, Mar. 5, 1981.

[6] J. W. Archer, "Millimeter wavelength frequency multipliers," *IEEE Trans. Microwave Theory Tech.*, vol. MTT-29, pp. 552–557, June 1981.

[7] D. H. Martin and E. Puplett, "Polarized interferometric spectroscopy for the millimeter and sub-millimeter spectrum," *Infrared Phys.*, vol. 10, pp. 105–109, 1969.

[8] A. R. Kerr and R. J. Mattauch, and J. Grange, "A new mixer design for 140–220 GHz," *IEEE Trans. Microwave Theory Tech.*, vol. MTT-25, pp. 399–401, May 1977.

[9] H. Cong, A. R. Kerr, and R. J. Mattauch, "The low noise 115-GHz receiver on the columbia-giss 4-foot radio telescope," *IEEE Trans. Microwave Theory Tech.*, vol. MTT-27, pp. 245–248, Mar. 1979.

[10] D. L. Fenstermacher, "A computer-aided analysis routine including optimization for microwave circuits and their noise," NRAO Electronics Div. Internal Rep. 217, July 1981.

[11] J. Payne, NRAO, Tucson, AZ, private communication.

[12] S. Weinreb, D. L. Fenstermacher, and R. Harris, "Ultra low-noise, 1.2–1.7 GHz cooled GaAsFET amplifiers," *IEEE Trans. Microwave Theory Tech.*, vol. MTT-30, pp. 849–853, June 1982.

[13] J. Ondria, "Wideband mechanically tunable and dual in-line radial mode *W*-Band CW GaAs gunn diode oscillators," in *Proc. 7th Biennial Cornell Elec. Eng. Conf.*, Cornell Univ., (Ithaca, NY,) Aug. 14–16, 1979, pp. 309–32.

Part VIII
Superconducting Junction Mixers

A FURTHER improvement in sensitivity of millimeter-wave mixers can be obtained using superconducting mixing elements cooled to a few degrees Kelvin. The superconductor–insulator–superconductor (SIS) mixer so far has shown the best results and is a very promising device for frequencies between about 30 GHz and a few hundred gigahertz. However, since further development concerning other types of superconducting mixers can be expected it is not settled yet which mixer types will survive in the longer perspective.

The first paper, by Dickman *et al.,* describes mixer results obtained with Schottky diodes where the metal contact is superconducting. However, it seems difficult to get a high cutoff frequency due to the diode parasitics, a problem which may limit the usefulness of the super-Schottky mixer to low millimeter-wave frequencies.

The Josephson mixer is a parametric down-converter and a promising device, in particular for frequencies of several hundred gigahertz. In the second paper Taur analyzes the Josephson mixer and points out what is required by the embedding circuit for a reasonable conversion gain. The next paper, by Taur and Kerr, describes an experimental device for 115 GHz showing a conversion loss of only 2.4 dB and a mixer noise temperature of 140 K. This noise temperature is not overwhelmingly low as compared to cooled Schottky diode mixers, but would of course be remarkable if it remains the same at several hundred gigahertz. The future will hopefully make this clear.

The most promising low noise millimeter-wave mixers are those based on SIS elements. The strength of the (resistive) nonlinearity makes it necessary to use quantum theory [1] for proper understanding, and as was shown in paper 10 of Part I, conversion gain can be achieved. An extensive analysis of the SIS mixer is given by Smith and Richards in the fourth paper of this part. Pan *et al.* report excellent experimental results at 115 GHz: a *receiver* noise temperature of 68 $\pm$ 3 K was obtained.

REFERENCES

[1] J. R. Tucker, "Quantum limited detection in tunnel junction mixers," *IEEE J. Quantum Electron.*, vol. QE-15, pp. 1234–1258, Nov. 1979.

Super-Schottky Mixer Performance at 92 GHz

ROBERT L. DICKMAN, WILLIAM J. WILSON, MEMBER, IEEE, AND GERALD G. BERRY

Abstract—As part of a program to explore the behavior of superconducting Schottky mixers at high frequencies ($\nu_{RF} \geqslant 90$ GHz), the mixing and video performance of several super-Schottky diodes have been tested at 92 GHz. The diodes used ($\sim$3-μm active diameter, doping concentration $\sim 2 \times 10^{19}$ cm^{-3}) were identical to those recently developed at Aerospace for use in a 31-GHz mixer. The WR-10 mixer mount, designed specifically for this experiment, utilizes a quartz stripline assembly for the diode, whisker, and IF choke, suspended across quarter-height RF waveguide.

At 92 GHz, video responsivities were typically $\sim$80 A/W (corrected for RF mismatch). Conversion loss (corrected for both RF and IF mismatches) was typically measured to be $\gtrsim$18 dB. As expected, T_{diode} was small (<5 K). Video responsivity and conversion loss were also measured at an RF frequency of 3.95 GHz. These data were used with the measured I–V characteristics of the diodes to compare theoretical predictions of diode performance at 92 GHz in both the video and mixing modes, with the high-frequency data.

Manuscript received July 28, 1980; revised February 24, 1981. This work was supported by the Aerospace Sponsored Research Programs.

R. L. Dickman was with the Electronics Research Laboratory, The Aerospace Corporation, Los Angeles, CA 90009. He is now with Five College Radio Astronomy Observatory, Amherst, MA 01003.

W. J. Wilson was with the Electronics Research Laboratory, The Aerospace Corporation, Los Angeles, CA 90009. He is now with the Jet Propulsion Laboratory, Pasadena, CA 91103.

G. G. Berry is with the Electronics Research Laboratory, The Aerospace Corporation, Los Angeles, CA 90009.

I. Introduction

THE DEVELOPMENT of increasingly sensitive receivers at millimeter wavelengths imposes stringent mixer performance requirements. The super-Schottky di-

Reprinted from *IEEE Trans. Microwave Theory Tech.*, vol. MTT-29, pp. 788–793, Aug. 1981.

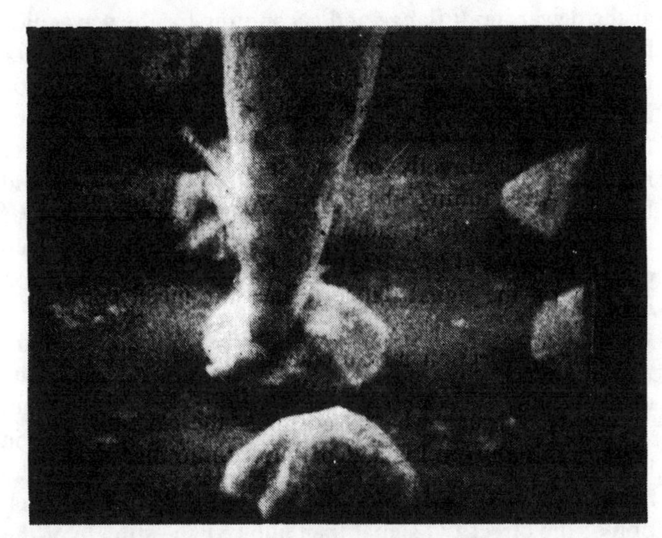

Fig. 1. Scanning-electron microscope (SEM) photograph of a super-Schottky diode array and whisker contact. The diameter of the diode mushrooms is 5 μm.

ode, a superconductor–semiconductor tunneling junction, has performed well as a mixer at both 9 and 31 GHz. At 9 GHz, the diode output temperature was measured to be 1.2 K with a single-sideband conversion loss of 7 dB [1]. At 31 GHz, use of an improved diode [2] also yielded an output temperature of 1.2 K; however, conversion loss increased slightly to 9 dB [3]. Given its low output noise temperature, the super-Schottky mixer appeared promising for low-noise receiver applications at short millimeter wavelengths. Uncertainties persisted, however, regarding the extent to which high-frequency (i.e., 90 GHz and above) performance would be degraded by junction parasitics and possible quantum-loss mechanisms [4]. Consequently, a program to study the performance of a super-Schottky mixer at ~90 GHz was undertaken. Results of the first stage of this program, that of testing available diodes originally fabricated [2] for use at 31 GHz, are reported here.

II. Diode and Mixer Mount

The super-Schottky diodes used in this experiment were provided by the Solid State Electronics Department of the Aerospace Corporation's Electronics Research Laboratory and were similar to those tested recently at 31 GHz [2], [3]. Arrays of 3-μm-diameter Pb dots were laid down by high-field pulse plating over 2×10^{19} cm^{-3} p-type gallium arsenide (GaAs) prepared as described in [2]. The Pb contacts were formed into mushroom shapes with ≈5-μm diameter to eliminate mechanical stress at the Pb/GaAs Schottky interface, which could destroy the extreme nonlinearity of the diode. A thin gold overplating was then applied to the contacts to inhibit oxide growth. Once fabricated, the large chip was sawed into 0.005×0.010-in squares for mounting. All diodes tested in this experiment were derived from the same parent chip. A scanning-electron microscope (SEM) photograph of a typical diode array (with a whiskered contact) is shown in Fig. 1.

The mixer mount design was based on previous uncooled and cryogenic millimeter-wave mixers [5]. A cross section of the mount is shown in Fig. 2. The signal enters via

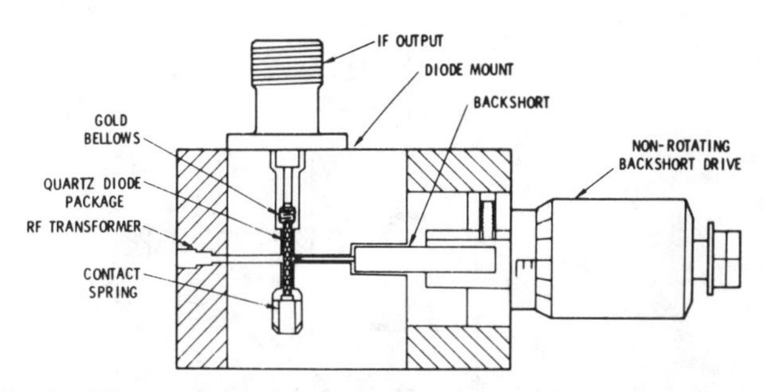

Fig. 2. Cross-sectional view of the 92-GHz super-Schottky mixer mount used in these measurements. The input waveguide is WR-10 (0.100× 0.050 in).

WR-10 waveguide, which is then stepped down to lower impedance quarter-height guide. A contacting short behind the mixer is used for tuning. The diode, whisker, and IF choke are mounted on a quartz stripline assembly across the reduced-height waveguide. Further details of the mount's construction and assembly are given in [6].

III. Measurement System and Test Procedures

A. Measurement System

A number of similar measurement system configurations were used in the course of this experiment. For clarity, we focus on the configuration used in the last series of tests, as shown in Fig. 3(a)–(c) in block diagram form. A convenient division of the test system into three subsystems—cold assembly, RF system, and IF system—facilitates discussion. Each is described below.

1) Cold Assembly: This subsystem incorporates the mixer, RF waveguide components, and IF coaxial components which reside in the dewar. The RF assembly consists of two WR-12 silver waveguide lines, with a section of thin-wall stainless-steel waveguide inserted in each run to increase thermal isolation. The signal in one line (usually the local oscillator (LO) signal) passes through a 20-dB metallized mica attenuator. The other waveguide run (used for calibration and tuning signals) couples into this line below the attenuator via a high-directivity (≳40 dB) 10-dB directional coupler.

Three identical 0.141-in-diameter semirigid coaxial lines carry output signals from the mixer, a 50-Ω termination, and a short circuit up through the header plate to a coaxial switch. These lines have a stainless outer jacket (for thermal isolation) and a beryllium–copper inner conductor (for relatively low IF loss). A fourth stainless coaxial line with a 50-Ω load can be used to terminate the IF receiver input circulator. The super-Schottky mixer and the IF lines mounted in the cold assembly are shown in Fig. 4.

Careful waveguide and coaxial-loss measurements were essential to derive desired mixer parameters. These loss measurements were repeated several times at 295 and 1 K in the course of the experiment.

2) RF System: As shown in Fig. 3(a), this subsystem incorporates the mixer, LO assembly, and klystron, calibration gas tube, and signal and RF reflection assembly with an auxiliary 92-GHz receiver. The LO power incident at the mixer was determined by measuring the klystron power

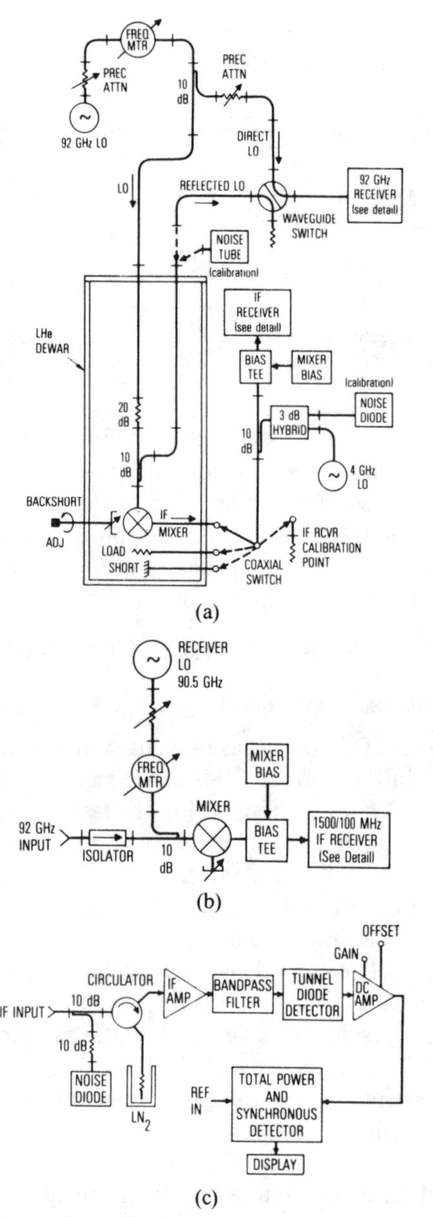

Fig. 3. (a) Block diagram of the super-Schottky measurement system. (b) Detail of RF reflectometer receiver. The IF receiver used is functionally identical to that shown in Fig. 3(c). (c) Block diagram of IF receivers used in this experiment. The two receivers used had center frequencies/bandwidth (MHz) of 1500/100 and 350/100.

Fig. 4. Photograph of the super-Schottky mixer and IF coaxial lines mounted in the dewar cold assembly. A carbon resistor temperature sensor is mounted on top of the mixer.

at the dewar input flange and accounting for the waveguide losses in the cold assembly. In most of our measurements, the mixer was tuned for optimum performance by connecting a gas tube noise source and synchronously modulated ferrite switch directly on the signal line at the dewar header. After tuning, the ferrite switch was removed and the noise tube, whose radiometric output temperature was measured to be $(12.7\pm0.6)\times10^3$ K at 90 GHz, was placed directly on the signal flange to measure mixer conversion loss.

For RF reflection measurements, the 92-GHz auxiliary was used to compare the reflected LO signal from the mixer (up through what is normally the calibration line) with the attenuated LO signal coupled into the receiver via the waveguide switch shown in Fig. 3(a). Knowing the round-trip losses in the dewar assembly and the attenuation of the direct LO signal then allowed an estimate of the high-level mixer return loss to be made. In subsequent discussions we assume this to be nearly identical to the small-signal RF return loss.[1]

3) IF System: Two IF receivers at 350 and 1500 MHz were used during the course of these measurements. The design of both is basically the same; a block diagram is shown in Fig. 3(c). The receivers incorporate a noise diode injected into a directional coupler (for IF reflection measurements), an input circulator with a cooled termination, low-noise amplifiers, a filter to define the bandpass, and a tunnel diode detector followed by a low noise dc amplifier to provide an output signal proportional to input power. Gain and offset controls are provided on the dc amplifier to scale the output to read directly in degrees Kelvin. A synchronous detector was also incorporated in all receivers and its output signal was used to optimize mixer performance when the input RF signal was synchronously modulated. The bulk of the measurements reported here were made with the 350-MHz (100-MHz bandwidth) radiometer because it generally gave the best mixer performance; possibly this was due to the lower IF loss in the mixer, less sideband gain imbalance, or lower calibration errors (since line losses were lowest at 350 MHz).

B. 4-GHz RF System

Because all diodes tested here are taken from the same chip, it was necessary to verify that the quality of the junctions tested was similar to those previously measured at lower frequencies. This was done by determining mixer output temperature, conversion loss, and video responsivity using an RF of 4 GHz, with a 350-MHz IF. Even without a backshort to assist in RF matching, it was expected that

[1] In general, the high- and low-level return losses may be different. However, at 3.95 GHz the corrections for RF mismatch applied to our data are small; moreover, a comparison of high- and low-level return losses for a super-Schottky mixer at 9 GHz [1] showed the two to differ only slightly (R. J. Pederson, private communication). At 92 GHz, the small-signal return loss was calculated from the impedance mismatch between the 1/4-height waveguide and the mixer, modeled following Lidholm [15]. The diode parameters were taken from the present set of measurements, whisker inductance was calculated from the whisker dimensions [6], and a purely reactive backshort was assumed. A minimum value (at optimum backshort setting) of $|\Gamma^2|\sim0.6$ was obtained, very close to that measured for the high-level LO.

similarity of these results to those obtained previously at X-band [2] would be a good indication of unimpaired junction quality (see Section IV-A). Further, as will be seen, the 4-GHz measurements assisted in analyzing the observed mixer and video behavior at 92 GHz.

The 4-GHz measurements were made by using a high-directivity coupler mounted at the IF input to inject LO and RF calibration signals along the common coaxial line to the mixer IF port (see Fig. 3(a)). A 4-GHz signal generator and a calibrated noise diode were combined in a 3-dB hybrid to provide the signals. Line losses to the dewar header were measured and added to the dewar coaxial losses at the respective RF and IF frequencies to calibrate the measurements.

C. *Video Responsivity Measurements*

Video responsivity was measured by frequency modulating the mixer LO ($\sim$100 Hz) at various incident power levels and synchronously detecting the video output voltage with a lock-in amplifier. At 90 GHz, the LO klystron was injected through the LO waveguide, while at 4 GHz, the signal was injected through the mixer IF port. The mixer was biased for optimum output response in these measurements using a constant-current bias supply.

D. *Mixer Measurements*

Mixer conversion loss, output-noise temperature, and IF reflection coefficient can be determined from mixer and short outputs when the calibration and IF noise diode signals are cycled [5]—provided that mixer output voltages are calibrated in degrees Kelvin at the IF port. In practice, IF radiometer output voltages were calibrated in degrees Kelvin at the coaxial switch using ambient and liquid-nitrogen-cooled terminations. Consequently, corrections were applied to the data, as detailed in [6], to account for the loss and emission of the dewar IF line, and IF receiver noise and mismatches when the noise diode was fired. IF data from the mixer, load, and short allow these contributions to be self-consistently determined assuming that coaxial-line mismatches occur within the IF receiver. This assumption is conservative, in the sense that mixer output-noise temperatures so derived are in fact upper limits.

IV. Results and Discussion

A. *Results*

Experimental results for the super-Schottky diodes are presented in Table I. Tabulated quantities are typical values selected from the larger set of many measurements. Estimated worst case standard errors are also listed in the table and were derived assuming all measurement uncertainties to be uncorrelated [6]. The tabulated uncertainties are believed to be conservative.

Several trends are evident in the data. Despite very low diode output-noise temperatures (expected for a super-Schottky junction [1]), mixer conversion efficiency at 92 GHz is poor; likewise, video responsivity is observed to be roughly two orders of magnitude worse at 92 GHz than at either 9 or 31 GHz [1], [3]. As noted in Section III-B, the mixer and video responsivity performance observed at 3.95 GHz suggests that the diode chip fabricated for this experiment is comparable to those previously tested at X-band [1]. Although the mixer RF reflection coefficient was found to be rather large (roughly 50 percent of the incident RF signal being lost by reflection), the conversion loss would only improve by $\sim$3 dB if a perfect match could be attained. Moreover, sideband imbalances can neither satisfactorily account for the generally poor conversion efficiency at 350-MHz IF (a sideband separation of 700 MHz), nor is it required to explain the marginally worse performance (just below our estimated uncertainty level) at the 1500-MHz IF.

TABLE I
Summary of Super-Schottky Results

			MIXER				DETECTOR	
ν_{LO} (GHz)	ν_{IF}/BW (MHz)	P_{LO} (nW)	SSB[a] L_{conv} (dB)	T_{diode} (K)	$\lvert\Gamma_{IF}\rvert^2$	$\lvert\Gamma_{RF}\rvert^2$	R_d (Ω)	**R** (A/W)
92	1500/100	100	25[b]	≤ 3[c]	0.20[d]	—	—	—
92	350/100	500	22[b]	≤ 3[c]	0.14[d]	0.63[g]	140	30
3.95	350/100	16	8.3[e]	≤ 5[f]	0.06	0.15[g]	156	2800

Notes: (a) Single-sideband conversion loss, corrected for IF mismatch
(b) Estimated worst-case error $\leq$ 2 dB (Ref. 6)
(c) Estimated worst-case error $\leq$ 80% (Ref. 6)
(d) Estimated worst-case error $\leq$ 10% (Ref. 6)
(e) Estimated worst-case error $\leq$ 1 dB (Ref. 6)
(f) Estimated worst-case error $\leq$ 4 K (Ref. 6)
(g) See § IIIA. 2.

B. *Discussion*

Previously developed theoretical models [1], [2] for super-Schottky diodes can be used to discuss the observed video and mixing performance of the junctions at both 4 and 92 GHz. Within the limitations of the theories, a comparison of data at both frequencies affords an important check of the experiment's internal self-consistency. Moreover, a good agreement between the experimental results and theoretical predictions would strongly argue for the basic validity of the theoretical models.

1) Video Responsivity: Consider first the video data at 4 GHz. Correcting the observed responsivity for RF mismatch yields

$$\boldsymbol{R}_{\text{cor}}(4) = \frac{\boldsymbol{R}_{\text{obs}}(4)}{(1-|\Gamma_{\text{RF}}|^2)} \sim 3294 \text{ A/W}. \tag{1}$$

Theoretically, $\boldsymbol{R}_{\text{cor}}$ is related to the diode's S-parameter (a measure of I–V nonlinearity—cf., below) and physical temperature T, by [7], [8]

$$\boldsymbol{R}_{\text{cor}} = \frac{S}{2L_p'}\left[\frac{\tanh(h\nu/2kT)}{(h\nu/2kT)}\right] \tag{2}$$

where L_p' is the diode's parasitic loss in the video mode, h and k are Planck's and Boltzmann's constants, respectively, and ν is the RF frequency. The factor in square brackets bridges the high- and low-frequency regimes of video operation. In the absence of parasitics, $\boldsymbol{R}_{\text{cor}}$ [1] corresponds to the energy per unit charge of electrons undergoing photon-assisted tunneling. At low frequencies, an energy $\sim(q/2kT)^{-1}$ is required to raise an electron above the

Fermi sea for tunneling; this corresponds to the absorption of many LO photons. At high frequencies, however, each electron can only absorb a single photon—hence, $\boldsymbol{R}_{\text{cor}}$ must have the asymptotic limit $\sim q/h\nu$. These limits are, therefore, in general applicable to all video detectors. The explicit $\tanh(x)/x$ form in (2) is a consequence of the (assumed) exponential I–V characteristic of the super-Schottky diodes around the bias voltage point. At $T=1$ K, the bracketed factor has the values 1.00 and 0.44 at 3.95 and 92 GHz, respectively.

The parasitic loss of the diode appearing in (2) can be expressed as [1]

$$L_p'=1+R_s/R_d+\omega^2C^2R_sR_d. \tag{3}$$

Here

- ω RF angular frequency, rad·s^{-1}
- R_s spreading resistance, Ω
- R_d dV/di dynamical diode resistance (Ω), from I–V curve (Table I)
- C junction capacitance, F.

At dc, the super-Schottky diodes tested here (with a doping concentration $N=2\times10^9$ cm^{-3}) have a spreading resistance [9] $R_s\cong 8.3$ Ω. At 92 GHz, skin-depth effects [11] can be expected to increase this value; for example, Held and Kerr [12] have measured an increase of ~ 2 Ω at 115 GHz in R_s for a 2.5-μm GaAs diode. In what follows, therefore, we shall assume $R_s\cong 8.3$ Ω at 4 GHz, and $R_s\cong 10$ Ω at 92 GHz, respectively.

The junction capacitance can be found from [10]

$$C=A\left[\frac{q^2\epsilon\epsilon_0 N}{2V_B}\right]^{1/2} \quad \text{F} \tag{4}$$

where

- A junction area, m^{-2} $=\pi D^2/4$
- q electron charge, 1.6×10^{-19} C
- ϵ dielectric constant of GaAs at 1 K, ~ 12.5
- ϵ_0 $=8.85\times10^{-12}$ F·m^{-1}
- N doping concentration $\cong 2\times10^{25}$ m^{-3}
- V_B barrier height near zero bias $\cong 0.4$ eV$=6.4\times10^{-20}$ J.

Hence $C\cong 149\times10^{-15}$ F.

Since $R_d=156$ Ω was measured at 4 GHz, one predicts $L_p'=1.07$ (0.3 dB) at that frequency. Relations (1) and (2) then imply $S\sim 7058$ V^{-1}. Now assuming a perfectly exponential I–V characteristic with no leakage to describe the I–V curve measured, one has $I=I_0\,[\exp(SV)-1]$, where I_0 is the saturation current. Using the I–V curve measured when the above data were taken, we obtain $S\sim 5600$ V^{-1}, a difference of only ~ 25 percent.

Using the measured value of S, the predicted video responsivity at 92 GHz using (2) is

$$\boldsymbol{R}_{\text{pred}}=\frac{5600}{2L_p'(92)}\times 0.44=\frac{1232}{L_p'(92)}\ \text{A/W}. \tag{5}$$

Evaluating $L_p'(92)$ using (3) and the measured value $R_d=$ 140 Ω then yields

$$L_p'(92)=11.42\ (10.6\ \text{dB}) \tag{6}$$

and hence

$$\boldsymbol{R}_{\text{pred}}(92)\cong 108\ \text{A/W}. \tag{7}$$

In fact, correcting $\boldsymbol{R}_{\text{obs}}(92)\sim 30$ A/W for the observed RF mismatch yields $\boldsymbol{R}_{\text{cor}}(92)\sim 81$ A/W. This is within about 30 percent of the predicted value and therefore in reasonable agreement.

TABLE II
PREDICTED VERSUS OBSERVED SUPER-SCHOTTKY CONVERSION LOSSES AT 4 AND 92 GHz

	CALCULATED			OBSERVED	
ν_{LO} (GHz)	L_p [1] (dB)	L_o [2] (dB)	L_{pred} [3] (dB)	L_{obs} [4] (dB)	$(L_{pred} - L_{obs})$ (dB)
3.95	0.5	6.7	7.2	7.6	-0.4
92	13.7	6.7	20.4	17.7	2.67

NOTES: 1 Parasitic loss (see text).
2 Intrinsic conversion loss from DC resistance ratio (see text).
3 $L_p + L_o$
4 Corrected for RF and IF mismatch; ν_{IF} = 350 MHz.

2) Conversion Loss: A comparison of mixer conversion loss at the two RF frequencies, while more uncertain than the foregoing, is also of interest. Neglecting ohmic losses, the single-sideband conversion loss of a broad-band mixer (corrected for RF and IF mismatches) is conventionally represented as a product of two terms

$$L=L_0L_p. \tag{8}$$

L_p is the parasitic loss of the mixer and is functionally similar to (3) [2]

$$L_p=(1+R_s/R_{\text{IF}})(1+R_s/R+\omega^2C^2R_sR). \tag{9}$$

Here R is the RF resistance of the nonlinear junction, R_{IF} its IF resistance, and the other terms have their previous meanings. Since R and R_{IF} were not directly measured, a straightforward evaluation of L_p is impossible. However, R_{IF} can be estimated from the slope of mixer I–V curve at optimum bias (with the LO applied). Since the bias point and LO drive levels for optimum mixing were similar to their best video response values, the approximation $R_{\text{IF}}=R_d$ may be used. This leads to a contribution ~ 0.3 dB to L_p from the first parenthetical term in (9). Further, to within a factor of ~ 2, the approximation $R=2R_{\text{IF}}$ is conventionally employed [11]. Noting that the uncertainty in this last assumption translates into <0.1 and $\sim$3-dB errors in L_p at 3.95 and 92 GHz, respectively, we obtain the results given in the Table II.

L_0 is the intrinsic loss associated with the frequency conversion process. It can be estimated from the following relations [1] using the measured dc resistance ratio $(R_{\max}/R_{\min})$ for the diode (≈ 140). Let

$$x=\tfrac{1}{2}\ln(R_{\max}/R_{\min}). \tag{10}$$

Then

$$L_0 = (2/\eta)\left[1+\sqrt{1-\eta}\right]^2 \tag{11}$$

where

$$\eta = \frac{2I_1(x)^2}{I_0(x)\left[I_0(x)+I_2(x)\right]} \tag{12}$$

and $I_j(x)$ is a modified Bessel function of the first kind of order j.

Several uncertainties are associated with this estimate. First, it implicitly assumes a Y-connected embedding network for the mixer (i.e., all harmonics of the RF and LO short-circuited), probably an adequate assumption given the large capacitance of the diode. Second, (9) assumes the LO drive level to encompass a full sweep of the diode conductance characteristic, a situation not likely to be fully attained. In both cases, deviations from the ideal will raise L_0. A final assumption implicit in the (classical) relations above is the neglect of collective quantum effects. These might either further degrade L_0 or improve it in a manner similar to the conversion gain predicted and observed in SIS mixers [13], [14]. Detailed theoretical modeling will be required to settle this issue.

Bearing these limitations in mind, a comparison of observed conversion losses at 4 and 92 GHz with those anticipated from the above relations (subject to the assumptions stated) was made. Results are given in Table II. The close agreement between measured and predicted losses may be fortuitous, given both the experimental and modeling uncertainties. However, the agreement between the experimental results at two widely differing RF frequencies and the theoretical models invoked here does suggest that these models are adequate for evaluating the potential of the super-Schottky diode at frequencies $\gtrsim$100 GHz. This being the case, it is clear that the major factor contributing to the poor diode performance observed at 92 GHz is junction parasitics. Therefore, an effort to reduce junction capacitance while maintaining spreading resistance values comparable to those here will be required, if substantially improved mixing and video performance at high frequencies is to be attained.

V. Summary

Both the video responsivity and mixing performance of heavily doped ($N=2\times10^{19}$ cm^{-3}) 3-μm active-diameter super-Schottky diodes have been measured at 92 GHz. A typical single-sideband conversion loss of $\sim$18 dB (corrected for RF and IF mismatches) was measured; a video responsivity of $\sim$80 A/W (likewise corrected for RF mismatch) was also observed. Conversion loss and responsivity were also measured at an RF frequency of 3.95 GHz, enabling the data at both frequencies to be compared using theoretical models. Agreement between the observations and model predictions is surprisingly good considering the various uncertainties (both theoretical and experimental) which enter into the comparison. This argues for both the internal consistency of the experiment performed here, as well as the basic validity of the theories developed to describe the super-Schottky diode. A major effort to reduce junction parasitics in the device will be required before mixing performance comparable to that observed at 10 and 31 GHz can be attained at frequencies $\gtrsim$100 GHz.

Acknowledgment

The authors wish to thank Dr. A. Silver for providing support, advice, and encouragement during this program. They gratefully acknowledge useful discussions with Dr. P. F. Goldsmith, Dr. A. R. Kerr, Dr. R. Pedersen, Dr. J. Tucker, and Dr. M. McColl. They thank A. B. Chase and R. Robertson for their assistance in providing the diodes tested.

References

[1] F. L. Vernon, Jr., M. F. Millea, M. F. Bottjer, A. H. Silver, R. L. Pedersen, and M. McColl, "The super Schottky diode," *IEEE Trans. Microwave Theory Tech.*, vol. MTT-25, pp. 286–294, Apr. 1977.

[2] M. McColl, M. F. Bottjer, A. B. Chase, R. J. Pedersen, A. H. Silver, and J. R. Tucker, "The super Schottky diode at 30 GHz," *IEEE Trans. Magn.*, vol. MAG-15, pp. 468–470, Jan. 1979.

[3] A. H. Silver, W. J. Wilson, R. J. Pedersen, M. McColl, and R. L. Dickman, "Superconducting low-noise receiver," The Aerospace Corp. Tech. Rep. ATR-80(8403)-1, Nov. 1979.

[4] J. R. Tucker, private communication.

[5] A. R. Kerr, "Low noise room-temperature and cryogenic mixers for 80–120 GHz," *IEEE Trans. Microwave Theory Tech.*, vol. MTT-23, pp. 781–787, Oct. 1975.

[6] R. L. Dickman, W. J. Wilson, and G. G. Berry, "Super-Schottky mixer performance at 92 GHz," The Aerospace Corp., Tech. Rep. (in publication), July 1981.

[7] J. R. Tucker and M. F. Millea, "Photon detection in non-linear tunneling devices," *Appl. Phys. Lett.*, vol. 33, no. 7, Oct. 1, 1978.

[8] ——, "Superconductive tunneling devices as millimeter wave photon detectors," *IEEE Trans. Magn.*, vol. MAG-15, no. 1, Jan. 1979.

[9] M. McColl, private communication.

[10] S. M. Sze, *Physics of Semiconductor Devices*. New York: Wiley, 1969, pp. 370–372.

[11] W. M. Kelly and G. T. Wrixon, "Conversion losses in Schottky-barrier diode mixers in the submillimeter region," *IEEE Trans. Microwave Theory Tech.*, vol. MTT-27, pp. 665–672, July 1979.

[12] D. N. Held and A. R. Kerr, "Conversion loss and noise of microwave and millimeter-wave mixers: Part I—Theory," *IEEE Trans. Microwave Theory Tech.*, vol. MTT-26, pp. 49–61, Feb. 1978.

[13] J. R. Tucker, "Predicted conversion gain in superconductor-insulator-superconductor quasiparticle mixers," *Appl. Phys. Lett.*, vol. 36, no. 6, pp. 477–479, Mar. 15, 1980.

[14] T. M. Shen, P. L. Richards, R. E. Harris, and F. L. Lloyd, "Conversion gain in millimeter-wave quasiparticle heterodyne mixers," *Appl. Phys. Lett.*, vol. 36, no. 9, pp. 777–779, May 1, 1980.

[15] S. Lidholm, "Low-noise mixers for 80–120 GHz," Onsala Space Observatory, Res. Rep. 129, Oct. 1977.

Josephson-Junction Mixer Analysis Using Frequency-Conversion and Noise-Correlation Matrices

YUAN TAUR

***Abstract*–A complete characterization and optimization have been carried out for an externally pumped Josephson-junction mixer. A noise-driven nonlinear pump equation is first solved in the time domain on a computer in order to obtain a conversion matrix and noise-correlation matrix for the small-signal current and voltage. A set of linear circuit equations formed by the matrices is then solved in the frequency domain for the mixer noise temperature and conversion efficiency. Finally, optimization is made with respect to circuit, bias, and junction parameters to find the ultimate theoretical performance.**

I. Introduction

RECENT experimental work has shown that point-contact Josephson junctions can make low-noise millimeter-wave mixers with high conversion efficiency (or gain) [1]-[3]. However, the observed noise is usually one to two orders of magnitude larger than thermal noise at the bath temperature. Although the excess noise has been attributed to a nonlinear process [4], [5], the conversion and noise correlation characteristics of a Josephson mixer are still not well understood quantitatively. This is partly because the nonlinear Josephson equations are difficult to analyze in the case of mixing. In addition, the high-order effect of noise in the junction plays such an important role that the nonlinear equation must be solved with a fluctuating term for thermal noise. This is in contrast to a classical mixer analysis [6] in which noise is simply treated as a small signal.

The first computer calculation on a Josephson mixer by Auracher and Van Duzer [7] predicted conversion gain from the modulation of I-V curve by an RF signal. However, the analysis was carried out for a current source configuration without a treatment of the nonlinear interactions between the mixer and its RF circuit. A subsequent generalization of the calculation [8] took the finite RF source impedance into account to arrive at coupling figures for conversion efficiency. But the approach lacked a noise analysis and was limited only to broad-band resistive circuits. A better understanding of the Josephson mixer was obtained from an electronic analog computer which modeled the narrow-band RF circuit more realistically [9], [10]. With a proper simulation of thermal noise, the analog computer can be used to evaluate mixer noise temperature given a set of parameters. However, analog simulation is inherently limited to one special case at a time, making it very tedious to cover all the parameter values of interest. Moreover, since a general formulation does not exist for the nonlinear conversion process, a systematic search for the optimum condition cannot be carried out.

Manuscript received February 21, 1980; revised April 18, 1980. This work was supported in part by Columbia Radiation Laboratory, Columbia University, New York, NY, and by Rockwell International Science Center, Thousand Oaks, CA.

The author was with NASA Goddard Institute for Space Studies, Goddard Space Flight Center, New York, NY 10025. He is now with Rockwell International Science Center, Thousand Oaks, CA 91360.

In this paper, we present a complete analysis to evaluate and optimize the noise temperature of a Josephson-junction mixer. The nonlinear Josephson equation, including both a large RF drive and a noise term, is first solved in the time domain using a digital computer. The results along with their fluctuations can be used to derive a small-signal impedance matrix and a noise-correlation matrix in the frequency domain. Following a general mixer analysis, we then solve a linear circuit equation for conversion efficiency and noise temperature to find out the optimum RF impedance. In this method, all the effects of up-conversion, image-conversion, and noise interaction are included. Furthermore, the advantage of image rejection (single-sideband mixer) can be easily investigated in the frequency-domain analysis. The computation is carried out for a variety of junction and bias parameters to determine the ultimate performance limits of the Josephson mixer.

II. Circuit Model

Since the experimental I-V curves of a low-capacitance point-contact Josephson junction are best described by the resistively shunted junction (RSJ) model [11], this formalism is used in our mixer analysis. In the RSJ model, a resistor of constant resistance R accounts for the quasi-particle current, and is in parallel with an ideal Josephson element which accounts for the superconducting pair current, as shown in the box of Fig. 1. The electrical characteristics of an ideal Josephson element are described by the Josephson relations [12]

$$I_J(t) = I_c \sin \phi(t) \tag{1}$$

$$V(t) = \frac{\hbar}{2e} \frac{d}{dt} \phi(t). \tag{2}$$

Here $\phi(t)$ is the superconducting phase difference across the junction, and I_c is a constant equal to the maximum supercurrent of the junction. Also shown in Fig. 1 is a current source representing Johnson noise having an autocorrelation function

$$\langle \delta I_n(t) \delta I_n(t') \rangle = \frac{2kT}{R} \delta(t - t') \tag{3}$$

Reprinted from *IEEE Tran. Electron Devices*, vol. ED-27, pp. 1921–1928, Oct. 1980.

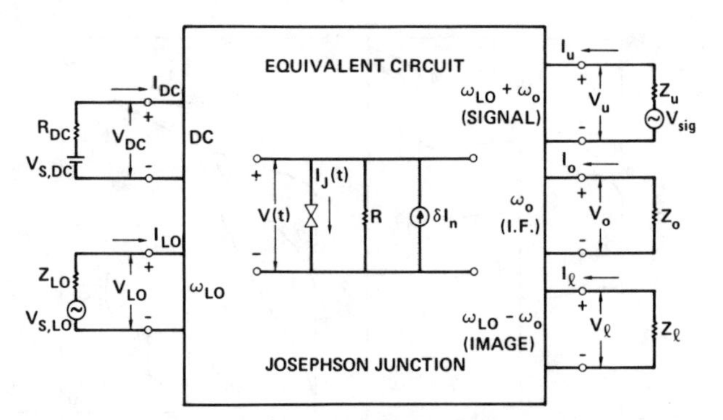

Fig. 1. Equivalent circuit of a Josephson-junction mixer with large-signal sources to the left and small-signal circuits to the right of the junction box. All the external circuits are assumed to be connected to the junction terminals but at different frequencies. The ac current and voltage are expressed in complex half amplitudes.

where T is the ambient temperature. Such a noise term must be included in the nonlinear pump equation as thermal noise cannot be regarded as a small signal.

If the junction is driven by a dc source and a local oscillator (LO) at frequency ω_{LO} as shown to the left of Fig. 1, the pump equation can be written in time domain as

$$\frac{\hbar}{2eR}\frac{d}{dt}\phi(t)+I_c\sin\phi(t)=I_{dc}+(I_{LO}e^{j\omega_{LO}t}+I_{LO}^*e^{-j\omega_{LO}t})+\delta I_n(t). \quad (4)$$

Here we use a complex notation for all ac amplitudes such as $\{V_{S,LO}\exp(j\omega_{LO}t)+V_{S,LO}^*\exp(-j\omega_{LO}t)\}$ for the LO generator voltage, and $\{I_{LO}\exp(j\omega_{LO}t)+I_{LO}^*\exp(-j\omega_{LO}t)\}$ for the LO current, etc. Equation (4) is valid under the assumption that the embedding circuit has a very high impedance at all frequencies except near dc and ω_{LO}, so that no external current can be generated at the harmonics of the LO frequency: $2\omega_{LO}, 3\omega_{LO}, 4\omega_{LO}, \cdots$, etc. This assumption has been verified from measurements on a scaled point-contact mixer model [13]. However, the impedance of the LO source is finite, which implies that the LO drive is not a constant current source. For a given $V_{S,LO}$, the magnitude of I_{LO} changes with dc bias, and the static *I-V* curve deviates significantly from that previously published for a constant current source [11].

One can choose a time origin such that I_{LO} is real, and express (4) in terms of dimensionless variables as follows:

$$\frac{d}{d\tau}\phi(\tau)+\sin\phi(\tau)=i_{dc}+2i_{LO}\cos\Omega_{LO}\tau+\delta i_n(\tau) \quad (5)$$

where $i_{dc}=I_{dc}/I_c$, $i_{LO}=I_{LO}/I_c$, and $\delta i_n=\delta I_n/I_c$ are normalized currents; $\tau=(2eRI_c/\hbar)t$ is normalized time; and $\Omega_{LO}=\hbar\omega_{LO}/2eRI_c$ is normalized LO frequency. A factor of two arises in the LO current due to the half-amplitude notation used here. The noise relationship becomes

$$\langle\delta i_n(\tau)\delta i_n(\tau')\rangle=2\Gamma\delta(\tau-\tau') \quad (6)$$

where $\Gamma=2ekT/\hbar I_c$ is a dimensionless noise parameter, equal to the thermal energy kT divided by the Josephson coupling energy $\hbar I_c/2e$. In dimensionless units, the junction is characterized by the parameters Ω_{LO} and Γ; while the bias conditions are represented by the parameters i_{dc} and i_{LO}. Both Ω_{LO} and Γ are important factors governing the mixer performance. It is desirable to keep $\Omega_{LO}<1$ since most of the RF current would then interact with the inductive Josephson element. The condition $\Gamma<1$ should also be satisfied, otherwise the Josephson nonlinearity would be smeared out by noise saturation. The region of interest in our analysis is, therefore, restricted to $\Omega_{LO}<1$ and $\Gamma<1$, which are satisfied in most experimental situations.

III. Method of Computation and *I-V* Curves

Given a junction and its bias, the pump equation (5) can be solved numerically for $\phi(\tau)$ on a digital computer (IBM 360/95). A good choice for the integration step is $H=\Delta\tau=0.25$. No discrepancy is found on a numerical test at half the step size. The noise $\delta i_n(\tau)$ is generated by calling a Gaussian-distributed (approximately) random number at each step of integration. The random numbers have a zero mean and a variance

$$\sigma^2=2\Gamma/H \quad (7)$$

depending on the step size. Due to discrete sampling, their spectrum is frequency independent (white) only below $\Omega_n\approx\pi/H(\gg 1)$. However, the junction noise beyond the cutoff does not have any significant effect on the mixer behavior, as has been confirmed in the half-step test.

The normalized junction voltage $v(\tau)=V(t)/RI_c=d\phi(\tau)/d\tau$ can be evaluated once ϕ is solved as a function of τ. It can be expressed in the frequency domain by taking a Fourier transform over a period P equal to ten LO cycles, i.e., $P=10(2\pi/\Omega_{LO})$

$$v(\tau)=v_{dc}+(v_{LO}e^{j\Omega_{LO}\tau}+v_{LO}^*e^{-j\Omega_{LO}\tau})+(\text{higher frequency components}) \quad (8)$$

where $v_{dc}=V_{dc}/RI_c$ is real but $v_{LO}=V_{LO}/RI_c$ is complex because there is a phase difference between V_{LO} and I_{LO}. Voltage components at harmonic frequencies can be ignored since they do not induce external currents. Both v_{dc} and v_{LO} contain fluctuations due to the noise at low frequency and Ω_{LO}

$$v_{dc}=\langle v_{dc}\rangle+\delta v_{dc} \quad (9)$$

$$v_{LO}=\langle v_{LO}\rangle+\delta v_{LO}. \quad (10)$$

In order to obtain the average as well as the spectral density of fluctuation, the Fourier transform is repeated to yield $v_{dc}^{(1)}, v_{LO}^{(1)}; v_{dc}^{(2)}, v_{LO}^{(2)}; v_{dc}^{(3)}, v_{LO}^{(3)}; \cdots; v_{dc}^{(K)}, v_{LO}^{(K)}$. Therefore, the total period of integration is KP, where $K=1000$ for an accuracy better than 5 percent. Then the auto- and crosscorrelations can be evaluated as follows:

$$\langle v_{dc}\rangle=\frac{1}{K}\sum_{i=1}^{K}v_{dc}^{(i)} \quad (11)$$

$$\langle v_{LO}\rangle=\frac{1}{K}\sum_{i=1}^{K}v_{LO}^{(i)} \quad (12)$$

$$\langle(\delta v_{dc})^2\rangle_P=\frac{1}{K}\sum_{i=1}^{K}(v_{dc}^{(i)})^2-\langle v_{dc}\rangle^2 \quad (13)$$

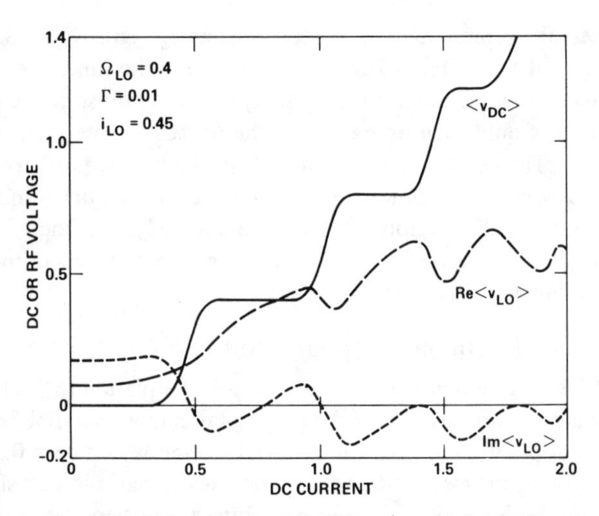

Fig. 2. Normalized dc voltage and in-phase/out-of-phase RF voltages versus dc current (normalized). One expects $\mathrm{Re}\langle v_{LO}\rangle \to i_{LO}$ and $\mathrm{Im}\langle v_{LO}\rangle \to 0$ at very large dc bias.

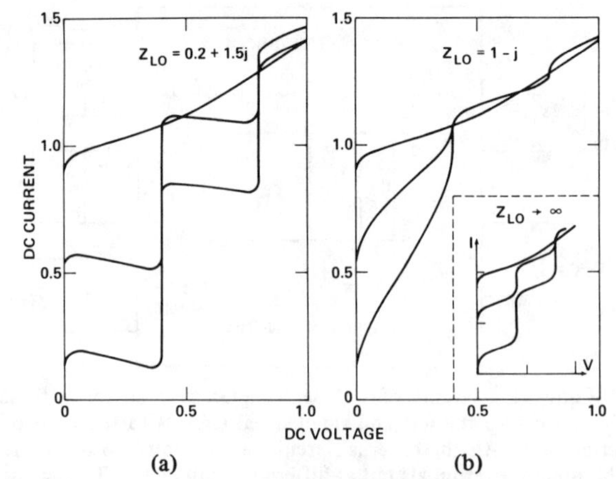

Fig. 3. Two series of I–V curves computed at Ω_{LO} = 0.4 and Γ = 0.01 for different LO impedances (normalized) shown. In each case, the LO power is zero for the top curve, then increases toward lower curves. The inset shows the case of constant LO current for comparison.

$$\langle|\delta v_{LO}|^2\rangle_P = \frac{1}{K}\sum_{i=1}^{K} |v_{LO}^{(i)}|^2 - |\langle v_{LO}\rangle|^2 \tag{14}$$

$$\langle(\delta v_{LO})^2\rangle_P = \frac{1}{K}\sum_{i=1}^{K} (v_{LO}^{(i)})^2 - \langle v_{LO}\rangle^2 \tag{15}$$

$$\langle(\delta v_{dc})(\delta v_{LO})\rangle_P = \frac{1}{K}\sum_{i=1}^{K} v_{dc}^{(i)} v_{LO}^{(i)} - \langle v_{dc}\rangle\langle v_{LO}\rangle. \tag{16}$$

Here (12), (15), (16) are complex. The subscript P indicates that the correlations depend on the period of Fourier transformation ($\propto 1/P$). These quantities are the ones needed for the mixer noise analysis.

For a given junction and LO frequency, $\langle v_{dc}\rangle$ and $\langle v_{LO}\rangle$ are functions of bias currents i_{dc}, i_{LO}. The dependence on dc bias is shown in Fig. 2, where the quasi-periodic variation is a result of RF-induced Josephson steps. It is found that the out-of-phase RF voltage or $\mathrm{Im}\langle v_{LO}\rangle$ does not vanish between steps, which differs from previously computed results in the absence of noise [14]. In fact, it changes from inductive to capacitive as the bias is increased from the first step to the second. This is due to noise rounding or partial RF synchronization of the ac Josephson current off the steps. Another related result is that the noise correlation of the dc and out-of-phase RF voltage or $\mathrm{Im}\langle(\delta v_{dc})(\delta v_{LO})\rangle_P$ is always near 100 percent between steps whenever $\Gamma < 0.1$. These effects show that the mixer characteristics are strongly influenced by the nonlinear interaction of noise.

Once the functions $\langle v_{dc}\rangle = f(i_{dc}, i_{LO})$ and $\langle v_{LO}\rangle = g_1(i_{dc}, i_{LO}) + jg_2(i_{dc}, i_{LO})$ are evaluated under various bias conditions, one can compute the modified I–V curves for a finite source impedance Z_{LO}. The load equations for the circuit on the left of Fig. 1 are

$$f(i_{dc}, i_{LO}) + r_{dc} i_{dc} = v_{S,dc} \tag{17}$$

$$g_1(i_{dc}, i_{LO}) + jg_2(i_{dc}, i_{LO}) + z_{LO} i_{LO} = v_{S,LO} = |v_{S,LO}|e^{j\theta} \tag{18}$$

where $r_{dc} = R_{dc}/R$, $z_{LO} = Z_{LO}/R$, and $v_{S,dc}$, $v_{S,LO}$ are normalized with respect to RI_c. For convenience, the phase angle θ of the LO generator is adjusted such that i_{LO} is real. Given r_{dc}, z_{LO}, $|v_{S,LO}|$, and $v_{S,dc}$, the algebraic equations (17), (18) can be solved graphically for θ, i_{dc}, i_{LO}, and, therefore $\langle v_{dc}\rangle = f(i_{dc}, i_{LO})$. The provides a point for the I–V curve. At a fixed LO power (constant $|v_{S,LO}|$), the entire I–V curve can be generated by repeating the process with different values of $v_{S,dc}$. The source resistance r_{dc} does not affect the shape of the I–V curve except for stability when there is a region of negative slope. Shown in Fig. 3 are two families of I–V curves computed at various LO power levels for two values of z_{LO}. The deviations from the well-known constant current case (inset) are obvious. They are in good agreement with experimental curves and analog simulator results [9]. When the RF termination is inductive as in Fig. 3(a), the current i_{LO} increases with dc bias between steps, resulting in a negative differential resistance. It can be stably biased only if r_{dc}^{-1} is larger than the magnitude of the slope. On the other hand, if the RF termination is capacitive as in Fig. 3(b), i_{LO} decreases as the bias voltage is increased. In this case, the dynamic resistance between steps becomes much lower, and the first step is reduced appreciably.

IV. Conversion and Noise-Correlation Matrices

Now we consider the mixer response to a small applied signal at a normalized frequency $\hbar\omega_u/2eRI_c = \Omega_u = \Omega_{LO} + \Omega_0$ (upper sideband or usb), where $\Omega_0 \ll \Omega_{LO}$. Based on the embedding circuit assumption, the current generated by the nonlinear Josephson element is only at intermediate frequency (IF) Ω_0 and image frequency $\Omega_I = \Omega_{LO} - \Omega_0$ (lower sideband or lsb), as shown on the right of Fig. 1. The small-signal current is then

$$i_{ss}(\tau) = (i_u e^{j\Omega_u\tau} + i_u^* e^{-j\Omega_u\tau}) + (i_0 e^{j\Omega_0\tau} + i_0^* e^{-j\Omega_0\tau}) + (i_I e^{j\Omega_I\tau} + i_I^* e^{-j\Omega_I\tau}) \tag{19}$$

in normalized units. If we add $i_{ss}(\tau)$ to (5) as a perturbation, the additional voltage (normalized) $d\phi_{ss}(\tau)/d\tau$ can be written as

$$v_{ss}(\tau) = (v_u e^{j\Omega_u \tau} + v_u^* e^{-j\Omega_u \tau}) + (v_0 e^{j\Omega_0 \tau} + v_0^* e^{-j\Omega_0 \tau}) + (v_l e^{j\Omega_l \tau} + v_l^* e^{-j\Omega_l \tau}) + \text{(higher sidebands)}. \quad (20)$$

Again we may ignore higher sideband voltage at $(2\Omega_{LO} \pm \Omega_0), \cdots$, etc. There are, of course, noise components at frequencies Ω_u, Ω_0, and Ω_l arising from fluctuations δv_{dc} and δv_{LO}. In the small-signal limit, the relationship between $v_{ss}(\tau)$ and $i_{ss}(\tau)$ is linear and can be described by a 3 X 3 impedance matrix

$$\begin{bmatrix} v_u \\ v_0 \\ v_l^* \end{bmatrix} = \begin{bmatrix} z_{uu} & z_{u0} & z_{ul} \\ z_{0u} & z_{00} & z_{0l} \\ z_{lu} & z_{l0} & z_{ll} \end{bmatrix} \begin{bmatrix} i_u \\ i_0 \\ i_l^* \end{bmatrix} + \text{(noise)} \quad (21)$$

or

$$\tilde{v}_s = \tilde{\tilde{Z}} \cdot \tilde{i}_s + \tilde{\delta v}_s. \quad (22)$$

All the matrix elements are normalized with respect to R. Since the IF is much lower than signal frequency in practice, the matrix can be determined from the pump solutions $\langle v_{dc} \rangle = f(i_{dc}, i_{LO})$, $\langle v_{LO} \rangle = g_1(i_{dc}, i_{LO}) + jg_2(i_{dc}, i_{LO})$ and their derivatives following a generalized mixer theory [15]

$$z_{uu} = \frac{1}{2}\left(\frac{\partial \langle v_{LO} \rangle}{\partial i_{LO}} + \frac{\langle v_{LO} \rangle}{i_{LO}}\right) \quad \text{(RF dynamic impedance)} \quad (23)$$

$$z_{u0} = \frac{\partial \langle v_{LO} \rangle}{\partial i_{dc}} \quad \text{(up-conversion)} \quad (24)$$

$$z_{ul} = \frac{1}{2}\left(\frac{\partial \langle v_{LO} \rangle}{\partial i_{LO}} - \frac{\langle v_{LO} \rangle}{i_{LO}}\right) \quad \text{(image conversion)} \quad (25)$$

$$z_{0u} = \frac{1}{2}\frac{\partial \langle v_{dc} \rangle}{\partial i_{LO}} \quad \text{(down-conversion)} \quad (26)$$

$$z_{00} = \frac{\partial \langle v_{dc} \rangle}{\partial i_{dc}} \quad \text{(dc dynamic resistance)} \quad (27)$$

and

$$z_{0l} = z_{0u}, \quad z_{lu} = z_{ul}^*, \quad z_{l0} = z_{u0}^*, \quad z_{ll} = z_{uu}^* \quad (28)$$

from symmetry. The second row of the matrix is real since i_{LO} has a zero phase. In contrast to resistive mixers, the calculated conversion matrix of a Josephson mixer is neither reciprocal nor passive; therefore, conversion gain is possible.

The impedance matrix depends on the junction parameters Ω_{LO}, Γ, as well as on the bias parameters i_{dc}, i_{LO}. (Without loss of generality, the bias currents can be used as a set of independent parameters instead of the dc and LO generator voltages.) A typical case is shown in Fig. 4, where the dc and RF dynamic impedance and the down-conversion impedance z_{0u} are plotted versus dc bias. All but the RF resistance show a symmetric shape between the zeroth and first induced steps. The mixer should be biased halfway between the steps for a maximum z_{0u}. If the noise parameter Γ is less than 0.1, z_{0u} is proportional to z_{00} as expected from earlier calculations [7]. In addition, the maximum z_{00} and z_{0u} are found to vary as $\Gamma^{-1/2}$ at a fixed Ω_{LO} [10], as shown in Fig. 5. As $\Gamma \to 1$, however, down-conversion is much less effective because of severe noise rounding of the steps.

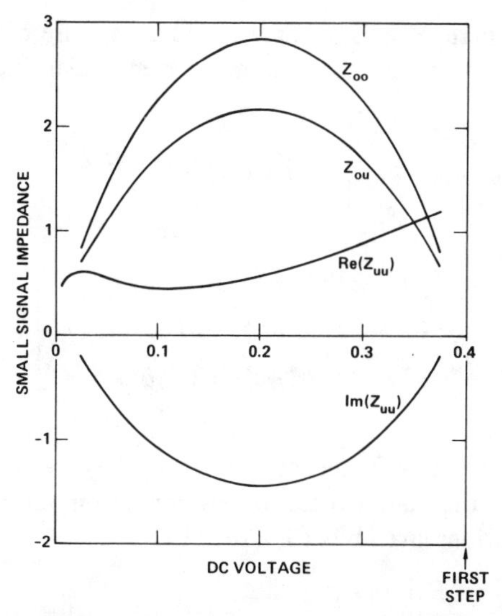

Fig. 4. Elements of the conversion matrix (normalized to R) as functions of bias voltage between the zeroth and first Josephson steps. They are computed for the same Ω_{LO}, Γ, and i_{LO} as those in Fig. 2, but for a more limited range of v_{dc}.

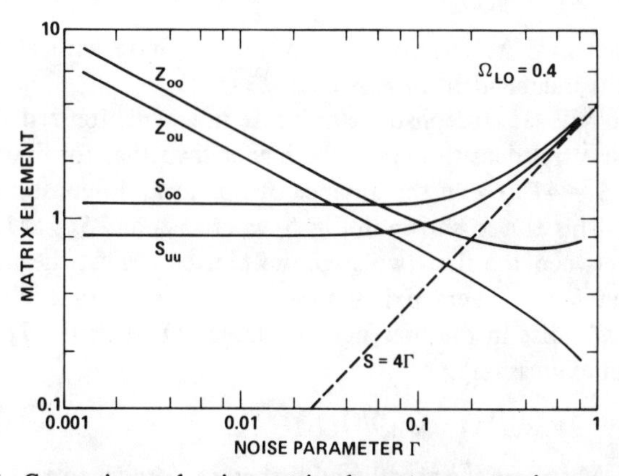

Fig. 5. Conversion and noise matrix elements versus noise parameter. All the quantities are taken at halfway between the first two steps, where the maximum values are insensitive to i_{LO}. The dashed line shows the spectral density of thermal noise without the ideal Josephson element.

In order to express the noise components in (21) explicitly, we consider a very-long observation time T (normalized) and expand δv_{dc}, δv_{LO} in Fourier series [16]

$$\delta v_{dc} = \sum_{n=-\infty}^{\infty} a_n e^{j(2n\pi/T)\tau}$$

$$\delta v_{LO} = \sum_{n=-\infty}^{\infty} b_n e^{j(2n\pi/T)\tau} \quad (29)$$

where $a_{-n} = a_n^*$ but $b_{-n} \neq b_n^*$. If a normalized bandwidth $\Delta f = 1/T$ is used for the mixer noise, we have

$$\tilde{\delta v}_s = \begin{bmatrix} b_N \\ a_N \\ b_{-N}^* \end{bmatrix} \quad (30)$$

for (22), where N satisfies $2\pi N/T = \Omega_0$. A noise correlation matrix can then be defined in terms of spectral densities

$$\tilde{\tilde{S}} = \begin{bmatrix} S_{uu} & S_{u0} & S_{ul} \\ S_{0u} & S_{00} & S_{0l} \\ S_{lu} & S_{l0} & S_{ll} \end{bmatrix} = 2T\langle(\tilde{\delta v}_s) \times (\tilde{\delta v}_s)^{T*}\rangle$$

$$= \begin{bmatrix} 2T\langle b_N b_N^* \rangle & 2T\langle b_N a_N^* \rangle & 2T\langle b_N b_{-N} \rangle \\ 2T\langle a_N b_N^* \rangle & 2T\langle a_N a_N^* \rangle & 2T\langle a_N b_{-N} \rangle \\ 2T\langle b_{-N}^* b_N^* \rangle & 2T\langle b_{-N}^* a_N^* \rangle & 2T\langle b_{-N}^* b_{-N} \rangle \end{bmatrix}. \tag{31}$$

Since the intermediate frequency Ω_0 is much smaller than Ω_{LO}, only the low-frequency spectra in (29) are of interest. In this case, the matrix $\tilde{\tilde{S}}$ can be evaluated from the auto- and crosscorrelations given by (13)-(16) [16]

$$S_{uu} = S_{ll} = 2P\langle|\delta v_{LO}|^2\rangle_P \tag{32}$$

$$S_{u0} = S_{0l} = S_{0u}^* = S_{l0}^* = 2P\langle(\delta v_{dc})(\delta v_{LO})\rangle_P \tag{33}$$

$$S_{ul} = S_{lu}^* = 2P\langle(\delta v_{LO})^2\rangle_P \tag{34}$$

$$S_{00} = 2P\langle(\delta v_{dc})^2\rangle_P. \tag{35}$$

The matrix $\tilde{\tilde{S}}$ is Hermitian and positive-definite with all the elements independent of P.

When the ac Josephson current is not synchronized, the noise spectral densities are much higher than that for thermal noise, $S = 4\Gamma$ (6), in the absence of nonlinear Josephson element. This is also shown in Fig. 5, where S_{00} and S_{uu} at halfway between the first two steps are plotted against the noise parameter Γ. There exists an analytic model for such an "excess" noise in the presence of a large LO current [17]. It gives an expression

$$S_{00} = 4\pi\langle v_{dc}\rangle(1 - \langle v_{dc}\rangle/\Omega_{LO}) \tag{36}$$

for $\langle v_{dc}\rangle$ between the zeroth and first steps. Similar to z_{0u}, S_{00} also has a maximum value ($=\pi\Omega_{LO}$) at $\langle v_{dc}\rangle = \Omega_{LO}/2$. Our computed results check out very well with (36) within the validity of the analytic theory in which $\Gamma < 0.1$. One notices that the voltage noise stays constant no matter how small the driving noise Γ is. This is consistent with the observed noise being much greater than thermal noise in a low-resistance, high critical-current junction [18]. The calculation also shows that the cross correlations given by the off-diagonal elements of $\tilde{\tilde{S}}$ are rather strong when $\Gamma \ll 1$ and $\Omega_{LO} < 1$.

V. Mixer Noise Temperature and its Optimization

Knowing the embedding circuit at signal, IF, and image frequencies, mixer conversion efficiency and noise temperature can be calculated from the conversion and noise correlation matrices [6]. The equations for the circuit on the right of Fig. 1 are

$$\begin{bmatrix} v_u \\ v_0 \\ v_l^* \end{bmatrix} + \begin{bmatrix} z_u & 0 & 0 \\ 0 & z_0 & 0 \\ 0 & 0 & z_l^* \end{bmatrix} \begin{bmatrix} i_u \\ i_0 \\ i_l^* \end{bmatrix} = \begin{bmatrix} v_{sig} \\ 0 \\ 0 \end{bmatrix} \tag{37}$$

or

$$\tilde{v}_s + \tilde{\tilde{Z}}_s \cdot \tilde{i}_s = \tilde{v}_{sig}. \tag{38}$$

Here $v_{sig} = V_{sig}/RI_c$, and z_u, z_0, z_l^* are normalized to R. Matrix equations (22) and (38) can be solved for the small-signal current

$$\tilde{i}_s = (\tilde{\tilde{Z}} + \tilde{\tilde{Z}}_s)^{-1} \cdot (\tilde{v}_{sig} - \tilde{\delta v}_s). \tag{39}$$

If we let

$$\tilde{\tilde{Y}} = (\tilde{\tilde{Z}} + \tilde{\tilde{Z}}_s)^{-1} = \begin{bmatrix} Y_{uu} & Y_{u0} & Y_{ul} \\ Y_{0u} & Y_{00} & Y_{0l} \\ Y_{lu} & Y_{l0} & Y_{ll} \end{bmatrix} \tag{40}$$

and

$$\tilde{Y}_0 = \begin{bmatrix} Y_{0u} \\ Y_{00} \\ Y_{0l} \end{bmatrix} \tag{41}$$

the IF current is simply given by the scalar equation

$$i_0 = \tilde{Y}_0^T \cdot (\tilde{v}_{sig} - \tilde{\delta v}_s) = Y_{0u}v_{sig} - \tilde{Y}_0^T \cdot \tilde{\delta v}_s. \tag{42}$$

The mean square IF current is, therefore,

$$\begin{aligned} 2\langle i_0 i_0^* \rangle &= 2|Y_{0u}|^2|v_{sig}|^2 + 2\tilde{Y}_0^T \cdot \langle(\tilde{\delta v}_s) \times (\tilde{\delta v}_s)^{T*}\rangle \cdot \tilde{Y}_0^* \\ &= 2|Y_{0u}|^2|v_{sig}|^2 + (\tilde{Y}_0^T \cdot \tilde{\tilde{S}} \cdot \tilde{Y}_0^*)\Delta f. \end{aligned} \tag{43}$$

Here (31) has been used for a noise bandwidth $\Delta f = 1/T$. Since the mixer conversion efficiency η is equal to the down-converted IF power divided by the available signal power, the signal term in (43) gives

$$\eta = \frac{2\,\mathrm{Re}\,(z_0)|Y_{0u}|^2|v_{sig}|^2}{2|v_{sig}|^2/4\,\mathrm{Re}\,(z_u)} = 4|Y_{0u}|^2\,\mathrm{Re}\,(z_u)\,\mathrm{Re}\,(z_0). \tag{44}$$

In order to obtain the mixer noise temperature T_M, we let the signal term equal the noise term in (43) and substitute $2|v_{sig}|^2$ with $4\,\mathrm{Re}\,(z_u)\Gamma_M\Delta f$ where $\Gamma_M = 2ekT_M/\hbar I_c$

$$\frac{T_M}{T} = \frac{\Gamma_M}{\Gamma} = \frac{(\tilde{Y}_0^T \cdot \tilde{\tilde{S}} \cdot \tilde{Y}_0^*)}{4\Gamma|Y_{0u}|^2\,\mathrm{Re}\,(z_u)}. \tag{45}$$

Another parameter of importance is the IF output impedance

$$z_{out} = Y_{00}^{-1} - z_0. \tag{46}$$

For a double-sideband (DSB) mixer, $z_u = z_l = z_{LO}$, it can be shown that z_{out} is real. For a single-sideband (SSB) mixer, $z_u \neq z_l$, and z_{out} is complex. In any case, z_{out} is independent of z_0. The IF load should, therefore, be conjugate matched for maximum conversion efficiency. That is, $z_0 = z_{out}^*$, provided that $\mathrm{Re}\,(z_{out}) > 0$. However, the output impedance may have a negative real part, such as the case in Fig. 3(a). Then the conversion efficiency is potentially unbounded. On the contrary, the mixer noise temperature is independent of z_0 and always remains finite and continuous. Therefore, it is the purpose of our optimization to find the minimum mixer noise temperature under a variety of conditions.

We first consider the dependence of T_M/T on RF impedance z_u, z_l for a given pair of matrices $\tilde{\tilde{Z}}, \tilde{\tilde{S}}$. An example is shown in Fig. 6, where constant T_M/T contours are plotted in a complex plane of signal impedance z_u. In both DSB and SSB

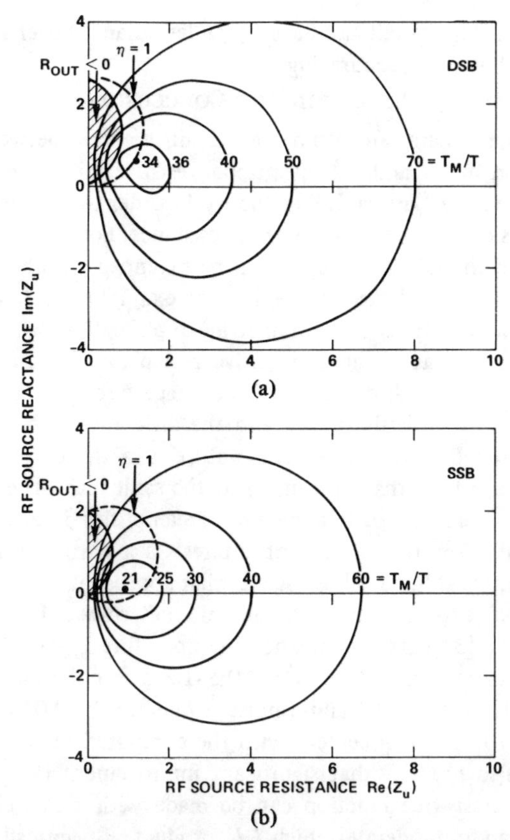

Fig. 6. Optimization of mixer noise temperature with respect to signal source impedance at a fixed bias, $\langle v_{dc} \rangle = 0.16$. Other parameters are the same as in Fig. 2. The dots indicate the signal impedances for minimum $T_M/T = 34$ (DSB) and $T_M/T = 21$ (SSB). For the SSB case in (b), the image impedance is fixed at its best value, $z_I = 4j$.

cases,[1] minimum T_M/T takes place outside a shaded region of negative output resistance. Also shown in Fig. 6 are two dashed curves for unity conversion efficiency assuming a matched IF load. Stable conversion gain is obtained inside two moon-shape regions. The lowest noise temperature for SSB is appreciably better than that for DSB, since the image impedance z_I can be varied independently in the former case. The best image termination is always reactive and close to an open circuit. A short circuit at the image frequency usually results in a large noise temperature.

Optimization with respect to RF terminations is repeated for different matrices to obtain minimum T_M/T as a function of dc bias between the first two steps. A typical case is shown in Fig. 7, where the corresponding conversion efficiency is also plotted. Here the mixer is very noisy near either step, and the best bias is slightly below the voltage midway between the steps. In general, the bias voltage for maximum conversion efficiency does not coincide with that for lowest mixer noise temperature. Bias beyond the first Josephson step has also been explored, but the result is not as good. For a given junction (fixed Ω_{LO} and Γ), the minimum T_M/T is rather insensitive to LO current (i_{LO}), provided that a significant fraction of the zero-voltage current is suppressed by the LO

[1]The terms DSB and SSB used in this paper are referring only to the input terminations of the mixer or receiver. All the conversion efficiency and noise temperature are for the detection of a narrow-band signal on one side of the LO, rather than for the figures inferred from a broad-band radiometric measurement.

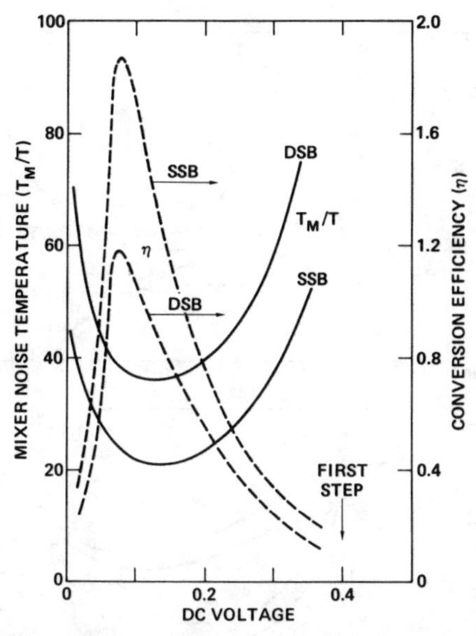

Fig. 7. Mixer noise temperature (solid curves) and corresponding conversion efficiency (dashed curves) versus dc bias. The optimum signal/image impedances for each bias are determined individually following a procedure similar to Fig. 6. The parameters $\Omega_{LO} = 0.4$, $\Gamma = 0.01$, $i_{LO} = 0.45$ are the same as in Fig. 2. At the bias voltage where T_M/T is lowest, the IF output impedances (normalized) are $z_{out} = 3.0$ for DSB, and $z_{out} = 1.0 + 1.6j$ for SSB.

power like the cases in Fig. 3. At a much higher LO power, however, the mixer performance does degrade.

The minimum T_M/T can be obtained under different optimum bias for a variety of junction parameters Γ and Ω_{LO}. The result is shown in Fig. 8 versus noise parameter Γ for two of the normalized frequencies studied. Also shown, in broken lines, are the corresponding conversion efficiencies. At a fixed normalized frequency, the mixer noise temperature divided by the ambient temperature is lowest around $\Gamma = 0.1$, where $\eta \approx 0.3$. It arises from two opposing effects: larger "excess" noise with respect to thermal noise at $\Gamma \ll 1$, and severe noise saturation when $\Gamma \rightarrow 1$. Such a minimum figure of T_M/T improves significantly toward lower values of Ω_{LO}. In particular, the theory predicts that a mixer noise temperature as low as five times the ambient temperature can be achieved in an SSB mixer at $\Omega_{LO} = 0.2$.

The sensitivity of a millimeter-wave heterodyne receiver is limited not only by the noise in the mixer but also by the noise in the following IF amplifier. The total receiver noise temperature with respect to the ambient temperature is

$$T_R/T = T_M/T + (1/\eta)(T_A/T) \tag{47}$$

where T_A is the amplifier noise temperature. Its contribution depends on the value of conversion efficiency. Since minimum T_M/T does not take place at maximum η, one must carry out the optimization for T_R/T also. Assuming a T_A/T equal to 5, which can be achieved by a cooled FET amplifier when $T \approx 4$ K, we obtain minimum receiver noise contours in an Ω_{LO}-Γ plane as shown in Fig. 9. The results are calculated for DSB and SSB receivers under the restriction $\text{Re}(z_{out}) > 0$ and with a matched IF load. For a fixed normalized frequency, lowest T_R/T is found at a noise parameter between 0.02 and 0.05. This is shifted from $\Gamma \approx 0.1$ for minimum T_M/T toward a higher conversion efficiency. A comparison

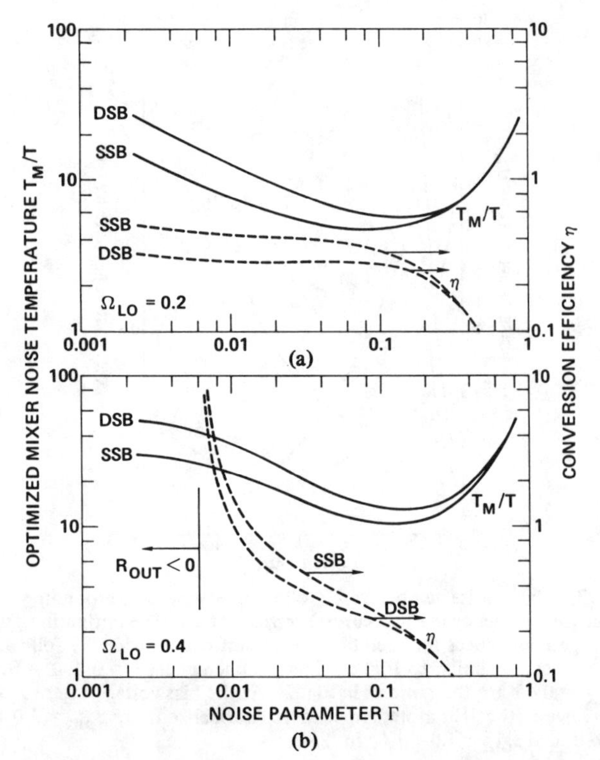

Fig. 8. DSB/SSB mixer noise temperature (solid) at best bias and corresponding conversion efficiency (dashed) versus noise parameter for two normalized frequencies shown. The output resistance becomes negative and η is unbounded below $\Gamma \equiv 2ekT/\hbar I_c \approx 0.006$ when $\Omega_{LO} \equiv \hbar\omega_{LO}/2eRI_c = 0.4$. There is no such divergence for $\Omega_{LO} = 0.2$ over the range of parameter studied.

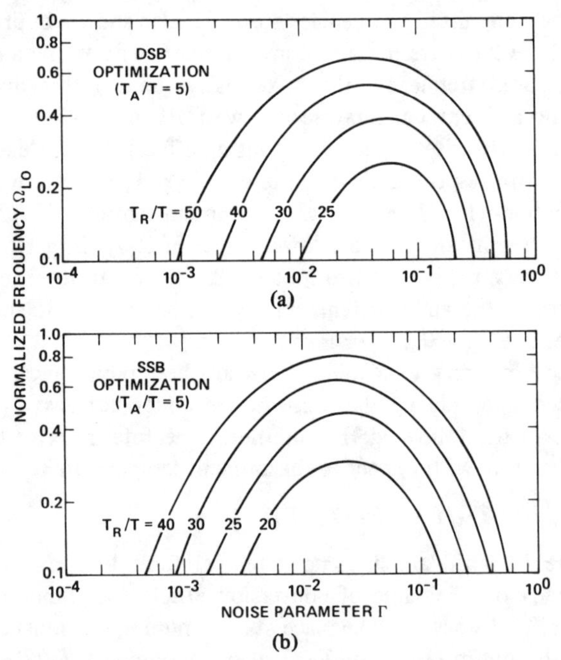

Fig. 9. Constant DSB and SSB receiver noise temperature contours in a plane of normalized frequency-noise parameter. Each number represents the best figure that can be achieved theoretically for a given junction. The IF output resistance is positive and matched throughout all cases. Here $\Gamma \equiv 2ekT/\hbar I_c$, and $\Omega_{LO} \equiv \hbar\omega_{LO}/2eRI_c$.

between Fig. 9(a) and (b) shows that an SSB receiver is better than a DSB receiver at the same Ω_{LO} and Γ by an average factor of one and a half. The prediction that an SSB Josephson receiver at $\Omega_{LO} = 0.2$ can have T_R/T less than 20 over a wide range of Γ is very encouraging.

VI. Discussion and Conclusion

Although there are fundamental differences between a Josephson mixer and a conventional resistive mixer, we have generalized the classical mixer theory in order to include such effects as conversion gain, negative resistance, noise saturation, and correlation. The analysis can be extended to other nonlinear Josephson devices as well. For example, we have examined a few cases for parametric amplification when the RF input impedance has a negative real part. The amplifier operates either with one idler at the image frequency or with two idlers at both the image and the difference frequencies. It is found that the noise temperature at a finite voltage is much worse than that of a mixer at the same bias. The range of parameters for such a parametric amplifier is also very restricted. Another device of interest is a harmonic mixer which can be studied following a similar approach.

Previous experimental data on millimeter-wave Josephson mixers [1]-[3] have shown a noise temperature $T_M/T = 20$-50 with a conversion efficiency $\eta = 0.5$-1.3 at a normalized frequency $\Omega_{LO} = 0.3$-0.4 and a noise parameter $\Gamma = 0.01$ or less. They are in good agreement with the computed results. The theory also predicts that significant improvement is possible if a hysteresis-free junction can be made with a low critical current and a moderately high RI_c product. Specifically, for an RI_c product equal to one-half of the niobium energy gap or 1 mV and an $I_c = 10\ \mu$A at $T = 4.2$ K, the noise temperature of an SSB Josephson receiver at 100 GHz would be as low as $T_R = 70$ K, provided that $T_A = 20$ K for the IF amplifier. A single-sideband configuration can be realized using a reactive narrow-band filter in front of the mixer. It has an additional advantage that the background noise contribution to the system is reduced by a factor of two.

In the beginning of the analysis, it is assumed that the spectral density of thermal noise is kT at all frequencies. The assumption breaks down in the quantum limit at a frequency higher than kT/h [10], or approximately 100 GHz at $T = 4.2$ K. In the limit when $\omega_{LO} \approx kT/\hbar$, the extra photon noise at frequencies beyond ω_{LO} does not have a very strong effect on the mixer since the noise components in Fig. 5 are practically independent of Γ. However, as the ambient temperature is reduced such that $kT \ll \hbar\omega_{LO}$, one can no longer expect T_M to improve with T. In order to explore such a transition into quantum limit, the noise term in (4) must satisfy a nonuniform spectrum for quantum fluctuations. This can be simulated in computer calculations with a series of properly correlated random numbers in the time domain.

In conclusion, we have carried out a complete numerical characterization as well as optimization for a Josephson-junction mixer. A nonlinear pump equation based on the RSJ model with thermal noise is solved in the time domain to obtain a conversion matrix and a noise correlation matrix. These matrices are then used in the frequency domain to calculate the noise temperature and conversion efficiency. Optimization is made in a multiparameter space to find the best mixer and receiver performance. The prediction from the analysis is very promising and may be realized using high-quality junctions.

ACKNOWLEDGMENT

The author would like to thank A. R. Kerr and J. H. Claassen for many fruitful discussions during the course of the work.

REFERENCES

[1] Y. Taur, J. H. Claassen, and P. L. Richards, "Conversion gain in a Josephson effect mixer," *Appl. Phys. Lett.*, vol. 24, pp. 101–103, Jan. 1974.
[2] J. H. Claassen and P. L. Richards," Point-contact Josephson mixers at 130 GHz," *J. Appl. Phys.*, vol. 49, pp. 4130–4140, July 1978.
[3] Y. Taur and A. R. Kerr, "Low-noise Josephson mixers at 115 GHz using recyclable point contacts," *Appl. Phys. Lett.*, vol. 32, pp. 775–777, June 1978.
[4] J. H. Claassen, Y. Taur, and P. L. Richards, "Noise in Josephson point contacts with and without RF bias," *Appl. Phys. Lett.*, vol. 25, pp. 759–761, Dec. 1974.
[5] Y. Taur, "Noise down-conversion in a pumped Josephson junction," *J. Phys.*, vol. 39-C6, pp. 575–576, Aug. 1978.
[6] D. N. Held and A. R. Kerr, "Conversion loss and noise of microwave and millimeter-wave mixers: Part 1–Theory," *IEEE Trans. Microwave Theory Tech.*, vol. MTT-26, pp. 49–55, Feb. 1978.
[7] F. Auracher and T. Van Duzer, "Numerical calculations of mixing with superconducting weak links," in *Proc. Appl. Superconductivity Conf.*, pp. 603–607, Sept. 1972.
[8] Y. Taur, "Josephson junctions as microwave heterodyne detectors," Ph.D dissertation, University of California, Berkeley, 1974.
[9] Y. Taur, J. H. Claassen, and P. L. Richards, "Conversion gain and noise in a Josephson mixer," *Rev. Phys. Appl.*, vol. 9, pp. 263–268, Jan. 1974.
[10] J. H. Claassen and P. L. Richards, "Performance limits of a Josephson-junction mixer," *J. Appl. Phys.*, vol. 49, pp. 4117–4129, July 1978.
[11] Y. Taur, P. L. Richards, and F. Auracher, "Application of the shunted junction model to point-contact Josephson junctions," in *Proc. 13th Conf. Low Temp. Phys.*, vol. 3, pp. 276–280, Aug. 1972.
[12] B. D. Josephson, "Possible new effects in superconductive tunneling," *Phys. Lett.*, vol. 1, pp. 251–253, July 1962.
[13] A. R. Kerr, private communication.
[14] F. Auracher and T. Van Duzer, "RF impedance of superconducting weak links," *J. Appl. Phys.*, vol. 44, pp. 848–851, Feb. 1973.
[15] H. C. Torrey and C. A. Whitmer, *Crystal Rectifiers.* New York: McGraw-Hill Book, 1948, ch. 5.
[16] A. Van der Ziel, *Noise; Sources, Characterization, Measurement.* Englewood Cliffs, NJ: Prentice-Hall, 1970, ch. 2.
[17] D. W. Peterson and Y. Taur, to be published.
[18] D. W. Peterson, "The unclamped and current clamped SUPARAMP: Studies of the unbiased Josephson junction parametric amplifier," Ph.D dissertation, University of California, Berkeley, 1978.

Low-noise Josephson mixers at 115 GHz using recyclable point contacts

Y. Taur[a)] and A. R. Kerr

NASA Goddard Institute for Space Studies, Goddard Space Flight Center, New York, New York 10025
(Received 3 February 1978; accepted for publication 9 March 1978)

Thermally recyclable Nb point-contact Josephson junctions are investigated as low-noise mixers with an external local oscillator at 115 GHz. The best single sideband mixer noise temperature achieved is 140 (±20) K with a (SSB) conversion loss of 2.4 (±0.5) dB. Such rugged junctions are suitable for use in practical receivers and should give unprecedented sensitivity at the shorter millimeter wavelengths.

PACS numbers: 74.50.+r, 07.62.+s, 85.25.+k

Millimeter-wave mixers, used as spectral line receivers, are finding an increasing number of applications in the fields of radio astronomy, atmospheric physics, and plasma diagnostics. Existing cooled Schottky diode mixers[1,2] have noise temperatures two orders of magnitude higher than the ideal photon noise limit. Since the efficiency of a heterodyne receiver is inversely proportional to the square of the system noise temperature,[3] the development of better low-noise mixers at millimeter wavelengths is clearly desirable.

It is well known that Josephson junctions have a great potential as efficient heterodyne mixers. This is due to their nonlinear response, which extends well into the submillimeter region,[4,5] and also their extremely small LO power requirement. Attempts at mixing with thin-film Josephson microbridges have resulted in very large conversion losses,[5-7] mostly due to poor coupling from the signal source to the very-low-impedance junctions. The coupling problem is more easily solved with high-impedance point-contact Josephson junctions, which have yielded very encouraging mixer performance at 36 and 130 GHz.[8-11] However, these point contacts all had the serious disadvantage of being mechanically unstable, and had therefore to be adjusted at low temperature. On the other hand, the recyclable point contacts which have been fabricated for SQUID applications[12,13] are not adequate for high-frequency mixing because they are either of the low-resistance type or shunted by a superconducting path. In this paper we report on Josephson mixers at 115 GHz using preadjusted recyclable high-impedance point contacts. They have excellent electrical and mechanical characteristics, and are therefore suitable for practical low-noise receivers.

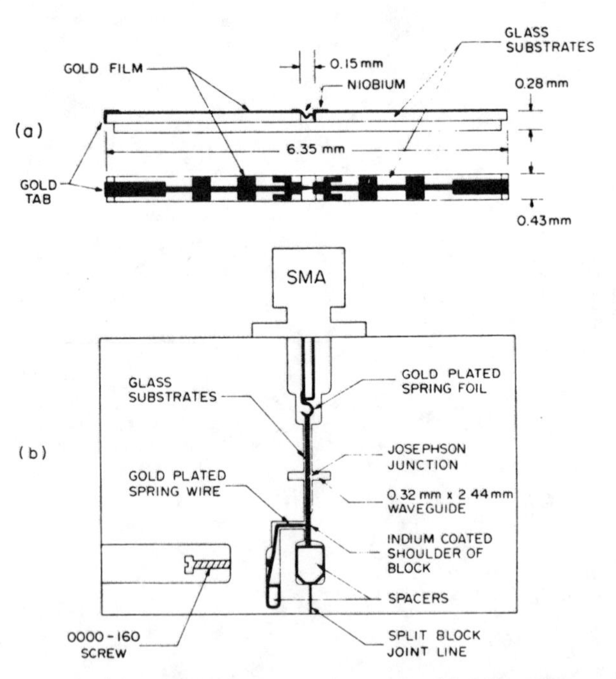

FIG. 1. (a) Side and top view of an assembled Josephson point contact. The substrates are glued together with Eastman 910 adhesive applied from the edges after alignment. (b) Mixer mount. The substrate lies sideways in a slot at the surface of the brass block, and is centered in the slot by the spring wire which also maintains the dc/i.f. ground contact between the rf choke and the block.

[a)]NAS–NRC Research Fellow, now with Columbia Radiation Laboratory, Columbia University, New York, N.Y. 10027.

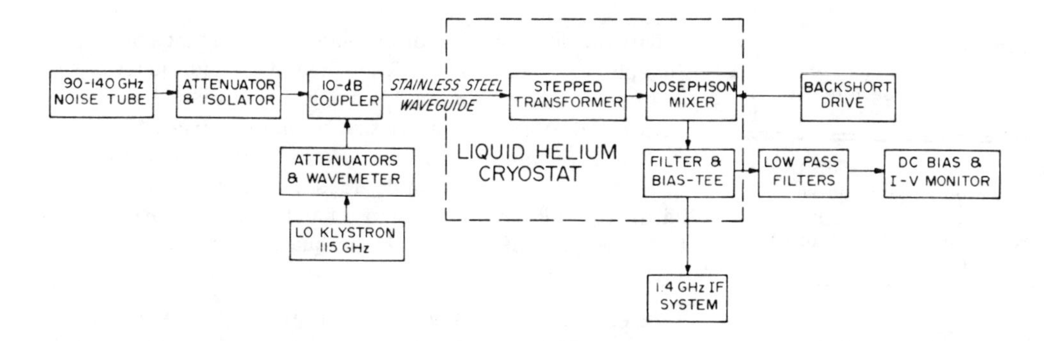

FIG. 2. Block diagram of the setup for testing Josephson mixers. Additional waveguide and coaxial cables for calibrating the attenuation and noise temperature of the input and output lines are not shown.

Our Nb point contacts are made between a Nb wire and foil, attached to metallized substrates which are glued to a third base substrate, as shown in Fig. 1(a). The configuration is similar to the scheme for cooled Schottky diode mixers,[2] except that Corning 8260 glass is used, instead of quartz, to match the expansion of Nb. Choke patterns on the substrates are fabricated by photolithographic techniques to reflect a proper reactive impedance for optimum coupling to the junction. A 12-μm-thick Nb bracket and an 18-μm-diam Nb wire are separately welded to the ends of the choke substrates. The tip of the Nb wire is etched electrochemically to a sharp point of radius < 0.25 μm. When the junction is assembled, the substrate with the Nb bracket is first glued to the base, then the substrate with the Nb wire is moved in with a differential micrometer. While making the contact, we monitor the junction resistance and adjust for a value of 20–30 Ω before gluing the whisker substrate to the base substrate.

The assembled junction is placed in a reduced-height waveguide mount as shown in Fig. 1(b). It is located, by an adjustable spring wire, in a slot machined in the surface of the mixer block. One choke section, on the substrate bearing the Nb foil, is pressed against the indium-coated shoulder to provide the ground contact. The intermediate frequency (i.f.) output and dc bias are connected to the SMA connector via a spring contact. Such a configuration avoids any rigid constraint on the glass structure, and allows it to expand and withstand mechanical shocks without deforming the junction.

The block diagram in Fig. 2 shows the setup for our mixer tests. Both the local oscillator (LO) and signal power at 115 ± 1.4 GHz are coupled to the junction via a stainless-steel waveguide (1.27 mm × 2.54 mm inside diameter) and a stepped transformer. A contacting backshort in the 0.32-mm waveguide behind the point contact can be adjusted for maximum coupling to the junction. The i.f. output at 1.4 GHz passes through a narrowband filter (bandwidth 200 MHz) to an i.f. measuring system which consists of an isolator, amplifier, and detector.[1] Leads for dc bias and I-V curve measurement are filtered and connected to the i.f. line with blocking capacitors and inductors.

The mixer block is sealed in He exchange gas before cooling. The junction resistance usually increases by ~15% from 300 to 4.2 K, probably because of a slight pressure change in the contact area. Such a resistance change, however, is reversible and reproducible, and the superconducting I-V characteristics stay the same between cycles. This can be observed in Fig. 3 which shows data taken from a junction which has been cycled between room and liquid-He temperatures more than ten times over a period of one month. In order to reduce thermal noise and to achieve a large resistance–critical current (RI_c) product, the junctions are operated at a temperature of 1.8 K. The RI_c product of our recyclable point contacts ranges from 600 to 800 μV, about one-third of the theoretical value. According to the mixing theory based on the resistively shunted Josephson junction (RSJ) model,[8,9] this gives a normalized signal frequency ($\hbar\omega_s/2eRI_c$) between 0.3 and 0.4, which is appropriate for making an efficient mixer at 115 GHz. Most of the junctions have no hysteresis after LO is applied.

The 1.4-GHz i.f. radiometer/reflectometer system[1] (50 MHz bandwidth) measures the output noise power from the junction in absolute temperature units, and also the match at the i.f. port of the mixer. Corrections for the coaxial cable loss (1.6 dB) and its effective temperature (135 K) are made, to refer the noise temperature $T_{i.f.}$ to the bandpass filter. The signal noise tube is calibrated at 115 GHz against hot and cold loads using a Schottky diode mixer.[2] It produces an input temperature change of $T_s = 250$ K after the loss of the stainless-steel waveguide (10 ± 0.25 dB, determined using a similar return guide) is taken into account. The mixer output noise temperature can be written

$$T_{i.f.} = (T_M + 2T_s + 2T_{in})/L, \qquad (1)$$

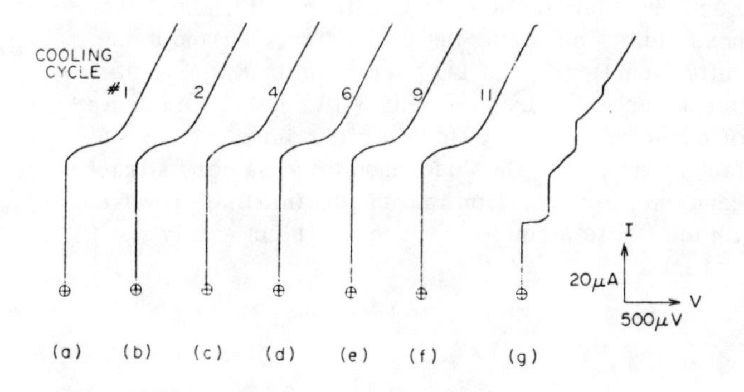

FIG. 3. (a)–(f) Static I-V characteristics of a niobium point contact in its first, second, fourth, sixth, ninth, and eleventh cooling cycles, respectively. (g) I-V curve of this junction showing Josephson steps induced by the applied LO at 115 GHz.

TABLE I. Summary of junction parameters and mixer performance. The conversion loss is measured from the waveguide transformer to the i.f. filter. All mixers have been tested for linear i.f. response.

Resistance (Ω)	Critical current (μA)	Normalized frequency	Conversion loss (± 0.5 dB) (dB)	SSB mixer noise temp. (± 20 K) (K)
26	25	0.37	5.0	180
38	19	0.33	1.4	200
22	34	0.32	2.4	140

where L is the single-sideband (SSB) conversion loss from the stepped transformer to the coaxial filter, T_M is the SSB equivalent input noise temperature of the mixer, and T_{in} is the input temperature when the noise tube is off. To determine the contribution of the waveguide thermal radiation to T_{in}, we used a series of small carbon resistors, mounted along the stainless-steel waveguide, as thermometers to measure the temperature distribution. We found $T_{in} = 70 \pm 5$ K, including the attenuated radiation from the room-temperature termination. Equal mixer response is assumed at the two sidebands, $f_{LO} \pm f_{i.f.}$.

The conversion loss of the mixer can be measured by observing the change in $T_{i.f.}$ as the noise tube is turned off and on ($T_s = 0$ and 250 K), and the value of T_M is then given by Eq. (1). The results from several junctions are summarized in Table I. In each case, the backshort is adjusted for maximum mixer response. As expected from the theory,[14] the best dc bias is always almost midway between the zeroth and first steps. In all cases, the i.f. mismatch between the junction and the 50-Ω system is only a few percent.

Both the conversion loss and mixer noise temperature represent a significant improvement over cooled Schottky diode mixers.[2] It is understood from the RSJ model that the excess i.f. noise is contributed by down-converted thermal noise at harmonics of the LO and Josephson frequencies.[10] Our results are within a factor of 2 of the theoretical prediction taking the conversion of high-frequency quantum noise into account.[11] Further reduction of T_M to less than 100 K should be possible by improving the choke structure and using better junctions. The LO power required to operate our Josephson mixers is only 3 nW, nearly six orders of magnitude less than for Schottky diode mixers—a fact which makes the Josephson mixer a very attractive candidate for extension into the submillimeter-wave region where strong LO sources are not easily available.

Saturation problems often play an important role in Josephson-type devices. The linearity of our junctions has been examined using a waveguide filter and a calibrated attenuator after the rf noise source. We find that such problems are less severe in our mixers than in parametric Josephson devices.[15,16] In general, no saturation occurs if the signal bandwidth is restricted to a few GHz for $T_{in} \approx 300$ K, in qualitative agreement with the RSJ model.[11]

In conclusion, we have made thermally recyclable point-contact Josephson junctions which have excellent mixing characteristics at 115 GHz. We plan to construct a practical receiver, using such a mixer, in the near future. The main differences between this receiver and our present laboratory system will be reduced input losses and the use of a low-noise cooled GaAs FET i.f. amplifier following the mixer. This receiver will be used on the Columbia/GISS Sky Survey Telescope at Columbia University, and will hopefully improve the observation efficiency by an order of magnitude.

The authors wish to thank J. Grange, I. Silverberg, and H. Miller for their valuable help in fabricating the junctions and mixers.

Note added in proof. In a recent experiment at 6 K, we have obtained a better figure $P_M = 120$ K (SSB) with $L = 0$ dB (SSB) at 115 GHz.

[1]S. Weinreb and A.R. Kerr, IEEE J. Solid-State Circuits SC-8, 58 (1973).
[2]A.R. Kerr, IEEE Trans. Microwave Theory Tech. MTT-23, 781 (1975).
[3]A. van der Ziel, *Noise: Sources, Characterization, Measurement* (Prentice-Hall, Englewood Cliffs, N.J., 1970).
[4]P.L. Richards, F. Auracher, and T. Van Duzer, Proc. IEEE 61, 36 (1973).
[5]F. Auracher, Ph.D. Thesis (University of California, Berkeley, 1973) (unpublished).
[6]L.K. Wang, A. Callegari, B.S. Deaver, D.W. Barr, and R.J. Mattauch, Appl. Phys. Lett. 31, 306 (1977).
[7]W. Howard and Y.H. Kao (private communication).
[8]Y. Taur, Ph.D. Thesis (University of California, Berkeley, 1974) (unpublished).
[9]Y. Taur, J.H. Claassen, and P.L. Richards, Appl. Phys. Lett. 24, 101 (1974).
[10]J.H. Claassen, Y. Taur, and P.L. Richards, Appl. Phys. Lett. 25, 759 (1974).
[11]J.H. Claassen and P.L. Richards (private communication).
[12]R.A. Buhrman, S.F. Strait, and W.W. Webb, The Material Science Center, Cornell University Technical Report No. 1555, 1971 (unpublished).
[13]J.E. Zimmerman, P. Thiene, and J.T. Harding, J. Appl. Phys. 41, 1572 (1970).
[14]Y. Taur, J.H. Claassen, and P.L. Richards, Rev. Phys. Appl. 9, 261 (1974).
[15]M.J. Feldman, P.T. Parrish, and R.Y. Chiao, J. Appl. Phys. 46, 4031 (1975).
[16]Y. Taur and P.L. Richards, J. Appl. Phys. 48, 1321 (1977).

Analytic solutions to superconductor-insulator-superconductor quantum mixer theory

A. D. Smith and P. L. Richards
Department of Physics, University of California, Berkeley, California 94720

(Received 14 September 1981; accepted for publication 5 January 1982)

Three-port quantum mixer theory for superconductor-insulator-superconductor tunnel junctions is solved for several simple cases of practical interest. Closed form solutions are derived for the conversion gain in the low local oscillator power limit, and also the general conditions for zero and infinite gain. These results are useful in understanding more general computer calculations of predicted performance. Practical limits to gain and dynamic range are analyzed both for series arrays and single junctions.

PACS numbers: 84.30.Qi, 74.50. + r, 85.25. + k

INTRODUCTION

The possibility of developing quantum noise limited, near millimeter wave receivers for radio astronomy has been suggested by recent progress in superconductor-insulator-superconductor (SIS) quasiparticle mixers. Photon assisted tunneling theory worked out by Tucker[1] quickly lead to predictions of conversion gain and low noise[2,3] when applied to the SIS tunnel junction mixer. Early experimental progress was reviewed by Richards and Shen.[4] rf-induced negative resistance[5,6] and output IF power in excess of rf mixer input power[7] are recent experimental verifications of these predictions of the quantum theory.

For small *signal* power, mixer response can be linearized and described by an admittance matrix Y which relates output currents to applied voltages for the important frequencies (ports) of the junction.[8,9] The Y matrix can be used to calculate mixer impedance and conversion efficiency. In three-port Y-mixer analysis, ports at harmonics of the signal frequency are assumed to be shorted (for example by junction capacitance), so that only the signal, intermediate frequency (IF), and image frequencies are considered. The elements of the 3 by 3 Y-matrix depend upon the I-V characteristic of the mixer element and on the amplitude of the local oscillator (LO). Quantum theory[1] evaluates the Y elements as a discrete convolution of the unpumped I-V curve. Unfortunately, the infinite sums of Bessel functions involved in the calculations make the predictions in the general case more suitable for computer calculations than direct analytic solution. Many of the most useful predictions (such as gain and negative resistance) were realized only after laborious computer simulations. Much can be understood, however, by examining analytic solutions for simple limiting cases. In this paper we develop a number of analytic solutions to the three-port Y mixer problem for SIS junctions with nearly ideal Bardeen-Cooper-Schrieffer (BCS) I-V curves. These include the limiting case of low LO power, which has been used by several researchers as a check on numerical calculations. The prediction of large mixer gain in these calculations is intimately linked to the availability of large (and even negative) dynamic resistance on the pumped I-V curve. An explanation of the origins of negative resistance is reviewed, showing the importance of proper rf tance G_s in obtaining gain. Finally, practical limitations for gain and dynamic range are discussed, both for single junctions and for series arrays of junctions.

I. LOW P_{LO} LIMIT

In the limit of very low applied local oscillator power, the admittance elements can be expanded in lowest order in the normalized rf voltage present across the junction $\alpha = eV_\omega/\hbar\omega$. An important feature of the quantum theory of a mixer biased with dc voltage V_0 is that the junction I-V curve enters the calculations only for voltages $V_0 + n\hbar\omega/e$ where $n = 0, \pm 1, \pm 2 \ldots$ which are called photon points. The I-V curve predicted by weak coupling BCS theory for an SIS junction at $T = 0$ has no current flow below the gap, so that for such ideal junctions $I_n = I_{dc}(V_0 + n\hbar\omega/e) = 0$ for $n \leqslant 0$. For our calculations we choose values of dc bias below the gap, so that all current is rf related, but sufficiently large that a single photon can supply enough energy to break a pair, i.e., $(2\Delta - \hbar\omega)/e < V_0 < 2\Delta/e$. For many practical situations this bias point on the first photon step below the gap gives maximum gain when the 3-port model is valid. In addition, it is assumed that $\omega_{IF} \ll \omega_{LO}$. In the low LO power limit, the elements of the Y matrix can be expressed in terms of the unpumped I-V curve at only a few photon points near V_0. To lowest order in α, the elements can be written in the form $Y_{mn} = G_{mn} + iB_{mn}$, where

$$G_{00} = \alpha^2 I'_1/4, \tag{1.1}$$

$$G_{10} = G_{-10} = \alpha I'_1/4,$$

$$G_{01} = G_{0-1} = \alpha I_1 e/2\hbar\omega,$$

$$G_{1-1} = G_{-11} = \alpha^2(I_2\text{-}2I_1)e/16\hbar\omega,$$

$$G_{11} = G_{-1-1} = I_1 e/2\hbar\omega,$$

$$B_{10} = -B_{-10} = \alpha\,\mathrm{Re}(j'_1 - 2j'_0 + j'_{-1})/4,$$

$$B_{11} = -B_{-1-1} = \mathrm{Re}(j_1 - 2j_0 + j_{-1})e/2\hbar\omega,$$

$$B_{1-1} = -B_{-11} = \alpha^2\,\mathrm{Re}(j_2 - 2j_0 + j_{-2})e/16\hbar\omega, \text{ and}$$

$$B_{00} = B_{01} = B_{0-1} = 0,$$

where prime denotes derivative with respect to V and j_n is the complete complex response function whose real part is the Kramers–Kronig transform of I_n. The subscripts -1,0, and 1 on G and B refer to the image, IF, and signal channels,

Reprinted with permission from *J. Appl. Phys.*, vol. 53, no. 5, pp. 3806–3812, May 5, 1982.

respectively.

The conversion efficiency of the mixer depends upon the Y elements and also on the external embedding admittances Y^{ext} presented at the output ports. The quantum rf susceptances B_{11} and B_{-1-1} can be canceled by external reactive elements

$$B_{11} + B_{1}^{\text{ext}} = B_{-1-1} + B_{1}^{\text{ext}} = 0. \quad (1.2)$$

Except at very high frequencies, this choice leads to nearly maximum gain.[10] The gain of the mixer is $G = 4G_S G_L |Z_{01}|^2$, where G_S is the rf conductance of the signal source, G_L is the conductance of the IF load, and $||Y_{mn} + \delta_{mn} Y_m^{\text{ext}}|| = ||Z||^{-1}$ is the augmented admittance matrix.[8] This augmented admittance matrix includes the effects of the embedding admittances. The IF conductance of the mixer Y_{IF} is Z_{00}^{-1}.[11] Expressing the signal and image conductances as $y_s G_{11}$ and $y_i G_{11}$, respectively, the mixer IF admittance is

$$Y_{\text{IF}} = R_{\text{Dyn}}^{-1} = \frac{G_{00}(y_s y_i - 1)}{(1 + y_s)(1 + y_i)}. \quad (1.3)$$

When $y_s y_i \gg 1$, the dynamic resistance R_{Dyn} is positive and finite. As $y_s y_i$ decreases, R_{Dyn} becomes arbitrarily large and then negative. For an IF-matched mixer, the gain is

$$G = 2\frac{I_1}{I_1' \hbar\omega/e} \frac{y_s(1 + y_i)}{(1 + y_s)(y_s y_i - 1)}. \quad (1.4)$$

The region of large R_{Dyn} has large available gain.

It is convenient to rewrite Eq. (1.4) in terms of simpler junction parameters. For bias points not too far below the gap, the I-V curve of the junction can be approximated as[12]

$$I_1 \approx \frac{\pi}{4} \frac{2\Delta}{e} G_N \quad (1.5)$$

and

$$I_1' \approx G_N,$$

where G_N is the junction normal state conductance. The gain can then be written in terms of convenient experimental parameters

$$G \approx \frac{4g_s[1 + 4g_i(\hbar\omega/\pi\Delta)]}{[1 + 4g_s(\hbar\omega/\pi\Delta)][16 g_s g_i(\hbar\omega/\pi\Delta)^2 - 1]}, \quad (1.6)$$

where $g_{s,i} = G_{S,I}/G_N$ are rf source conductances referred to the normal state junction conductance. In general, both gain and junction dynamic resistance diverge as $y_s y_i \to 1$ in the low LO limit. Thus for SIS junctions modeled as ideal BCS I-V curves with no subgap leakage current, quantum tunneling theory makes two surprising predictions. First, conversion gain is possible. Second, the conversion gain is possible for arbitrarily small applied local oscillator power P_{LO}. Although this limit is not approached in practical mixers for reasons discussed in Sec. II, it is a convenient one for analytic calculations.

A. Identical idler and signal termination

For many applications involving $\omega_{\text{IF}} \ll \omega_{\text{LO}}$, the terminations at the image and signal ports are nearly identical. Tucker[2] has calculated mixer performance in the special case that $G_S = G_I$. Tucker's mixer parameters can be evaluated to lowest significant order in α as

$$L_0 = I_1' \hbar\omega/eI_1, \quad (1.7)$$

$$\eta = 2,$$

$$\xi = 1,$$

and

$$\beta\gamma = \gamma^2 = 0,$$

Tucker's general formula for IF-matched three-port mixer gain G can be written as

$$G = \frac{\eta y_s[(\xi + y_s)^2 + \gamma^2]}{L_0 y_L[(\xi + y_s)(1 + y_s) - \gamma^2]^2}, \quad (1.8)$$

where the IF admittance $G_{00} y_L$ is equal to $G_{00}(y_s - 1)/(y_s + 1)$. For the low LO limit this reduces to a simple form equal to Eqs. (1.4) and (1.6) with y_i replaced by y_s.

$$G = 2\frac{I_1}{I_1' \hbar\omega/e} \frac{y_s}{y_s^2 - 1} \approx \frac{g_s \pi^2}{4[(g_s \hbar\omega/\Delta)^2 - \pi^2/16]}. \quad (1.9)$$

This result predicts unbounded gain as $y_s \to 1$ from above. For $y_s < 1$ negative resistance appears on the first photon step.

Several of the operating parameters can be calculated in the low LO limit. The rf conductance of the junction is given by $G_\omega = eI_1/2\hbar\omega$. This is equal to the value of the source conductance at the threshold of infinite gain. Thus mixer conversion efficiency in this low LO limit diverges for a source impedance equal to the rf input impedance. In general rf match does not insure maximum gain in three-port Y mixers.[9] The IF output impedance, however, should always be matched. The IF impedance of the mixer tends to infinity as α^{-2}. In practice, therefore, operating an SIS mixer near the low LO limit would require special care in IF impedance matching in order to couple the output power effectively.

B. Image rejection

The low LO limit calculations can also be used to study the case of a shorted image channel. Such an "image rejected" configuration is sometimes used to enhance signal gain in classical mixers. For a two-port mixer, with $y_i \to \infty$, gain and dynamic resistance are finite for all values of y_s.[13] This contrasts with the equal termination case, in which gain deverges for $y_s \to 1$. Applying Eq. (1.6) yields

$$G_{2\text{-}port} = 2\frac{I_1}{I_1' \hbar\omega/e} \frac{1}{(1 + y_s)} \approx \frac{1}{\hbar\omega/\pi\Delta + 4g_s(\hbar\omega/\pi\Delta)^2}. \quad (1.10)$$

The gain approaches the maximum value[14] of $G_{2\text{-}port} \to \pi\Delta/\hbar\omega$ as $g_s \to 0$. These two-port results are in agreement with calculations of Sollner.[13]

The low LO limit calculated here provides a useful test for numerical calculations and helps provide an intuitive understanding of more general calculations. It also gives interesting predictions for gain greater than the classical limits.[8] It is seen that, at least in the limit of low LO, equally terminated signal and image ports can produce larger gain than is the case with a shorted image port.

II. GAIN CONTOURS

Mixer efficiency is a complicated function of the complex source admittance, load admittance, LO power, frequency and junction characteristics. For a fixed LO frequency and an ideal BCS I-V characteristic, several of the device parameters can be readily chosen for optimum performance. As before, the rf source reactance can be used to cancel the junction capacitance and linear quantum capacitance. The load resistance should be equal to the dynamic resistance of the pumped I-V curve evaluated at the bias point $V_{0,\max} = (2\Delta - \hbar\omega/2)/e$. It remains, however, to choose an appropriate rf source conductance and an optimum LO power.

Figure 1 shows computer generated gain contours as a function of rf source conductance and applied $P_{\rm LO}$ for an ideal BCS I-V curve at zero temperature. The signal and image ports are terminated identically. Contours of this type have been published for a wide variety of mixer parameters.[3,7] Such contours are difficult to understand intuitively. Some insight can be obtained, however, from analytic calculations of specific limiting cases. The low LO limit for unity gain calculated above is shown as an arrow. In this section we will discuss a number of other approximate analytic calculations which help to explain the general features of these gain contours.

A. Zero conversion efficiency

While the general solution for moderate available local oscillator power $P_{\rm LO}$ is quite complicated, the right-hand side of the region of large gain in Fig. 1 is bounded by contours of zero conversion efficiency which can be calculated relatively easily.

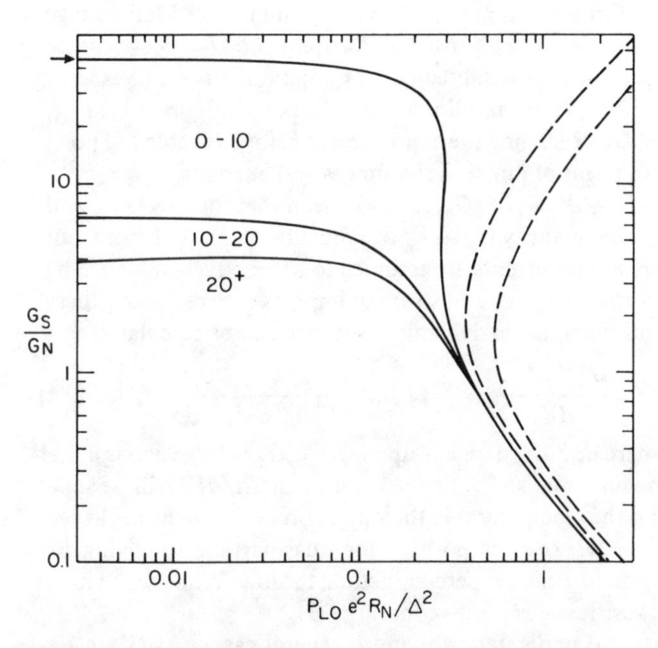

FIG. 1. Computer generated gain contours from three-port Y-mixer theory using an idealized BCS I-V curve at $T = 0$. Gain figures are in dB. The junction is biased at $V_{0,\max} = (2\Delta - \hbar\omega/2)/e$, and an IF match is assumed. The normalized LO frequency $\hbar\omega/2\Delta = 0.12$, which corresponds to 36 GHz for tin junctions. Gain contours are shown as a function of the normalized source conductance and normalized available LO power. The arrow indicates the low LO power limit for unity gain. The dashed lines are zero efficiency curves corresponding to $\alpha = 2.65$ and $\alpha = 3.49$.

The application of LO power introduces step structures on the dc I-V curve. For $\omega_{\rm sig} \approx \omega_{\rm LO}$, the signal can be thought of as a small increment to the applied LO voltage. The resultant change in the dc I-V curves represents the IF output signal. If the dc I-V curve is independent of $P_{\rm LO}$, then no mixing is possible. Therefore $\partial I_{\rm dc}(V, P_{LO}, Y_S, ...)/\partial P_{LO} = 0$ implies zero conversion efficiency. The dc current through the pumped junction can be written as a Bessel function-weighted average of photon points on the unpumped I-V curve $I_{\rm dc}(V)$[15]:

$$I(V_0, P_{\rm LO}) = \sum_{n=-\infty}^{\infty} J_n^2(\alpha) I_n, \tag{2.1}$$

where $\alpha = eV_\omega/\hbar\omega$ and $I_n = I_{\rm dc}(V_0 + n\hbar\omega/e)$ as before. Since the values of I_n are independent of $P_{\rm LO}$ the condition for zero conversion efficiency is simply

$$\sum \frac{\partial}{\partial \alpha} J_n^2(\alpha) I_n = 0. \tag{2.2}$$

It should be noted that this result is quite general. Any value of α satisfying Eq. (2.2) for a given I-V curve corresponds to zero conversion efficiency.

While Eq. (2.2) can be readily evaluated numerically for any given I-V curve, it is instructive to extract a closed form solution for the limit $\hbar\omega \ll 2\Delta$ with the junction biased on the first step below the gap. If there is no subgap leakage $I_n = 0$ for $n \leqslant 0$. Since only reasonably small values of n contribute significantly to the pumped current, I_n can be set equal to its value just above the gap.

$$I_n = \frac{\pi}{4} \frac{2\Delta}{e} G_N = I_+ \quad n > 0. \tag{2.3}$$

Using a Bessel function summation rule, the requirement of Eq. (2.2) may be reformulated without further loss of generality as

$$I_+ \sum_{n=1}^{\infty} \frac{\partial}{\partial \alpha} J_n^2(\alpha) = \frac{I_+}{2} \frac{\partial}{\partial \alpha} J_0^2(\alpha) = 0, \tag{2.4}$$

which has roots $\alpha_1 = 2.405$, $\alpha_2 = 3.832$, $\alpha_3 = 5.520$,.... Mixer performance is generally optimum for α less than the first root. For the more general cases of higher LO frequencies or nonideal I-V curves, solution of Eq. (2.2) can produce different numerical values for α.

As a measure of LO drive, α is convenient theoretically, but in practice α must be related to the available LO power $P_{\rm LO}$, which is the readily measured quantity. This can be done by modeling the LO power source as a current generator with amplitude I_S and admittance G_S The *available* power $P_{\rm LO}$ is defined as the power $I_S^2/(8G_S)$ dissipated by a matched load. Since $I_S = I_\omega + V_\omega G_S$,

$$P_{\rm LO} = (I_\omega + V_\omega G_S)^2/8G_S. \tag{2.5}$$

Of this amount, a quantity $I_\omega V_\omega/2$ is dissipated as heat in the junction, while the remainder of the available power is reflected. Junction currents out of phase with the rf voltage are generally small and have been ignored in this derivation. Using the calculated value of α_1, and computing the resultant I_ω, the required (available) $P_{\rm LO}$ can be expressed as

$$P_{LO}R_Ne^2/\Delta^2 = [1 + 4.81g_s(\hbar\omega/2\Delta)]^2 8g_s, \qquad (2.6)$$

where $g_s = G_S/G_N$ as before.

Contours of expected zero conversion efficiency are shown as dashed lines in Fig. 1 for roots α_1 and α_2 of Eq. (2.2). Note that for $G_S = I_\omega/V_\omega$ the *available* power required for a given LO amplitude α is minimized, while for other values of G_S more power is needed to maintain the same voltage amplitude across the junction.

B. Fixed IF load impedance; Real junction *I-V*

It is important to recognize mixer performance which is unrealistic due to limitations in available SIS junction quality or constraints in rf and IF matching. Several device parameter choices made for simplicity in the previous section do not correspond to common mixer configuration. In practice, the IF load impedance (IF amplifier input impedance) cannot usually be varied freely. In addition, real tunnel junctions show some subgap leakage and also some gap smearing. Incorporating these effects into numerical calculations makes significant changes in mixer efficiency, although large gain is still possible.[7]

Three-port numerical calculations corresponding to experiments using good quality tin junctions[7] are shown in Fig. 2. The observed unpumped *I-V* curve was used to obtain values of I_n. The junction normal state resistance was 23 Ω, and the IF amplifier was assumed to present a 50-Ω load to the mixer. The operating frequency was 36 GHz. In addition to predicted gain contours, a zero conversion efficiency curve is shown corresponding to the first root of Eq. (2.2) using the experimental *I-V* curve. This curve is nearly identical to the curve for α_1 in Fig. 1 which was calculated for an ideal *I-V* curve. The box gives an estimate of the experimental operating conditions which gave the best mixer efficiency (4.3-dB gain). Although the three-port model overestimates the available gain, the agreement with the theoretical values of G_S and P_{LO} is quite good.

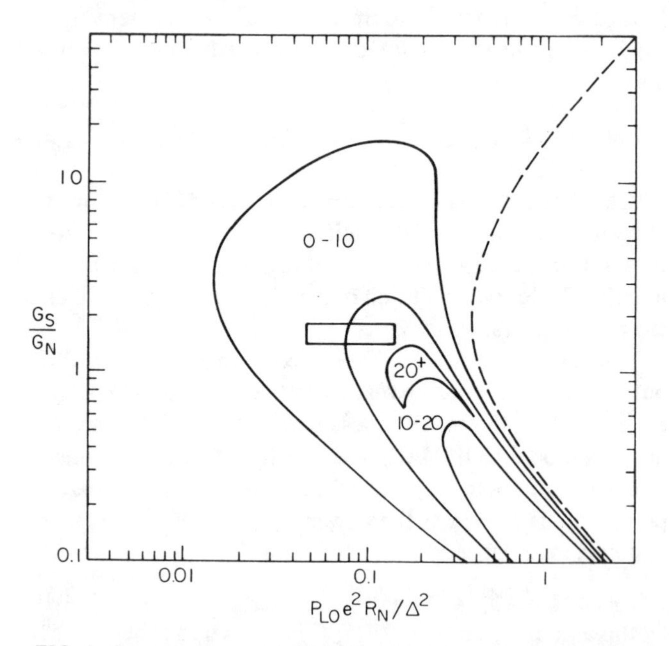

FIG. 2. Computer generated 36-GHz gain contours for the experimental tin *I-V* curves of Ref. 7. The junction impedance was $R_N = 23\ \Omega$ and the IF impedance was fixed at 50 Ω. The dashed line is a zero conversion efficiency curve corresponding to $\alpha_1 = 2.75$. The range of parameters for which large gain was observed experimentally are given by the box.

In summary, the general shape of Fig. 2 can be understood within the framework of the analytic calculations. Large gain is expected for a bounded region in the plotted parameter space. For larger P_{LO}, the zero conversion efficiency limit requires that the gain fall to zero. For smaller P_{LO}, the output current decreases and R_{Dyn} rises as α^{-2}. The resultant IF mismatch between the mixer and amplifier reduces the coupled gain. Impedance mismatch also plays a role if the source conductances becomes too small. If the source conductance becomes too large, the available IF power decreases due to the decrease in R_{Dyn}.

III. NEGATIVE RESISTANCE

The availability of high dynamic resistance and even negative dynamic resistance is intimately linked to gain in three-port SIS mixers. Computer calculations based on three-port theory predict that the dynamic resistance of the pumped *I-V* curve can be negative for certain values of rf source conductance.[2] In the absence of a simple physical interpretation, this prediction of negative resistance was somewhat mysterious, expecially since it was not seen in the extensive earlier investigations of photon assisted tunneling. Recently negative resistance in pumped SIS junctions has been observed.[5,6]. For completeness, we repeat here a simple model of Smith *et al.*[5] which explains the origins of the negative resistance.

For a junction with dc bias V_0 and induced RF voltage amplitude $\alpha = eV_\omega/\hbar\omega$, the low-frequency *I-V* curve can be expressed as a weighted sum of unpumped *I-V* curves as in Eq. (2.1). The rf amplitude variable α depends upon V_0, P_{LO}, and G_S. Recalling the expression (2.6) for available LO power in terms of junction rf voltages and currents, $P_{LO} = (I_\omega + V_\omega G_S)^2 8/G_S$, it can be seen that for fixed P_{LO} and G_S, the quantity $(I_\omega + V_\omega G_S)$ must be fixed. As before, out of phase rf currents are assumed to be reactively matched by external elements and so are unimportant here. Under these conditions the dc dynamic resistance can be calculated as

$$\frac{dI(V_0,P_{LO})}{dV_0} = \sum\left[J_n^2(\alpha)\frac{dI_n}{dV_0} + \frac{d\alpha}{dV_0}\frac{\partial(J_n^2)}{\partial\alpha}I_n\right], \qquad (3.1)$$

where $d\alpha/dV_0$ depends upon G_S. If G_S is large enough that the junction is rf voltage biased, then $d\alpha/dV_0$ will be small and the second term in the square brackets can be neglected. Since dI_n/dV_0 is positive for quasiparticle tunneling between identical superconductors, the low-frequency dynamic resistance is positive for large G_S.

To understand the more general case, $d\alpha/dV_0$ in Eq. (3.1) can be calculated. Again ignoring the (generally small) out-of-phase contributions, the rf current flowing through the junction I_ω can be calculated as[2]

$$I_\omega = \sum_{n=-\infty}^{\infty} J_n(J_{n+1} + J_{n-1})I_n. \qquad (3.2)$$

Taking the derivative with respect to dc voltage gives

$$I'_\omega = \sum J_n (J_{n+1} + J_{n-1}) I'_n + \frac{d\alpha}{dV}\frac{\partial}{\partial\alpha}[J_n(J_{n+1} + J_{n-1})] I_n, \tag{3.3}$$

The external rf biasing conductance constraint that $(I_\omega + V_\omega G_S)$ must be fixed requires

$$\frac{d\alpha}{dV_0} = -\frac{e}{\hbar\omega}\frac{I'_\omega}{G_S}. \tag{3.4}$$

Eliminating I'_ω from Eqs. (3.3) and (3.4) we find

$$\frac{d\alpha}{dV_0} = \frac{-\sum J_n(J_{n+1} + J_{n-1}) I'_n}{G_S\hbar\omega/e + \sum (d/d\alpha)[J_n(J_{n+1} + J_{n-1})] I_n}. \tag{3.5}$$

As expected, $d\alpha/dV_0$ tends to zero for rf voltage bias ($G_S \to \infty$). For rf current bias ($G_S \to 0$) $d\alpha/dV_0$ approaches a (negative) limiting value. This negative $d\alpha/dV_0$ is responsible for the observed negative resistance in pumped junctions.

In summary, the negative conductance for current-biased junctions can be understood from simple circuit-model arguments. An increase in dc bias voltage causes a decrease in the rf impedance of the junction. For high impedance bias, this decreases the rf voltage V_ω present across the junction (i.e., $d\alpha/dV_0 < 0$), which in turn reduces the rf-induced dc current. If this effect [second term in Eq. (3.1)] outweighs the leakage conductance and the pumped conductance of the junction which are given by the first term in Eq. (3.1), the observed resistance dV_0/dI will be negative.

IV. SATURATION

Calculations of mixer performance have been carried out assuming that the input signal power is small. The input signal is treated as a small perturbation on the applied LO drive voltage. The second- and higher-order terms are specifically ignored in Tucker's derivation of Y mixer theory. In this limit the IF output power is linearly related to the rf signal power presented to the mixer. For sufficiently large input powers, the IF signal will saturate and spurious responses can occur. The framework of Y mixer theory can be extended to predict the saturation point for SIS mixers. This section presents quite general estimates of the saturation power of mixers operated in the quantum limit. Resultant limits on SIS mixer dynamic range are derived. In addition, a simple analytic model is used to calculate the spurious harmonic response in saturated SIS mixers.

For junctions with $\hbar\omega$ slightly less than the gap spread, Rudner and Claeson[16] found little saturation for $P_{sig} \leqslant 10\% P_{LO}$. When quantum effects become important saturation occurs at much lower values of P_{LO}. Shen *et al.*[17] observed 3% gain compression for $P_{sig} = 0.3\% P_{LO}$. This corresponded to approximately 10-pW signal power.

A. Maximum IF output power

Sollner has pointed out[18] that in mixers with quantum response, saturation can occur first at the IF port. Nonlinear response can occur because the gain is a sensitive function of V_0. Gain, $G(V)$, has a local maximum at $V_{0,max} = (2\Delta - \hbar\omega/2)/e$, and falls to minima at $V_0 = (2\Delta - \hbar\omega)/e$ and $V_0 = 2\Delta/e$. Low-frequency IF output shifts the instantaneous bias voltage away from this optimum with a frequency ω_{IF}. The effective gain of the mixer is a weighted average of $G(V)$ over the IF voltage swing. To avoid gain compression, the amplitude of these oscillations δV must be significantly less than the step width, that is $\delta V = \gamma_0\hbar\omega/e$, where $\gamma_0 \ll 1$. For an IF-matched mixer this corresponds to an applied rf signal power of

$$P_{sig} \leqslant \frac{(\gamma_0\hbar\omega)^2}{2e^2 G R_{Dyn}}, \tag{4.1}$$

where R_{Dyn} is the dynamic resistance of the pumped junction, P_{sig} is the rf signal power, and G is the mixer gain. The coefficient γ_0 depends upon the allowable gain compression and can be estimated by inspection of the measured $G(V)$ of a mixer. For a 0.2-dB tolerance, $\gamma_0 \approx 0.1$ for the particular $G(V)$ curve measured by McGrath *et al.*[7] For a 36-GHz mixer, with $\gamma_0 = 0.1$, $G = 2$, and $R_{Dyn} = 50\ \Omega$, the corresponding limit to P_{sig} is 2 pW, in excellent agreement with the observation of McGrath *et al.*[7] who observed 0.2-dB gain compression at P_{sig} 1.5 pW.

There is a trade-off between gain and saturation power. Very general mixer arguments[8] lead to the relationship

$$G = \left[\frac{\partial I_{LO}}{\partial\sqrt{P}}\right]^2 R_{Dyn}, \tag{4.2}$$

which allows the saturation condition to be rewritten as

$$P_{sig} \leqslant \tfrac{1}{2}\left[\frac{\gamma_0\hbar\omega/e}{R_{Dyn}(\partial I_{LO}/\partial\sqrt{P})}\right]^2. \tag{4.3}$$

Provided γ_0 is roughly fixed, this shows that while gain can be improved by increasing R_{Dyn}, the saturation limit is simultaneously degraded as R^2_{Dyn}.

B. Maximum gain-bandwidth-dynamic range

The saturation in the IF output also provides a constraint on the gain-bandwidth product for a single mixer. Photon fluctuation noise over a bandwidth B produces input noise

$$P_{Noise} = \hbar\omega B. \tag{4.4}$$

If the maximum input signal is expressed as a dynamic range factor D times the inherent noise level, then Eq. (4.1) requires $D\hbar\omega BG \leqslant [(\gamma_0\hbar\omega/e)^2/2R_{Dyn}]$. The gain-bandwidth-dynamic range product is limited to

$$GBD \leqslant \frac{\gamma_0^2\hbar\omega}{2e^2 R_{Dyn}}. \tag{4.5}$$

For a junction at 36 GHz, taking $\gamma_0 = 0.1$ and assuming $R_{Dyn} = 50\ \Omega$, the GBD product is limited to 93 GHz. For this example, a mixer operated over a bandwidth of 0.5 GHz and with a gain of 2 could have a dynamic range of 93 (approximately 20 dB).

The IF dynamic range limit is quite general and may be compared with previously reported rf dynamic range limits. For a two-port mixer, large rf saturation power is predicted for bias far below the gap.[13] Under conditions otherwise similar to the above example, the requirement that $V_{sig} < V_{LO}$ places an upper bound of 79 dB on the mixer dynamic range.[13] While specifying a 0.2 dB allowable gain compression will impose a somewhat tighter rf constraint, it is clear that the 20-dB IF limit is the significant limiting factor in this case.

The existence of a GBD limit imposes an important design constraint for practical SIS receivers. Although large G may be available for an SIS mixer, the importance of mixer gain must be weighed against dynamic range and bandwidth requirements.

C. Third-order intermodulation

The two-tone, third-order intercept power is a conventional measure of the saturation point of a mixer. Intermodulation between two separate rf signals produces spurious responses within the IF bands of saturated mixers. An rf input consisting of two equal power, monochromatic signals, with angular frequencies $(\omega_1 + \omega_{LO})$ and $(\omega_2 + \omega_{LO})$, produces IF output at ω_1 and ω_2. A saturated mixer will also produce spurious responses, including an intermodulation signal with angular frequency $(2\omega_1 - \omega_2)$. In contrast to the IF outputs at ω_1 and ω_2, which are proportional to the individual rf powers P_{sig}, the $(2\omega_1 - \omega_2)$ signal is proportional to P_{sig}^3. As a measure of the power at which nonlinear effects are important, the intersection of the extrapolated P_{sig} and P_{sig}^3 responses computed at low power defines a third-order intermodulation intercept P_{int}.

In the limit of $\hbar\omega_{IF} \ll \Delta$, three-port mixer theory can be used to estimate the third-order intermodulation intercept of a quantum mixer. For rf power sufficient to induce IF voltage amplitude δV at the maximum gain bias point, the IF voltage waveform can be written

$$V(t) - V_{0,max} = f(V)\delta V\,[\cos(\omega_1 t) + \cos(\omega_2 t)], \quad (4.8)$$

where $f(V)$ describes the voltage dependence of the gain. As before, gain is maximum at $V_{0,max} = (2\Delta - \hbar\omega/2)/e$, and falls to minima at $V_0 = (2\Delta - \hbar\omega)/e$ and $V_0 = 2\Delta/e$. To model this general behavior, the gain function $f(V)$ can be approximated in the range $(2\Delta - \hbar\omega)/e < V_0 < 2\Delta/e$ as a simple parabola.

$$f(V) = 1 - 4(V - V_{0,max})^2 e^2/(\hbar\omega)^2. \quad (4.9)$$

Equations (4.8) and (4.9) can be solved self consistently to obtain $V(t)$. To lowest order in δV, the IF waveform is given by

$$V - V_{0,max} = \delta V\cos(\omega_1 t) + \delta V\cos(\omega_2 t) - 3(\delta V)^3 e^2/(\hbar\omega)^2 \cos((2\omega_1 - \omega_2)t) - \ldots, \quad (4.10)$$

where terms at frequencies $(2\omega_1 - \omega_2)$, $3\omega_1$, $3\omega_2$,...have not been explicitly written. The ω_1 term and the $(2\omega_1 - \omega_2)$ term have equal amplitudes for $\delta V = \hbar\omega/e\sqrt{3}$. This amplitude corresponds to a signal input power of

$$P_{int} = \frac{(\hbar\omega/e)^2}{6GR_{Dyn}}. \quad (4.11)$$

Note that this limit is quite similar to the gain saturation criterion of Eq. (4.1). While the exact value of the third-order intermodulation intercept depends on the functional dependence of $f(V)$, Eq. (4.11) provides a working estimate. For the parameters chosen in Sec. IV B, $P_{int.} = 30$ pW. In practice, for spurious signals to be kept 30 dB below desired linear signals, P_{sig} must be kept 10 dB below the calculated $P_{int.}$.

In summary, IF port saturation appears to be the dominant power limiting mechanism for quantum SIS mixers. The voltage dependence of the gain imposes constraints upon the amount of IF output power which can be obtained for a given allowable gain compression. These power constraints limit the gain-bandwidth-dynamic range which can be expected from a single junction SIS mixer. As signal powers approach the third-order intermodulation intercept, intermodulation between separate signals produces spurious response signals at IF frequencies.

V. SERIES ARRAYS

A significant feature of SIS quasiparticle mixers is their low characteristic power levels. The LO requirements for a single junction mixer are on the order of 1 nW, and saturation powers of ~1 pW are typical. For applications where larger saturation power is desirable, series-arrays of junctions have practical importance.[6,16,19]

In the design of an array of N identical junctions, the rf and IF matching constraints require that the series resistance of the junctions be comparable to that of a single junction in a corresponding mixer. This is accomplished by reducing the resistance per junction by a factor N. For the same operating frequency, the RC product for the junctions in the array should be the same as for the single junction. Thus the junction current density is held fixed, and the area of the individual junctions is increased to accomplish the reduction in individual resistances. Each junction is dc biased at $(2\Delta - \hbar\omega/2)/e$. In practice this can be done either by using junctions with nearly identical resistances, or possibly by individually biasing the different junctions with a low-pass biasing scheme. The resultant mixer array can be considered as an array of generators whose voltages add in series, producing a total output power which is proportional to N^2.

Table I summarizes the dependence of the properties of a series-array mixer upon the number of junctions. A distinc-

TABLE I. Dependence on N.

	Per Junction	Entire Array
$V_{0,IF,LO,...}$	1	N
$I_{0,IF,LO,...}$	N	N
$R_{0,IF,LO,...}$	N^{-1}	1
Capacitance	N	1
Mixer gain	1	1
LO power	N	N^2
Saturation power	N	N^2
Quantum noise voltage	$N^{-1/2}$	1
Quantum noise power	1	1
Shot noise power	1	1
Dynamic range	N	N^2
Direct detector responsivity	1	N^{-1}
Dimensions	$N^{1/2}$	

tion is made between noise attributable to shot noise caused by individual electron tunneling events, and quantum noise caused by zero-point photon fluctuation in the entire mixer.

The mixer gain must be calculated using an admittance matrix Y^{total} which characterizes the array. The voltage across junction i at port m resulting from an applied unit current at port n is given by the impedance matrix element $Z_{mn}^{(i)} = [Y^{(i)-1}]_{mn}$. Because the junctions are connected in series, the current through the individual junctions for a given port is the same. The resultant voltages across the junctions add, so that the total voltage can be given by an overall impedance matrix

$$Z^{\text{total}} = [Y^{\text{total}}]^{-1} = \sum_{i=0}^{N} [Y^{(i)}]^{-1}. \tag{5.1}$$

For identical junctions with resistances scaled as above, Y^{total} is therefore independent of N. Equation (5.1) can also be used for heterogeneous arrays provided the $Y^{(i)}$ are calculated for fixed LO and dc current through the junctions. It has been assumed that the length of the array is much less than relevant electrical wavelengths, so that phase shifts across the junctions can be neglected.

Several researchers have noted the quantum noise level of arrays should be independent of the number of junctions.[20] This prediction can be understood in a quasiclassical circuit model. Zero-point fluctuation on the individual junctions each produce IF power $\hbar\omega BG$, with IF impedance R_{IF}/N. This corresponds to a rms IF voltage noise of $(\hbar\omega BGR_{\text{IF}}/N)^{1/2}$. Provided the electrostatic coupling energy from junction to junction $(e^2/2C)$ is small compared to the thermal energy (kT_{bath}), the noise in the junctions is uncorrelated. The total array voltage noise is found by a sum-of-squares addition, yielding $(\hbar\omega BGR_{\text{IF}})^{1/2}$, which is independent of N. Thus, while each junction contributes IF power $\hbar\omega BG$, only a fraction $1/N$ of this power is coupled to the external IF load. The same analysis holds true for shot noise.

Junction biasing is critical for series-array mixers operated with high gain. Because the gain of the individual junctions depends strongly on the bias voltage, each junction must be operated as close as possible to the optimum voltage $V_{0,\text{max}}$. For biasing schemes where the same dc current flows through each junction, this requires uniform junction resistances. Rudner *et al.*[19] found a 5%–10% spread in junction resistances caused significant departures from predicted pumped I-V characteristics. Uniformity is most critical for high gain mixers, which encorporate high dynamic resistance R_{Dyn}. For these mixers small normal state resistance variations lead to large voltage differences between the junctions. It appears likely that applications requiring high gain will tend to use a small number of junctions, while uses which put a premium on high saturation power will use larger N.

VI. CONCLUSIONS

In summary, analytic solutions to three-port Y mixer theory applied to SIS tunnel junctions predict negative resistance and large gain. The conversion efficiency of the mixers can be described by gain contours as a function of the available power P_{LO} and the source impedance G_S. While exact quantitative descriptions of the gain contours requires extensive numerical calculations, the qualitative behavior can be understood by examining special limiting cases where analytic solutions are possible. The cases of low P_{LO} and also zero conversion efficiency have been derived for SIS tunnel junctions. The possibility of IF saturation imposes a constraint on the maximum gain-bandwidth-dynamic range product which may be achieved using single junction SIS quasiparticle mixers. Both gain compression and signal intermodulation occur for signals close to the saturation power level. Larger saturation powers are available using series-array mixers, although at the cost of increased biasing problems.

ACKNOWLEDGMENTS

We are grateful for helpful discussions with M. J. Feldman, K. E. Irwin, W. R. McGrath, J. Peterson, and G. Sollner. Calculation of noise in arrays was done in collaboration with D. Scalapino. A portion of this research was supported by the U. S. Office of Naval Research.

[1] J. R. Tucker, IEEE J. Quantum Electron. **QE-15**, 1234 (1979).
[2] J. R. Tucker, Appl. Phys. Lett. **36**, 477 (1980).
[3] T-M. Shen and P. L. Richards, IEEE Trans. Magn. **MAG-17**, 677 (1981).
[4] P. L. Richards and T-M. Shen, IEEE Trans. Electron Devices **ED-27**, 1909 (1980).
[5] A. D. Smith, W. R. McGrath, P. L. Richards, H. van Kempen, D. Prober, and P. Santhanam, Physica B + C **108**, 1367 (1981).
[6] A. R. Kerr, S.-K. Pan, M. J. Feldman, and A. Davidson, Physica B + C **108**, 1369 (1981).
[7] W. R. McGrath, P. L. Richards, A. D. Smith, H. van Kempen, R. A. Batchelor, D. E. Prober, and P. Santhanam, Appl. Phys. Lett. **39**, 655 (1981).
[8] H. C. Torrey and C. A. Whitmer, *Crystal Rectifier*, M. I. T. Radiat. Lab. Ser. **15** (McGraw-Hill, New York, 1948).
[9] A. A. M. Saleh, *Theory of Resistive Mixers* (M. I. T., Cambridge, MA, 1971).
[10] M. Wengler (private communication).
[11] It is valid and covenient to calculate Y_{IF} with $Y_0^{\text{ext}} = 0$.
[12] M. Tinkham, *Introduction to Superconductivity* (McGraw-Hill, New York, 1975).
[13] T. C. L. G. Sollner, Physica B + C **108**, 1365 (1981).
[14] The limit of $g_i \to \infty$ and $g_s \to 0$ is taken with $g_s g_i \gg 1$.
[15] P. K. Tien and J. P. Gordon, Phys. Rev. **129**, 647 (1963).
[16] S. Rudner and T. Claeson, Appl. Phys. Lett. **34**, 711 (1979).
[17] T-M. Shen, P. L. Richards, R. E. Harris, and F. L. Lloyd, Appl. Phys. Lett. **36**, 777 (1980).
[18] T. C. L. G. Sollner (private communication).
[19] S. Rudner, M. J. Feldman, E. Kollberg, and T. Claeson, IEEE Trans. Magn. **MAG-17**, 690 (1981).
[20] M. J. Feldman and S. Rudner, *Reviews in Infrared and Millimeter Waves*, edited by K. J. Button (Academic, New York, 1982), Vol. 2.

Low-noise 115-GHz receiver using superconducting tunnel junctions

S. -K. Pan,[a] M. J. Feldman,[a] and A. R. Kerr
NASA Goddard Institute for Space Studies, New York, New York 10025

P. Timbie
Physics Department, Princeton University, Princeton, New Jersey 08544

(Received 24 June 1983; accepted for publication 28 July 1983)

A 110–118-GHz receiver based on a superconducting quasiparticle tunnel junction mixer is described. The single-sideband noise temperature is as low as 68 ± 3 K. This is nearly twice the sensitivity of any other receiver at this frequency. The receiver was designed using a low-frequency scale model in conjunction with the quantum mixer theory. A scaled version of the receiver for operation at 46 GHz has a single-sideband noise temperature of 55 K. The factors leading to the success of this design are discussed.

PACS numbers: 07.62. + s, 84.30.Qi, 85.25. + k

The superconductor-insulator-superconductor (SIS) quasiparticle tunnel junction mixer[1,2] has recently been used as a first stage in millimeter-wave receivers for radio astronomy. First at 115 GHz,[3] and then at frequencies from 45 to 250 GHz,[4–6] these receivers have rivaled the best conventional (i.e., Schottky diode) receivers. Now we have constructed an SIS receiver for 115 GHz which is almost twice as sensitive as any previously reported.[7] This letter describes the design of the receiver, its performance, and the factors leading to its success. These results may influence the choice of receiver for a number of millimeter-wave astronomy facilities now being planned.

The mixer employs a series pair of Pb(InAu)-oxide-Pb(Bi) tunnel junctions. The dc *I-V* curve of these junctions at 2.5 K is shown in Fig. 1(a). The differential resistance R, measured at 8 mV, is 94 Ω. An estimate of the capacitance C gives $\omega RC \sim 7$. The junctions were deposited using a photoresist liftoff process at NBS-Boulder. Each junction is $1.8 \times 2.5\ \mu$m, defined by an SiO window.

A schematic drawing of the mixer is shown in Fig. 2. The fused quartz SIS chip, $0.010 \times 0.005 \times 0.010$ in. thick, is mounted on a 0.003-in.-thick fused quartz substrate with a gold stripline circuit pattern. The substrate is placed across a quarter-height WR-10 waveguide. The transition to the reduced height waveguide is accomplished using a broadband channel-waveguide transformer.[8] (The waveguide width is actually 0.096 in., slightly narrower than standard, to permit operation beyond the normal operating band without evanescent mode resonances.) Two sliding shorts of the contacting spring-finger type serve as adjustable tuning elements. One sits in the main waveguide behind the substrate, while the other is in a secondary waveguide, parallel to the first. The two waveguides are coupled through the suspended stripline as shown in the figure. Compared to conventional mixer blocks with only a single backshort, this configuration has much greater tunability at a given frequency, but the tuning is a relatively sharp function of frequency.

The potential performance of the mixer was analyzed by applying the quantum mixer theory[9,10] to the unpumped *I-V* curve, Fig. 1(a). A conversion loss as low as 1.5 dB was predicted if the signal and image ports were terminated with appropriate, equal, impedances. But our relatively sharp tuning assured that only one sideband could be well terminated, while the other was approximately short circuited (by the large junction capacitance). The best conversion loss with the image short circuited was predicted to be 4.3 dB. The suspended-substrate printed circuit pattern was designed to aid in achieving the requisite signal termination. This was accomplished as in Ref. 11 by using a $40 \times$ scale model of the mixer block, which included the sliding shorts, the substrate, and the printed circuitry. Calculations based upon the scale model results indicated that the ohmic circuit losses would add 1–3 dB to the mixer's conversion loss in the vicinity of the best operating point, giving an expected overall conversion loss of 5–8 dB. We also calculated the shot noise of the mixer, as predicted in Ref. 9, and the ambient 2.5-K thermal noise reflected at the IF port of the mixer. In the vicinity of the best operating point, the theoretical mixer *output* temperature is 3.9 ± 0.4 K.

The performance of the mixer, tuned for single-sideband operation, was measured in a laboratory dewar with a monochromatic signal at 115.3 GHz. The best conversion loss is 6.9 ± 0.5 dB. The mixer's instantaneous bandwidth is

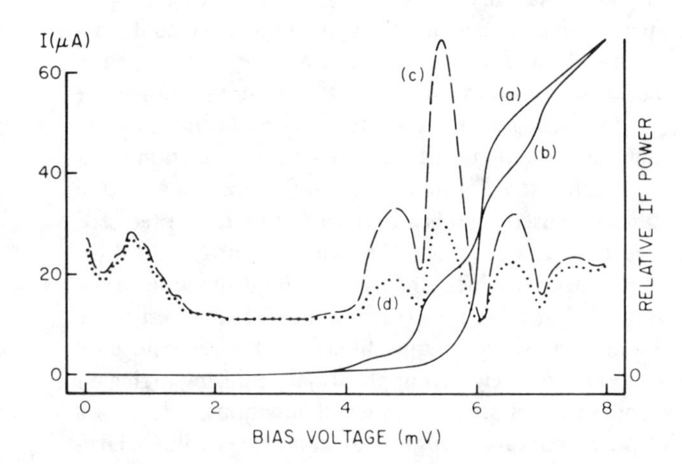

FIG. 1. (a) dc *I-V* curve of an unpumped series pair of SIS junctions, at 2.5 K. (b) dc *I–V* curve when LO power of ~250 nW at 113.9 GHz was applied to the mixer. The receiver is tuned for a lower-sideband rejection > 25 dB. The receiver's output noise power at the 1.4 GHz IF, in response to room temperature and liquid nitrogen loads, is shown in curves (c) and (d), respectively.

[a] Also Physics Department, Columbia University, New York, NY 10027.

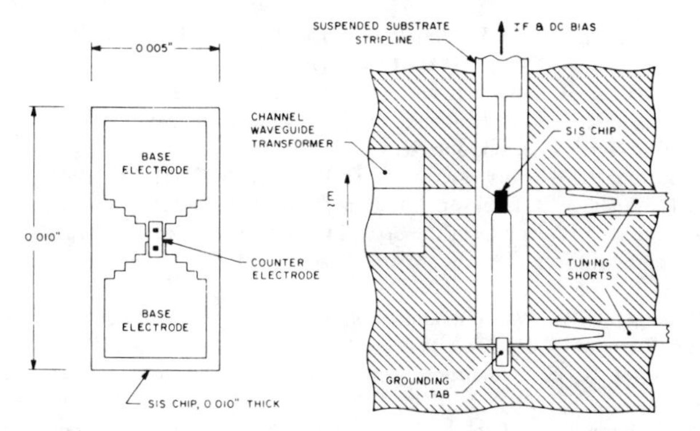

FIG. 2. Schematic drawing of the SIS chip and a cross-sectional view of the mixer block.

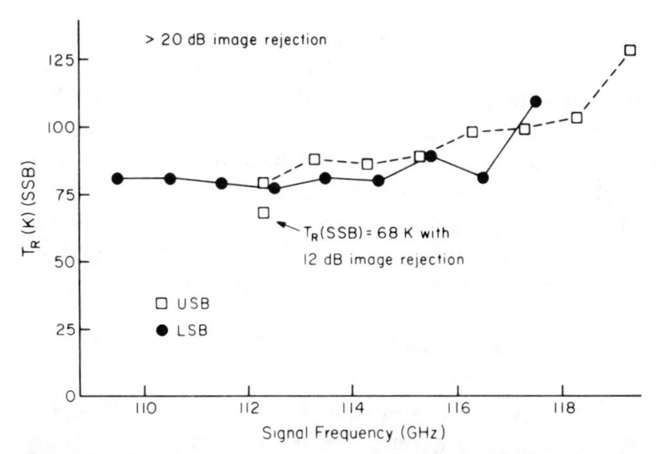

FIG. 3. Receiver's noise temperature as a function of signal frequency, for both upper-sideband (□) and lower-sideband (●) operation. The receiver is tuned for > 20 dB image rejection. By relaxing the image rejection it was possible to obtain somewhat better results, e.g., a single-sideband $T_R = 68$ K at 112.3 GHz. The uncertainty in T_R is ± 3 K.

~350 MHz, and the 1-dB gain compression point is typically 4 nW.

In the receiver dewar, the mixer is connected through a cross-guide LO injection coupler (coupling = 20 dB) and a stainless steel waveguide to a room-temperature scalar horn. The loss between the mouth of the horn and the mixer is 0.6 dB and contributes 12 K to the receiver noise temperature. The 1.2–1.6-GHz output from the mixer passes through a bias tee and an isolator[12] to a GaAs field-effect transistor amplifier,[13] all cooled to 2.5 K. The noise temperature of the IF section is 10.5 ± 1.0 K, referred to the output port of the mixer. This quantity was measured by using the shot noise in the unpumped SIS junctions, biased with a dc current, as an IF noise source. If a series array of N junctions is biased with a current I_{dc} at a differential resistance R_d, then the available shot noise power per unit bandwidth is $eI_{dc}\ R_d/2N$ (see Ref. 14).

Figure 1(b) shows the dc I-V curve when LO power of ~250 nW at 113.9 GHz was applied to the mixer. The mixer was tuned to maximize the conversion, while maintaining a large image rejection (> 25 dB). This was accomplished using a monochromatic source. Under certain other tuning conditions it is possible to obtain a region of negative differential resistance on the pumped I-V curve, as in Ref. 15. This does not, however, imply a negative output resistance at the 1.4 GHz IF, because of the relatively sharp tuning of the mixer. Figures 1(c) and 1(d) show the receiver's IF output noise power, in a 50-MHz bandwidth, in response to room temperature and liquid nitrogen loads, respectively. At the dc bias voltage of 5.45 mV the receiver noise temperature is 89 ± 3 K at the 115.3-GHz signal frequency.

The experimental data given above can be used to infer that the noise temperature of the mixer itself at 115.3 GHz is $T_m = 15$ K, with a maximum uncertainty of ± 14 K. The theoretical value of T_m (referred to the mixer input using the experimental conversion loss) is 19 ± 4 K. Thus it appears that the main contributions to our mixer's noise are well understood.

Figure 3 shows the receiver's noise temperature as a function of signal frequency, tuned for > 20 dB image rejection. Note that $T_R \simeq 80$ K over a broad range of frequencies. By relaxing the image rejection somewhat it is possible to reduce the receiver noise temperature further. For example, with 12-dB image rejection it is possible to obtain (single sideband) $T_R = 68$ K at 112.3 GHz. The rise of T_R at higher frequencies is a result of the LO coupler, which has an evanescent mode resonance at 118 GHz. The receiver's noise temperature changes by less than 10 K within a 270-MHz instantaneous bandwidth.

No magnetic field was applied to the junctions in these experiments. Nevertheless, the Josephson critical current in Fig. 1(a) is completely suppressed, whereas on other cool downs various amounts of Josephson critical current were apparent. We ascribe this variation to different amounts of magnetic flux trapped in the junctions. Whenever the Josephson critical current was seen, a large IF noise was generated by the pumped mixer at bias voltages below about 2.5 mV.[16] The Josephson currents have no discernible effect at larger bias voltages.

Although the initial optimization of the receiver for each frequency was time consuming, the settings could be noted and quickly regained on subsequent cool downs. There was no perceptible change in the junctions' behavior through 10 cool downs over four months, during which time the junctions were stored in a dry nitrogen atmosphere at a temperature of 0 °C. The long term stability remains to be checked. This receiver is presently in use on the Columbia/GISS CO Sky Survey telescope.

A scaled version of this receiver has also been built for observations of the 3-K cosmic blackbody radiation at 46 GHz. A larger pair of junctions, with $R = 34\,\Omega$ (for the pair) and $\omega RC \sim 2.5$, was used. The single-sideband receiver noise temperature was 55 K. This is comparable to the noise performance of maser receivers at this frequency. More detailed information will be reported elsewhere.

Why are these receivers more sensitive than previous[3–6] SIS receivers? The I-V curve of the junctions, Fig. 1(a), is not particularly sharp compared to other SIS junctions which have been used. Rather, we believe that two other factors are important. First is the relatively large ωRC product. The significance of this is discussed in Ref. 14. It can be inferred from Ref. 16, Fig. 9, that $\omega RC \gtrsim 4$ is required for the mixer's conversion to reach the value predicted using the three-frequency model, at a bias voltage of $N\hbar\omega/2e$ below the energy gap. Second is the superior tuning capability of our mixers. This factor is especially important in light of the large junction capacitance, which cannot be precisely controlled during fabrication.

In conclusion, we have demonstrated an SIS receiver for 115 GHz which is almost twice as sensitive as any competitor. The design has been scaled to 46 GHz with comparable success. These results should firmly establish the SIS mixer as the first choice for ultralow-noise millimeter-wave receivers.

The authors wish to thank F. L. Lloyd and C. A. Hamilton of the National Bureau of Standards for their assistance during the junction fabrication, J. A. Grange for developing the mixer assembly techniques, E. S. Palmer for designing the overall receiver, and P. Thaddeus and D. Wilkinson for their continuing support of this work.

[1]P. L. Richards, T.-M. Shen, R. E. Harris, and F. L. Lloyd, Appl. Phys. Lett. **34**, 345 (1979).
[2]G. J. Dolan, T. G. Phillips, and D. P. Woody, Appl. Phys. Lett. **34**, 347 (1979).
[3]T. G. Phillips and D. P. Woody, Ann. Rev. Astron. Astrophys. **20**, 285 (1982); T. G. Phillips, D. P. Woody, G. J. Dolan, R. E. Miller, and R. A. Linke, IEEE Trans. Magn. **MAG-17**, 684 (1981).
[4]R. Blundell, K. H. Gundlach, and E. J. Blum (unpublished).
[5]L. Olsson, S. Rudner, E. Kollberg, and C. O. Lindström, Int. J. Infrared and Millimeter Waves (unpublished).
[6]E. C. Sutton, IEEE Trans. Microwave Theory Tech. **MTT-31**, 589 (1983).
[7]A. V. Räisänen, N. R. Erickson, J. L. R. Marreno, P. F. Goldsmith, and C. R. Predmore, Proceedings of the Sixth International Conference on Infrared and Millimeter Waves, Miami Beach, 1981 (IEEE Cat. No. 81 CH1645-1 MTT).
[8]P. H. Siegel, D. W. Peterson, and A. R. Kerr, IEEE Trans. Microwave Theory Tech. **MTT-31**, 473 (1983).
[9]J. R. Tucker, IEEE J. Quantum Electron. **QE-15**, 1234 (1979).
[10]J. R. Tucker, Appl. Phys. Lett. **36**, 477 (1980).
[11]M. J. Feldman, S. -K. Pan, A. R. Kerr, and A. Davidson, IEEE Trans. Magn. **MAG-19**, 494 (1983).
[12]Passive Microwave Technology, 8030 Remmit Ave., Canoga Park, Ca 91304, cryogenic *L*-band isolator, model 1102.
[13]S. Weinreb, D. L. Fenstermacher, and R. W. Harris, IEEE Trans. Microwave Theory Tech. **MTT-30**, 849 (1982).
[14]M. J. Feldman and S. Rudner, in *Reviews of Infrared and Millimeter Waves*, Volume 1, edited by K. J. Button (Plenum, New York, 1983), pp. 47–75.
[15]A. R. Kerr, S. -K. Pan, M. J. Feldman, and A. Davidson, Physica B **108**, 1369 (1981).
[16]S. Rudner, M. J. Feldman, E. Kollberg, and T. Claeson, J. Appl. Phys. **52**, 6366 (1981).

Part IX
FET Mixers

MIXERS using MESFET's as the nonlinear element can be designed so that conversion gain is obtained. The signal and the local oscillator can be connected to the terminal of the FET in different ways implying that there are different modes of operation available [1].

The most common type of FET mixer is the gate mixer in which the signal and the local oscillator are both connected between the gate and the source of the FET. In the first paper Pucel *et al.* describe the gate mixer in considerable detail, both from the theoretical and the experimental point of view. A more complete theoretical discussion of the gate mixer is presented in papers 4 and 5 of this part. Begemann and Hecht analyze the conversion gain and the stability and discuss essential constraints on the embedding circuit. Tie and Aitchison present a more complete noise theory and show that the theory agrees well with experiments. They find that although the RF output and IF input circuits of a MESFET mixer are important for high gain performance, their effects upon the mixer noise performance are less significant.

In another mode of operation the signal is still applied between the gate and the source, but the local oscillator is applied between the drain and the source. In the second paper of this part Bura and Dikshit describe experimental gate and drain mixers and also discuss their relative merits. In paper 6 Begemann and Jacob discuss the theory of the drain mixer. They include some numerical examples.

The dual-gate MESFET offers some advantages over the ordinary single-gate MESFET in mixer applications. The local oscillator circuit can be isolated from the signal circuit by connecting them to separate gate electrodes. In the third paper, by Cripps *et al.*, single-gate and dual-gate mixers are evaluated and compared. In the final paper Tsironis *et al.* show how one of the gates can be used for creating oscillations, so that an external local oscillator source can be avoided.

References

[1] R. S. Pengelly, *Microwave Field-Effect Transistors—Theory, Design, and Applications.* Chichester, England: Research Studies, 1982.

Performance of GaAs MESFET Mixers at X Band

ROBERT A. PUCEL, SENIOR MEMBER, IEEE, DANIEL MASSÉ, MEMBER, IEEE, AND RICHARD BERA

***Abstract*—A theoretical analysis and experimental verification of the signal properties of the GaAs MESFET mixer are presented. Experimental techniques for evaluating some of the mixer parameters are described.**

Experiments performed on GaAs MESFET mixers at X band show that good noise performance and large dynamic range can be achieved with conversion gain. A conversion gain over 6 dB is measured at 7.8 GHz. Noise figures as low as 7.4 dB and output third-order intermodulation intercepts of +18 dBm have been obtained at 8 GHz with a balanced MESFET mixer.

I. Introduction

THE low-noise performance of GaAs Schottky barrier gate field-effect transistors (MESFET's) as high-gain linear amplifiers in the high microwave band (C–X band) has been demonstrated by many laboratories [1].[1] In this paper we shall show that these transistors also have the potential for low-noise operation as microwave mixers with gain and high dynamic range. As such, they combine the best features of the tunnel diode and the Schottky barrier diode mixer.

We also show in this paper how certain of the FET mixer parameters can be measured. Using the measured parameter values in the theory developed here, we obtain good agreement with experiment.

Our experiments were performed at X band with single-gate GaAs MESFET's. These studies, a natural extension of our investigations of the GaAs FET linear amplifier, were intended to assess the suitability of the GaAs FET as a building block for the three basic active components of an integrated front end at X band, namely the RF amplifier, local oscillator, and mixer. The devices used in the mixer studies were the same as those used in the low-noise amplifier experiments. That is to say, no attempt was made to optimize them for mixer applications.

II. Frequency Conversion in an FET

The small-signal equivalent circuit of the unpackaged FET valid at frequencies up to X band and higher is illustrated in Fig. 1. Additional parasitic elements, such as lead inductances and interelectrode capacitances, must be added for a device mounted in a package. (Our experiments were made with unpackaged FET's with beam leads.) In a well-designed GaAs MESFET the parasitic contact resistances R_{gm}, R_s, and R_{dr} in the gate, source, and drain leads, respectively, are small compared to the drain resistance R_d.

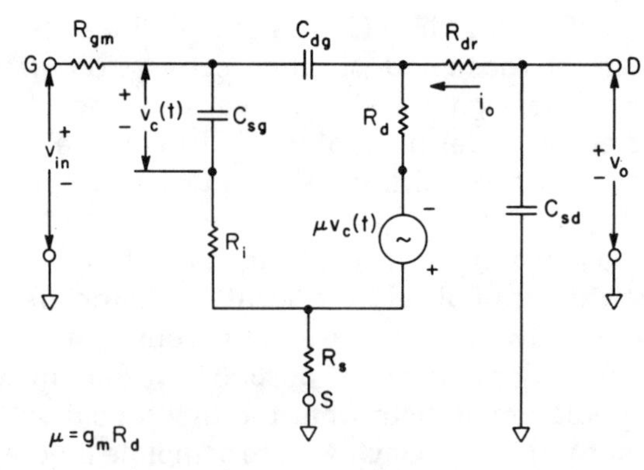

Fig. 1. Small-signal equivalent circuit of the GaAs field-effect transistor.

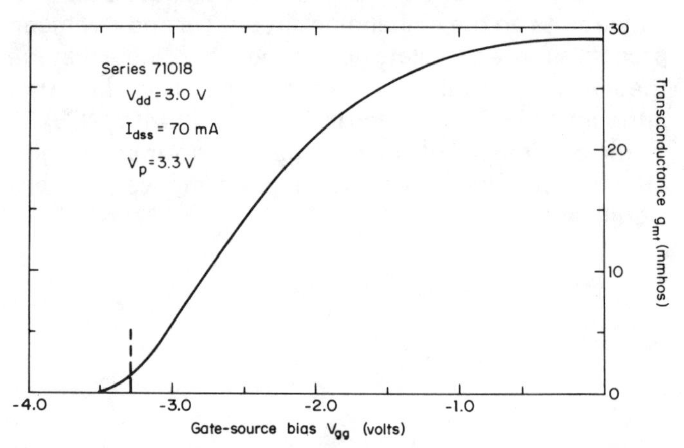

Fig. 2. Measured terminal transconductance as a function of gate bias of a GaAs MESFET used in mixer experiments.

However, R_{gm} and R_s are the principal sources of extrinsic noise [2].

Mixing occurs in an FET when the small-signal elements representing the FET are varied at a periodic rate by a large local oscillator signal impressed between a pair of the device terminals, usually the gate-source terminals. In a GaAs MESFET the strongest gate-bias dependence is exhibited by the transconductance g_m. The mixing products attributable to parametric "pumping" of the source-gate capacitance C_{sg} and its charging resistance R_i are negligible. The drain resistance also shows a strong gate-bias dependence. However, since it is not integral to the gain mechanism of the FET, we have used only its time-averaged value in our theory. The theory can be generalized to remove this simplification, but our experimental results did not require this, at least for the signal properties of the mixer.

Fig. 2 is a graph of the measured transconductance of a

Manuscript received October 3, 1975; revised December 17, 1975.
The authors are with the Research Division, the Raytheon Company, Waltham, MA 02154.

[1] Since in this paper we shall consider only GaAs Schottky barrier gate field-effect transistors, we use the terms GaAs MESFET, GaAs FET, and FET interchangeably.

Reprinted from *IEEE Trans. Microwave Theory Tech.*, vol. MTT-24, pp. 351–360, June 1976.

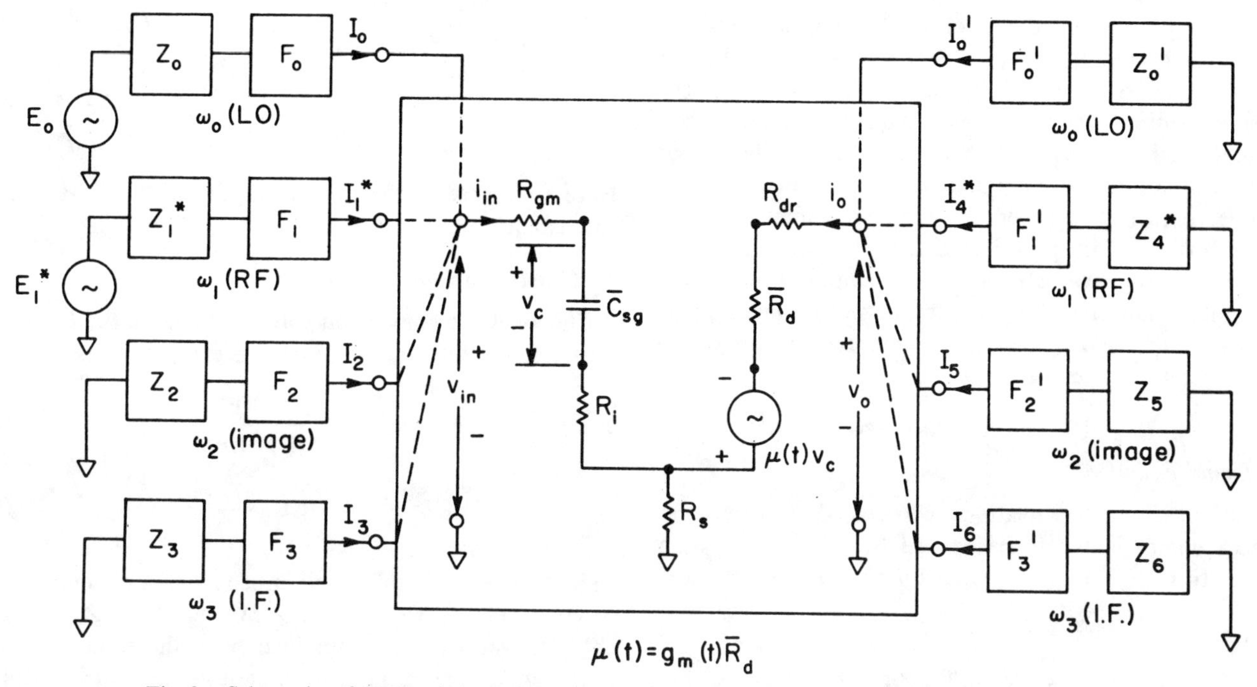

Fig. 3. Schematic of FET mixer, including signal, image, and IF circuits, used in signal analysis.

typical microwave FET as a function of the potential difference V_{gg} between the gate and source terminals when the drain-source bias voltage V_{dd} has a value somewhere above the knee of the I–V characteristic, in this case 3.0 V. The functional dependence for any other value of drain voltage above the knee is similar.

Assume that for a fixed value of gate-source bias, a large LO signal is superimposed on the gate-source terminals. The transconductance becomes a time-varying function $g_m(t)$ with a period equal to that of the LO. If ω_0 denotes the oscillator frequency, we may write

$$g_m(t) = \sum_{k=-\infty}^{\infty} g_k e^{jk\omega_0 t} \tag{1}$$

where

$$g_k = \frac{1}{2\pi}\int_0^{2\pi} g(t)e^{-jk\omega_0 t}d(\omega_0 t). \tag{2}$$

All harmonic amplitudes can be assumed to be real by proper choice of the time reference. In accordance with our earlier stated assumption, we will neglect all harmonic components of $R_d(t)$, so that the time-varying voltage amplification factor $\mu(t)$ can be written as $\mu(t) \approx \bar{R}_d g_m(t)$, where $\bar{R}_d = R_0$ is the time-averaged component of the drain resistance.

Let a small-signal $v_c(t)$ of frequency $\omega_1 \neq \omega_0$ be impressed on the gate capacitance, as indicated in Fig. 1. By the mixing action of the time-varying transconductance, a voltage $\mu(t)v_c(t)$ is generated in the drain circuit. This signal has side-band frequency components $|n\omega_0 \pm \omega_1|$, where n takes on all integer values.

An analysis of a practical FET mixer must include side-band components in both the gate (input) and drain (output) circuits. In our analysis we shall focus attention on the down-converter, where the output or intermediate (IF) frequency $\omega_3 = |\omega_0 - \omega_1|$ is less than the signal frequency ω_1. Then the only remaining frequency component of first-order importance is the image frequency ω_2, where $|\omega_2 - \omega_1| = 2\omega_3$. For convenience we shall assume $\omega_1 < \omega_0 < \omega_2$.

Let V_1, V_2, V_3 and I_1, I_2, I_3 represent the complex voltage and current amplitudes of the signal, image, and IF components in the gate circuit, and V_4, V_5, V_6 and I_4, I_5, I_6 the corresponding voltage and current amplitudes, respectively, of the components in the drain circuit. Suppose the input signal at ω_1 is driven by a voltage source E_1 of internal impedance Z_1 and all other side-band components, including the desired output signal at frequency ω_3, are terminated in complex impedances. Then the boundary conditions at the gate source and drain-source terminals, relating the voltage and current amplitudes V_k and I_k are

$$V_k = E_k - I_k Z_k, \qquad (k = 1,2,\cdots,6) \tag{3}$$

where $E_k = 0$ for $k \neq 1$.

The equivalent circuit of the FET mixer, based on Fig. 1, representing (3) is shown in Fig. 3. Because we have chosen the incoming signal frequency to be the "lower side band," i.e., $\omega_1 < \omega_0$, we must use the complex conjugate component of the signal. This will be denoted by the asterisk (*). In our representation the boxes labeled F_k and F_k' are fictitious ideal (lossless) filters which have zero impedance at the desired frequency and infinite impedance at all the remaining side-band components. In practice, of course, these filters are not ideal and are not necessarily separable in a physical sense. Neither are the set of input ports, or output ports, physically separable. The source-drain capacitance does not appear in Fig. 3 since we have included it as part of the filter-termination circuit of each output port.

We have also included a set of input and output oscillator ports. The oscillator signal is assumed to be injected at the gate terminal as shown in Fig. 3. It could also be inserted in series with the source terminal, but this is not a convenient method at high frequencies. The oscillator port at the output is necessary because a large current component at this frequency is generated by the "pumping" of the drain current. This port is usually terminated in a low impedance, for example, a short-circuiting transmission line stub. When the IF is low, the output termination for the oscillator, signal, and image components can be the same circuit element.

III. Circuit Analysis

A. Signal Equations

The linear circuit relations between the various frequency components can be derived by a loop analysis of Fig. 3. In matrix notation these equations are written as

$$[E] = [V] + [Z_t][I] \tag{4a}$$

$$= [Z_m][I] + [Z_t][I] \tag{4b}$$

where

$$[E] = \begin{bmatrix} E_1^* \\ 0 \\ 0 \\ 0 \\ 0 \\ 0 \end{bmatrix} \quad [V] = \begin{bmatrix} V_1^* \\ V_2 \\ V_3 \\ V_4^* \\ V_5 \\ V_6 \end{bmatrix} \quad [I] = \begin{bmatrix} I_1^* \\ I_2 \\ I_3 \\ I_4^* \\ I_5 \\ I_6 \end{bmatrix}$$

and $[Z_m]$ and $[Z_t]$ are, respectively, the matrices representing the mixer proper and its terminations. These are given by

$$[Z_m] = \begin{bmatrix} Z_{11}^* & 0 & 0 & Z_{14}^* & 0 & 0 \\ 0 & Z_{22} & 0 & 0 & Z_{25} & 0 \\ 0 & 0 & Z_{33} & 0 & 0 & Z_{36} \\ Z_{41}^* & 0 & Z_{43} & Z_{44}^* & 0 & 0 \\ 0 & Z_{52} & Z_{53} & 0 & Z_{55} & 0 \\ Z_{61}^* & Z_{62} & Z_{63} & 0 & 0 & Z_{66} \end{bmatrix}$$

$$[Z_t] = \begin{bmatrix} Z_1^* & 0 & 0 & \cdot & \cdot & \cdot \\ 0 & Z_2 & & & & \\ 0 & & Z_3 & & & \\ \cdot & & & Z_4^* & & \\ \cdot & & & & Z_5 & \\ \cdot & & & & & Z_6 \end{bmatrix}.$$

If we neglect mixing by harmonics of $g_m(t)$ higher than the first, the matrix elements are

$$Z_{kk}(\omega_k) = R_{gm} + R_i + R_s + \frac{1}{j\omega_k \bar{C}}, \qquad (k = 1,2,3)$$

$$= R_{dr} + \bar{R}_d + R_s, \qquad (k = 4,5,6) \tag{5a}$$

$$Z_{14} = Z_{25} = Z_{36} = R_s \tag{5b}$$

$$Z_{41} = \frac{-g_0\bar{R}_d}{j\omega_1\bar{C}} + R_s \qquad Z_{61} = \frac{-g_1\bar{R}_d}{j\omega_1\bar{C}}$$

$$Z_{52} = \frac{-g_0\bar{R}_d}{j\omega_2\bar{C}} + R_s \qquad Z_{62} = \frac{-g_1\bar{R}_d}{j\omega_2\bar{C}}$$

$$Z_{63} = \frac{-g_0\bar{R}_d}{j\omega_3\bar{C}} + R_s \qquad Z_{43} = Z_{53} = \frac{-g_1\bar{R}_d}{j\omega_3\bar{C}}. \tag{5c}$$

Here $\bar{C}$ represents the time-averaged value of the source-gate capacitance.

B. Conversion Gain

The available conversion gain G_{av} between the RF input, port 1, and the IF output, port 6, is expressible as

$$G_{av} = \frac{|I_6|^2 \operatorname{Re} Z_6}{|E_1|^2/4 \operatorname{Re} Z_1} \tag{6a}$$

$$= 4R_g R_L \left|\frac{I_6}{E_1}\right|^2 \tag{6b}$$

where the source and load impedances are defined, respectively, as $Z_1 \equiv Z_g = R_g + jX_g$ and $Z_6 \equiv Z_L = R_L + jX_L$. The ratio I_6/E_1 is obtained from the solution of (4). If we let Δ_z denote the determinant of the matrix $[Z_m] + [Z_t]$, and Δ the determinant of the matrix obtained by deletion of the first row and sixth column, then $I_6/E_1 = -\Delta/\Delta_z$.

The expression for G_{av} is a complicated function of the terminations on each port. The derivation is given in the Appendix. However, when the intermediate frequency is small compared to the input signal frequency, drastic simplifications are introduced. Not only does the image at the input side "see" the same termination as the source, but the gain becomes insensitive to the IF termination at the input side and the image and RF terminations at the output side. We have verified this insensitivity to terminations in our experiments since our IF was 30 MHz and the RF was 8 GHz.

We find from the Appendix that the gain expression for this case simplifies to

$$G_{av} = \left(\frac{2g_1\bar{R}_d}{\omega_1\bar{C}}\right)^2 \frac{R_g}{(R_g + R_{in})^2 + \left(X_g - \dfrac{1}{\omega_1\bar{C}}\right)^2} \cdot \frac{R_L}{(\bar{R}_d + R_L)^2 + X_L^2} \tag{7}$$

where $R_{in} = R_{gm} + R_i + R_s$ is the input resistance. Note that the gain has a bandpass shape factor for the input and output circuits, as one might expect. The gain is maximum at band center when the source and load are conjugately matched to the FET, that is, for $R_g = R_{in}$, $X_g = (\omega_1\bar{C})^{-1}$; $R_L = \bar{R}_d$, $X_L = 0$. Defining $G_c = G_{av,\max}$, we find

$$G_c = \frac{g_1^2}{4\omega_1^2\bar{C}^2} \frac{\bar{R}_d}{R_{in}}. \tag{8}$$

We shall hereafter refer to the maximum available mixer gain as the conversion gain.

It is not surprising that this expression for conversion gain is of the same form as that for the maximum available amplifier gain G_a

$$G_a = \frac{g_m^{\,2}}{4\omega_1^{\,2}C^2}\frac{R_d}{R_{in}} \tag{9}$$

where the time-averaged quantities g_1, $\bar{C}$, and $\bar{R}_d$ are replaced by the values pertaining to a specific bias condition. The ratio of these two gains, corresponding to the same signal input frequency,

$$\frac{G_c}{G_a} = \left(\frac{g_1}{g_m}\right)^2 \left(\frac{C}{\bar{C}}\right)^2 \frac{\bar{R}_d}{R_d} \tag{10}$$

can be larger than unity. That is, the conversion gain can exceed the amplifier gain. We shall demonstrate this later. Even though $g_1/g_m < 1$, the ratios $C/\bar{C}$ and $\bar{R}_d/R_d$ are greater than unity since for maximum conversion gain the device is biased near pinchoff, in contrast to the amplifier for which the gate is operated at or near zero bias.

C. Bias Dependence of Mixer Gain Parameters

The conversion gain is a strong function of the gate bias and LO drive. The quantity most strongly dependent is the conversion conductance g_1. One might expect g_1 to be greatest for a gate bias near pinchoff, since it is here that the transconductance is most sensitive to bias modulation by the LO. This is true. To demonstrate this, let the instantaneous voltage between the gate-source terminals be represented as

$$V_{sg}(t) = V_{gg} + V_0 \cos \omega_0 t \tag{11}$$

where V_{gg} is the dc bias and V_0 the peak RF amplitude of the LO drive. This modulation wave is shown in the inset of Fig. 4(a).

Also illustrated in this figure is a plot of the theoretical conversion conductance as a function of dc bias when the LO amplitude is chosen to be the maximum possible value. This is taken to correspond to the onset of forward conduction of the Schottky barrier gate (approximately 0.5-V forward bias). Excessive forward conduction requires unnecessary LO power. The curve was calculated by a numerical Fourier analysis of $g_m(t)$ for the experimental curve shown in Fig. 2. Notice that a broad maximum in the curve occurs at approximately $V_{gg} = -3.2$ V, near the pinchoff point for g_m (Fig. 2).

Observe from Figs. 2 and 4(a) that the maximum value of g_1 is approximately one-third of the maximum (zero-bias) value of g_m. This ratio is close to the $1/\pi$ ratio obtained for the "ideal" case when g_m is a step function of gate bias.

For LO drive amplitudes below the maximum level, curves similar to Fig. 4(a) would be obtained, but of progressively lower value. We show this dependence explicitly for the gate bias corresponding to pinchoff [Fig. 4(b)]. The steeper the g_m curve is near pinchoff, the steeper the g_1 curve is near zero LO drive. In fact, for an ideal g_m versus V_{gg} dependence, the g_1 curve in Fig. 4(b) would also be a step function, with maximum g_1 occurring for vanishing LO drive.

Gate capacitance exhibits a much milder dependence on gate bias. The experimental points in Fig. 5 were obtained for the same FET whose transconductance curves were shown previously. Beyond the pinchoff bias, C_{sg} continues to decrease. Because of the mild variation of C_{sg}, we have approximated the experimental data by a linear function of bias, as shown by the solid line. For other devices, quadratic and higher order terms may be needed.

In general, if the gate bias dependence of C_{sg} is highly nonlinear, the average capacitance $\bar{C}$ "seen" by the small signals differs from the average capacitance $\bar{C}_0$ "seen" by

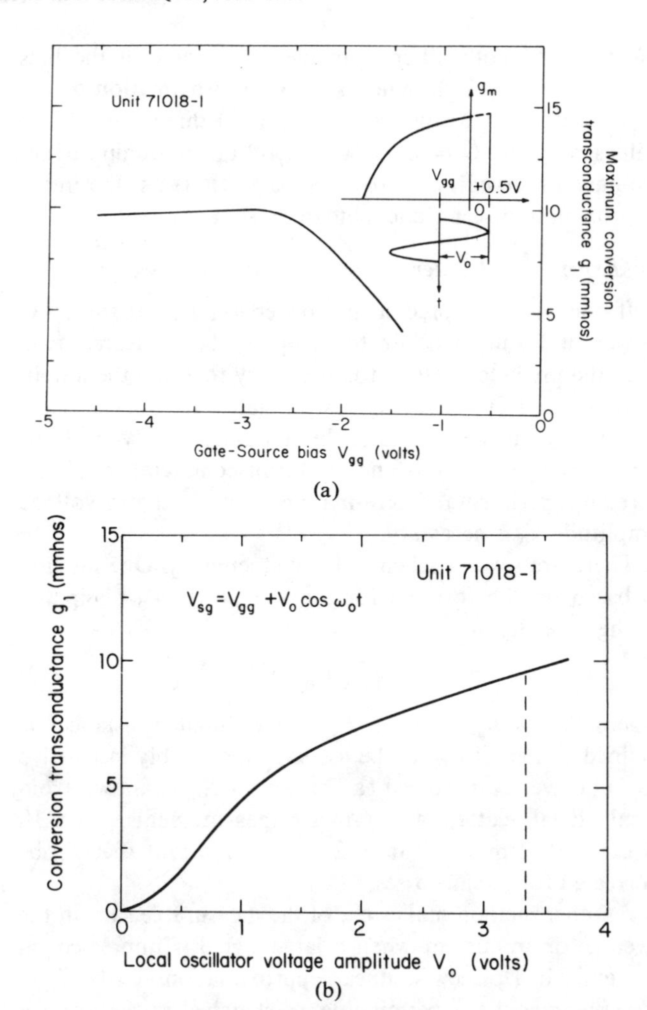

Fig. 4. Calculated (a) conversion transconductance at maximum LO drive as a function of gate bias and (b) conversion transconductance at pinchoff bias as a function of LO drive.

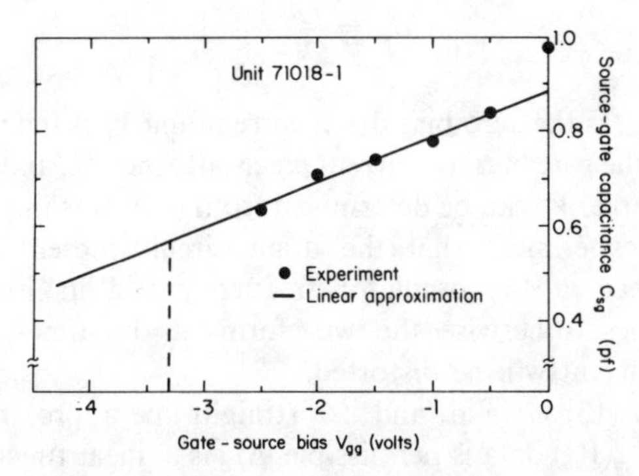

Fig. 5. Measured source-gate capacitance as a function of gate bias and its linear approximation.

the LO, and both differ from the static value at the bias point $C_{sg} = C_{sg}(V_{gg})$. When a linear approximation to the $C_{sg}(V_{gg})$ data is possible, as in Fig. 5, all three capacitance values coincide, $\bar{C} = \bar{C}_0 = C_{sg}(V_{gg})$. For the nonlinearities we have observed in microwave GaAs MESFET's, the linear approximation seems adequate in most cases.

D. Experimental Determination of Mixer Parameters

If one wishes to predict the conversion gain at some LO power and gate bias, or to compare the measured gain with the predicted value, it is necessary to know the amplitude of the LO voltage across the gate capacitance. Then g_1 and $\bar{R}_d$ can be calculated by a Fourier analysis of the time-varying drain resistance and transconductance. Therefore an experimental determination of the oscillator voltage amplitude V_0 is necessary.

There are several schemes for deducing V_0. One method is based on the formula for the LO power P_0 dissipated in the gate circuit

$$P_0 = \tfrac{1}{2}(\omega_0 \bar{C}_0 V_0)^2 R_{in} \tag{12}$$

where $R_{in} = R_{gm} + R_i + R_s$ is the input resistance as defined earlier. If $\bar{C}_0$ can be assumed reasonably insensitive to LO drive, as discussed earlier, and if R_{in} is known from small-signal scattering parameter measurements, then V_0 is easily determined from (12). It is important that P_0 be corrected for circuit losses.

Another method makes use of the measurable shift in the average drain current when a large signal is impressed on the gate. If a piecewise linear approximation to the $I_d - V_{sg}$ characteristic is permissible, as sketched in the inset of Fig. 6, then when biased at pinchoff, the shift in average drain current is linearly related to the oscillator voltage

$$\Delta \bar{I}_d = \frac{I_{dss}}{\pi} \frac{V_0}{V_p} \tag{13}$$

where I_{dss} is the zero bias drain current and V_p is the magnitude of the gate bias for current pinchoff. Since I_{dss} and V_p are measurable, V_0 can be determined from (13). It is important in the experiment that the drain circuit present a low impedance at the oscillator frequency and its first few harmonics, otherwise the waveform of the time-varying drain current will be distorted.

When (13) is valid, and if a straight-line approximation of the $C_{sg}(V_{sg})$ data is permissible, $\Delta \bar{I}_d$ is a linear function of $\sqrt{P_0}$ by virtue of (12). The data in Fig. 6 are an example of this linear relationship. The slope of the line

$$S = \frac{1}{\pi} \frac{I_{dss}}{V_p} \sqrt{\frac{2}{\omega_0{}^2 \bar{C}_0{}^2 R_{in}}} \tag{14}$$

allows one to determine the product $\bar{C}_0{}^2 R_{in}$ without knowledge of R_{in}. If $\bar{C}_0$ is known from capacitance data, one may determine R_{in} without recourse to high-frequency small-signal measurements. Furthermore, since $\bar{C} \approx \bar{C}_0$, this R–C product can be inserted directly into the denominator of the conversion gain expression (8).

The determination of $\bar{R}_d$ by Fourier analysis is not recommended since the low-frequency drain resistance often shows an erratic dependence on gate bias which is not evident at higher frequencies. It is preferable to obtain $\bar{R}_d$ by measurement of the resistance presented to the FET by the IF circuit when the IF load is adjusted for maximum conversion gain at each LO drive. This was the method used by us.

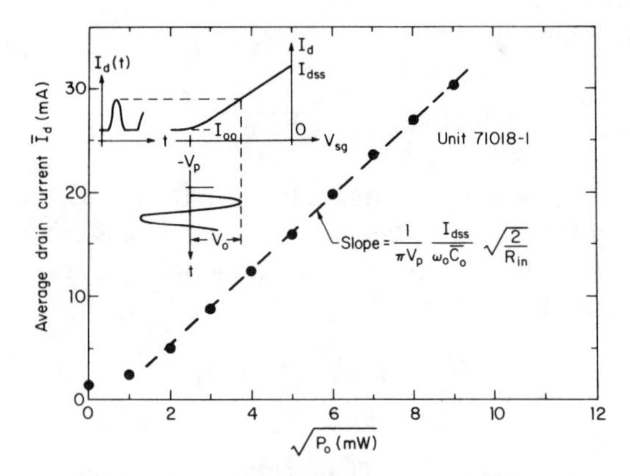

Fig. 6. Increase in average drain current as a function of LO drive when FET is biased near pinchoff.

E. Noise

Noise in a microwave FET is produced by sources intrinsic to the device; by thermal sources associated with the parasitic resistances, i.e., the gate metallization and source and drain contact resistances, and by extraneous sources arising from defects in the semiconductor material such as traps. The spectrum of the instrinsic and thermal sources is "flat," i.e., white noise, extending well beyond the microwave band. The trap noise, generally, shows a rapid drop with frequency, often exhibiting a $1/f$-like character. As such it is more pronounced in the IF frequency band than in the signal band.

Experiments performed at this laboratory show that the $1/f$ component of the noise spectrum in the drain current extends at least up to 100 MHz. The experimental results reported here give evidence that the $1/f$ noise may be responsible for the degradation of the mixer noise performance at the 30 MHz IF.

The theoretical analysis of noise in a GaAs MESFET mixer is complicated because the correlation between the instrinsic drain noise and the induced gate noise cannot be neglected and is also a time-varying function. An analysis of the noise performance of a GaAs MESFET mixer is in progress.

IV. Experimental Results

A. Introduction

In this section we shall describe experimental results obtained with two different series of GaAs MESFET's. The first FET, series 71018, has a gate length of approximately 2.5 μm, a gate width of 500 μm, a terminal pinchoff voltage $V_p \approx 3.3$ V,[2] and a channel doping of 8×10^{16} cm^{-3}.

[2] By terminal pinchoff we mean the intrinsic pinchoff voltage W_{00} less the built-in barrier potential ϕ of the gate, where $\phi \approx 0.9$ V.

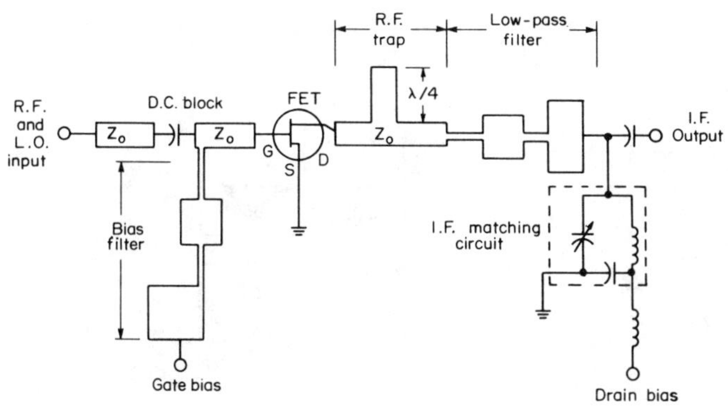

Fig. 7. Circuit configuration of integrated GaAs MESFET mixer used in X-band experiments.

The data shown in Figs. 2, 4–6 pertain to the 71018 series. The second FET, series 40713, has the aforementioned gate width, a gate length of 1.4 μm, a terminal pinchoff voltage of 2.1 V, and a channel doping of 10^{17} cm^{-3}. The f_{max} value of the latter is in the range 25–30 GHz. For the 71018 series it is lower, but was not measured.

All experiments were performed with the mixer circuit in integrated form on a 20-mil-thick alumina substrate. All conductors were gold. The RF frequency was in the vicinity of 8 GHz, the IF frequency was maintained at 30 MHz.

A schematic of the mixer circuit is shown in Fig. 7. The alumina substrate consists of two parts which lie on a common metal base containing a ridge. The ridge divides the base into two contiguous sections. The two alumina wafers butt against the ridge and connect to the gate and drain terminals of the beam-leaded FET chip which is mounted, grounded source, on the ridge. One of the wafers contains the input or RF circuitry which appears to the "left" of the FET in the schematic (Fig. 7). The other wafer contains the IF circuitry. The IF impedance matching network shown in Fig. 7 was made of lumped elements and was contained in a separate chassis because of the low frequency.

The blocking capacitor in series with the RF line was in interdigitated form. The bias filter in the gate circuit provided a low impedance termination at the IF. On the IF side a quarter-wave stub acts as a short-circuit termination for the RF, image, and LO signals. The IF impedance at 30 MHz is high, approximately 1500–2000 Ω. It is matched to 50 Ω by adjustment of the air capacitors in the matching circuit. A low-pass filter precedes the matching circuit.

We have measured only the gain characteristics of the 71018 series to convince ourselves of the validity of the theoretical small-signal model. Following this, we then changed to the higher frequency 40713 series which we also used to measure gain. We then arranged a pair of these single-ended mixers in a balanced configuration to minimize LO-introduced noise. With this balanced mixer configuration we again measured gain and, in addition, the gain compression level, intermodulation products, and noise performance.

A block diagram of the balanced mixer configuration is shown in Fig. 8. Notice the presence of phase shifters in each

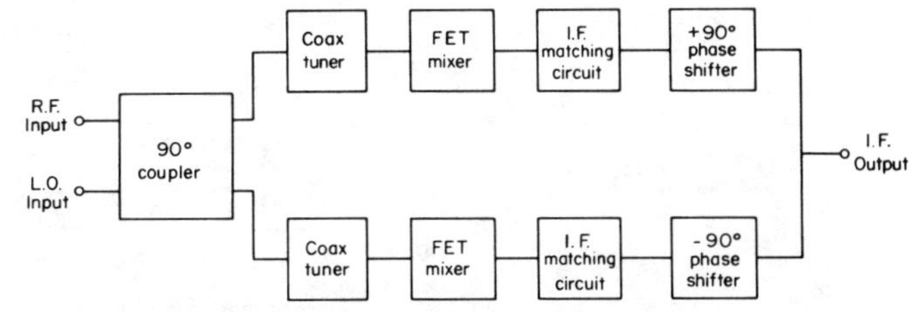

Fig. 8. Block diagram of balanced FET mixer showing phase shifters in output circuit for cancellation of LO noise.

IF branch. In diode mixers cancellation of the LO-introduced noise can be achieved by reversing the terminals of one of the diodes. Obviously, this cannot be done with FET's. The 180° phase shift between the two IF branches can be accomplished by use of leading and lagging phase shifters as shown in Fig. 8. These phase shifters are constructed of lumped elements and can be incorporated into the IF matching network.

Before we display our results, a brief description will be given of the microwave test setup for measuring the gain and noise figure.

B. Experimental Test Arrangement

The experimental data were obtained with the setup described in Fig. 9. The LO power was supplied by a klystron (X13) capable of delivering up to 80 mW at 8 GHz. The signal generator's output was in excess of +6 dBm. The measurements at the 30-MHz IF frequency were made with an AIL receiver. Because of its narrow bandwidth, we found it necessary to lock the signal 30 MHz away from the LO frequency. A second mixer and a 30-MHz discriminator were used for this purpose.

1) Gain Measurements: The LO and signal power were carefully calibrated with a power meter. The levels could be determined easily with the precision attenuators of the setup. The 30-MHz output power was measured on the receiver, which was calibrated with a separate 30-MHz generator and a power meter. By using a precision IF attenuator at the input, the receiver was kept at a constant level, thus eliminating errors due to nonlinearities.

2) Noise Measurements: The noise figure of the mixer was measured by the so-called Y factor method with the IF precision attenuator and the 30-MHz receiver. The noise source was solid state with an ENR of 15.5 dB, and was broad band. The noise figure is the double channel value.[3]

3) Intermodulation Products: The third-order intermodulation products were measured with the help of an HP 8553B low-frequency spectrum analyzer. The outputs of the two signal generators, 1 MHz apart, were adjusted to have equal amplitude. They were combined and the resultant level adjusted independently. The magnitude of the intermodulation products could be measured directly on the calibrated screen of the analyzer.

For the measurement of gain compression the level of

[3] Simultaneous excitation at signal and image channels.

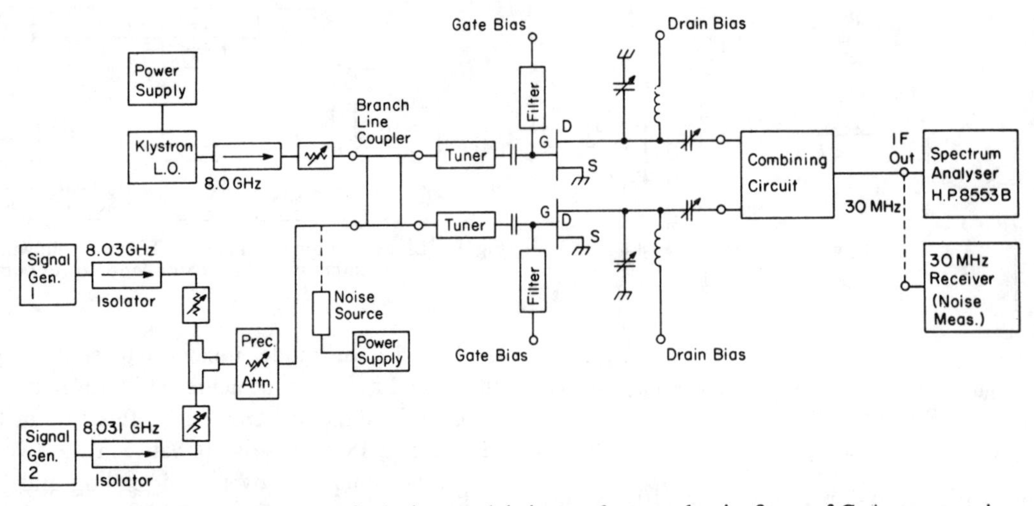

Fig. 9. Test setup for measurement of gain, intermodulation products, and noise figure of GaAs MESFET mixer.

LO power and one signal generator was set, while the other signal level was raised until a decrease of 1 dB was observed at the IF output corresponding to the fixed signal.

4) Corrections: Our results have been corrected to take into account the ohmic losses of the input circuit. The microstrip 3-dB hybrid coupler and the double-stub tuners present losses which attenuate the LO and signal powers and affect the noise figure measurement.

The input circuit was tuned for best operation of the mixer. Its attenuation was determined by measuring the insertion loss of each component separately. The 3-dB directional coupler had an insertion loss of 0.8 dB and the stub tuners 1.2 dB, giving a total loss of 2.0 dB for the input circuit.

The LO and signal powers were corrected by this amount. Thus the gain reported here is the conversion gain of the FET devices.

The noise figure measured includes the noise contributions of the input matching network preceding the mixer and the receiver preamplifier following the mixer. If we attribute the insertion loss of the hybrid coupler and double-stub tuners to dissipation only, then the noise figure of the input circuit is just equal to its insertion loss. Letting the insertion loss be L and the noise figure of the receiver preamplifier be F_r, then by the cascade noise formula the measured noise figure F_m is given by

$$F_m = L + (F - 1)L + \frac{(F_r - 1)L}{G_c} \tag{15}$$

where F is the noise figure of the mixer and G_c is its conversion gain. From the preceding equation, the noise figure of the mixer is

$$F = \frac{F_m}{L} - \frac{F_r - 1}{G_c}. \tag{16}$$

The corrected noise figure F is the one reported here.

C. Conversion Gain

The measured conversion gain as a function of LO power for a 71018 unit is illustrated in Fig. 10 [3]. The device was operated near pinchoff. Also shown is the theoretical gain calculated by the method described in Section II. Note the excellent agreement. Observe that conversion gain is obtained with a LO drive level as low as 3 mW. It is interesting that the 6.4-dB maximum conversion gain exceeds the maximum linear amplifier gain of 4.7 dB at 7.8 GHz predicted on the basis of the measured S parameters.

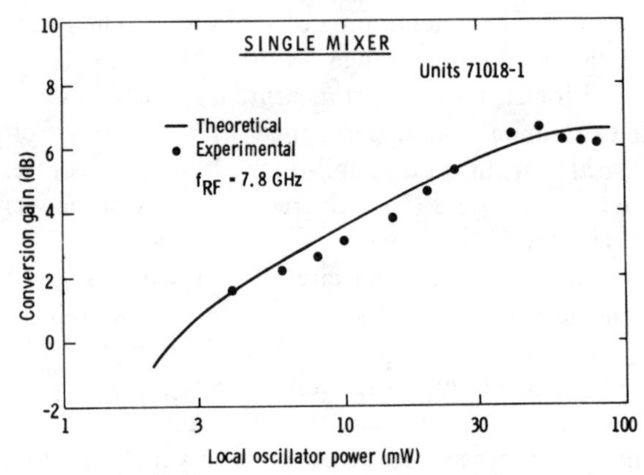

Fig. 10. Measured and calculated conversion gain of a single-ended MESFET mixer showing good agreement between theory and experiment.

The LO power required for maximum conversion gain can be reduced substantially by decrease of the pinchoff voltage. We have verified this with the 40713 series. The measured conversion gain is shown in Fig. 11(a) [4]. Note that nearly the same 6-dB maximum gain is obtained as before, but with approximately 50 percent of the LO power. By a further 50-percent reduction of the LO power to 8 mW, only 1-dB reduction in gain results.

Fig. 11(b) illustrates the excellent signal-handling property of the FET mixer. Note that unlike the diode mixer a higher LO power does not necessarily imply a higher 1-dB gain compression point. Indeed, our measurements indicate that there exists an optimum LO power to give a maximum 1-dB gain compression point. It should be pointed out that the 1-dB gain compression level at the lower LO power

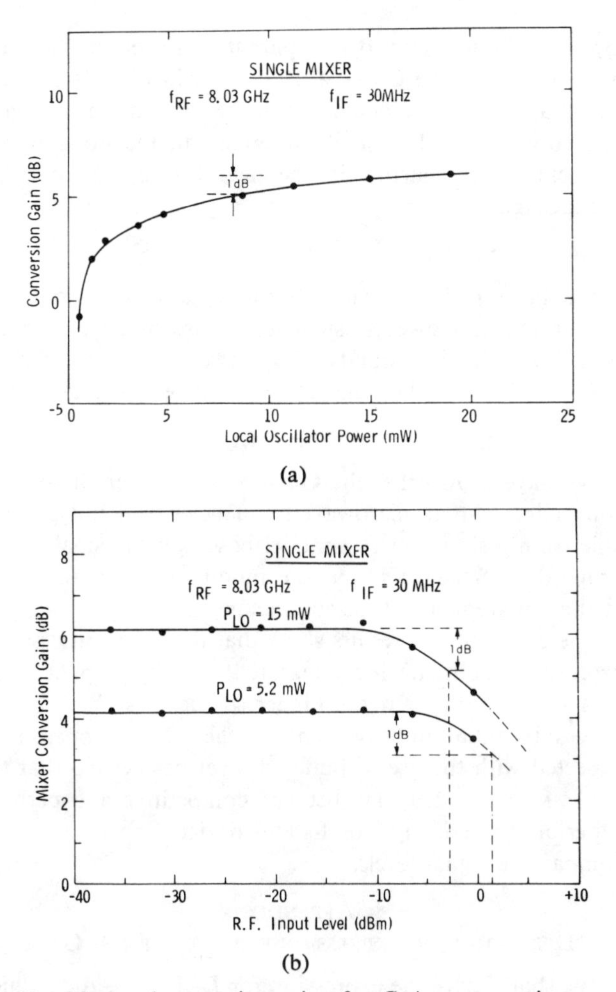

Fig. 11. Measured conversion gain of a GaAs MESFET mixer as a function of (a) LO power and (b) RF input signal level.

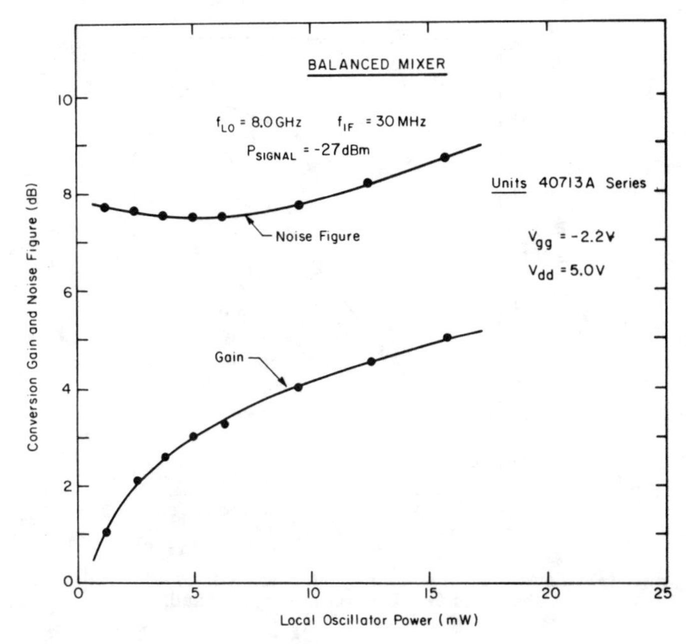

Fig. 12. Measured conversion gain and double channel noise figure of a balanced GaAs MESFET mixer at X band as a function of LO drive.

occurs at an IF power level exceeding +5 dBm. This is about 10 dB higher than for low-level (Class I) commercially available balanced diode mixers [5].

D. Noise Figure

A balanced mixer was assembled using two separate single mixer modules, and the gain, distortion, and noise characteristics were measured. Correcting for losses in the coaxial tuners and the 90° hybrid coupler (totaling about 2 dB), we obtained the very promising results shown in Fig. 12.

We believe that the minimum noise figure, 7.4 dB, is the lowest reported for an FET mixer operating at this high an RF frequency. Note that at the minimum noise point, the total LO power is only 6 mW, well within the range of an FET oscillator. At this operating point the gain is still in excess of 3 dB, nearly 10 dB higher than for a diode mixer—thus eliminating the need of a preamplifier.

The mild dependence of the noise figure on LO power suggests that there may be a considerable 30-MHz component of $1/f$ base-band noise being amplified. There is some evidence that this noise can be reduced considerably by using a high-resistivity buffer layer between the channel epitaxial layer and the substrate interface. For example, noise figures of buffered-layer FET amplifiers show a significant improvement in noise performance [1]. Sitch and Robson have also obtained appreciable noise reduction with buffered-layer FET mixers operating at S band [6].

Noise degradation by the LO source also is possible because of a mismatch in either the conversion gain or phase shift of the two single-ended mixers. To eliminate this possibility, we have analyzed the noise contribution of the LO source.

Let $(N/C)_{dsb}$ denote the double side-band AM noise-to-carrier ratio of the LO in a band B, 30 MHz from the carrier as seen by the mixer. If P_{0t} is the total LO power delivered to the balanced mixer gate terminals, then the degradation in the noise figure attributable to a phase imbalance $\Delta\phi$ in the two mixer IF ports is

$$\Delta F_{\phi} = \left(\frac{N}{C}\right)_{dsb} \frac{P_{0t}}{kT_0B} (\tan \Delta\phi/2)^2 \qquad (17)$$

where $k = 1.38 \times 10^{-23}$ J/K and $T_0 = 290$ K. We define the phase imbalance as $\Delta\phi = 180° - |\phi_1 - \phi_2|$, where ϕ_1 and ϕ_2 are the phase shifts in either IF arm, with due regard for algebraic sign. Expression (17) does not include gain imbalance.

The degradation from gain imbalance is expressible as

$$\Delta F_g = \left[\frac{\sqrt{\frac{G_2}{G_1}} - 1}{\sqrt{\frac{G_2}{G_1}} + 1}\right]^2 \left(\frac{N}{C}\right)_{dsb} \frac{P_{0t}}{2kT_0B}. \qquad (18)$$

The measured phase imbalance of the phase-shifting network in a 10-MHz band centered at 30 MHz was less than 12°. The gain imbalance was approximately 1 dB. The measured $(N/C)_{dsb} = -179$ dB in a 1-Hz band. For a total oscillator power of 6 mW (the value of the noise minimum

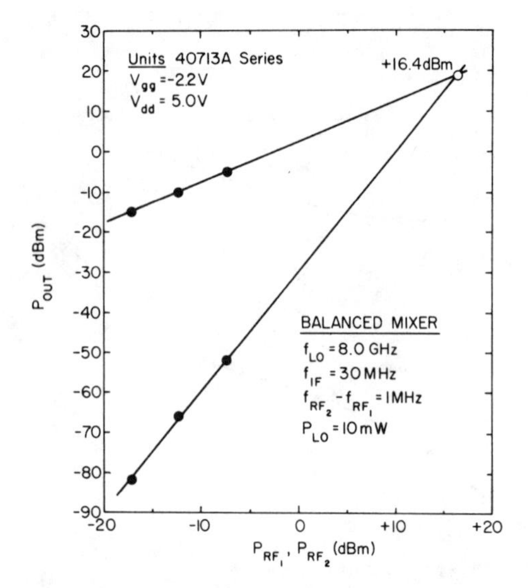

Fig. 13. Third-order two-tone modulation curves obtained with balanced GaAs MESFET mixer at X band.

Mixer	Maximum gain	Minimum noise figure	Output * 3rd - order IM intercept	Output 1dB gain compression level
GaAs FET	+ 6 dB	7.4 dB	+ 20 dBm	+ 5.5 dBm
Diode † (low - level)	-5 dB	5-7 dB	+ 5 dBm	-6 to -1 dBm

* at P_{LO} = + 8 dBm † Cheadle, MICROWAVES (Dec. 1973)

Fig. 14. Table comparing noise performance and signal-handling capabilities of the GaAs MESFET mixer and low-level diode mixer.

in Fig. 11), the calculated degradation in the mixer noise figure is $\Delta F_\phi = 2.1 \times 10^{-2}$ for the phase imbalance and $\Delta F_g = 3.1 \times 10^{-3}$ for the gain imbalance. Therefore the increase in the noise figure at the minimum is only 0.02 dB, a negligible quantity.

E. Intermodulation Distortion

Fig. 13 displays our results obtained in a two-tone intermodulation experiment. Note that the intersection corresponds to a third-order intermodulation level of +16.4 dBm at the input to the mixer, or +20 dBm at the IF output! These values are consistent with those reported by Sitch and Robson for an S-band GaAs MESFET mixer [6]. By comparison the output IM point typical of low-level balanced diode mixers is in the range of −6 to −1 dBm, over 10 dB lower. Furthermore, the output level corresponding to the 1-dB gain compression point is also some 10 dB lower than for the FET mixer.

Fig. 14 is a table comparing the signal-handling capabilities and noise performance of the GaAs FET and the low-level diode mixer. It is apparent that even at this stage of development the GaAs MESFET mixer has nearly a 10-dB advantage over the diode mixer in the gain and in the signal distortion levels. The small difference in the noise figure, we believe, will narrow in the near future, and possibly change sign.

F. Burn-Out Level

We have not performed burn-out tests on the GaAs MESFET mixer. However, such tests were made on similar FET's in amplifier circuits. CW power levels in excess of 1 W were impressed on the gate without permanent damage.

V. Summary

We have shown that the GaAs MESFET mixer can exhibit conversion gain at microwave frequencies, which is predictable from a simple circuit model based on the small-signal properties of the FET and the modulation characteristics of the low-frequency transconductance.

The experimental results show that the GaAs MESFET is a promising candidate for integrated front-end applications at X band, and possibly at higher frequencies.

A substantial improvement in the noise properties is expected with the use of buffered-layer devices, so that the GaAs MESFET mixer may become competitive and perhaps superior to existing solid-state devices operating with comparable signal levels.

Appendix
Derivation of Expression for Conversion Gain

We shall derive the expression for $I_6/E_1 = -\Delta/\Delta_z$ which appears in the gain expression (6a). For convenience let $Z_{kk} + Z_k = Z_{kk}'$. The determinant Δ_z is given by

$$\Delta_z = \begin{vmatrix} Z_{11}'^* & 0 & 0 & Z_{14}^* & 0 & 0 \\ 0 & Z_{22}' & 0 & 0 & Z_{25} & 0 \\ 0 & 0 & Z_{33}' & 0 & 0 & Z_{36} \\ Z_{41}^* & 0 & Z_{43} & Z_{44}'^* & 0 & 0 \\ 0 & Z_{52} & Z_{53} & 0 & Z_{55}' & 0 \\ Z_{61}^* & Z_{62} & Z_{63} & 0 & 0 & Z_{66}' \end{vmatrix} \tag{A1}$$

and Δ is obtained from Δ_z by deleting the first row and sixth column.

We now assume that $\bar{R}_d \gg R_s, R_{dr}$, and $g_0 R_s \ll 1$, both of which are true for well-designed GaAs MESFET's. Then one may show that

$$\frac{I_6}{E_1} = +j\left(\frac{g_1 \bar{R}_d}{\omega_1 \bar{C}}\right) \frac{1}{(Z_{11}^* + Z_1^*)(Z_{66} + Z_6) + \delta} \tag{A2}$$

where δ is given by

$$\delta = \frac{1}{\omega_3}\left(\frac{g_1 R_s \bar{R}_d}{\bar{C}}\right)^2 \frac{\omega_2^{-1}(Z_{11}^* + Z_1^*)(Z_{44}^* + Z_4^*) - \omega_1^{-1}(Z_{22} + Z_2)(Z_{55} + Z_5)}{(Z_{22} + Z_2)(Z_{33} + Z_3)(Z_{44}^* + Z_4^*)(Z_{55} + Z_5)}. \tag{A3}$$

Notice that the first term in the denominator of (A2) involves only the input RF circuit and the output IF circuit. The terminations at the remaining ports only enter in δ.

When $\omega_3 \ll \omega_1$ the image loop impedances can be approximated by the first two terms of the Taylor expansion of the signal impedances,

$$Z_{22} + Z_2 = Z_{11} + Z_1 + 2\frac{\partial(Z_{11} + Z_1)}{\partial\omega_1}\omega_3 + \cdots \tag{A4}$$

$$Z_{55} + Z_5 = Z_{44} + Z_4 + 2\frac{\partial(Z_{55} + Z_5)}{\partial\omega_1}\omega_3 + \cdots. \tag{A5}$$

Inserting these expansions into (A3), noting that $\omega_2^{-1} \cong \omega_{-1}(1 - 2\omega_3/\omega_1)$ and $Z_{33} + Z_3 \sim 1/\omega_3\bar{C}$, one obtains, after neglecting all terms of order ω_3/ω_1, the result

$$\delta = \frac{(g_1R_s\bar{R}_d)^2}{\omega_1\bar{C}}\,\mathrm{O}\left[\frac{1}{(Z_{44} + Z_4)}\right] \tag{A6}$$

where the symbol O denotes "the order of" the argument in parentheses. But $Z_{44} + Z_4 \geq \bar{R}_d$, so

$$\delta < (g_1R_s)^2\frac{\bar{R}_d}{\omega_1\bar{C}}.$$

To compare δ with the first term in the denominator of (A2), we note that

$$(Z_{11} + Z_1)^*(Z_{66} + Z_6) = \mathrm{O}(R_{\mathrm{in}}\bar{R}_d).$$

Thus the ratio of δ to this term is of the order of $(g_1R_s)^2/\omega_1\bar{C}R_{\mathrm{in}}$. Since $\omega_1\bar{C}R_{\mathrm{in}} = \mathrm{O}(0.5)$ for a well-designed FET, whereas $g_1R_s = \mathrm{O}(0.05)$, the δ term in the denominator of (A2) is less than 1 percent of the first term and can be neglected. Thus the terminations at the ports other than the signal input and IF output are not of critical importance. Therefore the available conversion gain is given to good accuracy by

$$G_{\mathrm{av}} = 4R_gR_L\left(\frac{g_1\bar{R}_d}{\omega_1\bar{C}}\right)^2\frac{1}{|Z_{11} + Z_1|^2|Z_{66} + Z_6|^2} \tag{A7a}$$

$$= \left(\frac{2g_1\bar{R}_d}{\omega_1\bar{C}}\right)^2\frac{R_gR_L}{|Z_{11} + Z_1|^2|Z_{66} + Z_6|^2}. \tag{A7b}$$

When the substitutions $Z_1 = Z_g = R_g + jX_g$, $Z_6 = R_L + jX_L$, $Z_{66} = \bar{R}_d$, and $Z_{11} = R_{\mathrm{in}} + (j\omega_1\bar{C})^{-1}$ are made, (7) of the text is obtained.

Acknowledgment

The authors wish to thank Dr. C. F. Krumm for fabricating the excellent beam-lead devices, S. R. Steele for the high-quality epitaxial material, and R. W. Bierig for his constant encouragement throughout this study.

References

[1] R. A. Pucel, D. Massé, and C. F. Krumm, "Noise performance of gallium arsenide field-effect transistors," *Proc. Fifth Biennial Cornell Electrical Engineering Conf.*, Cornell University, 1975.

[2] R. A. Pucel, H. A. Haus, and H. Statz, *Advances in Electronics and Electron Physics 38*, "Signal and noise properties of gallium arsenide field-effect transistors." New York: Academic, 1975, pp. 195–265.

[3] R. A. Pucel, R. Bera, and D. Massé, "An evaluation of GaAs FET oscillators and mixers for integrated front-end applications," *Digest of Technical Papers, 1975 IEEE Int. Solid-State Circuits Conf.*, Philadelphia, PA, pp. 62–63.

[4] R. A. Pucel, D. Massé, and R. Bera, "Integrated GaAs FET mixer performance at X-band," *Electronics Letters*, vol. 11, no. 9, pp. 199–200, May 1, 1975.

[5] D. Cheadle, "Selecting mixers for best intermodulation performance," *Microwaves*, Part I, pp. 48–52, Nov. 1973; Part II, pp. 58–62, Dec. 1973.

[6] J. E. Sitch and P. N. Robson, "The performance of GaAs field-effect transistors as microwave mixers," *Proc. IEEE*, vol. 61, pp. 399–400, Mar. 1973.

FET MIXERS FOR COMMUNICATION SATELLITE TRANSPONDERS

P. Bura and R. Dikshit

RCA Limited
Ste. Anne de Bellevue
Quebec, Canada

Abstract

Two different types of FET mixer circuits have been developed for 6/4 GHz frequency translation for communications satellite transponder.

Gate mixer uses non-linear I_d - V_g characteristic, with LO being injected into the gate circuit, while in the drain mixer the LO is injected into the drain circuit and non-linear I_d - V_d characteristic is utilized.

Both circuits yielded conversion gains up to 3 dB over 500 MHz bandwidth. However, the noise figure of the drain mixer as low as 4 dB was measured compared with 5.7 dB for the gate mixer.

When a sufficiently strong LO signal is applied to the gate of an FET biased close to pinch-off, the drain current will be modulated between zero and the saturation value I_{dss}. Simultaneously FET transconductance, gm, will also vary be between zero and its peak value. Since gm remains fairly constant down to small values of the drain current, the gm waveform can be regarded as approximating a square-wave. For this limiting condition, the intrinsic mixer conversion gain is given by:

$$G_c = \frac{1}{4} \left(\frac{gm}{\pi \omega \overline{C_g}} \right)^2 \frac{R_D}{R_g}$$

where, $\overline{C_g}$ - is the time average of the gate-to-source capacitance,

R_D, R_g - drain and gate resistances.

For a typical 1 μ gate FET this yields a conversion gain of 7 dB at 6 GHz.

In the case of the drain mixer, the LO is injected into the drain circuit with resultant modulation of the drain resistance between a low and high value corresponding to the saturated electron flow in the channel. An accurate estimate of the intrinsic conversion gain is more difficult in this case as both the resistance and the transconduction are modulated.

Both types of the MIC mixer circuit are developed for communications satellite transponder application. The input frequency range is from 5.925 to 6.425 GHz and the output frequency is 3.7 to 4.2 GHz with LO frequency being 2.225 GHz.

The gate mixer circuit is shown in Figure 1. A diplexer circuit, consisting of a bandpass filter for the signal and a lowpass filter for the LO is used in the common input to the gate. A broadband signal matching circuit then follows while the LO matching is done before the lowpass filter. The broadband output circuit matches the drain impedance to 50 ohm output. In the actual transponder receiver, additional filtering is used to further suppress the LO and its second harmonic in the output.

The drain mixer is shown in Figure 2. It consists of the input and output matching circuits with a lowpass filter for the LO injection to the drain. LO rejection in the output is parpartially accomplished by means of a quarter-wave open-circuited stub.

The conversion gain of the two circuits is shown in Figure 3. Although conversion gain of 6 dB was measured when optimized for narrow-band performance, in good agreement with the calculated intrinsic value, a reduction in the gain had to be accepted when the circuits were broad-banded over the required frequency range. The filter insertion losses are not included in the curves shown. Typically, the conversion gain will be reduced by 1 dB due to the insertion loss of the input and output filter.

The noise figure of the two mixers is shown in Figure 4. The minimum noise figure for the drain mixer of 4 dB is only 1.3 dB higher than the noise figure of the FET in a 6 GHz amplifier circuit.

It is interesting to note that the drain mixer has an appreciably lower noise figure than the gate mixer. This is probably due to the parametric upconversion of the low frequency noise in the FET to the output frequency, in the case of the gate mixer. With the gate Schottky-barrier remaining in the reverse bias condition over the whole of LO cycle, the variation of the depletion layer width results in the corresponding capacitance modulation which leads to noise upconversion. In the drain mixer, very little gate capacitance modulation takes place.

The third-order intermodulation distortion intercept point of 16 dBm was measured. The group delay ripple, including the contribution from the filters, was $\pm$ 0.5 nsec. The inband spurious product (2 f_s - 4 f_{LO}) was $\overline{70}$ dB below carrier at the output.

Reprinted from *IEEE MTT-S Int. Microwave Symp. Dig.*, 1976, pp. 90–92.

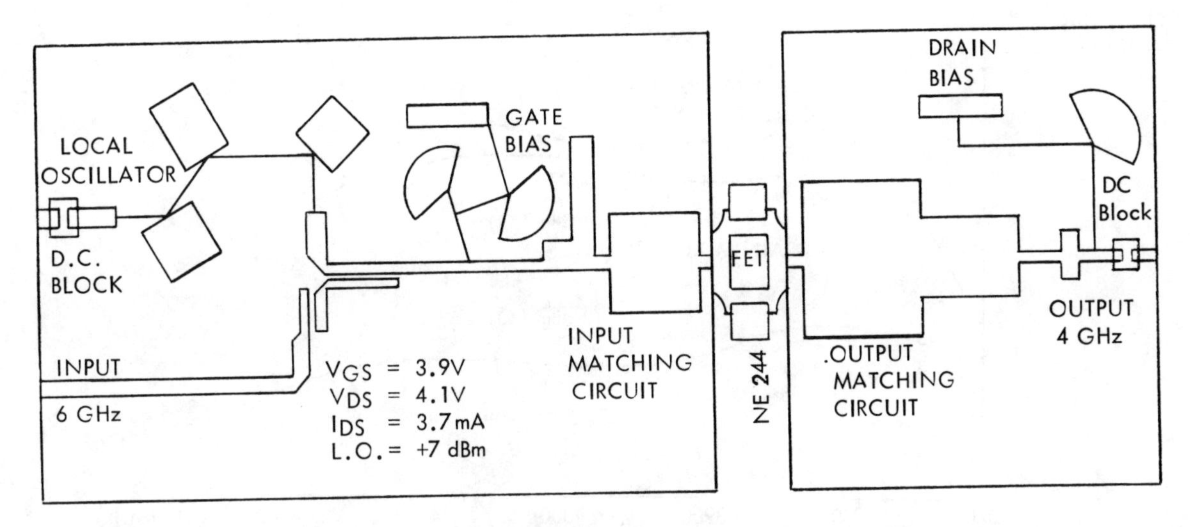

FIGURE 1. GATE MIXER.

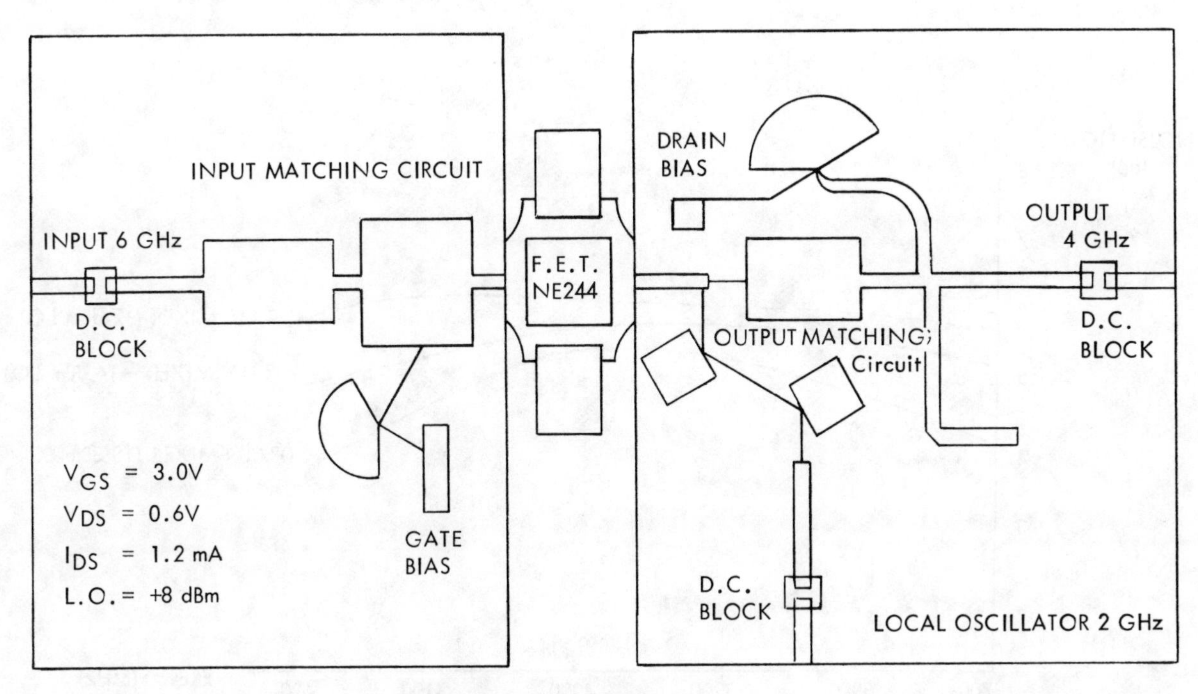

FIGURE 2. DRAIN MIXER.

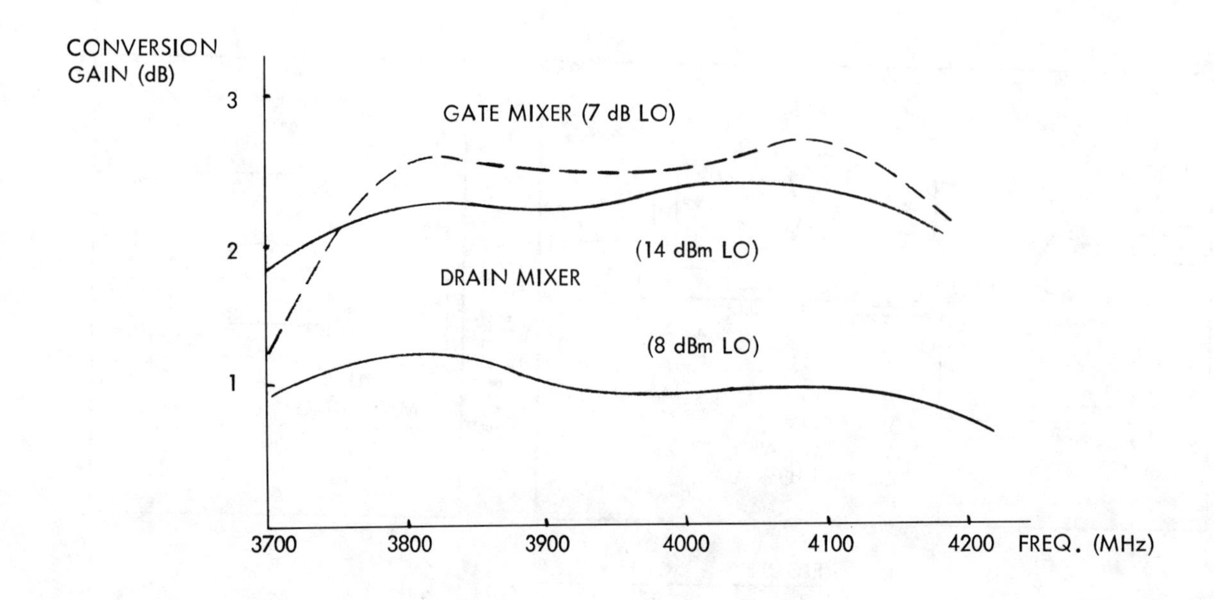

FIGURE 3. MIXER CONVERSION GAIN.

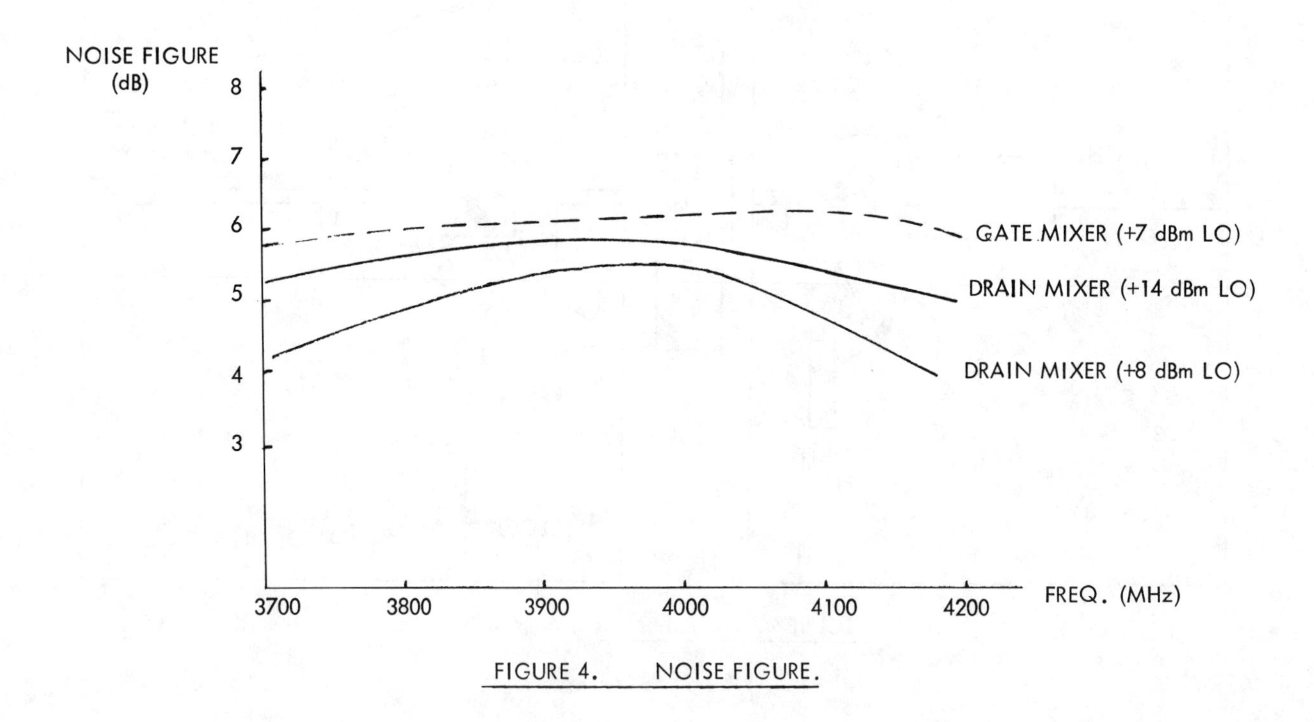

FIGURE 4. NOISE FIGURE.

AN EXPERIMENTAL EVALUATION OF X-BAND GaAs FET MIXERS USING SINGLE AND DUAL-GATE DEVICES

S. C. Cripps, O. Nielsen, D. Parker, J. A. Turner
Allen Clark Research Centre,
The Plessey Company Limited,
Caswell, Towcester, Northants.,
U.K.

Abstract

Experimental results are presented for X-band GaAs FET mixers.

Two circuits using commercially available single-gate devices have yielded good conversion gains at 10 GHz, and a specially developed dual-gate device in a simple microwave circuit has yielded 11 dB conversion gain and 6.5 dB noise figure (D.S.B.), at 10 GHz.

Introduction

The use of a GaAs FET as a microwave mixer which can combine low noise performance and conversion gain has been an attractive possibility since the first appearance of the device. Unlike low-noise amplifier applications, however, the FET mixer has a well established solid-state competitor in the Schottky diode mixer, which forms a standard against which results must be compared. It is important, however, in making such comparisons of electrical performance, that due account is taken of other factors such as circuit complexity and size.

This paper will begin with an assessment of suitable devices and circuit configurations for mixer applications, and then present experimental results from different circuit configurations using commercially available 1 micron single-gate devices. A dual-gate 1 micron FET will then be described, which has been specifically developed for mixer applications. Experimental results are presented for this device in a simple microwave circuit which has yielded 11 dB conversion gain with an associated DSB noise figure of 6.5 dB at 10 GHz, using a 30 MHz I.F.

Device Selection

In order to use a FET as a mixing element, it is necessary to cause the local oscillator signal to modulate the transconductance in a linear manner and with the greatest possible swing. Early FET devices, which were fabricated on unbuffered epilayers, had a substantial square-law region, over which the transconductance varied in a linear manner, over a significant range (fig. 1a). These devices had poor noise and gain characteristics, however, and it has been shown[1] that much improved performance can be obtained using devices fabricated on material with a semi-insulating buffer layer, with an abrupt transition to the n-type active layer. These devices have a distinctive D.C. characteristic, in which the transconductance remains nearly constant with applied gate voltage right down to pinch-off, as shown in fig. 1b. However, it is clear that these devices would give poor mixer conversion gain, unless operated in a 'switching' mode, whereby the device is D.C. biased to a point near, or just beyond, pinch-off, and the local oscillator voltage swings the operating point into the constant transconductance region.

Such a 'switching mode' mixer suffers from the drawback that the transconductance variation with time is non-sinusoidal, and ideal second order mixing is not achievable. The dual-gate device, described in a later section, offers a way of utilising the desirable low-noise properties of a device displaying this very linear gm characteristic, without need to operate in a switching mode, thereby reducing local oscillator power requirements and improving both noise performance and mixing linearity.

Single Gate Mixer Circuits

The initial problem in designing a mixer circuit around a single gate device is the requirement to excite the gate with both signal and local oscillator voltages. One convenient way to overcome this problem is to use a balanced arrangement, as shown in Fig. 2. The 90° 3 dB hybrid was of the interdigital Lange type, which had been previously developed for balanced amplifier applications. The I.F. signals appearing in the drain circuits are in antiphase, and were combined using a centre-tapped tuned transformer with a bifilar wound primary. Fig. 3 shows a picture of the practical circuit, which used commercial Plessey GAT5 devices. The low-pass filters in the drain circuit were realised with networks of high impedance lines and circular capacitive elements. The signal frequencies at the drain were terminated with a 50 ohm load in an attempt to improve overall stability. The correct terminations of the devices at signal frequencies has been found to be a critical factor in achieving good mixer performance.

All results to date have been obtained using an I.F. of 30 MHz, since this appears to be the most useful in potential systems applications. Similar circuits are, however, being constructed for a 1 GHz I.F., in an attempt to resolve the question as to whether I/F noise at 30 MHz causes degraded noise figure performance from FET mixers, as suggested by other works.[2,3]

The circuit shown in Fig. 3 has given 6 dB of conversion gain with a signal frequency of 10 GHz, using simple 'disc' tuning on the input lines to achieve a good match at the signal frequency. The associated noise figure was 8.5 dB (D.S.B.) which includes a contribution from a 3 dB noise figure I.F. amplifier. These results were obtained using 10 dBm L.O. power, obtained from a Gunn diode source.

An alternative circuit has also been evaluated in which two devices are connected in series, L.O. power being applied to one gate and signal to the other. This arrangement has yielded high conversion gains (12 dB at 10 GHz) but inferior noise figures (14 dB).

Dual-Gate FET

Fig. 4 shows the structure of a new Plessey dual-gate GaAs FET. The structure of source and drain is similar to that of the GAT4/5 series, but the source drain spacing has been increased to allow two 1 micron gates to be deposited over the channel region. The bonding pads for the two gates are in a convenient symmetrical arrangement.

The action of a second gate is to control the transconductance of the first gate, and D.C. characteristics of this device shown that this variation is linear over nearly the whole range of gm.

Reprinted from *IEEE MTT-S Int. Microwave Symp. Dig.*, 1977, pp. 285–287.

Thus by applying a local oscillator signal to the second gate, which is D.C. biased to a mid-point in the gm range, very linear mixing action can be obtained. (Fig. 5).

A 10 GHz Dual-gate FET mixer has been evaluated experimentally, using a very simple microwave circuit, with matching at each gate input, and a similar LPF arrangement in the drain as used in the single-gate circuits, with the exception that no bifilar transformer is necessary (Fig. 6). This circuit has given 11 dB of conversion gain, with 6.5 dB noise figure (DSB). L.O. power was 10 mW.

The simplicity and compactness of the circuit are particularly significant, since to obtain a comparable noise figure using a Schottky diode mixer would usually require a complicated double-balanced arrangement, using several hybrids.

An image concellation circuit is presently being constructed, using two dual-gate FETs, which will hopefully give a significant improvement in performance.

Measurements of dynamic range and intermodulation performance for the dual-gate mixer compare very favourably with Schottky diode mixers.

Conclusions

In this paper, it is shown that modern low-noise single-gate FETs are not well suited to mixer applications unless used in switching mode circuits. Two such circuits have yielded good conversion gains at 10 GHz, but the best results have been obtained from a dual-gate 1 micron device which has been developed for mixer applications. This device, in a simple microwave circuit, has given considerably better conversion gain and noise figure at 10 GHz than any other GaAs FET mixer results reported to date.

References

1. R.S. Butlin, et al., Proc. 1976 European. Microw. Conf., p606.

2. Pucel et al., IEEE Trans., MTT-24, p351, 1976.

3. B. Loriou, J.C. Leost, Elect. Lett., 12, p373, 1976.

FIG. 1

GaAs F.E.T. TRANSFER CHARACTERISTICS

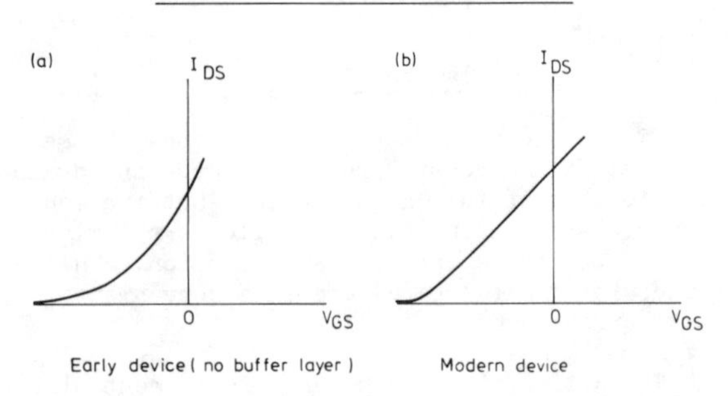

FIG. 2

BALANCED F.E.T. MIXER

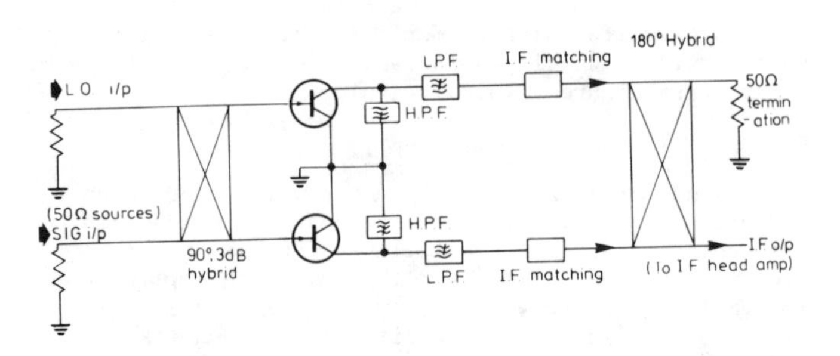

FIG. 3

DUAL-GATE MIXER CIRCUIT

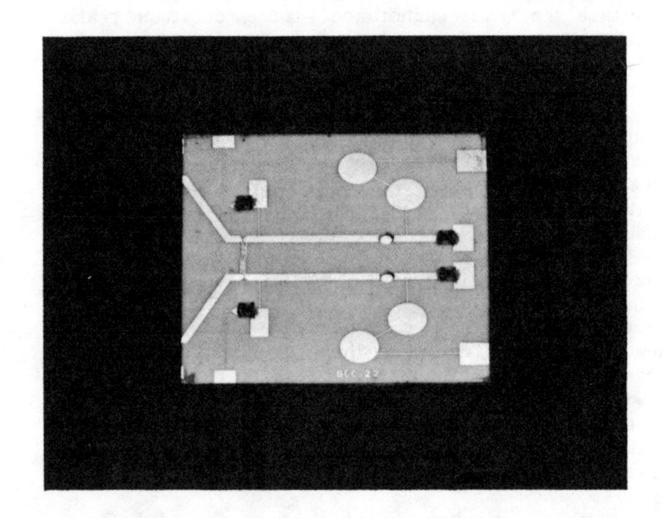

FIG. 4

PLESSEY DUAL GATE DEVICE

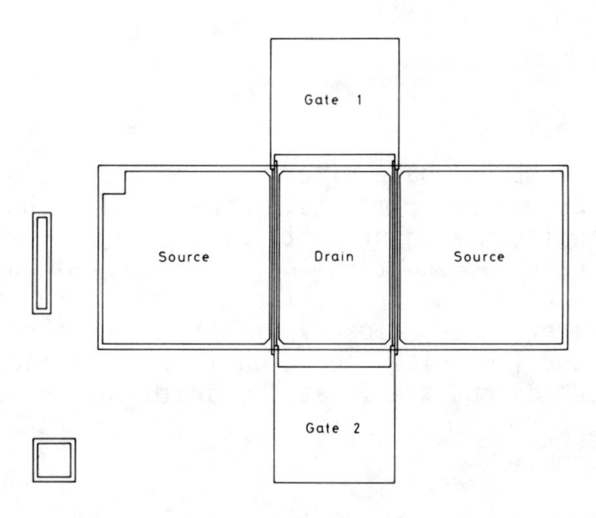

FIG 5

DUAL GATE F.E.T.

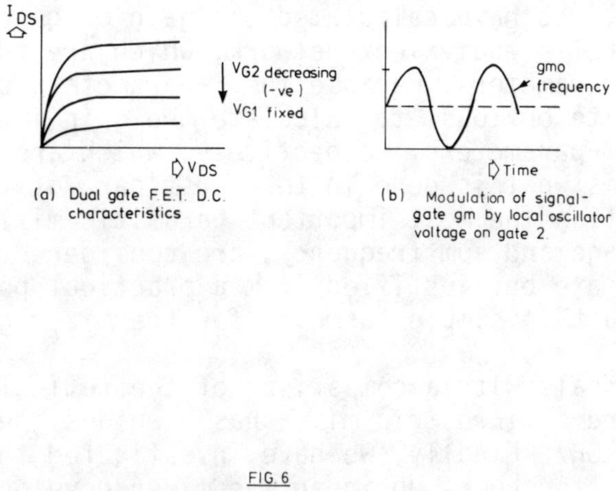

I_{DS}

V_{G2} decreasing (-ve)

V_{G1} fixed

V_{DS}

(a) Dual gate F.E.T. D.C. characteristics

gmo frequency

Time

(b) Modulation of signal-gate gm by local oscillator voltage on gate 2

FIG 6

DUAL-GATE F.E.T. MIXER

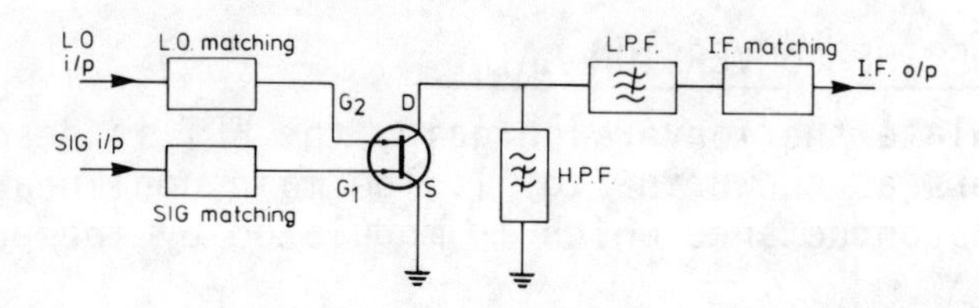

THE CONVERSION GAIN AND STABILITY OF MESFET GATE MIXERS

Günther Begemann*, Andreas Hecht**

ABSTRACT

The conversion gain of MESFET gate mixers has been calculated based on the directly measurable S- und Y-parameters, respectively. The analytical results show that parasitic mixing products can be neglected and that the reactions can be suppressed by shorting the IF at the gate and the signal, image, and sum frequency at the drain. The gain which has been measured at a 3.8 GHZ microstrip mixer agrees closely with the one which has been calculated. For a stable and low noise operation it is essential to short all the mixing frequencies except the IF at the drain and the signal at the gate.

INTRODUCTION

Along with the rapid improvement of the GaAs technology there are now single gate MESFET's available working well as active nonlinear elements in mixers up to signal frequencies of more than 10 GHz. According to the significance of active mixers a lot of papers have been published throughout the last few years dealing with different MESFET mixer configurations. However, most of these publications deal with the practical aspect of the problem and till now there is a lack of theoretical investigations which give direct informations according to which mixers can be designed best.

On the field of gate mixers the most profound work has been published by Pucel et al. /1/, who have calculated the gain of gate mixers based on the elements of the FET's equivalent network, which are but difficult to determine from the S-parameters. Because the S-parameters can be measured directly, it is quite obvious, to calculate the gain immediately from the S-parameters or Y-parameters, respectively, which are related by analytical formulas. This is what is done in the first part of the paper. Furthermore, in this contribution the most important parasitic mixing frequencies, that is to say the image and sum frequency, are considered. Under certain assumptions, which are but justified from a practical point of view, the investigations result in simple formulas for the mixer gain.

The second part deals with a comparison of theoretical and practical results. To this end a microstrip mixer has been designed working at a signal frequency of 3.8 GHz. Finally, we have investigated the phenomenon of instability which often comes up in active mixer development. Instabilities occuring throughout the measurements could be explained theorectically.

THE CALCULATION OF THE CONVERSION GAIN

In order to calculate the conversion gain, the FET is described by its admittance parameters as shown in Fig. 1. The main nonlinearity in a gate mixer is the transconductance which is modulated by the pump signal at the

* Institut für Hochfrequenztechnik, TU Braunschweig, Postfach 3329, D-3300 Braunschweig

** Rhode u. Schwarz, Postfach 80 14 69, D-8000 München 80

Reprinted with permission from *Conf. Proc. 9th Euro. Microwave Conf.*, 1979, pp. 316–320.
Published by Microwave Exhibitions and Publishers, Ltd.

gate-source diode. All of the other parameters are replaced by their time-average values. At the input the FET is loaded by

$$Y_{kk} = Y_k + y_{11}, \qquad k = 1 \ldots 4,$$

and at the output by

$$Y_{kk} = Y_k + y_{22}, \qquad k = 5 \ldots 8,$$

where the index k holds for the signal (k=1), image (k=2), sum (k=3), and the intermediate frequency (k=4) at the input, and for the signal (k=5), image (k=6), sum (k=7), and the intermediate frequency (k=8) at the output. All of the other mixing products are assumed to be shorted. The filters F_k and F'_k are lossless and have zero impedance at the desired frequency and infinite impedance at all the remaining side-bands.

The transconductance $y_{21}(t)$ is developed into a Fourier expansion

$$y_{21}(t)\Big|_{\omega_k} = y_{21}^{(0)} + 2\, y_{21}^{(1)} \cos \omega_k t + 2\, y_{21}^{(2)} \cos 2\, \omega_k t\,, \qquad (1)$$

where ω_k are pump circle frequencies with k = 1 ... 4 denoted as above. Because the transconductance is a function of frequency /1/, one has to consider its Fourier coefficients at the corresponding mixing frequency. Fourier coefficients of higher order have been neglected because in the particular case of the FET transconductance they are small and do only contribute on indirect ways to the intermediate frequency mixing product.

With the Fourier coefficients known one can set up a conversion matrix, which relates the currents and voltages at the FET

$$\begin{bmatrix} I_1 \\ 0 \\ 0 \\ 0 \\ 0 \\ 0 \\ 0 \\ 0 \end{bmatrix} = \begin{bmatrix} Y_{11} & 0 & 0 & 0 & Y_{15} & 0 & 0 & 0 \\ 0 & Y_{22}^* & 0 & 0 & 0 & Y_{26}^* & 0 & 0 \\ 0 & 0 & Y_{33} & 0 & 0 & 0 & Y_{37} & 0 \\ 0 & 0 & 0 & Y_{44} & 0 & 0 & 0 & Y_{48} \\ Y_{51} & Y_{52} & Y_{53}^* & Y_{54} & Y_{55} & 0 & 0 & 0 \\ Y_{61}^* & Y_{62}^* & 0 & Y_{64}^* & 0 & Y_{66} & 0 & 0 \\ Y_{71} & 0 & Y_{73} & Y_{74} & 0 & 0 & Y_{77} & 0 \\ Y_{81}^* & Y_{82} & Y_{83}^* & Y_{84} & 0 & 0 & 0 & Y_{88} \end{bmatrix} \cdot \begin{bmatrix} U_1 \\ U_2 \\ U_3 \\ U_4 \\ U_5 \\ U_6 \\ U_7 \\ U_8 \end{bmatrix}, \qquad (2)$$

where I_1 is the signal current,

U_k, k = 1 ... 8, are the voltages at the input and output,

Y_{ki}, k = 1 ... 4, i = 5 ... 8, are the reactions and

Y_{ki}, k = 5 ... 8, i = 1 ... 4, are the Fourier coefficients of the transconductance:

$$Y_{51} = y_{21}^{⓪}(\omega_1) \quad Y_{52} = y_{21}^{②}(\omega_2) \quad Y_{53} = y_{21}^{①}(\omega_3) \quad Y_{54} = y_{21}^{①}(\omega_4)$$

$$Y_{61} = y_{21}^{②}(\omega_1) \quad Y_{62} = y_{21}^{⓪}(\omega_2) \quad Y_{63} = 0 \quad Y_{64} = y_{21}^{①}(\omega_4)$$

$$Y_{71} = y_{21}^{①}(\omega_1) \quad Y_{72} = 0 \quad Y_{73} = y_{21}^{⓪}(\omega_3) \quad Y_{74} = y_{21}^{②}(\omega_4)$$

$$Y_{81} = y_{21}^{①}(\omega_1) \quad Y_{82} = y_{21}^{①}(\omega_2) \quad Y_{83} = y_{21}^{②}(\omega_3) \quad Y_{84} = y_{21}^{⓪}(\omega_4) \; .$$

The asterix indicates a conjugate complex value. Applying eq. (2), the conversion gain

$$G_M = 4\, G_G\, G_L\, \frac{U_8^2}{I_1} \qquad (3)$$

results in the approximation

$$G_M = 4\, G_G\, G_L \left| \frac{Y_{81}^*}{Y_{11} Y_{88}} \frac{1}{1 - \frac{Y_{37} Y_{83}^*}{Y_{33} Y_{77}} - \frac{Y_{26}^* Y_{61}^*}{Y_{22}^* Y_{66}^*} - \frac{Y_{15} Y_{51}}{Y_{11} Y_{55}} - \frac{Y_{48} Y_{84}}{Y_{44} Y_{88}}} \right|^2 , \qquad (4)$$

where G_G is the real part of the signal generator admittance and G_L the real part of the intermediate frequency load.

The very small products of reactions have been neglected. For large values of Y_{kk}, $k = 2 \ldots 7$, and conjugate matching at the input and output, eq. (4) simplifies to

$$G_M = \frac{| Y_{81}^* |^2}{4\, \bar{g}_{11}\, \bar{g}_{22}} , \qquad (5)$$

where $\bar{g}_{11}$ and $\bar{g}_{22}$ are the average values of the real parts of y_{11} and y_{22} at the signal and intermediate frequency, respectively. Eq. (5) shows that

1. the reactions are suppressed by large Y_{kk}, $k = 2 \ldots 7$ and
2. only the knowledge of the transconductance and of the time average value of y_{11} at the signal frequency and of the time average value of y_{22} at the intermediate frequency are necessary in order to calculate the conversion gain.

Eq. (5) equals to an expression given in /2/, if the transconductance is a step function and if

$$\bar{g}_{11} = \omega^2\, \bar{C}_g^2\, R_g \quad \text{and} \quad \bar{g}_{22} = \frac{1}{R_D} ,$$

where $\bar{C}_g$ is the time-average value of the gate-source capacitance and R_D, R_g are the drain and gate resistances.

MEASUREMENTS

In order to prove eq. (5) the Y-parameters of the GaAs MESFET NE 24483* have been measured at an IF of 100 MHz and an RF of 3.8 GHz. Calculating the gain gave a 10.5 dB maximum with a bias near to pinch off (Fig.2). Measurements at a microstrip-mixer similar to the one of /1/ at the same bias have yielded a gain of 10.7 dB. Fig. 2 shows the close agreement of theoretical and practical results. Biasing the FET more to negative gate voltages gave slightly higher gains but required higher pump powers. As shown in Fig. 3 the SSB noise figure was measured to be 4.5 dB minimum with an image frequency gain about 10 dB less than the RF gain.

STABILITY

Because FET's are active devices, some attention has to be given to the stability of a mixer. Eq. (4) shows that the gain increases with decreasing Y_{kk}, k = 2 ... 7. As the measurements have shown, it is most important to short the signal frequency close to the output of the FET. This is illustrated in Fig. 4 for different real parts of the external signal frequency load Y_5. By choosing a proper imaginary part of Y_5 the gain increases rapidly. We have measured a shape of gain vs. Y_5 similar to the one denoted by G_5 = 1.5 mS. In order to ensure stability the signal frequency load has finally been chosen to have a capacitive component greater than 5 mS. To increase the gain by decreasing Y_5 is unattractive from a practical point of view, because the noise figure increases rapidly, too.

CONCLUSIONS

The conversion gain of MESFET gate mixers has been calculated based on the directly measurable S- and Y-parameters, respectively. Shorting all the mixing products except of the IF at the drain and the signal frequency at the gate, which is an easy task in most of the practical cases, is of advantage in many respects:

1. the parasitiv mixing products can be neglected,
2. the reactions are suppressed,
3. the expression for the gain becomes simple and handy,
4. the mixer operates stable.

The theoretical results have been confirmed by measurements.

REFERENCES

/1/ Pucel, R. A., Masse, D., Bera, R.: Performance of GaAs MESFET at X-Band. IEEE Trans. on Microwave Theory and Techniques, vol. MTT-24, no. 6, 351 - 360, June 1976.

/2/ Bera, R., Dikshit, R.: FET Mixers for Communication Satellite Transponders. Proc. of the 1976 IEEE MTT-S Microwave Symposium, 90 - 92.

* by NEC

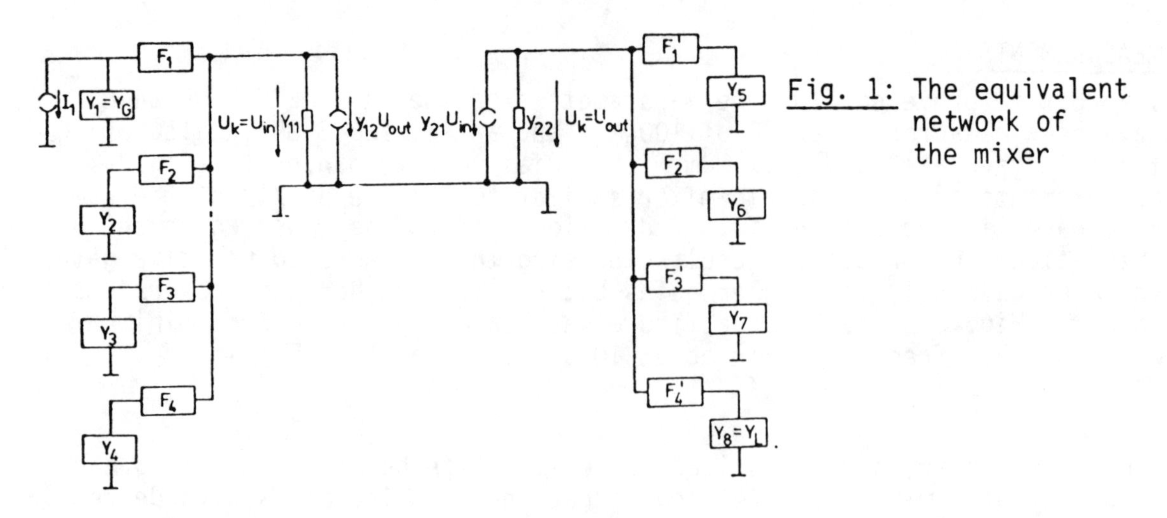

Fig. 1: The equivalent network of the mixer

Fig. 2: The calculated and measured mixer gain vs. local oscillator power P_{LO}
Parameters:
f_{RF} = 3.8 GHz
f_{IF} = 100 MHz
Gate-source bias
U_{gs} = - 2.2 V

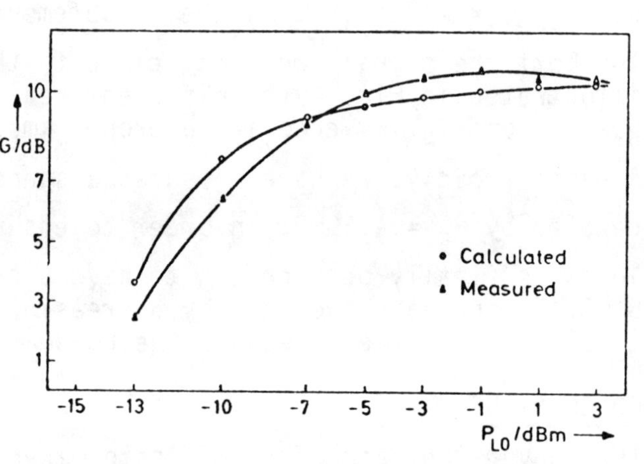

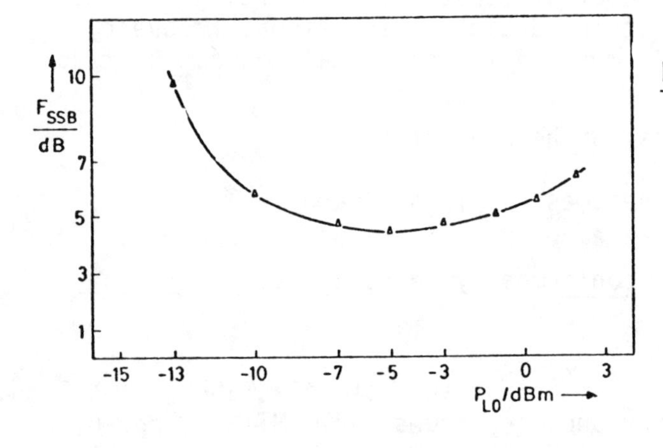

Fig. 3: The measured SSB noise figure F_{SSB} vs. local oscillator power P_{LO}. Parameters as in Fig. 2.

Fig. 4: The mixer gain as a function of the external signal frequency load at the output $Y_5 = G_5 + jB_5$

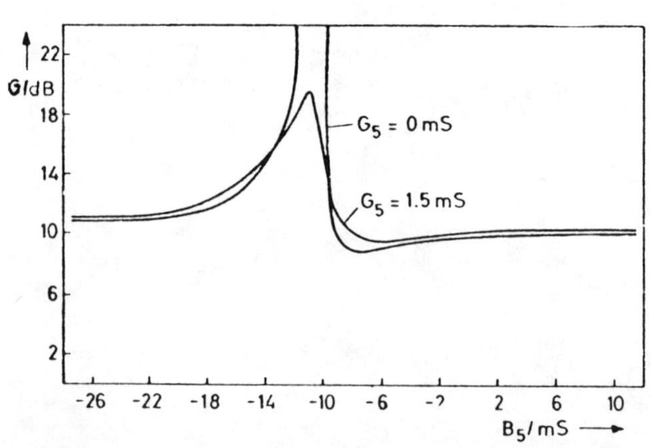

NOISE FIGURE AND ASSOCIATED CONVERSION GAIN OF A MICROWAVE MESFET GATE MIXER

G.K. Tie* and C.S. Aitchison*

ABSTRACT

The noise figure of a microwave MESFET mixer has been calculated for the first time. This paper presents the theoretical and experimental results of the noise figure and conversion gain of a microwave MESFET mixer. The effects of the RF circuit at the mixer's output side and the IF circuit at its input side upon its noise and gain performance are also presented.

INTRODUCTION

In recent years the application of MESFET as a microwave mixer has attracted much attention because of its advantages over the diode mixer. Amongst these advantages is the possibility of obtaining high conversion gain with low noise figure, and this has been demonstrated in the recent papers [1], [2]. Although many observations on the noise figure performance of the MESFET mixer have been reported in the recent years [1]-[4], there is still a lack of satisfactory literature for the analysis of its noise performance as a mixer at the present time. This paper outlines the derivation of expressions for the MESFET mixer noise figure and conversion gain. These expressions enable the noise figure and conversion gain of the mixer to be calculated for different circuit conditions. The device used in this study is the Plessey GAT-6 chip form GaAs MESFET. The LO signal frequency, RF and IF chosen are 8.03GHz, 8GHz and 30MHz, respectively.

THEORETICAL ANALYSIS

It has been widely accepted that the noise properties of a GaAs MESFET can be described by an equivalent circuit as shown in Fig. 1 [5]. The current generator i_d in the drain circuit represents the device drain current noise, the i_g in the gate circuit represents its induced-gate current noise. The voltage generators e_{gm}, e_s and e_{dr} represent the thermal noise source due to the device parasitic contact resistances R_{gm}, R_s and R_{dr} in the gate, source and drain leads, respectively. The mean square values of the drain current noise source $\overline{i_d^2}$, that of the induced-gate current noise source $\overline{i_g^2}$ and the correlation coefficient C between the i_d and i_g are given by

$$\overline{i_d^2} = 4k\,T\,B\,g_m\,P; \quad \overline{i_g^2} = 4k\,T\,B\,\frac{\omega^2 C_{gs}^2}{g_m}R; \quad C = \frac{\overline{i_g^* i_d}}{\sqrt{\overline{i_g^2}\;\overline{i_d^2}}}$$

where k is Boltzmann's constant, T the absolute temperature, B the frequency band of operation, ω the radian frequency of the signal, g_m the device static transconductance, C_{gs} its static gate-source capacitance, P and R

*Department of Electronics, Chelsea College (University of London), Pulton Place, London SW6 5PR, U.K.

Reprinted with permission from *Conf. Proc. 13th Euro. Microwave Conf.*, 1983, pp. 579–584. Published by Microwave Exhibitions and Publishers, Ltd.

are dimensionless noise parameters of the device, and the asterisk (*) represents the complex conjugate.

From the noise equivalent circuit of the MESFET of Fig. 1, the noise equivalent circuit of a MESFET mixer as shown in Fig. 2 can be obtained. Note that in Fig. 2 the time-averaged values of the device gate-source capacitance $\overline{C_{gs}}$ and drain resistance $\overline{R_d}$ are used. Let ω_o denote the LO radian frequency, then the device time-varying transconductance $g_m(t)$ can be expressed as

$$g_m(t) = \sum_{k=-\infty}^{\infty} g_k e^{jk\omega_o t}$$

where
$$g_k = \frac{1}{2\pi} \int_{-\pi}^{\pi} g_m(t) e^{-jk\omega_o t} d(\omega_o t)$$

The noise sources in the mixer's gate circuits i_{ng1} and i_{ng2} are functions of the device induced-gate noise source at the RF i_{g1} and the IF i_{g2} respectively. The mean square values of the i_{g1} and i_{g2} are

$$\overline{i_{gm}^2} = 4k\,T\,B \frac{(\omega_m \overline{C_{gs}})^2}{g_o} \overline{R}, \quad \text{where } m = 1,2$$

Similarly, the current generators i_{nd1} and i_{nd2} of the mixer are functions of the drain noise source i_d, where the mean square value of i_d is $\overline{i_d^2} = 4k\,T\,B\,g_o\overline{P}$. The $\overline{P}$ and $\overline{R}$ are the time-averaged values of the device noise parameter P and R, g_o is the time-averaged value of $g_m(t)$. The Z_1 in Fig. 2 is the mixer's RF input matching impedance, Z_4 the IF output matching impedance, the Z_2 and Z_3 are termination loads at the mixer's IF input and RF output ports. The mean square values of the voltage generators representing the thermal noise sources in the various sideband circuits are

$$\overline{e_1^2} = 4k\,TBR_1, \quad \overline{e_g^2} = 4k\,TBR_{gm}, \quad \overline{e_2^2} = 4k\,TB[R_{gm} + \mathrm{Re}(Z_2)],$$

$$\overline{e_3^2} = 4k\,TB[R_{dr} + \mathrm{Re}(Z_3)], \quad \overline{e_{dr}^2} = 4k\,TBR_{dr}, \quad \overline{e_s^2} = 4k\,TBR_s$$

A plot of the device's transconductance versus its gate-source voltage used in our mixer model is shown in Fig. 3. The fall of the g_m at high V_{Gs} values is the main reason for the drop in mixer's conversion gain and the rise in its noise figure at large LO power.

From Fig. 2 it can be shown that the noise figure F of the mixer is

$$F = \frac{\left| I_{no1} + I_{no2} + I_{no3} + i_{nd2} + (4k\,TB/R_{dr})^{\frac{1}{2}} \right|^2}{\left| I_{no} \right|^2}$$

where I_{no1}, I_{no2} and I_{no3} are the noise current components at the mixer IF output port from the noise sources in the RF input, IF input and RF output circuits, respectively. The i_{nd2} and $(4k\,TB/R_{dr})^{\frac{1}{2}}$ are the noise current components generated within the IF drain circuit itself. I_{no} is the noise current component in the mixer IF output circuit engendered by its RF input matching impedance. Let Y_{t1}, Y_{t2} and Y_{t3} be the transfer admittances and G_{fc1}, G_{fc2}, G_{fc3} the current gains from the mixer's RF input, IF input and RF output ports to its IF output port, respectively. Then the mixer noise figure F can be expressed as

$$F = 1 + \frac{R_{gm}+R_s}{R_1} + \frac{|Y_{t2}|^2}{|Y_{t1}|^2}\left(\frac{R_{gm}+R_s+Re(Z_2)}{R_1}\right)$$

$$+ \frac{|Y_{t3}|^2}{|Y_{t1}|^2}\left(\frac{R_{dr}+Re(Z_3)}{R_1}\right) + \frac{1}{|Y_{t1}|^2 R_1 R_{dr}}$$

$$+ \frac{|G_{fc1}|^2}{|Y_{t1}|^2} \frac{\overline{|i^*_{ng1}|^2}}{4k\ T\ B\ R_1} + \frac{|G_{fc3}|^2}{|Y_{t1}|^2} \frac{\overline{|i^*_{nd1}|^2}}{4k\ T\ B\ R_1}$$

$$+ \frac{|G_{fc2}|^2}{|Y_{t1}|^2} \frac{\overline{|i_{ng2}|^2}}{4k\ TB\ R_1} + \frac{\overline{|i_{nd2}|^2}}{|Y_{t1}|^2\ 4k\ TBR_1}$$

$$+ \frac{2(\overline{|i^*_{ng1}|^2}\ \overline{|i^*_{nd1}|^2})^{\frac{1}{2}}}{|Y_{t1}|^2\ 4k\ T\ B\ R_1} Re(G^*_{fc1}\ G_{fc3}\ \bar{C}_{r1})$$

$$+ \frac{2(\overline{|i_{ng2}|^2}\ \overline{|i_{nd2}|^2})^{\frac{1}{2}}}{|Y_{t1}|^2\ 4k\ T\ R\ R_1} Re(G^*_{fc2}\ \bar{C}_{r2})$$

where $\bar{C}_{r1}$ and $\bar{C}_{r2}$ are the time-averaged values of the correlation coefficients between the i^*_{ng1} and i^*_{nd1} and the i_{ng2} and i_{nd2} respectively.

When the values of the mixer's IF input load Z_2 and the RF output load Z^*_3 are known, its conversion gain G can be evaluated using the definition

$$|G| = \frac{|I_4|^2\ \ Re(Z_4)}{|E^*_1|^2 /4\ Re(Z^*_1)} = 4\ Re(Z^*_1)\ Re(Z_4)\left|\frac{I_4}{E^*_1}\right|^2$$

where I_4 is the IF output current, E_1 the E.M.F. of the RF input generator Z_4 the IF output matching impedance and Z^*_1 the RF input matching impedance. The Z_4 of the mixer is usually the complex conjugate value of its IF output impedance, and two types of optimum Z^*_1 are possible. For maximum gain performance Z^*_1 is the complex conjugate of the mixer RF input impedance, and the other is for minimum noise performance where the value of Z^*_1 can be computed from the mixer noise figure expression.

PRACTICAL RESULTS

To confirm the foregoing theory experimentally three MESFET mixers with different RF output and IF input load combinations were constructed. Mixer 1 has short circuit IF input and RF output termination loads. Mixer 2 has a high impedance IF input load and a short circuit IF input and RF output termination loads. Mixer 2 has a high impedance IF input load and a short circuit RF output load. Mixer 3 uses high impedance loads for both its IF input and RF output termination. The schematic diagrams of Mixer 1 and Mixer 3 and the photographs of their microwave parts can be seen in Fig. 4, Fig. 5, Fig. 6 and Fig. 7. All the three mixers were designed for maximum gain performance, i.e. $Z^*_1 = (Z^*_{in})^*$ and $Z_4 = Z^*_{out}$. The theoretical and experimental results of the noise figure and conversion gain performance of the three mixers are shown in Fig.8, 9 and 10, respectively. The noise figure and conversion gain in these figures are expressed as functions of the LO power level. The agreement between the calculated and measures conversion gain results for all the three mixers is good. The results obtained show that the IF input circuit and the RF output circuit of a MESFET do have significant influence on its gain performance. This is, however, not the case for the mixer's noise figure performance. It can be seen from the results that the effects of the mixer IF input and RF output circuits upon its noise figure performance are much less significant. Although there is an observable discrepancy in the LO power level at which minimum noise figure occurs between the calculated and measured results, the agreement in their minimum noise figure values and the slopes of the curves have been good. This discrepancy in LO power level at minimum noise is thought to be due to discrepancy between the shapes of the published and actual variation of P, R and C with normalised drain current used in our calculations.

CONCLUSION

The noise figure and conversion gain analyses of a microwave MESFET gate mixer have been presented. Useful agreement between the calculated and measured results for mixers with different IF input and RF output circuit combinations has been obtained. It is found that although the RF output and IF input circuits of a MESFET mixer are important for high gain performance, their effects upon the mixer noise performance are not significant.

REFERENCES

1. R.A. Pucel, R. Bera and D. Masse. "An evaluation of GaAs FET oscillators and mixers for integrated front-end applications" Digest of Technical Papers, 1975 IEEE International Solid-State Circuits Conference, Philadelphia, PA, pp.62-63.

2. B. Loriou and J.C. Leost. "GaAs F.E.T. mixer operation with high intermediate frequencies" Electronics Letters, Vol 12, pp.373-375, July 1976.

3. R.A. Pucel, D. Masse and R. Bera. "Performance of GaAs MESFET mixers at X-band" IEEE Trans., Microwave Theory & Techniques, Vol MTT-24, pp.351-360, June 1976.

4. G. Begemann and A. Hecht. "The conversion gain and stability of MESFET gate mixer" Proc. of the 9th European Microwave Conference, pp.316-320, 1979.

5. W. Baechtold. "Noise behaviour of Schottky barrier gate field-effect transistors at microwave frequencies" IEEE Trans., Electron Devices, Vol ED-18, pp.97-104, Feb. 1971.

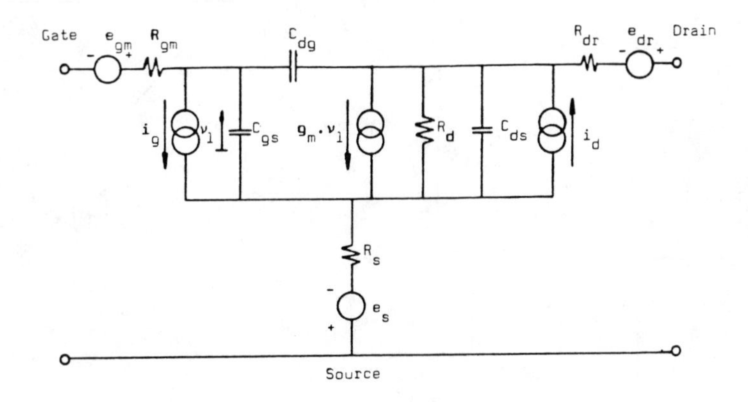

Fig. 1. The Noisy MESFET equivalent circuit.

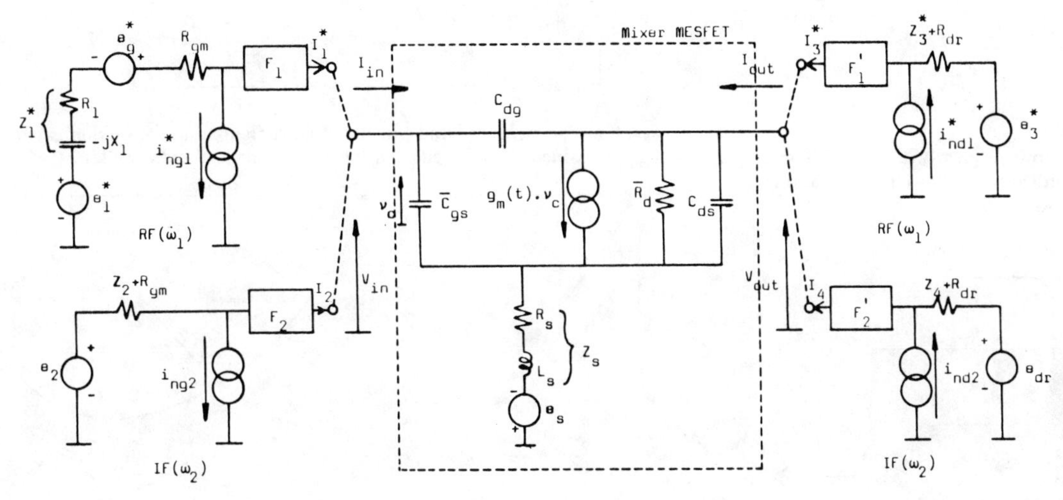

Fig. 2. The MESFET mixer circuit including the IF input and RF output circuits for noise analysis.

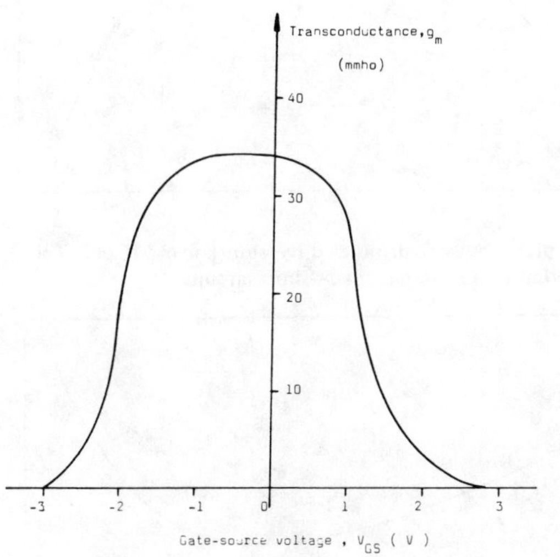

Fig. 3. The device's transconductance versus its gate-source voltage characteristic used in mixer model.

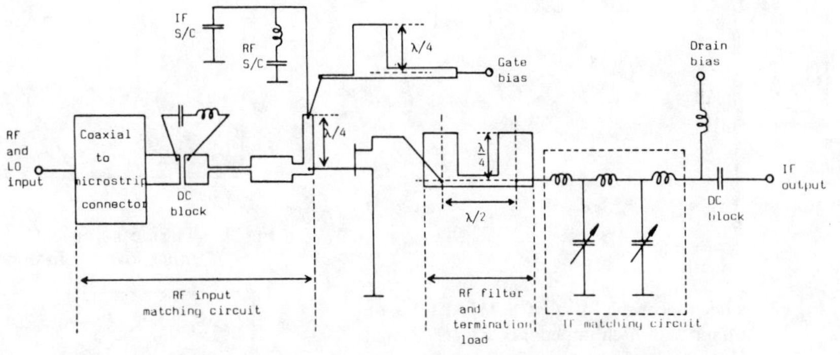

Fig. 4. Schematic of the Mixer 1 used in measurements.

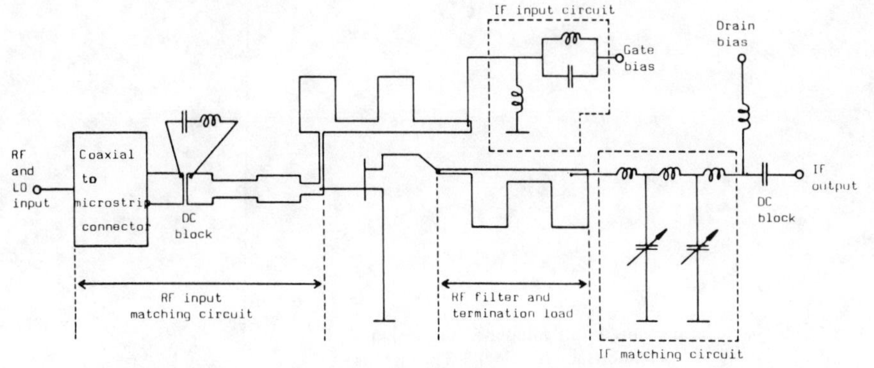

Fig. 5. Schematic of the Mixer 3 used in measurements.

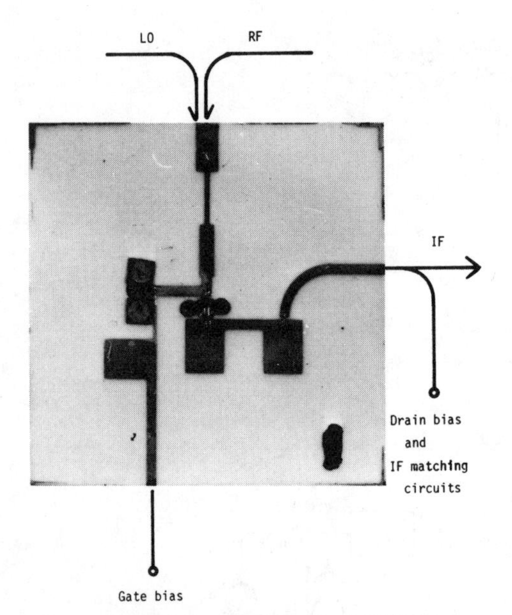

Fig. 6. The microwave part of a MESFET gate mixer with short circuit RF (O/P) and IF (I/P) loads.

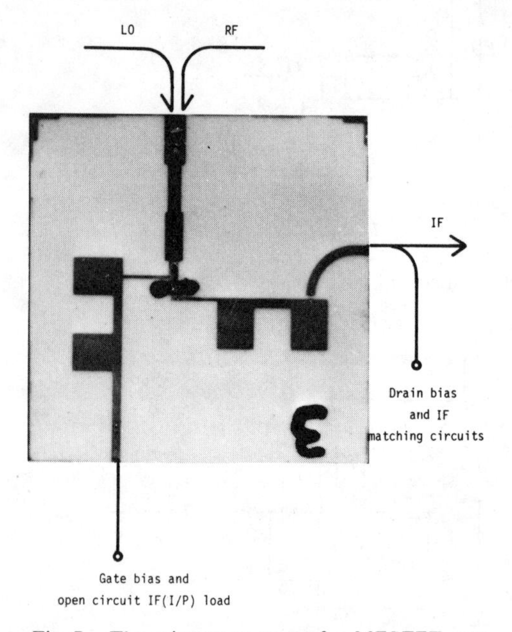

Fig. 7. The microwave part of a MESFET gate mixer with high impedance RF (O/P) and IF (I/P) loads.

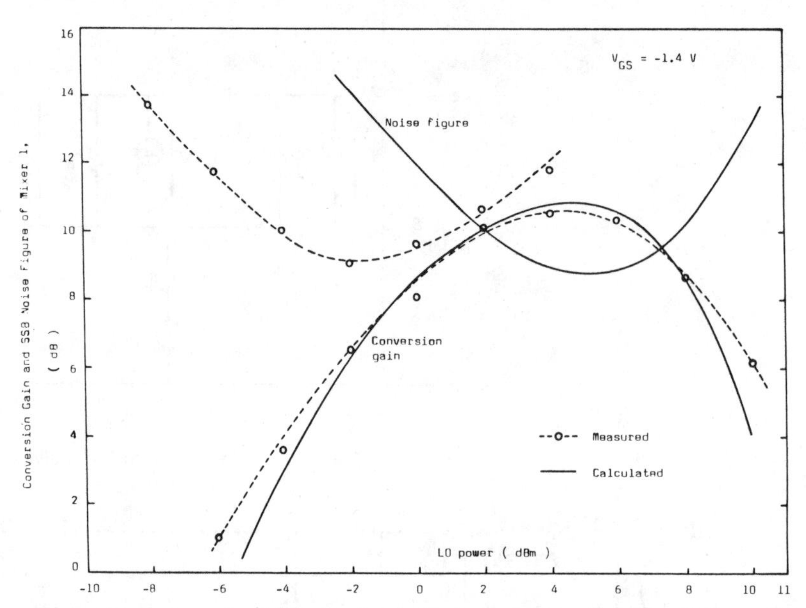

Fig. 8. The measured and calculated noise figure and conversion gain of Mixer 1, in which the IF input and RF output loads are short circuits.

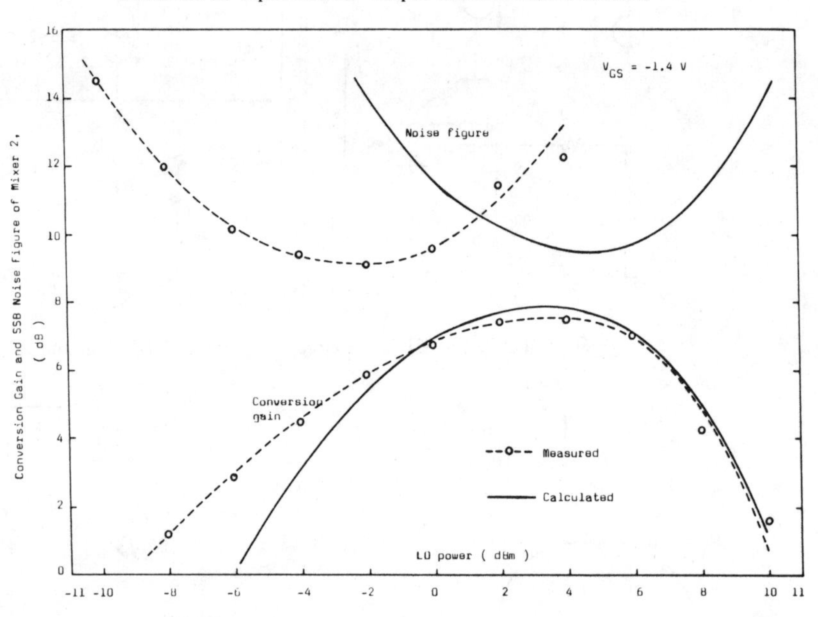

Fig. 9. The measured and calculated noise figure and conversion gain of Mixer 2. The IF input load=high impedance, RF output load=short circuit.

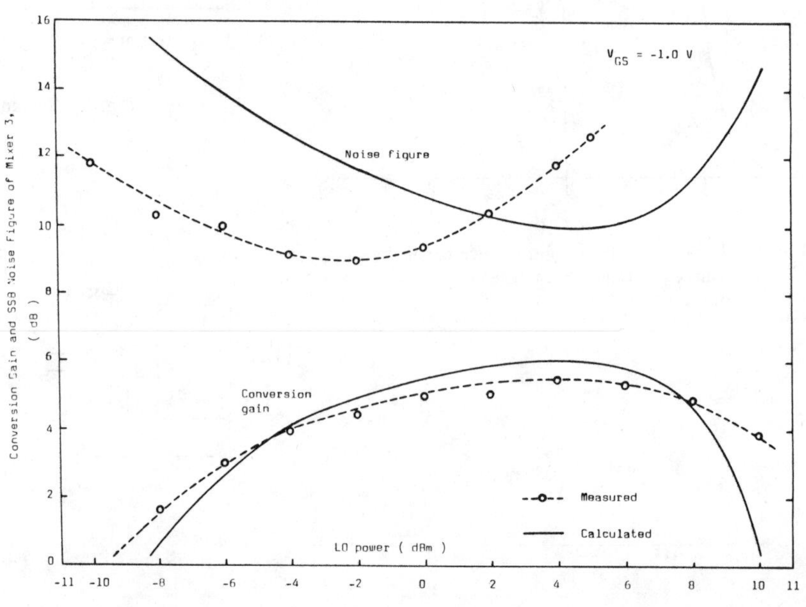

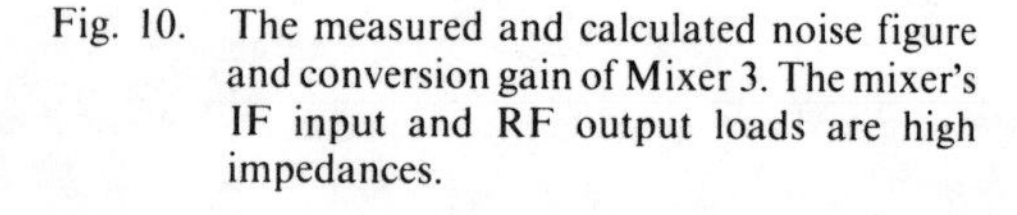

Fig. 10. The measured and calculated noise figure and conversion gain of Mixer 3. The mixer's IF input and RF output loads are high impedances.

CONVERSION GAIN OF M.E.S.F.E.T. DRAIN MIXERS

Indexing terms: Field-effect transistor circuits, Mixers, Semiconductor device models, Schottky-barrier field-effect circuits

A theoretical analysis of the gain properties of m.e.s.f.e.t. drain mixers is presented. The m.e.s.f.e.t. model includes the nonlinearity of both the transconductance and the drain resistance. For a special case, a simple analytical expression for the gain is given. Numerical results for a typical example are briefly presented as an illustration.

Based on the analytical model for gate mixers[1] the gain of drain mixers has been calculated. The investigations start from a m.e.s.f.e.t. equivalent network (Fig. 1) which describes the main physical effects with sufficient accuracy. The equivalent network includes the main nonlinearities for the application in drain mixers, i.e. the transconductance g_m and the drain resistance R_d. All of the other elements of the equivalent network, i.e. the parasitic resistances R_{gm}, R_s, and R_{dr}, the gate resistance R_i, the gate-source capacitance C_{sg}, and the gate-drain capacitance C_{gd}, which are partly voltage dependent, are assumed to be constant.

Applying a large local oscillator (l.o.) signal of frequency ω_0 between the drain and source terminals, the transconductance g_m and the drain resistance R_d become time-varying functions with a period equal to that of the l.o. Thus the voltage amplification factor $\mu = g_m R_d$ also becomes a time-varying function. By superimposing a small r.f. signal of frequency ω_1 between the gate and source terminals, mixing occurs in the f.e.t. by the action of both the nonlinearities $\mu(t)$ and $R_d(t)$. Among the mixing products are the intermediate frequency (i.f.) $\omega_3 = \omega_1 - \omega_0$ and the image frequency $\omega_2 = 2\omega_0 - \omega_1$. Only these frequencies are considered in the complete mixer equivalent network in Fig. 1. All of the other mixing products are assumed to be suppressed by the filters F_k, $k = 1 \ldots 6$, where the index k holds for the r.f. ($k = 1$), image frequency ($k = 2$) and i.f. at the input (gate), and for the r.f. ($k = 4$), image frequency ($k = 5$) and i.f. at the output (drain). The f.e.t. is loaded by complex impedances Z_k, $k = 1 \ldots 6$, with k denoted as above.

With the Fourier coefficients μ_n and R_{dn} known, where n is an integer, $-\infty < n < +\infty$, one can set up a conversion matrix, which relates the currents and voltages at the f.e.t. When the feedback capacitance C_{gd} is equal to zero, i.e. when the dashed line in Fig. 1 is open, one obtains

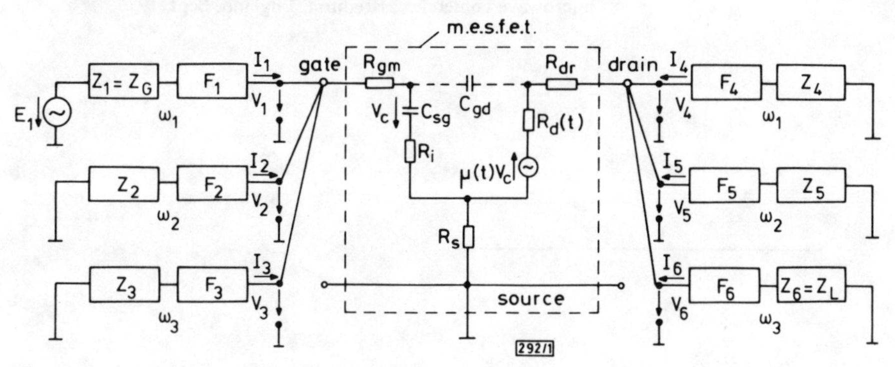

Fig. 1 *Drain mixer equivalent network based on a simple m.e.s.f.e.t. model*

$$\begin{bmatrix} E_1 \\ 0 \\ 0 \\ 0 \\ 0 \\ 0 \end{bmatrix} = \begin{bmatrix} Z_{11} & 0 & 0 & Z_{41} & 0 & 0 \\ 0 & Z_{22}^* & 0 & 0 & Z_{25}^* & 0 \\ 0 & 0 & Z_{33} & 0 & 0 & Z_{36} \\ Z_{41} & Z_{42}^* & Z_{43} & Z_{44} & Z_{45}^* & Z_{46} \\ Z_{51} & Z_{52}^* & Z_{53} & Z_{54} & Z_{55}^* & Z_{56} \\ Z_{61} & Z_{62}^* & Z_{63} & Z_{64} & Z_{65}^* & Z_{66} \end{bmatrix} \begin{bmatrix} I_1 \\ I_2^* \\ I_3 \\ I_4 \\ I_5^* \\ I_6 \end{bmatrix} \quad (1)$$

where E_1 is the signal generator voltage and I_k, $k = 1 \ldots 6$, are the currents at the input and output. The asterisks indicate complex conjugate values. The matrix elements are:

$$Z_{kk} = Z_k + R_{gm} + R_i + R_s + \frac{1}{j\omega_k C_{sg}}$$

$$Z_{k+3,k+3} = Z_{k+3} + R_{dr} + R_s + R_{d0}$$

$$Z_{k,k+3} = R_s, \qquad Z_{k+3,k} = R_s - \frac{\mu_0}{j\omega_k C_{sg}}$$

$$k = 1, 2, 3$$

$$Z_{43} = -\frac{\mu_1}{j\omega_3 C_{sg}}, \qquad Z_{53} = -\frac{\mu_1^*}{j\omega_3 C_{sg}},$$

$$Z_{61} = -\frac{\mu_1^*}{j\omega_1 C_{sg}}$$

$$Z_{62} = -\frac{\mu_1^*}{j\omega_2 C_{sg}}, \qquad Z_{46} = R_{d1},$$

$$Z_{56} = Z_{64} = Z_{65} = R_{d1}^*$$

$$Z_{42} = -\frac{\mu_2^*}{j\omega_3 C_{sg}}, \qquad Z_{51} = -\frac{\mu_2^*}{j\omega_1 C_{sg}},$$

$$Z_{45} = Z_{54} = R_{d2}^*$$

Applying eqn. 1, the conversion gain

$$G_M = 4R_G R_L \cdot \left| \frac{I_6}{E_1} \right|^2 \quad (2)$$

where R_G is the real part of the signal generator impedance and R_L the real part of the i.f. load, becomes a complicated function of the terminations at each port. For a simple case, however, i.e. for large values of Z_3, Z_4, and Z_5 and conjugate matching at the signal input and i.f. output, eqn. 2 becomes

$$G_M = \frac{|\mu_1|^2}{4\omega_1^2 C_{sg}^2 (R_{gm} + R_i + R_s)(R_{dr} + R_s + R_{d0})} \quad (3)$$

Note, that for this case, not only are the reactions suppressed (the f.e.t. becomes unilateral), but also the mixing action by the drain resistance. The nonlinearity of R_d only appears in the first harmonic component μ_1 of the voltage amplification factor, in contrast to the gain expression for a gate mixer as developed in References 1, 2, where R_d was supposed to be constant.

For $C_{gd} \neq 0$ a feedback current through C_{gd} must be taken into account. It has sideband components at all of the frequencies occurring during the mixing process. For the sake of simplicity we took into account only components at ω_1, ω_2, and ω_3. Nevertheless, the conversion matrix and hence the gain expression becomes so unwieldy that we only present the numerical results of this analysis.

To calculate the mixer gain, the transconductance g_m and the drain resistance R_d of the f.e.t. NE 24483 by NEC have been measured as a function of the gate (V_{gs}) and the drain (V_{ds}) bias. The remaining elements of the equivalent network have been taken from the data sheet of the f.e.t. The g_m and R_d characteristics have been numerically approximated by polynomial functions. Finally, the gain has been calculated under assistance of eqn. 3. Simultaneously we assumed the drain-source diode to be driven by the instantaneous voltage

$$v_{ds}(t) = V_{ds0} + \hat{V} \cos \omega_{0t} \beta,$$

where V_{ds0} is the d.c. bias, $\hat{V}$ is the peak local-oscillator amplitude and ω_0 is the l.o. angular frequency.

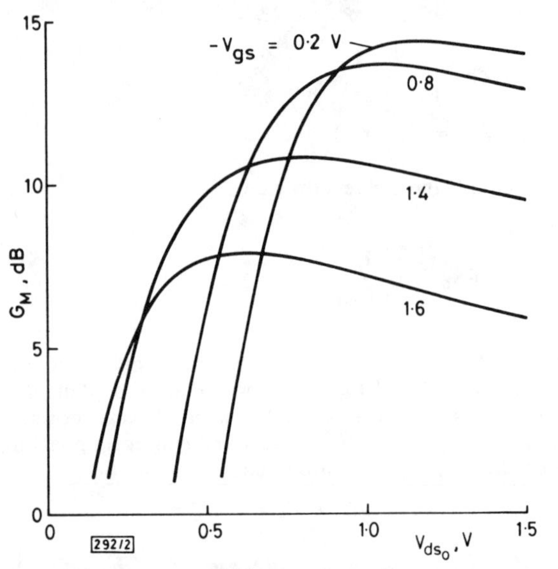

Fig. 2 *Available gain G_M against the drain bias V_{ds0} with the gate bias V_{gs} as a parameter; l.o. amplitude $\hat{V} = V_{ds0}$*

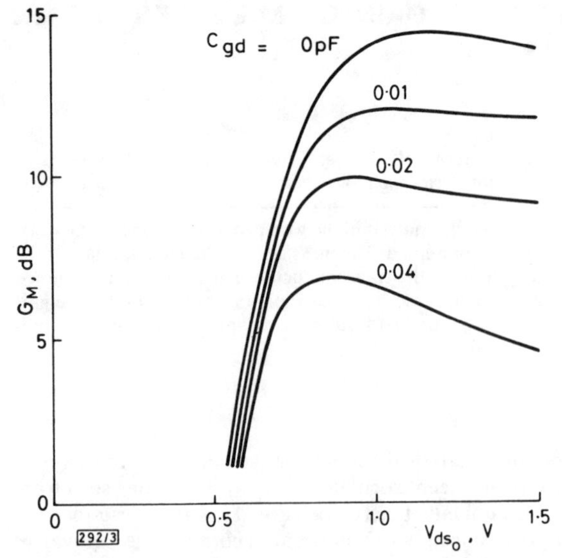

Fig. 3 *Available gain G_M against the drain bias V_{ds0} with the feedback capacitance C_{gd} as a parameter; gate bias $V_{gs} = -0{\cdot}2$ V; l.o. amplitude $\hat{V} = V_{ds0}$*

Fig. 2 shows the gain as a function of the drain bias at a maximum l.o. drive $\hat{V} = V_{ds0}$ for several gate bias points. The signal and intermediate frequencies were chosen to be 4·1 GHz and 0·1 GHz, respectively. The gain decreases with increasing negative values of the gate bias. The maximum available gain is 14·3 dB at $V_{ds0} = 1{\cdot}2$ V and $V_{gs} = -0{\cdot}2$ V. With decreasing l.o. amplitude the gain decreases as well.

Fig. 3 shows the influence of C_{gd}. For higher values of C_{gd}, the feedback current through C_{gd} becomes stronger and leads to more and more important losses in the mixer. It seems that C_{gd} cannot be neglected in an accurate mixer analysis.

GÜNTHER BEGEMANN
ARNE JACOB

Institut für Hochfrequenztechnik
Technische Universität Braunschweig
Postfach 3329, D-3300 Braunschweig, W. Germany

16th July 1979

References

1 PUCEL, R. A., MASSÉ, D., and BERA, R.: 'Performance of GaAs MESFET mixers at X-band', *IEEE Trans.*, 1976, **MTT-24**, pp. 351–360

2 BEGEMANN, G., and HECHT, A.: 'The conversion gain and stability of MESFET gate mixers'. To be presented at the 9th European microwave conference, Brighton, England, Sept. 1979

A SELF-OSCILLATING DUAL GATE MESFET X-BAND MIXER WITH 12 DB CONVERSION GAIN.

Christos TSIRONIS, Rainer STAHLMANN, Frederik PONSE[+]

ABSTRACT
An x-band receiver stage including preamplifier, mixer and local oscillator has been realized by a dual gate GaAs MESFET in common source configuration. The conversion gain for a signal frequency of 10 GHz and an I.F. of 1GHz was 12 dB by appropriate matching the input (gate 1) and output (drain) port. A variable short, connected to gate 2 controlled the L.O. x-band oscillation. The isolation between drain and gate 1 was 16 dB. Using disc and $BaTi_4O_9$ resonator matching on Al_2O_3 substrate with only I.F. external tuning enabled conversion gain of 5dB. The tuner matched circuit had DSB noise figure of 5.5dB and associated conversion gain of 4dB at an I.F. of 1GHz whereas the best noise figure achieved by the disc-tuned circuit was 7.7dB at an I.F. of 300MHz and associated conversion gain of -1.5dB.

I. INTRODUCTION

GaAs MESFETs have been used as x-band mixers either as dual gate FETs using an external local oscillator /1/ or as single gate selfoscillating FETs /2/. The disadvantage is in the first case the need for an external oscillator, whereas in the second case the input port is not sufficiently isolated from the oscillator signal, which then is fed into the generator (or the antenna). Using a dual gate GaAs MESFET as a selfoscillating (s.o.) mixer a) there is no need of an external oscillator, b) there is sufficient isolation between input and output port, since the oscillation is controlled by the second gate and c) the oscillator power needed for optimum conversion conditions is lower, because of the intrinsic amplification mechanism /3/. We realized s. o. mixer circuits using a) external slug tuner and b) disc and $BaTi_4O_9$ resonator matching on the alumina substrate with comparable performance. The $BaTi_4O_9$ has an ε_r of 37-38 and a Q factor of 3 000 - 4 000.

II. DUAL GATE MESFET MIXING

The dual gate (D.G.) GaAs MESFETs fabricated on a $1 \cdot 10^{17} cm^{-3}$ doped unbuffered epilayer were designed for mixer applications: The gate 1, near source, was 1 µm long for sufficient x-band gain and the gate 2, near drain, was 2 µm long for higher power capability and life time. The drain to source gap was 8 µm and the FET width 200 µm. The saturation currents I_{sat} were 20-40 mA for $V_{G1S}=V_{G2S}=0$, the pinchoff voltages 3-3.5V and the intrinsic transconductance g_m of G1 had a maximum of 10-15mS for V_{G1S} between -1 and -2V (fig.1b), important for low noise applications. The MESFET chips were bonded in 50Ω microstrips on 0.25 mm thick Al_2O_3 substrates (fig.3).

The mixing mechanism can be described as follows: The input signal at ω_{SIG}, which is fed into G1 is amplified in the first (common source) MESFET (fig.2) and flows then through the channel of the second (common gate) MESFET. The transconductance of the whole FET is modulated at ω_{LO} by the local oscillator voltage generated by oscillations of the second FET. The output current contains therefore a spectral component of the angular frequency $\omega_{IF}=|\omega_{SIG}-\omega_{LO}|$. The input capacitance C_{G1S} and the output resistance R_d vary with V_{G2S} between +1.5 and -3V less than 1% and 4% resp., as s-parameter measurements at 500MHz showed. Fig. 1 demonstrates the mixing mechanism

[+]Institute of Semiconductor Electronics, Technical University of Aachen, Templergraben 55, D 5100 Aachen, F.R.G.

Reprinted with permission from *Conf. Proc. 9th Euro. Microwave Conf.,* 1979, pp. 321-325.
Published by Microwave Exhibitions and Publishers, Ltd.

and the conversion transconductance g_1, calculated by Fourier analysis of $g_m(t)$ in fig.1a. The optimum conversion gain is achieved by biasing the gate

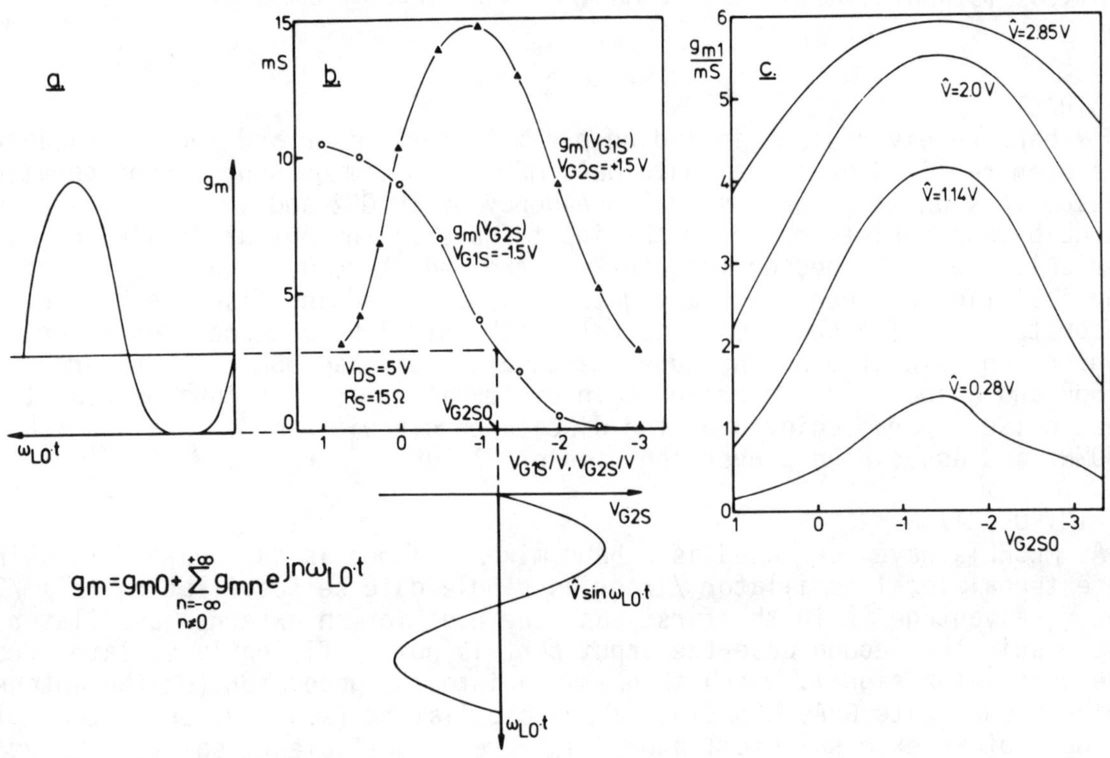

Fig. 1: Dual Gate MESFET mixing mechanism. Device E1. R_S: Source resistance.

2 at -1.5V and choosing a maximum local oscillator amplitude. The maximum conversion gain G_{cmax} can be higher than the maximum available gain G_{max}. After /4/

$$\frac{G_{cmax}}{G_{max}} = \left(\frac{g_{m1}}{g_m}\right)^2 \cdot \left(\frac{C_{gs}}{\overline{C_{gs}}}\right)^2 \cdot \frac{\overline{R_d}}{R_d} \tag{1}$$

In the case of dual gate MESFETs not only the quotients $C_{gs}/\overline{C_{gs}}$, $\overline{R_d}/R_d$ can be larger than unity but also g_{m1}/g_m as can be seen from fig.1b and 1c for $V_{G2S} = -1.5V$. The transconductance g_m has a maximum for $V_{G1S} \lessapprox -1.0V$ and shifts to lower V_{G1S} for lower V_{G2S}.

III. EXPERIMENTAL SETUP AND TEST CIRCUITS

Fig.2 gives the MESFET-mixer experimental test setup:

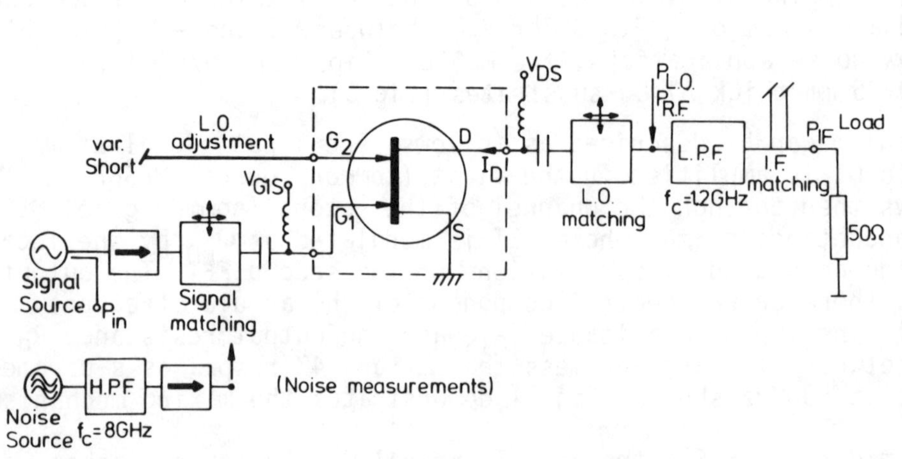

Fig.2: Dual gate selfoscillating MESFET-mixer test circuit.

The variable short at G2 causes the device to oscillate between 7.5 and 12 GHz. The oscillation frequency and magnitude can be adjusted by this short and shifted by the drain R.F. slug tuner over 50-100MHz.For optimum conversion gain an additional tuning at G1 and I.F. behind the low pass filter (L.P.F.) is necessary. Because of the isolation between G1 and G2 this tuning only slightly affects the L.O. frequency, which can be restored by retuning the drain R.F. slug tuner. For noise measurements the signal source was replaced by a solid state noise source and a high pass filter (fig.2). Behind the L.P.F. a broadband low noise preamplifier was used.
Testing the disc and BT_4 resonator-tuned circuit shown in fig. 3, all tuner matching of the test setup in fig. 2, except the I.F. double stub tuner, were omitted. The circuit oscillates at x-band frequencies depending on the ion of the BT_4 resonator and the drain- discs. The L.O. frequency can be adjusted over 30 - 50MHz by a massive earth contact on the top of the BT_4 resonator as reported in /5/. G1 is directly connected to the 50Ω generator.

Fig.3: Disc and BT_4-tuned s.o. MESFET mixer circuit.

IV. RESULTS

Experimental results, measured on tuner and disc-matched circuits, are given in figures 4 - 8 and table I. The conversion gain for x-band frequencies is given on fig. 4. The upper curve corresponds to an external tuner matched and the lower curve to the disc and BT_4 tuned circuit. The gain decrease with V_{DS} is caused by the oscillation magnitude drop in both cases as can be seen in fig. 5. The L.O. power varies over 18 dB for $3.75V \leq V_{DS} \leq 6.5V$ whereas the frequency pushing amounts to 8.2MHz or 0.1% per volt. The linearity of the disc-tuned circuit is given in fig. 6. The 1dB compression point is at an input power of -5dBm. The dual gate s.o. mixer circuit promisses good linearity behaviour due to the needed low L.O.power /4/. The gain-decrease is a consequence of the L.O. power decrease with P_{in}, as can be seen by comparison of the I_D-curve with that one of fig. 5. The temperature dependence of the L.O. power and frequency is shown on fig. 7. The disc tuned and BT_4 resonator-stabilized circuit has a frequency shifting of 12.5ppm/K, without BT_4 resonator the frequency variation was ca. 50ppm/K, for chip holder temperatures of 25 to 70^0 C. The L.O. power shows a tempe-

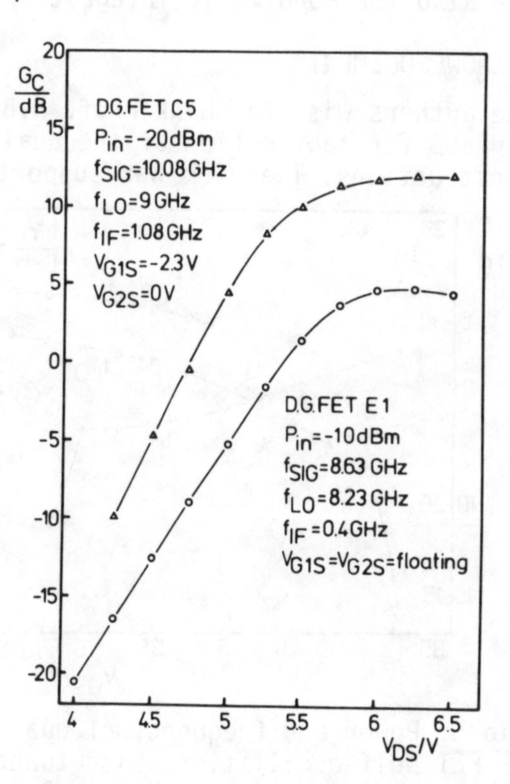

Fig. 4: Conversion gain of s.o. MESFET mixers.

rature hysteresis effect higher than reported in /5/. Fig. 8 finally gives the isolation properties of the disc-tuned circuit. P_{LO} is here the oscillation power measured at the drain. The L.O. power at the "oscillating" gate 2 is throughout 10 to 20 dB higher than at the input gate 1 confirming the assumption that the oscillation is originated by the second FET.
Measurements of the noise figure confirmed the reported dependence on the I.F. /6/. At an I.F. of 30MHz (table I) the low frequency excess noise deteriorates the whole noise figure of these unbuffered devices. Though noise figure increases slightly with V_{DS}, in these circuits we had to find a compromise between local oscillator power (V_{DS}), oscillating conditions (V_{G1S}, V_{G2S}) and input matching. This iterative process, inspite of the desired effect of the maximum of g_m for $V_{G1S} \lessapprox -1$ V (see also ch.II), has not been optimized yet. The results of noise measurements are given on table I:

circuit ⟶	slug tuner matched		disc + BT_4 reson. tuned
$f_{SIG}/f_{LO}/f_{IF}$ (GHz)	10/9.97/0.03	10.08/9/1.08	8.5/8.2/0.3
$V_{DS}/V_{G1S}/V_{G2S}$ (V)	5/-2.4/0	5/-2.3/0	6/float/float
DSB NF (dB)	10.7	5.5	7.7
G_c assoc. (dB)	3.5	4.0	-1.5

Table I: Noise performance of s.o. dual gate MESFET mixer.

CONCLUSION

The dual gate GaAs MESFET enabled x-band single device receiver circuits in self-oscillating common source configuration. A disc- and $BaTi_4O_9$ resonator tuned circuit on alumina substrate gave conversion gain of 5dB for mixing from 8.6GHz to 0.4GHz. The DSB noise figure of the same circuit was 7.7dB with associated conversion gain of -1.5dB. The circuit seems to be appropriate also for monolithic integration.

ACKNOWLEDGEMENT

The authors wish to thank Prof. H.Beneking for his encouragement, Miss G. Neuhauß for fabrication of the dual gate FETs and J.Wenn for the mechanical constructions. The work was supported by the SFB 56 -Festkörperelektronik.

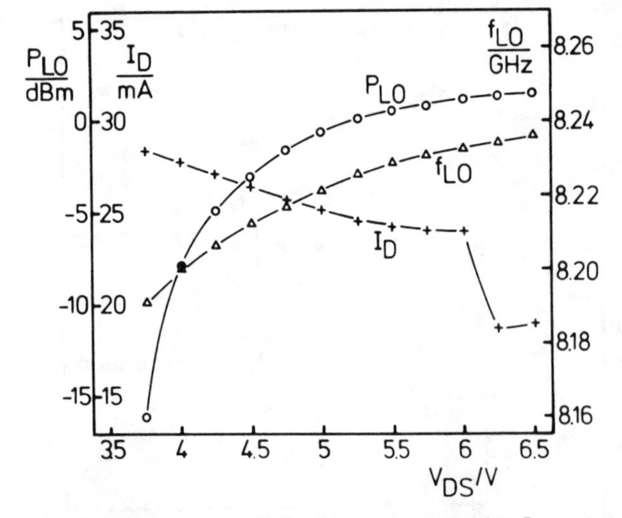

Fig.5: Power and frequency of dual gate MESFET self-oscillation. Disc-tuned and BT_4 stabilized circuit. V_{G1S} = 0 V.

REFERENCES

/1/ S.C.Cripps et al.,"An experimental evaluation of x-band mixers using dual gate GaAs MESFETs", Proc. 7th Europ. Micr.Conf.,Copenhagen, 1977, p.101.
/2/ Y.Tajima, "GaAs FET applications for injections-locked oscillators and self-oscillating mixers", Proc. MTT-S, Ottawa,1978, p. 303.
/3/ P.Harrop et al.,"Performance of some GaAs MESFET mixers", Proc. 6th Europ.Micr.Conf.,Rome,1976, p.8.
/4/ R.A.Pucel et al.,"Performance of GaAs MESFET mixers at x-band", IEEE Trans. MTT-24,1976, p.351.

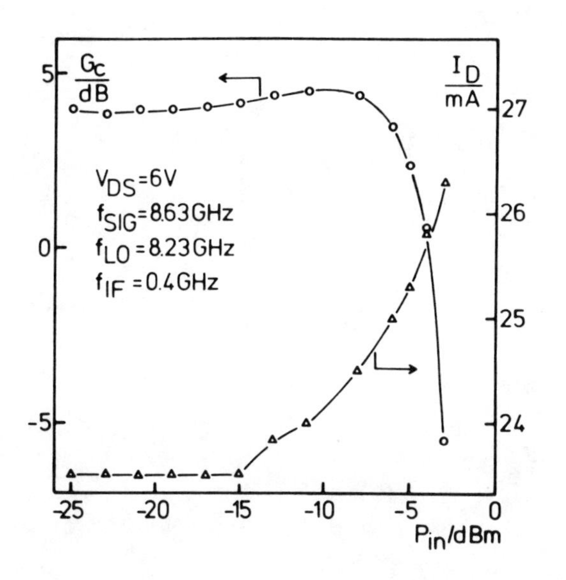

/5/ P.Lesartre et al.,"Stable FET local oscillator at 11 GHz with electronic amplitude control", Proc. 8th Europ. Micr. Conf., Paris 1978, p. 264.

/6/ B.Loriou and J.C.Leost,"Design and performance of low noise c- and x-band mixers", Proc.7th Europ. Micr. Conf. Copenhagen, 1977, p.95.

Fig. 6:Conversion gain as a function of input power. Disc-tuned circuit. V_{G1S} = 0 V.

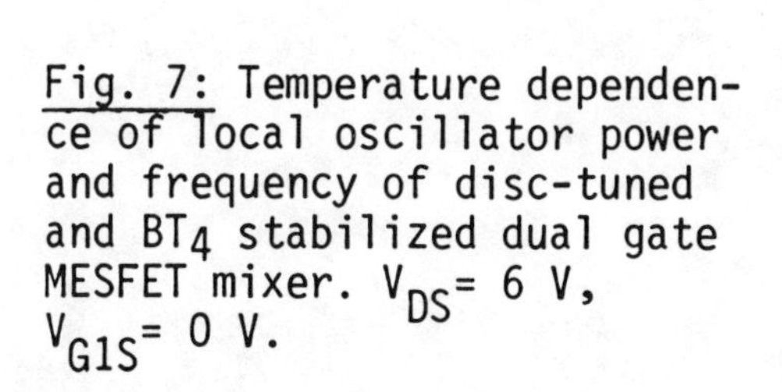
Fig. 7: Temperature dependence of local oscillator power and frequency of disc-tuned and BT_4 stabilized dual gate MESFET mixer. V_{DS}= 6 V, V_{G1S}= 0 V.

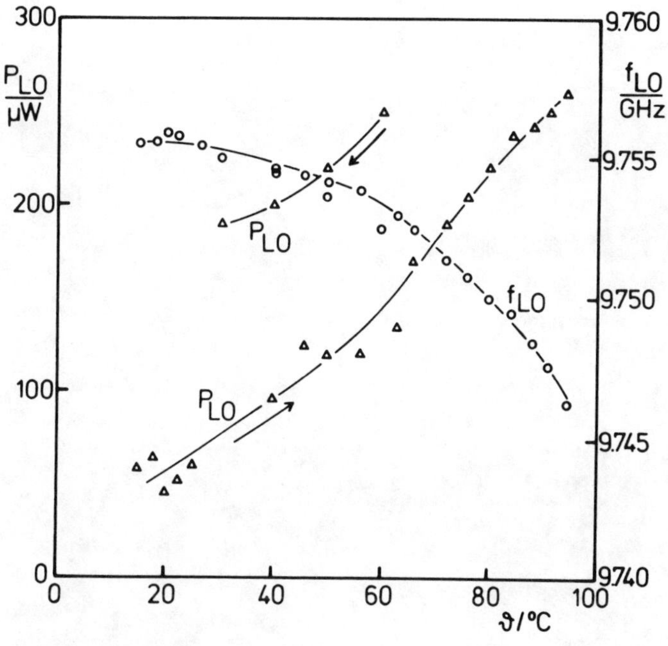

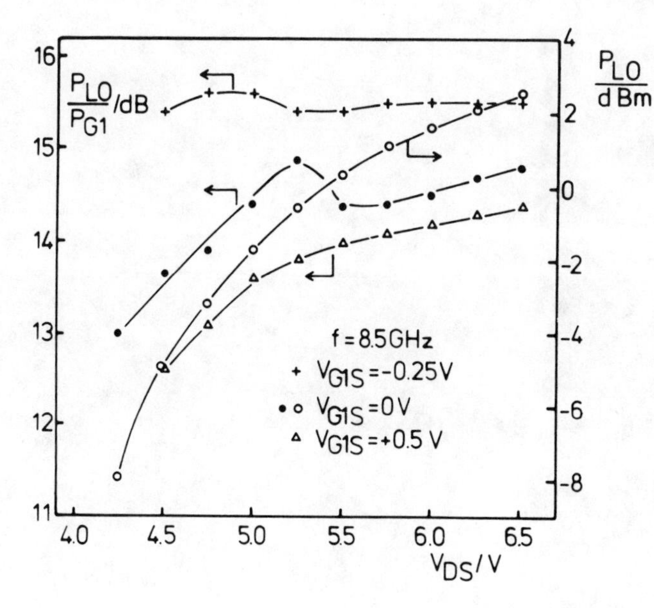

Fig.8: Isolation of input port (G 1). Disc-tuned s.o. MESFET mixer.

Part X
Further Aspects on Microwave and Millimeter-Wave Mixers

IN this last part we have collected papers on various important topics concerning mixers that have not been discussed in the previous sections. The first paper concerns a very important and rapidly developing technology: monolithic integrated circuits (MIC). Chao *et al.* describe MIC's developed on semi-insulating GaAs substrates for millimeter-wave (~34 GHz) balanced mixers. In this mixer the GaAs chip is used as a suspended stripline in a crossbar mixer circuit. The next paper, by Clifton, gives an overview of receiver technology for the millimeter- and submillimeter-wave regions. The extension of waveguide mixers into the submillimeter region and various types of quasioptical receivers are discussed. Receiver performance is reviewed and the potential impact of MIC technology is considered.

The last four papers in this volume describe mixers using other types of nonlinear elements not discussed in the previous papers. In paper 3 Kim discusses tunnel diode mixers. Dixon and Jacobs describe indium phosphide Gunn diodes used in self-mixing oscillators for 60 GHz. An experimental investigation of conversion properties and the minimum detectable signal is presented. In the fifth paper Förg and Freyer investigate Ka-band self-oscillating mixers using Schottky baritt diodes. As in the two previous papers, conversion gain is obtained.

In the last paper another cryogenically cooled mixing element is used: indium antimonide bolometers. Very low noise temperature can be obtained at millimeter- and submillimeter-waves [1]. However, due to the long time constant, only about 2 MHz IF can be used.

References

[1] T. G. Phillips and D. P. Woody, "Millimeter- and submillimeter-wave receivers," *Amer. Rev. Astron. Astrophys.*, vol. 20, pp. 285-321, 1982.

Ka-Band Monolithic GaAs Balanced Mixers

CHENTE CHAO, MEMBER, IEEE, A. CONTOLATIS, STEPHEN A. JAMISON, AND PAUL E. BAUHAHN, MEMBER, IEEE

Abstract—**Monolithic integrated circuits have been developed on semi-insulating GaAs substrates for millimeter-wave balanced mixers. The GaAs chip is used as a suspended stripline in a cross-bar mixer circuit. A double sideband noise figure of 4.5 dB has been achieved with a monolithic GaAs balanced mixer filter chip over a 30- to 32-GHz frequency range. A monolithic GaAs balanced mixer chip has also been optimized and combined with a hybrid MIC IF preamplifier in a planar package with significant improvement in RF bandwidth and reduction in chip size. A double sideband noise figure of less than 6 dB has been achieved over a 31- to 39-GHz frequency range with a GaAs chip size of only 0.5×0.43 in. This includes the contribution of a 1.5-dB noise figure due to IF preamplifier (5–500 MHz).**

I. Introduction

The performance of GaAs devices has been steadily improved with recent advances in material, process, and device technology. Monolithic integration of passive elements and active devices on GaAs substrates becomes increasingly attractive for use at millimeter-wave frequencies as opposed to the more conventional MIC hybrid approach where the effects of parasitics are difficult to control. Considerable attention has been given to the development of millimeter-wave monolithic balanced mixers which serve as an important building block for a number of potential systems. Recently, a monolithic mixer IF preamplifier using a microstrip circuit approach has been demonstrated with good performance at millimeter-wave frequencies. [1] The circuit requires an on-chip coupler to combine the local oscillator and signal frequencies increasing the chip size, but has the advantage of simple interfaces. The present paper describes a monolithic balanced mixer using a suspended stripline circuit approach in a cross-bar mixer circuit [2] which does not require an on-chip coupler. The monolithic mixer has also been integrated with a hybrid MIC IF amplifier in a wafer-type waveguide package. In this approach the only passive elements on the monolithic GaAs chip are the coupling circuits for millimeter-wave frequencies. The low-pass filter is incorporated with the bipolar IF preamplifier on a hybrid MIC.

II. Circuit Design

The basic configuration of the monolithic balanced mixer is illustrated in Fig. 1. The GaAs chip is used as a suspended stripline and is coupled to local oscillator power and RF signal via two full-height waveguide ports. It consists of two planar mixer diodes and matching circuits for RF waveguide, LO coupling, and IF filter. There are several unique features of the mixer that significantly reduce the complexity and take advantage of the monolithic circuit technique. The diodes are electrically in series with respect to the RF signal. The RF impedance of a single mixer is usually much lower than that of waveguide. With two diodes in series, the RF input impedance of the chip at the signal port can be matched easily with straight full-height waveguide. Furthermore, if a reduced height or ridged waveguide is used to match the impedances of circuit and chip over the desired frequency range, then the RF bandwidth of this mixer could be increased to nearly full waveguide bandwidth.

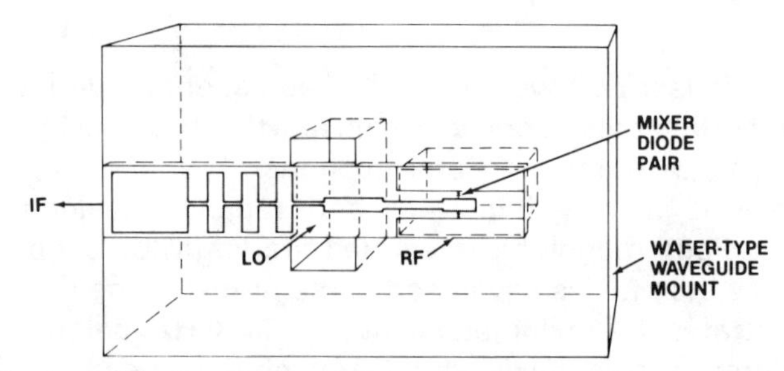

Fig. 1. Millimeter-wave integrated mixer chip design containing two planar mixer diodes and matching circuits for RF waveguide, LO coupling, and IF filter.

The diodes are electrically in parallel with respect to the local oscillator and are in a direction such that the induced LO currents in the diodes are out of phase. This eliminates the need for a magic tee to cancel the noise contributed by the local oscillator. The conventional waveguide magic tee is inherently a narrow-band component and is very expensive to make, particularly at millimeter-wave frequencies. Furthermore, the two diodes are physically close to each other and can have nearly identical parameters because they are monolithically fabricated. Such a well-matched diode pair gives excellent LO noise suppression.

The isolation between the local oscillator and signal ports can be very high over a wide range of frequencies since the dominant TE_{10} mode in the RF waveguide is orthogonal to the quasi-TEM LO input.

III. Chip Layout and Fabrication

The monolithic mixer described above has been fabricated with advanced semiconductor technology available in our laboratory. The monolithic chip design and the key

Manuscript received May 1, 1982.

The authors are with Honeywell Corporate Technology Center, Bloomington, MN 55420.

Reprinted from *IEEE Trans. Microwave Theory Tech.*, vol. MTT-31, pp. 11–15, Jan. 1983.

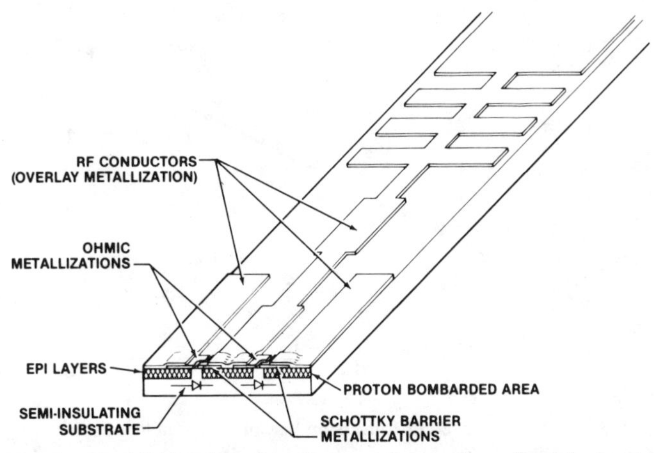

Fig. 2. Monolithic mixer chip shown with two planar Schottky-barrier mixer diodes isolated parasitically by a proton ion-implantation technique and RF matched by overlay metallization patterns.

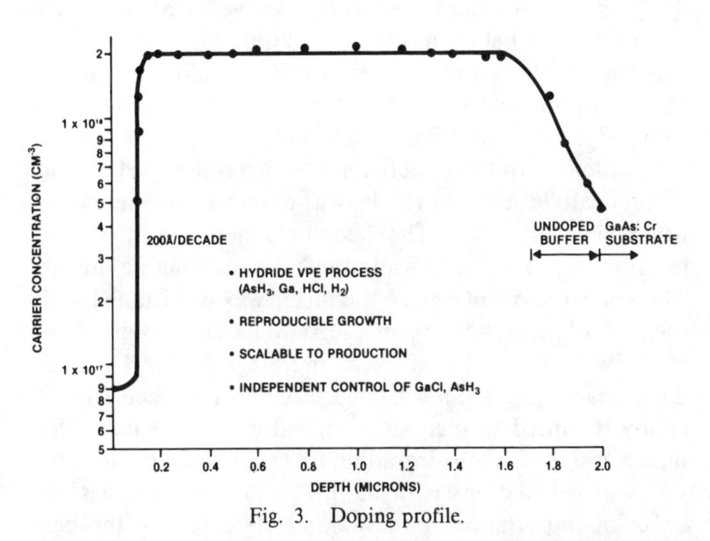

Fig. 3. Doping profile.

technology needed for its fabrication are illustrated in Fig. 2. The diodes are fabricated with VPE-grown n-n^+ layers on a semi-insulating substrate isolated by proton bombardment [3]. High-quality n-n^+ layers are grown on 10-mil semi-insulating GaAs:Cr substrates which have been qualified by our established qualification test procedure. The n^+ layer has a doping density of 2×10^{18} cm^{-3} and is 2 μm thick and the n layer which is on top of n^+ layer has a doping density of 9×10^{16} cm^{-3} and is 0.1 μm thick. These layers have been grown by our vapor-phase epitaxial reactor with the hydride process (AsH_3, HC1, Ga, H_2). Carrier concentration as a function of depth from the surface of the n-n^+ epi-layer is shown in Fig. 3. Schottky-barrier (TiW/Au) and contact (AuGe/Ni/TiW/Au) metallizations are then deposited for the planar mixer diodes. The chip is bombarded by high energy protons everywhere except at the diodes which are protected by a thick layer of photoresist and gold metal. This proton bombardment process isolates the diodes from the circuit parasitics. After the protective layers are removed, overlay metallization is deposited for the RF matching circuits. Individual chips are cut from the large wafer and mounted in the wafer-type waveguide package as shown in Fig. 4.

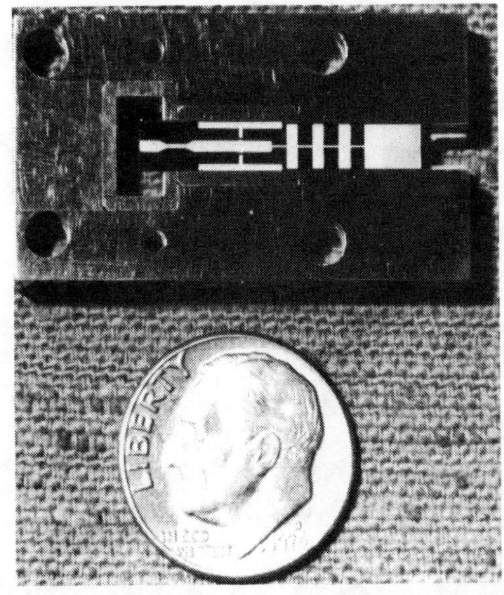

Fig. 4. Monolithic mixer chip in a wafer-type mount.

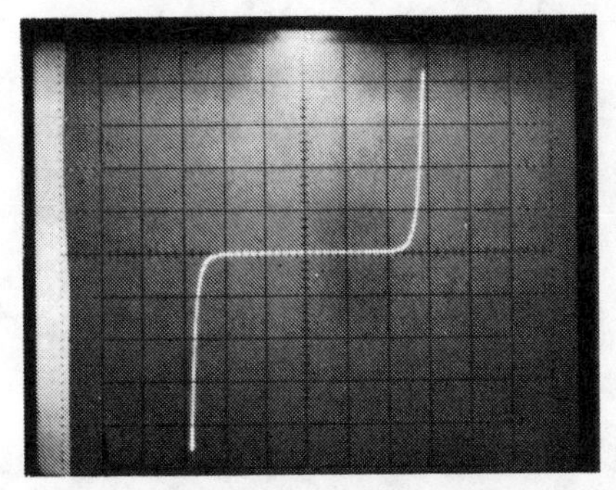

Fig. 5. $I-V$ characteristics of the monolithic diode pair. Horizontal: 0.2 V/div; Vertical: 1 μA/div.

Diodes with a zero biased cutoff frequency of better than 600 GHz have been achieved. Fig. 5 shows the $I-V$ characteristics of a monolithic diode pair.

IV. Monolithic Balanced Mixer-Filter

Fig. 6 shows the double sideband noise figure of the mixer filter chip as a function of the local oscillator power for the three different LO frequencies. A double sideband noise figure of 4.5 dB has been achieved with the monolithic GaAs balanced mixer chip over a 30- to 32-GHz frequency range at a LO power of approximately 10 mW. This includes the contribution of a 1.5-dB noise figure due to an IF preamplifier which has a bandwidth of 5–500 MHz.

This performance is quite competitive with the best conventional mixers at this frequency. Measurements of noise figure as a function of LO frequency for three different mechanical tuning positions are also plotted in Fig. 7. This indicates that some tuning is possible with the monolithic balanced mixer but the current chip configuration works best around 31 GHz with a low noise IF bandwidth of about 500 MHz.

The isolation between the local oscillator and signal

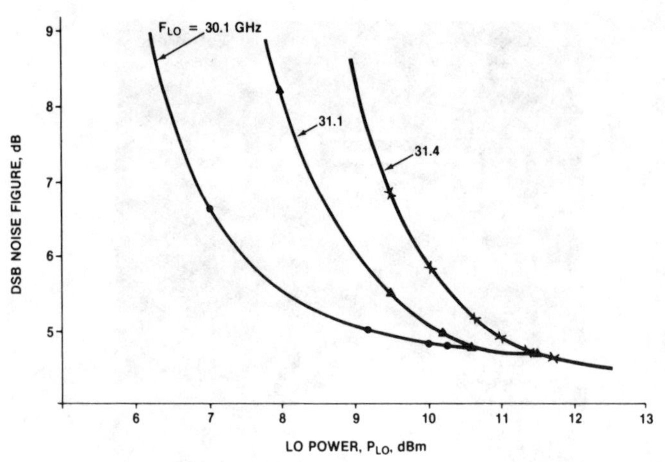

Fig. 6. Noise figure versus P_{LO} of the monolithic mixer.

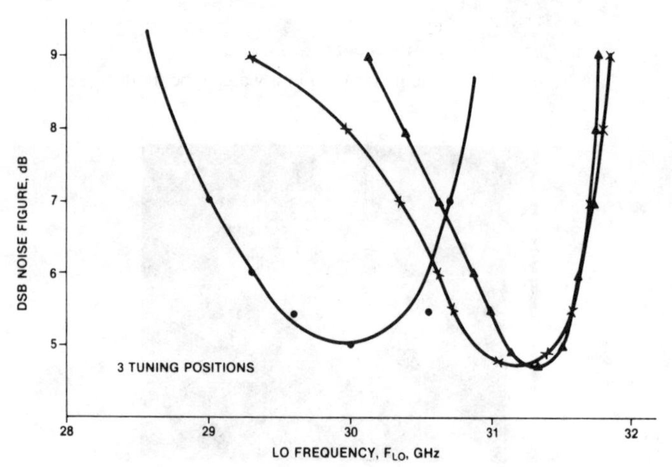

Fig. 7. Noise figure versus F_{LO} of the monolithic mixer.

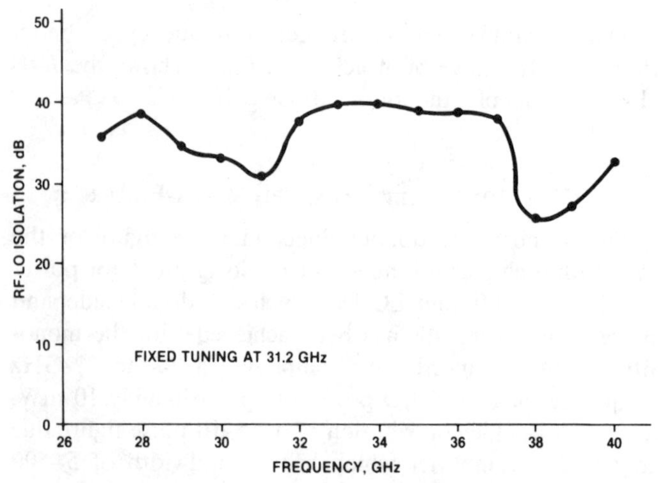

Fig. 8. Isolation versus frequency of the monolithic mixer.

ports of the mixer is excellent as indicated in Fig. 8. An isolation of better than 30 dB has been achieved over a frequency range of 27–37 GHz when the mixer is tuned at 31 GHz. This excellent isolation property is a direct consequence of the mixer circuit design approach, i.e., the LO and RF ports are decoupled because the E-fields of the dominant modes are orthogonal to each other.

The single-sideband conversion loss of the mixer is about

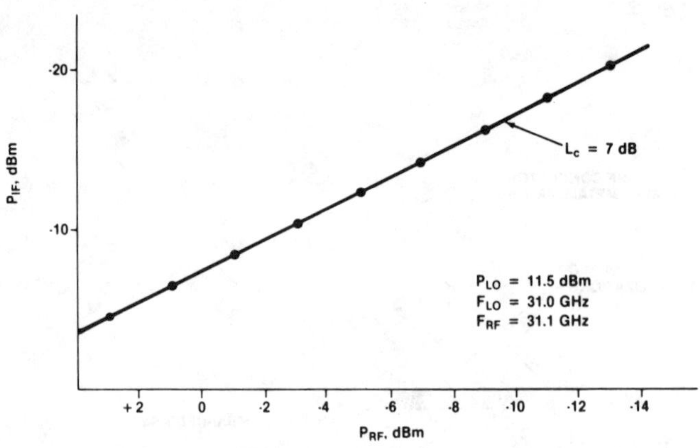

Fig. 9. Single sideband conversion loss of the monolithic mixer.

7 dB as shown in Fig. 9. The corresponding single sideband noise figure is 7.6 dB. This means that the noise due to the diode is very low and the diodes are well matched. This confirmed our belief that the monolithic diode pair fabricated in close proximity to each other would have nearly identical parameters and therefore the LO noise suppression property of the mixer is excellent.

Excellent correlation between the experimental data and theoretically calculated results was obtained for the monolithic mixer filter chip. The theoretical model has also been used to aid the optimization of the monolithic mixer. Theoretical performance of the mixer was determined with the aid of a very complete mixer analysis program [4] which was obtained from Kerr and Siegel at NASA Goddard Space Flight Center. This program has been significantly modified to increase its speed and convenience for mixer design without degrading its accuracy. An extensive series of calculations with the program indicates that the order of importance of the parameters affecting the performance of a mixer diode are the series resistance, capacitances, and then the series inductance. The optimum inductance value is a function of the circuit impedance level and other factors.

Using realistic parameters for the various diode characteristics, the performance of the monolithic mixer indicated in Fig. 10(a) was obtained. The equivalent circuit is given in Fig. 10(b). The performance predicted for the series inductance estimated from the diode lead geometry for the monolithic mixer is indicated. The measured performance is essentially identical. It is clear that a slight adjustment of the circuit could give considerably better noise performance by improved impedance matching between the diode and waveguide.

V. Integration of Monolithic Mixer and Hybrid IF Preamplifier

A monolithic GaAs balanced mixer has been combined with a hybrid IF preamplifier in a planar waveguide package. Fig. 11 is a schematic of the circuit configuration and Fig. 12 is a photograph of the package. The size of the monolithic balanced mixer has been minimized by incorporating only those circuit elements which are critical for matching of devices at millimeter-wave frequencies on the

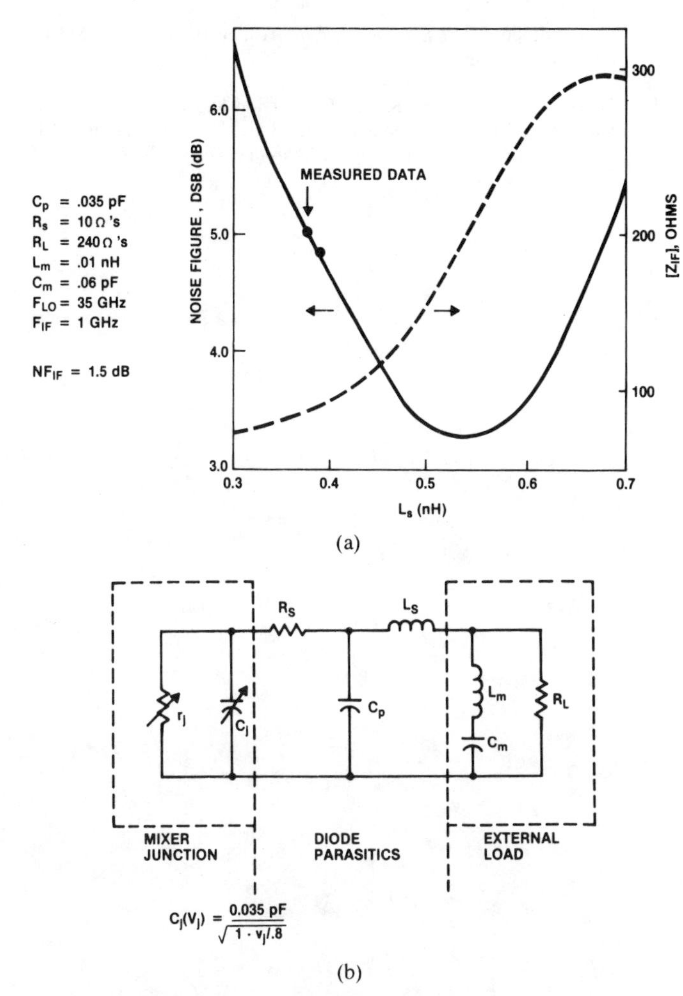

Fig. 10. (a) Comparison of measured data and calculated performance of the monolithic mixer filter chip. (b) Equivalent circuit used for mixer junction embedding impedance calculation.

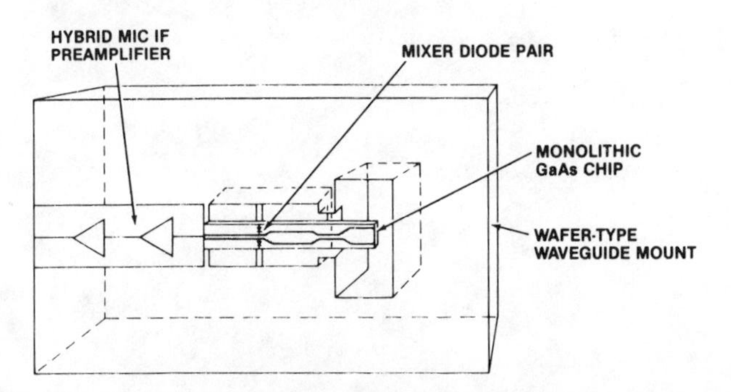

Fig. 11. Packaging technique for combining a monolithic balanced mixer and a hybrid IF preamplifier.

Fig. 12. *Ka*-band monolithic GaAs balanced mixer chip integrated with a hybrid MIC IF preamplifier on a planar package.

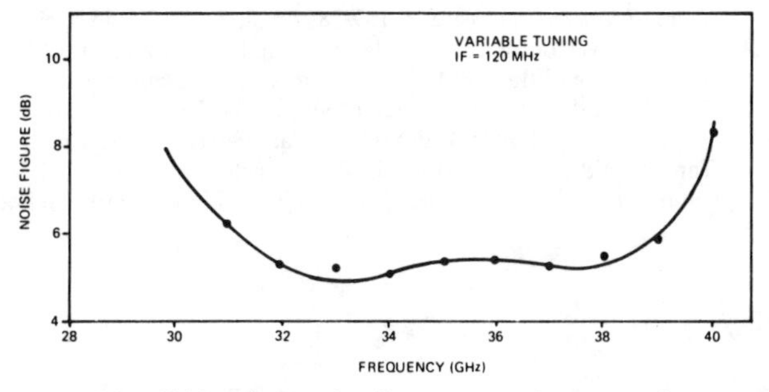

Fig. 13. Monolithic mixer noise figure versus frequency demonstrates wide tunable operating frequency range for flexible system applications (includes 1.5-dB IF preamplifier noise).

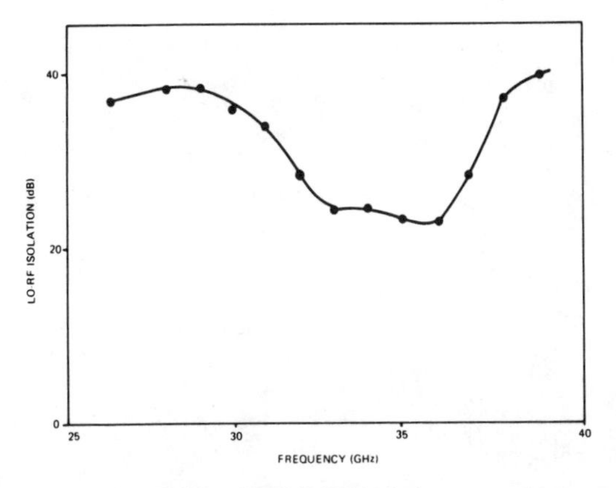

Fig. 14. Monolithic mixer LO-RF isolation versus frequency.

GaAs substrate. Significant reduction in chip size and improvement in RF bandwidth have been achieved.

Fig. 13 shows noise figure versus frequency of a monolithic chip from 30 to 40 GHz. A double sideband noise figure of less than 6 dB has been achieved over an 8-GHz bandwidth with a GaAs chip size of only 0.05×0.43 in. This includes the contribution of a 1.5-dB noise figure from IF preamplifier which has a bandwidth of 5–500 MHz. This performance gives an improvement of RF bandwidth by a factor of 4 and a reduction of chip size by a factor of 5 with respect to a result reported previously [2]. The isolation between the local oscillator and signal ports of the mixer is very good as indicated in Fig. 14. An isolation better than 20 dB over 26–40 GHz has been achieved.

VI. Conclusions

A monolithic GaAs balanced mixer chip with a minimum chip size can be combined with a hybrid MIC IF preamplifier in a unique circuit configuration to achieve high-performance and potentially cost-effective components for millimeter-wave receiver applications.

Acknowledgment

The authors would like to express their appreciation to T. Peck and J. Abrokwah for supplying VPE material and to D. Hickman for technical assistance.

References

[1] A. Chu, W. E. Courtney, and R. W. Sudbury, "A 31-GHz monolithic GaAs mixer/preamplifier circuit for receiver applications," *IEEE*

Trans. Electron Devices, vol. ED-28, no. 2, pp. 149–154, Feb. 1981.

[2] C. Chao, A. Contolatis, S. A. Jamison, and E. S. Johnson, "Millimeter-wave monolithic GaAs balanced mixers," presented at 1980 Gallium Arsenide Integrated Circuit Symp., Las Vegas, NV, paper 32.

[3] R. A. Murphy, C. O. Bozler, C. D. Parker, H. R. Fetterman, P. E. Tannenwald, B. J. Clifton, J. P. Donnelly, and W. T. Lindley, "Submillimeter heterodyne detection with planar GaAs Schottky-barrier diodes," *IEEE Trans. Microwave Theory Tech.*, vol. MTT-25, pp. 494–495, June 1977.

[4] D. N. Held and A. R. Kerr, "Conversion loss and noise of microwave and millimeter-wave mixers: Part I—Theory," and "Part II—Experiment," *IEEE Trans. Microwave Theory Tech.*, vol. MTT-26, pp. 49–61, Feb. 1978.

Receiver technology for the millimeter and submillimeter wave regions

Brian J. Clifton
Lincoln Laboratory, Massachusetts Institute of Technology
Lexington, Massachusetts 02173

Abstract

Increased interest in the millimeter and submillimeter wavelength regions during the past decade has stimulated the development of sensitive receivers for a wide range of applications. The extension of waveguide mixers into the submillimeter region and the development of various types of quasi-optical receivers are reviewed. The development of novel GaAs integrated circuit mixers for the millimeter and submillimeter regions is discussed. Receiver performance is reviewed and the potential impact of monolithic receiver technology on systems applications is considered.

Introduction

The past decade has seen an increased interest in the millimeter and submillimeter regions of the spectrum and in the related receiver technology.[1] The development of heterodyne receivers with improved sensitivity has resulted in a wide range of applications in radio astronomy, plasma physics, frequency standards and spectroscopy, radar, aeronomy and in satellite-based radiometry. The diversity of these applications with vastly different requirements generates the need for continuing development of heterodyne receivers with lower noise, improved reliability and at lower cost.

Although there are a number of different devices that are used as the non-linear mixing elements in heterodyne receivers, the most common device is the Schottky-barrier diode. Excellent results have been obtained with whisker-contacted diodes in waveguide and in quasi-optical mixer mounts, but it is considerably more difficult to make contact to the smaller-diameter diodes used at the higher frequencies and the reliability of the resulting diodes is inferior to that obtained at lower frequencies. The assembly of whisker-contacted diodes in mixer mounts is very labor intensive, time consuming and expensive. Naturally, there has been considerable interest in replacing the whisker-contact to the Schottky diode with a photolithographically-fabricated contact on the surface of the device. At microwave frequencies, beam-lead diodes have been integrated in a hybrid fashion with microstrip, stripline or dielectric-guide circuit elements. Similar techniques have been extended into the millimeter wave region with some success. The recent renewed interest in GaAs monolithic integrated circuits for the microwave region of the spectrum and the development of monolithic integrated circuit mixers for the millimeter and submillimeter regions offer the systems designer the possibility of totally integrated monolithic receivers with improved performance and reliability, and the potential for low-cost mass production. More difficult, but very exciting, is the possibility of building monolithic imaging arrays in the millimeter and submillimeter.

Receiver design concepts

The purpose of a heterodyne receiver is to convert a high-frequency signal to a lower frequency signal, preserving the phase information and adding a minimum of noise or degradation to the signal in the conversion process. A block diagram of a typical millimeter or submillimeter-wave heterodyne receiver is shown in Figure 1. The heart of any heterodyne receiver is a non-linear mixing element that combines a low-level radio frequency (RF) input signal with a local oscillator (LO) signal to produce a signal at the difference or intermediate freqeuncy (IF). The receiver elements after the IF amplifier are essentially independent of the frequency of the input signal. They depend mainly on the signal processing performed on the information in the IF signal and are not dealt with here.

The LO signal is usually many orders of magnitude larger than the RF signal and gives rise to the non-linearity of the mixer by driving or switching the mixer diode over some portion of its non-linear range. A typical millimeter or submillimeter mixer requires 1 to 10 mW of LO power, which is close to the maximum output power of typical present-day LO sources in the submillimeter region. Thus, in contrast to the microwave frequency range where there is usually abundant LO power available, the millimeter and submillimeter region requires that nearly all the available LO power be delivered to the mixer element.

The function of the diplexer is to combine the LO and RF signals and to send both signals to the mixer with minimum loss. In the microwave or lower-millimeter wave range, the diplexer could be a hybrid, magic T or even a directional coupler with, in all cases, the RF and LO energy propagating as guided waves confined within the component. At higher frequencies where quasi-optical techniques are more appropriate, the diplexer will be a quasi-optical component and the RF and LO energy will propagate in free-space beam modes within the diplexer.

The primary antenna receives the RF signal as a free-space plane wave and focuses the energy into a beam of appropriate diameter to send to the diplexer with minimum reflection and loss. The RF and LO energy exiting from the diplexer in overlapping beams must be coupled to a secondary antenna whose purpose is to convert the free-space energy propagating in beam modes into guided or confined wave energy that can be coupled to the Schottky-barrier junction where it appears as current through and voltages across the junction. Martin & Lesurf[2] provide a comprehensive discussion of the problems of applying classical geometrical optics to the design of submillimeter-wave optics and indicate how beam-mode concepts can be applied.

The function of all the elements preceding the mixer is to collect, focus and convert the energy of the RF and LO waves, which are propagating as free-space electromagnetic waves, into electrical current flowing in the non-linear mixing element. In the case of a Schottky-barrier diode, the actual diode has a diameter of the order of a few micrometers whereas the RF and LO free-space waves can have beam diameters of the order of centimeters. This implies considerable concentration of beam energy density in coupling the free-space energy to the diode junction. The coupling of electromagnetic energy must be achieved efficiently with minimum loss and little or no reflection. In general the impedance of the Schottky-barrier junction is quite different from that of free space so that the coupling function involves matching between impedance levels.

At microwave frequencies the RF and LO waves are guided and coupled to the diode by coaxial, stripline, microstrip, waveguide or other transmission-line media. Since the dimensions of the guiding media must be of the order of a wavelength to prevent energy propagation in undesirable modes, an upper frequency limit is imposed by problems in fabricating the guiding structures, by difficulties in mounting signal processing components within the guide structures, and by guide attenuation that increases rapidly as dimensions are reduced and frequency is increased.

For most practical purposes, mechanical fabrication is limited to components with dimensions greater than about 0.1 mm, although tolerances and surface finish can be held to much smaller dimensions. Typical matching and coupling structures in the microwave region have dimensions from a quarter to one-tenth wavelength. Thus, if similar structures are to be scaled to operate in the millimeter and submillimeter regions, present-day mechanical fabrication techniques will restrict fabrication of coupling and matching components to frequencies below 400 GHz. However, photolithographic techniques used in semiconductor device fabriction allow dimension control down to about one micrometer, and x-ray or electron beam lithography can extend dimension control down to below 0.1 μm. Clearly, it is attractive to consider using these techniques to build coupling and matching elements for the frequency range above 400 GHz where mechanical fabrication techniques are extremely difficult to apply. Thus, there is considerable interest in the fabrication of monolithic integrated circuits in which the elements of Figure 1, from the secondary antenna up to and possibly including the IF amplifier, are fabricated on a single piece of GaAs. Critical circuit elements can be located close to the diode junction with the possibility of much lower coupling loss.

Discrete diode mixers

The fundamental waveguide mixer is used extensively throughout the millimeter region with considerable success (Figure 2). A whisker-contacted axial diode is usually mounted in reduced-height waveguide in order to provide an optimum impedance environment to the diode. A discrete diode chip mounted flush with or on one broad wall of the waveguide, and a contact whisker centered across the waveguide and contacting one of many Schottky-barrier junctions on the diode chip comprise the non-linear mixing element. An impedance transformer is required to match the full-height waveguide impedance to the reduced-height mixer mount. At the higher millimeter-wave frequencies, the additional fabrication complexity and loss introduced by the transformer and reduced height waveguide usually favor mounting the diode in full-height guide. A movable or fixed backshort located approximately a quarter wavelength behind the diode presents a high impedance at the whisker-diode location and provides reactive tuning of the whisker and chip parasitics so that the diode junction is presented with a real impedance at the signal frequency. The backshort is a source of loss and both contacting and non-contacting shorts have been used. Non-contacting shorts give more repeatable results but are difficult to fabricate at the higher frequencies and have slightly more loss than the best contacting shorts. The coaxial RF choke in the IF output port can be difficult to fabricate at higher frequencies and is replaced in many designs by a photolithographically formed filter on suspended stripline.[3]

A number of workers[3-10] have reported excellent results with fundamental waveguide mixers operating both at room temperature and at cryogenic temperatures. Particularly noteworthy are papers by Vizard et al.[6] and by Keen et al.[7-9] which report results obtained with Mott barriers in which the n layer is fully depleted at zero bias. Keen et al.[8] reported a single side-band (SSB) mixer noise temperature of 98K and conversion loss of 5.0 dB at 115 GHz for a fundamental waveguide mixer cooled to 42 K. At higher frequencies, Erickson[10] reported a fundamental-mode waveguide mixer at 318 GHz having a SSB mixer noise temperature of 3100K and a conversion loss of 9.3 dB.

At lower millimeter wave frequencies discrete beam lead diodes have been integrated successfully with printed circuit mixer structures fabricated in various transmission line media.[11,12] The mixer reported by Cardiasmenos and Parrish[12] uses advanced GaAs beam-lead diodes mounted onto a fused silica suspended stripline mixer circuit. This is an impressive example of production-type mixer at 94 GHz with an uncooled SSB mixer noise temperature of 760 K and conversion loss of 6.0 dB. Unfortunately, the parasitics associated with the discrete beam-lead diodes will probably limit their applications to frequencies below 200 GHz.

Above 400 GHz it becomes exceedingly difficult to fabricate fundamental waveguide mixers. As a result mixers have been built in a quasi-optical free-space environment in which the propagating energy is not constrained to a single mode. In the simplest mounts a Schottky diode is located at the focus of a spherical reflector with a whisker contact used as a simple wire antenna. Gustincic[13-15] developed a quasi-optical biconical mixer mount, shown in Figure 3, in which a diode and whisker contact wire are mounted between the apexes of the cone-shaped pieces forming the biconical antenna. An interesting feature of this mount is that the characteristic impedance is essentially independent of frequency and is determined by the cone angle. Thus optimum RF matching to the diode should be possible simply by selecting the appropriate cone angle. GaAs Schottky-barrier diodes 2.5 μm in diameter were used in several versions of this mixer at frequencies up to 671 GHz.[16]

Lincoln Laboratory[17] adopted a long-wire antenna approach in a number of corner-reflector mixer mounts covering the 120 μm to 1mm wavelength range. The corner-reflector diode mount (Figure 4) uses a 4-wavelength long antenna wire located 1.2 wavelengths from the corner of a 90° corner reflector. The diode chip is mounted flush with the surface of the ground plane. Diodes having diameters in the range 1 μm to 2 μm are contacted by lowering the antenna whisker into contact with a diode, while observing the contacting operation through a high power optical microscope. The corner reflector mixer has an almost circular beam cross-section with a 14° beamwidth at the 3 dB points. The corner reflector mixers have been used successfully to detect interstellar carbon monoxide at 434 μm[17,18] in the Orion molecular cloud, and have also been used for fusion plasma diagnostics at 400 μm and 120 μm. The results of blackbody radiometric measurements using a standard Y-factor method interpretation of the data are shown in Table I.

TABLE I

QUASI-OPTICAL CORNER REFLECTOR MIXER PERFORMANCE

λ (μm)	FREQUENCY (GHz)	TOTAL SYSTEMS NOISE TEMPERATURE (DSB) K	MIXER TEMPERATURE (K)	CONVERSION LOSS (dB)
434	692	3,000	2,600	8.6
184	1630	19,000	≈17,000	15.6
119	2521	32,000	≈28,000	17.8

Monolithic Integrated Circuit Receivers

The rapid growth of interest in monolithic microwave integrated circuits is evidenced by the recent special issue of IEEE Transactions on Electron Devices[19] devoted to this area. At microwave and low millimeter wave frequencies, a considerable amount of GaAs surface area will be occupied by passive circuit elements, and the active device area to overall chip area ratio will be very small. Thus material cost, processing cost and active device yield tend to make a monolithic approach less cost effective at these lower frequencies than a hybrid approach in which active device and passive circuit technology are separated. Overall chip area can be reduced in some cases by replacing and simulating passive circuit functions with active devices. However, the penalty is usually increased noise and power dissipation.

The earliest reported work on millimeter wave integrated circuits on GaAs was by Texas Instruments[20,21] in 1968 in which single-ended and balanced mixers were fabricated for use at 94 GHz. The passive integrated circuit components were microstrip lines deposited on 100 μm thick semi-insulating GaAs substrates. More recently Chu et al.[22] have fabricated a 31 GHz monolithic receiver front end, as shown in Figure 5, using microstrip circuits on 175 μm thick semi-insulating GaAs. The balanced mixer-preamplifier combinations typically exhibit a conversion gain of 4 dB and a single-sideband noise figure of 11.5 dB.

Clifton et al.[23,24] have reported the development of a novel GaAs monolithic integrated circuit mixer (Figure 6) which is impedance matched to fundamental waveguide. It consists of a slot coupler, coplanar transmission line, surface-oriented Schottky-barrier diode, and RF bypass capacitor monolithically integrated on the GaAs surface. In this mixer the radiation propagates through the GaAs to a slot coupler fabricated photolithographically in a metallic ground plane on the surface of the GaAs, in contrast to monolithic mixer structures using microstrip circuits in which the radiation is guided along the surface of the GaAs dielectric. The bulk dielectric is actually part of the impedance matching circuit. The slot coupler is connected to a diode by an appropriate section of coplanar line, and an integrated bypass capacitor completes the mixer circuit providing a short circuit to millimeter wave frequencies and an open circuit at the IF. At 110 GHz, a monolithic mixer module mounted in the end of a waveguide horn has an uncooled double-side-band (DSB) mixer noise temperature of 339 K and a conversion loss of 3.8 dB. The same mixer module cooled to 77 K has a DSB mixer noise temperature of 50 K and a conversion loss of 4.5 dB. A similar monolithic mixer module designed to operate at 350 GHz has an uncooled DSB mixer noise temperature of 6500 K.

Monolithic Receiver Fabrication Considerations

At microwave and lower millimeter wave frequencies microstrip techniques can be applied very effectively to build passive circuit components on semi-insulating GaAs. However, in the submillimeter regime it is more appropriate to use waveguide and quasi-optical techniques. Thus, beam waveguide, focusing optics, antenna structures and dielectric waveguides are concepts that can be applied to the design of receiver circuits in the millimeter and submillimeter. The monolithic integrated circuit mixer (Figure 6) developed at Lincoln Laboratory combines antenna concepts with properties of the bulk GaAs dielectric to produce a very efficient coupling structure at 110 GHz. We believe that these concepts can be applied throughout the millimeter and a large portion of the submillimeter region.

In considering the Schottky-barrier diode, similar design rules apply to monolithic mixers as to conventional whisker-contacted diode mixers. However, since the millimeter wave circuit is connected directly to the Schottky diode, diode lead parasitics can be eliminated in a correctly designed monolithic mixer circuit. Diode junction capacitance and series resistance are parasitic elements that degrade mixer performance and should be minimized. A Schottky-barrier diode designed for use at frequencies near 100 GHz would have an equivalent circular diameter of about 2 μm on GaAs material with a carrier concentration in the low 10^{17} cm^{-3} range. Figure 7, which is a detailed drawing of the surface-oriented diode portion of the Lincoln Laboratory monolithic mixer, illustrates the main fabrication and topological details of the device. The n layer should be thin to minimize the contribution of the undepleted epitaxial layer to diode series resistance, and will typically be 0.1 to 0.2 μm thick with a carrier concentration of 1-2 x 10^{17} cm^{-3}. If the n layer is fully depleted at zero bias, the resulting Mott barrier not only has lower series resistance than a Schottky barrier but also requires lower local oscillator drive power and gives lower noise performance in cooled mixer applications. Such devices require very uniform epitaxial layers with a very sharp transition between the n^+ and n layers and an n layer thickness of less than 0.1 μm. The Schottky barrier can be formed using standard metallization techniques. The particular choice of Schottky-barrier metal may be influenced, however, by the temperature stability of the barrier during subsequent device fabrication. The most critical and difficult aspect of device fabrication is the location of the Schottky-barrier junction at one edge of the conducting pocket of GaAs. This must be accomplished with a registration accuracy of better than ± 0.5 μm since this directly determines the resulting junction capacitance. A stripe geometry with the long axis perpendicular to the pocket edge is superior to a circular geometry since registration error has less effect on junction capacitance and, in addition, spreading resistance is lower. Reduction of spreading resistance requires an ohmic contact of low resistivity in close proximity to the Schottky barrier and an n^+ epitaxial layer with a thickness of several skin depths at the operating frequency. However, if proton bombardment is used to isolate the conducting pocket of GaAs, the total epitaxial layer thickness that can be converted to semi-insulating is determined by the proton energy available. For example, 400 keV protons will reliably convert up to 3 μm of n^+ GaAs with a carrier concentration of 2×10^{18} cm-3.

The diode ohmic contact region can be defined on the GaAs surface by a variety of means, such as by ion implantation or by etching away the n layer and alloying an evaporated contact metallization into the exposed n^+ layer. Specific contact resistance as low as 2×10^{-8} ohm-cm^2 has been achieved in the Lincoln Laboratory diode by alloying an evaporated Ni/Au-Ge/W/Au metallization into the n^+ layer. Other workers have reported extremely low specific contact resistance for non-alloyed ohmic contacts formed by direct metallization onto an epitaxial Ge layer grown on top of the n^+ layer. Similar results might be expected for non-alloyed metallizations onto n^+ layers with carrier concentration in the 10^{19} cm^{-3} range.

The technique selected to isolate and define the conducting area of the diode depends on many factors. Although proton bombardment has proven to be a convenient and successful technique, it has certain limitations. Thicker epitaxial layers, higher n^+ carrier concentration, or dopant tails extending into the semi-insulating substrate may not be reliably converted by the available proton energy. This can result in an undesirable buried conducting layer which attenuates electromagnetic radiation and degrades mixer performance. In addition, subsequent processing steps must be limited to temperatures below 300°C to prevent annealing of the proton damage. This could place severe restrictions on the fabrication sequence for more complex monolithic integration. Mesa etching might be an alternative, but it results in a non-planar surface which in most cases precludes subsequent high-resolution photolithography.

As we move from the integration of single components such as mixers or amplifiers to a completely monolithic receiver front end, we find that material that is optimum for mixer diodes is not suited to MESFET devices. Selective epitaxy might prove to be the most cost effective technique to provide such starting material, since this technique offers the possibility of tailoring the material characteristics within a particular pocket to suit a specific active device. Of course, ion implantation and proton bombardment will still be valuable techniques to modify material characteristics during fabrication. Although selective epitaxy into etched pockets in a wafer of semi-insulating GaAs has been used by many workers to produce isolated regions of conducting GaAs, problems of non-planar surfaces, moat-like voids at the periphery of the pockets, and quality of the epitaxial layers within the pockets have delayed the use of these techniques for monolithic mixers in the millimeter and submillimeter. The recent development[25,26] of ion beam assisted etching of semi-insulating GaAs to create pockets with vertical walls and smooth surfaces appears to offer an ideal pocket structure into which high-quality n on n^+ layers can be grown by MBE. The resulting void-free planar wafers should be ideal for the fabrication of monolithic mixers and receivers.

Monolithic mixer circuits can be scaled to operate at higher frequencies simply by scaling the circuit dimensions smaller in the ratio of the higher to lower frequency. However, contact bonding pads on the millimeter monolithic mixers must have minimum linear dimensions of about 50 μm if conventional ribbon or wire bonding techniques are used to connect to the device. As the devices are scaled into the submillimeter wave regime, contact pads will inevitably become too small and alternative techniques must be considered to interface to the device. In addition, the module thickness is reduced by the same scale factor. Techniques have been developed at Lincoln Laboratory that allow controlled thinning and separation of individual modules using front-to-back alignment and a cantilevered extended ground plane and coplanar IF lead as shown in Figure 8. Such devices are easy to integrate with subsequent low-frequency circuits and provide an integrated package that can be conveniently handled and interfaced with high-frequency quasi-optical components.

Conclusions

The state-of-the-art in low noise receiver technology in the 100-1000 GHz frequency range is summarized in Figure 9. Monolithic integrated circuit mixers appear to be competitive with the best waveguide mixers at 110 GHz. Above 400 GHz quasi-optical discrete diode mixers still dominate and significant improvement in performance has been obtained with corner reflector mixer mounts used for radio astronomy at 700 GHz.

Considerable progress has been made in the past few years in the development of monolithic integrated circuits for the microwave, millimeter and submillimeter wave regions of the spectrum. Material quality has improved and work is underway to provide high-quality large-area semi-insulating substrates. More work is still required to provide high-quality selective epitaxy for advanced monolithic integrated circuits. Device fabrication technology is well developed and given low-cost high-quality GaAs wafers there is no reason why monolithic receivers should not dominate the millimeter and submillimeter wave spectrum during the next decade. Integration of a low-noise submicrometer-gate FET IF-amplifier with the monolithic mixer offers the intriguing possibility of a cooled monolithic receiver at 3mm with a total system DSB noise temperature of 100 K or less. The monolithic mixers with appropriate integrated packaging concepts should be scalable to 3 THz. More difficult, but very exciting, is the possibility of building monolithic imaging arrays in the millimeter and submillimeter.

Acknowledgment

The author would like to acknowledge the contributions of many workers who have helped develop receiver technology to its present state. In particular, I wish to thank my colleagues in the Solid State Division at Lincoln Laboratory whose work, comments, and encouragement have contributed significantly to the preparation of this paper. The preparation of this paper was supported by the Department of the Air Force and the U.S. Army Research Office.

References

1. Clifton, B. J., "Schottky Diode Receivers for Operation in the 100-1000 GHz Region," Radio Electron Eng., Vol. 49, pp. 333-346, 1979.
2. Martin, D. H., and Lesurf, J., "Submillimeter Wave Optics", Infrared Physics, Vol. 18, pp. 405-412, 1978.
3. Wrixon, G. T., "Low Noise Diodes and Mixers for the 1-2 mm Wavelength Region," IEEE Trans. Microwave Theory Tech., Vol. MTT-22, pp. 1159-1165, 1974.
4. Kerr, A. R., "Low-noise Room-temperature and Cryogenic Mixers for 80-120 GHz, IEEE Trans. Microwave Theory Tech., Vol. MTT-23, pp. 781-787, 1975.
5. Kerr, A. R., Mattauch, R. J., and Grange, J. A., "A New Mixer Design for 140-220 GHz", IEEE Trans. Microwave Theory Tech. Vol. MTT-25, pp. 399-401, 1977.
6. Vizard, D. R., Keen, N. J., Kelly, W. M., and Wrixon, G. T., "Low Noise Millimeter Wave Schottky Barrier Diodes with Extremely Low Local Oscillator Power Requirements," 1979 IEEE-MTT-S International Microwave Symposium Digest, IEEE Cat. No. 79CH1439-9MTT-S, pp. 81-83, 1979.
7. Keen, N. J., Haas, R. W., and Perchtold, E., "A Very Low Noise Mixer at 115 GHz, using a Mott Diode Cooled to 20K", Electron Lett., Vol. 14, pp. 825-826, 1978.
8. Keen, N. J., Kelly, W. M., and Wrixon, G. T., "Pumped Schottky Diodes with Noise Temperatures of less than 100K at 115 GHz," Electron Lett., Vol. 15, pp. 689-690, 1979.
9. Keen, N. J., "Low-Noise Millimeter-Wave Mixer Diodes: Results and Evaluation of a Test Program," IEE Proc., Vol. 127, Pt.I, pp. 188-198, 1980.
10. Erickson, N. R., "A 0.9 mm Heterodyne Receiver for Astronomical Observations," 1978 IEEE-MTT-S International Microwave Symposium Digest, IEEE Cat. No. 78CH1355-7 MTT, pp. 438-439, 1978.
11. Meier, P. J., "Printed-Circuit Balanced Mixer for the 4- and 5-mm Bands," 1979 IEEE-MTT-S International Microwave Sysmposium Digest, IEEE Cat No. 79CH1439-9MTT-S, pp. 84-86, 1979.
12. Cardiasmenos, A. G., and Parrish, P. T., "A 94 GHz Balanced Mixer using Suspended Substrate Technology", 1979 IEEE-MTT-S International Microwave Symposium Digest, IEEE Cat. No. 79CH1439-9MTT-S, pp. 22-24, 1979.
13. Gustincic, J. J., "A Quasi-optical Radiometer", Second International Conference and Winter School on Submillimeter Waves and their Applications Digest, IEEE Cat. No. 76CH1152-8MTT, pp. 106-107, 1976.
14. Gustincic, J. J., Receiver Design Principles", Proc. Soc. Phot-Opt. Instrum. Engrs., Vol. 105, pp. 40-43, 1977.
15. Gustincic, J. J. "A Quasi-optical Receiver Design", 1977 IEEE-MTT-S International Microwave Symposium Digest, IEEE Cat. No. 77CH1219-5MTT, pp. 99-101, 1977.

16. Gustincic, J. J., DeGraauw, T. R., Hodges, D. T., and Luhmann, Jr. N. C., "Extension of Schottky Diode Receivers into the Submillimeter Region," Unpublished report 1977.

17. Goldsmith, P. F., Erickson, N. R., Fetterman, H. R., Clifton, B. J., Peck, D. D., Tannenwald, P. E., Koepf, G. A., Buhl, D., and McAvoy, N., "Detection of the J=6→5 Transition of Carbon Monoxide," Astrophys. J., Vol. 243, pp. L79-L82, 1981.

18. Fetterman, H. R., Koepf, G. A., Goldsmith, P. F., Clifton, B. J., Buhl, D., Erickson, N. R., Peck, D. D., McAvoy, N., and Tannenwald, P. E., "Submillimeter Heterodyne Detection of Interstellar Carbon Monoxide at 434 Micrometers", Science, Vol. 221, pp. 580-582, 1981.

19. IEEE Tans. Electron Devices, Vol. ED-28, No. 2, 1981.

20. Mehal, E. W., and Wacker, R. W., "GaAs Integrated Microwave Circuits," IEEE Trans. Electron Devices, Vol. ED-15, pp. 513-516, 1968.

21. Mao, S., Jones, S., and Vendelin, G. D., "Millimeter-Wave Integrated Circuits," IEEE Trans. Electron Devices, Vol. ED-15, pp. 517-523, 1968.

22. Chu, A., Courtney, W. E., and Sudbury R. W., "A 31 GHz Monolithic GaAs Mixer/Preamplifier Circuit for Receiver Applications," IEEE Trans. Electron Devices, Vol. ED 28, pp. 149-154, 1981.

23. Clifton, B. J., Alley, G. D., Murphy, R. A., and Mroczkowski, I. H., "High Performance Quasi-Optical GaAs Monolithic Mixer at 110 GHz." IEEE Trans. Electron Devices, Vol. ED-28, p. 155-157, 1981.

24. Clifton, B. J., Alley, G. D., Murphy, R. A., Piacentini, W. J., Mroczkowski, I. H., and Macropoulos, W., "Cooled Low Noise GaAs Monolithic Mixers at 110 GHz," 1981 IEEE-MTT-S International Microwave Symposium Digest, IEEE Cat. No. 81CH1592-5MTT, pp. 444-446, 1981.

25. Geis, M. W., Efremow, N. N., and Lincoln, G. A., "A Novel Dry Etching Technique." (Submitted to the Journal of Vacuum Science & Technology).

26. Geis, M. W. Lincoln, G. A., and Efremow, N. N., "A Novel Dry Etching Technique," Proceedings of the 16th symposium on Electron, Ion, and Photon Beam Technology, 1981.

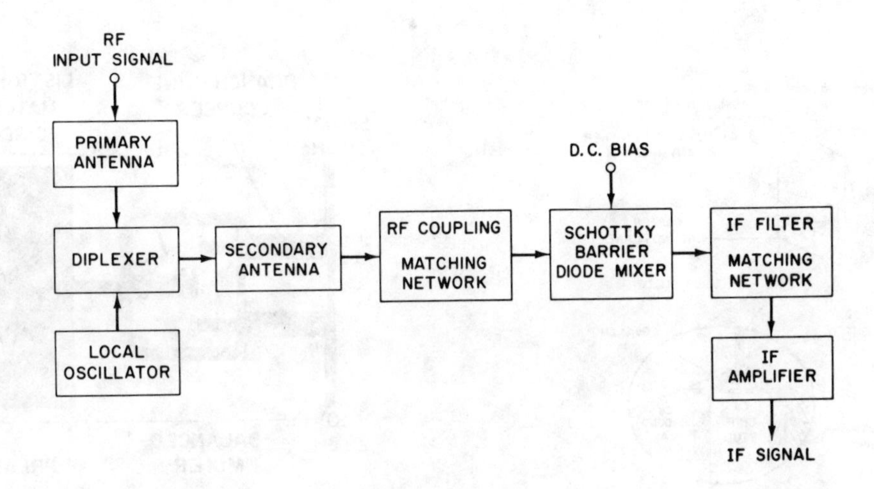

Figure 1. Block diagram of a typical millimeter or submillimeter-wave heterodyne receiver.

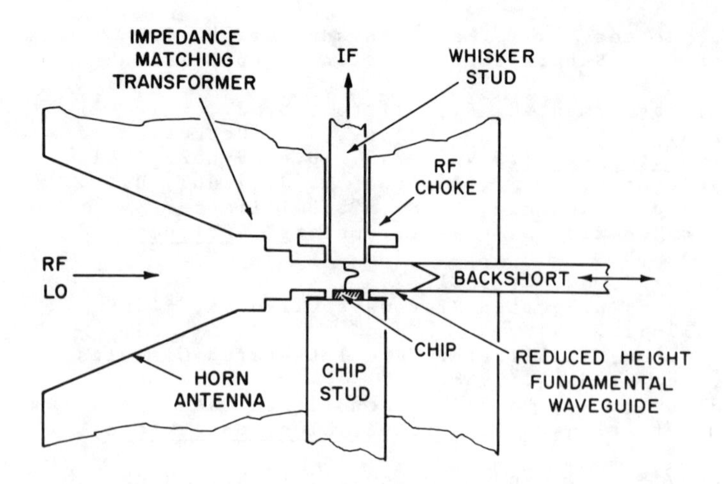

Figure 2. Cross-sectional detail of a typical millimeter wave fundamental waveguide mixer.

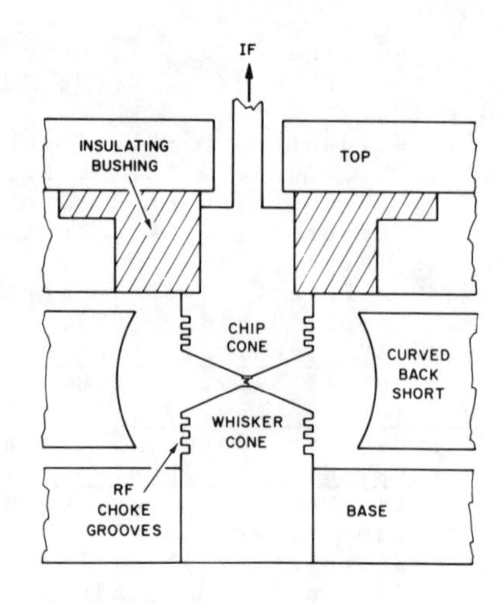

Figure 3. Quasi-optical biconical mixer mount. (Gustincic)

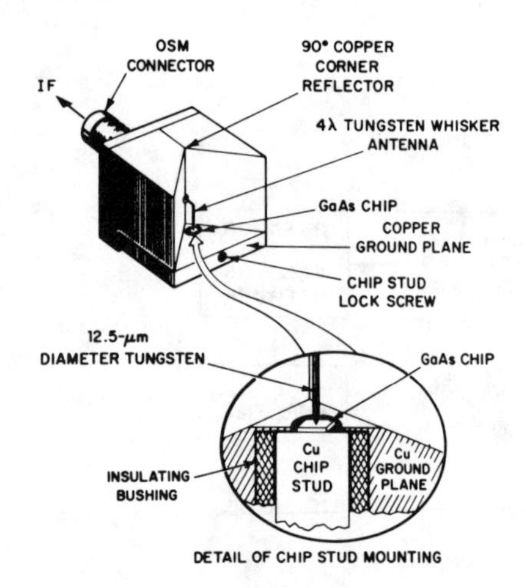

Figure 4. Corner-reflector mixer mount.

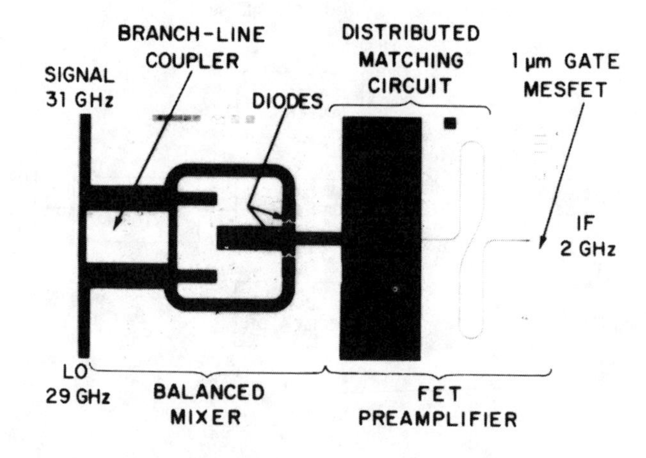

Figure 5. 31-GHz GaAs monolithic receiver showing a balanced mixer and MESFET amplifier. (Chu et al.)

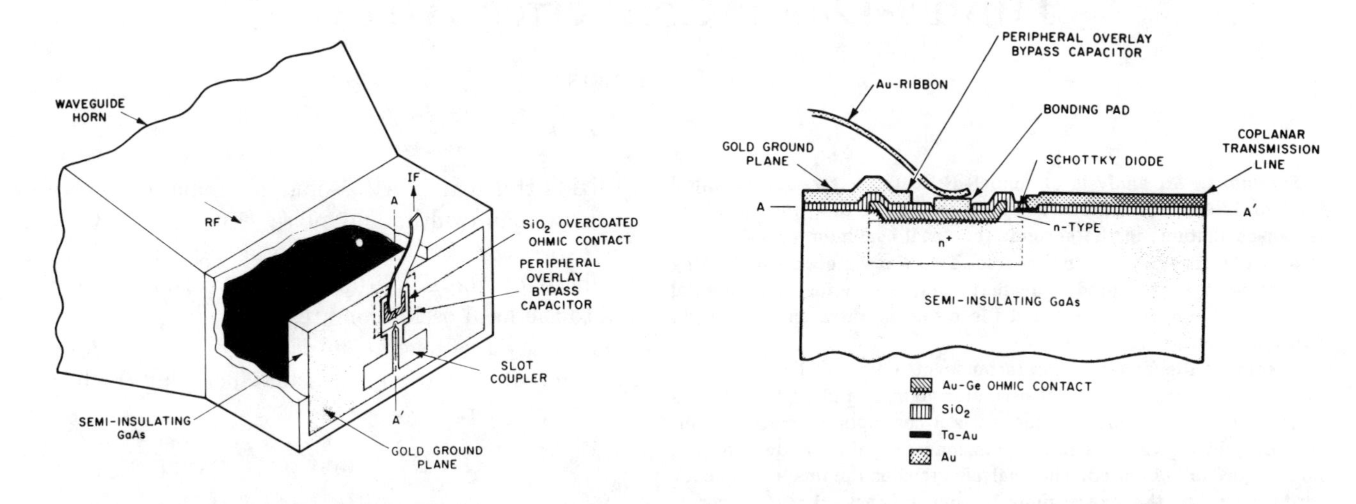

Figure 6. Monolithic integrated-circuit mixer mounted in TE_{10} waveguide horn.

Figure 7. Cross-section through A-A' showing fabrication details.

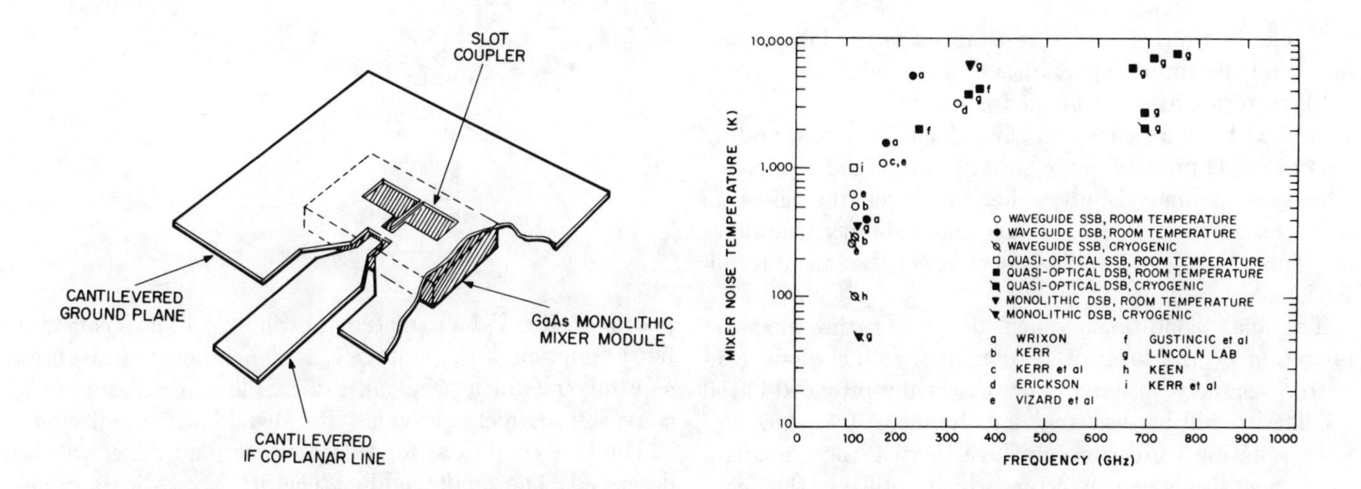

Figure 8. Integrated mixer module for the submillimeter with cantilevered ground plane and IF coplanar line.

Figure 9. State-of-the-art- in low noise receiver technology in the 100-1000 GHz range.

Tunnel-Diode Converter Analysis*

C. S. KIM†, MEMBER, IRE

Summary—**An analysis of tunnel-diode converters is presented for two different conditions of operation. In the first, the converter is self-oscillating; in the second, the local oscillator (LO) voltage is provided from an external source. Two classes of self-oscillating converters are presented. The first operates on the fundamental of the oscillation frequency and the other operates on the second harmonic.**

Assuming the dynamic conductance of a tunnel diode changes only as a function of the local oscillator voltage, and expressing this conductance in a Fourier series, a 3 x 3 conductance matrix is obtained. Utilizing this matrix, expressions for gain, bandwidth, and noise figure are obtained. The analysis includes the image frequency termination. In the expressions for noise figure, the cross-correlation terms, produced by the amplitude change of the dc equivalent shot-noise current due to the LO voltage, are also included.

I. Introduction

Considerable work[1,2] on tunnel-diode converters with both external and internal (self-excited) local oscillation has been carried out. In these works, the problems were very much simplified by introducing assumptions and using special conditions such as:

1) elimination of series resistance, series inductance, and the shunt capacitance of a tunnel diode;
2) no termination at image frequency.

1) and 2) have an effect on gain and noise figure and, in particular, 1) provides large error at very high frequencies. The general analysis given here includes the effect of series resistance, series inductance, and shunt capacitance of a tunnel diode, as well as the effect of the image termination.

The bias conditions which determine the mode of operation of the local oscillation in a tunnel-diode converter, namely internal- or externally-provided local oscillation, will be discussed. Furthermore, two classes of self-oscillating converters are investigated. The first is a fundamental-mode converter which utilizes the conductance change of the diode at the oscillation frequency. The second is the harmonic-mode converter which utilizes the conductance change of the diode at twice the oscillation frequency.

Assuming the tunnel-diode conductance only changes with the local oscillator voltage (*i.e.*, it is independent of the signal voltage) and expressing the conductance in a Fourier series, a 3 × 3 conductance matrix is obtained. Utilizing this matrix with some manipulation, expressions for gain, bandwidth, and noise figure are obtained. In the expression for noise figure, cross-correlation terms produced by the amplitude change of shot-noise current due to the local oscillation are included.

The analysis is general and can be used for both external and internal oscillation converters including the harmonic mode converter.

II. Dynamic Conductance

As a short-circuit stable device,[3] the dynamic conductance g of a tunnel diode has the characteristic shown in Fig. 1, where g is negative between the peak and valley

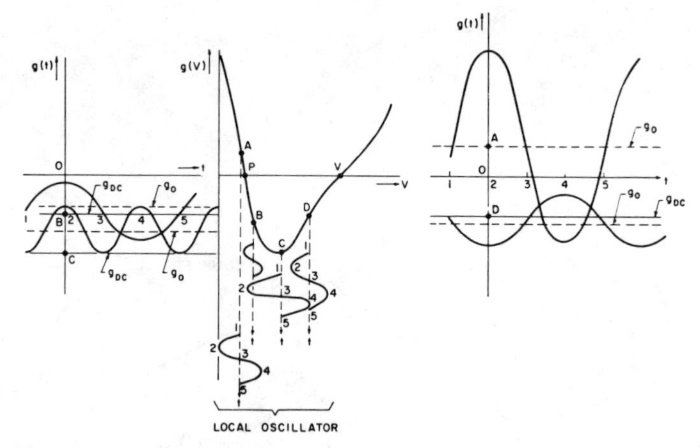

Fig. 1—Various bias points.

points of the V-I characteristic. Since g is not constant but a function of the voltage v, a tunnel diode can be used as a mixer (which needs an external local oscillator) or a converter (which provides its own local oscillation).

The bias conditions for the converter and mixer will be described. The diode conductance $g(t)$ varies instantaneously with the local oscillator voltage and has an average value g_0. To prevent oscillation regardless of termination, g must be positive. This condition is fulfilled by bias point A. On the other hand, in a self-excited converter, g_0 must be negative in order to provide oscillation at the local oscillator frequency. The possible bias points for the self-excited mode are several (B, C, and D).

It is assumed that the tunnel diode's dynamic conductance $g(t)$ varies only with the local oscillation voltage and is independent of the signal; then the conductance variation as a function of the local oscillation voltage for various bias points is given in Fig. 1. Expressing the

* Received by the PGED, February 7, 1961; revised manuscript received, April 26, 1961.

† Electronics Lab., General Electric Co., Syracuse, N. Y.

[1] K. K. Chang, *et al.*, "Low-noise tunnel-diode down converter having conversion gain," Proc. IRE, vol. 48, pp. 854–858; May, 1960.

[2] D. I. Breitzer, "Noise figure of tunnel diode mixer," Proc. IRE (Correspondence), vol. 48, pp. 935–936; May, 1960.

[3] H. W. Bode, "Network Analysis and Feedback Amplifier Design," D. Van Nostrand Co., Inc., New York, N. Y.

Reprinted from *IRE Trans. Electron Devices,* vol. ED-8, pp. 394–405, Sept. 1961.

conductance $g(t)$ in terms of a Fourier series yields

$$g(t) = \sum_{n=-\infty}^{n=\infty} g_n e^{jn\omega_0 t}, \tag{1}$$

where ω_0 is the angular frequency of the local oscillator. The time reference is placed such that $g(t)$ has mirror symmetry with respect to time at $t = 0$. g_n then becomes real and

$$g_n = g_{-n} \tag{2}$$

as seen from Fig. 1. It should be noted that the first term g_0 of $g(t)$ of (1) is positive for the external local oscillator mode (bias point A) and is negative for the internal local oscillator mode (bias points B, C, and D). The various values of g_0 are different from the corresponding values of conductance g_{DC} at various bias points as shown in Fig. 1. Furthermore, if the reference is chosen such that g_1 of the bias D point is made a negative real value, then g_1 at bias points A and B or g_2 at bias points C become a positive real value.

For the fundamental mode of operation, the bias point should be placed in such a region that g is nearly linear. This is the case for points A, B, and D. On the other hand, for the second-harmonic mode, g should be a square function of the local oscillator voltage as in the region of the point C.

As will be seen later, for a large gain and small noise figure it is desirable to make g_1 and g_2 large for the fundamental and harmonic modes, respectively. However, there is a limit for g_1 and g_2 as seen from Fig. 1. In the external local oscillator mode, the maximum voltage is confined to the A-C range since a local-oscillator voltage larger than the voltage difference between point C and A will not increase g_1. While in the internal local oscillator modes, the operation is more or less limited to the range P-C for the bias point B, to the range C-V for the bias point D, and to the range P-V for the bias point C. These limits result from the requirement for self-oscillation. If the oscillation frequency is low, it is possible to extend the local oscillator voltage into the positive region of g, and the local oscillation becomes a relaxation oscillation. However, for higher frequencies, the oscillation becomes more or less sinusoidal.

The slope of g in the region A-C is steeper than the slope of region C-V. Therefore, a smaller value of local oscillator voltage is needed in region A-C for the same magnitude of g_1. Since it is more difficult to obtain the larger voltage at high frequencies, the operation at point D is restricted to lower frequencies.

III. Circuit Requirements

A. External Local Oscillator Mode

The operational principal of a tunnel-diode mixer (external local oscillator) is similar to that of an ordinary diode mixer. The main differences are:

1) the conversion gain of a tunnel-diode mixer can be greater than unity,

2) the conversion gain of an ordinary diode mixer is always smaller than unity.

The above difference results from the fact that

1) in a tunnel-diode mixer, $|g_1|$ can be made larger than g_0

$$|g_1| > g_0, \tag{3}$$

2) in an ordinary mixer, g_1 is always smaller than g_0

$$g_1 < g_0. \tag{4}$$

B. Internal Local Oscillator Modes

The operation of a self-excited tunnel-diode converter is more complicated than that of a tunnel-diode mixer. In a self-excited converter, the following functions must be performed simultaneously in a single diode:

1) oscillation at the desired local oscillator frequency
2) amplification at RF
3) mixing
4) amplification at IF.

In order to satisfy the preceding conditions, it is necessary to provide a circuit that presents the proper admittance Y across the diode[4] at each of the three frequencies involved. For the generalized tunnel-diode circuit shown in Fig. 2, the conditions for oscillation are as follows:

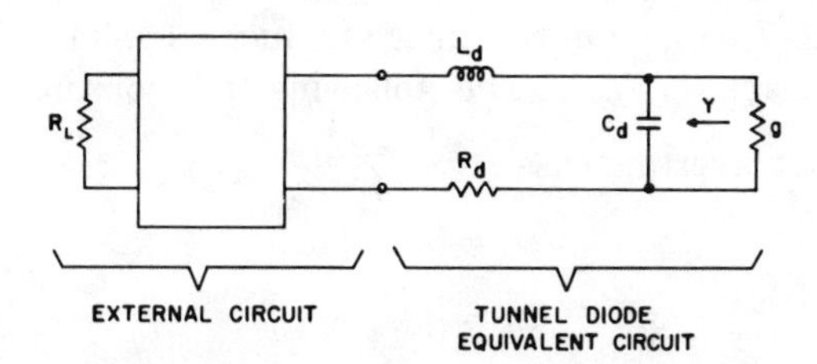

Fig. 2—Tunnel diode with general external circuit.

1) $Re\{Y\} \leq |g|$ at the oscillation frequency,

2) $Im\{Y\} = 0$ at the oscillation frequency. (6)

If these conditions are not satisfied, the circuit will not oscillate. Conversely, stability can be achieved by the following combinations of conditions:

1) $Re\{Y\} > |g_0|$ over entire frequency range or

2) In the frequency range where $Re\{Y\} \leq g$, $Im\{Y\} \neq 0$.

For the converter, then, the following conditions are desirable:

$$\left.\begin{aligned} 1)\ Re\{Y\} &\leq |g| \\ Im\{Y\} &= 0 \end{aligned}\right\} \text{ at local oscillator frequency,} \tag{7}$$

[4] "Across the diode" implies here across the negative-conductance element of the diode, as opposed to across the diode's external terminals which include parasitic elements.

2) $Re\{Y\} > |g_0|$, $Im\{Y\} = 0$ at RF,

3) $Re\{Y\} > |g_0|$, $Im\{Y\} = 0$ at IF.

The frequency relationship for fundamental-mode and harmonic-mode converters are given by:

for fundamental mode, $\omega_s = \omega_0 \pm \omega_i$

for harmonic mode, $\omega_s = n\omega_0 \pm \omega_i$ (8)

+ for noninverting case
− for inverting case
$n = 2, 3, 4, \cdots,$

where

ω_s = RF signal angular frequency
ω_0 = LO angular frequency
ω_i = IF angular frequency.

IV. Equivalent Circuit

When two ac sources (*i.e.*, a local oscillator voltage and the RF signal) drive a mixer or converter, four different voltages and currents of interest appear in the circuit, provided that the appropriate tank circuits are present. These voltages and currents are the local oscillator voltage $V_0(\omega_0)$ and current $I_0(\omega_0)$, the RF signal voltage $V_s(\omega_s)$ and current $I_s(\omega_s)$, the IF signal voltage $V_i(\omega_i)$ and current $I_i(\omega_i)$, and the image frequency voltage $V_k(\omega_k)$ and current $I_k(\omega_k)$. The following relationship exists:

1) noninverting case

$$\begin{aligned} \omega_s &= n\omega_0 + \omega_i \\ \omega_k &= n\omega_0 - \omega_i \end{aligned} \tag{9}$$

2) inverting case

$$\begin{aligned} \omega_s &= n\omega_0 - \omega_i \\ \omega_k &= n\omega_0 + \omega_i \end{aligned} \tag{10}$$

where $n = 1, 2, 3, \cdots$.

The local oscillator voltage in a tunnel-diode converter may be provided by self-oscillation or by an external pump. The analysis presented here holds for both cases. If the local oscillator voltage is large compared with the RF input signal, the instantaneous dynamic conductance g is considered as a function of this voltage only, as shown in (1) and (2). Consider the admittance $Y(\omega)$ facing the negative conductance of the diode. This admittance includes all the parasitic elements of the diode structure. If $Y(\omega)$ is expressed by $Y_s(\omega_s)$, $Y_i(\omega_i)$ and $Y_k(\omega_k)$ at ω_s, ω_i and ω_k, respectively, Fig. 2 can be represented by Fig. 3. The dynamic conductance g is expressed in a Fourier series in terms of the local oscillator voltage, and the termination at ω_0 is not required. The impedances $Z_s''(\omega_s)$, $Z_i''(\omega_i)$, and $Z_k''(\omega_k)$ are the actual external terminal impedances across the tunnel diode at ω_s, ω_i and ω_k, respectively.

Expressing Fig. 3 in the form of Fig. 4, the following relationships are obtained:

$$\left.\begin{aligned} Y_s(\omega_s) &= Y_s' + j\omega_s C_d \\ Y_i(\omega_i) &= Y_i' + j\omega_i C_d \\ Y_k(\omega_k) &= Y_k' + j\omega_k C_d \\ I_g(\omega_s) &= E_g Y_s'(\omega_s) \end{aligned}\right\}, \tag{11}$$

where

$$Y_s'(\omega_s) = \frac{1}{Z_s'' + j\omega_s L_d + r_g + R_d}$$

$$Y_k'(\omega_i) = \frac{1}{Z_i'' + j\omega_i L_d + R_d}$$

$$Y_k'(\omega_k) = \frac{1}{Z_k'' + j\omega_k L_d + R_d}.$$

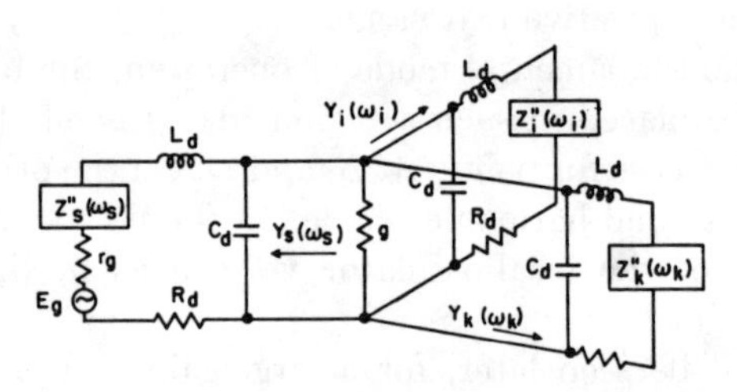

Fig. 3—Equivalent circuit of tunnel-diode converter with actual terminations.

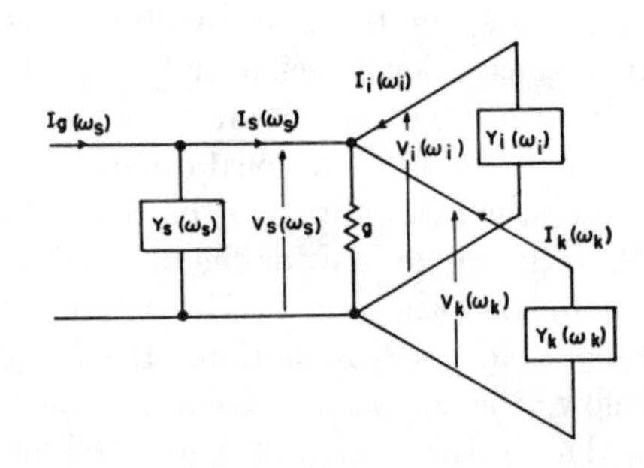

Fig. 4—Final equivalent circuit of tunnel-diode converter.

The immittance transformation given in (11) makes it possible to use a 3 × 3 matrix of g with simple terminations. Furthermore, it will give simple expressions for gain and noise figure compared with other methods.

The total voltage across the conductance g is

$$v_T(t) = v_s + v_i + v_k, \tag{12}$$

and the current through the conductance g is

$$i_T(t) = i_s + i_i + i_k. \tag{13}$$

Expressing v_T and i_T as complex variables, and knowing the relationship of

$$i_T = g(t)v_T, \tag{14}$$

and (2) and (3), the following matrix relationships are obtained:

$$\begin{bmatrix} I_g \\ 0 \\ 0 \end{bmatrix} = [Y] \begin{bmatrix} V_s \\ V_i \\ V_k^* \end{bmatrix} \quad \text{(noninverting case)}, \tag{15}$$

where I_g is the equivalent current source at RF termination

$$[Y] = \begin{bmatrix} Y_s + g_0 & g_1 & g_2 \\ g_1 & Y_i + g_0 & g_1 \\ g_2 & g_1 & Y_k^* + g_0 \end{bmatrix} \tag{16}$$

and

$$\begin{bmatrix} I_g \\ 0 \\ 0 \end{bmatrix} = [Y] \begin{bmatrix} V_s \\ V_i^* \\ V_k^* \end{bmatrix} \quad \text{(inverting case)}, \tag{17}$$

where

$$[Y] = \begin{bmatrix} Y_s + g_0 & g_1 & g_2 \\ g_1 & Y_i^* + g_0 & g_1 \\ g_2 & g_1 & Y_k^* + g_0 \end{bmatrix}. \tag{18}$$

If g_1 and g_2 of the fundamental mode are replaced by g_2 and g_4, respectively, the above relationships are satisfied for the second-harmonic mode.

V. Gain and Bandwidth

A. Gain and Bandwidth

The expressions for the voltage and current relationships of (15) and (17) are the same for the fundamental- and second-harmonic modes, except that g_1 and g_2 of the fundamental mode are replaced by g_2 and g_4 of the second-harmonic mode, respectively. Therefore, only the fundamental mode expressions will be developed.

Solving (15) for V_i, the following expression is obtained:

$$V_i(\omega_i) = \frac{-g_1(Y_k^* + g_0 - g_2)I_g}{(Y_s + g_0)(Y_i + g_0)(Y_k^* + g_0) - (Y_s^* + Y_k^* + 2g_0 - 2g_2)g_1^2 - (Y_i + g_0)g_2^2} \tag{19}$$

$$V_i(\omega_i)/I_g = H_{is}$$

for noninverting case.

For the inverting case, $V_i(\omega_i)$ and Y_i in (19) should be replaced by $V_i^*(\omega_i)$ and $Y_i^*(\omega_i)$, respectively.

Referring to (11) and Figs. 4 and 5, the available power gain P_{Ga} is given by

$$P_{Ga} = \frac{|V_i|^2 g_i}{|E_g|^2/4r_g} = \frac{|V_i|^2 g_i |\gamma|^2}{|I_g|^2/4g_g}, \tag{20}$$

where

E_g = voltage of the signal generator
r_g = $1/g_g$ = internal resistance of the generator.
γ = Y_s'/g_g

Here it is assumed that g_i is the load conductance (since the effect of R_d and $\omega_i L_1$ are negligible at ω_i). Assume the signal angular frequency ω_s varies an increment $\Delta\omega$ from its center angular frequency ω_{s0},

$$\omega_s = \omega_{s0} + \Delta\omega. \tag{21}$$

The corresponding angular frequency change in ω_i and ω_k can be expressed by

$$\omega_i = \omega_{i0} \pm \Delta\omega$$
$$\omega_k = \omega_{k0} - (\pm\Delta\omega) \tag{22}$$

+ for noninverting case
− for inverting case.

The bandwidth of the converter can be defined as the frequency range $B = f_2 - f_1$ where at f_2 and f_1

$$\frac{P_{Ga}}{[P_{Ga}]_{\Delta\omega=0}} = \tfrac{1}{2}. \tag{23}$$

B. Three-Tuned Circuits

Referring to Fig. 5 for three-tuned circuits, assume that

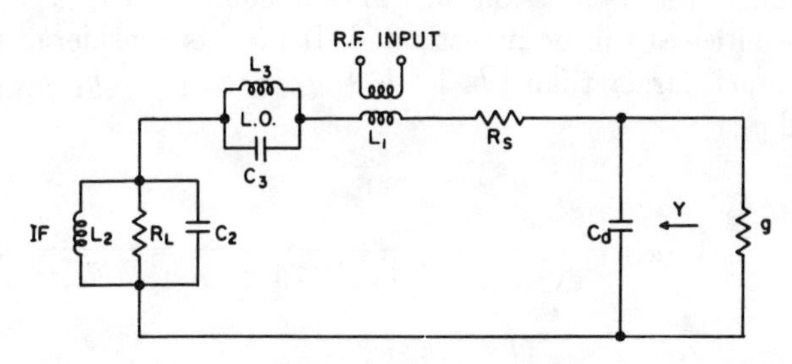

Fig. 5—Tunnel-diode converter with three-tuned circuits.

C_2 and C_3 are short circuits at ω_s, C_2 and L_1 are short circuits at ω_0, and L_1 and L_3 are short circuits at ω_i. Furthermore, C_d has negligible effect at ω_i and ω_0. If the Q of the converter circuit at various resonance frequencies is considered to be relatively high, it is possible to express the terminal admittances as follows:

$$Y_s = g_s + jb_s \doteqdot g_s + j\left(\omega_s C_d - \frac{1}{\omega_s L_1}\right) \tag{24}$$
$$= g_s(1 + j2\,\delta_s Q_s),$$

where

$$\delta_s = \frac{\Delta\omega}{\omega_{s0}}; \quad Q_s = \frac{\omega_{s0} C_d}{g_s}; \quad \omega_{s0}^2 = \frac{1}{L_1 C_d} \tag{25}$$

$$Y_i = g_i + jb_s \doteqdot g_i + j\left(\omega_i C_2 - \frac{1}{\omega_i L_2}\right)$$
$$= g_s(1 + j2\,\delta_i Q_i),$$

where

$$\delta_i = \frac{\Delta\omega}{\omega_{i0}}; \qquad Q_i = \frac{\omega_{i0}C_2}{g_i}; \qquad \omega_{s0}^2 = \frac{1}{L_2C_2}$$

$$Y_k = g_k - jb_k. \tag{26}$$

Substituting these equations into (19) gives

$$V_i(\omega_i) = \frac{-g_1 I_g}{\dfrac{(\bar{g}_s - g_2)(\bar{g}_k - g_2) + b_s b_k}{(\bar{g}_k - g_2)^2 + b_k^2}(\bar{g}_i g_2 - g_1^2) + \dfrac{b_i g_2\{b_k(\bar{g}_s - g_2) - b_s(\bar{g}_k - g_2)\}}{(\bar{g}_k - g_2)^2 + b_k^2}}$$
$$+ \bar{g}_s\bar{g}_i - b_s b_i - g_1^2 + j\left\{\frac{b_s(\bar{g}_k - g_2) - b_k(\bar{g}_s - g_2)}{(\bar{g}_k - g_2)^2 + b_k^2}(\bar{g}_i g_2 - g_1^2)\right.$$
$$\left. + \frac{[(\bar{g}_s - g_2)(\bar{g}_k - g_2) + b_s b_k]}{(\bar{g}_k - g_2)^2 + b_k^2} b_i g_2 + b_s\bar{g}_i + b_i\bar{g}_s\right\}, \tag{27}$$

where

$$g_s + g_0 = \bar{g}_s$$
$$g_k + g_0 = \bar{g}_k$$
$$g_i + g_0 = \bar{g}_i.$$

Since the expression of (27) is complicated, a few special cases will be investigated. If $|b_k|$ is considered to be much larger than $|b_s|$, $|b_i|$, $\bar{g}_s - g_2$, $\bar{g}_k - g_2$, $\bar{g}_i$, g_1 and g_2,

$$V_i(\omega_i) = \frac{-g_1 I_g}{(Y_i + g_0)(Y_s + g_0) - g_1^2} = \frac{-g_1 I_g}{\bar{g}_s\bar{g}_i - g_1^2 + 4g_i g_s\,\delta_i^2 Q_i Q_s \omega_i/\omega_s + j2\,\delta_i(\bar{g}_i g_s Q_s \omega_i/\omega_s + \bar{g}_s g_i Q_i)}. \tag{28}$$

Referring to Fig. 5 where, in order to make $b_s = Im\{Y_s\} = 0$ at ω_{s0}, $Im\{Y_s'\}$ must be negative. Referring to (11), this implies that

$$L_d + \frac{Im\{Z_s''(\omega_s)\}}{j\omega_s} = L_1 > 0. \tag{29}$$

In this case, if

$$\frac{\Delta\omega}{\omega_s} \ll 1, \tag{30}$$

it can be assumed that

$$Y_s'(\omega_s) = Y_s'(\omega_{s0}). \tag{31}$$

Furthermore, according to (20), if

$$(R_d + r_g)^2 \ll \omega_s^2 L_1^2,$$

then

$$|\gamma|^2 = \frac{1}{\omega_s^2 L_1^2 g_g^2} \doteqdot \frac{C_d}{L_1} r_g^2 \doteqdot \frac{g_s}{g_g}. \tag{32}$$

The available gain for this case can be expressed for both noninverting and inverting case by

$$P_{Ga} = \frac{4g_s g_i g_1^2}{(\bar{g}_s\bar{g}_i - g_1^2 + 4g_i g_s\,\delta_i^2 Q_i Q_s \omega_i/\omega_s)^2 + 4\,\delta_i^2(\bar{g}_i g_s Q_s \omega_i/\omega_s + \bar{g}_s g_i Q_i)^2}. \tag{33}$$

The available gain at the center frequency, *i.e.*, $\Delta\omega = 0$ or $\delta_i = 0$, is given by

$$P_{Ga}]_{\Delta\omega=0} = \frac{4g_s g_i g_1^2}{(\bar{g}_s\bar{g}_i - g_1^2)^2}. \tag{34}$$

As mentioned in Section II, in order to make $P_G]_{\Delta\omega=0}$ large, it is desirable to make $|g_1|$ large for a given value of $(\bar{g}_s\bar{g}_i - g_1^2)^2$. Of course, by making $\bar{g}_s\bar{g}_i - g_1^2$ small, further $P_G]_{\Delta\omega=0}$ can be increased. Small $\bar{g}_s = g_s + g_0$, and $\bar{g}_i = g_i + g_0$ will make the circuit less stable and this should be avoided.

In the case of external local oscillation, $\bar{g}_s = g_s + g_0$ and $\bar{g}_i = g_i + g_0$ can not be made small since g_0 is positive. Therefore, in order to make $P_G|_{\Delta\omega=0}$ large, g_1 must be made large so that $\bar{g}_s\bar{g}_i - g_1^2$ becomes small.

The stability condition for (34) can be given as follows:

1) $g_s + g_0 > 0$
2) $g_i + g_0 > 0$
3) $(g_s + g_0)(g_i + g_0) - g_1^2 > 0.$

The normalized bandwidth $B_n = 2\delta_i$, where δ_i is obtained by solving (23) for this case, can be expressed by

$$B_n = \sqrt{\frac{b}{a}\left(\sqrt{1 + \frac{a}{b^2}}\right) - 1}, \tag{35}$$

where

$$\sqrt{a} = \frac{4g_i g_s Q_i Q_s \omega_i/\omega_s}{\bar{g}_s\bar{g}_i - g_1^2}$$

$$b = \sqrt{a} + 2\left(\frac{\bar{g}_i\bar{g}_s Q_s \omega_i/\omega_s + \bar{g}_s\bar{g}_i Q_i}{\bar{g}_s\bar{g}_i - g_1^2}\right)^2.$$

It should be noted that as $\bar{g}_s\bar{g}_i - g_1^2$ decreases, the gain of (34) increases, and the normalized bandwidth of (35) decreases.

C. Two-Tuned Circuits

If the difference between f_s and f_0 is small, referring to Fig. 6, the following conditions can be satisfied[5]

$$\bar{g}_s^2 = (g_s + g_0)^2 \ll b_k^2 = b_s^2$$
$$\bar{g}_k^2 = (g_k + g_0)^2 \ll b_k^2 = b_s^2. \tag{36}$$

Since the second harmonic is small for the fundamental mode, the effect of g_2 is smaller than that of $b_s = b_k$. Eq. (27) can be then be simplified

$$V_i(\omega_i) = \frac{-g_1 I_g}{\bar{g}_i(\bar{g}_s + g_2) - b_i\left\{g_2(g_k - g_s)\frac{1}{b_s} + b_s\right\} - 2g_1^2 + j\left\{b_i(\bar{g}_s + g_2) + (g_k - g_s)(\bar{g}_i g_2 - g_1^2)\frac{1}{b_s} + b_s \bar{g}_i\right\}}. \tag{37}$$

The expression of (37) is still rather complicated. Inspecting (37), it is desirable to make the imaginary part of $V_i(\omega_i)$ zero for $\Delta\omega = 0$ in order to increase the gain; the factor b_i is zero, but $b_s(\omega_s)$ is not zero. However, usually $\Delta\omega/\omega_s \ll 1$ and $b_s(\omega_s)$ can be considered to be independent of $\Delta\omega$ changes, *i.e.*,

$$b_s(\omega_s) = b_s(\omega_{s0}). \tag{38}$$

The imaginary term will then be

$$b_s^2 = \frac{(g_1^2 - \bar{g}_i g_2)(g_k - g_s)}{\bar{g}_i}. \tag{39}$$

Furthermore, one can assume from the previous approximation that

$$g_2(g_k - g_s) \ll b_s^2. \tag{40}$$

Then (37) can be reduced to

$$V_i(\omega_i) = \frac{-g_1 I_g}{\bar{g}_i(\bar{g}_s + g_2) - b_i b_s - 2g_1^2 + j b_i(\bar{g}_s + g_2)}. \tag{41}$$

Of course, b_s in (41) must satisfy the condition in (39) for $\Delta\omega = 0$. Using (41), the available gain[6] can be expressed for both the noninverting and the inverting case by

$$P_{Ga} = \frac{4 g_s g_i g_1^2}{\{\bar{g}_i(\bar{g}_s + g_2) - 2g_1^2 - 2g_i \delta_i Q_i b_s\}^2 + 4\delta_i^2 Q_i^2(\bar{g}_s + g_2)^2 g_i^2}. \tag{42}$$

The available gain $P_{Ga}]_{\Delta\omega=0}$ at the center frequency is given by

$$P_{Ga}]_{\Delta\omega=0} = \frac{4 g_s g_i g_1^2}{\{\bar{g}_i(\bar{g}_s + g_2) - 2g_1^2\}^2}. \tag{43}$$

Solving $P_{Ga}/P_{Ga}]_{\Delta\omega=0} = \frac{1}{2}$ for δ_i gives

$$\delta_i = a \pm \sqrt{2a^2}, \tag{44}$$

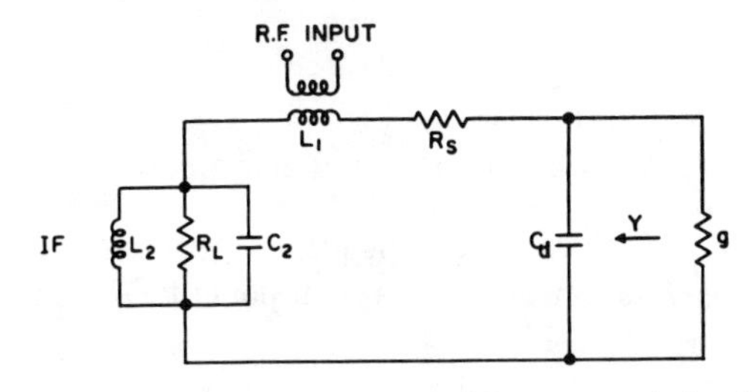

Fig. 6—Tunnel-diode converter with two-tuned circuits.

where

$$a = \frac{b_s[\bar{g}_i(\bar{g}_s + g_2) - 2g_1^2]}{2Q_i g_i[b_s^2 + (\bar{g}_s + g_2)^2]} \doteqdot \frac{\bar{g}_i(\bar{g}_s + g_2) - 2g_1^2}{2Q_i g_i b_s}. \tag{45}$$

It should be noted that δ_i which satisfies (45) has two values, δ_{i1} and δ_{i2}. This results from the symmetry term $2\, g_i \delta_i Q_i b_s$ in (42). The normalized bandwidth B_n is then defined[7] by

$$B_n = |\,\delta_{i2} - \delta_{i1}\,| = 2\sqrt{2a^2}. \tag{46}$$

VI. Noise Figure

A. Noise Figure Definition

The noise figure can be defined as follows:[8]

$$F = \frac{N_0/S_0}{N_i/S_i} = 1 + \frac{N_e}{P_G N_i}, \tag{47}$$

where

N_0 = actual noise power delivered at the output
S_0 = actual signal power delivered at the output
N_i = actual noise power delivered to the input
S_i = actual signal power delivered to the input
N_e = excess noise power (noise power delivered at the output due to internal sources in the network)
$P_G = S_0/S_i$ = actual power gain.

This definition is frequently difficult to apply to negative-resistance devices since the power delivered to the input can not easily be determined. It can be shown that the difficulty is easily overcome by making use of the following equality

$$P_{Ga} \cdot N_a = P_G \cdot N_i, \tag{48}$$

[5] Note that the imaginary part of Y_s and Y_k have opposite signs.
[6] Here, too, it is assumed that (32) is satisfied.

[7] In (45), b_s is positive for the inverting converter and negative for the noninverting converter. In order to make B_n always positive, A is used in (46).
[8] C. S. Kim, "Four terminal equivalent circuits of parametric and diodes," 1959 IRE WESCON Convention Record, pt. 2, pp. 91–101.

where

P_{Ga} = available power gain
$N_a = KT\Delta f$ = available input noise power.

Here S_i and N_i are always positive even when the input immittance has a negative real component. Making use of the above relationship

$$F = 1 + \frac{N_e}{P_G N_i} = 1 + \frac{N_e}{P_{Ga} N_a}. \tag{49}$$

B. Excess Noise

There are two sources of excess noise in tunnel-diode networks, amplifiers and converters. One results from the shot noise[9,10] generated by tunnel currents, and the other results from thermal noise produced by the internal series resistance.

There are two tunneling currents, the forward and backward. The dc current of a tunnel diode is the arithmetic sum of two tunneling currents which flow in opposite directions. However, the shot noise generated by the two currents does not necessarily cancel, but usually adds on, and rms basis since the current flows are usually uncorrelated. The total shot-noise current i_s can then be expressed by

$$\overline{i_s^2} = 2e(I_f + |I_b|)\,\Delta f = 2eI_{eg}\,\Delta f, \tag{50}$$

where

I_f = forward tunnel current
I_b = backward tunnel current
I_{eg} = equivalent dc current.

It can be shown that with zero bias and when thermal equilibrium is maintained, $\overline{i_s^2}$ is equivalent to the thermal noise $\overline{i_T^2}$[9] expressed by $\overline{i_s^2} = \overline{i_T^2} = 4kTG_{eg}\Delta f$, where

$$\overline{i_s^2} = 2eI_{eg}\,\Delta f = 4eI_f\,\Delta f = 4e\,|I_b|\,\Delta f$$
$$|I_f| = |I_b|, \qquad G_{eg} = \left.\frac{\partial I}{\partial V}\right|_{v=0}. \tag{51}$$

In the negative-conductance region, I_b is almost negligible compared with I_f, and the dc current I_{dc} and I_{eq} can be considered to be equal.

$$\overline{i_s^2} = 2eI_{dc}\,\Delta f. \tag{52}$$

As seen from (51) at the origin of the I-V curve, $\overline{i_T^2} = \overline{i_s^2}$ must be zero at $T = 0°\text{K}$, and I_f and I_b become zero. This indicates that I_{eq} is temperature dependent. This can be understood from the fact that both I_f and I_b are functions of the Fermi distribution, and hence are temperature sensitive. Typical curves of I_{eq} for room temperature and a lower temperature are given in Fig. 7.

[9] J. J. Tiemann, "Shot noise in tunnel diode amplifiers," Proc. IRE, vol. 48, pp. 1418–1423; August, 1960.

[10] E. G. Nielsen, "Noise in tunnel diodes circuits," *Proc. Natl. Electronics Conf.*, Chicago, Ill., October, 1960, pp. 785–790.

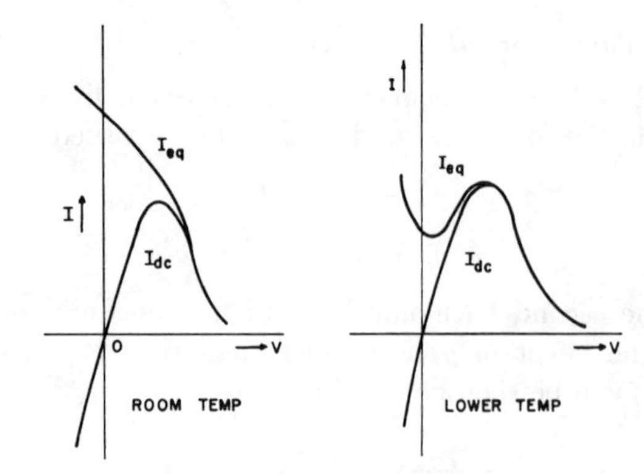

Fig. 7—Tunnel-diode characteristics and equivalent shot-noise dc current at different temperatures.

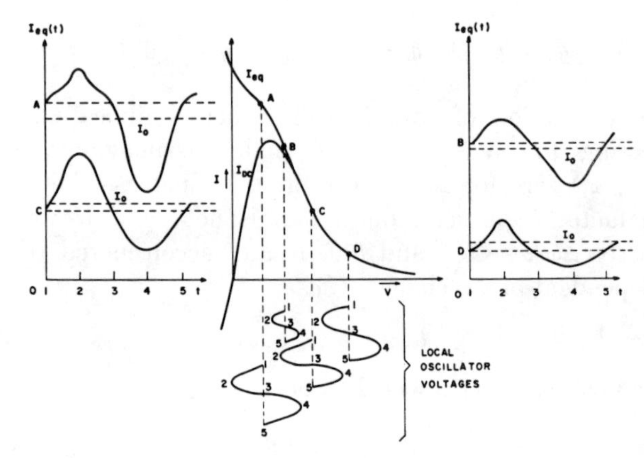

Fig. 8—Equivalent shot-noise current variation at various bias points.

The thermal noise produced by the internal series resistance can be expressed by the usual equation

$$\overline{e_t^2} = 4kT_dR_d\,\Delta f, \tag{53}$$

and it is assumed to be uncorrelated with $\overline{i_s^2}$.

C. Variations of dc Equivalent Shot-Noise Current

The dc equivalent shot-noise current I_{eq} shown in Fig. 8 is fixed for a given bias voltage. However, in a converter circuit, since a large local oscillator voltage is superimposed on the dc bias, I_{eq} varies with the local oscillator voltage. Fig. 8 shows the I_{eq} variations for the four bias points corresponding to Fig. 1. The I_{eq} characteristics are at room temperature. With the same reference used in Fig. 1, $I_{eq}(t)$ can be expressed as a Fourier series as follows:

$$I_{eq}(t) = I_0 + 2I_1 \cos \omega_0 t + 2I_2 \cos 2\omega_0 t + \cdots. \tag{54}$$

Of course, the average shot-noise equivalent current I_0 is positive, and the value of I_0 decreases in the order of bias points A, B, C, and D.[11] Therefore, the bias point A produces more excess noise than B, etc. As will be seen later, the shot-noise contribution in the noise figure expression not only comes from the average shot-noise equivalent current I_0, but also from the fundamental

[11] In Chang,[1] the dc equivalent shot-noise current at point A is defined to be equal to I_{dc}. This is in error; it should be I_0 of (54).

and harmonic terms I_1 and I_2. These contributions are brought out by correlation terms.

By reducing the diode's temperature, the magnitude of I_{eq} in the region of the origin and bias point A is reduced. This reduces the corresponding value of I_0 for bias point A. However, as shown in Fig. 7, I_{eq} is not decreased in the negative-conductance region. This implies that the reduction of the shot-noise contribution by cooling the diode cannot be achieved for the operation at the bias points B, C, and D.

D. Shot Noise in Tunnel-Diode Converters or Mixer

The shot-noise effect due to the periodic change of the dc equivalent shot-noise current can be analyzed using a method similar to that of Strutt.[12]

It will be shown in the Appendix that all frequencies except ω_s, ω_k, and ω_i are removed by the filter. As a result, the instantaneous shot-noise current $i_n(t)$ will be composed of three components, namely the RF noise current i_{ns}, the IF noise current i_{ni}, and the image frequency noise current i_{nk}. Then, the shot-noise equivalent-converter circuit can be represented by Fig. 9.

If IF voltages across the load due to i_{ns}, i_{ni}, and i_{nk} can be given by $h_{is}i_{ns}$, $h_{ii}i_{ni}$ and $h_{ik}i_{nk}$, where $h_{is}(t)$, $h_{ii}(t)$ and $h_{ik}(t)$ are the respective transfer or impedance functions, then the total IF noise voltage $v_{ni}(t)$ can be expressed by

$$v_{ni}(t) = h_{is}i_{ns} + h_{ii}i_{ni} + h_{ik}i_{nk}. \tag{55}$$

It is shown in the Appendix that the time average of $V_{ni}^2(t)$ can be expressed by the following relation from a knowledge of the correlation properties of phase angles of i_{ns}, i_{ni}, i_{nk}, and (54):

$$\begin{aligned}\overline{V_{ni}^2}(t) = 2e\,\Delta f[(|H_{is}|^2 + |H_{ii}|^2 + |H_{ik}|^2)\,I_0 \\ + \{H_{ii}(H_{is}^* + H_{ik}^*) + H_{ii}^*(H_{is} + H_{ik})\}I_1 \\ + (H_{is}H_{ik}^* + H_{is}^*H_{ik})I_2]\end{aligned} \tag{56}$$

where H_{is} is the Laplace transform of h_{is} and the asterisk represents the complex conjugate.

The above expression holds for fundamental mode operation. The first term of (56) is noise-power contributions due to I_0. The second term represents noise correlation between RF and IF, and between IF and image frequencies. The third term represents the noise correlation between RF and image frequencies. It can be shown that the result identical to (56) can be obtained by extending the analysis given by van der Ziel and Watters.[13]

E. Thermal Noise

The excess noise due to the series resistance R_d is given by (53). Since this noise has a white Gaussian spectrum and R_d is considered a constant for the given

[12] M. J. O. Strutt, "Noise figure reduction in mixer stages," Proc. IRE, vol. 34, pp. 942–950; December, 1946.

[13] A van der Ziel and R. L. Watters, "Noise in mixer tubes," Proc. IRE (Correspondence), vol. 46, pp. 1426–1427; July, 1958.

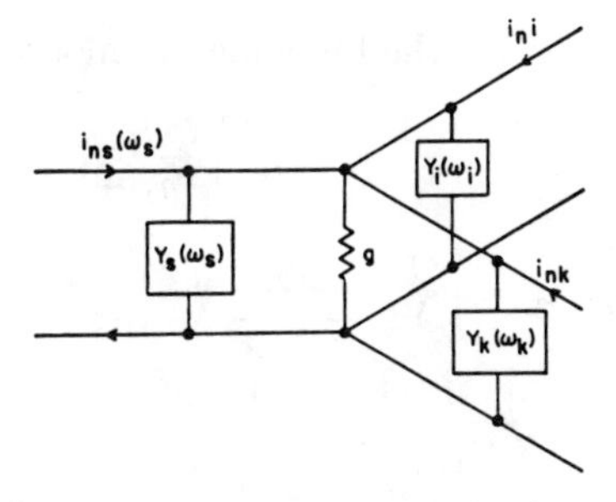

Fig. 9—Noise equivalent circuit.

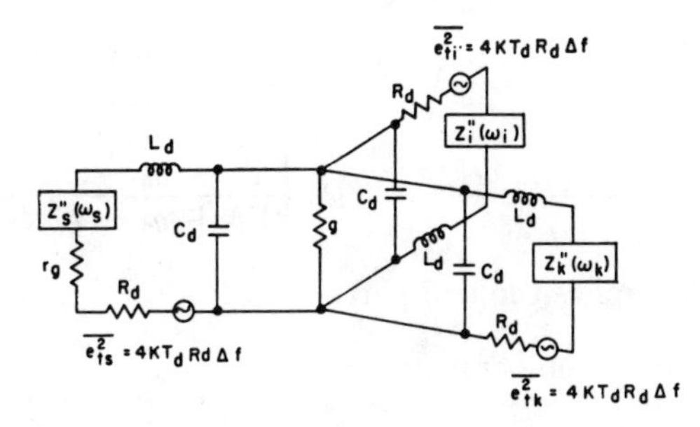

Fig. 10—Noise equivalent circuit for series resistance R_d.

diode, the equivalent noise circuit in this case can be represented (refer to Fig. 3) by Fig. 10.

By transforming the voltage generators e_{ts}, e_{ti}, and e_{tk} at the RF, IF, and image frequencies, respectively, into corresponding current generators i_{ts}, i_{ti} and i_{tk}, and referring to (11), the following relationships between the voltage and current generators exist:

$$\begin{aligned}\overline{i_{ts}^2} &= \overline{e_{ts}^2}\,|Y_s'(\omega_s)|^2 \\ \overline{i_{ti}^2} &= \overline{e_{ti}^2}\,|Y_i'(\omega_i)|^2 \\ \overline{i_{tk}^2} &= \overline{e_{tk}^2}\,|Y_k(\omega_k)|^2.\end{aligned} \tag{57}$$

Then the excess noise output across the load at IF due to thermal noise in R_d is

$$\begin{aligned}\overline{v_{tn}^2}\cdot g_i &= \{\overline{i_{ts}^2}\,|H_{is}|^2 + \overline{i_{ti}^2}\,|H_{ii}|^2 + \overline{i_{tk}^2}\,|H_{ik}|^2\}g_i \\ &= 4_kT_dR_dg_i\,\Delta f\{|Y_s'|^2\,|H_{is}|^2 \\ &\quad + |Y_i'|^2\,|H_{ii}|^2 + |Y_k'|^2\,|H_{ik}|^2\}.\end{aligned} \tag{58}$$

F. Converter Noise Figure

From (49), the noise figure of the fundamental mode converter is, referring to (20), (56), and (58),

$$F = 1 + \frac{N_e}{P_{Ga}N_a} = 1 + \frac{\overline{v_{ni}^2} + \overline{v_{tn}^2}}{P_{Ga}N_a}\,g_i, \tag{59}$$

where

$$N_e = (\overline{v_{ni}^2} + \overline{v_{tn}^2})g_i.$$

As matter a convenience, define

$$\begin{aligned}2eI_0\,\Delta f &= 4kT_dG_0\,\Delta f \\ 2eI_1\,\Delta f &= 4kT_dG_1\,\Delta f \\ 2eI_2\,\Delta f &= 4kT_dG_2\,\Delta f \\ 2eI_4\,\Delta f &= 4kT_dG_4\,\Delta f.\end{aligned} \tag{60}$$

Referring to (17) for the fundamental mode

$$\frac{|H_{ii}|^2}{|H_{is}|^2} = \left|\frac{(Y_s + g_0)(Y_k^* + g_0) - g_2^2}{g_1(Y_k^* + g_0 - g_2)}\right|^2$$
$$\frac{|H_{ik}|^2}{|H_{is}|} = \left|\frac{Y_s + g_0 - g_2}{Y_k^* + g_0 - g_2}\right|^2 \tag{61}$$

$$\frac{H_{ii}(H_{is}^* + H_{ik}^*) + H_{ii}^*(H_{is} + H_{ik})}{|H_{is}|^2} = -2\, Re\left\{\frac{(Y_s + g_0)(Y_k^* + g_0) - g_2^2}{g_1(Y_k^* + g_0 - g_2)} \cdot \left(1 + \frac{Y_s^* + g_0 - g_2}{Y_k + g_0 - g_2}\right)\right\}$$

$$\frac{H_{is}H_{ik}^* + H_{is}^*H_{ik}}{|H_{is}|^2} = 2\, Re\left\{\frac{Y_s + g_0 - g_2}{Y_k^* + g_0 - g_2}\right\},$$

where Re implies the real part.

T_d = diode temperature

T = generator temperature.

The expression for F is given for the fundamental mode by

$$F = 1 + \frac{T_d |Z_s'|^2}{Tr_g}\Bigg[(G_0 + R_d |Y_s'|^2) + \left|\frac{(Y_s + g_0)(Y_k^* + g_0) - g_2^2}{g_1(Y_k^* + g_0 - g_2)}\right|^2 (G_0 + R_d |Y_i'|^2) + \left|\frac{Y_s + g_0 - g_2}{Y_k^* + g_0 - g_2}\right|^2 (G_0 + R_d |Y_k'|^2) - 2\, Re\left\{\frac{(Y_s + g_0)(Y_k^* + g_0) - g_2^2}{g_1(Y_k^* + g_0 - g_2)} \cdot \left(1 + \frac{Y_s + g_0 - g_2}{Y_k + g_0 - g_2}\right)\right\} G_1 + 2\, Re\left\{\frac{Y_s + g_0 - g_2}{Y_k^* + g_0 - g_2}\right\} G_2\Bigg]. \tag{62}$$

For the second-harmonic mode, $g_1 g_2$, G_1 and G_2 in (62) should be replaced by g_2, g_4, G_2 and G_4, respectively. The noise figure expressions derived here are very general. The only approximation used is the small-signal approximation. These expressions are valid for self-excited or externally-excited converters, for both the noninverting and inverting cases.

If the image frequency termination is not a pure reactance, $R_d |Y_k'|^2$ in the fourth term (the image frequency term) of (62) should be replaced by $(R_d + r_k) |Y_k'|^2$, where r_k is the real part of $Z_k''(\omega_k)$ (refer to Fig. 3). This allows the noise of r_k to be included. Furthermore, if the noise contribution from the load g_i is needed,[14] $G_0 + R |Y_i'|^2$ in the third term of (62) should be changed to $G_0 + g_i + R |Y_i'|^2$.

[14] The noise figure is often considered to be independent of the noise from the load.

It should be noted that $|Z_s'|^2/r_g$ in (62) represents the equivalent parallel resistance of r_g and is equal to g_s as shown in (32). Therefore, as a first step in making the noise figure F small,

$$|Z_s'|^2 G_0/r_g = \frac{G_0}{g_s} \quad \text{or} \quad \frac{I_0}{g_s}$$

should be made as small as possible. Also it should be pointed out that the noise contribution of the series resistance R_d can be made small if

$$G_0 \gg R_d |Y_s'|^2$$
$$G_0 \gg R_d |Y_k'|^2 \tag{63}$$
$$G_0 \gg R_d |Y_i'|^2.$$

The cutoff frequency of the diode ω_c is defined as the frequency above which the diode no longer presents a negative resistance across its terminals.

$$\omega_c = \frac{|g|}{C_d}\sqrt{\frac{1}{|g| R_d} - 1}. \tag{64}$$

It can be shown that the condition indicated by (63) requires a diode with a high cutoff frequency (*i.e.*, $\omega_c \gg \omega_s$ and $\omega_c \gg \omega_k$). The fourth term, the noise contribution from image frequency, can be made small by terminating Y_s and Y_k such that

$$\left|\frac{Y_s + g_0 - g_2}{Y_k^* + g_0 - g_2}\right|^2 \ll 1. \tag{65}$$

The above condition also reduces the noise contribution of the RF and image frequency correlation term [last term in (62)]. By making

$$\left|\frac{(Y_s + g_0)(Y_k^* + g_0) - g_2^2}{g_1(Y_k^* + g_0 - g_2)}\right|^2 \ll 1, \tag{66}$$

the noise contribution from the IF can be made small.

As was mentioned previously, g_1 for bias points A and B is positive, and g_1 for bias points D is negative for fundamental modes. The term g_2 for bias points C corresponding to the second-harmonic mode can be considered as positive. For positive g_1 (fundamental) and g_2 (second-harmonic mode), the fifth term in (62) is positive. This term describes the correlation between the RF, IF, and image frequencies. Therefore, this term will reduce the noise figure since there is a negative sign preceding it. In this case, the larger the terms

$$|Y_s + g_0 - g_2/Y_k^* + g_0 - g_c|^2$$

$$\text{and} \quad |(Y_s + g_0)(Y_k^* + g_0) - g_2^2/g_1(Y_k^* + g_0 - g_2)|^2,$$

the larger the correlation term. Therefore for the g_1 positive cases, in order to obtain the optimum noise figure, some compromise between the third and fourth terms should be made. Of course, for the g_1 negative case, both (65) and (66) are satisfied so that the correlation terms also become small and tend to reduce F.

As a special case, if $Im\{Y_k\}$ becomes very large, which corresponds to the case of (28) and (33), and if the noise contribution from R_d is negligible, referring to (32), (62) can be reduced to, including the noise contribution from the load g_i,

$$F = 1 + \frac{T_d G_0}{T g_s}\left[1 + \frac{(g_s + g_0)^2}{g_1^2}\left(1 + \frac{g_i}{G_0}\right) - 2\frac{(g_s + g_0)G_1}{g_1 G_0}\right], \quad (67)$$

where $b_s = 0$ is assumed.

It should be re-emphasized that $I_0 \propto G_0$, which is a function of the bias point as shown in Fig. 8. Optimization of the noise figure for the various cases involves complicated procedures. However, from the above analysis, one may conclude that the operations at bias points C and D are more favorable for low noise figures. This is true because not only G_0 is reduced, but also g_0 is negative. $g_s + g_0$ can be made very small as can be seen in (67).

The third term in (67) can be written as

$$\frac{(g_s + g_0)(g_i + g_0)}{g_1^2} \cdot \frac{(g_s + g_0)}{(g_i + g_0)} \cdot \left(1 + \frac{g_i}{G_0}\right). \quad (68)$$

It can be seen from (34) that the value $(g_s + g_0)(gi + g_0)/g_1^2$ should be slightly larger than unity to obtain reasonable power gain and to fulfill the stability condition. Therefore, in order to make the contribution of (68) small, the value of $g_s + g_0$ must be very small compared with the value of $g_i + g_0$, provided that g_i/G_0 is not very large. This means that a much larger amplification is required at the RF frequency than at the IF frequency. A low value of $g_s + g_0$ also makes the fourth term (correlation) of (67) small. As a limiting case, the third and fourth term will be zero and the noise figure expression of (67) becomes

$$F = 1 + \frac{T_0 G_0}{T g_s}. \quad (69)$$

This is nothing but the noise figure expression for the ideal amplifier at RF.

An ideal low-noise tunnel diode for a converter operation should have a sharp triangular shape of the V-I characteristic as shown in Fig. 11(a). The corresponding dynamic-conductance $g(V)$ is shown in Fig. 11(b). The sharper the triangle (*i.e.*, the limit will be a δ-function), the larger the magnitude of the negative conductance will be. Furthermore, it is desirable to have a linear slope in region B and D to increase $|g_1|$. Since G_0/g_s can be made very small when operating in the D region, the noise figure will be low in an ideal case.

In conclusion, the noise figure can be made relatively low. A noise figure of 3 db in a converter may be obtained having a low value of $g_s + g_0$ with presently available devices by operating at a frequency much lower than the cut-off frequency, and by making G_0/g_s smaller than unity.

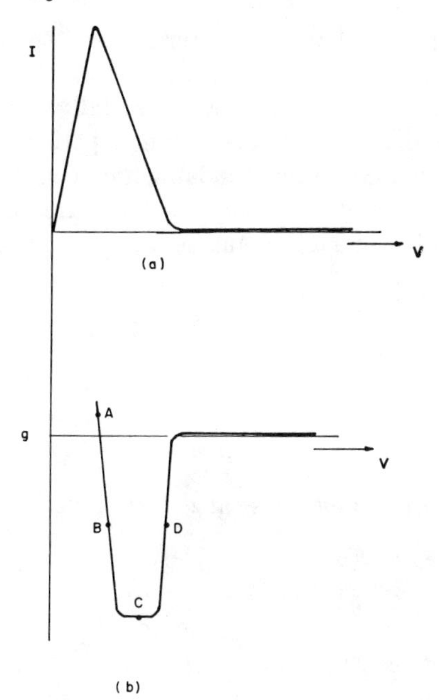

Fig. 11—Ideal tunnel-diode characteristic for low-noise converter.

Appendix

A. Shot Noise in Tunnel-Diode Converters or Mixers

Using a method similar to that of Strutt,[12] the shot-noise effect due to the periodic change of the dc equivalent shot-noise current will be investigated.

The instantaneous shot-noise current $i_n(t)$ can be expressed as a Fourier series

$$i_n(t) = \sum_{n=1}^{\infty} B_n \cos(\omega_n t + \theta_n). \quad (70)$$

If the shot noise is white Gaussian, and the amplitude B_n of (70) are periodic functions of LO voltage, B_n can be written as

$$B_n(t) = B(t) = B_0 + 2B_1 \cos\omega_0 t + 2B_2 \cos 2\omega_0 t + 2B_3 \cos 3\omega_0 t + \cdots. \quad (71)$$

The distribution of random phase angles θ_n in (70) is Gaussian, and θ_m and θ_n are not correlated for $m \neq n$.

By combining (70) and (71), it can be shown that $i_n(t)$ is an amplitude- and phase-modulated noise current. Expanding the amplitude-modulation terms gives

$$\begin{aligned} i_n(t) &= \sum_{n=1}^{\infty} (B_0 + 2B_1 \cos\omega_0 t + 2B_2 \cos 2\omega_0 t + 2B_3 \cos 3\omega_0 t + \cdots)\cos(\omega_n t + \theta_n) \\ &= \sum_{n=1}^{\infty}\Big[B_0 \cos(\omega_n t + \theta_n) + \sum_{r=1}^{\infty} B_r\{\cos[(r\omega_0 - \omega_n)t + \theta_n] + \cos[(r\omega_0 + \omega_n)t + \theta_n]\}\Big]. \end{aligned} \quad (72)$$

The frequency bands of interest are the incremental bands $\Delta\omega$ centered around ω_s, ω_i, and ω_k. All frequencies ω_n, $(r\omega_0 - \omega_n)$, and $(r\omega_0 + \omega_n)$, that fall within the $\Delta\omega$'s at ω_s, ω_i and ω_k, contribute to noise performance. Consider the noninverting fundamental converter where $\omega_s = \omega_0 + \omega_i$ and $\omega_k = \omega_0 - \omega_i$. The frequencies ω_n of (72), which become ω_s, ω_i and ω_k, are:

$$\begin{aligned} \omega_n &= \omega_s \\ \omega_n &= \omega_i \\ \omega_n &= \omega_k \\ \omega_n &= r\omega_0 \pm \omega_i \quad \text{where} \quad r = 2, 3, 4 \cdots . \end{aligned} \tag{73}$$

The corresponding phase angles are given by

$$\begin{aligned} \theta_n &= \theta_s \\ \theta_n &= \theta_i \\ \theta_n &= \theta_k \\ \theta_n &= \theta_{\pm r}, \quad \text{where} \quad r = 2, 3, 4 \cdots . \end{aligned} \tag{74}$$

Since all frequencies except ω_s, ω_k and ω_i are filtered out, the noise current $i_n(t)$ consists of three currents, namely the RF noise current i_{ns}, the IF noise current i_{ni}, and the image frequency noise current i_{nk}, *i.e.*,

$$i_n(t) = i_{ns} + i_{ni} + i_{nk}, \tag{75}$$

where

$$\begin{aligned} i_{ns} = B_0 \cos (\omega_s t + \theta_s) & \\ + B_1\{\cos (\omega_s t + \theta_i) &+ \cos (\omega_s t + \theta_2)\} \\ + B_2\{\cos (\omega_s t + \theta_k) &+ \cos (\omega_s t + \theta_3)\} \\ + B_3\{\cos (\omega_s t + \theta_{-2}) &+ \cos (\omega_s t + \theta_4)\} \\ + B_4\{\cos (\omega_s t + \theta_{-3}) &+ \cos (\omega_s t + \theta_5)\} \\ + B_5\{\cos (\omega_s t + \theta_{-4}) &+ \cos (\omega_s t + \theta_6)\} \\ + \cdots\cdots\cdots\cdots & \end{aligned} \tag{76}$$

$$\begin{aligned} i_{ni} = B_0 \cos (\omega_i t + \theta_i) & \\ + B_1\{\cos (\omega_i t + \theta_s) &+ \cos (\omega_i t + \theta_k)\} \\ + B_2\{\cos (\omega_i t + \theta_2) &+ \cos (\omega_i t + \theta_{-2})\} \\ + B_3\{\cos (\omega_i t + \theta_3) &+ \cos (\omega_i t + \theta_{-3})\} \\ + B_4\{\cos (\omega_i t + \theta_4) &+ \cos (\omega_i t + \theta_{-4})\} \\ + \cdots\cdots\cdots\cdots & \end{aligned} \tag{77}$$

$$\begin{aligned} i_{nk} = B_0 \cos (\omega_k t + \theta_k) & \\ + B_1\{\cos (\omega_k t + \theta_i) &+ \cos (\omega_k t + \theta_{-1})\} \\ + B_2\{\cos (\omega_k t + \theta_s) &+ \cos (\omega_k t + \theta_{-3})\} \\ + B_3\{\cos (\omega_k t + \theta_2) &+ \cos (\omega_k t + \theta_{-4})\} \\ + B_4\{\cos (\omega_k t + \theta_3) &+ \cos (\omega_k t + \theta_{-5})\} \\ + B_5\{\cos (\omega_k t + \theta_4) &+ \cos (\omega_k t + \theta_{-6})\} \\ + \cdots\cdots\cdots\cdots &. \end{aligned} \tag{78}$$

It should be noted that random phase angles θ_s, θ_i, θ_k and $\theta_{\pm r}$ in (76), (77), and (78) were not changed in the conversion process.

The shot-noise equivalent converter circuit is shown in Fig. 9.

The IF voltage across the load due to i_{ns}, i_{ni} and i_{nk} can be given by $h_{is}i_{ns}$, $h_{ii}i_{ni}$ and $h_{ik}i_{nk}$ where $h_{is}(t)$, $h_{ii}(t)$ and $h_{ik}(t)$ are the respective transfer or impedance functions. Then the total IF noise voltage $v_{ni}(t)$ can be expressed by

$$v_{ni}(t) = h_{is}i_{ns} + h_{ii}i_{ni} + h_{ik}i_{nk}. \tag{79}$$

Taking the Laplace transform of (79) gives

$$\begin{aligned} V_{ni}(\omega_i) &= H_{is}I_{ns} + H_{ii}I_{ni} + H_{ik}I_{nk} \\ &= (H_{is}B_0 + H_{ii}B_1 + H_{ik}B_2)e^{j\theta_s} \\ &\quad + (H_{is}B_1 + H_{ii}B_0 + H_{ik}B_1)e^{j\theta_i} \\ &\quad + (H_{is}B_2 + H_{ii}B_1 + H_{ik}B_0)e^{j\theta_k} \\ &\quad + (H_{is}B_1 + H_{ii}B_2 + H_{ik}B_3)e^{j\theta_2} \\ &\quad + (H_{is}B_3 + H_{ii}B_2 + H_{ik}B_1)e^{j\theta_{-2}} \\ &\quad + (H_{is}B_2 + H_{ii}B_3 + H_{ik}B_4)e^{j\theta_3} \\ &\quad + (H_{is}B_4 + H_{ii}B_3 + H_{ik}B_2)e^{j\theta_{-3}} \\ &\quad + (H_{is}B_3 + H_{ii}B_4 + H_{ik}B_5)e^{j\theta_4} \\ &\quad + (H_{is}B_5 + H_{ii}B_4 + H_{ik}B_3)e^{j\theta_{-4}} \\ &\quad + (H_{is}B_4 + H_{ii}B_5 + H_{ik}B_6)e^{j\theta_5} \\ &\quad + (H_{is}B_6 + H_{ii}B_5 + H_{ik}B_4)e^{j\theta_{-5}} \\ &\quad + \cdots\cdots\cdots\cdots . \end{aligned} \tag{80}$$

It should be noted that the random phase angles θ_s, θ_i, θ_k and $\theta_{\pm r}$ $(r = 2, 3, 4 \cdots)$ are uncorrelated. Therefore, if the phase average of the quadratic form of $V_{ni}(\omega_i)$ is taken, the following result is obtained:[15]

$$\begin{aligned} |\widetilde{V_{ni}(\omega_i)}|^2 = \tfrac{1}{2}\{(|H_{is}|^2 + |H_{ii}|^2 + |H_{ik}|^2)A_0 & \\ + \{H_{ii}(H_{is}^* + H_{ik}^*) + H_{ii}^*(H_{is} + H_{ik})\}A_1 & \\ + (H_{is}H_{ik}^* + H_{is}^*H_{ik})A_2\}, & \end{aligned} \tag{81}$$

where

$$\begin{aligned} A_0 &= B_0^2 + 2B_1^2 + 2B_2^2 + 2B_3^2 + \cdots \\ A_1 &= 2(B_0B_1 + B_1B_2 + B_2B_3 + \cdots) \\ A_2 &= 2\left(\frac{B_1^2}{2} + B_0B_2 + B_1B_3 + B_2B_4 + \cdots\right), \end{aligned} \tag{82}$$

and the asterisk denotes complex conjugate.

[15] This can be obtained by multiplying each term in (80) with its complex conjugate and adding all terms.

Squaring $B(t)$ of (71), the following relation is obtained:

$$\begin{aligned} B^2(t) &= A(t) \\ &= (B_0 + 2B_1 \cos \omega_0 t + 2B_2 \cos 2\omega_0 t \\ &\quad + 2B_3 \cos 3\omega_0 t + \cdots)^2 \\ &= A_0 + 2A_1 \cos \omega_0 t + 2A_2 \cos 2\omega_0 t \\ &\quad + 2A_3 \cos 3\omega_0 t + \cdots), \end{aligned} \quad (83)$$

where

$$\begin{aligned} A_0 &= B_0^2 + 2B_1^2 + 2B_2^2 + 2B_3^2 + \cdots \\ A_1 &= 2(B_0B_1 + B_1B_2 + B_2B_3 + \cdots) \\ A_2 &= 2\left(\frac{B_1^2}{2} + B_0B_2 + B_1B_3 + B_2B_4 + \cdots\right). \end{aligned} \quad (84)$$

Note that (82) and (84) are identical.

It can be shown[16] that the time average of the square of the shot-noise current $i_s(t)$ for the case where no amplitude modulation is given in the interval Δf by

$$\overline{i_s^2(t)} = 2eI_0 \, \Delta f, \quad (85)$$

where I_0 is the dc current. It can be shown also that the time average of $i_n^2(t)$ in the interval Δf at a frequency f_n is given by

$$\overline{i_n^2(t)} = \frac{\overline{B^2(t)}}{2} \Delta f = \frac{A_0}{2} \Delta f. \quad (86)$$

[16] A. van der Zeil, "Noise," Prentice-Hall, Inc., New York N. Y., p. 91; 1954.

In the case of tunnel diodes, I_0 of (85) is the averaged value of the dc equivalent shot-current I_{eq} as shown in (54). Therefore, $A_0/2$ should be equal to $2eI_0$, and

$$\overline{i_n^2(t)} = \frac{A_0}{2} \Delta f = 2eI_0 \, \Delta f, \quad (87)$$

where

$$\frac{A_0}{2} = 2eI_0. \quad (88)$$

Similarly, the following identity can be obtained

$$\begin{aligned} \frac{A_1}{2} &= 2eI_1 \\ \frac{A_2}{2} &= 2eI_2, \end{aligned} \quad (89)$$

where I_1 and I_2 are the second and third terms of $I_{eq}(t)$ in (54).

Then, referring to (79) and (81), the expression for the time average of $v_{ni}^2(t)$ can be given by

$$\begin{aligned} \overline{v_{ni}^2(t)} = 2e \, \Delta f[(\,|H_{is}|^2 + |H_{ii}|^2 + |H_{ik}|^2)I_0 \\ + \{H_{ii}(H_{is}^* + H_{ik}^*) + H_{ii}^*(H_{is} + H_{ik})\}I_1 \\ + (H_{is}H_{ik}^* + H_{is}^*H_{ik})I_2]. \end{aligned} \quad (90)$$

Acknowledgment

The author wishes to acknowledge the assistance and encouragement given by A. Brandli, E. G. Nielsen, and J. J. Suran at the Electronics Laboratory, General Electric Company, Syracuse, N. Y.

Millimeter-Wave InP Image Line Self-Mixing Gunn Oscillator

SAMUEL DIXON, JR., SENIOR MEMBER, IEEE, AND HAROLD JACOBS, FELLOW, IEEE

Abstract—**Indium Phosphide Gunn Diodes have been used to develop self-mixing oscillators in the 60-GHz region using dielectric image line techniques. Experiments have been performed to measure conversion characteristics and minimum detectable signal. Applications are suggested based on minimum detectable signal for systems requiring small size and low cost.**

I. Introduction

THE millimeter-wave image guide self-mixing oscillator is potentially a key element for future low-cost receiver front end designs. The main reason for this is that the self-mixing approach embodies simplifications in circuitry for the entire receiver system. Schottky barrier and other rectifier diodes suffer from the disadvantage of having a separate local oscillator and low burnout power limit. Bulk-effect Gunn self-oscillating mixers offer less sensitivity at 60 GHz, but have the attractive alternative of a high-power handling capability.

In conventional mixers, there usually exists a signal frequency, mixer diodes of rectifier type and a separate local oscillator. In the self-oscillating mixer, the mixer diode is eliminated. The Gunn diode will serve both as a local oscillator, and because nonlinearities are always present in an oscillator, as a mixing element. With the Gunn diode oscillator serving both these functions, the integrated receiver front-end design using the dielectric image guide approach becomes extremely compact and simplified. In the latter arrangement, the signal is fed directly into the oscillator and a suitable IF probe will remove the IF power for use in subsequent amplifier stages.

One of the objectives of this paper was the design of self-oscillating mixers with considerable simplifications, and hence, reduction in cost. In the quest for lower cost, the image guide technology was applied using a Gunn diode in a simply constructed dielectric cavity. The InP Gunn diode was imbedded in an aperture which was cut in a high resistivity aluminum oxide (Al_2O_3) ceramic waveguide. The significance of the image line technology is that active devices, as well as passive components, can be developed and integrated into circuit modules to construct functional subsystems.

Manuscript received November 24, 1980; revised February 18, 1981.

The authors are with the Millimeter Wave Devices and Circuits Team, U.S. Army Electronics Technology and Devices Laboratory, Fort Monmouth, NJ 07703.

II. Device Design

The Gunn oscillator cavity design is based on an image-line concept first formulated by Marcatili [1] and later modified for millimeter waves [2]. The fundamental electromagnetic wave propagating in an image guide is the E_{11}^{y} mode, a hybrid mode which propagates when correctly launched. Application of theoretical considerations indicated that the image guide for proper operation should be on the order of one wavelength in the medium in width, and less than one-half wavelength in height. At 60 GHz, cross-sectional dimensions of the image guides were oversized, i.e., slightly greater than 1 mm in height and about 2 mm in width. Experiments indicated that in this oversized condition, the E_{11}^{y} mode dominated. The resonant section in back of the Gunn diode was approximately $(2N+1)\lambda/2$ in length. A diagram of the image guide self-oscillating mixer is shown in Fig. 1. This simplified schematic indicates the manner of coupling to the metal waveguide showing one end of the resonant cavity being tapered. This taper can effect a low-loss match to the metal waveguide by sliding the tapered end into the metal waveguide for maximum power transfer. This matched condition also yielded optimum IF output when the RF input signal was introduced. Fig. 2 shows a more detailed cutaway view of the device investigated. As can be seen, the signal power propagates down the dielectric from the left with the IF output being extracted out the top of the image guide. A metal disk is being utilized as a tuning element for the Gunn diode. Fig. 3 shows an exploded view of the 60-GHz image waveguide self-oscillating mixer which utilizes a tunable short to optimize performance. The metal housing was designed to support the biasing structure and the dimensions were such that it just covered the immediate area of the Gunn diode. The dielectric waveguide was exposed for most of its length, and experiments were not made on the unit without the dielectric. The Gunn diode is mounted flush with the bottom of the metal structure. The Al_2O_3 image waveguide with tapered front end was bonded to the metal housing, using a modified silver conductive adhesive, in such a way that the Gunn diode top protruded up into the dielectric. The dimensions of the image dielectric guide was 0.060 by 0.120 in. A 0.045-in hole in the dielectric allowed the IF and bias voltage post to come down and make a pressure contact with the top of the Gunn diode. This method of applying the bias voltage

Reprinted from *IEEE Trans. Microwave Theory Tech.*, vol. MTT-29, pp. 958–961, Sept. 1981.

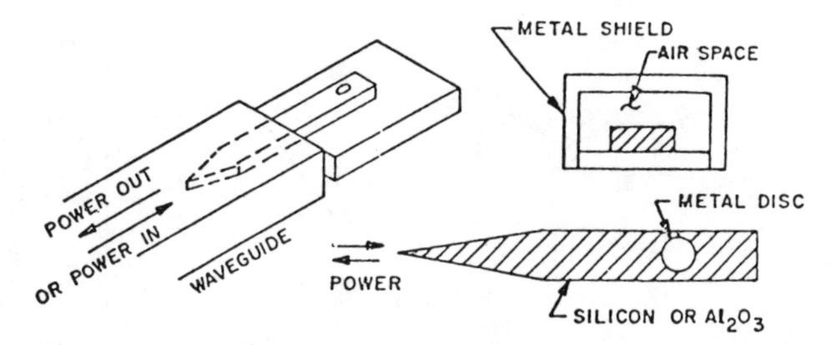

Fig. 1. Image guide self-oscillating mixer coupling to waveguide.

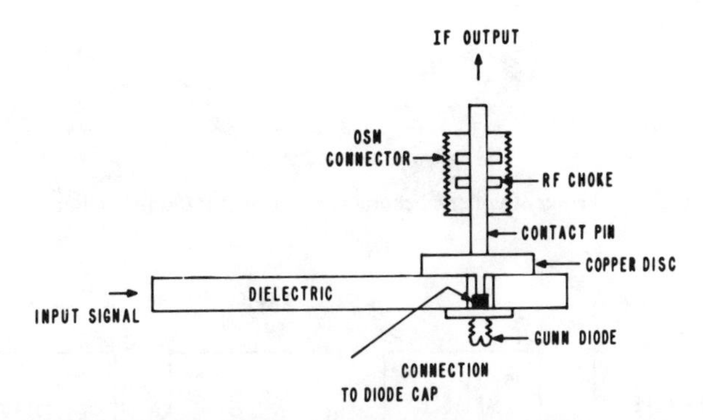

Fig. 2. Cutaway view of image guide self-oscillating mixer.

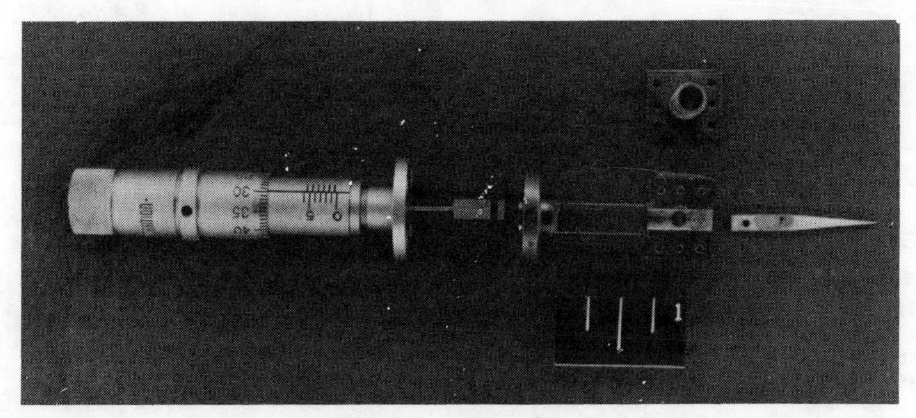

Fig. 3. Exploded view of the image guide self-oscillating mixer.

made it possible to mount a tuning short behind the image guide.

III. Experimental Results

Fig. 4 shows the output power and frequency characteristics of the InP Gunn diode self-oscillating mixer as a function of the bias voltage. Fig. 5 shows the circuit used in the evaluating of the InP self-oscillating mixer. A wide-band (50 to 75 GHz) backward wave oscillator in a manual tuning mode was used as a signal source. The single frequency outputs from the source was stable within ± 0.001 percent with nonharmonic spurious signals that are 40 dB down. This stable signal source was tuned 300 MHz above or below the InP Gunn oscillator to produce an IF falling into the filter passband. All measurements were made in an IF noise bandwidth of 100 MHz. The amplitude of the IF power in dBm was compared with measured values of input signal power for conversion gain or loss measurements. Fig. 6 shows the conversion loss or gain as a function of the bias voltage. The minimum detectable signal is the principle parameter for determining the sensitivity of the self-mixing oscillator, that is, how weak a signal the device can detect. To determine the minimum detectable signal (MDS), the IF power, as viewed on the wide-band scope (500-MHz bandwidth), is decreased by placing attenuation in the signal channel. This IF energy was set equal to the noise level. At this point, the signal level is equal to the noise level and this signal power is defined as the minimum detectable signal measured in decibels referred to 1.0 mW (dBm). Fig. 7 shows the minimum detectable signal in a 100-MHz bandpass as a function of bias voltage. As can be seen, the InP Gunn diodes MDS varies from -79 to -55 dBm as the bias voltage is varied from 7.3 to 8.1 V. The MDS and conver-

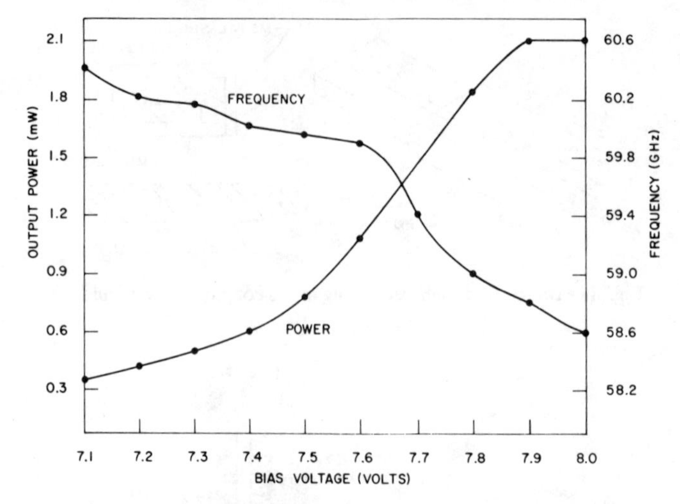

Fig. 4. Power/frequency characteristics of InP Gunn oscillator.

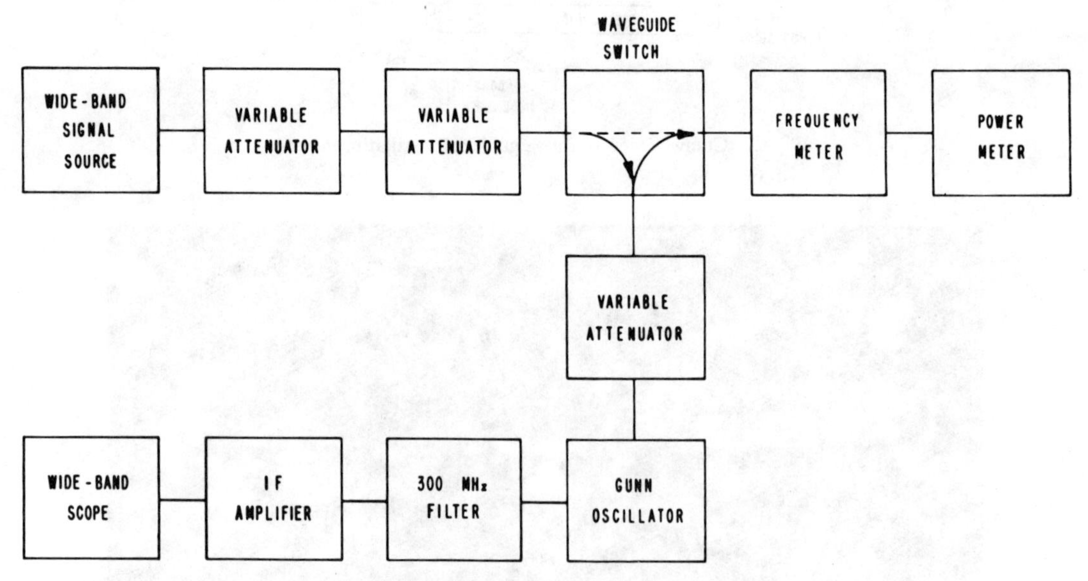

Fig. 5. Experimental setup for Gunn characterization.

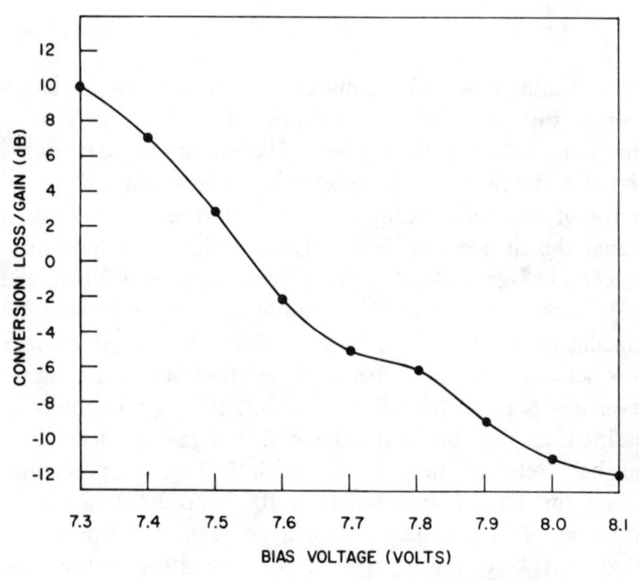

Fig. 6. Self-mixing oscillator conversion characteristics.

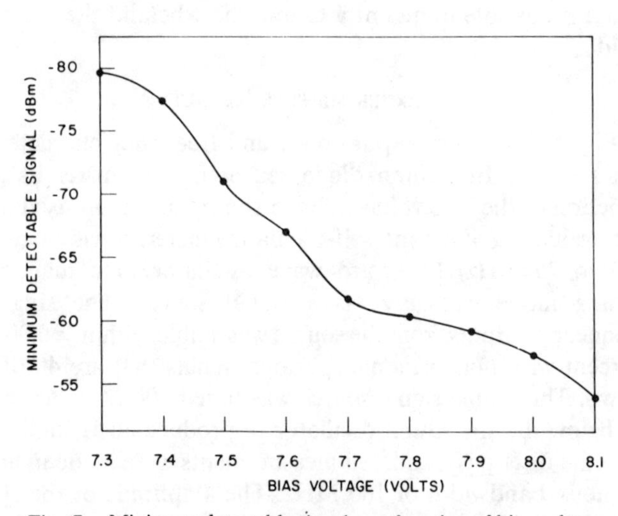

Fig. 7. Minimum detectable signal as a function of bias voltage.

sion characteristic peak at a voltage just above threshold for oscillation. The conversion gain characteristic tends to be slightly unstable at this point and better operation can be attained at a higher bias voltage. This behavior is consistent with information reported previously by investigators at lower frequencies of operating using GaAs Gunn diodes in waveguide cavities [3]–[5].

The InP Gunn diode has emerged as a potentially lower noise device, and hence higher sensitivity self-mixing oscillator, than those utilizing GaAs material. The larger peak-to-valley ratio of InP indicates higher electron velocity for a given field and leads to greater operating efficiency than GaAs. It is believed that the various scattering mechanisms which limit the high-frequency performance of transferred electron devices predict higher frequency operations for InP than GaAs [6].

IV. Conclusion

It has been shown that, self-oscillating mixers using InP Gunn diodes can be successfully designed with low cost and simplified construction techniques. These devices are very lightweight and lend themselves to applications in integrated circuit modules and subassemblies. Experimental data indicate a peak sensitivity in the order of −79 dBm which is lower than a conventional Schottky barrier balanced mixer. However, the image waveguide device has the advantages of having a simplified construction with a high-signal power burnout level coupled with low unit cost. These characteristics make the image waveguide self-mixing oscillator, a viable device in low-cost receivers, expendable EW sensors, and short range terminal guidance. In addition, when compared with GaAs Gunn diodes, the InP self-mixing device has greater potential in the higher millimeter-wave frequency region (above 100 GHz) due to InP having a higher effective transit velocity and fast intervalley scattering [6], [7].

References

[1] E. J. Marcatili, "Dielectric rectangular waveguide and directional couplers for integrated optics," *Bell Syst. Tech. J.*, vol. 48, pp. 2017–2102, Sept. 1979.

[2] H. Jacobs, G. Novick, C. M. LoCasia, and M. M. Chrepta, "Measurement of guide wavelength in dielectric rectangular waveguide," *Proc. IEEE*, vol. MTT-24, pp. 812–815, Nov. 1976.

[3] M. J. Lazarus, S. Novak, and E. D. Bullimore, "New millimeter-wave receiver using self-oscillating Gunn diode mixers," *Microwave J.*, vol. 14, no. 7, pp. 43–45, July 1971.

[4] M. J. Lazarus and S. Novak, "Millimeter-wave Gunn mixer with −90 dBm sensitivity, using a MOSFET/bipolar AFC circuit," *Proc. IEEE*, vol. 60, p. 747, June 1972.

[5] M. Kotani and S. Mitsui, "Self-mixing effect of Gunn oscillator," *Electron. Comm. Jap.*, vol. 55-B, no. 12, pp. 60–67, 1972.

[6] F. A. Myers, "Efficient InP Gunn diodes shrink power requirements," *Microwaves*, May 1980.

[7] J. D. Crowley, S. B. Hyder, J. J. Sowers, "Millimeter-wave indium phosphide Gunn devices," R&D Tech. Rep. DELET-TR-2940, June 1979.

Ka-BAND SELF-OSCILLATING MIXERS WITH SCHOTTKY BARITT DIODES

Indexing terms: Baritt diodes, Solid-state microwave devices, Microwave mixers

The fabrication of Pt n-p^+ silicon Baritt diodes is described which can deliver more than 4 mW at Ka-band frequencies. The diodes were investigated as self-oscillating mixers. A minimum detectable signal of −154 dBm as well as a conversion gain of 26 dB could be realised.

Introduction: Recent investigations on Baritt diodes as self-oscillating mixers have shown their superior application as Doppler sensitive detectors at X-band frequencies compared to Impatt diodes or Gunn devices.[1] The high mixing sensitivity results from the well known low noise behaviour of Baritt diodes[2,3] in addition to a considerable conversion gain[4,5] at low output power of the Baritt diode oscillator. Higher frequencies make Doppler system applications more interesting because of the increasing velocity resolution and the decreasing dimensions, e.g. the antenna. Therefore, efforts are being made to develop Baritt diodes which can deliver sufficient output power at Ka-band frequencies. These diodes are investigated as self-oscillating mixers and the dependence of the minimum detectable signal and the conversion gain versus oscillator output power at low Doppler frequencies is shown.

Device characterisation: For the following investigations Ka-band Pt n-p^+ silicon Baritt diodes are fabricated following design rules published earlier.[6] The resistivity of the n-epitaxial layer is 0·69 Ωcm and the length of the active layer is 1·8 μm leading to optimum r.f. performance at about 30 GHz. The diodes are fabricated by a special etching step[7] to reduce drastically the thickness of the remaining p^+-substrate to less than 1·5 μm and thus reducing the losses induced by ohmic series resistance and skin effect. The total device thickness is 3 μm. The area of the mesa-type diode is chosen as 5×10^{-5} cm^2 for optimum matching of the diode impedance to the resonator, which consists of a waveguide cap structure. Because of the very low negative resistance of Baritt diodes, the reduction of the series resistance and a proper r.f. impedance matching mainly determines the power output. At 29 GHz maximum output power of 4·2 mW could be attained with a diode voltage of 27·8 V and a bias current of 35 mA.

Measurements: (a) Minimum detectable signal: The minimum detectable signal *MDS*, that is, the weakest signal level which can be distinguished from the noise level, is a principal criterion to determine the sensitivity of a self-oscillating mixer system. *MDS* depends on the noise of the oscillator itself, on the detection bandwidth, and on the mixer characteristics, and is defined as:

$$MDS = 10 \log (kT \,.\, B \,.\, NF) \quad \text{decibels} \qquad (1)$$

where k is Boltzmann's constant, T absolute temperature, B detection bandwidth and NF the noise figure of the mixer system. The measurement set-up for the determination of *MDS* simulating a Doppler system is shown in Fig. 1. The output power of the Baritt diode is attenuated by precision attenuators and fed through a circulator and a phase shifter to adjust a proper phase relation between the oscillator and the Doppler signal; the ferrite modulator generates an amplitude modulated signal (two sidebands) which, after passing again through the circulator, is further attenuated and then mixed with the oscillator signal. The Doppler frequency is determined by the square wave generator and the Doppler signal is measured with a selective voltage meter in the bias circuit. For Doppler measurements, only one sideband has to be taken into account.

The behaviour of *MDS* as a function of the output power is investigated for signal frequencies near the carrier. *MDS* normalised to 1 Hz detection bandwidth versus oscillator output power is shown for three Doppler frequencies in Fig. 2. As can be seen, *MDS* decreases with decreasing output power, i.e. the mixing sensitivity increases. For a 1 kHz Doppler frequency a minimum detectable signal of −154 dBm could be measured at an output power of −26 dBm. *MDS* increases with decreasing Doppler frequency due to the increasing oscillator noise near the carrier.

(b) Conversion gain: Besides the minimum detectable signal, the conversion gain also describes the mixing properties of a self-oscillating mixer system. Passive mixers show a conversion loss, whereas active mixers can generate a conversion gain. The conversion gain is defined as:

$$G = 10 \log (P_D/P_S) \quad \text{decibels} \quad (G > 0) \qquad (2)$$

where P_D is the power at the detected low frequency (Doppler frequency) and P_S is the power at the signal frequency.

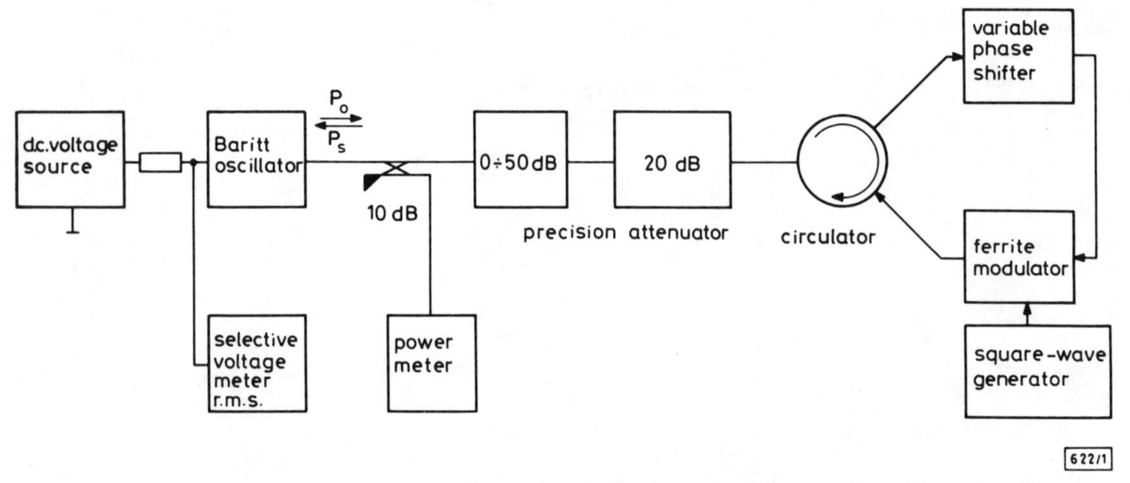

Fig. 1 *Measurement set-up for Baritt diode self-oscillating mixer for determination of minimum detectable signal and conversion gain*

Reprinted with permission from *Electron. Lett.*, vol. 16, pp. 827–829, Oct. 23, 1980.

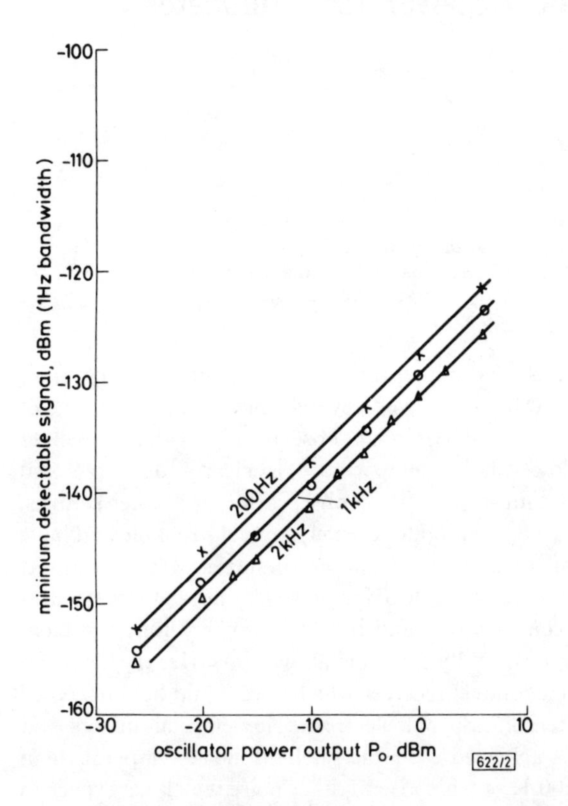

Fig. 2 *Minimum detectable signal versus oscillator output power. Doppler frequency is parameter; $f_0 = 29$ GHz*

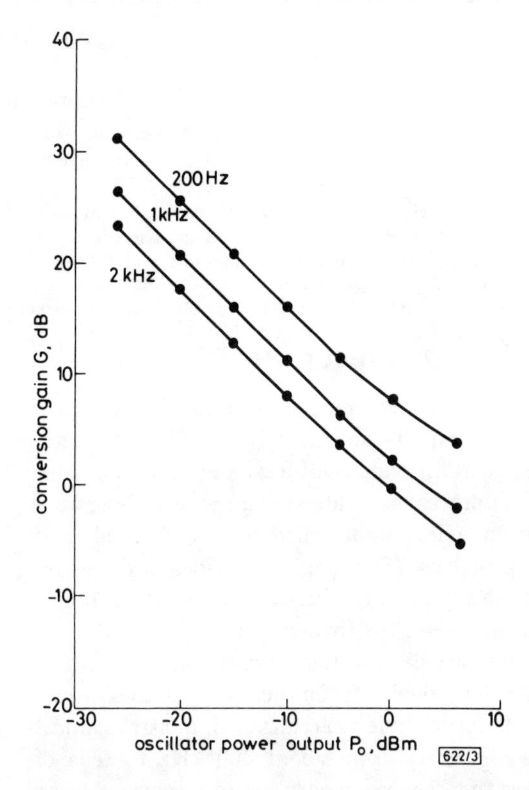

Fig. 3 *Conversion gain versus oscillator output power. Doppler frequency is parameter; $f_0 = 29$ GHz*

The measurement set-up for the conversion gain is the same as for *MDS*. The value of P_D can be determined by measuring the diode impedance (in terms of the Doppler frequency and the oscillator output power) and the root mean square voltage of the detected Doppler signal in the bias circuit. The signal power follows from the oscillator power P_o attenuated by the overall attenuation. Fig. 3 shows the measured conversion-gain versus oscillator output power for three Doppler frequencies. A conversion gain of 26 dB could be measured at -26 dBm oscillator output power and 1 kHz Doppler frequency. The conversion gain decreases with increasing output power and reaches a value of -2 dBm at maximum output power. As can be seen, the Baritt diode shows a conversion loss at this power level and acts in the same way as a passive mixer.

Conclusion: It has been shown that Baritt diodes can deliver sufficient output power of more than 4 mW at Ka-band frequencies. A minimum detectable signal of -154 dBm at -26 dBm oscillator output power could be measured for 1 kHz Doppler frequency. The increasing mixing sensitivity with decreasing oscillator output power and increasing Doppler frequencies has been shown. At -26 dBm output power a conversion gain of 26 dB could be attained. This demonstrates the very high mixing sensitivity of Baritt diodes as self-oscillating Doppler detectors at Ka-band frequencies.

Acknowledgments: The authors gratefully acknowledge the financial support of the Deutsche Forschungsgemeinschaft and thank Prof. Dr. W. Harth and Prof. Dr. M. Claassen for helpful discussions.

P. N. FÖRG
J. FREYER

17th September 1980

Lehrstuhl für Allgemeine Elektrotechnik und Angewandte Elektronik
Technische Universität München
D-8000 München 2, Arcisstrasse 21, W. Germany

References

1 EAST, J. R., NGUYEN-BA, H., and HADDAD, G. I.: 'Design, fabrication, and evaluation of Baritt devices for doppler system applications', *IEEE Trans.*, 1976, **MTT-24**, pp. 943–948

2 FREYER, J.: 'Baritt-dioden für X- und K_u-band frequenzen', *Mikrowellen-Magazin*, 1977, **6**, pp. 504–507

3 FREYER, J., AHMAD, S., and HARTH, W.: 'High power low noise Pt Schottky Baritt diodes'. Presented at 6th European Microwave Conference, Rome, Sept. 1976

4 VANOVERSCHELDE, A., SALMER, G., RAMAUT, J., and MEIGNANT, D.: 'The use of punch-through diodes in self-oscillating mixers', *J. Phys. D*, 1975, **8**, pp. 1108–1114

5 HARTH, W.: 'Conversion gain of self-oscillating Baritt diode mixers', *AEÜ*, to be published

6 FREYER, J., and FÖRG, P. N.: 'Baritt diodes for K_a-band frequencies', *IEE Proc. I, Solid-State & Electron Devices*, 1980, **127**, (2), pp. 78–80

7 FREYER, J.: 'Investigations of the fabrication of thin silicon films for microwave semiconductor transit time devices', *J. Electrochem. Soc.*, 1975, **122**, pp. 1238–1240

A Low Temperature Bolometer Heterodyne Receiver for Millimeter Wave Astronomy

T. G. Phillips and K. B. Jefferts
Bell Laboratories, Murray Hill, New Jersey 07974
(Received 12 March 1973)

Liquid helium cooled InSb hot electron bolometers are used in a balanced mixer configuration as detectors for an imageless microwave receiver. The system is designed for mounting at the prime focus of the National Radio Astronomy Observatory (NRAO) 11 m antenna at Kitt Peak, Arizona, and is suitable for the study of rotational line spectra of interstellar gas molecules. Currently the operating frequency is in the 90–140 GHz band where the double sideband system noise temperature is 250 K.

INTRODUCTION

The study of microwave emission spectra of interstellar gas molecules[1] is a rapidly advancing field of astrophysics, which is expected to provide considerable information concerning chemical processes in interstellar space and possibly give information relevant to current cosmological and astrophysical problems. This paper describes a detection scheme designed to improve the sensitivity of such observations and to extend the frequency range.

Fundamental molecular rotational transitions vary in frequency from a few gigahertz for heavy molecules up to about 1000 GHz for hydride molecules and, in astronomical sources, linewidths vary from about 100 kHz to tens of MHz. To cover this range several types of receivers will be required using various observational platforms. Currently, observations are made using ground based reflecting telescopes with low noise microwave receivers at the prime focus. For further information the reader is referred to review articles by Buhl and Snyder,[2] and Penzias and Burrus.[3]

The effective emission strength of typical astronomical sources is often of the order of only 1 K in terms of equivalent black body radiation temperature (T_s) so that great care must be taken to reduce the noise temperature of the receiving equipment (T_n). We may assess the requirements from the following relation for the signal to noise ratio for such a system:

$$\frac{S}{N}=\frac{T_s}{T_n}(\Delta\nu\Delta t)^{\frac{1}{2}}, \tag{1}$$

where $\Delta\nu$ is the receiving channel bandwidth and Δt is the integration time. It is customary to use a multiplexing system in which several independent channels cover the frequency range of interest, $\Delta\nu$ being fixed at somewhat less than the spectral linewidth. Taking typical values of one hour for Δt and 250 kHz for $\Delta\nu$, Eq. (1) shows that a system noise temperature of 3000 K is required to reach a signal-to-noise ratio of 10. This is approximately the performance level of Schottky barrier diode mixer receivers[2,3] operating at about 100 GHz. [*Note added in proof*. The most recent receivers of this type have shown improvements of a factor of 2.] In fact, it is usually necessary to frequency switch the receiver on and off the emission line and to position switch the telescope on and off the source to cancel systematic effects from the receiver, telescope and atmosphere. This means that the proposed one hour of integration will take 16 h to achieve,[4] severely limiting the amount of information obtained in an observing period. A receiver with a lower noise temperature is clearly desirable and would, of course, permit the observation of weaker sources.

Schottky barrier diode receivers are very reliable and have proved of great value in the early stages of this field. At frequencies above 100 GHz, however, noise temperatures increase considerably, and it seems unlikely that the technique can be usefully extended above 300 GHz. In this paper we describe a novel receiver which uses liquid helium cooled bulk material InSb hot electron bolometers as mixers and which has achieved a double sideband noise temperature of 250 K (100 K of which is amplifier noise which we expect to eliminate) at 120 GHz with a useful bandwidth of about 4 MHz. This represents a significant improvement over current systems. The receiver was developed for use on the NRAO 11 meter antenna at Kitt Peak, Arizona, and has been successfully used for the observation of interstellar gas emission lines. We believe that the techniques described here can be extended to at least 500 GHz.

THE InSb BOLOMETER MIXER

The InSb hot electron bolometer was initially developed as a microwave and far infrared direct detector.[5,6] When used (in conjunction with a klystron local oscillator) as a microwave mixer we find that it displays characteristics which are uniquely suited to certain experiments, in particular, line spectroscopy. In this section we describe the mode of operation of the bolometer mixer and indicate the methods used to measure the detection characteristics and parameters.

At liquid helium temperatures high purity InSb ($N_D-N_A \approx 3\times10^{13}$ cm^{-3}) exhibits hot electron effects.[5] That is to say, the temperature of the partially degenerate conduction electron gas (T_e) can be easily raised with respect to the lattice temperature (T_0). The electrons interact rapidly among themselves to allow one to define an electron gas temperature, but the gas is only weakly coupled to the lattice. For an input power (P) to the electronic system the temperature rise (ΔT) is given by Kogan's[7] theory. However, the analytic results are complicated and it is simpler to consider small temperature changes only ($\Delta T \ll T_0$). This is sufficient to display most of the bolometer characteristics, and we can write

$$P=K\Delta T, \tag{2}$$

Reprinted with permission from *Rev. Sci. Instr.*, vol. 44, no. 8, pp. 1009–1014, Aug. 1973.

where K is the coefficient of thermal conductance (considered temperature independent) and is determined by the processes contributing to the electron–phonon coupling.

The temperature rise of the electron gas causes an increase in the electron mobility (μ), because in this temperature range μ is limited by ionized impurity scattering, which is a process depending strongly on electron velocities ($\mu\alpha T_e^{\frac{3}{2}}$). Consequently, under constant current (I_0) operation the voltage change across the bolometer for a temperature change ΔT is

$$\Delta V = I_0 \Delta R = -(3/2) I_0 R_0 (\Delta T/T_0), \qquad (3)$$

where R_0 is the bolometer resistance at the lattice temperature. Equations (2) and (3) can now be used to determine the bolometer sensitivity to radiation input power.[8] However, we are interested in both the frequency dependence and the behavior in the presence of a local oscillator field, i.e., in the mixer properties.[9] In the case of a bolometer element mounted in a waveguide where the bolometer dimensions are small compared with the guide wavelength (see Fig. 1), the incident power is described by

$$P \approx I_0^2 R + P_{lo} + 2(P_{lo} P_s)^{\frac{1}{2}} e^{i\omega t}, \qquad (4)$$

where the three terms are dc power, local oscillator power, and the mixing term at angular frequency ω (the difference between local oscillator and signal frequencies). P_s is the small signal power. As the bolometer resistance changes so does the input power, in a way that depends on the bias conditions. For constant current conditions[10]

$$P = P^0\left(1 - \frac{3}{2}\frac{\Delta T}{T_0}\right),$$

as in Eq. (3), and the mixer equation is

$$[I_0^2 R_0 + P_{lo}^0 + 2(P_{lo}^0 P_s^0 e^{i\omega t})^{\frac{1}{2}}]\left(1 - \frac{3}{2}\frac{\Delta T}{T_0}\right) = K\Delta T + \frac{C d(\Delta T)}{dt}, \qquad (5)$$

where C is the thermal capacity of the conduction electron gas. In conjunction with Eq. (3) this can be solved to give

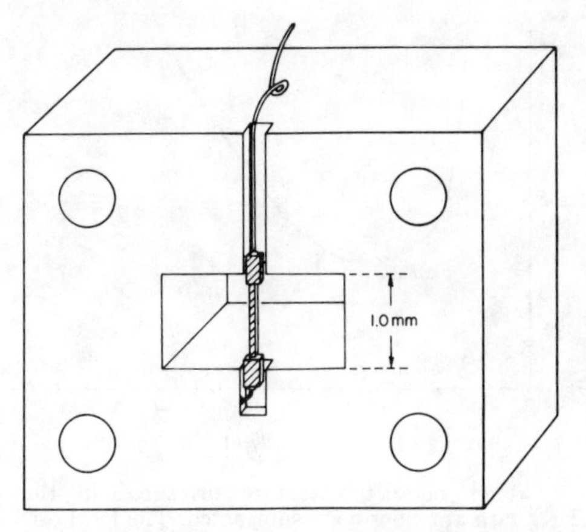

FIG. 1. InSb bolometer mounted in RG-138 waveguide. The active region of the bolometer is etched to dimensions of about 1×0.1×0.1 mm. Ohmic contacts are formed using sulphur-doped indium solder.

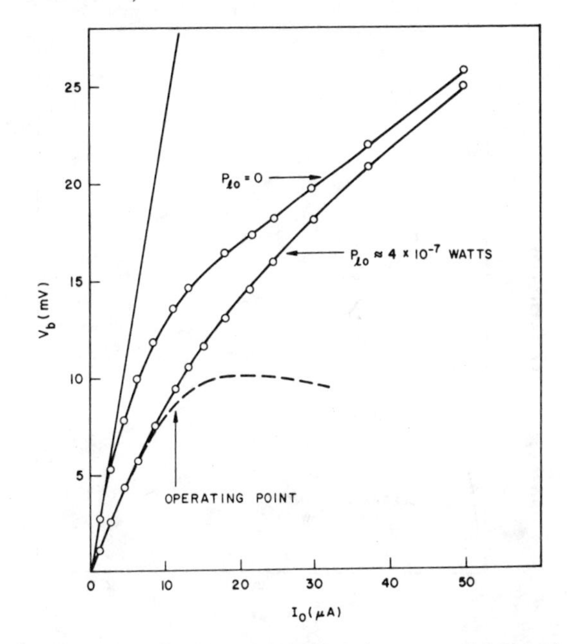

FIG. 2. V–I characteristics of an InSb bolometer at 4.2 K with and without local oscillator bias power showing deviation from Ohm's law. Also shown is the result from the linear theory (dashed curve) using values of $KT_0 = 5.9\times10^{-7}$ W and $P_{lo} = 4.0\times10^{-7}$ W. The operating point corresponds to the peak sensitivity.

the voltage across the bolometer,

$$V_b = I_0 R_0 \left[1 - \frac{3}{2}\frac{(I_0^2 R_0 + P_{lo}^0)}{K'T_0} - \frac{3(P_{lo}^0 P_s^0)^{\frac{1}{2}}}{K'T_0}\frac{e^{i\omega t}}{1 + i\omega\tau'}\right], \qquad (6)$$

where $K' = K + (3/2)(I_0^2 R_0 + P_{lo}^0)/T_0$ and $\tau' = C/K'$. The last term in Eq. (6) is the signal voltage at the difference frequency. The bolometer does not respond to the microwave frequency itself since $\tau' = C/K'$ is of the order of 10^{-7} sec.

Figure 2 shows the bolometer V–I characteristics with and without local oscillator power. The deviation from Ohm's law is clearly observable, but at high power levels is not as great as the predictions of the linear theory Eq. (6). Figure 3 shows the signal voltage as a function of I_0 for fixed local oscillator and signal power. The signal rises initially linearly with I_0, reaching a maximum at $I_0^2 R_0 \approx P_{lo}^0$, but again deviating from the linear theory at high powers. In practice maximum signal is found with $I_0^2 R_0 \approx P_{lo}^0 \approx KT_0 \approx 5\times10^{-7}$ W, for the 120 GHz bolometers. The bolometer dimensions are typically 1×0.1×0.1 mm giving a dynamic resistance of about 500 Ω at $T_0 = 4.2$ K. The operating point is determined mainly by the peak sensitivity and the linear theory gives a reasonable description of the bolometer characteristics up to this point. However, this is approximately the power level at which nonlinear effects become important and the bolometer is in fact operating with $\Delta T \sim 1$ K.

The parameter τ' determines the bandwidth of the mixer. This differs from the intrinsic hot electron relaxation time $\tau = C/K$ and indicates that the system bandwidth should increase with increasing input power. This is displayed in Fig. 4 where it is seen that the bandwidth in the region of maximum sensitivity is approximately 2 MHz.

We now turn to the laboratory values of the mixer noise temperature as measured by the scheme of Fig. 5. The bolometer is mounted across RG-138 waveguide with a

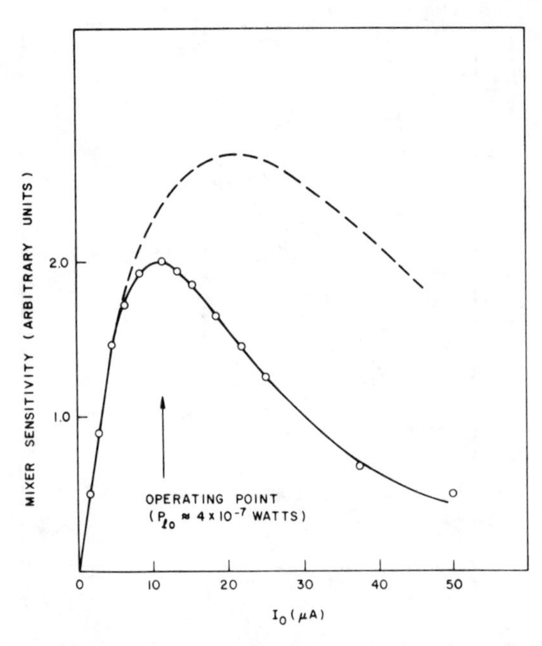

FIG. 3. Mixer sensitivity at 4.2 K as a function of bias current for fixed local oscillator power of 4.0×10^{-7} W. The dashed curve represents the linear theory result for the same values as in Fig. 2.

backing plunger for matching purposes. Ohmic contacts[5] are provided for the dc bias, and microwave bias power is from a 100 mW klystron operating at a frequency $f=119$ GHz and feeding a -10 dB coupler. In order to facilitate setting of microwave and dc bias conditions for optimum sensitivity a homodyne scheme is used in which a small signal, derived from the same klystron at -140 dB, and amplitude modulated at 100 kHz is fed to the bolometer. The magnitude of the 100 kHz mixer output is now optimized as in Fig. 3.

With the 100 kHz signal removed the input signal is provided solely by the room temperature load (a correction must be made for the room temperature local oscillator line coupled at -10 dB). The bolometer noise power is now measured with the aid of a low noise amplifier and rectification scheme.

The noise consists of several parts: (a) Johnson noise, (b) current noise, (c) amplifier noise, (d) local oscillator noise,

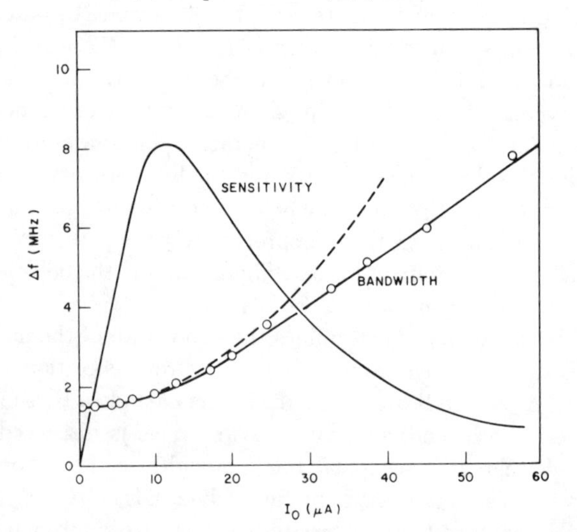

FIG. 4. Bandwidth (Δf) as a function of bias current for fixed local oscillator power of 4.0×10^{-7} W. The dashed curve represents the linear theory result for the same values as in Fig. 2, normalized at $I_0=0$.

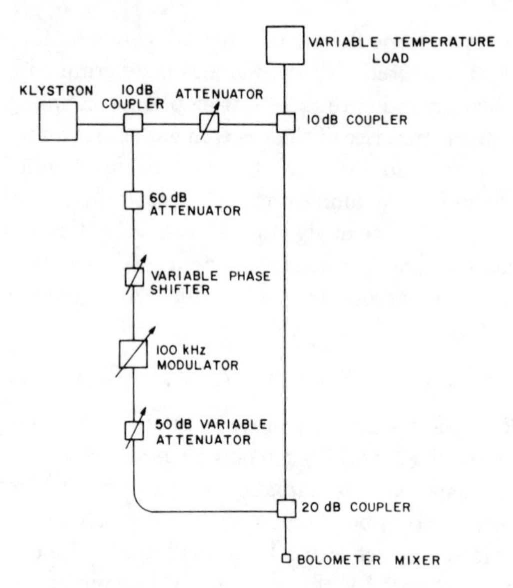

FIG. 5. A block diagram of the scheme used to measure the bolometer mixer noise temperature.

and (e) signal noise from the room temperature load. It is possible to separate these components, but here we subtract out only the amplifier contribution. We now determine the over-all bolometer noise temperature by taking values at two load temperatures. It is easily shown that provided $hf \ll k_B T_{1,2}$ the system noise temperature is given by

$$T_N=\frac{1}{2}\left[\left(\frac{NP_1+NP_2}{NP_1-NP_2}\right)(T_1-T_2)-(T_1+T_2)\right], \tag{7}$$

where $NP_{1,2}$ is the noise power for load temperature $T_{1,2}$. Here $T_{1,2}$ were ambient and liquid nitrogen temperatures. Figure 6 shows the system noise temperature as a function of dc bias, optimizing the local oscillator power at each point. The minimum noise temperature is found at a bias current slightly below that of maximum sensitivity (Fig. 3) due to the presence of current noise.

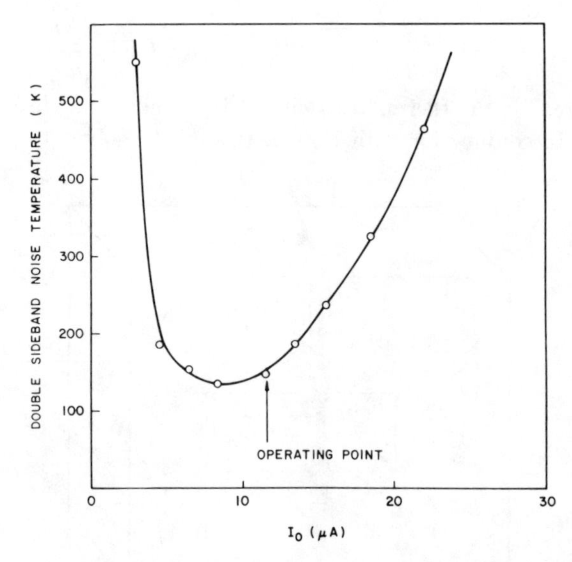

FIG. 6. Mixer noise temperature measured in the laboratory ($T_0=4.2$ K) with amplifier noise subtracted. The local oscillator power is adjusted for minimum noise temperature at each point and has a value of about 4×10^{-7} W at the operating point. This measurement is of the double sideband temperature.

It must finally be pointed out that the noise temperatures obtained are relevant to double sideband operation, since both 2 MHz wide sidebands (one above and one below the l.o. frequency) are available for detection. In a single bolometer detection scheme the two sidebands are superimposed. For use as a spectroscopic detector it is necessary to separate information in these sidebands. To accomplish this we use two mixers in an imageless processor configuration, obtaining two independent sidebands. The single sideband noise temperatures which are now appropriate are found by multiplying the temperatures of Fig. 6 by a factor of 2.

THE ASTRONOMICAL RECEIVER

The apparatus that fits at the prime focus of the Kitt Peak telescope consists of a Dewar containing the bolometer receiver; a klystron and microwave components of the local oscillator system; an electronic phase lock system for oscillator frequency control; amplifiers and imageless processor circuitry. This entire package is mounted in a space frame (Fig. 7) which bolts to a support ring at the focus.

The helium Dewar contains a gas cooled radiation shield to reduce the servicing requirements. Although it is filled from the top it operates satisfactorily to an angle of 15° from horizontal allowing full use of the elevation performance of the telescope. The bottom of the helium space is a copper cold plate which acts as a heat sink for the receiver components which are mounted in vacuum at the bottom of the Dewar. A Mylar window in the Dewar bottom plate room temperature vacuum wall permits the $f/0.8$ radiation from the telescope mirror to enter the receiver. There are no optical or infrared filters in the system since the waveguide components perform this function adequately.

A block diagram of the receiver in imageless processor form is shown in Fig. 8. Attached to the helium cold plate are two InSb detector elements mounted across the center of RG-138 waveguide and each backed by a tuning plunger, micrometer driven from the outside of the Dewar. The bolometers are symmetrically mounted with respect to a

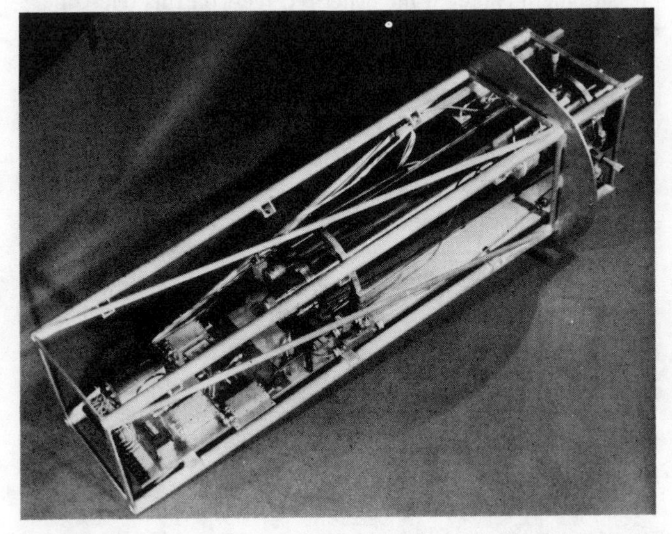

FIG. 7. Space frame package which is bolted to a support ring at the prime focus of the NRAO Kitt Peak 11 m antenna. The local oscillator phase lock frequency control system has been removed in this view. It essentially fills the remaining space in the rear of the package.

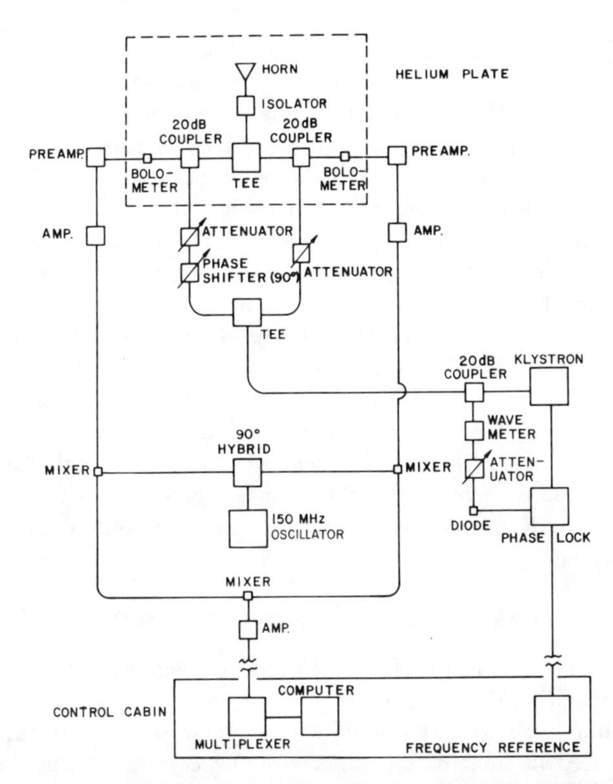

FIG. 8. Block diagram of the receiver. Components within the dashed box are attached to the helium plate. Those within the solid box are in the control cabin and are supplied by NRAO.

hybrid T which divides the input power equally. The signal is received by a microwave horn and fed through a Faraday rotation isolator to the H port of the T. The E port of the T can be matched, but in fact is used to aid in precise balancing of the mixer. Each bolometer receives the local oscillator power via a cross-guide coupler and the amplitude and phase difference of the two local oscillator signals is adjustable, the imageless processor requiring 90° of phase shift. Figure 9 is a photograph of the receiver front end showing the cold plate and microwave circuit. The isolator is used to prevent local oscillator leakage from the feed horn into the telescope structure. If the leakage can be sufficiently reduced by matching and balancing the bolometers then the isolator is not required. Currently the system is run without the isolator for frequencies near the design value of 115 GHz.

The bolometers are biased in the constant current mode with a battery and large series resistors, and the signal output is via short thin wires to specially constructed room temperature preamplifiers. Since the bolometer noise voltage is only that developed by a 500 Ω resistance at about 4 K (or about 7 Ω at room temperature) it is very difficult to make the amplifier contribution negligible. The multiple transistor amplifier currently in use contributes less than 50% excess noise to the system and has a 5 MHz bandwidth. Cooling of the preamplifier was rejected in view of the undesirability of heat loading the cryostat and also because of the temperature stability required to eliminate gain variation.

After further amplification the signals are converted to the filter bank processing frequency of 150 MHz by mixing with a second local oscillator. A second 90° phase shift is introduced between the signals at this point, and it can be shown that, after adding of the signals, the two 2 MHz wide sidebands will be independently displayed above and below the 150 MHz second local oscillator frequency. The total bandwidth of the system is then 4 MHz. The signal is now fed down from the focus of the telescope to a set of filters which form the frequency channels and then to the multiplexer and computer in the control cabin of the telescope, output being in the form of a set of 250 kHz (or 100 kHz) channels. Also in the control room is a 100 MHz crystal oscillator which provides a reference frequency which is multiplied to 4 GHz. To this is added a variable frequency in the range 100–250 MHz and the sum is then transmitted to the focus. A phase lock system maintains the klystron frequency to a precise value by locking to a high harmonic (say 30th) of the reference 4 GHz by means of a harmonic mixer crystal diode which is driven by the klystron coupled about −20 dB.

PERFORMANCE ON THE TELESCOPE

The noise temperature of the receiver itself is obtained by comparing the noise power observed in the channels of the filter bank when two different temperature absorbers are placed in front of the window of the cryostat. As in the laboratory, the two temperatures are ambient and liquid nitrogen and Eq. (7) is used. A double sideband temperature of 250 K is obtained. This is inferior to the InSb laboratory performance since there is a 100 K contribution from the amplifiers, also some attenuation from the front end microwave components, and to some extent the side lobes of the horn see the room temperature wall of the cryostat. The noise temperature of the complete system including the telescope is somewhat inferior again due to side lobes of the horn which illuminate the ground outside the reflector. Further degradation is caused by telescope aperture blockage and atmospheric radiation (sky noise). Such effects have been studied previously,[3] and if we take a typical value of 70 K for excess antenna noise plus sky noise at 30° elevation we can calibrate the receiver again by comparing the system noise (horn illuminating dish) with the noise when

FIG. 9. Front end of the receiver with vacuum wall and radiation shields removed.

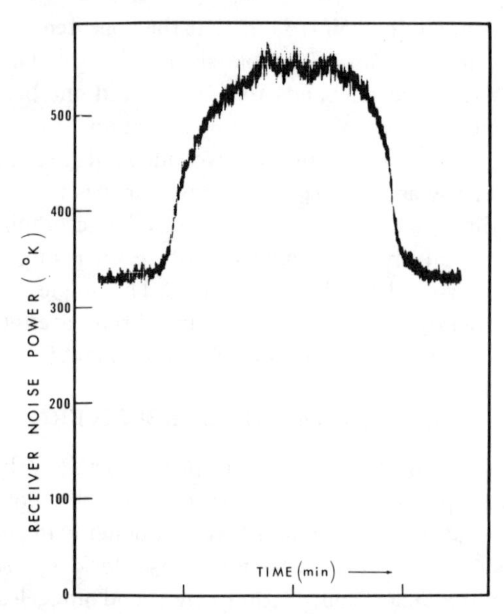

FIG. 10. A drift scan across the moon at an elevation of 30°. The vertical axis represents the total receiver noise power and is equivalent to 320 K (250 K receiver noise +70 K combined sky and side lobe noise) when off the moon, and 570 K on the point of maximum moon temperature.

the ambient temperature absorber (chopper) is in front of the feed horn. Again we find a value of 250 K.

Receiver checks which involve absorbers in front of the feed horn might be suspect since it is possible for local oscillator leakage power to be reflected back into the receiver by a nonperfect absorber altering the bias condition of the bolometers. However, a further check is possible by observing an astronomical object such as the moon. Such an observation is shown in Fig. 10. This is a plot of the receiver total power (4 MHz bandwidth) during a drift scan across the moon (not full) at an elevation of 30°. Values of 250 K for receiver temperature and 70 K for sky plus side lobe temperature lead to a moon antenna temperature of 250 K, giving a true temperature of about 380 K when the antenna efficiency is taken into account.[11] This is in agreement with measurements made using other receivers on this telescope,[11] and thus confirms the switched absorber measurements.

The receiver has been used to observe emission lines of carbon monoxide in the sources DR21, Cloud 4 and the Orion Nebula. Information is gathered suzciently rapidly that it is possible to map such sources by the method of drift scans. Figure 11 shows a set of such scans at varying declinations across the Orion Nebula taken on January 22, 1973. The receiver frequency was 115.2547 GHz (the $^{12}C^{16}O$ frequency) and the full 4 MHz receiver bandwidth was used to obtain an integral over the CO radiation.[12] Each scan takes 3 min (local standard time) and the integration time is 10 sec implying a slight distortion of intensity profiles.

Several problems have been encountered which are not necessarily specific to this receiver, but which may affect any low noise system. The two major problems are concerned with the local oscillator noise. It is very important with a low intermediate frequency system that the frequency reference for the local oscillator lock does not introduce any noise. In view of the large harmonic multiplication factors

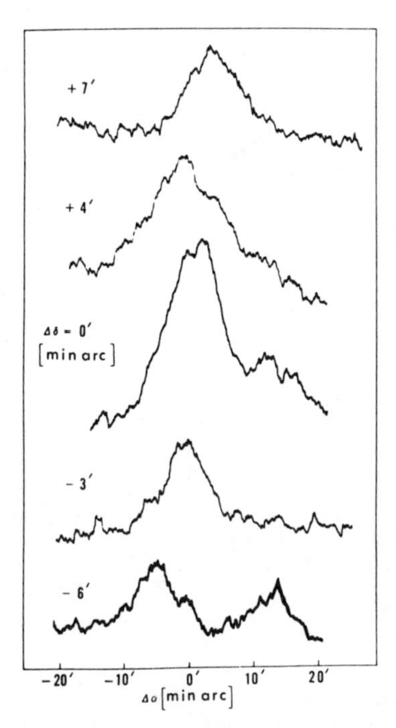

FIG. 11. A set of drift scans across the Orion Nebula in carbon monoxide ($^{12}C^{16}O$) at a variety of declination settings. The receiver frequency was 115.2547 GHz with a total bandwidth of 4 MHz. The horizontal axis represents local standard time converted to minutes of arc and the antenna beam width is about 1′. The plot is nominally centered in both declination and right ascension at the Orion A molecular peak. The peak of the central scan represents an antenna temperature of (30±5) K. Atmospheric fluctuations are responsible for the majority of the noise.

involved (∼1000) this requires considerable spectral purity from the initial frequency source. The phase lock scheme has been sufficiently troublesome from this point of view that we are currently considering replacing it wlth a frequency lock scheme. Also it is essential that there be no sizeable local oscillator leakage from the feed horn since there are multiple modes in the telescope and dome structure which are excited by the main lobe and side lobes of the horn and telescope. These modes interfere with the frequency discrimination of the receiver, giving rise to baseline variation and setting a lower limit for observable signals.

The results obtained so far indicate that this receiver will be useful at the present frequency for rapid mapping of interstellar gas clouds and possibly in searches for low intensity molecular lines. However, the most significant use will probably be at higher frequencies where the bulk material nature of the detector permits controlled manufacture and handling in spite of the small waveguide dimension. Also the local oscillator power requirements are sufficiently low that use can be made of harmonic generation techniques.

ACKNOWLEDGMENTS

We would like to thank the National Radio Astronomy Observatory and the staff at Kitt Peak for extending to us their facilities and also for their assistance. We also thank A. A. Penzias and R. W. Wilson for helpful discussions, R. E. Miller for technical assistance in preparing the equipment, and P. Wannier for assistance in taking data.

[1]D. M. Rank, C. H. Townes, and W. J. Welch, Science **174,** 1083 (1971); and P. M. Solomon, Physics Today, March (1973).

[2]D. Buhl and L. E. Snyder, Nat. Phys. Sci. **232,** 161 (1971).

[3]A. A. Penzias and C. A. Burrus, Annual Review of Astronomy and Astrophysics (to be published).

[4]This may be reduced to 8 h if the frequency switching mode is one which does not shift the signal completely out of the receiver bandpass, but presents it in a different set of channels.

[5]M. A. Kinch and B. V. Rollin, Br. J. Appl. Phys. **14,** 672 (1963).

[6]E. H. Putley, Appl. Opt. **4,** 649 (1965).

[7]Sh. M. Kogan, Sov. Phys.-Solid State **4,** 1813 (1963).

[8]For a complete discussion of direct radiation bolometer detectors see R. C. Jones, J. Opt. Soc. Am. **43,** 1 (1953).

[9]For a description of bolometer mixing terms of i.f. conversion loss see F. Arams, C. Allen, B. Peyton and E. Sard, Proc. IEEE **54,** 612 (1966); and J. J. Whalen and C. R. Westgate, *Proceedings of the Symposium on Submillimeter Waves* (Polytechnic Press, New York, 1970).

[10]The microwave bias, in general, will not correspond precisely to either constant current or constant voltage conditions. This may be accounted for in Eq. (5) and leads to minor changes in Eq. (6).

[11]E. K. Conklin (private communication).

[12]For a discussion of CO observations using diode receivers see A. A. Penzias, K. B. Jefferts, and R. W. Wilson, Astrophys. J. **165,** 229 (1971) and A. A. Penzias, K. B. Jefferts, R. W. Wilson, H. S. Liszt, and P. N. Solomon, Astrophys. J. (to be published).

Author Index

Subject Index

N

Editor's Biography

Erik L. Kollberg (M'83–SM'83) was born in Stockholm, Sweden. He received the M.Sc. degree in electrical engineering in 1961, the Tekn. Licentiat degree in 1965, and the Teknologie Doktor degree in 1970, all from Chalmers University of Technology, Göteborg, Sweden.

He is now a Professor at Chalmers University of Technology and is responsible for development of low-noise receivers for the Onsala Space Observatory. From 1963 to 1976 he worked on low-noise maser amplifiers. He has developed eight masers for the Onsala Space Observatory, ranging in frequency from 1 to 36 GHz. Partially because of this pioneering work, for many years the observatory has been in the forefront of the world in radio astronomy. Since 1972 he has also been working on low-noise millimeter-wave mixers, and his group has successfully developed cryogenically cooled low-noise mixer receivers for frequencies of up to 150 GHz. These mixers have been used for radio astronomy observations at Onsala since 1979. More recently, he has been working on extremely low-noise superconducting (quasiparticle) mixers, and experimental receivers are now in use at the Onsala Observatory. He and his coworkers have also started development work on GaAs millimeter-wave components. In the fields mentioned above, he has authored about 60 scientific papers. At the 1982 12th European Microwave Conference in Helsinki, he and his coauthors were rewarded the Microwave Prize. Aside from his research and development work he also lectures at the University.

Dr. Kollberg is Chairman of the Microwave Theory and Techniques Chapter of the IEEE Swedish Section. He is also a member of the Swedish branch of URSI.